Highlights in Theoretical Chemistry

Vol. 9

Series Editors: Christopher J. Cramer • Donald G. Truhlar

More information about this series at http://www.springer.com/series/11166

Ron Shepard • Russell M. Pitzer • Thom Dunning
Volume Editors

Isaiah Shavitt

A Memorial Festschrift from Theoretical Chemistry Accounts

With contributions from

Aleksey B. Alekseyev • Ibon Alkorta • Shigeko Asai • George B. Bacskay
Mario Barbatti • Rodney J. Bartlett • Charles W. Bauschlicher Jr.
Michael J. Bearpark • Scott R. Brozell • Robert J. Buenker
Joachim Burgdörfer • Petr Čársky • Daniel M. Chipman • Roman Čurík
Rathawat Daengngern • Janet E. Del Bene • Thom H. Dunning Jr.
José Elguero • Walter C. Ermler • Nathan D. Erxleben • Peter Funke
Gergely Gidofalvi • Friedrich Grein • Yundong Guo • Supa Hannongbua
Yanjun Hao • Frank E. Harris • James F. Harrison • Miho Hatanaka
So Hirata • Mark R. Hoffmann • Shawn Horn • Amir Karton
Jacek Karwowski • Gary S. Kedziora • Khanittha Kerdpol
Manoj K. Kesharwani • Zhuangfei Kou • Stefan Krebs • Nawee Kungwan
Benjamin G. Levine • Xiangzhu Li • Florian Libisch
Heinz-Peter Liebermann • Wenjian Liu • Hans Lischka • Victor F. Lotrich
Rahul Maitra • Jan M.L. Martin • Todd J. Martínez • Spiridoula Matsika
Rei Matsuzaki • David A. Mazziotti • Stephen L. Meloni
David Mendive-Tapia • Robert W. Molt Jr. • Debashis Mukherjee
Thomas Müller • Josef Paldus • Ajith Perera • Robin A. Phillips
R.M. Pitzer • Felix Plasser • Alessandra Ricca • Michael A. Robb
Nicholas C. Rubin • Henry F. Schaefer • Sangita Sen • Ron Shepard
Debalina Sinha • Tyler Y. Takeshita • Jeffrey L. Tilson • Iogann Tolbatov
Joseph J. Urban • Morgane Vacher • Aaron M. Virshup
C. William McCurdy • Henryk A. Witek • Yaoming Xie • Lu T. Xu
Satoshi Yabushita • Li-Juan Yu • Meiling Zhang • Zhiyong Zhang
S.C. Zimmerman

Springer

Volume Editors
Ron Shepard
Chemical Sciences and Engineering Division
Argonne National Laboratory
Argonne, IL, USA

Russell M. Pitzer
Ohio State University
Columbus, OH, USA

Thom Dunning
National Center for Supercomputing Applications (NCSA)
Urbana, IL, USA

Originally published in Theor Chem Acc, Volume 133 (2014),

ISSN 2194-8666 ISSN 2194-8674 (electronic)
Highlights in Theoretical Chemistry
ISBN 978-3-662-48147-9 ISBN 978-3-662-48148-6 (eBook)
DOI 10.1007/978-3-662-48148-6

Library of Congress Control Number: 2015953006

Springer Heidelberg New York Dordrecht London

Printed on acid-free paper

Springer-Verlag GmbH Berlin Heidelberg is part of Springer Science+Business Media (www.springer.com)

Contents

Theor Chem Acc (2014) 133:1488
DOI 10.1007/s00214-014-1488-3

OVERVIEW

Isaiah Shavitt: Computational chemistry pioneer

S. C. Zimmerman · R. M. Pitzer

Received: 24 March 2014 / Accepted: 28 March 2014 / Published online: 27 April 2014

Abstract Isaiah (Shi) Shavitt (1925–2012) was a pioneer in using digital computers to study chemical problems. From the days of vacuum tube computers with no floating point arithmetic to the days of massively powerful computers, he showed how we could solve otherwise intractable chemical problems by making use of computers. He started with a statistical mechanical problem and soon switched to quantum mechanical problems. He and his associates showed how the configuration interaction method worked, both in terms of advantages and difficulties. An early problem was the effect of tunneling on kinetics calculations on an H_3 potential energy surface. Later problems included the π-electron excited states of benzene and the lowest excited state of methylene. He showed how spin-eigenfunctions could be used efficiently in configuration interaction calculations instead of Slater determinants. His leadership led to the Columbus suite of programs put together with many collaborators.

Keywords Computational chemistry history · Isaiah Shavitt · Configuration interaction · Electron correlation · Graphical Unitary Group Approach · COLUMBUS Programs

Published as part of the special collection of articles dedicated to the memory of Professor Isaiah Shavitt and celebrating his many contributions.

S. C. Zimmerman
Department of Chemistry, University of Illinois, 600 S. Mathews Ave., Urbana, IL 61801, USA

R. M. Pitzer (✉)
Department of Chemistry and Biochemistry, The Ohio State University, 100 W. 18th Ave., Columbus, OH 43210, USA
e-mail: pitzer.3@osu.edu

1 Introduction

Isaiah (Shi) Shavitt was born in Poland in 1925, immigrated to Palestine at a very early age, and was raised in Haifa. He was educated at the Technion and Cambridge University, earning a PhD in 1957. As a graduate student, he learned to use digital computers. He soon set an example by using high standards in computational work and in writing about and evaluating such work. He performed very careful calculations to extend and evaluate the method of configuration interaction for molecules. He wrote many extensive review articles on this method and its evaluation and co-authored a book on electron correlation.

2 Family

Isaiah (Shi) Shavitt (né Kruk) was born to Zvi and Chava Kruk on July 29, 1925, in Kutno, Poland. In 1926, his father immigrated to Palestine to establish a new life for his family in Haifa. In 1929, his wife and son joined him there.

Later, Shi changed his last name to Shavitt, which in Hebrew means comet.

In 1957, Shi married Vera Neuwirth. Shi often said that the best decision of his life was marrying Vera. They had a daughter, Sharon, who married Steve Zimmerman. Both Sharon and Steve have faculty positions at the University of Illinois, starting in 1987 and 1985, respectively. They have two daughters, Arielle and Ellie.

3 Career

Shi studied chemical engineering at the Technion (Israel Institute of Technology) during 1945–1951. His studies were interrupted by his service during 1948–1949 in Israel's War of Independence. Shi completed his BSc in 1950 and a Diploma in Engineering a year later, both degrees in Chemical Engineering. He then decided to change to theoretical chemistry and was awarded a British Council Scholarship for Study in Britain to do so. He earned his PhD at University of Cambridge in 1957 and then returned to Israel. In his graduate work, he learned computing on the EDSAC I, a primitive vacuum tube computer that frequently required repeating calculations just to ensure they were correct.

Short-term positions followed as a postdoctoral fellow at the University of Wisconsin, a teaching position at Brandeis University, and a staff position at IBM Watson Laboratory.

Long-term positions ensued. In 1962, he accepted a faculty position at Technion, where he established his research program and worked with several graduate students. Then in 1967 he accepted a position as Research Leader in the newly formed theoretical chemistry group at Battelle Memorial Institute in Columbus, Ohio, adding an adjunct professor position in the Chemistry Department at Ohio State University a year later. After several years in Columbus, Shi and Vera became US citizens. In 1981, he left Battelle and became full time on the Ohio State University faculty. He retired at the end of 1994.

4 Postretirement science

In September 1995, Shi and the memory of his PhD advisor, S. F. Boys, were jointly recognized by a conference in their names at St. Catharine's College at Cambridge University. Papers from this conference were gathered in a festschrift [1], which included a paper describing the Shavitt research group's work up until that time. Some of the attributes of this work were described [1] as follows, "In an often skeptical community, his results have the highest credibility. It is well-known that questions of numerical accuracy and stability are always addressed in his work and that he has instilled his standards in his students and colleagues." He was also noted [1] for his many carefully done review articles, particularly nos. 11, 39, 66, and 83 in his list of publications in the "Appendix."

He continued his work in retirement, including detailed studies of the energy and dipole-moment surfaces of the water molecule (no. 90), studies of the solvated proton in water (nos. 92, 93), a description of the COLUMBUS programs (many collaborators, no. 94), describing the computer program system inspired by his unitary-group work, studies of DNA base-pair binding (no. 100. 103), and finally, a book with R. J. Bartlett (no. 104), describing the theory and use of many-body methods in describing electron correlation in molecules. He spent several years working on this book, during which he first sought equation-writing and diagram-producing software that met his exacting standards for the manuscript.

In summarizing one component of Shi's scholarly impact, T. H. Dunning, Director of the National Center for Supercomputer Applications (NCSA), recently noted, "It was Shi and his generation that laid the foundation for the use of digital computers to obtain rigorous solutions of the electronic Schroedinger equation."

5 Final years

With their daughter Sharon and her husband Steve Zimmerman both having faculty positions at the University of Illinois, the Shavitts decided to obtain a second house near them in order to increase their family enjoyment and support. After several years of splitting their time between the locations, the Shavitts sold their Columbus house in 2001 and stayed full time in Illinois. Shi was appointed Adjunct Professor of Chemistry at the University of Illinois in 1996. Shi enjoyed completing the *New York Times* crossword puzzle daily, loved traveling, limericks, and the poems of Ogden Nash. He was an extraordinary scholar and intellect, but humble, generous, and kind in his interactions with students, colleagues, and anyone he met. He is remembered as a loving and supportive husband and father, and a doting grandfather. After fighting disease for several years, Shi Shavitt died December 8, 2012.

Appendix

Publications of Isaiah Shavitt

1. S. F. Boys, G. B. Cook, C. M. Reeves, and I. Shavitt, Automatic Fundamental Calculation of Molecular Structure, *Nature*, **178,** 1207–1209 (1956).

2. I. Shavitt and S. F. Boys, A General Expression for Intermolecular Potentials, *Nature*, **178**, 1340 (1956).
3. S. F. Boys, I. Jones, and I. Shavitt, Quelques Problèmes et Résultats de Prévision de Propriétés Moléculaires Fondamentales par le Calcul Automatique, in *Calcul des Fonctions d'Onde Moléculaire (Colloques Internationaux du Centre National de la Recherche Scientifique)* (R. Daudel, editor), **82**, C.N.R.S., Paris (1958), pp. 253–261.
4. S. F. Boys and I. Shavitt, A Fundamental Calculation of the Energy Surface for the System of Three Hydrogen Atoms, University of Wisconsin Naval Research Laboratory Report WIS-AF-13 (1959).
5. I. Shavitt, The Tunnel Effect Correction to the Rates of Reactions with Parabolic and Eckart Barriers, University of Wisconsin Theoretical Chemistry Laboratory Report WIS-AEC-23 (1959).
6. I. Shavitt, A Calculation of the Rates of the Ortho-Para Conversions and Isotope Exchanges in Hydrogen, J. *Chem. Phys.* **31**, 1359–1367 (1959).
7. S. F. Boys and I. Shavitt, Intermolecular Forces and Properties of Fluids. I. The Automatic Calculation of Higher Virial Coefficients and Some Values of the Fourth Coefficient for the Lennard-Jones Potential, *Proc. Roy. Soc. London*, **A254,** 487–498 (1960).
8. S. F. Boys and I. Shavitt, Intermolecular Forces and Properties of Fluids. II. A General Functional Representation of Intermolecular Potentials and Some Values of the 2nd, 3rd, and 4th Virial Coefficients for Systematically Differing Potentials, *Proc. Roy. Soc. London*, **A254,** 499–506 (1960).
9. I. Shavitt and M. Karplus, Multicenter Integrals in Molecular Quantum Mechanics, *J. Chem. Phys.* **36,** 550–551 (1962).
10. M. Karplus and I. Shavitt, Multicenter Pi-Electron Integrals for the Benzene Molecule, *J. Chem. Phys.* **38,** 1256–1257 (1963).
11. I. Shavitt, The Gaussian Function in Calculations of Statistical Mechanics and Quantum Mechanics, *Meth. Comput. Phys.* **2,** 1–45 (1963).
12. I. Shavitt and M. Karplus, Gaussian Transform Method for Molecular Integrals. I. Formulation for Energy Integrals, *J. Chem. Phys.* **43**, *398–414* (1965).
13. U. Kaldor and I. Shavitt, LCAO-SCF Computations for Hydrogen Peroxide, *J. Chem. Phys.* **44**, 1823–1829 (1966).
14. U. Kaldor and I. Shavitt, LCAO-SCF Computations for Ammonia, *J. Chem. Phys.* **45,** 888–895 (1966).
15. U. Kaldor and I. Shavitt, LCAO-SCF Computations for Ethylene, *J. Chem. Phys.* **48**, 191–203 (1968).
16. Z. Gershgorn and I. Shavitt, The Construction of Symmetry Adapted Functions in Configuration Interaction Calculations for Molecules with One Principal Axis of Symmetry, *Int. J. Quantum Chem. Symp.* **1**, 403–417 (1967).
17. I. Shavitt, R. M. Stevens, F. L. Minn, and M. Karplus, Potential Energy Surface for H_3, *J. Chem. Phys.* **48**, 2700–2713 (1968).
18. Z. Gershgorn and I. Shavitt, An Application of Perturbation Theory Ideas in Configuration Interaction Calculations, *Int. J. Quantum Chem.* **2**, 751–759 (1968).
19. A. Pipano and I. Shavitt, Convergence Studies and the Selection for Configuration Functions in Configuration Interaction Calculations, *Int. J. Quantum Chem.* **2,** 741–749 (1968).
20. I. Shavitt, A Correlation of Experimental Rate Constants of the Hydrogen Exchange Reactions with a Theoretical H_3 Potential Surface, Using Transition State Theory, *J. Chem. Phys.* **49**, 4048–4056 (1968).
21. M. Rubinstein and I. Shavitt, Theoretical Study of the Potential Surface for the H_4 System by Double-Zeta Configuration Interaction Calculations, *J. Chem. Phys.* **51**, 2014–2024 (1969).
22. A. Pipano, R. R. Gilman, C. F. Bender, and I. Shavitt, *Ab Initio* Calculation of the Inversion Barrier in Ammonia, *Chem. Phys. Lett.* **5,** 583–584 (1970).
23. A. Pipano, R. R. Gilman, and I. Shavitt, Invariance of Inner Shell Correlation Energy with Geometry Changes in a Polyatomic Molecule, *Chem. Phys. Lett.* **5,** 285–287 (1970).
24. I. Shavitt, Modification of Nesbet's Algorithm for the Iterative Evaluation of Eigenvalues and Eigenvectors of Large Matrices, *J. Comput. Phys.* **6**, 124–130 (1970).
25. C. F. Bender and I. Shavitt, An Iterative Procedure for the Calculation of the Lowest Real Eigenvalue and Eigenvector of a Non-Symmetric Matrix, *J. Comput. Phys.* **6,** 146–149 (1970).
26. R. P. Hosteny, R. R. Gilman, T. H. Dunning, Jr., A. Pipano, and I. Shavitt, Comparison of Slater and Contracted Guassian Basis Sets in SCF and CI Calculations of H_2O, *Chem. Phys. Lett.* **7,** 325–328 (1970).
27. J. Paldus, J. Čížek, and I. Shavitt, Correlation Problems in Atomic and Molecular Systems, IV. Extended Coupled-Pair Many-Electron Theory and Its Application to the BH_3 Molecule, *Phys. Rev. A.* **5,** 50–67 (1972).
28. I. Shavitt, C. F. Bender, A. Pipano, and R. P. Hosteny, The Iterative Calculation of Several of the Lowest or Highest Eigenvalues and Corresponding Eigenvectors of Very Large Symmetric Matrices, *J. Comput Phys.* **11**, 90–108 (1973).

29. T. H. Dunning, Jr., R. P. Hosteny, and I. Shavitt, Low-Lying π-Electron States of *trans*-Butadiene, *J. Am. Chem. Soc.* **95,** 5067–5068 (1973).
30. P. J. Hay and I. Shavitt, Large-Scale Configuration Interaction Calculations on the π-Electron States of Benzene, *Chem. Phys. Lett.* **22**, 33–36 (1973).
31. P. J. Hay and I. Shavitt, *Ab Initio* Configuration Interaction Studies of the π-Electron States of Benzene, *J. Chem. Phys.* **60**, 2865–2877 (1974).
32. L. R. Kahn, P. J. Hay, and I. Shavitt, Theoretical Study of Curve Crossing: *Ab Initio* Calculations on the Four Lowest $^1\Sigma^+$ States of LiF, *J. Chem. Phys.* **61**, 3530–3546 (1974).
33. R. P. Hosteny, T. H. Dunning, Jr., R. R. Gilman, A. Pipano, and I. Shavitt, *Ab Initio* Study of the π-Electron States of *trans*-Butadiene, *J. Chem. Phys.* **62,** 4764–4779 (1975).
34. B. J. Rosenberg and I. Shavitt, *Ab Initio* SCF and CI Studies on the Ground State of the Water Molecule. I. Comparison of CGTO and STO Basis Sets near the Hartree–Fock Limit, *J. Chem. Phys.* **63**, 2162–2174 (1975).
35. I. Shavitt, B. J. Rosenberg, and S. Palalikit, Comparison of Configuration Interaction Expansions Based on Different Orbital Transitions, *Int. J. Quantum Chem. Symp.* **10,** 33–46 (1976); erratum, *ibid.* **11**, 651 (1977).
36. I. Shavitt, Computers and Quantum Chemistry, in *Computer Science and Scientific Computing* (J. M. Ortega, editor), Academic Press (1976), pp. 227–253.
37. B. J. Rosenberg, W. C. Ermler, and I. Shavitt, *Ab Initio* SCF and CI Studies on the Ground State of the Water Molecule. II. Potential Energy and Property Surfaces, *J. Chem. Phys.* **65,** 4072–4080 (1976).
38. R. C. Raffenetti, K. Hsu, and I. Shavitt, Selection of Terms for a CI Wavefunction to Preserve Potential Surface Features, *Theor. Chim. Acta* **45**, 33–44 (1977).
39. I. Shavitt, The Method of Configuration Interaction, in *Methods of Electronic Structure Theory (Modern Theoretical Chemistry, Vol 3)* (H. F. Schaefer III, editor), Plenum Press, New York (1977), pp. 189–275.
40. I. Shavitt, Graph Theoretical Concepts for the Unitary Group Approach to the Many-Electron Correlation Problem, *Int. J. Quantum Chem. Symp.* **11**, 131–148 (1977).
41. R. J. Bartlett and I. Shavitt, Determination of the Size-Consistency Error in the Single and Double Excitation Configuration Interaction Model, *Int. J. Quantum Chem. Symp.* **11**, 165–173 (1977); erratum, *ibid.* **12** 543–544 (1978).
42. R. J. Bartlett and I. Shavitt, Comparison of High-Order Many-Body Perturbation Theory and Configuration Interaction for H_2O, *Chem. Phys. Lett.* **50**, 190–198 (1977); erratum, *ibid.* **57,** 157–158 (1978).
43. C. W. Bauschlicher and I. Shavitt, Accurate ab initio Calculations on the Singlet–Triplet Separation in Methylene, *J. Am. Chem. Soc.* **100,** 739–743 (1978).
44. I. Shavitt, Matrix Element Evaluation in the Unitary Group Approach to the Electron Correlation Problem, *Int. J. Quantum Chem. Symp.* **12**, 5–32 (1978).
45. I. Shavitt, The Utilization of Abelian Point Group Symmetry in the Graphical Unitary Group Approach to the Calculation of Correlated Electronic Wavefunctions, *Chem. Phys. Lett.* **63**, 421–427 (1979).
46. R. J. Bartlett, I. Shavitt, and G. D. Purvis, The Quartic Force Field of H_2O Determined by Many-Body Methods that Include Quadruple Excitation Effects, *J. Chem. Phys.* **71,** 281–291 (1979).
47. I. Shavitt, The Graphical Unitary Group Approach (GUGA) to the Electron Correlation Problem-Survey and Recent Advances, in *Electron Correlation: Proceedings of the Daresbury Study Weekend, 17–18 November, 1979* (M. F. Guest and S. Wilson, editors), Science Research Council, Daresbury Laboratory, Warringon, England (1980), pp. 60–68.
48. C. W. Bauschlicher Jr. and I. Shavitt, A Low-Energy Passage for $C^+ + H_2 \rightarrow CH^+$ ($^1\Sigma^+$) + H, *Chem. Phys. Lett.* **75**, 62–65 (1980).
49. I. Shavitt and L. T. Redmon, Quasidegenerate Perturbation Theories. A Canonical van Vleck Formalism and Its Relationship to Other Approaches. *J. Chem. Phys.* **73,** 5711–5717 (1980).
50. C. F. Jackels and I. Shavitt, Accuracy of Energy Extrapolation in Multireference Configuration Interaction Calculations, *Theor. Chim. Acta* **58,** 81–96 (1981).
51. I. Shavitt, The Graphical Unitary Group Approach and Its Application to Direct Configuration Interaction Calculations, in *The Unitary Group for the Evaluation of Electronic Energy Matrix Elements (Lectures Notes in Chemistry* **22)** (J. Hinze, editor), Springer Verlag, Berlin (1981), pp. 51–99.
52. M. J. Redmon, R. J. Bartlett, B. C. Garrett, G. D. Purvis III, P. M. Saatzer, G. C. Schatz, and I. Shavitt, Collisional Excitation of H_2O by O-Atom Impact : Classical Dynamics on an Accurate ab initio Potential Energy Surface, in *Potential Energy Surfaces and Dynamics Calculations* (D. G. Truhlar, editor), Plenum, New York (1981), pp. 771–803.
53. H. Lischka, R. Shepard, F. B. Brown, and I. Shavitt, New Implementation of the Graphical Unitary Group Approach for Multireference Direct Configuration

Interaction Calculations, *Int. J. Quantum Chem. Symp.* **15**, 91–100 (1981).
54. P. G. Lykos and I. Shavitt, editors, *Supercomputers and Chemistry (ACS Symposium Series* **173**), American Chemical Society, Washington, D. C. (1981), 278 pp.
55. R. Shepard, I. Shavitt, and J. Simons, Comparison of Convergence Characteristics of Some Iterative Wave Function Optimization Methods, *J. Chem. Phys.* **76**, 543–557 (1982).
56. G. Born and I. Shavitt, A Unitary Group Formulation of Open-Shell Electron Propagator Theory, *J. Chem. Phys.* **76**, 558–567 (1982).
57. P. Čársky, M. Svrček, I. Hubač, F. B. Brown, and I. Shavitt, Correlation Energy in Triplet States. Comparison of Many-Body Perturbation Theory and Configuration Interaction for CH_2 and O_2, *Chem. Phys. Lett.* **85,** 17–20 (1982).
58. I. Shavitt, The Unitary Group and the Electron Correlation Problem, in *New Horizons of Quantum Chemistry* (P.-O. Löwdin and B. Pullman, editors), Reidel, Dordrecht (1983), pp. 279–293.
59. F. B. Brown, I. Shavitt, and R. Shepard, Multireference Configuration Interaction Treatment of Potential Energy Surfaces: Symmetric Dissociation of H_2O in a Double-Zeta Basis, *Chem. Phys. Lett.***105**, 363–369 (1984).
60. I. Shavitt, The Treatment of Electron Correlation: Where Do We Go From Here? in *Advanced Theories and Computational Approaches the Electronic Structure of Molecules* (C. E. Dykstra, editor), Reidel, Dordrecht (1984). pp. 185–196.
61. I. Shavitt, Geometry and Singlet-Triplet Energy Gap in Methylene: A Critical Review of Experimental and Theoretical Determinations, *Tetrahedron* **41**, 1531–1542 (1985).
62. W. C. Ermler, B. J. Rosenberg, and I. Shavitt, *Ab Initio* SCF and CI Studies on the Ground State of the Water Molecule. III. Vibrational Analysis of Potential Energy and Property Surfaces, in *Comparison of Ab Initio Quantum Chemistry with Experiment: State of the Art* (R. J. Bartlett, editor), Reidel, Dordrecht (1985), pp. 171–215.
63. D. C. Comeau, R. J. Zellmer, and I. Shavitt, The Location and Characterization of Stationary Points on Molecular Potential Energy Surfaces, in *Geometrical Derivatives of Energy Surfaces and Molecular Properties* (P. Jørgensen and J. Simons, editors), Reidel, Dordrecht (1986), pp. 243–251.
64. I. Shavitt, F. B. Brown, and P. G. Burton, Configuration Selection and Extrapolation in Multireference Configuration Interaction Calculations: The $(H_2)_2$ van der Waals Complex as a Benchmark Example, *Int. J. Quantum Chem.* **31**, 507–520 (1987).
65. R. J. Barlett, S. J. Cole, G. D. Purvis, W. C. Ermler, H. C. Hsieh, and I. Shavitt, The Quartic Force Field of H_2O Determined by Many-Body Methods. II. Effects of Triple Excitations, *J. Chem. Phys.* **87**, 6579–6591 (1987).
66. I. Shavitt, Unitary Group Approach to Configuration Interaction Calculations of the Electronic Structure of Atoms and Molecules, in *Mathematical Frontiers in Computational Chemical Physics* (D. G. Truhlar, editor), Springer-Verlag, Berlin (1988), pp. 300–349.
67. R. Ernenwein, M. Benard, and I. Shavitt, Vectorizing a Sequence of Conditional Branches: the Calculation of the Class Index of Two-Electron Repulsion Integrals on Cray Computers, *Comput. Phys. Commun.* **48**, 175–180 (1988).
68. R. Shepard, I. Shavitt, R. M. Pitzer, D. C. Comeau, M. Pepper, H. Lischka, P. G. Szalay, R. Ahlrichs, F. B. Brown, and J.–G. Zhao, A Progress Report on the Status of the COLUMBUS MRCI Program System, *Int. J. Quantum Chem. Symp.* **22**, 149–165 (1988).
69. D. C. Comeau, I. Shavitt, P. Jensen, and P. R. Bunker, An ab initio Determination of the Potential Energy Surfaces and Rotation-Vibration Energy Levels of Methylene in the Lowest Triplet and Singlet States and the Single-Triplet Splitting, *J. Chem. Phys.* **90**, 6491–6500 (1989).
70. J. E. Del Bene and I. Shavitt, Comparison of Methods for Determining the Correlation Contributions to Hydrogen Bond Energies, *Int. J. Quantum Chem. Symp.* **23**, 445–452 (1989).
71. J. E. Del Bene and I. Shavitt, Comparison of Theoretical Methods for the Determination of the Protonation and Deprotonation Energies of NH_3, H_2O, HF, PH_3, H_2S, HCl, and HCN, *J. Phys. Chem.* **94,** 5514–5518 (1990).
72. J. E. Del Bene, E. A. Stahlberg, and I. Shavitt, A Theoretical Study of the Complexes of N_2O with H^+, Li^+, and HF Using Various Correlation Methods, *Int. J. Quantum Chem. Symp.* **24**, 455–466 (1990).
73. J. E. Del Bene and I. Shavitt, Comparison of Theoretical Methods for the Determination of the Li^+ Affinities of Neutral and Anionic First- and Second-Row Bases, *Int. J. Quantum Chem. Symp.* **24**, 365–373 (1990).
74. J. E. Del Bene, K. Kim, and I. Shavitt, An ab initio Study of Symmetry Breaking in Calculations on the First Excited Singlet State of N_2H_2, *Can. J. Chem.* **69**, 246–250, (1991).
75. J. E. Del Bene, H. D. Mettee, and I. Shavitt, The Structure, Binding Energy, and Vibrational Frequencies of $CH_3CN\ldots HCl$, *J. Phys. Chem.* **95**, 5387–5388 (1991).

76. J. E. Del Bene and I. Shavitt, A Theoretical Study of the Neutral, Protonated, and Deprotonated Trimers of HF and HCl, *J. Mol. Struct. (THEOCHEM)* **234**, 499–508 (1991).
77. I. Shavitt, J. E. Del Bene, and D. W. Ewing, *Ab Initio* Study of Spiropentadiene, C_5H_4, *J. Am. Chem. Soc.* **113**, 9389–9391 (1991).
78. R. D. Kent, M. Schlesinger, and I. Shavitt, Graphical Unitary Group Approach to Spin–Spin Interaction, *Int. J. Quantum Chem.* **41,** 89–103 (1992).
79. J. E. Del Bene, D. H. Aue, and I. Shavitt, Stabilities of Hydrocarbons and Carbocations. I. A Comparison of Augmented 6-31G, 6-311G, and Correlation-Consistent Basis Sets, *J. Am. Chem. Soc.* **114,** 1631–1640 (1992).
80. I. Shavitt, The A_k and B_k Approximate CI Methods. Comment on a Paper by Maynau and Heully, *Chem. Phys. Lett.* **192**, 135–137 (1992).
81. K. Kim, I. Shavitt, and J. E. Del Bene, Theoretical Study of the Di-imide (N_2H_2) Molecule in Ground and $n \rightarrow \pi^*$ Excited States, *J. Chem. Phys.* **96**, 7573–7579 (1992).
82. D. W. Ewing and I. Shavitt, Double Zeta Basis Sets in Carbon Cluster Calculations, in *Physics and Chemistry of Finite Systems: From Clusters to Crystals,* Vol. I. (P. Jena, S. N. Khanna, and B. K. Rao, editors), Kluwers, Dordrecht, Holland (1992), pp. 561–567.
83. I. Shavitt, The History and Evolution of Gaussian Basis Sets, *Israel J. Chem.* **33**, 357–368 (1993).
84. J. E. Del Bene and I. Shavitt, Basis-Set Effects on Computer Acid–Base Interaction Energies Using the Dunning Correlation-Consistent Polarized Split-Valence Basis Sets, *J. Mol. Struct, (THEOCHEM)* **307**, 27–34 (1994).
85. J. E. Del Bene and I. Shavitt, An ab initio Study of the Complexes of HCl with the Chloromethanes, *J. Mol. Struct.* (THEOCHEM) **314**, 9–17 (1994).
86. M. J. M. Pepper, I. Shavitt, P. v. R. Schleyer, M. N. Glukhovstev, R. Janoschek, and M. Quack, Is the Stereomutation of Methane Possible? *J. Comput. Chem.* **16**, 207–225 (1995).
87. L. Ojamäe, I. Shavitt, and S. J. Singer, Potential Energy Surfaces and Vibrational Spectra of $H_5O_2^+$ and Larger Hydrated Proton Complexes, *Int. J. Quantum Chem. Symp.* **29**, 657–668 (1995).
88. N. C. Handy, J. A. Pople, and I. Shavitt, Samuel Francis Boys, *J. Phys. Chem.* **100**, 6007–6016 (1996).
89. G. S. Kedziora and I. Shavitt, Calculation and Fitting of Potential Energy and Dipole Moment Surfaces for the Water Molecule: Fully ab initio Determination of Vibrational Transition Energies and Band Intensities, *J. Chem. Phys.* **106**, 8733–8745 (1997).
90. J. E. Del Bene and I. Shavitt, The Quest for Reliability in Calculated Properties of Hydrogen-bonded Complexes, in *Molecular Interactions, From van der Waals to Strongly Bound Complexes*, (Wiley Tutorial Series in Theoretical Chemistry) (S. Scheiner, editor), Wiley: Chichester (1997), pp. 157–179.
91. I. Shavitt, The History and Evolution of Configuration Interaction, *Mol. Phys.* **94**, 3–17 (1998).
92. L. Ojamäe, I. Shavitt, and S. J. Singer, Potential Models for Simulations of the Solvated Proton in Water, *J. Chem. Phys.* **109**, 5547–5564 (1998).
93. C. V. Ciobanu, L. Ojamäe, I. Shavitt, and S. J. Singer, Structure and Vibrational Spectra of $H^+(H_2O)_8$: Is the Excess Proton in a Symmetrical Hydrogen Bond? *J. Chem. Phys.* **113**, 5321–5330.
94. H. Lischka, R. Shepard, R. M. Pitzer, I. Shavitt, M. Dallos, T. Müller, P. G. Szalay, M. Seth, G. S. Kedziora, S. Yabushita, and Z. Zhang, High-Level Multireference Methods in the Quantum-Chemistry Program System COLUMBUS: Analytic MR-CISD and MR-AQCC Gradients and MR-AQCC-LRT for Excited States, GUGA Spin–Orbit CI and Parallel CI Density, *Phys. Chem. Chem. Phys.* **3**, 664–673 (2001).
95. R. Shepard, I. Shavitt, and H. Lischka, Software News and Updates Reducing I/O Costs for the Eigenvalue Procedure in Large-Scale Configuration Interaction Calculations, *J. Comp. Chem.* **23**, 1121–1125 (2002).
96. I. Shavitt, Multi-State Multireference Rayleigh-Schroedinger Perturbation Theory for Mixed Electronic States: Second and Third Order, *Int. J. Mol. Sci.* **3**, 639–655.
97. J.-L. Kuo, C.V. Ciobanu, L. Ojamäe, I. Shavitt, and S. J. Singer, Short H-Bonds and Spontaneous Self-dissociation in $(H_2O)_{20}$: Effects of H-Bond Topology, *J. Chem. Phys.* **118**, 3583–3588.
98. J. A. Karwowski and I. Shavitt, Configuration Interaction, in *Handbook of Molecular Physics and Quantum Chemistry* (S. Wilson, editor) **2,** Wiley: Chichester (2003), pp. 227–271.
99. I. Shavitt, Are Exponential-Type Basis Sets Preferable to Gaussians? *Int. J. Quantum Chem.* **100**, 105–108 (2004).
100. J. R. Quinn, S. C. Zimmerman, J. E. Del Bene, and I. Shavitt, Does the A-T or G-C Base-Pair Possess Enhanced Stability? Quantifying the Effects of CH···O Interactions and Secondary Interactions on Base-Pair Stability Using a Phenomenological Analysis and ab initio Calculations, *J. Am. Chem. Soc.* **129**, 934–941 (2007).

101. R. Shepard, G. S. Kedziora, H. Lischka, I. Shavitt, T. Müller, P. G. Szalay, M. Kállay, and M. Seth, The Accuracy of Molecular Bond Lengths Computed by Multireference Electronic Structure Methods, *Chem. Phys.* **349**, 37–57 (2008).
102. I. Shavitt, Tribute to Russell M. Pitzer, *J. Phys. Chem. A* **113**, 12339–12342 (2009).
103. J. R. Quinn, S. C. Zimmerman, J. E. Del Bene, and I. Shavitt, Prebiotic Selection of the AT Base Pair? A Physical Organic Approach to Understanding AT Base-Pair Stability Indicates Special Stability, in *ACS Symposium Series* (2009), **1025** (*Chemical Evolution* **II**), pp. 95–107.
104. I. Shavitt and R. J. Bartlett, *Many-Body Methods in Chemistry and Physics: MBPT and Coupled-Cluster Theory*, Cambridge University Press: Cambridge, (2009), 532 pp.
105. H. Lischka, T. Müller, P. G. Szalay, I. Shavitt, R. M. Pitzer, and R. Shepard, COLUMBUS-A Program System for Advanced Multireference Theory Calculations, in *Wiley Interdisciplinary Reviews: Computational Molecular Science* **1**, (2011), pp. 191–199.
106. I. Shavitt, Perspective: Björn Roos and Direct Configuration Interaction, *Int. J. Quantum Chem.* **111**, 3263–3266 (2011).

Reference

1. Kaldor U, Pitzer RM (1996) Isaiah Shavitt. J Phys Chem 100:6017–6022

Theor Chem Acc (2014) 133:1512
DOI 10.1007/s00214-014-1512-7

REGULAR ARTICLE

Wave function analysis with Shavitt graph density in the graphically contracted function method

Gergely Gidofalvi · Scott R. Brozell · Ron Shepard

Received: 11 February 2014 / Accepted: 24 May 2014 / Published online: 15 July 2014

Abstract The goals of electronic structure theory are to make quantitative predictions of molecular properties and to provide qualitative insight into bonding as well as features of potential energy surfaces. Oftentimes, the two goals are at odds as an accurate treatment requires a complicated wave function that obscures chemical insight. The multifacet graphically contracted function (MFGCF) method offers a new approach that allows both goals to be addressed simultaneously. The recursive product structure of the MFGCF wave function reduces the exponential scaling of the exact wave function and allows the computation of molecular properties with polynomial scaling with respect to system size. Additionally, the graph density concept provides an intuitive tool for visualizing and analyzing the qualitative features of the wave function. In this work, the graph densities for model systems are examined to demonstrate their utility in analyzing the changes in wave function character along potential energy surfaces and near avoided crossings. Finally, we demonstrate that the graph density exposes the structure of the exact wave function for a system of noninteracting molecules as a product of the fragment wave functions.

Keywords Multifacet graphically contracted function method · Configuration interaction · Graphical unitary group approach · Shavitt graph · Wave function analysis · Graph density · Node density · Arc density

Dedicated to the memory of Professor Isaiah Shavitt and published as part of the special collection of articles celebrating his many contributions.

G. Gidofalvi (✉)
Department of Chemistry and Biochemistry, Gonzaga University, 502 E. Boone Ave., Spokane, WA 99258-0102, USA
e-mail: gidofalvi@gonzaga.edu

S. R. Brozell · R. Shepard
Chemical Sciences and Engineering Division, Argonne National Laboratory, Argonne, IL 60439, USA
e-mail: srb@osc.edu

R. Shepard
e-mail: shepard@tcg.anl.gov

1 Introduction

At the heart of electronic structure theory lies the electronic Schrödinger equation, and the ultimate goal of quantum chemistry is the accurate and efficient computation of its solutions: the energies and associated wave functions of the electronic states of atoms and molecules. The graphical unitary group approach (GUGA) of Shavitt [1–5] establishes a convenient representation for expansions of electronic wave functions in terms of spin-adapted configuration state functions (CSFs). GUGA is based on the unitary group approach in which an integer triple (a, b, c) of a Paldus ABC tableau [6–9] defines the number of orbitals $n = a + b + c$, the number of electrons $N = 2a + b$, and the spin quantum number $b = 2S$. In GUGA, each node of a Shavitt graph corresponds to a unique integer triple. The Shavitt graph for a three-electron three-orbital doublet full-CI wave function is shown in Fig. 1a. The tail node $(0, 0, 0)$ represents the physical vacuum, and the head node represents the n, N, and S of the wave function. The vertical levels in a Shavitt graph coincide with a specific sequence of orbitals. The nodes between adjacent levels are connected by arcs or steps, and the four possible step numbers are $d = 0$, 1, 2, and 3, each with $(\Delta a, \Delta b, \Delta c)$ values of $(0, 0, 1)$, $(0, 1, 0)$, $(1, -1, 1)$, and $(1, 0, 0)$ respectively. Step numbers $d = 0$ and 3 denote empty and doubly occupied

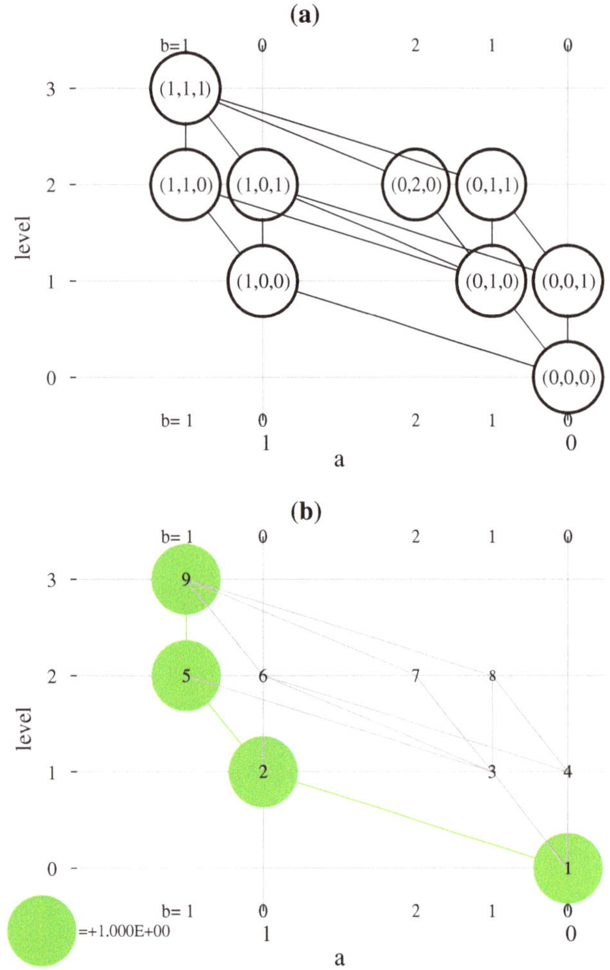

Fig. 1 **a** Full-CI Shavitt graph for a three-electron three-orbital doublet state. Nodes of the graph are labeled according to the integer triples (a_j, b_j, c_j). The slant of the arcs away from vertical increases with the step number; $d = 0$ is vertical, whereas $d = 3$ has the largest slant. **b** The graph density plot for the one-configuration SCF wave function $|\phi_1^2\phi_2^1\phi_3^0\rangle$. The orbital levels are ordered according to orbital energy and the nodes are labeled according to the contiguous indexing scheme $j = 1, \ldots, N_{row}$, where N_{row} is the number of nodes in the Shavitt graph. Arc(node) densities with appreciable values are *green*, while *gray* indicates arcs(nodes) with negligible densities

orbitals, respectively, while step number $d = 1(2)$ corresponds to a singly occupied orbital such that the cumulative spin is increased(decreased) by $\Delta b = \pm 1$. A walk along arcs from the vacuum node to the head node defines a spin-adapted CSF expansion term. The GUGA formulation has been employed in several electronic structure packages for configuration interaction (CI) and multiconfiguration self-consistent field (MCSCF) calculations, such as GAMESS [10, 11] and MOLCAS [12], and notably in the COLUMBUS Program System of Shavitt and coworkers [3, 13–15] within which the present work has been performed.

The energy (E) is of central importance in variational electronic structure methods as a criterion for optimizing the wave function parameters. In GUGA CI, the CSF expansion coefficients ($\mathbf{x}$) are obtained by solving the eigenvalue equation $\mathbf{Hx} = E\ \mathbf{x}$ using iterative diagonalization techniques [16] that rely on the efficient evaluation and use of the Hamiltonian ($\mathbf{H}$) matrix elements between CSFs [4]. Although the energy and its geometrical derivatives are useful in predicting equilibrium and transition state structures as well as energies of reactions and barrier heights, the wave function itself is equally important as it provides insight into the chemical and physical properties; unfortunately, this utility of the wave function is oftentimes obscured by the (*possibly very*) large number of parameters. Thus, several techniques have been developed to allow the efficient interpretation and visualization of electronic wave functions. Most notably, the one-electron reduced density matrix (1-RDM) may be used in population analysis methods, such as those proposed by Mulliken [17] and Löwdin [18], to estimate gross atomic charges and bond orders [19]. Relative to the canonical molecular orbitals, the eigenvectors of the 1-RDM, the natural orbitals, are more appropriate in some circumstances for correlated calculations and have been used in various CI [20–24] and coupled-cluster [25–29] methods. Although changes in wave function character can be observed by examination of the 1-RDM matrix elements, the information contained in the 1-RDM is highly contracted, and the subtle details associated with spin-coupling in the wave function are obscured.

In previous work, we introduced [30] and implemented [31–33] a novel wave function analysis technique based on Shavitt graphs. It is denoted as the graph density concept, and it enables wave functions to be characterized and compared in terms of nodes and arcs of the Shavitt graph. The graph density technique arose in the context of our graphically contracted function (GCF) method [30–41], which opens up the possibility to compute astronomically large CI expansions. For such wave functions, traditional analysis approaches based on individual CSF coefficients become intractable. However, the graph density concept is generally applicable to any expansion length or form that can be represented on a Shavitt graph. Thus, the graph density concept offers a tool that complements existing wave function analysis methods. Node densities contain information related to both orbital occupations and spin-coupling within the molecule, whereas the occupation numbers do not contain information about spin-coupling. For example, a singlet and triplet wave function can, in principle, have the same natural orbitals and occupations, whereas the graph densities of the two wave functions would be distinct. Arc densities also contain information that can aid the determination of the seniority quantum number of the wave function [42]. Additionally, arc densities are related to the one-orbital reduced density matrices introduced to study orbital interaction in density matrix renormalization group theory [43] using quantum information theory [44, 45]. These one-orbital reduced density matrices may be considered as contractions of the arc densities and provide useful insight into changes in electronic structure along the potential energy surface [46] and aid in the characterization of chemical bonds [47].

After a brief review of the GCF method and the computation of graph densities in that context, we present wave function analyses via graph densities along the potential energy curves for the C_{2v} insertion pathway for BeH_2, the dissociation of C_2, and D_{2h} and D_{4h} dissociation paths of H_8. The purpose of the present work is to develop the necessary background for interpreting graph densities and to demonstrate their utility in analyzing the wave function. Thus, to minimize the complexity, each example presented herein considers valence-only correlation and employs small basis sets that capture the essential features of the wave function. Nonetheless, we emphasize that this analysis approach is equally applicable to arbitrary wave functions.

2 Method

Within the GCF approach, the wave function is expressed as

$$|\psi\rangle = \sum_{M=1}^{N_{GCF}} c_M |M\rangle, \tag{1}$$

where each GCF $|M\rangle$ is a multiconfigurational function formally defined as a linear combination of all the CSFs $|\tilde{m}\rangle$ comprising the underlying CI space

$$|M\rangle = \sum_{m=1}^{N_{CSF}} x_m^M |\tilde{m}\rangle, \tag{2}$$

and the CSF expansion coefficients x_m^M are defined as ordered products of arc factors $\boldsymbol{\alpha}^M$

$$x_m^M = \boldsymbol{\alpha}_{j_0,j_1}^{M,d_{j_0 j_1}^m} \boldsymbol{\alpha}_{j_1,j_2}^{M,d_{j_1 j_2}^m} \ldots \boldsymbol{\alpha}_{j_{n-1} j_n}^{M,d_{j_{n-1} j_n}^m} = \prod_{p=1}^{n} \boldsymbol{\alpha}_{j_{p-1} j_p}^{M,d_{j_{p-1} j_p}^m}. \tag{3}$$

In Eq. (3), $\boldsymbol{\alpha}_{j_{p-1} j_p}^{M,d_{j_{p-1} j_p}}$ is a real $f_{j_{p-1}}$ by f_{j_p} matrix in which $f_{j_{p-1}} (f_{j_p})$ is the number of facets associated with node $j_{p-1} (j_p)$ in level $p-1(p)$. The superscript $d_{j_{p-1} j_p}^m$ indicates the step d that connects nodes j_{p-1} and j_p that are touched by the CSF $|\tilde{m}\rangle$. In this way, the Shavitt graph plays a central role in the mathematical structure of the resulting wave function and in its dependence on the nonlinear arc factor parameters. It is also important to recognize that an arc of the Shavitt graph is shared by a possibly large number of CSFs. Thus, whereas the number of linear CSF coefficients scales *exponentially* ($\mathcal{O}(n^N)$) with system size, the number of parameters in a GCF is proportional to the number of arcs in the Shavitt graph and scales *polynomially* ($\mathcal{O}(N^2 n)$) with system size [30]. Because of the product structure, the GCF basis function is also seen to be a matrix product state [48] and is similar to the wave function Ansatz employed in spin-adapted density matrix renormalization group methods [49, 50].

The linear coefficients $\mathbf{c}$ are obtained by solving the generalized eigenvalue equation $\mathbf{Hc} = E\,\mathbf{Sc}$, where $\mathbf{H}$ and $\mathbf{S}$ are the Hamiltonian and overlap matrices evaluated in the GCF basis [34]. The arc factors $\boldsymbol{\alpha}$ are optimized to minimize the energy of a single state, or the weighted average of the energies of several states [32]. Efficient routines for computing overlap and Hamiltonian matrix elements and evaluating derivatives of the state-specific and state-averaged energies have been developed [32]. In addition to enabling the efficient computation of the quantities needed during the optimization, the GCF energy is variational and size extensive in a localized orbital basis. The variational flexibility of the GCF wave function may be improved by increasing the size of the GCF expansion basis in Eq. (1), increasing the number of facets associated with the nodes of the Shavitt graph, or increasing both. In our initial work every node was restricted to a single facet, but preliminary results indicate that varying the number of facets to form a multifacet GCF (MFGCF) while restricting N_{GCF} to be equal to the number of electronic states of interest results in more flexible basis functions and in better convergence [32].

The character of the wave function may be analyzed through the graph density concept introduced in Ref. [30]. Formally, the node density of a particular node of the Shavitt graph is equal to the sum of the squares of the CI coefficients for those CSFs that pass through that node. Thus, it is a measure of the importance of that node in describing the wave function. Using an analogous argument, the arc densities quantify the importance of individual arcs in the wave function. In fact, since the node density is the sum of the arc densities leading to that node, node densities may be considered to be contractions over arc densities. From the formal definition of the node and arc densities, it readily follows that the sum of the node densities within any level of the Shavitt graph and the sum of the arc densities connecting any two adjacent levels of the Shavitt graph are both equal to unity for a normalized wave function. The graphical representation of the node and arc densities using graph density plots provides an intuitive tool for analyzing the wave function. Within these graphs, the magnitude of the node density is proportional to the area of the circle representing the node. Thus, the head and tail nodes are represented by circles with unit area. Arcs with "appreciable" densities above some predefined threshold are represented by green lines, and all other arcs with small densities are represented by gray lines. The threshold is proportional to the maximum arc density in the normalized wave function. In this work the thresholds were 0.001 for all graph density plots. To illustrate the graph density concept, consider the SCF wave function for a three-electron three-orbital molecule that consists of a single configuration with the first orbital doubly occupied, the second orbital singly occupied, and the highest energy orbital empty. The graph density plot for this wave function is shown in Fig. 1b. For such a wave function, only those three arcs that are consistent with this occupation pattern have appreciable arc densities and all remaining arc densities are zero. Thus, since the node densities are contractions over the connecting arc densities, it is readily seen that, at each level of the Shavitt graph, there is a single node with unit node density. Note that for wave functions with more than one dominant configuration, several nodes and arcs within each level of the Shavitt graph may acquire significant densities. Thus, a significant "spread" in node and arc density within a level indicates the multiconfigurational character of the wave function.

For a GCF expansion, the density contributions can be determined directly from the partial overlap matrices and the arc factors. This allows graph density analysis of wave function expansions that are much larger than could be considered with conventional approaches that rely on explicit summations of CSF contributions. For a normalized GCF wave function, the node density D_k may be computed as [30, 32]

$$D_k = \sum_{M,N}^{N_{GCF}} c_M c_N D_k^{MN} = \sum_{M,N}^{N_{GCF}} c_M c_N \mathrm{Tr}\left[\left(\gamma_k^{MN}\right)^T \bar{\gamma}_k^{MN}\right]. \quad (4)$$

The partial lower-walk overlap matrix γ_k^{MN} measures the overlap of all those CSFs that lead from the vacuum node of the Shavitt graph to node k, while the partial upper-walk overlap matrix $\bar{\gamma}_k^{MN}$ determines the overlap of all those CSFs that lead from the head node of the Shavitt graph to node k. Since the CSF coefficients are expressed as products of the arc factor matrices in Eq. (3), the partial overlap matrices may be computed recursively [30]. Assuming that every node of the Shavitt graph is characterized by f_{max} facets, the recursive computation of *all* matrices $\{\gamma_k^{MN}\}$ and $\{\bar{\gamma}_k^{MN}\}$ would scale as $\mathcal{O}(N_{row} f_{max}^3)$, where N_{row} is the number of nodes in the Shavitt graph and scales as $\mathcal{O}(N^2 n)$ for full-CI wave functions. Thus, computing the node densities for a GCF wave function with N_{GCF} expansion GCFs scales linearly with the number of orbitals $\mathcal{O}(N_{GCF}^2 N^2 f_{max}^3 n)$. A similar analysis in terms of *transition arc densities* reveals that evaluating arc densities with GCF wave functions has a similar scaling. It is noteworthy that the matrix product structure of the CSF coefficients in Eq. (3) is not required for computing graph densities and only facilitates their construction. Thus, the graph density concept is equally useful in the interpretation of traditional CI wave functions. However, in contrast to these traditional CI expansions for which the computational cost of analysis increases exponentially with the system size, the GCF wave function may be analyzed at polynomial cost. Although the examples presented in this work are small enough so that examining the qualitative features of the wave functions from both the CSF and graph density perspectives is tractable, the graph density concept will afford the analysis of future GCF wave functions for which the number of CSFs would be prohibitively large. Graph densities of a given electronic state at various points on a potential energy surface (PES) and densities of various electronic states can be compared in order to determine the essential differences in the wave functions. Graph densities can be monitored along a reaction path in order to characterize the changes in the wave function associated with barriers, wells, and other PES features.

3 Results

3.1 BeH_2 C_{2v} insertion pathway

The equilibrium geometry for collinear BeH_2 and H_2, as well as the saddle point geometry are optimized at the complete active-space self-consistent field (CASSCF) level using the $1a_1^2(2-4a_1, 1b_1, 1-2b_2)^4$ active space. All calculations employ Dunning's cc-pVTZ basis sets [51, 52], and C_{2v} symmetry is imposed by restricting the line segment connecting the center of mass of H_2 and Be to be perpendicular to the H_2 bond axis (Fig. 2). In terms of (R_{Be-H_2}, R_{H-H}), where R_{Be-H_2} is the distance between the beryllium atom and the center of mass of molecular hydrogen and R_{H-H} is the separation between the two hydrogen atoms, the equilibrium geometry of collinear BeH_2 is $(0.0a_0, 5.1142a_0)$ and the transition state geometry

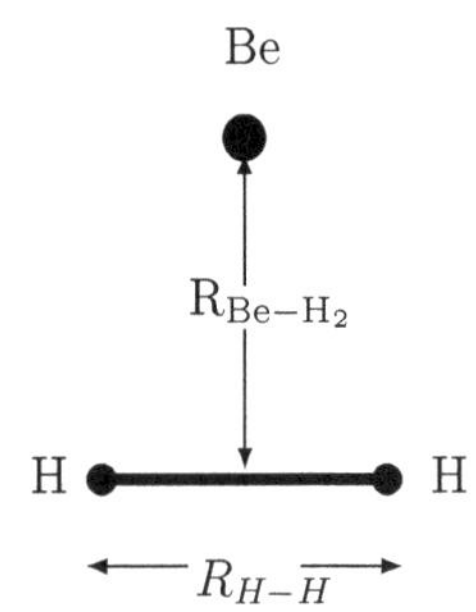

Fig. 2 Geometrical parameters used to describe the potential energy curve for the C_{2v} insertion pathway in BeH_2. The drawing is to scale for the transition state geometry

is given by ($2.5024a_0$, $2.1507a_0$). The asymptotic H_2 equilibrium bond distance is $1.4272a_0$. As in Ref. [53], the reaction path is defined as two straight-line segments in terms of R_{Be-H_2} and R_{H-H}. The first line segment passes through the collinear equilibrium geometry, the transition state geometry, and ($R^*_{Be-H_2}$, $1.4272a_0$), where $R^*_{Be-H_2} = 3.1133a_0$ is the intersection between the two line segments. For $R_{Be-H_2} \geq R^*_{Be-H_2}$, the second line segment is the horizontal line (R_{Be-H_2}, $1.4272a_0$).

The reaction $Be(^1S) + H_2\left(^1\Sigma_g^+\right) \rightarrow BeH_2(^1A_1)$ along the C_{2v} path is interesting because the 1^1A_1 ground state involves the avoided crossing between the two closed-shell configurations $1a_1^2 2a_1^2 3a_1^2$ and $1a_1^2 2a_1^2 1b_2^2$. The former configuration dominates at large Be – H_2 separations and corresponds to the ground-state fragments $Be(^1S)$ and $H_2\left(^1\Sigma_g^+\right)$, while the latter configuration is dominant near the collinear equilibrium geometry. The 2^1A_1 PES corresponds to the reaction $Be(^1P) + H_2\left(^1\Sigma_g^+\right) \rightarrow BeH_2(^1A_1)$. The shape of the 2^1A_1 PES is more complex as it involves three avoided crossings between the aforementioned closed-shell configurations as well as the two open-shell configurations $1a_1^2 2a_1 3a_1 1b_2^2$ and $1a_1^2 2a_1^2 3a_1 4a_1$. Thus, a flexible wave function expansion is necessary to describe the qualitative features of the PESs. State-averaged full-valence CASSCF calculations were used to optimize the $1a_1$ core orbital and the $(2-4a_1, 1b_1, 1-2b_2)$ active orbitals. Natural orbital resolution is imposed on the active orbital space. The subsequent state-averaged MFGCF calculations with $N_{GCF} = 2$ are carried out using the valence orbitals and employ the 6^4 full-CI Shavitt graph. The convergence of the MFGCF energy was examined by a sequence of calculations with a fixed maximum number of facets, f_{max}, for every node of the Shavitt graph. Future developments will explore the possibility of dynamically adjusting the number of facets for individual nodes of the Shavitt graph in order to satisfy overall accuracy requirements.

For both PESs, the MFGCF energies with $f_{max} = 2$ are within 2.1 mE_h of the CASSCF results, and the MFGCF energies with $f_{max} = 4$ (Fig. 3) are exact. The graph density plots in Fig. 4 demonstrate the changes in the electronic structure of the two lowest energy $^1A_1$ states. The core $1a_1$ orbital is excluded from the graphs. For the ground state, the graph densities at the collinear equilibrium geometry, $R_{Be-H_2} = 0a_0$, and near the asymptotic limit, $R_{Be-H_2} = 4.5a_0$, show that the wave function is dominated by a single configuration at either limit. It is also evident that there is a significant change in the wave function character, and the graph density at $R_{Be-H_2} = 2.5a_0$ displays the mixed nature of the wave function. The graph density

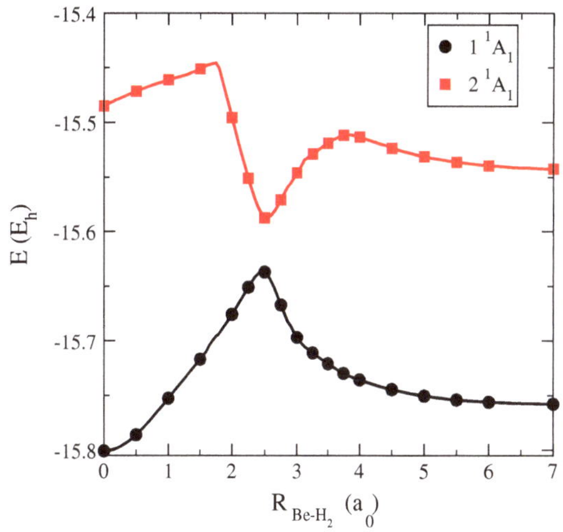

Fig. 3 Potential energy surfaces for the 1^1A_1 and 2^1A_1 states of BeH_2 along the C_{2v} insertion path. *Symbols* indicate the MFGCF results ($N_{GCF} = 2$ and $f_{max} = 4$) and *lines* denote the CASSCF energies

plots for the excited state show a more complicated structure. Near the collinear geometry the open-shell configuration $1a_1^2 2a_1 3a_1 1b_2^2$ dominates; however, for $R_{Be-H_2} \geq 1.5a_0$, the closed-shell configuration $1a_1^2 2a_1^2 3a_1^2$ becomes important. The graph densities for the 1^1A_1 and 2^1A_1 electronic states are essentially identical near the transition state geometry at $R_{Be-H_2} = 2.5a_0$. This is because both states are dominated by the same closed-shell configurations with nearly identical coefficient magnitudes at that geometry (Table 1). After the avoided crossing between the two closed-shell configurations, the $1a_1^2 2a_1^2 1b_2^2$ configuration is prominent in the 2^1A_1 state until $R_{Be-H_2} \leq 4.0a_0$. For longer R_{Be-H_2}, the wave function character is dominated by the open-shell configuration $1a_1^2 2a_1^2 3a_1 4a_1$, which corresponds to the $Be(^1P)$ and $H_2\left(\Sigma_g^+\right)$ dissociation fragments.

3.2 C_2 potential energy surface

The carbon dimer has received considerable attention due to its importance in interstellar chemistry and combustion processes [54]. From a theoretical point of view, the carbon dimer is interesting because of its unusual bonding, its complicated electronic structure even near the equilibrium geometry, and the presence of several low-lying excited states [55–59]. To assess the robustness of the state-averaged MFGCF approach, the PESs for the three lowest energy singlet electronic states of C_2 were calculated on the interval $1.5a_0 \leq R_{CC} \leq 6.0a_0$, where R_{CC} is the distance between the two carbon atoms. State-averaged CASSCF

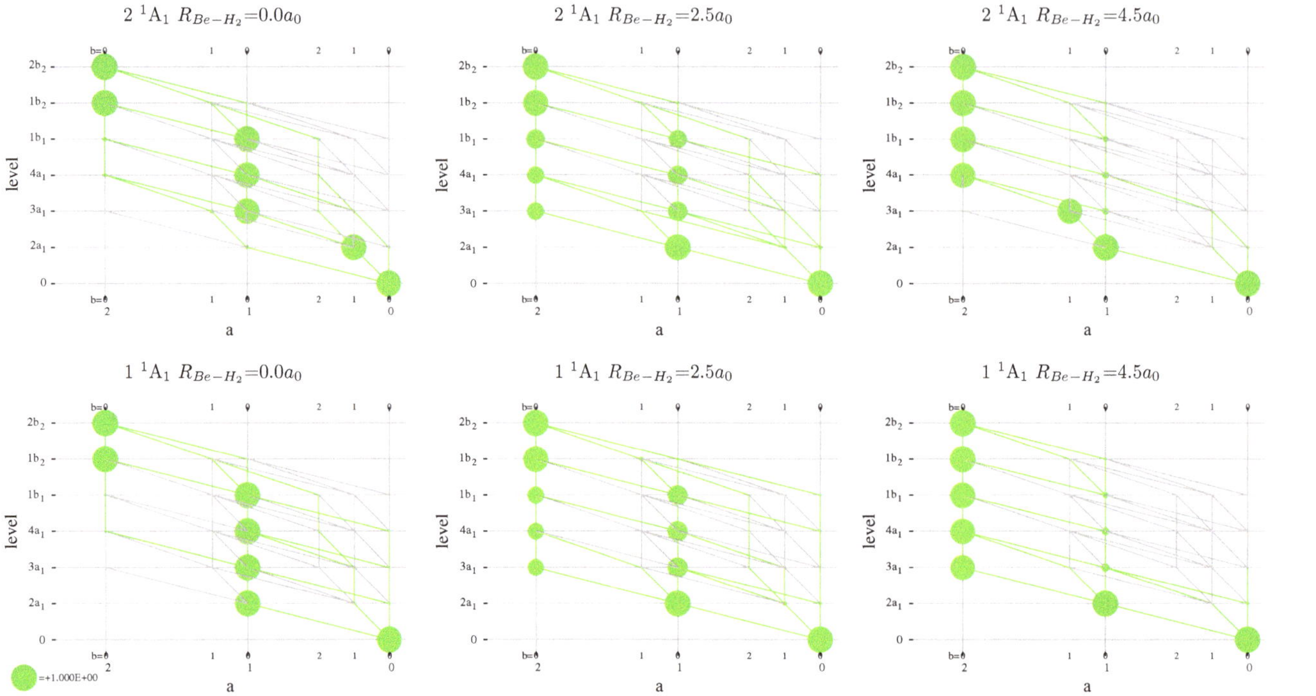

Fig. 4 Graph density plots for the 1^1A_1 and 2^1A_1 states of BeH_2 along the C_{2v} insertion reaction path. R_{Be-H_2} is defined as the distance between Be and H_2. Graph densities are computed from an MFGCF wave function with $N_{GCF} = 2$ and $f_{max} = 4$

Table 1 BeH_2 coefficients

Configuration	$R_{Be-H_2} = 0.0a_0$		$R_{Be-H_2} = 2.5a_0$		$R_{Be-H_2} = 4.5a_0$	
	1^1A_1	2^1A_1	1^1A_1	2^1A_1	1^1A_1	2^1A_1
$1a_1^2 2a_1^2 3a_1^2$	0.001	−0.000	−0.633	0.666	−0.947	0.024
$1a_1^2 2a_1^2 1b_2^2$	0.985	−0.000	0.727	0.647	0.186	−0.184
$1a_1^2 2a_1 3a_1 1b_2^2$	0.000	0.974	0.004	−0.044	−0.021	0.010
$1a_1^2 2a_1^2 3a_1 4a_1$	0.000	−0.153	0.031	−0.047	−0.087	−0.960

Coefficients of the two dominant closed-shell and the two open-shell configurations in the 1^1A_1 and 2^1A_1 electronic states of BeH_2 at selected R_{Be-H_2}. The coefficients are extracted from the $N_{GCF} = 2$ and $f_{max} = 4$ MFGCF wave function. −0.000 indicates a small negative number

calculations with equal weights for the three lowest energy ^1A_g states were used to optimize the core $1\sigma_g$ and $1\sigma_u$ and the $(2-3\sigma_g,\ 1\pi_{u,x},\ 1\pi_{u,y},\ 2-3\sigma_u,\ 1\pi_{g,x},\ 1\pi_{g,y})$ active orbitals within the cc-pVTZ basis set [51]. Natural orbital resolution is imposed on the active orbital space. State-averaged full-valence MFGCF calculations with equal weights for all states are performed with $N_{GCF} = 3$, and the convergence of the MFGCF wave function is examined through a series of calculations with $2 \leq f_{max} \leq 5$. The lowest energy $1\sigma_g$ and $1\sigma_u$ orbitals are frozen in all MFGCF calculations. Although dynamic correlation is neglected, previous work has demonstrated that a full-valence CASSCF approach captures the important qualitative features of the PESs for the three states considered here [57].

In terms of the nonparallelity error (NPE), Fig. 5 shows the accuracy of the MFGCF wave function as the maximum number of facets per node is systematically increased. The NPE is defined as the difference between the maximum and minimum deviations of the MFGCF energy relative to the state-averaged CASSCF results. For the three lowest energy singlet states, the CASSCF results are reproduced with mE_h accuracy for $f_{max} = 3$, and the NPE is <10 μE_h for $f_{max} = 5$. For all computed PESs, the maximum errors occur in the region $3.0a_0 \leq R_{CC} \leq 4.5a_0$ and are at least an order of magnitude larger than the minimum errors that occur near the dissociation region. This is in accord with the general observations from earlier studies [32] using the MFGCF method. The results for the $B^1\Delta_g$ state are consistently more accurate than those for the $^1\Sigma_g^+$

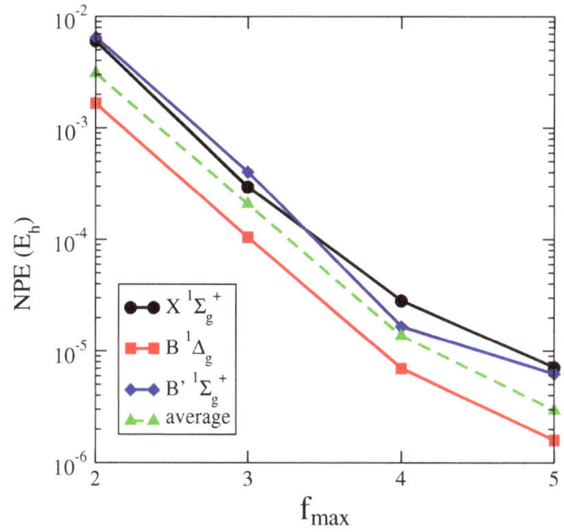

Fig. 5 The nonparallelity error in the MFGCF PESs for C_2 with $N_{GCF} = 3$ and $2 \leq f_{max} \leq 5$

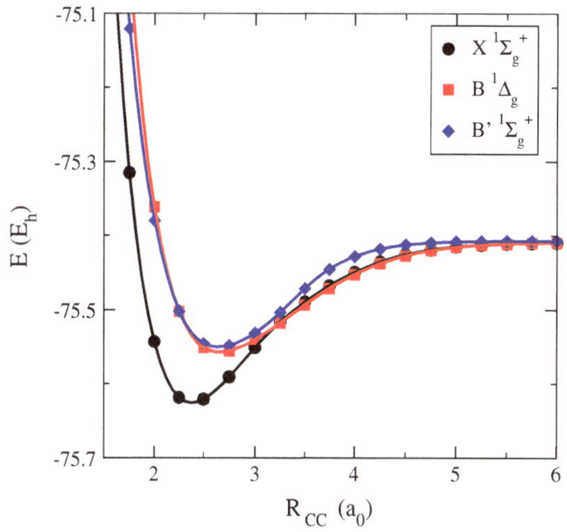

Fig. 6 Potential energy surfaces for the $X^1\Sigma_g^+$, $B'^1\Sigma_g^+$, and $B^1\Delta_g$ states of C_2. *Symbols* indicate the MFGCF results ($N_{GCF} = 3$ and $f_{max} = 5$) and *lines* denote the CASSCF energies

states. Although the decrease in the NPE between $f_{max} = 4$ and 5 for the $B'^1\Sigma_g^+$ state is smaller than expected, the NPE for the state-averaged energy follows the typical trend. Based on this, we believe that the unexpected shape of the NPE for the $B'^1\Sigma_g^+$ state is not likely to be the result of the optimization being trapped in a local minimum, but rather, it is a consequence of the state-averaged approach for computing excited-state energies.

The PESs obtained from the MFGCF wave function with $f_{max} = 5$ are shown in Fig. 6. For short bond distances, the $B'^1\Sigma_g^+$ state is lower in energy than the $B^1\Delta_g$ state; however, by $R_{CC} = 2.25a_0$, the energies of the two states switch. By $R_{CC} = 3.25a_0$, the energy of the $B^1\Delta_g$ state is below the energy of the $X^1\Sigma_g^+$ state. It is also at this point that the qualitative features of the PESs for the two $^1\Sigma_g^+$ states start to display unusual features. As a result of an avoided crossing, the energy of the $X^1\Sigma_g^+$ state flattens out while the PES of the $B'^1\Sigma_g^+$ state starts to increase more steeply. As reported by Boschen et al. [59], this avoided crossing results in a relatively inaccurate even-tempered Gaussian fit to the PES. For $R_{CC} > 5.5a_0$, the $X^1\Sigma_g^+$ state, in these state-averaged calculations, is once again below the $B^1\Delta_g$ state, and all three states dissociate to the ground-state $C(^3P)$ fragments.

The complicated electronic structure of the $X^1\Sigma_g^+$ and $B'^1\Sigma_g^+$ states at selected bond lengths in terms of graph densities is shown in Fig. 7. Contrary to the BeH_2 system where the ground-state wave function is dominated by a single configuration near the collinear equilibrium geometry, there are several configurations that are important in describing the $X^1\Sigma_g^+$ state at $R_{CC} = 2.25a_0$. In addition to the restricted Hartree–Fock (RHF) configuration $2\sigma_g^2 2\sigma_u^2 1\pi_{u,x}^2 1\pi_{u,y}^2$ (coefficient of 0.84), the coefficient of the doubly excited configuration $2\sigma_g^2 1\pi_{u,x}^2 1\pi_{u,y}^2 3\sigma_g^2$ in the MFGCF wave function for the $X^1\Sigma_g^+$ state is 0.39. Such a large coefficient for a molecule near its equilibrium geometry is rather unusual. Although the coefficient magnitudes of the two doubly excited configurations $2\sigma_g^2 2\sigma_u^2 1\pi_{u,x}^2 3\sigma_g^2$ and $2\sigma_g^2 2\sigma_u^2 1\pi_{u,y}^2 3\sigma_g^2$ are negligible in the $X^1\Sigma_g^+$ wave function, these configurations account for almost 80 % of the character of the $B'^1\Sigma_g^+$ wave function. The difference in the character of the $^1\Sigma_g^+$ states is clearly reflected by the wider distribution of significant node densities for the $B'^1\Sigma_g^+$ state. For $R_{CC} \geq 3.25a_0$, the appearance of the graph densities of the two $^1\Sigma_g^+$ states reverse, indicating the drastic change in the wave function character associated with the avoided crossing. Also note that for both states, the densities for nodes with $b \neq 0$ or 2 are negligible near the equilibrium bond distance, and the increased magnitude of these node densities for $R_{CC} = 6.0a_0$ indicate the open-shell fragments are near the dissociation limit. At the $R_{CC} = 3.50a_0$ and $R_{CC} = 6.0a_0$ geometries, the coefficient in the $X^1\Sigma_g^+$ state of the $2\sigma_g^2 2\sigma_u^2 1\pi_{u,x}^2 1\pi_{u,y}^2$ configuration, which is dominant at $R_{CC} = 2.25a_0$, is 0.29 and 0.19, respectively. This small contribution is evident from the small (4,0,1) node density in Fig. 7. Single reference methods have difficulty describing accurately this ground-state dissociation because of this drastic change in the nature of the wave function.

For the $B^1\Delta_g$ state, symmetry requires that the coefficient of the $2\sigma_g^2 2\sigma_u^2 1\pi_{u,x}^2 1\pi_{u,y}^2$ configuration be zero. The

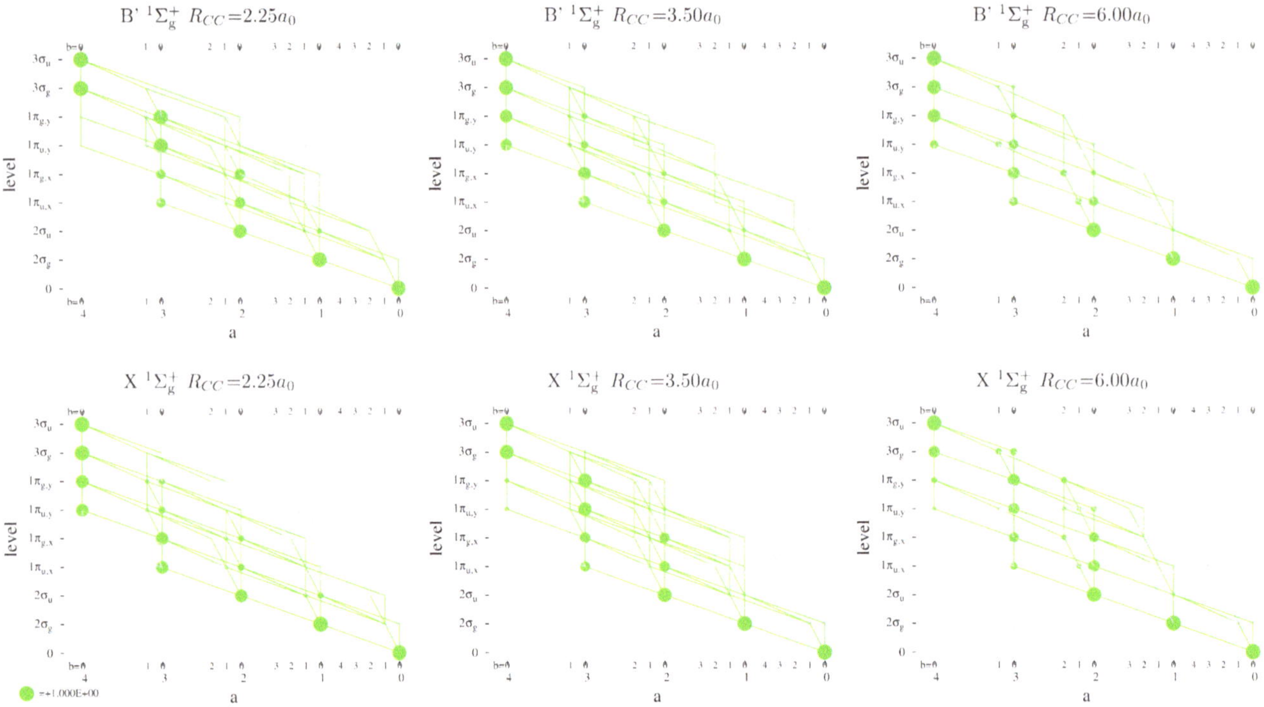

Fig. 7 Graph density plots for the $X^1\Sigma_g^+$ and $B'^1\Sigma_g^+$ states of C_2 at $R_{CC} = 2.25a_0$, $3.50a_0$, and $6.0a_0$. Graph densities are computed from an MFGCF wave function with $N_{GCF} = 3$ and $f_{max} = 5$

vanishing densities of nodes (4, 0, 0) and (4, 0, 1) in the graph density of the $B^1\Delta_g$ state at $R_{CC} = 2.25a_0$ shown in Fig. 8 are consistent with this restriction. Although the difference in character of the $B^1\Delta_g$ and $X^1\Sigma_g^+$ states is apparent by inspection, this difference is less obvious upon comparing the density plot of the $B^1\Delta_g$ state to that of the $B'^1\Sigma_g^+$ state since, at this geometry, the $2\sigma_g^2 2\sigma_u^2 1\pi_{u,x}^2 1\pi_{u,y}^2$ configuration is negligible in both states. Indeed, distinguishing these two states requires the inspection of the coefficients for the $2\sigma_g^2 2\sigma_u^2 1\pi_{u,x}^2 3\sigma_g^2$ and $2\sigma_g^2 2\sigma_u^2 1\pi_{u,y}^2 3\sigma_g^2$ configurations in the MFGCF wave function. Within both states, the magnitude of the coefficients should be the same; however, the signs should be the same for the $^1\Sigma_g^+$ states and opposite for the $B^1\Delta_g$ state. In addition to affording the identification of states with different symmetries, these coefficients can also serve as a measure of the symmetry contamination in the approximate MFGCF wave function. The magnitude of the symmetry error defined by $2(|c_1| - |c_2|)/(|c_1| + |c_2|)$, where c_1 and c_2 are the coefficients of the aforementioned configurations, does not exceed 10^{-3} for the entire $B'^1\Sigma_g^+$ PES. Although the symmetry errors for the $X^1\Sigma_g^+$ and $B^1\Delta_g$ states remain below this threshold for most of the PESs, the symmetry errors peak near $R_{CC} = 3.25a_0$ and $5.75a_0$. Not too surprisingly, these bond distances correspond to points along the PESs where the energies of the two states cross.

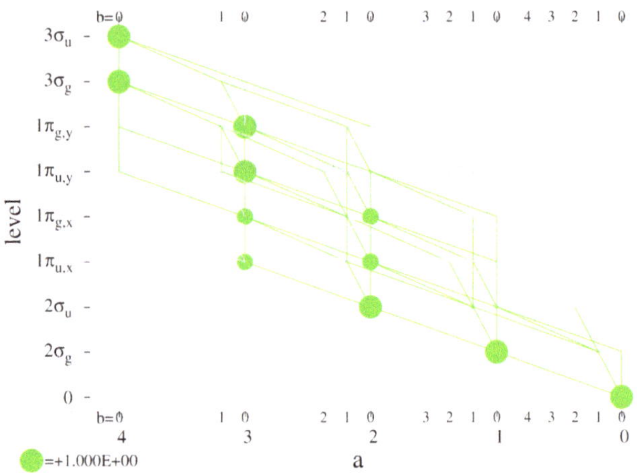

Fig. 8 Graph density plot for the $B^1\Delta_g$ state of C_2 at $R_{CC} = 2.25a_0$. The graph density is computed from an MFGCF wave function with $N_{GCF} = 3$ and $f_{max} = 5$

3.3 H_8 model system

The H_8 model of Jankowski et al. [60] consists of four interacting stretched H_2 molecules arranged in an elongated octagon. All four molecular bond distances are fixed at $2.0a_0$. Symmetric displacement of exactly two H_2 molecules from opposite sides distorts the limiting D_{8h} regular octagonal configuration into various D_{2h} elongated octagonal configurations. The model parameter α measures the displacement as shown in Fig. 9. This system has been used extensively for benchmarking multireference coupled-cluster methods [61–64]. The minimal basis set (MBS) of Huzinaga [65] is employed in most studies and is used herein. Investigations performed with double zeta and double zeta plus polarization basis sets have concluded that the MBS model captures all the essential features [66].

For the range $0 < \alpha \leq 1.0a_0$ the eight molecular orbitals (MOs) in ascending energy order are $1a_{1g}, 1b_{2u}, 1b_{3u}, 2a_{1g}, 1b_{1g}, 2b_{3u}, 2b_{2u}$, and $2b_{1g}$. Throughout this range the $^1A_{1g}$ ground-state RHF wave function is $1a_{1g}^2 1b_{2u}^2 1b_{3u}^2 2a_{1g}^2$. This configuration dominates in the exact wave function for $0.1a_0 \leq \alpha$; its coefficient is 0.83 at $\alpha = 0.1a_0$ and 0.94 at $\alpha = 1.0a_0$. However, as α approaches 0 the $1b_{2u}$ and $1b_{3u}$ MOs approach degeneracy, the highest occupied MO (HOMO) $2a_{1g}$ and the lowest unoccupied MO (LUMO) $1b_{1g}$ become increasingly quasidegenerate, and the configuration with a double excitation from the HOMO to the LUMO $1a_{1g}^2 1b_{2u}^2 1b_{3u}^2 1b_{1g}^2$ becomes increasingly important. The exact wave function is predominantly a two-configuration wave function at $\alpha = 0$ with coefficients of magnitude 0.67 and opposite signs for $1a_{1g}^2 1b_{2u}^2 1b_{3u}^2 2a_{1g}^2$ and $1a_{1g}^2 1b_{2u}^2 1b_{3u}^2 1b_{1g}^2$.

Figure 10 contains a sequence of graph density plots computed from MFGCF expansions that have been converged to the full-CI limit. The order of levels in these Shavitt graphs is the same as the MO order above. Of particular interest are the node densities at the $2a_{1g}$ level.

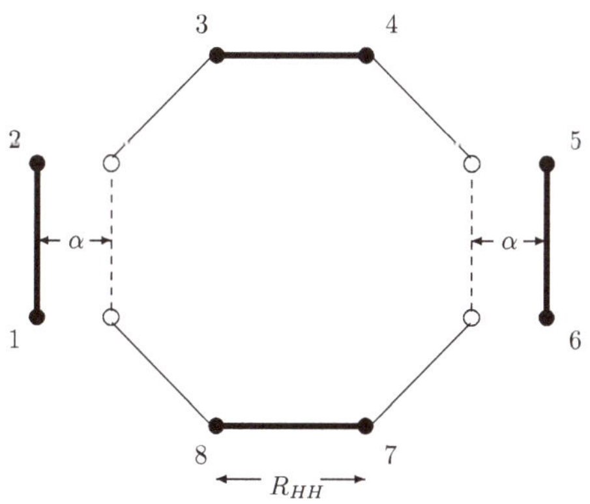

Fig. 9 The H_8 model nuclear configurations and parameter definitions. All four H_2 molecular bond distances R_{HH} are fixed at $2.0a_0$. α is the displacement from a regular octagon to an elongated octagon of D_{2h} symmetry. The drawing is to scale for α equal to $1.0a_0$. In the extended H_8 model, all four H_2 fragments are displaced symmetrically from the center of mass along the D_{4h} dissociation path

At $\alpha = 0.0001a_0$ the densities of nodes $(4, 0, 0)$ and $(3, 0, 1)$ are approximately equal showing the two-configuration character. At $\alpha = 0.1a_0$ density has shifted from $(3, 0, 1)$ to $(4, 0, 0)$, and at $\alpha = 1.0a_0$ the density at $(4, 0, 0)$ is approaching unity indicating the single-configuration character. Thus, the progression of node densities in the

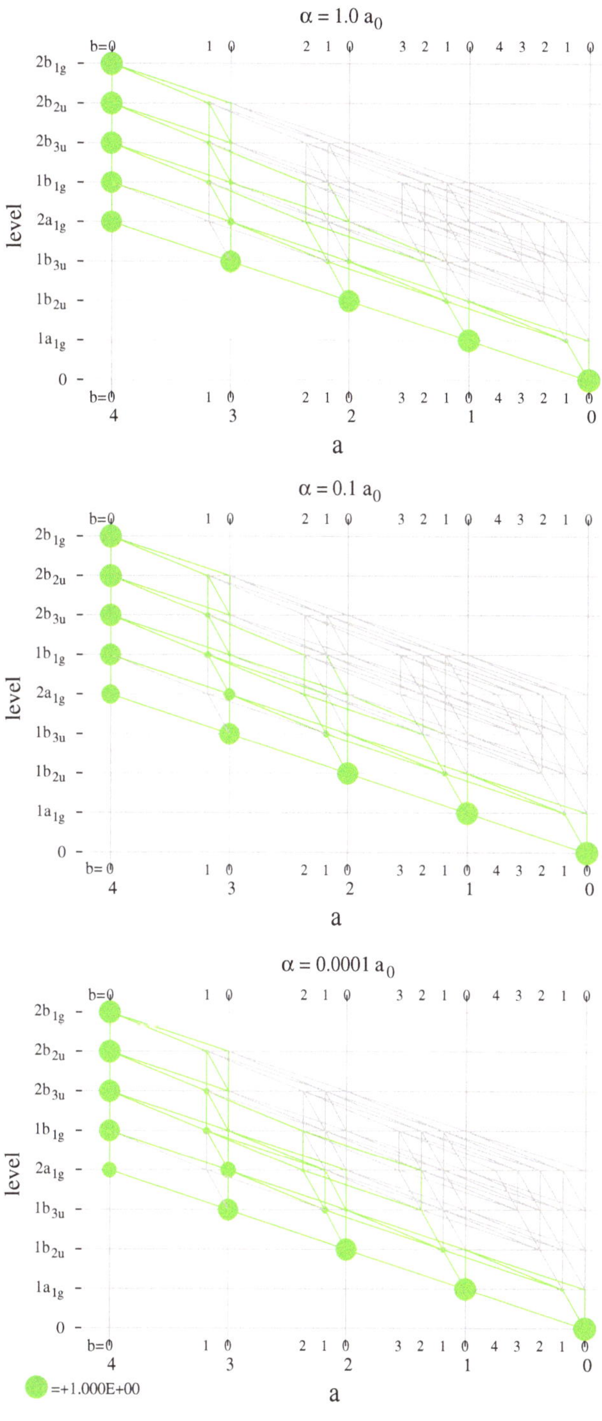

Fig. 10 Graph density plots for the ground state of the H_8 model at $\alpha = 0.0001, 0.1$, and $1.0a_0$

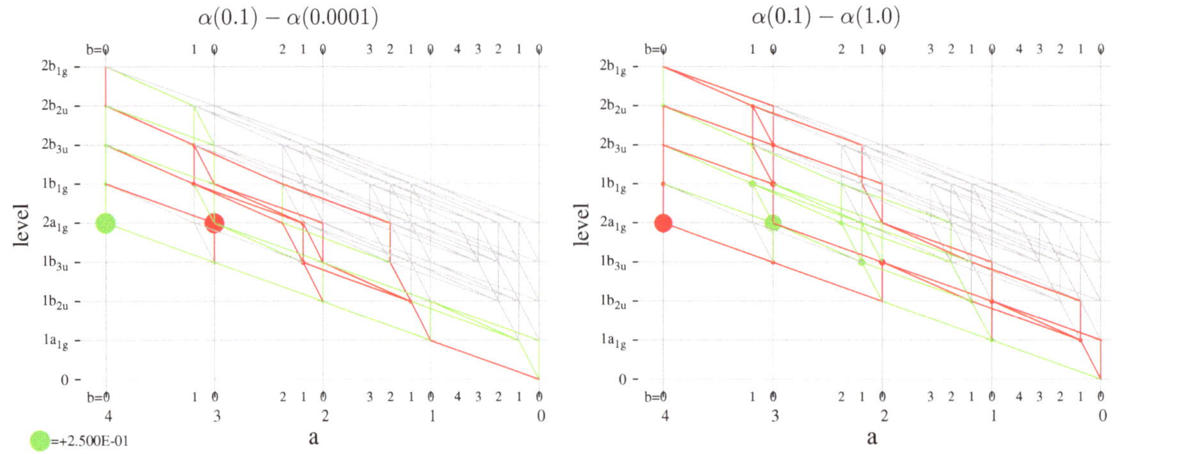

Fig. 11 Graph density difference plots for the ground state of the H_8 model with respect to the $\alpha = 0.1a_0$ geometry. Positive density differences are *green* and negative density differences are *red*

figure reveals the fundamental change in the character of the wave function. The graph density difference plots in Fig. 11 highlight the shift in density. In the left-hand plot the $\alpha = 0.0001a_0$ density is subtracted from the $\alpha = 0.1a_0$ density, and in the right-hand plot the $\alpha = 1.0a_0$ density is subtracted from the $\alpha = 0.1a_0$ density. Significant positive density differences are indicated in green, significant negative density differences are indicated in red, and small differences of any sign are gray. The sum of the density differences at any level equals zero. Thus, relatively greater density for $\alpha = 0.1a_0$ is green in both plots. The left-hand plot shows density shifting to node $(3, 0, 1)$ in the $\alpha = 0.0001a_0$ wave function relative to the $\alpha = 0.1a_0$ wave function which indicates an increasing contribution of the $1a_{1g}^2 1b_{2u}^2 1b_{3u}^2 1b_{1g}^2$ configuration. In contrast, the right-hand plot shows density shifting to node $(4, 0, 0)$ in the $\alpha = 1.0a_0$ wave function relative to the $\alpha = 0.1a_0$ wave function which indicates an increasing contribution of the $1a_{1g}^2 1b_{2u}^2 1b_{3u}^2 2a_{1g}^2$ configuration.

3.4 Extended H_8 model system

A straightforward extension of the H_8 model of Jankowski et al. [60] (see Sect. 3.3) is displacement of all four H_2 molecules [64]. In particular to study size extensivity, all H_2 molecules are symmetrically displaced to infinity or as a practical matter in this work to the D_{4h} dissociation geometry of $\alpha = 994a_0$. A C_1 localized orbital (LO) set computed from the symmetrically orthonormalized atomic orbitals [18] is employed. This set corresponds to the orthonormal MOs that are closest in the least squares sense to the atomic orbitals. Thus, these molecular calculations are performed without a preliminary SCF or MCSCF orbital optimization step. The order of levels in the Shavitt graph is such that the LOs in each H_2 are adjacent. Given these conditions—noninteracting singlet fragments, a LO basis set, and a fragment ordered Shavitt graph—a single facet single GCF is formally size extensive. Figure 12 shows the graph density plot computed from a GCF expansion with $N_{GCF} = 1$ and $f_{max} = 1$ for the ground state of one H_2 in its LO set, and Fig. 13 contains the corresponding plot from an analogous GCF expansion for the ground state of H_8 in this extended model with its LO set and noninteracting H_2 molecules. The H_8 plot is clearly composed of four H_2 plots joined head to tail demonstrating the product nature of the wave function, the gray nodes in Fig. 13 have zero density verifying that the H_2 fragments are noninteracting, and for these GCF expansions the energy of the composite system H_8 is $-4.411215155407E_h$ exactly four times that of the fragment H_2 energy calculated at $-1.102803788852E_h$.

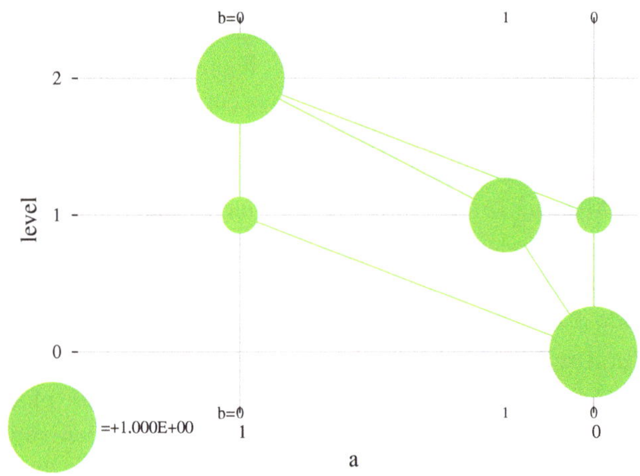

Fig. 12 Graph density plot for the ground state of a single H_2 in the extended H_8 model using a localized orbital basis

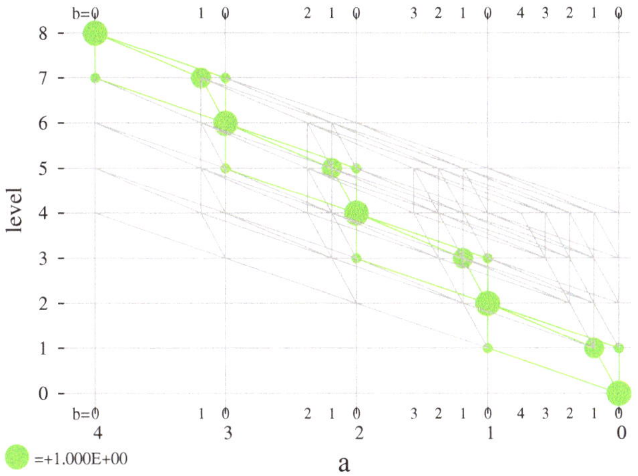

Fig. 13 Graph density plot for the ground state of H_8 in the extended model using a localized orbital basis and where the H_2 molecules are well separated and thus noninteracting

4 Conclusion

Among the goals of electronic structure theory are the prediction of energies of reactions, relative energies of conformers, and energies of electronic states, as well as other molecular properties. As a result of the instantaneous repulsion between electrons, accurate predictions demand a complicated wave function structure that often obfuscates an equally important aspect of electronic structure theory to provide qualitative chemical insight into bonding and potential energy surface features. The MFGCF method allows for both goals to be addressed simultaneously.

The MFGCF method, in addition to offering the possibility of reducing the exponential scaling of the exact wave function with system size, also affords the efficient manipulation and analysis of the wave function. Although the MFGCF wave function has a quite complicated recursive structure, the underlying Shavitt graph coupled with the graph density concept provides an intuitive approach for analyzing the character of the wave function. Since graph densities retain important information related to the spin-coupling structure of the wave function, they complement existing analysis methods based on the density matrix or Mulliken and Löwdin populations. Even though this concept may be employed with traditional wave function methods, it is particularly useful in the context of the MFGCF wave function as the recursive construction of the graph densities scales polynomially with system size.

In this work, the MFGCF method was applied to model systems for studying the changes in wave function character from a graph density perspective. The graphical nature of the analysis allows a very intuitive avenue for visualizing the complex changes in the electronic structure. To wit, for the extended H_8 model, in which the four H_2 molecule fragments are noninteracting, the product structure of the wave function in terms of fragment wave functions is readily apparent and demonstrates the correct behavior of the MFGCF wave function with system size. For the systems studied, the sizes of the wave functions are small enough so that the more detailed analyses of the wave functions in terms of CSF coefficients are tractable. The qualitative features inferred from the node and arc densities are in good agreement with the quantitative analyses based on CSF coefficients. Nonetheless, it should be recognized that the graph density concept is not always sufficient for a detailed analysis of a wave function. Since the graph density is formally based on the magnitude of the CSF coefficients, potentially important phase information is lost. Near avoided crossings, such as in the BeH_2 system examined herein, two states may have nearly indistinguishable graph densities. To distinguish the $X^1\Sigma_g^+$ and $B'^1\Sigma_g^+$ states of C_2 from the $B^1\Delta_g$ states also requires a more quantitative analysis of the wave function. In both these cases, the actual CSF coefficients (including sign) are needed. For larger systems, should the use of CSF or Slater determinant coefficients be desirable, efficient routines for extracting the coefficients of user-specified CSFs or Slater determinants have been developed [32, 37].

The concept of graph density is not only a complementary wave function analysis tool, but also offers possibilities for improving the optimization of the MFGCF wave function parameters. In the current approach, the number of facets for the individual nodes is restricted to not exceed some user-specified value f_{max}. Clearly, this is not the optimal approach as nodes that contribute significantly to the wave function or that have a more multiconfigurational nature may require more facets than other nodes in order to achieve a specified wave function accuracy. A method for identifying these nodes during the optimization is needed. Algorithms similar to the decimation procedure in density matrix renormalization group methods [48–50] or concepts related to node densities, namely natural facet densities, may be used to develop reliable tools for identifying these important nodes. Developments along these lines and the implementation of an optimization method with dynamic facet counts will be explored in future work.

Acknowledgments This work was supported by the Office of Basic Energy Sciences, Division of Chemical Sciences, Geosciences, and Biosciences, U.S. Department of Energy, under contract DE-AC02-06CH11357. G.G. was supported by an award from the Research Corporation for Science Advancement and a grant to Gonzaga University from the Howard Hughes Medical Institute through the Undergraduate Science Education Program. S.R.B. acknowledges the use of computational facilities at the Ohio Supercomputer Center.

References

1. Shavitt I (1977) Int J Quantum Chem S11:131
2. Shavitt I (1978) Int J Quantum Chem S12:5
3. Lischka H, Shepard R, Brown FB, Shavitt I (1981) Int J Quantum Chem S15:91
4. Shavitt I (1981) In: Hinze J (ed) The unitary group for the evaluation of electronic energy matrix elements, lecture notes in chemistry, vol 22. Springer, New York, pp 51–99
5. Shavitt I (1988) In: Truhlar DG (ed) Mathematical frontiers in computational chemical physics, the IMA volumes in mathematics and its application. Springer, New York, pp 300–349
6. Paldus J (1974) J Chem Phys 61:5321
7. Paldus J, Boyle MJ (1980) Phys Scr 21:295
8. Paldus J (1981) In: Hinze J (ed) The unitary group for the evaluation of electronic energy matrix elements, lecture notes in chemistry. Springer, New York, pp 1–50
9. Paldus J (1988) In: Truhlar DG (ed) Mathematical frontiers in computational chemical physics, the IMA volumes in mathematics and its application. Springer, New York, pp 262–299
10. Brooks BR, Schaefer HF III (1979) J Chem Phys 70:5092
11. Brooks BR, Laidig WD, Saxe P, Handy NC, Schaefer HF III (1980) Phys Scr 21:312
12. Aquilante F, Vico LD, Ferré N, Ghigo G, Malmqvist PÅ, Neogrády P, Pedersen TB, Pitonak M, Reiher M, Roos BO, Serrano-Andrés L, Urban M, Veryazov V, Lindh R (2010) J Comput Chem 31:224
13. Shepard R, Shavitt I, Pitzer RM, Comeau DC, Pepper M, Lischka H, Szalay PG, Ahlrichs R, Brown FB, Zhao JG (1988) Int J Quantum Chem 22:149
14. Lischka H, Shepard R, Pitzer RM, Shavitt I, Dallos M, Müller T, Szalay PG, Seth M, Kedziora GS, Yabushita S, Zhang Z (2001) Phys Chem Chem Phys 3:664
15. Shepard R (1994) In: Malli GL (ed) Relativistic and electron correlation effects in molecules and solids, NATO Advanced Science Institutes. Plenum Press, New York, pp 447–460
16. Davidson ER (1975) J Comput Phys 17:87
17. Mulliken RS (1955) J Chem Phys 23:1833
18. Löwdin PO (1950) J Chem Phys 18:365
19. Weinhold F (1998) In: Schleyer PVR, Allinger NL, Clark T, Gasteiger J, Kollman PA, Schaefer HF III, Schreiner PR (eds) Encyclopedia of computational chemistry. Wiley, Chichester, pp 1792–1811
20. Ivanic J, Ruedenberg K (2002) Theor Chem Acc 107:220
21. Buenker RJ, Peyerimhoff SD (1974) Theor Chim Acta 35:33
22. Abrams ML, Sherrill DC (2002) J Chem Phys 118:1604
23. Shavitt I, Rosenberg BJ, Palalikit S (1976) Int J Quantum Chem Symp 10:33
24. Barr TL, Davidson ER (1970) Phys Rev A 1:644
25. Klopper W, Noga J, Koch H, Helgaker T (1997) Theor Chem Acc 97:164
26. Sosa C, Geersten J, Trucks GW, Barlett RJ, Franz JA (1989) Chem Phys Lett 159:148
27. Taube AG, Bartlett RJ (2005) Collect Czechoslov Chem Commun 70:837
28. Landau A, Khistyaev K, Dolgikh S, Krylov AI (2010) J Chem Phys 132:014109
29. DePrince AE, Sherrill DC (2013) J Chem Theory Comput 9:293
30. Shepard R (2005) J Phys Chem A 109:11629
31. Shepard R, Minkoff M, Brozell SR (2007) Int J Quantum Chem 107:3203
32. Shepard R, Gidofalvi G, Brozell SR (2014) The multifacet graphically contracted function method: I. Formulation and implementation. J Chem Phys (in press)
33. Shepard R, Gidofalvi G, Brozell SR (2014) The multifacet graphically contracted function method: II. A general procedure for the parameterization of orthogonal matrices and its application to arc factors. J Chem Phys (in press)
34. Shepard R (2006) J Phys Chem A 110:8880
35. Shepard R, Minkoff M (2006) Int J Quantum Chem 106:3190
36. Brozell SR, Shepard R, Zhang Z (2007) Int J Quantum Chem 107:3191
37. Gidofalvi G, Shepard R (2009) J Comput Chem 30:2414
38. Gidofalvi G, Shepard R (2009) Int J Quantum Chem 109:3552
39. Brozell SR, Shepard R (2009) J Phys Chem A 113:12741
40. Shepard R, Gidofalvi G, Hovland PD (2010) Int J Quantum Chem 110:2938
41. Gidofalvi G, Shepard R (2010) Mol Phys 108:2717
42. Judd BR (1963) Operator techniques in atomic spectroscopy. McGraw-Hill, New York
43. White SR (1992) Phys Rev Lett 69:2863
44. Rissler J, Noack RM, White SR (2006) Chem Phys 323:519
45. Legeza Ö, Sólyom J (2003) Phys Rev B 67:195116
46. Boguslawski K, Tecmer P, Barcza G, Legeza Ö, Reiher M (2013) J Chem Theory Comput 9:2959
47. Kurashige Y, Chan GKL, Yanai T (2013) Nat Chem 5:660
48. Schollwöck U (2011) Ann Phys 326:96
49. Sharma S, Chan GKL (2012) J Chem Phys 136:124121
50. Wouters S, Limacher PA, Neck DV, Ayers PW (2012) J Chem Phys 136:134110
51. Dunning TH Jr (1989) J Chem Phys 90:1007
52. Dunning TH Jr (1994) J Chem Phys 100:2975
53. Purvis GD III, Shepard R, Brown FB, Bartlett RJ (1983) Int J Quantum Chem 23:835
54. Orden AV, Saykally RJ (1998) Chem Rev 98:2313
55. Hoffmann R (1995) Am Sci 83:309
56. Abrams ML, Sherrill CD (2004) J Chem Phys 121:9211
57. Sherrill CD, Piecuch P (2005) J Chem Phys 122:124104
58. Booth GH, Cleland D, Thom AJW, Alavi A (2011) J Chem Phys 135:084104
59. Boschen JS, Theis D, Ruedenberg K, Windus TL (2014) Theor Chem Acc 133:1425
60. Jankowski K, Meissner L, Wasilewski J (1985) Int J Quantum Chem 28:931
61. Jankowski K, Paldus J, Wasilewski J (1991) J Chem Phys 95:3549
62. Piecuch P, Adamowicz L (1994) J Chem Phys 100:5792
63. Evangelista FA, Allen WD, Schaefer HF III (2006) J Chem Phys 125:154113:1
64. Pape D, Hanrath M (2012) Chem Phys 401:157
65. Huzinaga S (1965) J Chem Phys 42:1293
66. Kucharski SA, Balková A, Szalay PG, Bartlett RJ (1992) J Chem Phys 97:4289

Theor Chem Acc (2014) 133:1522
DOI 10.1007/s00214-014-1522-5

REGULAR ARTICLE

Aspects of size extensivity in unitary group adapted multi-reference coupled cluster theories: the role of cumulant decomposition of spin-free reduced density matrices

Rahul Maitra · Debalina Sinha · Sangita Sen · Debashis Mukherjee

Received: 9 February 2014 / Accepted: 14 June 2014 / Published online: 11 July 2014

Abstract We present in this paper a comprehensive study of the various aspects of size extensivity of a set of unitary group adapted multi-reference coupled cluster (UGA-MRCC) theories recently developed by us. All these theories utilize a Jeziorski–Monkhorst (JM) inspired spin-free cluster Ansatz of the form $|\Psi\rangle = \sum_\mu \Omega_\mu|\phi_\mu\rangle c_\mu$ with $\Omega_\mu = \{\exp(T_\mu)\}$, where T_μ is expressed in terms of spin-free generators of the unitary group $U(n)$ for n-orbitals with the associated cluster amplitudes. $\{\ldots\}$ indicates normal ordering with respect to the common closed shell *core* part, $|0\rangle$, of the model functions, $\{\phi_\mu\}$ which is taken as the vacuum. We argue and emphasize in the paper that maintaining size extensivity of the associated theories is consequent upon (a) connectivity of the composites, G_μ, containing the Hamiltonian H and the various powers of T connected to it, (b) proving the connectivity of the MRCC equations which involve not only G_μs but also the associated connected components of the spin-free reduced density matrices (RDMs) obtained via their cumulant decomposition and (c) showing the extensivity of the cluster amplitudes for non-interacting groups of orbitals and eventually of the size-consistency of the theories in the fragmentation limits. While we will discuss the aspect (a) above rather briefly, since this was amply covered in our earlier papers, the aspect (b) and (c), not covered in detail hitherto, will be covered extensively in this paper. The UGA-MRCC theories dealt with in this paper are the spin-free analogs of the state-specific and state-universal MRCC developed and applied by us recently.We will explain the unfolding of the proof of extensivity by analyzing the algebraic structure of the working equations, decomposed into two factors, one containing the composite G_μ that is connected with the products of cumulants arising out of the cumulant decomposition of the RDMs and the second term containing some RDMs which is disconnected from the first and can be factored out and removed. This factorization ultimately leads to a set of connected MRCC equations. Establishing the extensivity and size-consistency of the theories requires careful separation of truly extensive cumulants from the ones which are a measure of spin correlation and are thus connected but not extensive. We have discussed in detail, using diagrams, the factorization procedure and have used suitable example diagrams to amplify the meanings of the various algebraic quantities of any diagram. We conclude the paper by summarizing our findings and commenting on further developments in the future.

Keywords Unitary group adaptation · Size-extensive SS and SU MRCC · Spin-free RDM · Spin-free cumulants · Connectivity of cumulants · Spin correlation of cumulants

Rahul Maitra and Debalina Sinha have contributed equally to this work.

Dedicated to the memory of Professor Isaiah Shavitt and published as part of the special collection of articles celebrating his many contributions.

R. Maitra · D. Sinha · S. Sen · D. Mukherjee (✉)
Raman Center for Atomic, Molecular and Optical Sciences, Indian Association for the Cultivation of Science, Kolkata 700 032, India
e-mail: pcdm@iacs.res.in

Present Address:
R. Maitra
Frick Chemistry Laboratory, Department of Chemistry, Princeton University, Princeton, NJ 08544, USA

1 Introduction

Since a non-relativistic Hamiltonian, H, for the many-electron problem is spin-free, it's eigenstates are all characterized by fixed eigenvalues for the operator S^2. This is why it is conceptually and computationally convenient to describe approximate eigenstates of H as a combination of basis which are anti-symmetric functions with a fixed eigenvalue of S^2. There is an enormous literature describing the various spin-adaptation procedures for a many-electron function. Among them, three methods, all group theoretic in origin but using different groups, stand out as: (a) permutation group (b) SU2 group and (c) unitary group.

Approaches based on these groups have been studied extensively in the context of electron correlation problems, in particular for configuration interaction (CI)-based formalisms. In spinorbital-based formalisms, the permutation group [1–7] or SU2 [8] group adaptations fit in naturally in a CI framework which requires spin adaptation of the CI coefficients appearing in a linear manner in the wave function. Both these approaches have found wide use. It is rather difficult, however, to extend these spin adaptations to a many-body formalism in a spinorbital basis using nonlinear cluster Ansatz of the wave operator such as $\exp(T)$. Even though spin adaptation can be achieved by using proper coupling coefficients for the operators T, their powers still lead to spin contamination. The Unitary Group Approach (UGA), on the other hand, has the property that a spin-free operator can be written in terms of spin-free generators of the unitary group whose powers also remain spin free. Hence, a coupled cluster representation of the wave operator for a spin-free T is spin free. For developing spin-free many-body theories, a unitary group approach (UGA) is thus the most natural choice. We may mention here that a UGA path to other nonlinear Ansätze for a wave function has also been explored in recent years [9–12]. From now on, we will refer to the spin-free generators of the unitary group as generators.

The application of unitary group to nuclear many-fermion problem involving Hamiltonians which are spin and iso-spin independent was developed by Moshinsky and Seligman [13–15]. The spin-free formulation for the spin-independent many-electron Hamiltonian was considered earlier by Matsen [16, 17]. But it is only after the advent of group theoretic simplifications for a Hamiltonian which is spin independent (and having no additional quantum number like iso-spin) via a unitary group approach by Paldus [18], the concomitant graphical representation by Shavitt [19, 20] of the unitary group adapted spatial Gel'fand states and of the associated matrix elements of generators between two Gel'fand states that the true revolution in compact and efficient unitary group-based spin-free theories was achieved. Subsequently, more efficient and general formulae for matrix element evaluation were developed, comprehensive accounts of which can be found in Refs. [21, 22]. The advantages of the unitary group formulation lie in the efficient evaluation of the possible matrix elements involving generators for which, at the current stage of development of quantum chemistry, there are already very efficient computer codes [23–29] which any new many-electron theory can embed in its own code for its own unitary spin-adaptation strategy. UGA theories have been utilized for CI [23], MCSCF [30–32] and spin-free single CSF-based UGA CC [33, 34]. There was also a prescient proposal for UGA-SUMRCC [33] using the Jeziorski–Monkhorst (JM)-type multi-exponential Ansatz.

It is important to mention here that for any practical implementation of a UGA scheme, one need not demand that the n-electron CSF's be adapted to $U(n)$. Rather, it is expedient to demand that the CSF's be adapted to the subgroup of $U(n), \boldsymbol{u}(n) = U(n_c) \otimes U(n_a) \otimes U(n_v)$, where n_c, n_a, and n_v are, respectively, the number of core, active, and virtual orbitals. This is simply due to the physical requirement that the maximum invariance of an approximate function that one may practically impose is the invariance of the function and the energy with respect to separate unitary transformation among core, active, and virtual orbitals. This was indeed done in the UGA CC papers by Paldus and others [33, 34]. In particular, both these papers have used the $\boldsymbol{u}(n)$-adapted scalar tensor generators in their choice of excitation operators. We also point out two references [35, 36] in this context where useful discussions of other UGA-based approaches and their interrelation can be found. Reference [36] has also presented in considerable detail a number of approximately spin-adapted CC approaches and their relationship with UGA CC. We also point out that Li and Paldus have applied the UGA CC method to many problems (see e.g., Refs. 39–48 in [37]).

As emphasized in the first para of our introduction, a spin-adapted formulation of any coupled cluster formalism using spinorbital basis is beset by the problem of spin contamination of the function for terms containing nonlinear powers of the cluster operators acting on the reference function. Nevertheless, progress was made in the development of such spin adapted formalisms by imposing on the wave function the constraint that the final function be an eigen function of the total spin with a fixed eigenvalue [38, 39]. An approximate spin-adapted formulation where the expectation value of S^2 is constrained to a fixed value corresponding to a given spin was considered somewhat earlier [40]. Nooijen and Bartlett have proposed a unitary group-based spin-adapted open-shell coupled cluster for a single reference function [41]. We should also mention the spin-adapted cluster (SAC) expansion formalism of Nakatsuji [42, 43] in this context. Here also, the linear term of the cluster expansion of the wave function is spin adapted although the full wave function is not.

Unfortunately, all these developments are rather difficult to generalize for a full exponential type of wave operator.

Recently, we have published the development and implementation of unitary group adapted state specific (SS) and state-universal (SU) multi-reference coupled cluster theories which are called the UGA-SSMRCC [44, 45] and UGA-SUMRCC theories [44, 46–48], respectively. This paper will focus on an exhaustive treatment of the proof of connectivity and eventually of extensivity of the MRCC equations of the spin-free formulations of state-specific (SS) [44, 45] and the state-universal (SU) [44, 46–48] variety. We therefore mention in this context that our recent spinorbital-based parent formulations are the most relevant, and we would not dwell in this paper on other MRCC approaches. The SUMRCC for complete active spaces in the spinorbital basis was first formulated by Jeziorski and Monkhorst [51]. Generalization to encompass incomplete active spaces was formulated later by Mukherjee [52, 53], Mukhopadhyay and Mukherjee [54] and by Bartlett et al. [55, 56]. The state-specific MRCC using the JM Ansatz in the spinorbital-based theory was formulated by Mukherjee et al. [57–59], which has been rather extensively used [60–62]. Study of importance of higher-rank excitations [63–65], parallelization strategies [66, 67] as well as analytic gradients was done subsequently by others [68]. There have also been multi-reference equation of motion (MREOM) versions for the state-specific theory [69–73]. For the sake of completeness, we should mention that internally contracted MRCC theories, using a single-wave operator, were also suggested [74–79] all of which were formulated in spinorbital basis and thus were not spin adapted. We will not discuss these formulations any further since we are not going to discuss the spin adaptations of these theories in this paper.

In the spin-free formulations of the UGA-MRCC theories, the use of CSFs entails that both the MRCC equations are in matrix form and the associated effective Hamiltonians will involve various n-body spin-free reduced density matrices (n-RDMs). n-RDMs are product separable and hence not size-extensive. From now on, we will refer to the spin-free RDMs as simply the RDMs. When spinorbital-based RDMs are discussed, we will explicitly indicate this. So, no confusions should arise. It is non-trivial to establish the extensivity of both the cluster operators and the effective Hamiltonian in spite of the occurrence of these n-RDMs. This paper will briefly review the formulation of the UGA-MRCC theories mentioned above and will present a comprehensive account of the aspects of connectivity which leads to extensivity. Although in some of our earlier papers [47] we sketched how size extensivity emerges after the cumulant decomposition of the n-RDMs, we will present here a detailed and thorough analysis of the underlying issues.

It will turn out that a proper understanding of the consequence of connectivity in a spin-free formalism requires a careful conceptual separation of three inter-related but distinct properties: (a) Connected entities which are not spatially product separable as happens for cumulants describing spin correlation between spatially separable fragments. (b) Connected entities which are spatially separable (i.e., matrix element of operators with mixed indices, simultaneously involving indices of orbitals belonging to different groups), which vanish if the inter-group interactions in the Hamiltonian are switched off. They are necessarily size-extensive at any point on the potential energy surface (PES). (c) Matrix elements of operators which vanish when the system separates asymptotically into non-interacting fragments; these are the size-consistent operators when the orbitals are localized on different fragments. As mentioned above, the connected entities with property (a) can arise due to the spin-free nature of the formalisms. If the system dissociates into non-interacting fragments A and B, the total wave function Ψ becomes an anti-symmetrized product of the form $A \sim \mathcal{A}_{AB}[\Psi_A^{s_A}\Psi_B^{s_B}]^s$, where $\Psi_A^{s_A}$ and $\Psi_B^{s_B}$ are anti-symmetric fragment functions of given spins s_A and s_B targeted by us as the corresponding fragment functions. Due to the constraint that the fragment functions have to be coupled to a given spin s, one introduces spin correlation between the fragment functions which only depends on the coupled spin s. The associated cumulants do not have any dependence on the inter-fragment separation. These cumulants have mixed orbital indices on different fragments but appearing in a permuted fashion like $\Lambda_{u_A v_B}^{v_B u_A}$, say, for a spin-free two-body cumulant. This is a connected entity by construction in the sense that such entities are non-factorizable into factors having no orbitals in common. However, they will remain non-vanishing even for inifinite inter-fragment separation, even if they are connected. In contrast, operators mentioned in (b) above are both connected and spatially separable which makes them vanishing and are the extensive entities we are after. For entities of type mentioned in (b) above with orbitals on different fragments, one gets the property (c), and their vanishing nature for non-interacting fragments will spell size-consistency.

The connecting threads in our developments will be as follows:

1. Use of unitary group algebra to choose the model space CSFs as Gel'fand states and compute the various matrix elements of the generators between them.
2. Reduction of the expressions for the MRCC equations to explicitly connected ones via the decomposition of the n-RDMs into products of spin-free one-body RDMs and spin-free cumulants [80–82] and establish

the connectivity of the projection equations of our MRCC formalisms.

3. Subsequent proof of size extensivity and consistency of the theories.

The paper is organized as follows: In Sect. 2, we begin with a brief résumé of the UGA-based MRCC theories developed by us and show the generic similarity in the algebraic structure of their working equations. Section 2.1 presents certain basic elements of the unitary group approach. This serves two purposes: both to introduce the notations and also to indicate the essential building blocks in a UGA-based MRCC theory. Section 2.2 presents brief discussions on the various UGA-MRCC theories developed by us, the working equations and highlights their differences and similarities. Moreover, we will emphasize here how the working equations of the various theories can all be subsumed under one set of equations which would facilitate the analysis of connectivity and extensivity. Section 3 analyzes the aspects of connectivity of the MRCC equations. In particular, we emphasize that a complete proof of connectivity is predicated by the connectivity of the operator composites generated from the Hamiltonian and the cluster operators from the associated wave operator as well as a reduction of the MRCC equations involving the above composites and spin-free cumulants, which necessitates extensive algebraic manipulation via cumulant decomposition of the n-RDMs. Section 3.1 begins with a brief review of the aspects of the connectedness of the operator composites, referring to our previous papers [44–48] for the details. This subsection is one of the central portions of our paper since it will also present the details of the connectivity of the working equations brought out by the cumulant decompositions of the n-RDMs. Section 3.2 addresses the other major issue., viz. the proof of extensivity and consistency of our two UGA-MRCC theories. Since our theories are not invariant with respect to the transformation of active orbitals, it is imperative to use active orbitals localized on different fragments to arrive at size-consistency from a size-extensive theory. Section 4 presents the summary of our presentation, the conclusions, and also the avenues for future development.

2 UGA-SUMRCC and UGA-SSMRCC formalism

2.1 The use of the unitary group in MRCC

Any spin-independent m-body operator can always be expressed in terms of m-products of one-body generators of the unitary group $U(n)$. With the spinorbital creation and annihilation operators for orbitals p and q and the spin state σ, the one-body generators E^p_q can be defined as

$$E^p_q = \sum_\sigma a^\dagger_{p_\sigma} a_{q_\sigma} \tag{1}$$

Using UGA, the spin-free part of an N-electron function with n-orbitals can be chosen as a basis of a particular irreducible representation (irrep) of the unitary group, $U(n)$. However, as emphasized in references [33, 34], it is much more convenient to adapt the spin-free part of an N-electron function to the irreducible representations of the subgroup $\boldsymbol{u}(n)$. Because of the availability of very efficient softwares (like GAMESS [29]), we have chosen model function CSFs ϕ_μ as Gel'fand states adapted to $U(n_a)$, since for a complete active space the doubly occupied core function $|0\rangle$ belongs trivially to the identity representation of $U(n_c)$. For a one-body generator, the matrix elements like, $\langle\phi_\mu|E^p_q|\phi_\nu\rangle$, can be conveniently computed via a Shavitt graph [19, 20]. In a similar manner, any m-body generator can be evaluated.

The Gel'fand functions, ϕ_μ, can be formally generated by the action of 'Gel'fand creators, $\{Y_{\mu_k}^{\Gamma}\}$ on $|0\rangle$, taken as the 'vacuum' for the development:

$$\begin{aligned} |\phi_\mu\rangle &= \{Y_{\mu_k}^{\Gamma}\}|0\rangle \\ &\equiv |\phi_{\mu_k}^{\Gamma}\rangle \end{aligned} \tag{2}$$

The indices Γ and k, indicate to which component k of the Gel'fand basis for irreducible representation, Γ, of the unitary group $U(n_a)$, the function ϕ_μ belongs. From now on, we will drop the indices Γ and k from our functions ϕ_μ for notational simplicity.

For our formulation, we will take all the generators involved in our formulation as *normal ordered* with respect to the core taken as the vacuum and are to be henceforth generically denoted as $\{E^p_q\}$, $\{E^{pr}_{qs}\}$, etc., for one-, two-body,... generators where the curly bracket indicates normal ordering with respect to the core. Here, p, q, etc., are arbitrary orbital indices. The rewriting of the products of normal-ordered spin-free generators of the unitary group can be done using a spin-free version of the Wick's theorem (WT) whose content was summarized by Kutzelnigg [49, 50].

The Hamiltonian and the cluster operators in our formulation will also be written in normal order and hence will involve normal-ordered generators. The Hamiltonian, normal ordered with respect to a core $|0\rangle$, has the structure:

$$H = \langle 0|H|0\rangle + \{F_c\} + \{V\} \tag{3}$$

where $\langle 0|H|0\rangle$ is the unperturbed core energy, $\{F_c\}$ is the core Fock operator:

$$\{F_c\} = \sum_{p,q} \langle p|f_c|q\rangle\{E^p_q\} \equiv \sum_{p,q} \{f_c\}^p_q\{E^p_q\} \tag{4}$$

whose elements are defined as

$$\{f_c\}^p_q = h^p_q + \sum_{i\in|0\rangle}\left(2v^{pi}_{qi} - v^{ip}_{qi}\right) \tag{5}$$

$\{V\}$ is the two-body Coulomb operator in normal order and is of the form:

$$V = \frac{1}{2}\sum_{pqrs} v_{qs}^{pr}\{E_q^p E_s^r\} \equiv \frac{1}{2}\sum_{pqrs} v_{qs}^{pr}\{E_{qs}^{pr}\} \quad (6)$$

The matrix element of the Hamiltonian, H, taken between two Gel'fand states, ϕ_μ and ϕ_ν can be written as

$$\langle\phi_\mu|H|\phi_\nu\rangle = \langle 0|H|0\rangle\delta_{\mu\nu} + \sum_{pq} f_{c_q}^{\ p}\langle\phi_\mu|\{E_q^p\}|\phi_\nu\rangle + \frac{1}{2}\sum_{pqrs} v_{qs}^{pr}\langle\phi_\mu|\{E_{qs}^{pr}\}|\phi_\nu\rangle \quad (7)$$

It is clear that the matrix elements of the one- and two-body generators between Gel'fand states ϕ_μ and ϕ_ν generate one- and two-particle transition density matrix elements respectively: $\gamma^{\mu\nu}$ and $\Gamma^{\mu\nu}$ with appropriate indices:

$$\gamma^{\mu\nu p}_{\ \ \ q} = \langle\phi_\mu|\{E_q^p\}|\phi_\nu\rangle \quad (8)$$

$$\Gamma^{\mu\nu pr}_{\ \ \ qs} = \langle\phi_\mu|\{E_{qs}^{pr}\}|\phi_\nu\rangle \quad (9)$$

Of course, if we know the transition density matrices for the one-body generators, for all (μ,ν) and all (p,q), then not only the two-body transition matrices but also the higher-body transition density matrices can be derived by resolution of identity over all intermediate states λ and using simple matrix multiplication. For such a situation to hold, obviously the space spanned by the Gel'fand states, λ, must be complete. In an MRCC setting, we choose an active (or a model) space which may or may not be complete with respect to distribution of active electrons in active orbitals, which are partially occupied in some ϕ_μ but not in others. We have used in all our applications hitherto the program system GAMESS [29] for getting the one-particle transition density elements obtained using the Graphical Unitary Group Approach (GUGA).

We conclude this subsection by summarizing our notation for the sections to come. We will denote the generic labels of orbitals, as mentioned above, by $p, q, \ldots$, etc., the occupied by $i, j, \ldots$ the active by $u, v, w, \ldots$, and the virtuals by $a, b, \ldots$, etc.

2.2 Working equations for UGA-MRCC theories

The Ansatz we have chosen for our unitary group-based MRCC methods [45–48] is designed to closely mimic the Jeziorski–Monkhorst (JM) Ansatz [51, 83] in order to follow quite closely the developments in the analogous non-spin-adapted theories. As mentioned, we choose a set of Gel'fand states, ϕ_μ, to denote the model functions. We next introduce our spin-free JM-inspired Ansatz in Eq. (10) for the wave operator Ω_μ acting on ϕ_μs. Our choice differs in two aspects from the traditional spinorbital-based JM Ansatz. The T_μs are written in terms of generators normal ordered with respect to a suitable closed shell vacuum $|0\rangle$. The first difference is operationally manifest as T_μs defined using spatial orbitals which do not commute with each other. The second difference arises due to our desire to have naturally truncating working equations after a finite power of T. This is accomplished by choosing Ω_μ to be of the following normal-ordered exponential form:

$$\Omega_\mu = \{\exp(T_\mu)\} \quad (10)$$

Although we feel that a combinatoric spin-free cluster expansion [84, 85] developed earlier in our group uses a wave operator that incorporates all the essential aspects of the analogous spinorbital-based JM Ansatz in a more complete way, we have chosen to work with our normal-ordered Ansatz which yields simpler equations and yet has been found to be very good. This is despite the fact that it misses some additional clusterings due to this normal ordering of the Ansatz. We have used the Ansatz in Eq. (10) for Ω_μ for both our unitary group adapted state-universal MRCC (UGA-SUMRCC) [46] and the corresponding state-specific MRCC (UGA-SSMRCC, also known as UGA-MkMRCC) [45].

The parametrization for the wave function is thus:

$$|\Psi_k\rangle = \sum_\mu \{\exp(T_\mu)\}|\phi_\mu\rangle c_{\mu k} \quad (11)$$

In the SS formalism, we target only one function $|\Psi\rangle$, and hence, the coefficients $\{c_{\mu k}\}$ have no k dependence, and thus, we drop the state index k. In the SU formalism, all $|\Psi_k\rangle$ states are targeted but the equations for obtaining the amplitudes with whose connectedness we are concerned have no dependence on $\{c_{\mu k}\}$.

We will use a shorthand notation $\{\varepsilon_\mu^l\}$ for a normal-ordered generator inducing transition of $|\phi_\mu\rangle$ to a virtual function $|\chi_\mu^l\rangle$. The operator $\{E\}$ is labeled by string of operators vacated and created, respectively. If a string of orbital indices, collectively labeled as I_μ, is replaced by another string of indices A_μ to convert ϕ_μ to χ_l, the excitation operator $\{\varepsilon_\mu^l\}$ is really nothing but the normal-ordered product $\{E_{I_\mu}^{A_\mu}\}$, where the associated string is explicitly specified in an abstract manner. The operators $\{E_{I_\mu}^{A_\mu}\}$ are neither linearly independent nor orthogonal in the sense of producing linearly independent or orthogonal $|\chi_\mu^l\rangle$s. We, however, want for a particular truncation strategy of the rank of the operator T_μ only those operators that can be classified into complete sets of I_μ of upto m_c core orbitals and m_a active orbitals vacated and in A_μ of upto m_v virtual orbitals and m_a' active orbitals created. Obviously, $m_c + m_a = m_v + m_a'$. In principle, one should choose for a fixed choice of m_c and m_v all those operators needed to span the full scalar tensor space of $\boldsymbol{u}(n)$. This fixes the highest values of m_a and m'_a, as $m_a^{\max}$, and $m'^{\max}_a$, given a fixed m_v. In practice, however, one does not necessarily have to include all operators that are needed to span the full tensor subspace corresponding to the replacement of m_c core with m_v virtual orbitals. It is enough to chose all excitation up to a certain m_a and m'_a which are less than $m_a^{\max}$ and $m'^{\max}_a$, respectively. We will use these excitation operators in our T_μ. They are still generally linearly dependent and one can extract the corresponding linearly independent scalar tensor generators of $\boldsymbol{u}(n)$ by group theoretic techniques as for example what was done in references [33, 34]. In our present formulation, in conformity with our earlier developments, we will use the set of operators I_μ and A_μ containing up to some fixed m_a and m'_a. This will correspond to a given truncation strategy of the rank of the cluster operators of T_μ. Neither the proof of connectivity nor of extensivity, however, depends on the rank of truncation.

In order to eliminate the redundancy of the operators $\{E_{I_\mu}^{A_\mu}\}$, we would introduce extra cluster operators as new variables and would provide equations for them to resolve the redundancy. The proof of the connectivity and extensivity of the theory is simpler if such χ_μ^ls are used. After the conclusion of the proof, we would indicate how we could have developed the corresponding UGA-based MRCC theories with only linearly independent excitations in T_μ. In fact, once the size extensivity of our theories with linearly dependent operators is established, we will discuss some possible choice of the linearly independent scalar tensor generators of $\boldsymbol{u}(n)$. The principal goal of this paper is not affected by whether we use linearly dependent or independent operators. We will also indicate how the analysis of the size extensivity provides us with some insight into what could be a natural choice for the excited state manifold and hence, the excitation operators, $\{\varepsilon_\mu^l\}$.

Although the starting points for the UGA-SUMRCC and UGA-SSMRCC are different, the equations ultimately assume the same structure which we shall generally treat to demonstrate both the connectivity and the size extensivity.

2.2.1 UGA-SUMRCC

The UGA-SUMRCC is a multi-root formalism, based on effective Hamiltonians, starting from the Bloch equation. The Bloch equation for our model space, with the projector P, for the wave operator, Ω, is of the form:

$$H\Omega P = \Omega P H_{\text{eff}} P \tag{12}$$

where, Ω is chosen as

$$\Omega = \sum_\mu \{\exp T_\mu\} |\phi_\mu\rangle\langle\phi_\mu| \tag{13}$$

H_{eff} is given by,

$$H_{\text{eff}} = [P\Omega P]^{-1} PH\Omega P \tag{14}$$

If we choose a complete active space (CAS), the intermediate normalization (IN), $P\Omega P = P$ is used and as a consequence, $H_{\text{eff}} = H\Omega P$. For incomplete model spaces, $P\Omega P \neq P$ and the expression of H_{eff} in Eq. (14) must be used [52–54]. Using our Ansatz in the Bloch equation, we arrive at the working equation:

$$\langle\chi_\mu^l|\{\overline{H}_\mu\}_{ex}|\phi_\mu\rangle$$

$$- \langle\chi_\mu^l|\sum_\nu\{\exp-(T_\mu)\exp(T_\nu)\overbrace{\exp(T_\nu)W_{\nu\mu}}\}_{ex}|\phi_\mu\rangle = 0 \tag{15}$$

where $W_{\nu\mu}$ is a closed operator which scatters from ϕ_μ to ϕ_ν. $\{\overline{H}_\mu\}_{ex}$ is the excitation component of the similarity transformed Hamiltonian $\{\exp(-T_\mu)H\exp(T_\mu)\}$ which features in what is called the '*direct*' term and occurs identically in the UGA-SSMRCC theory to follow. The second so-called coupling term controls the mixing of the different model functions in the T_μ-determining equations $\forall\mu$.

We write Eq. (15) in a compact notation as

$$\langle\chi_\mu^l|G_\mu^{SU}|\phi_\mu\rangle = 0 \tag{16}$$

where G_μ^{SU} is a composite connecting ϕ_μ and χ_μ^l containing both the direct and coupling term:

$$G_\mu^{SU} = \{\overline{H}_\mu\}_{ex} - \sum_\nu\{\exp-(T_\mu)\exp(T_\nu)\overbrace{\exp(T_\nu)W_{\nu\mu}}\}_{ex} \tag{17}$$

The superscript '*SU*' has been used to distinguish the structure of G_μ for state-universal (SU) MRCC vis a vis the state-specific (SS) MRCC theories to follow. For the details of the derivation, we refer to our paper [46]. In our first formulation [46–48], we wanted to apply the theory to study excitation energies or excited states relative to a closed shell ground state. For this, we chose the active space to consist of active holes and particles and all possible h-p excited CSFs were generated for this choice. In the terminology of Lindgren [86], such model spaces are called quasi–complete model spaces (QMS), and it was shown long ago by Mukherjee that the size extensivity of an effective Hamiltonian theory with an incomplete model space (IMS) including QMS requires that the customary intermediate normalization (IN) be abandoned [52–54]. In the general situation where a CAS is used, $H_{\text{eff}} = PH\Omega P$, and the equations are considerably simplified. Since we are mainly concerned with the issue of size extensivity of the theories, the modifications necessary to ensure size extensivity, by abandoning IN for Ω, are minor in nature and are not particularly germane for our later discussions of size extensivity.

2.2.2 UGA-SSMRCC

The UGA-SSMRCC is a single-root intruder-free MRCC starting from the Schrödinger equation. In a way exactly analogous to what was derived for the spinorbital-based SSMRCC [57–59], there will be the necessity of invoking sufficiency conditions to eliminate redundancy of the number of cluster amplitudes, and there will appear both a '*direct*' and a set of '*coupling*' terms. On account of the normal ordering of the UGA Ansatz, the direct term will terminate exactly at the quartic power, while the highest power of the coupling term will depend on the level of truncation of the cluster operators and the rank of $\tilde{H}_{\mu\nu}$.

We have shown in our parent UGA-SSMRCC paper [45] that there are two 'natural' ways of choosing the sufficiency conditions, and each leads to a set of UGA-SSMRCC equations for the cluster amplitudes which is inequivalent with the other. However, a study of the numerical performance of both the variants leads us to the conclusion that they produce very close results and no criteria for choosing one over the other can be established [44]. Since the version A involves fewer terms and is simpler in structure, our subsequent applications focused mainly on this alternative.

The target state $|\Psi\rangle$ in our UGA-SSMRCC formulation is sought to be obtained by the action of the wave operator Ω on to the reference function $|\Psi_0\rangle = \sum_\mu |\phi_\mu\rangle c_\mu$:

$$|\Psi\rangle = \Omega|\Psi_0\rangle = \sum_\mu \Omega_\mu|\phi_\mu\rangle c_\mu \tag{18}$$

The Schrödinger equation for $|\Psi\rangle$ is given by

$$H|\Psi\rangle = E|\Psi\rangle \tag{19}$$

where H is written in normal order with respect to the core function $|0\rangle$ as the vacuum, viz. Eq. 3.

Our working equations for the cluster amplitudes of the set $\{T_\mu\}$ will be obtained by substituting the expression, of Eq. (18) in Eq. (19):

$$\{H\}\sum_\mu\{\exp(T_\mu)\}|\phi_\mu\rangle c_\mu = E\sum_\mu\{\exp(T_\mu)\}|\phi_\mu\rangle c_\mu \tag{20}$$

The normal ordering is with respect to the core $|0\rangle$, taken as the vacuum. At the stage of invoking sufficiency conditions, analogous to the spinorbital-based SSMRCC, two inequivalent paths present itself leading to what we call sufficiency A and sufficiency B. These two alternatives, then lead to the versions A and B alluded to above. The reader is referred to our paper [45] for details leading to the following working equations for the amplitudes of the cluster operator $\{T_\mu\}$:

$$\langle\chi_\mu^l|\{\overline{H_\mu}\}_{ex}|\phi_\mu\rangle c_\mu + \sum_\nu\langle\chi_\mu^l|\{\exp(-T_\mu + T_\nu)\}_{ex}|\phi_\mu\rangle\tilde{H}_{\mu\nu}c_\nu - \sum_\nu\langle\chi_\mu^l|\{(\exp(\overbrace{T_\mu))_{ex}W_{\nu\mu}}\}_{ex}|\phi_\mu\rangle c_\mu = 0, \forall\mu, l \tag{21}$$

for the version A where $\tilde{H}_{\mu\nu} = \langle\phi_\mu|\{\overbrace{H\exp(T_\mu)}\}_{cl}|\phi_\nu\rangle$, and

$$\langle\chi_\mu^l|\{\overline{H}_\mu\}_{ex}|\phi_\mu\rangle c_\mu + \sum_\nu\langle\chi_\mu^l|\{\exp(-T_\mu + T_\nu)\}_{ex}|\phi_\mu\rangle\tilde{H}_{\mu\nu}c_\nu - \sum_\nu\langle\chi_\mu^l|\{\exp(-T_\mu + T_\nu)(\exp(\overbrace{T_\nu))W_{\mu\nu}}\}_{ex}|\phi_\nu\rangle c_\nu = 0\forall\mu, l \tag{22}$$

for the version B. We will henceforth refer to the methods as the UGA-SSMRCC-A and UGA-SSMRCC-B. Both these Eqs. (21) and (22) can also be generally written as

$$\langle\chi_\mu^l|G_\mu^{SS}|\phi_\mu\rangle = 0 \tag{23}$$

where, G_μ^{SS} is either G_μ^A or G_μ^B.

$$G_\mu^A = \{\bar{H}_\mu\}_{ex} + \sum_\nu \tilde{H}_{\mu\nu}\{\exp(-T_\mu + T_\nu)\}_{ex}\frac{c_\nu}{c_\mu} - \sum_\nu\{(\exp(\overbrace{T_\mu))\,W_{\nu\mu}}\}_{ex} \tag{24}$$

$$G_\mu^B = \{\bar{H}_\mu\}_{ex} + \sum_\nu[\tilde{H}_{\mu\nu}\{\exp(-T_\mu + T_\nu)\}_{ex} - \{\exp(-T_\mu + T_\nu)(\exp(\overbrace{T_\nu))W_{\mu\nu}}\}_{ex}]\frac{c_\nu}{c_\mu} \tag{25}$$

Unlike in the spinorbital-based formulation, the UGA-based MRCC theories use spin-free Gel'fand states ϕ_μ as the model functions and generate projection equations leading from each model function ϕ_μ to virtual functions χ_μ^l reachable by cluster operators T_μ in a given truncation scheme.

Depending on the theory, the operator G_μ^{SU} or G_μ^{SS} sandwiched between χ_μ^l and ϕ_μ are different, but in each of the versions, the matrix elements M_μ^l of the corresponding operator between χ_μ^l and ϕ_μ, written as $\langle\phi_\mu|\{\varepsilon_\mu^l\}^\dagger G_\mu|\phi_\mu\rangle$, turns out to be a sum of certain amplitude terms (as introduced in Eqs. (17), (24) and (25), involving composites of H and powers of $\{T_\mu\}$, $\{T_\nu\}$, etc., along with certain RDMs of different ranks. Since the RDMs are factorizable quantities, which can lead to disconnected terms in the working equations, its not obvious at first sight that the working equations are connected [45], and hence, the theory does not manifestly appear to be size-extensive (the energies of individual non-interacting fragments are not additively separable). We will present here a comprehensive analysis to prove the extensivity of the working equations.

3 Aspects of connectivity, size extensivity, and size-consistency in the UGA-MRCC theories

Although we have mentioned in the earlier sections about two distinct but inter-related aspects of size extensivity, it is essential that we present these two separately. We will then have a proper theoretical underpinning of these two aspects. Since our working equations have the generic structure:

$$M^l_\mu = \langle \chi^l_\mu | G_\mu | \phi_\mu \rangle \equiv \langle \phi_\mu | \{\varepsilon^l_\mu\}^\dagger G_\mu | \phi_\mu \rangle = 0 \qquad (26)$$

,they involve ultimately a sum of composites generated by the application of Wick's theorem (WT) on the product $\{\varepsilon^l_\mu\}^\dagger G_\mu$ to generate various terms in normal order [87]. Unless the operators left uncontracted in such a term have only active labels, the expectation value $\langle \phi_\mu | \{...\} | \phi_\mu \rangle$ will vanish, and henceforth, we will consider only those terms where the composites contain only active left uncontracted. If a particular composite contains k number of active indices destroyed and the same number of active indices created, the expectation value $\langle \phi_\mu | \{...\} | \phi_\mu \rangle$ will generate a k-RDM $\Gamma^{(k)}_{\mu\mu}$ containing k-pairs of active orbitals with the appropriate lower and upper active indices. These k-RDMs are quasi-diagonal in the sense that for a fixed set of lower indices, the upper indices can at most be the permutation of the lower indices.

The proof of size extensivity of each term of M^l_μ generated by WT is somewhat harder than all the previous MRCC formalisms, since each term containing some k-RDMs turns out to be a product of the specific RDMs and one or other of the amplitudes g_μ of a particular term of G_μ. From Eqs. (15), (21) and (22), the amplitudes of the composites G_μ are connected if T_μs are connected. The labels on the corresponding k-RDMs, Γ_k, are not necessarily connected in the composite, M^l_μ, with the labels of the G-block. The uncontracted active lines thus cannot come entirely from the active orbitals of the G-block and they may not have common orbital labels with the uncontracted active operators in the G-block. Unlike the amplitudes g_μ, the matrix elements of Γ_k are not connected since these are multiplicatively separable operators. Hence, the entire composite may not be connected.

As explained above, for proving the extensivity, we need to proceed in two steps. In Sect. 3.1, we will establish that the MRCC equations are connected if the cluster operators and G_μs are connected. In other words, we would first show that the connectivity of the composite G_μ, the cluster amplitudes T_μ and the MRCC equations are mutually compatible. Of course, we need finally to prove the more stringent property that the cluster operators are not only connected but also spatially extensive in the sense that the cluster amplitudes labeled by a set of orbitals will all go to zero if we keep these orbitals in two arbitrary groups and switch off interaction between these groups (which amounts to switching off the Hamiltonian matrix elements with mixed inter-group indices). This issue is non-trivial since several connected terms in the MRCC equations would contain inter-group cumulants with mixed indices which do not vanish if the interaction between these groups is turned off. As we will discuss in Sect. 3.2, a certain class of mixed cumulants, which we will call as inter-fragment 'exchange' cumulants, are nonzero due to spin correlations only and they have no spatial dependence. These are the connected composites of type (a) introduced in Sect. 1 which are connected but not spatially extensive. Since they appear as a multiplicative factor with the matrix elements of G_μ, connectivity of the MRCC equations does not necessarily guarantee extensivity.

In the next step, we analyze the structure of the various terms generated after the application of the WT to the matrix element M^l_μ in our working equations and establish that we can systematically eliminate the disconnected portion of M^l_μ, if we keep track of which components of the composites containing Γ_k and G_μ are connected. This particular analysis requires the concept of cumulant decomposition [75, 80, 88, 89] of the density matrix elements of Γ_k for various ranks k. Since the final working equations are connected after the elimination of the disconnected terms, the cluster amplitudes of $\{T_\mu\}$ are connected and are compatible with the connectivity of G_μ.

In the final step, we will prove that the cluster operators T_μ, obtained from the solution of the MRCC equations, are extensive as well which would establish the extensivity of the formalisms. Since these theories are not invariant with respect to the transformation of active orbitals, we need to use orbitals localized on fragments to show their size-consistency.

3.1 Connectivity of the operator composites and of the MRCC equations

We briefly discuss first the issue of the connectivity of the composite G_μ for the various UGA-MRCC theories. Since the connectedness of the G_μ for the various theories has been proven in detail in our papers on UGA-SUMRCC [46] and UGA-SSMRCC formalisms [44], we would dwell on this aspect rather briefly. On account of the underlying perturbative structure of these CC equations, the t_μ-amplitudes are taken as connected and remain so provided the working equations are connected. The connectivity of the G_μs can be discussed in two part: a) the 'direct' term and b) the 'coupling' term. Needless to say the 'direct' term is identical for G_μ^{SU}, G_μ^A and G_μ^B. The expression is explicitly connected because H and the powers of the operator $\{T_\mu\}$ are contracted to form a connected composite. Thus, if T_μ is a connected operator, the composite from the 'direct' term is also connected.

It is the 'coupling' term which is different for the three G_μs discussed above. Notably, the coefficients c_μ feature in $G_\mu^{A/B}$ but not in G_μ^{SU}. We briefly touch upon the simplest situation of G_μ^{SU} first and then go on to elucidate the additional considerations arising from the occurrence of c_μs. For the two schemes A and B, the connectivity of $G_\mu^{A/B}$ can be inferred from their operator structures. The proof is in line with what was invoked in the paper on spinorbital-based SSMRCC theory [57–59].

Connectivity of the coupling term for G_μ^{SU} for the UGA-SUMRCC hinges on several different aspects of connectivity. For the coupling term, $\overline{\exp(T_\nu)W_{\nu\mu}}$ we note that there are two parts of which $X_{\nu\mu} = \overline{\exp(T_\nu)W_{\nu\mu}}$ is explicitly connected. The connectedness of $\{\exp -(T_\mu - T_\nu)X_{\nu\mu}\}$ requires some deliberation. ϕ_ν and ϕ_μ may differ by at least one orbital or ϕ_μ and ϕ_ν may have the same orbital occupancy and either $\phi_\mu = \phi_\nu$ or they differ in the spin coupling scheme of the active orbitals. The details have been explained in our earlier paper [47].

For showing the connectivity of coupling term for G_μ^A for the UGA-SSMRCC-A, we recall from Eq. (24) the expression for G_μ^A:

$$G_\mu^A = \{\bar{H}_\mu\}_{ex} - \sum_\nu \{(\overline{\exp(T_\mu))_{ex}W_{\nu\mu}}\}_{ex} + \sum_\nu \{\exp(-T_\mu + T_\nu)\}_{ex}|\phi_\mu\rangle\langle\phi_\mu|\{\overline{H\exp(T_\mu)}\}_{cl}|\phi_\nu\rangle\frac{c_\nu}{c_\mu} \quad (27)$$

(27)

where we have interchanged now the second and third term of Eq. (24) for the ease of further discussions.

As in UGA-SUMRCC, we consider only the coupling terms (ie. the second and third terms). The second term in G_μ^A is connected if the T_μs are connected. The third term can also be proven to be connected as follows. Since we are dealing with a CAS and the coefficients $\{c_\lambda\}$ are obtained as eigenvectors of the connected matrix $\tilde{H}_{\mu\nu} = \langle\phi_\mu|\{\overline{H\exp(T_\mu)}\}_{cl}|\phi_\nu\rangle$ in the CAS, the ratio $[\frac{c_\nu}{c_\mu}]$ can always be expressed as

$$\frac{c_\nu}{c_\mu} = \langle\phi_\nu|\{\exp(\sigma_\mu)\}|\phi_\mu\rangle \quad (28)$$

where, the σ_μs are all connected operators. This ratio thus consists of products of amplitudes of σ_μ which can induce transitions from ϕ_μ to ϕ_ν. The difference $T_\nu(\mu) - T_\mu$ in the third term will be functionally dependent on all the active orbitals by which ϕ_μ and ϕ_ν differ, or when the orbital occupancies are the same, they will be dependent on all the orbitals involved in the respective spin couplings. The term $T_\nu(\mu)$ indicates those T_ν whose action on ϕ_μ is nonzero. In either case, $\{\exp(-T_\mu + T_\nu)\}_{ex}|\phi_\mu\rangle$ will have common orbital labels with the connected term $\langle\phi_\mu|\{\overline{H\exp(T_\mu)}_{cl}\}|\phi_\nu\rangle\frac{c_\nu}{c_\mu}$. Hence, the operator G_μ^A is connected if the Ts are connected. Essentially the same argument holds good for the scheme B and we will not elaborate on this further.

We now go on to show the connectivity of the working equations which will follow from the cumulant decomposition of the n-RDMs. To begin with, we would explicitly write out the general structure of the working equations, Eq. (26), by indicating the strings of operators involved in M_μ^l. For this purpose, we introduce a couple of new and more general notations to explain our procedure and illustrate, with example diagrams, the significance of the various notations describing the entities entering the components of M_μ^l. Every $|\chi_\mu^l\rangle = \{\varepsilon_\mu^l\}|\phi_\mu\rangle$ can be uniquely defined in terms of substitutions of a set of orbitals of the CSF ϕ_μ, involving at least one inactive orbital, hole, or particle by another set. For any specific function χ_μ^l, let us assume $\{\varepsilon_\mu^l\}$ of this form:

$$\{\varepsilon_\mu^l\} \equiv \{E_{I_\mu^n}^{A_\mu^n}\} \quad (29)$$

when χ_μ^l is obtained from ϕ_μ by replacing a string of ordered n operators denoted by I_μ^n with a similar ordered set A_μ^n. Both the strings I_μ and A_μ cannot simultaneously contain only active labels. The operator G_μ also consists of a sum of various excitation operators of ranks m which can be explicitly written as

$$G_\mu = \sum_{m,B_\mu,J_\mu} g_{\mu J_\mu^m}^{m B_\mu^m}\{E_{J_\mu^m}^{B_\mu^m}\} \quad (30)$$

Then, the expression of M_μ^l can be written in long hand as

$$M_\mu^l = \sum_{m,B_\mu,J_\mu} g_{\mu J_\mu^m}^{m B_\mu^m}\langle\phi_\mu|\{E_{I_\mu^n}^{A_\mu^n}\}^\dagger\{E_{J_\mu^m}^{B_\mu^m}\}|\phi_\mu\rangle \quad (31)$$

If we now write the matrix element $\langle\phi_\mu|\{E_{I_\mu^n}^{A_\mu^n}\}^\dagger\{E_{J_\mu^m}^{B_\mu^m}\}|\phi_\mu\rangle$ in normal order using WT and keep only the nonzero entities, then we can write

$$\langle\phi_\mu|\{E^{A^n_\mu}_{I^n_\mu}\}^\dagger\{E^{B^m_\mu}_{J^m_\mu}\}|\phi_\mu\rangle = \langle\phi_\mu|\{E^{I^n_\mu}_{A^n_\mu}\}\{E^{B^m_\mu}_{J^m_\mu}\}|\phi_\mu\rangle \equiv \sum_{r,s,k,p,q} \kappa^{r I^n_\mu B^m_\mu}_{s A^n_\mu J^m_\mu} \Gamma^{(k)p}_{\ \ q} \tag{32}$$

where, the term $\kappa^{r I^n_\mu B^m_\mu}_{s A^n_\mu J^m_\mu}$ consists of various contractions, r in number, contracting operators of $\{E^{A^n_\mu}_{I^n_\mu}\}$ with those in $\{E^{B^m_\mu}_{J^m_\mu}\}$. The rank k of Γ depends on the actual number of active excitation/de-excitation operators in $\{E^{A^n_\mu}_{I^n_\mu}\}$ and $\{E^{B^m_\mu}_{J^m_\mu}\}$, but the maximum rank of k can be $(m+n-r)$. We note here that all the inactive lines of $\{E^{A^n_\mu}_{I^n_\mu}\}$ and $\{E^{B^m_\mu}_{J^m_\mu}\}$ must be contracted among themselves to have a non-vanishing contribution of the expectation value $\langle\phi_\mu|\{E^{I^n_\mu}_{A^n_\mu}\}\{E^{B^m_\mu}_{J^m_\mu}\}|\phi_\mu\rangle$. In addition, several contractions with active lines connecting the two operators may appear. For a given number of contractions r, there would generally be terms of various types depending on the number of particle, hole, active particle, and active hole contractions. Also, for a fixed number of such contractions, the connectivity generated by the contractions would generally have different topologies. The various topologies are denoted by the symbol s and the overall factor associated with all these contractions for a given r and s is denoted by κ^r_s. From now on, we suppress the indices '${}^{I^n_\mu B^m_\mu}_{A^n_\mu J^m_\mu}$' from κ^r_s for brevity. We remember that the uncontracted operators will only have to be labeled by active indices in M^l_μ. Depending on the indices involved in κ^r_s, the strings of the lower and upper entries in Γ would be different which we have indicated by the symbols q and p, respectively, in Eq. (32).

Introducing the notation,

$$\bar{M}^{r I^n_\mu B^m_\mu}_{s A^n_\mu J^m_\mu} = \kappa^r_s g^{m B^m_\mu}_{\mu J^m_\mu} \tag{33}$$

then for a choice of the set of indices J^m_μ and B^m_μ we can write

$$M^l_\mu = \sum_{r,s,p,q,m,J^m_\mu,B^m_\mu} P^r_s \kappa^r_s g^{m B^m_\mu}_{\mu J^m_\mu} \Gamma^p_q = \sum_{r,s,p,q,m,J^m_\mu,B^m_\mu} \bar{M}^{r I^n_\mu B^m_\mu}_{s A^n_\mu J^m_\mu} \Gamma^p_q \tag{34}$$

Here, P^r_s involves a sum over identical permutations among the upper and lower indices of $\bar{M}^{r I^n_\mu B^m_\mu}_{s A^n_\mu J^m_\mu}$ with the same r and s but involving in effect, different labels of contracted operators. From now on, we will omit the permutation operator P^r_s for the sake of simplicity of expression, but—unless otherwise specified—such a summation over different selection of contracted labels will always be implied.

As an illustration of the various entities introduced above, we take specific terms appearing in the projection involving $\{\varepsilon^l_\mu\}^\dagger \equiv \{E^l_\mu\}^\dagger = \{E^\mu_l\}$ as shown in Fig. 1a–c, i.e., we focus on one specific term $\bar{M}^{r I^n_\mu B^m_\mu}_{s A^n_\mu J^m_\mu}\Gamma^p_q$ in the sum of the right side of Eq. (34).

The blue box on the left of each diagram is the de-excitation operator $\{\varepsilon^l_\mu\}^\dagger = \{\varepsilon^\mu_l\}$. The gray box denotes a specific term of G_μ of a given rank and a set of orbital labels. The gray box is the specific amplitude, g, and the entering and the leaving lines consist of the lower and upper indices of the operator accompanying it. The composites are the various terms obtained after the application of the WT with some possible uncontracted operators which must all be active lines. In Fig. 1a, the $G^{(m)}$ is of the form $G^{ax}_{iy} = g^{ax}_{iy}\{E\}^{ax}_{iy}$. The contribution of Fig. 1a to M^l_μ is given by $\sum_{xy} 2g^{ax}_{iy}\gamma^x_y$, where the factor 2 is the entity κ^r_s, and in this case, there is only one possible contraction and the value is 2 because of the loop. The uncontracted operators contribute to the matrix element γ^x_y. Similarly, Fig. 1b, c contribute $\sum_{xy} 2g^{ax}_{iy}\gamma^u_y\delta_{vx}$ for projections involving de-excitation $\{E^{iu}_{av}\}$, κ^r_s being $2\delta_{vx}$ and $\sum_{xy} 2g^{ax}_{iy}\Gamma^{xu}_{yv}$ for de-excitation $\{E^{iu}_{av}\}$, κ^r_s being 2. In our subsequent discussions, we may often omit the distinguishing indices s, p, q for the sake of brevity, particularly when we focus on one or more specific terms.

We now motivate toward the cumulant decomposition of the RDM Γ^p_q to discern which of the factors in the various terms after cumulant decomposition are connected to $\bar{M}^r_s$. Let us look at the connectivity of the three diagrams in Fig. 1. In Fig. 1a, the lines labeled by i and a connect the de-excitation and the G-block, and hence, the portion of the Fig. 1a with the pair i and a is connected. The RDM $\gamma^x_y\delta_{xy}$ resides entirely on the G-block and remains connected because the G-block is connected. Hence, Fig. 1a is connected. We call the connectivity where all pairs of lines on the de-excitation part is completely contracted with those coming from the G-block as connectivity of type (A). In Fig. 1b, because of the diagonality of the density matrix involving the labels u and y, they are equal and this is a connected quantity too. We would henceforth call such types of connectivity to be of type (B) where though the indices are not completely contracted, they form density matrices with active lines on the two vertices entering and leaving separately on the de-excitation vertex and G block, the term is connected because of the quasi-diagonality of the density matrices. The diagram in Fig. 1c, on the other hand, contains a two-body density matrix Γ^{ux}_{vy} and—although x and y are attached to the G-block—neither u nor v are so attached and it is conceivable that this matrix element of Γ will have a nonzero contribution even if the pairs (x, y) and (u, v) belong to different molecular fragments. Thus, this is a disconnected term. It is here that one

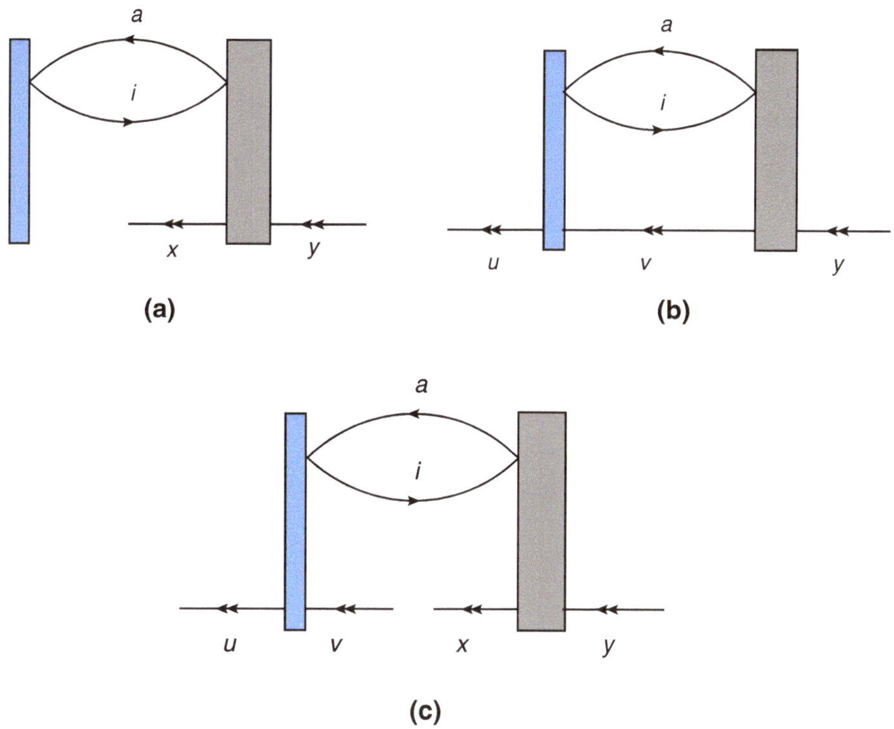

Fig. 1 A few typical terms contributing to the projection equation corresponding to the de-excitation operators ε_a^i (**a**) and ε_{av}^{iu} (**b, c**)

needs the cumulant decomposition of the RDMs. The cumulants [90] being non-product separable, are connected which would allow us to discern connectivity or otherwise of the resultant terms.

The general structure of the cumulant decomposition of a spin-free RDM of given rank is discussed in some detail in the Appendix. In our discussions here, we will make use of the expressions explained therein and for the sake of continuity of reading refer to the appendix for further details. The cumulant decomposition of Γ_{vy}^{ux}, a 2-body RDM, is given by Eq. (35)

$$\Gamma_{vy}^{ux} = \gamma_v^u \gamma_y^x - \frac{1}{2}\gamma_v^x \gamma_y^u + \Lambda_{vy}^{ux} \tag{35}$$

where Λ_{vy}^{ux} is the spin-free cumulant, and the first two terms are the direct and exchange factorized portions of Γ_{vy}^{ux}, respectively. The three terms generated from Fig. 1c generated after the cumulant decomposition are shown in Fig. 2.

The diagram of Fig. 2a is the 'direct' term coming from the factorization of Γ and the density matrices γ_y^x and γ_v^u are indicated by joining the appropriate labels (x, y) and (u, v). As mentioned above, the diagonality of the γs imply $x = y$ and $u = v$. Since there are no common labels between the indices of the two γs, such a term is disconnected and the density matrix element γ_v^u factors out from the rest of the connected composite. The diagram of Fig. 2b containing the 'exchange' product is also similarly indicated. Again, due to diagonality of γ's, $x = v$ and $u = y$ must be satisfied. Since u and v on the de-excitation operator have common indices with x and y on G, the term is connected due to the connectivity of G-block. The diagram in Fig. 2c contains the cumulant where the pair of active lines joined by the solid red vertical rectangle indicates the two body Λ. Since Λ is connected, Fig. 2c is also connected. Clearly, the connectivity of Fig. 2b, c stems from a common origin: (1) due to factorization in the exchange mode or (2) connected through a Λ. In general, such connectivities where various pairs of active lines joining the two vertices either via exchange factorization or by higher-body cumulants or both will be henceforth called as connectivity of type (C). Our task is to show that the disconnected term in Fig. 2a does not contribute to the projection equation involving $\{E_{av}^{iu}\}$ while the terms coming from Fig. 2b, c do. We note that the connected composite of Fig. 2a involves a de-excitation operator of the form $\{E_a^i\}$ which is a subset of the entire de-excitation operator $\{E_{av}^{iu}\}$. We can interpret the connected portion of the de-excitation $\{E_{iu}^{av}\}$ in Fig. 2a, as $\{E_a^i\}_c$. Similarly, since the entire composites of Fig. 2b, c are connected, we will call the corresponding de-excitation operator of the connected composite as $\{E_{av}^{iu}\}_c$.

We amplify the factorization of any diagram after cumulant decomposition into a disconnected factor on the de-excitation block and a connected composite by another example with a diagram containing three pairs of active lines forming a three-body RDM as shown in Fig. 3. Using the expression in Eq. (86) and representing the various terms diagrammatically using the convention introduced in

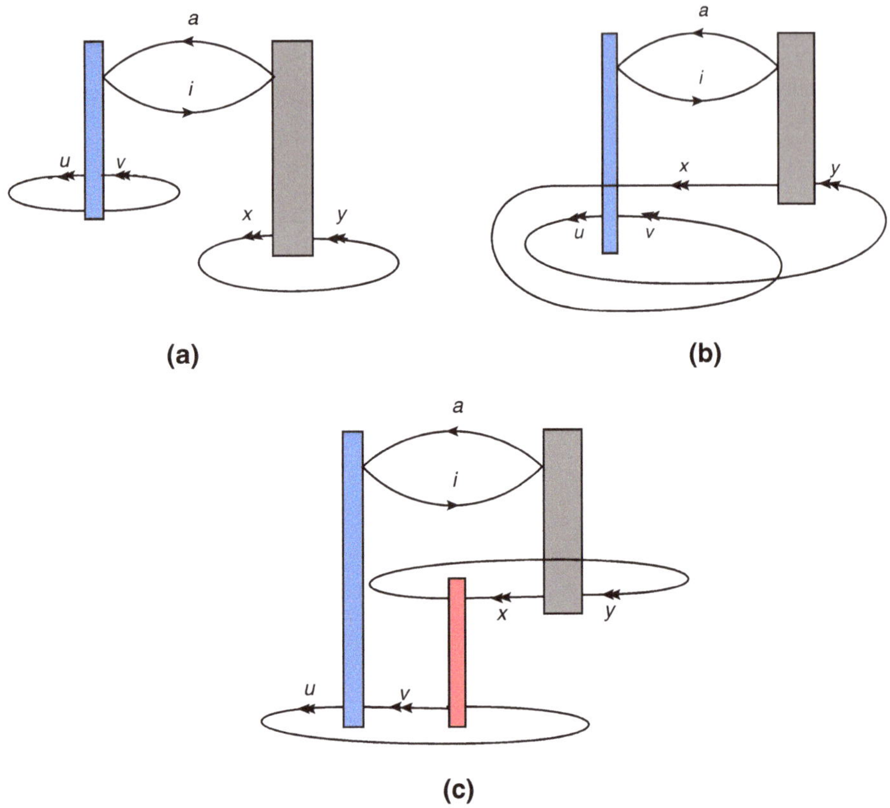

Fig. 2 Pictorial representation of the diagrams generated from Fig. 1c by the cumulant decomposition of two-body RDM. **b**, **c** Belong to the connectivity class C, as introduced in the text, while **a** does not

Fig. 1, we present a set of diagrams some of which are shown in Fig. 4. A common feature of all these diagrams is the connectedness of the de-excitation vertex containing the i and a with the G-block because of the connectivity of the type (A). Figure 4a–c is the diagram in which active lines of the pair on the de-excitation operator are contracted among themselves, and the two pairs residing entirely on the G-block are contracted among themselves in the three possible ways: Fig. 4a has two direct contractions, Fig. 4b has the exchange contractions, and Fig. 4c contains Λ_2. The interesting aspect of these three diagrams is that the various factorizations of the two pairs of active lines residing entirely on the G-block can be regrouped to generate a factorized term like $\gamma^u_v\Gamma^{xw}_{yz}$. Figure 4a–c thus has the disconnected part γ^u_v since this contraction resides entirely on the de-excitation and the rest of the composite is connected. Figure 4d, on the other hand, is an exchange diagram involving the pairs (u, v) and (x, y) which are, respectively, coming from the de-excitation and the G-block and the term then is connected because this spells equality of the indices u and y and x and v, respectively. The contraction of the pair w, z residing entirely on the G-block is irrelevant for connectivity as long as the other two pairs come from the de-excitation and the G-block, respectively. Figure 4e is another connected term containing ${\Lambda_2}^{ux}_{vy}$. Since the Λ connects the pair (u, v) and (x, y) residing on, respectively, the de-excitation and the G-block, it is a connected term. Finally, we show another connected diagram, Fig. 4f, containing Λ_3,viz. ${\Lambda_3}^{uxw}_{vyz}$. All the diagrams in Fig. 4d–f are completely connected with one-type (A) connectivity involving the indices i and a and one-type (C) connectivity.

For the connected composites of Fig. 4, we have the corresponding de-excitation operators $\{E^i_a\}_c$ for Fig. 4a–c and $\{E^{iu}_{av}\}_c$ for Fig. 4d–f.

In a similar vein, if several, m_1 pairs (which includes the case m_1=1) of active lines appearing after the cumulant decomposition come entirely from the de-excitation operator, then all such terms of different types of contractions among these pairs for every term of the rest of the operators contracted in the same way can be regrouped to form an m_1-RDM with the respective active labels residing entirely on the de-excitation operator.*This is the disconnected part which we want to factor out.* For each distinct matrix element of m_1 RDM with fixed labels, the rest of the diagram is connected. In addition, there may also be n_1 pairs of active lines residing entirely on the G-block, and connected among themselves in all possible ways, which can be summed back again to generate an n1-RDM residing entirely on the G-block, and hence, they are always connected due to their being on the G-block.

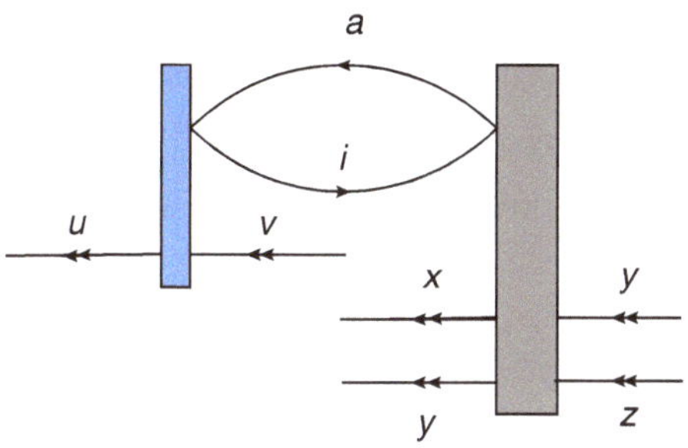

Fig. 3 A typical diagram containing three pairs of active lines

We are now in a position to rewrite the expression of $\bar{M}^{I^n_\mu B^m_\mu}_{r\,s\,A^n_\mu J^m_\mu} \Gamma^p_q$ in terms of products of one-body density matrices and cumulants obtained after cumulant decomposition and classified according to various partitions as described in the Appendix. A given Γ^p_q of rank k with ordered upper and lower strings of length k denoted as p and q is a sum of all terms where the upper and lower strings are grouped in several 'partitions' N_i of lengths k_i, $\sum_i k_i = k$ and for each partition there is a definite contribution of each term in the partition. For a given connectivity of each term, there is a prefactor which depends only on the connectivity. For a detailed description of the various entities entering the partition we refer to the Appendix and in particular the discussions following Eq. (84).

For each term of $\bar{M}^{I^n_\mu B^m_\mu}_{r\,s\,A^n_\mu J^m_\mu} \Gamma^p_q$, we now label the outgoing lines of active indices on the de-excitation block, which forms the disconnected piece as $u_1 \ldots u_{k_d}$ and the corresponding lines connected to the indices $v_1 \ldots v_{k_d}$ for the lines entering from the right. Similarly, the rest of the indices leaving and entering the composite are labeled as $u_{k_d} + 1 \ldots$ and $v_{k_d} + 1 \ldots$. The contribution of the disconnected entity is really a sum of various terms each of which is a product of various partitions with appropriate selection from the upper and lower indices from the strings $u_1 \ldots u_{k_d}$ and $v_1 \ldots v_{k_d}$. Using Eq. (86) of Appendix, the disconnected entity will have to be of the form:

$$\sum_{cl_{k_d}} P_{cl_{k_d}} \Delta^{cl}_{k_d} \equiv \Gamma^{(k_d)p_q}_{\quad\ q_d} \tag{36}$$

Similarly, the rest of the term in $\bar{M}^{I^n_\mu B^m_\mu}_{r\,s\,A^n_\mu J^m_\mu} \Gamma^p_q$ containing the product of one-body density matrices and cumulants which are connected, due to the connectivities of the form (A), (B), and (C) can also be written as

$$\sum_{cl_{k_c}} P_{cl_{k_c}} \Delta^{cl}_{k_c} \equiv \Gamma^{(k_c)p_c}_{\quad\ q_c} \tag{37}$$

Since all the labels $\{u\}$ and $\{v\}$ can be put in the outgoing indices of Γ in many different ways, the entire expression

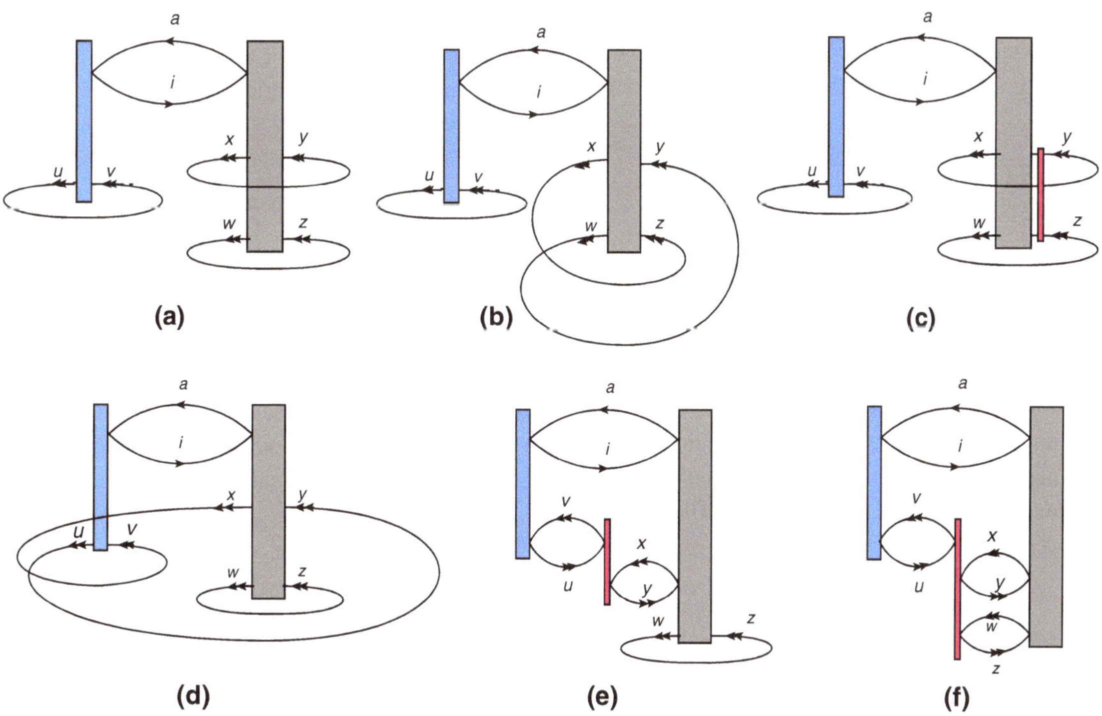

Fig. 4 Pictorial representation of a few terms obtained by cumulant decomposition of three-body RDM as shown in Fig. 3. Please see the text for details

for $\bar{M}^{r I^n_\mu B^m_\mu}_{s A^n_\mu J^m_\mu} \Gamma^p_q$ for various labellings on Γ can be compactly written as

$$\bar{M}^{r I^n_\mu B^m_\mu}_{s A^n_\mu J^m_\mu} \Gamma^p_q = \sum_{k_d,k_c} P_{k_c,k_d} \left[\kappa^r_s g^m \sum_{cl_{k_c}} P_{cl_{k_c}} \Delta^{cl}_{k_c} \right] \left[\sum_{cl_{k_d}} P_{cl_{k_d}} \Delta^{cl}_{k_d} \right] \tag{38}$$

where P_{k_c,k_d} interchanges groups of labels in $u_1 \ldots u_{k_d}$ and $u_{k_d}+1 \ldots$. in all possible manner and exactly the same interchange in $v_1 \ldots v_{k_d}$ and $v_{k_d}+1 \ldots$.. Clearly, after the action of P_{k_c,k_d}, $\bar{M} \ldots$ can be written as

$$\bar{M}^{r I^n_\mu B^m_\mu}_{s A^n_\mu J^m_\mu} \Gamma^p_q = \sum_{k_d,k_c,p_d,q_d,p_c,q_c} \left[\kappa^r_s g^m \Gamma^{(k_c)p_c}_{q_c} \right] \left[\Gamma^{(k_d)p_d}_{q_d} \right] \tag{39}$$

$$\equiv \sum_{k_d,k_c,p_d,q_d} R^{r I^n_\mu B^m_\mu}_{s A^n_\mu J^m_\mu} \Gamma^{(k_d)p_d}_{q_d} \tag{40}$$

In Eq. (39) above, the result of the action of P_{k_c,k_d} has generated different strings (p_c, q_c) and (p_d, q_d) after the interchange of the connected and disconnected labels. Obviously, we can now reduce the number of labels on the upper and lower strings of R by omitting the strings p_d and q_d since they do not appear in the connected composite R. In fact the entity R involves only those subsets of I^n_μ and A^n_μ which together with the contractions κ^r_s and Λ and $g^{(m)}$ constitutes a connected composite due to the connectivities discussed in point (A), (B), and (C) above. If we introduce symbols $I^{n_c}_\mu(c)$ and $A^{n_c}_\mu(c)$ as the subsets of I_μ and A_μ involved in the connected portion $R^{r(c)}_s$ with the connectivity implied by the labels in the set (c) and of string length n_c, then the expression M^l_μ can be written as

$$M^l_\mu = \sum_{r,s,p_d,q_d,m,k_d,n_c \in n, J^m_\mu, B^m_\mu} R^{r I^{n_c}_\mu(c) B^m_\mu}_{s A^{n_c}_\mu(c) J^m_\mu} \Gamma^{(k_d)p_d}_{q_d} \tag{41}$$

where, the sum runs over s, all possible contractions between the de-excitation operators $(I^{n_c}_\mu(c), A^{n_c}_\mu(c))$ with those in (J^m_μ, B^m_μ).

We note here that because of the contractions r, the set of lower and upper indices in R coming from J^m_μ and B^m_μ would take on values dictated by the contractions. However, the strings thus modified would still be of length m and we may denote them as $J^m_\mu(\rho)$ and $B^m_\mu(\rho)$. Thus, we can rewrite Eq. (41) as

$$M^l_\mu = \sum_{r,s,p_d,q_d,m,k_d,n_c \in n, J^m_\mu(\rho), B^m_\mu(\rho)} R^{r I^{n_c}_\mu(c) B^m_\mu(\rho)}_{s A^{n_c}_\mu(c) J^m_\mu(\rho)} \Gamma^{(k_d)p_d}_{q_d} \tag{42}$$

The matrix elements $\Gamma^{(k_d)p_d}_{q_d}$ obviously involves active strings p_d and q_d left uncontracted from the set (I_μ, A_μ). Since R depends entirely on the de-excitation operators $I^{n_c}_\mu(c)$ and $A^{n_c}_\mu(c)$, we no longer need the mnemonic l in its definition. But instead, we can interpret R as a matrix element involving a virtual CSF $\langle \chi^{l_\mu(c)}_\mu | = \langle \phi_\mu | \{ \varepsilon^{I^{n_c}_\mu(c)}_{A^{n_c}_\mu(c)} \}$ of rank necessarily lower than that of l_μ and the model CSF ϕ_μ of the type:

$$\langle \chi^{l_\mu(c)}_\mu | = \langle \phi_\mu | \left\{ \varepsilon^{I^{n_c}_\mu(c)}_{A^{n_c}_\mu(c)} \right\} \tag{43}$$

so that, we can write,

$$M^{l_\mu}_\mu = \sum_{p_d,q_d,k_d,c,l_\mu(c) \in l_\mu} R^{l_\mu(c)} \Gamma^{(k_d)p_d}_{q_d} \tag{44}$$

We now see that $R^{(c)}$ involves only connected quantities involving a state $l_\mu(c)$ as a de-excitation operator:

$$\begin{aligned} R^{l_\mu(c)} &= \sum_m \langle \phi_\mu | [\{E^\mu_{l_\mu(c)}\}\{E^m_\mu\}] | \phi_\mu \rangle_c g_m \\ &\equiv \sum_m \langle \phi_\mu | [\{E^\mu_{l_\mu}\}\{E^m_\mu\}] | \phi_\mu \rangle_c g_m \end{aligned} \tag{45}$$

where the difference between an ordinary matrix element $M^{l_\mu}_\mu$ and a connected matrix element $R^{l_\mu(c)}$ should be noted. While $M^{l_\mu}_\mu$ contains both connected and disconnected matrix elements, $R^{l_\mu(c)}$ contains connected matrix elements only.

Introducing a connected matrix D with elements D_{lm} defined as $D_{lm} = \langle \phi_\mu | [\{E^\mu_{l_\mu(c)}\}\{E^m_\mu\}] | \phi_\mu \rangle_c$, the entities $R^{l_\mu(c)}$ written as a column $\underline{R}$ can be expressed as

$$\underline{R} = D\underline{g}, \tag{46}$$

where $\underline{g}$ is the column of the entries g_m.

We now start hierarchically upwards for all excitations generating the various l_μs by singles, doubles, ... excitations. Then for the single excitations l^1_μ, the corresponding Ms would have no disconnected components $\Gamma^{(k_d)}$. Hence, it follows that:

$$M^{l^1_\mu}_\mu \equiv R^{l^1_\mu(c)} = 0 \quad \forall l^1_\mu, \mu \tag{47}$$

Similarly, for all two-, three-body, ... excitations involving only inactive orbitals, there cannot be any factorized Γ, since there are only inactive orbitals in the pair (I_μ, A_μ). Hence, for all inactive excitations, of excitation rank i, we can write the matrix equation to be solved as

$$M^{l^i_\mu}_\mu = R^{l^i_\mu(c)} = 0 \quad \forall l^i_\mu, \mu \tag{48}$$

We now consider the double, triples,...etc., excitations involving active orbitals. Thus, the pair (I_μ, A_μ) will contain one or more active orbitals. Then, the quantity M in such cases will be a sum of products of a one body R_1 and a Γ_1 and a connected two-body quantity R_2. Since all the one-body Rs which equal the corresponding Ms have already been proved to be zero at the converged solution, it follows that:

$$M_{\mu}^{l_{2\mu}^{i}} \equiv R_{\mu}^{l_{2\mu}^{a}(c)} = 0 \qquad \forall l_{2\mu}^{a} \tag{49}$$

Proceeding hierarchically upwards, it then follows that:

$$M_{\mu}^{l^{a}} = R_{\mu}^{l_{\mu}^{a}(c)} = 0 \qquad \forall l_{mu}^{a}, \mu \tag{50}$$

which shows that our working equations are connected. In a somewhat expanded form involving the matrix D, the entire set of Eq. (50) can be written as

$$\underline{R} = D\underline{g} = 0. \tag{51}$$

Incidentally, there is a more straightforward and a rather naive way to show the connectivity of the working equations which throws additional light on the ranks of cumulants that need to appear while using the cumulant decomposition of the Γ-matrices. We simply substitute the lower-body equations successively to write the product of disconnected Γs on the de-excitation and the rest, where the latter corresponds to a lower-body equation whose value vanishes. For a typical case where the active to active scattering in the projection operator is concentrated on one fragment and the active to active scattering in a block is concentrated on another fragment, we will present the analysis of size extensivity by replacing the lower-body equation in the higher-body equation in greater detail here.

Let us consider the block $\{^{aw}_{ix}\}$

Equation for $ix \rightarrow aw$

$$\begin{aligned} M_{ix}^{aw} \equiv\ & 2G_{i}^{a}\gamma_{w}^{x} + 2\sum_{uv} G_{iu}^{av}\Gamma_{uw}^{vx} - \sum_{uv} G_{ui}^{av}\Gamma_{uw}^{vx} + 2\sum_{uv} G_{iu}^{av}\delta_{v}^{w}\gamma_{u}^{x} \\ & - \sum_{uv} G_{ui}^{av}\delta_{v}^{w}\gamma_{u}^{x} + 2\sum_{uvyz} G_{iuy}^{avz}\delta_{v}^{w}\Gamma_{uy}^{xz} + 2\sum_{uvyz} G_{iuy}^{avz}\Gamma_{uwy}^{vxz} \\ & - \sum_{uvyz} G_{iuy}^{vaz}\delta_{v}^{w}\Gamma_{uy}^{xz} - \sum_{uvyz} G_{iuy}^{vaz}\delta_{z}^{w}\Gamma_{uy}^{vx} - \sum_{uvyz} G_{iuy}^{vaz}\Gamma_{uyw}^{vzx} = 0 \end{aligned} \tag{52}$$

Equations for $i \rightarrow a$

$$\begin{aligned} M_{i}^{a} = R_{i}^{a} \equiv\ & 2G_{i}^{a} + \sum_{uv} 2G_{iu}^{av}\gamma_{u}^{v} - \sum_{uv} G_{iu}^{va}\gamma_{u}^{v} + 2\sum_{uvyz} G_{iuy}^{avz}\Gamma_{uy}^{vz} \\ & - \sum_{uvyz} G_{iuy}^{vaz}\Gamma_{uy}^{vz} = 0 \end{aligned} \tag{53}$$

Performing cumulant decomposition on some of the terms of Eq. (52), we get:

$$\begin{aligned} M_{ix}^{aw} \equiv\ & 2G_{i}^{a}\gamma_{w}^{x} + 2\sum_{uv} G_{iu}^{av}[\gamma_{u}^{v}\gamma_{w}^{x} - (1/2)\gamma_{u}^{x}\gamma_{w}^{v} + \Lambda_{uw}^{vx}] \\ & - \sum_{uv} G_{ui}^{av}[\gamma_{u}^{v}\gamma_{w}^{x} - (1/2)\gamma_{u}^{x}\gamma_{w}^{v} + \Lambda_{uw}^{vx}] + 2\sum_{uv} G_{iu}^{av}\delta_{v}^{w}\gamma_{u}^{x} + \\ & - \sum_{uv} G_{ui}^{av}\delta_{v}^{w}\gamma_{u}^{x} + 2\sum_{uvyz} G_{iuy}^{avz}\delta_{v}^{w}\Gamma_{uy}^{xz} + 2\sum_{uvyz} G_{iuy}^{avz}[\gamma_{u}^{v}\gamma_{y}^{z}\gamma_{w}^{x} \\ & - 1/2\gamma_{u}^{z}\gamma_{y}^{v}\gamma_{w}^{x} - 1/2\gamma_{u}^{x}\gamma_{w}^{v}\gamma_{y}^{z} - 1/2\gamma_{y}^{x}\gamma_{w}^{z}\gamma_{u}^{v} + 1/4\gamma_{u}^{z}\gamma_{y}^{x}\gamma_{w}^{v} \\ & + 1/4\gamma_{u}^{x}\gamma_{w}^{z}\gamma_{y}^{v} + \Lambda_{uy}^{vz}\gamma_{w}^{x} + \Lambda_{yw}^{zx}\gamma_{u}^{v} + \Lambda_{uw}^{vx}\gamma_{y}^{z} - 1/2\Lambda_{yw}^{vx}\gamma_{u}^{z} \\ & - 1/2\Lambda_{yw}^{zv}\gamma_{u}^{x} - 1/2\Lambda_{uw}^{zx}\gamma_{y}^{v} - 1/2\Lambda_{uw}^{vz}\gamma_{y}^{x} - 1/2\Lambda_{yu}^{zx}\gamma_{w}^{v} \\ & - 1/2\Lambda_{uy}^{vx}\gamma_{w}^{z} + \Lambda_{uyw}^{vzx}] - \sum_{uvyz} G_{iuy}^{vaz}\delta_{v}^{w}\Gamma_{uy}^{xz} - \sum_{uvyz} G_{iuy}^{vaz}\delta_{z}^{w}\Gamma_{uy}^{vx} \\ & - \sum_{uvyz} G_{iuy}^{vaz}[\gamma_{u}^{v}\gamma_{y}^{z}\gamma_{w}^{x} - 1/2\gamma_{u}^{z}\gamma_{y}^{v}\gamma_{w}^{x} - 1/2\gamma_{u}^{x}\gamma_{w}^{v}\gamma_{y}^{z} \\ & - 1/2\gamma_{y}^{x}\gamma_{w}^{z}\gamma_{u}^{v} + 1/4\gamma_{u}^{z}\gamma_{y}^{x}\gamma_{w}^{v} + 1/4\gamma_{u}^{x}\gamma_{w}^{z}\gamma_{y}^{v} \\ & + \Lambda_{uy}^{vz}\gamma_{w}^{x} + \Lambda_{yw}^{zx}\gamma_{u}^{v} + \Lambda_{uw}^{vx}\gamma_{y}^{z} - 1/2\Lambda_{yw}^{vx}\gamma_{u}^{z} \end{aligned} \tag{54}$$

which can be reorganized as

$$\begin{aligned} M_{ix}^{aw} \equiv\ & \gamma_{w}^{x}[2G_{i}^{a} + 2\sum_{uv} G_{iu}^{av}\gamma_{u}^{v} - \sum_{uv} G_{ui}^{av}\gamma_{u}^{v} + 2\sum_{uvyz} G_{iuy}^{avz}\Gamma_{uy}^{vz} \\ & - \sum_{uvyz} G_{iuy}^{vaz}\Gamma_{uy}^{vz}] + 2\sum_{uv} G_{iu}^{av}[-(1/2)\gamma_{u}^{x}\gamma_{w}^{v} + \Lambda_{uw}^{vx}] \\ & - \sum_{uv} G_{ui}^{av}[-(1/2)\gamma_{u}^{x}\gamma_{w}^{v} + \Lambda_{uw}^{vx}] + 2\sum_{uv} G_{iu}^{av}\delta_{v}^{w}\gamma_{u}^{x} \\ & - \sum_{uv} G_{ui}^{av}\delta_{v}^{w}\gamma_{u}^{x} + 2\sum_{uvyz} G_{iuy}^{avz}\delta_{v}^{w}\Gamma_{uy}^{xz} 2\sum_{uvyz} G_{iuy}^{avz}[-1/2\gamma_{u}^{x}\gamma_{w}^{v}\gamma_{y}^{z} \\ & - 1/2\gamma_{y}^{x}\gamma_{w}^{z}\gamma_{u}^{v} + 1/4\gamma_{u}^{z}\gamma_{y}^{x}\gamma_{w}^{v} + 1/4\gamma_{u}^{x}\gamma_{w}^{z}\gamma_{y}^{v} \\ & + \Lambda_{yw}^{zx}\gamma_{u}^{v} + \Lambda_{uw}^{vx}\gamma_{y}^{z} - 1/2\Lambda_{yw}^{vx}\gamma_{u}^{z} \\ & - 1/2\Lambda_{yw}^{zv}\gamma_{u}^{x} - 1/2\Lambda_{uw}^{zx}\gamma_{y}^{v} - 1/2\Lambda_{uw}^{vz}\gamma_{y}^{x} - 1/2\Lambda_{yu}^{zx}\gamma_{w}^{v} \\ & - 1/2\Lambda_{uy}^{vx}\gamma_{w}^{z} + \Lambda_{uyw}^{vzx}] \\ & - \sum_{uvyz} G_{iuy}^{vaz}\delta_{v}^{w}\Gamma_{uy}^{xz} - \sum_{uvyz} G_{iuy}^{vaz}\delta_{z}^{w}\Gamma_{uy}^{vx} \\ & - \sum_{uvyz} G_{iuy}^{vaz}[-1/2\gamma_{u}^{x}\gamma_{w}^{v}\gamma_{y}^{z} \\ & - 1/2\gamma_{y}^{x}\gamma_{w}^{z}\gamma_{u}^{v} + 1/4\gamma_{u}^{z}\gamma_{y}^{x}\gamma_{w}^{v} + 1/4\gamma_{u}^{x}\gamma_{w}^{z}\gamma_{y}^{v} \\ & + \Lambda_{yw}^{zx}\gamma_{u}^{v} + \Lambda_{uw}^{vx}\gamma_{y}^{z} - 1/2\Lambda_{yw}^{vx}\gamma_{u}^{z} \\ & - 1/2\Lambda_{yw}^{zv}\gamma_{u}^{x} - 1/2\Lambda_{uw}^{zx}\gamma_{y}^{v} - 1/2\Lambda_{uw}^{vz}\gamma_{y}^{x} - 1/2\Lambda_{yu}^{zx}\gamma_{w}^{v} \\ & - 1/2\Lambda_{uy}^{vx}\gamma_{w}^{z} + \Lambda_{uyw}^{vzx}] = 0 \end{aligned} \tag{55}$$

Replacing Eq. (53) in Eq. (55), we get:

$$
\begin{aligned}
M_{ix}^{aw} \equiv\ & 2\sum_{uv} G_{iu}^{av}[-(1/2)\gamma_u^x\gamma_w^v + \Lambda_{uw}^{vx}] \\
& - \sum_{uv} G_{ui}^{av}[-(1/2)\gamma_u^x\gamma_w^v + \Lambda_{uw}^{vx}] + 2\sum_{uv} G_{iu}^{av}\delta_v^w\gamma_u^x \\
& - \sum_{uv} G_{ui}^{av}\delta_v^w\gamma_u^x + 2\sum_{uvyz} G_{iuy}^{avz}\delta_v^w\Gamma_{uy}^{xz} \\
& + 2\sum_{uvyz} G_{iuy}^{avz}[-1/2\gamma_u^x\gamma_w^v\gamma_y^z \\
& - 1/2\gamma_y^x\gamma_w^z\gamma_u^v + 1/4\gamma_u^z\gamma_y^x\gamma_w^v + 1/4\gamma_u^x\gamma_w^z\gamma_y^v + \Lambda_{yw}^{zx}\gamma_u^v \\
& + \Lambda_{uw}^{vx}\gamma_y^z - 1/2\Lambda_{yw}^{vx}\gamma_u^z \\
& - 1/2\Lambda_{yw}^{zv}\gamma_u^x - 1/2\Lambda_{uw}^{zx}\gamma_y^v - 1/2\Lambda_{uw}^{vz}\gamma_y^x - 1/2\Lambda_{yu}^{zx}\gamma_w^v \\
& - 1/2\Lambda_{uy}^{vx}\gamma_w^z + \Lambda_{uyw}^{vzx}] \\
& - \sum_{uvyz} G_{iuy}^{vaz}\delta_v^w\Gamma_{uy}^{xz} - \sum_{uvyz} G_{iuy}^{vaz}\delta_z^w\Gamma_{uy}^{vx} \\
& - \sum_{uvyz} G_{iuy}^{vaz}[-1/2\gamma_u^x\gamma_w^v\gamma_y^z \\
& - 1/2\gamma_y^x\gamma_w^z\gamma_u^v + 1/4\gamma_u^z\gamma_y^x\gamma_w^v + 1/4\gamma_u^x\gamma_w^z\gamma_y^v \\
& + \Lambda_{yw}^{zx}\gamma_u^v + \Lambda_{uw}^{vx}\gamma_y^z - 1/2\Lambda_{yw}^{vx}\gamma_u^z
\end{aligned}
\tag{56}
$$

The terms in Eq. (56) are connected as it involves terms of two types, viz. those containing cumulants which are connected by definition and those containing exchange densities involving the labels on the de-excitation as well as those on the G-block. This implies

$$
M_{ix}^{aw} = R_{ix}^{aw(c)} = 0. \tag{57}
$$

This concludes the proof of our assertion that the working equations Eqs. (21) and (22) are connected.

We note here that there is a subtle but significant aspect of any connected formulation where the connectivity is predicated by the connectedness of cumulants. In order to effect cancelation of the disconnected $\Gamma_{q_d}^{p_d}$ and the lower-body M which would invariably appear, it is obviously necessary to include certain higher-body densities up to rank $m_{\max} + n_{\max}$ in a given truncation scheme where a maximum of $n_{\max}$ number of active lines enter the G-blocks from the right and arbitrary pairs m of active pairs of de-excitation operators up to some $m_{\max}$. It then can so happen that for an $n_{\max}$ active electron situation, all the density matrices of rank greater than $n_{\max}$ are zero by the Pauli exclusion principle. This essentially implies that the numerical cancelation of the product of $\Gamma_{q_d}^{p_d}$ and the lower-body equations will implicitly generate Λs of rank up to $m_{\max} + n_{\max}$. As is known, connectivity of cumulants imparts the bizarre property of having non-vanishing values even when its rank exceeds the number of active electrons. Thus, it is remarkable that while applying our UGA-MRCC to the case of two active electrons, some of the projection equations involving at least a pair of active orbitals in the de-excitation would generate Γ_3 which are zero. However, if we include all G-blocks containing up to two valence lines entering the G-block, then the numerical cancelation of the disconnected terms will automatically imply inclusion of Λ_3. This is clearly manifest in Eq. (57), even for the case of two active electrons. Thus, our analysis above provides the vitally important insight that *the Λ s of rank greater than the number of active electrons* are necessary for maintaining size extensivity, in case we decide to work with equations containing Γ's, rather than product of Λs. In particular, if we look for applications for states with very few electrons, say two, then connectivity as well as accuracy demands that we keep terms containing Λ_3 (although the corresponding Γ_3 matrix elements are all zero).

A strategy, which we have adopted in practice in our first papers [44–46], works with the parent UGA-MRCC equations themselves. Although they contain γ and Γs of various ranks, we carefully choose to include in the equations only those terms in the higher-body equations, which on cumulant decomposition of the Γ, would precisely yield terms of the form $\Gamma_{q_d}^{p_d}$ times a connected M with all the terms of M contained in the lower-body connected equation. Thus, essentially we allow the higher-body working equations to take care of the product of $\Gamma_{q_d}^{p_d}$ and of the M not to contribute because of the lower-body equation set to zero. As an example, if we include in every equation all the G-blocks containing up to some fixed $n_{\max}$ number of active lines entering the G-blocks from the right, then it automatically follows that, after the factorization of terms like $\Gamma_{q_d}^{p_d}$, the lower-body connected factors M would also contain the G-blocks where the numbers of active lines entering the G-blocks can go up to $n_{\max}$. This ensures numerical cancelation of all the disconnected terms in the parent working equations. As long as the constraint on the number of active lines entering the G-blocks is maintained, we can take the liberty of including in our working equations for the highest rank of the Γs, m arbitrary pairs of active de-excitation operators up to some $m_{\max}$ which are not connected to G in anyway. This is precisely the way we have programed our working equations in our actual applications [44–46].

It has not escaped our attention that instead of using the de-excitation operators $\{E_{\mu}^{l_\mu}\}^{\dagger}$ if we use a modified operator $\{\tilde{E}_{\mu}^{l_\mu}\}^{\dagger}$ defined by:

$$
\{\tilde{E}_{\mu}^{l_\mu^n}\}^{\dagger} = \{E_{\mu}^{l_\mu^n}\}^{\dagger} - \sum_{k_d}\{E_{\mu}^{l_\mu^{n-k_d}}\}^{\dagger}\Gamma^{(k_d)} \tag{58}
$$

then the operator $\{\tilde{E}_{\mu}^{l_\mu}\}^{\dagger}$, in effect, subtracts from $\{E_{\mu}^{l_\mu^n}\}^{\dagger}$ all subsets of k_d pairs of active to active scatterings. Clearly, if instead of projecting on to $\{E_{\mu}^{l_\mu}\}^{\dagger}$ had we projected by

$\{\tilde{E}^{l_\mu}_\mu\}^\dagger$, then all the disconnected density matrices of all ranks residing entirely on the de-excitation operators would have canceled out. This implies that it is more convenient to choose the cluster operators in terms of amplitudes $\tilde{t}^l_\mu$ and the operators $\{\tilde{E}^{l_\mu}_\mu\}$. Interestingly, an operator $\{E^{l_\mu}_\mu\}_0$ defined in generalized normal ordering (GNO) with respect to ϕ_μ—introduced in a different context—has exactly the same expression as $\{\tilde{E}^{l_\mu}_\mu\}$. Cluster operators in GNO [75, 80, 88, 91, 92] with respect to CSFs and multi-reference functions were introduced much earlier to generate explicitly connected internally contracted multi-reference (ICMRCC) theories [76–79, 88, 89, 92]. Our analysis above implies that the UGA-MRCC theories would also benefit greatly by a similar choice of the cluster operators. In particular, it leads to a compact choice of linearly independent excitation operators adapted to the direct product group $\boldsymbol{u}$ for each fixed number m_c and m_v orbitals involved in the excitation. We hope to return to this soon in a forthcoming publication.

3.2 Proof of the extensivity and size-consistency of the UGA-MRCC theories

In order to establish the extensivity of the UGA-MRCC theories, it is necessary to prove the extensivity of the cluster operators T_μ. To prove this, we arbitrarily group all the orbitals into two groups having some fixed number of electrons in each group and equate the matrix elements of Hamiltonian containing mixed inter-group indices as zero. In such a situation, the total CAS function Ψ^0_{CAS} becomes an anti-symmetrized product $\Psi^0_{CAS} = \sum \mathcal{A}_{AB}[\Psi^{0_{S_A}}_A \Psi^{0_{S_B}}_B]^s$. The corresponding spinorbital n-RDMs for such a function can always be written as

$$\gamma^{(n)}_{AB} = \mathcal{A}_{AB}[\gamma^{(n_A)}_A \gamma^{(n_B)}_B]^s \tag{59}$$

where $\gamma^{(n_A)}_A$ and $\gamma^{(n_B)}_B$ are each anti-symmetric with respect to their own spinorbital indices. By summing over the spin indices, the analogous decomposition of the corresponding spin-free n-RDMs follows.

$$\Gamma^{(n)}_{AB} = \mathcal{P}_{AB}[\Gamma^{(n_A)}_A \Gamma^{(n_B)}_B]^s, \tag{60}$$

where $\mathcal{P}_{AB}$ ia a sum of all possible permutations between all the upper orbital indices of group A and group B accompanied with prefactors for each permutation that depends on the spin s. $\Gamma^{(n_A)}_A$ and $\Gamma^{(n_B)}_B$ are the RDMs corresponding to $\Psi^{0_{S_A}}_A$ and $\Psi^{0_{S_B}}_B$, respectively. Using cumulant decomposition of the entire density matrix, the $\Gamma^{(n)}_{AB}$ can be written as a sum of products of cumulants with each cumulant labeled by its own group indices and another sum where each term will contain at least one cumulant with mixed group indices. Of course, since the subsystems $\Psi^0{}_A$ and $\Psi^0{}_B$ contain fixed number of electrons, the number of indices in the lower and upper entries of each RDM should not only be the same but also would contain indices of the same set, i.e., they are quasi-diagonal. It is a straightforward matter to show that mixed cumulants with the same entries of the indices in the upper and lower positions, the so-called mixed direct cumulants, are zero from their very definition. As an example, for a model space with two active electrons coupled to a spin s and with orbitals u_A and v_B in different subgroups $\Gamma^{u_A v_B}_{u_A v_B} = 1$. The corresponding direct cumulant is zero:

$$\begin{aligned}\Lambda^{u_A v_B}_{u_A v_B} &= \Gamma^{u_A v_B}_{u_A v_B} - \gamma^{u_A}_{u_A}\gamma^{v_B}_{v_B} \\ &= 1 - 1 = 0\end{aligned} \tag{61}$$

In sharp contrast, the inter-group 'exchange' cumulants $\Lambda^{u_A v_B}_{v_B u_A}$ would have a dramatically different property since the corresponding 2-RDM $\Gamma^{u_A v_B}_{v_B u_A}$ has the value $(-1)^s$ for $s = 0, 1$. We find in a straightforward manner that:

$$\begin{aligned}\Lambda^{u_A v_B}_{v_B u_A} &= \Gamma^{u_A v_B}_{v_B u_A} + \frac{1}{2}\gamma^{u_A}_{u_A}\gamma^{v}_{B}{}_{v_B} \\ &= (-1)^s + \frac{1}{2} \\ &= \frac{2(-1)^s + 1}{2}\end{aligned} \tag{62}$$

which is nonzero for $s = 0, 1$. Thus, these mixed exchange cumulants are connected but not extensive and arises because of the constraint imposed on the overall function Ψ^0 to be of a given spin s. This is also reflected on the value of the cumulant which is explicitly dependent on the value of s.

We utilize the above property of the inter-group direct and exchange cumulants to prove the size extensivity of the cluster amplitudes of T_μ. We recall from Eq. (45) that each term in $R^{l^a_\mu(c)}_\mu$ is summed over m, with excitations leading from $|\phi_\mu\rangle$ to $|\chi^m_\mu\rangle$, and an accompanying amplitude g_m of G_μ:

$$R^{l^a_\mu(c)}_\mu = \sum_m \langle\phi_\mu|\{E^{l_a}_\mu\}^\dagger\{E^m_\mu\}|\phi_\mu\rangle_c g^m_\mu \tag{63}$$

The entity $\langle\phi_\mu|...|\phi_\mu\rangle_c$ is a very special connected scalar product where there are no disconnected products of cumulants residing entirely on the de-excitation operator. Assuming that T_μs are connected, the cumulants residing entirely on the G_μ block form a connected composite. The only cumulants that can make a term in $R^{l^a_\mu(c)}_\mu$ size-inextensive is where a mixed exchange cumulant appears connecting the operators $\{E^{l_a}_\mu\}^\dagger$ and $\{E^m_\mu\}$.

As a result, if we separate the orbitals into two groups in the limit of no inter-group interaction, then the only non-zero terms are those in which either all orbitals in the term are on A or on B or they have mixed exchange inter-group cumulants. The matrix D of Eq. (46) in the non-interacting limit has thus the following structure

$$\begin{bmatrix} D^{AA}_{AA} & 0 & 0 & 0 \\ 0 & D^{BB}_{BB} & 0 & 0 \\ 0 & 0 & 0 & D^{AB}_{BA} \\ 0 & 0 & D^{BA}_{AB} & 0 \end{bmatrix}$$

This indicates that the set of MRCC equation (46) in the limit of vanishing inter-group interaction decouples into set of equations on each fragment only, denoted by sub-columns $\underline{g}_{AA}$ and $\underline{g}_{BB}$. There is also a set where size-inextensive terms appear only because the corresponding D^{AB}_{BA} and D^{BA}_{AB} are nonzero. The matrix elements of D^{AB}_{BA}/D^{BA}_{AB} are all nonzero for every set of mixed indices in AB in D^{AB}_{BA}/D^{BA}_{AB} since they contain at least one mixed exchange cumulant. Interestingly, the matrix $\bar{D}_{AB}$,

$$\overline{D}_{AB} = \begin{bmatrix} 0 & D^{AB}_{BA} \\ D^{BA}_{AB} & 0 \end{bmatrix}$$

is non-singular since its determinant is given by:

$$\det\bar{D}_{AB} = -\det[D^{BA}_{AB} D^{AB}_{BA}] = -\det[D^{\dagger AB}_{\ BA} D^{AB}_{BA}] \tag{64}$$

so that the possible set of trouble-makers, leading to non-zero g_{AB}/g_{BA}, are of the form:

$$\begin{bmatrix} 0 & D^{AB}_{BA} \\ D^{BA}_{AB} & 0 \end{bmatrix} \begin{bmatrix} g_{AB} \\ g_{BA} \end{bmatrix} = 0 \tag{65}$$

The non-singularity of $\overline{D}_{AB}$ would imply

$$g_{AB} = 0 \tag{66}$$

$$g_{BA} = 0 \tag{67}$$

In effect set, g_{AB} is the same as the set g_{BA}. We have shown them separately to facilitate the analysis. Equation (66) can be taken as the set of equations determining T_{AB}. The corresponding equations for subgroup A and subgroup B, respectively, satisfy:

$$D^{AA}_{AA} g_{AA} = 0 \tag{68}$$

and

$$D^{BB}_{BB} g_{BB} = 0 \tag{69}$$

We now prove that the cluster amplitudes of T_μ are not only connected but also extensive. To begin with, we of course assume that there are three sets of T: T_{AA}, T_{BB} and T_{AB}, and all appear in a coupled manner in all the three Eqs. (66), (68), and (69). The extensivity of T follows by a close inspection of the algebraic structure of these equations.

Let us first look at the nature of connectivity of the matrix element of T_{AB} as obtained from the solution of Eq. (66). We note that the amplitudes g_{AB} have no explicit dependence on mixed cumulants and, indeed, of any cumulants at all since cumulants appear only in the matrix D. However, they may implicitly contain a cumulant since Eqs. (66), (68) and (69) have couplings between the operators T_{AA}, T_{BB} and T_{AB}, though they contain intra-group cumulants, Λ_{AA} and Λ_{BB}, which are harmless. In the expression of amplitude g_{AB}, this so-called leading term—H_{AB} having no T—is zero in the limit of vanishing inter-group interaction. There is also a term in the equation for the amplitude t^{AB}_l, say, of the form $\Delta E^{AB}_l t^{AB}_l$. The rest of the terms collected in $g'^{AB}_{\ l}$ will consist of all connected terms involving H_{AA} and all the Ts and also H_{BB} and all the Ts. For $g'^{AB}_{\ l}$ to be nonzero, there has to be one T_{AB} amplitude since H cannot lead to inter-group coupling in the no interaction limit. We now write the equation for the amplitude t^{AB}_l using the above consideration as

$$\Delta E^{AB}_l t^{AB}_l + g'^{AB}_{\ l} = 0 \tag{70}$$

leading to,

$$t^{AB}_l = -\frac{g'^{AB}_{\ l}}{\Delta E^{AB}_l} \tag{71}$$

We can generate the amplitude via an iterative procedure of the type

$$t^{AB(p+1)}_l = -\frac{g'^{AB(p)}_{\ l}}{\Delta E^{AB}_l} \tag{72}$$

where $g'^{AB(p)}_{\ l}$ is computed using the t amplitude of the p-th iteration step. Since there is no first order t^{AB}_l, H_{AB} being zero, we start out with zero amplitudes for all T^{AB}_m for all m. Since the possible nonzero contribution to $g'^{AB(p)}_{\ l}$ can come only from the contribution of $t^{AB(p)}_l$, it follows that t^{AB}_l remains zero in the successive iterations. It hence follows that all the cluster amplitudes with mixed indices are zero for vanishing inter-group interaction, and hence, T_{AB} is zero.

We now consider the solution of Eq. (68) which can be written in longhand as

$$D^{AA}_{l,l} H^{AA}_l + D^{AA}_{l,l} \Delta E^{AA}_l t^{AA}_l + \sum_m D^{AA}_{l,m} g'^{AA}_{\ m} = 0 \tag{73}$$

so that we can set up the following iterative procedure as

$$t^{AA(p+1)}_l = -\frac{H^{AA}_l}{\Delta E^{AA}_l} - \frac{\sum_m D^{AA}_{l,m} g'^{AA(p)}_{\ m}}{D^{AA}_{l,l} \Delta E^{AA}_l} \tag{74}$$

Now $g'^{AA(p)}_{\ m}$ is a connected composite containing either H_{AA}

or H_{BB} with all T_{AA} and T_{BB} (T_{AB} being zero), it follows that g'^{AA}_{m} can come only from composite of H_{AA} and powers of T_{AA} since no T_{BB} can connect to H_{AA} and no term containing H_{BB} can contribute to amplitudes of type AA. Thus, all the cluster amplitudes T_{AA} are localized on A and cannot see the fragment B. The same analysis holds for the amplitudes of T_{BB}. We have thus proved that the cluster operator T is extensive and the effective operator $\tilde{H}$ satisfies

$$\tilde{H} = \tilde{H}_{AA} + \tilde{H}_{BB} \tag{75}$$

in the vanishing inter-group interaction limit.

Since our UGA-MRCC theories are not invariant with respect to transformation among active orbitals, we need to localize the orbitals on our target fragments A and B, having fixed number of electrons n_A and n_B and fragment spins s_A and s_B, in order to prove size-consistency. In that case, all the arguments used above to prove the size extensivity carries through to the situation of vanishing inter-fragment interaction and shows that Ts with mixed inter-fragment indices are all zero. Since with localized representation the Hamiltonian is also additively separable, it follows in a straightforward manner that the effective matrix $\tilde{H}$ simply becomes a sum of $\tilde{H}_A$ and $\tilde{H}_B$:

$$\tilde{H} = \tilde{H}_A + \tilde{H}_B \tag{76}$$

which indicates $E = E_A + E_B$. There are two remarkable consequence of the proof presented above. One is the realization that at the fragmentation limit the MRCC equations for each fragment are just the amplitude equations, Eq. (66). These equations do not depend on the value of total spin s since the total spin enters the MRCC equations only via the mixed exchange cumulants. They, however, depend on s_A and s_B since the intra-group cumulants Λ_{AA} and Λ_{BB} depend on s_A and s_B, respectively. This implies that at the fragmentation limit into two fragments with n_A and n_B electrons and spins s_A and s_B, respectively, the total fragment asymptote energy E depends only on s_A and s_B and not on the total spin s. Thus, all non-interacting composite functions $\mathcal{A}_{AB}[\Psi_A^{s_A}\Psi_B^{s_B}]^s$ for all values of s ranging from $|s_A - s_B|$ to $|s_A + s_B|$ will be exactly degenerate. The second consequence is that at the asymptotic limit, the amplitude equations, for example, Eq. (66) and the projection equations Eqs. (68) and (69), are equivalent.

We have verified by numerical applications along the lines of [93] that both the consequences above hold good. For ethylene, using orbitals localized on the two triplet methylene fragments, the total energy at the fragmentation limit is just the sum of two methylene triplet energies for the three PES of singlet, triplet, and quintet ethylene with errors of the order of a few μH. We will present all such results in a comprehensive manner in a correspondence in the near future.

We conclude this subsection with a brief discussion of the strategy one may use for choosing only linearly independent cluster operators in T_μ. It is always better in this case to use $U(n)$ generators written in GNO [75, 80, 88, 91, 92], denoted as $\{E^l_\mu\}_0$. For excitation classes involving m_c core and m_a active destructions and m_v virtuals, and m'_a active creations, one may follow the strategy used by Li and Paldus [33] for example. Another possibility is to take pure excitation operators first with no spectator active orbitals which are all linearly independent. The higher-rank operators with spectator excitations can be chosen by successive orthogonalization at each rank to those in the lower rank starting with the pure excitations. Eventually, the linearly independent excitation operators in GNO span the same space when acting on the function $|\phi_\mu\rangle$. Since our analysis of extensivity relies on connectivity of MRCC equations and of the special property of mixed exchange cumulants, it is not imperative for us to resort to any group theoretic argument of the type invoked by Jankowski et. al. [34].

4 Summary and future outlook

In this paper, we have presented several aspects of the unitary group adapted MRCCs of the state-specific and state-universal type:

(a) We get rather compact working equations of MRCC via the use of a Jeziorski–Monkhorst inspired normal-ordered cluster Ansatz for the component $\Omega_\mu = \{\exp T_\mu\}$ of the wave operator Ω which acts on a spin-free model function ϕ_μ where the $\{...\}$ denotes normal ordering with respect to the common core part $|0\rangle$ of the CSFs.

(b) The attendant MRCC equations are obtained as matrix equations of a connected operator G_μ between the virtual functions χ^l_μ and ϕ_μ containing a sum of terms involving the amplitudes of a connected composite containing the Hamiltonian, H, and the powers of the cluster operator, T_μ multiplied by the n-RDMs defined in the active space. The connectivity of the cluster amplitudes then hinges on factorization of the MRCC equations via a cumulant decomposition of the n-RDMs into a connected part containing the composite and certain density matrices. This eventually leads us to a set of modified MRCC equations containing connected composites only. As a consequence, the cluster operators are all connected.

(c) We distinguish between cumulants which are connected entities but not necessarily size extensive (in the sense that they do not vanish if interactions between the groups of orbitals entering simultaneously in a term are switched off) and size-extensive entities which vanish in such situations.
(d) We show that exchange type of inter-group cumulants are connected though not extensive and the rest are extensive.
(e) Finally, we prove that in the limit of vanishing interaction between two groups A and B of orbitals, the MRCC equations for the mixed Ts reduce to just the amplitude equations for connected composites denoted as G_μ^{AB}. Since G_μ^{AB}s vanish for non-interacting fragments, the amplitudes with mixed indices are also zero. This leads to the extensivity of the theories. For orbitals in localized fragments, this also proves the size-consistencies of the theories.
(f) We also provide additional insights into reasonable truncation strategies for the various spin-free n-body cumulants, which we glean from an alternative way of looking at the connectivity problem.

The aspects of size extensivity depend essentially on the factorization of the MRCC equations into connected composites multiplied by disconnected densities. The very systematic nature of such factorization implies that suitably modified excitation operators in defining the cluster operators T_μ might be chosen in such a way that the disconnected densities will not appear at all. It is heartening to see that such modified excitation operators are nothing but some operators written in generalized normal ordering (GNO) introduced quite some time ago in the context of Internally Contracted MRCC [88, 89, 92] where the genesis of the concept of normal ordering came from a different but related motivation of an attempt to find operators in some special normal ordering such that the average of such normal-ordered operators have vanishing expectation value with respect to a general multireference function. We will use this unifying concept in all our future formulations both for the multi-component wave operator-based theories with Ω_μ and for the internally contracted state-specific MRCC with judicious use of normal ordering. We hope to come back soon with such considerations in our forthcoming publications.

Acknowledgments We thank Marcel Nooijen for very constructive critical remarks on the paper and for his insistence that we clearly distinguish between the asymptotic behavior of cumulants and extensive operators. Thanks are also due to an anonymous referee for in-depth and highly informative comments, which led us to improve our presentation by removing certain ambiguities and inserting more references. DM thanks Andreas Savin and Werner Kutzelnigg for very stimulating discussions on the properties of cumulants. R.M. and D.S. thank Department of Science and Technology (New Delhi) for research fellowships. S.S. thanks Centre of Scientific and Industrial Research, (New Delhi) for SPM fellowship. D.M. thanks IFCPAR/CEFIPRA and the Ind-Nor bilateral programs for research Grants. D.M. also acknowledges generous financial support of the Science and Engineering Research Board (New Delhi) for the J. C. Bose National Fellowship. We also thank Avijit Shee, Pradipta Kumar Samanta, and Avijit Sen for their unstinting help and for critical comments on the manuscript. DM reminisces the lavish, warm, and stimulating interactions with Isaiah Shavitt and dedicates this paper in his memory.

Appendix: Definition of spin-free RDMs and the associated spin-free cumulants

The n-particle density matrices are not size-extensive, but are instead product separable, just like the wave operator. However, one can factorize any n-body density matrix in spinorbital basis as anti-symmetrized products of 1-particle density matrices and '*cumulants*' which can be recursively defined as the rank n of γ_n increases. For now, we denote by $u, v....$, etc., the active spinorbitals. For a CASSCF function Ψ_0, all the core spinorbitals are fully filled in each ϕ_μ of Ψ_0, and hence, all the γ_ns corresponding to a ϕ_μ will factor out as anti-symmetrized product of γ_{n_a} and γ_{n_c} with $(n_a + n_c) = n$, where n_a and n_c are the number of valence and core occupancies:

$$\gamma_n = \frac{1}{n_a!n_c!}\mathcal{A}[\gamma_{n_a}\gamma_{n_c}] \tag{77}$$

Clearly, the indices in γ_{n_a} are all active whereas those in γ_{n_c} are all hole spinorbitals. γ_{n_c} can always be written as an anti-symmetric product of the n_c 1-body density matrix elements with hole labels:

$$\gamma_n = \mathcal{A}[\gamma_1 \otimes \gamma_1 \otimes \gamma_1 \otimes ... n_c \text{ times}] \tag{78}$$

with appropriate hole orbital indices. We thus see that γ_{n_c} is completely product separable but for a general ϕ_μ, which is a multi-determinant function, γ_{n_a} is not necessarily so. One can separate the product separable portion of γ_{n_a} via cumulant decomposition where the cumulants are a measure of the extent of correlation beyond what is contained in the factorizable portion including Fermi correlation. One can introduce cumulants recursively in the following manner:

$$\begin{aligned}\gamma_{2wx}^{uv} &= \gamma_w^u\gamma_x^v - \gamma_x^u\gamma_w^v + \lambda_{wx}^{uv} \\ &= \mathcal{A}[\gamma_w^u\gamma_x^v] + \lambda_{wx}^{uv}\end{aligned} \tag{79}$$

$$\gamma_{3xyz}^{uvw} = \mathcal{A}[\gamma_x^u\gamma_y^v\gamma_z^w] + \mathcal{A}[\gamma_x^u\lambda_{yz}^{vw}] + \lambda_{xyz}^{uvw} \tag{80}$$

and so on [80].

The spin-free form of the above cumulant decomposition requires careful handling. One may imagine that a straightforward spin-free form of the cumulant decomposition should be possible by an appropriate summation over spin indices for the same set of orbitals in the upper and the lower array at the same place in the string. However, some reflection indicates that a cumulant decomposition involving only spin-free one-body reduced densities and appropriate spin-free cumulants is not generally possible for a non-singlet ϕ_μ, with a fixed M_s value. Unless some spin-density matrix elements are introduced, generation of a spin-free Γ is not possible. To circumvent this difficulty, Kutzelnigg and Mukherjee defined an M_s averaged ensemble density matrix [80]:

$$\Gamma_{n\,v_1v_2\ldots v_n}^{\;u_1u_2\ldots u_n} = \frac{1}{2S+1}\sum_{M_s=-S,\sigma_1\ldots\sigma_n}^{S} {}^{M_s}\gamma_{n\,v_{1\sigma_1}v_{2\sigma_2}\ldots v_{n\sigma_n}}^{\;u_{1\sigma_1}u_{2\sigma_2}\ldots u_{n\sigma_n}} \tag{81}$$

where ${}^{M_s}\gamma_n$ is the spinorbital n-body-reduced density matrix for a given Γ with a given value of M_s. Since this is an equally weighted sum of the n-RDMs for all M_s values, it does not depend on M_s at all and is invariant under rotation of the quantization axis. Kutzelnigg and Mukherjee demonstrated that, with such spin-free density matrices, a cumulant decomposition is possible which does not involve spin densities.

A consequence of this definition above is that the spin-free γ_v^u can be defined as a sum of two spinorbital densities $\gamma_{v_\uparrow}^{u_\uparrow}$ and $\gamma_{v_\downarrow}^{u_\downarrow}$ and can be written as

$$\gamma_{v_\sigma}^{u_\sigma} = \frac{1}{2S+1}\sum_{M_s=-1/2}^{1/2} {}^{M_s}\gamma_{v_\sigma}^{u_\sigma} \tag{82}$$

For such M_s-averaged quantities, $\gamma_{v_\sigma}^{u_\sigma}$ for both up and down spins are equal and hence each equals $1/2\gamma_v^u$. With such simplifications, and introducing spin-free cumulants Λ_n in a similar manner:

$$\begin{aligned}\Lambda_{v_1v_2\ldots v_n}^{u_1u_2\ldots u_n} &= \frac{1}{2S+1}\sum_{M_s=-s,\sigma_1\ldots\sigma_n}^{s} {}^{M_s}\lambda_{v_{1\sigma_1}v_{2\sigma_2}\ldots v_{n\sigma_n}}^{u_{1\sigma_1}u_{2\sigma_2}\ldots u_{n\sigma_n}}\\ &\equiv \sum_{\sigma_1\ldots\sigma_n} \lambda_{v_{1\sigma_1}v_{2\sigma_2}\ldots v_{n\sigma_n}}^{u_{1\sigma_1}u_{2\sigma_2}\ldots u_{n\sigma_n}}\end{aligned} \tag{83}$$

It follows that:

$$\Gamma_{2\,wx}^{\;uv} \equiv \Gamma_{wx}^{uv} = \gamma_w^u\gamma_x^v - \frac{1}{2}\gamma_x^u\gamma_w^v + \Lambda_{wx}^{uv} \tag{84}$$

From now on, we would take the indices $u, v, \ldots$, etc., as *orbitals* and *not spinorbitals*.

In the general case of a k-body spin-free RDM, the cumulant decomposition would require the concept of partitions i of k labels with N_i as the length of the upper and lower strings appearing in the products of Λs in this partition. For the uniformity of notation, we would introduce the notation Λ_1 for γ. The product of Λs are all connected in each partition i due to connectivity of type (C). All the terms contributing to the partition are connected among themselves in a sense which is best explained by an example. Let us consider a partition of length 4 with labels as shown: $\binom{u_1u_2u_3u_4}{v_1v_2v_3v_4}$. Here the upper indices are labeled by the index u and lower indices are labeled by the index v and the numeral in the subscript indicates the column to which each index belongs. If two terms from all the products of cumulants generated have the connectivities as shown in Figs. 5 and 6, then the connectivities of these two terms can be inferred by following the type of pairings of the upper and lower indices.

In Fig. 5, u_1 and v_2 are paired which indicates that u_2 and v_1 are also connected being the corresponding indices. Columns 1 and 2 are thus connected. Further, since u_2 and v_3, u_3 and v_4 and u_4 and v_1 are also connected, the entire composite is connected. The overall contribution of this term is $\frac{1}{8}\Lambda_{v_2}^{u_1}\Lambda_{v_3}^{u_2}\Lambda_{v_4}^{u_3}\Lambda_{v_1}^{u_4}$. The prefactor $1/8$ *depends entirely* on the type of connectivity of cross-contractions between the four Λs. In Fig. 6, the pairing of u_1 and v_2 and u_2 and v_3 produce two Λs of rank 1, thereby connecting columns 1,2 and 3, and u_3 and u_4 shown in the box are a part of a two body Λ, automatically making all four indices connected. The overall factor of this term in given by$\frac{1}{2}\Lambda_{v_2}^{u_1}\Lambda_{v_3}^{u_2}\Lambda_{v_1v_4}^{u_3u_4}$ The accompanying factor $1/2$ again depends on connectivity and rank of Λs.

Let us use indices $j_i = 1$ and $j_i = 2$ to indicate the different patterns of connectivity belonging to the same partition i. The factors, accompanying every connectivity symbolized as $\kappa_{j_i}^i$, are $\kappa_1^i = -1/8$ and $\kappa_2^i = -1/2$,

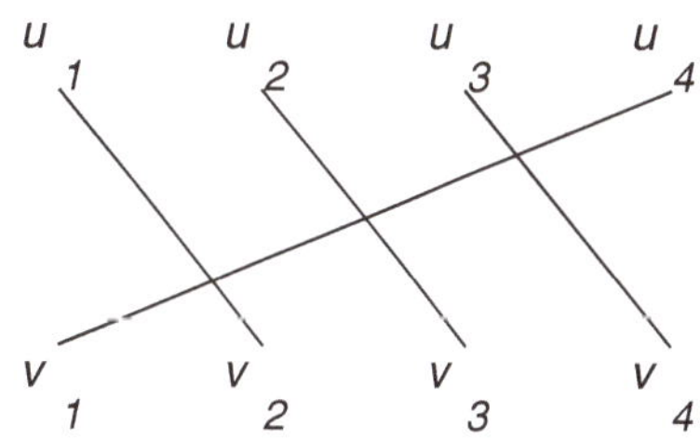

Fig. 5 One mode of connectivity corresponding to partition i and length $N_i = 4$ with four γ matrix elements

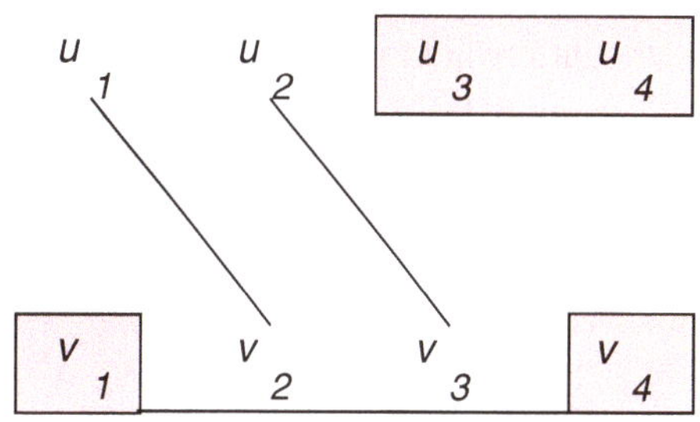

Fig. 6 Another mode of connectivity corresponding to partition i and length $N_i = 4$ where the pairs (u_3, u_4) and (v_1, v_4) form the upper indices of a Λ_2

Table 1 List of connectivities and the factors $\kappa^i_{j_i}$ for partitions N_i of ranks 2, 3, and 4

Partition (i)	j_i	Connectivity	$\kappa^i_{j_i}$
2	1	[(1 2)–(1 2)]	1
	2	[1–2, 2–1]	$-\frac{1}{2}$
3	1	[(1 2 3)–(1 2 3)]	1
	2	[1–2, (2 3)–(1 3)]	$-\frac{1}{2}$
	3	[1–2, 2–3, 3–1]	$-\frac{1}{4}$
4	1	[(1 2 3 4)–(1 2 3 4)]	1
	2	[1–2, (2 3 4)–(1 3 4)]	$-\frac{1}{2}$
	3	[(1 2)–(1 3), (3 4)–(2 4)]	$-\frac{1}{2}$
	4	[1–2, 2–3, (3 4)–(1 4)]	$\frac{1}{4}$
	5	[1–3, 2–4, (3 4)–(1 2)]	$\frac{1}{4}$
	6	[(1 2)–(3 4), (3 4)–(1 2)]	$\frac{1}{3}$
	7	[(1 2)–(3 4), (3 4)–(2 1)]	$\frac{1}{6}$
	8	[(1 2)–(4 3), (3 4)–(1 2)]	$\frac{1}{6}$
	9	[(1 2)–(4 3), (3 4)–(2 1)]	$\frac{1}{3}$
	10	[1–2, 2–3, 3–4, 4–1]	$-\frac{1}{8}$

respectively. *The factor* $\kappa^i_{j_i}$ *is unique to every* j_i *and does not depend on the connectivity of other partitions.* We have collected in Table 1 the quantities j_i and κ_{j_i} for partitions of length 2, 3, and 4. Instead of the diagrams like Figs. 5 and 6, we have used a compact algebraic notation to describe the connectivity. In this notation, the connections involving 1−body RDM are denoted by $u_P - v_Q$ to indicate the connection of a u in the Pth column with a v in the Qth column. Similarly, an n-tuple connection is indicated by $(u_{k_1}, u_{k_2}, \ldots, u_{k_n}) - (v_{l_1}, v_{l_2}, \ldots, v_{l_n})$. This particular n-tuple will generate a Λ^n where for an upper index k_1 the matching is with lower index l_1. Using this notation, Fig. 5 has the connectivity [1–2, 2–3, 3–4, 4–1] and Fig. 6 has the connectivity [1–2, 2–3, (3 4)–(4 1)].

Generally speaking, the total contribution Δ^{N_i} of a string of operators of length N_i can be compactly written as

$$\Delta^{N_i} = \sum_{j_i = \text{pattern}\mathrm{N}_{k_i}} \Pi_{k_i(\sum_{k_i} N_{k_i} = N_i)} \Lambda^{N_{k_i}} \kappa^i_{j_i} \tag{85}$$

It is now straightforward to write the cumulant decomposition of a k−body RDM. Let us assume that $\Gamma^{(k)}$ has a set of upper and lower indices $u_1 \ldots u_k$ and $v_1 \ldots v_k$, respectively. We may then generate all possible terms of the cumulant decomposition by first partitioning the set of u and v in all possible manner with the restriction $N_1 \geq N_2 \geq N_3 \ldots$ starting with $\binom{u_1}{v_1}$. The value of the partition N_i, Δ^{N_i}, will be labeled by unique ordered set of u and v and the overall contribution of the first term of the cumulant will be $\Pi_i \Delta^{N_i}$, where the upper indices $u_1 \ldots u_k$ and lower indices $v_1 \ldots v_k$ appear consecutively in the Γ with all possible connectivity in a given partition. Every pattern characterized by the set $\{N_i\}$ will be given a class symbol cl. The overall contribution for every class cl will be generated by simultaneous interchange of the same upper and lower indices in all possible ways between the different partitions. Clearly the possible permutations depend on the class cl and contribution of all the terms of the class cl can be compactly written as $P_{cl}\Delta^{cl}$. The overall contribution of the $\Gamma^{(k)}$ can be then compactly written as

$$\Gamma^{(k)} = \sum_{cl} P_{cl} \Delta^{cl} \tag{86}$$

with the identity permutation of P_{cl} as the unity operator given by $\mathbb{1}$.

References

1. Paldus J, Wormer PES (1978) Phys Rev A 18:827
2. Ruedenberg K (1971) Phys Rev Lett 27:1105
3. Ruedenberg K, Poshusta RD (1972) In: Löwdin PO (ed) Advances in Quantum Chemistry, vol 6. Academic Press, London, p 267
4. Salmon WI, Ruedenberg K (1972) J Chem Phys 57:2776
5. Pauncz R (1995) The symmetric group in quantum chemistry. CRC Press, Boca Raton
6. Kaplan IG (1975) Symmetry of many electron systems. Academic Press, New York
7. Drake GWF, Schlesinger M (1977) Phys Rev A 15:1990
8. Wormer PES, Paldus J (2006) Adv Quantum Chem 51:59
9. Shepard R (2005) J Phys Chem A 109:11629
10. Shepard R (2006) J Phys Chem A 110:8880
11. Shepard R, Minkoff Michael (2006) Int J Quantum Chem 106:3190
12. Shepard R, Wagner AF, Gray SK (2006) J Phys Conf Ser 46:239
13. Moshinsky M (1967) Group theory and the many body problem. Gordon and Breach, New York
14. Moshinsky M, Seligman TH (1971) Ann Phys 66:311
15. Seligman TH (1981) Lecture Notes in Chemistry 22:362
16. Matsen FA (1987) Int J Quantum Chem 32:71
17. Matsen FA, Cantu AA (1968) J Phys Chem 72:21
18. Paldus J (1974) J Chem Phys 61:5321
19. Shavitt I (1977) Int J Quantum Chem 12(S11):131
20. Shavitt I (1978) Int J Quantum Chem S12:5
21. Paldus J and Shavitt I (1988) In: Mathematical Frontiers in Computational Chemical Physics: IMA Series (Springer, Berlin) 15:262
22. Paldus J and Shavitt I (1988) In: Mathematical frontiers in computational chemical physics: IMA Series, 15. Springer, Berlin, p 300
23. Brooks BR, Schaefer HF (1979) J Chem Phys 70:5092
24. Lischka H, Shepard R, Brown FB, Shavitt I (1981) Int J Quantum Chem Quantum Chem Symp 15:91
25. Shepard R, Shavitt I, Pitzer RM, Comeau DC, Pepper M, Lischka H, Szalay PG, Ahlrichs R, Brown FB (1988) Int J Quantum Chem Quantum Chem Symp 22:149
26. Lischka H, Shepard R, Pitzer RM, Shavitt I, Dallos M, Mller Th, Szalay PG, Seth M, Kedziora GS, Yabushita S, Zhang Z (2001) Phys Chem Chem Phys 3:664

27. Lischka H, Mller Th, Szalay PG, Shavitt I, Pitzer RM, Shepard R (2011) WIREs 1:191
28. Lischka H, Shepard R, Shavitt I, Pitzer RM, Dallos M, Mller Th, Szalay PG, Brown FB, Ahlrichs R, Bhm HJ, Chang A, Comeau DC, Gdanitz R, Dachsel H, Ehrhardt C, Ernzerhof M, Hchtl P, Irle S, Kedziora G, Kovar T, Parasuk V, Pepper MJM, Scharf P, Schiffer H, Schindler M, Schler M, Seth M, Stahlberg EA, Zhao JG, Yabushita S, Zhang Z, Barbatti M, Matsika S, Schuurmann M, Yarkony DR, Brozell SR, Beck EV, and Blaudeau JP, Ruckenbauer M, Sellner B, Plasser F, and Szymczak JJ (2012) COLUMBUS, an ab initio electronic structure program, release 7.0.
29. GAMESS (1993) J Comput Chem 14:1347
30. Brooks BR, Laidig WD, Saxe P, Schaefer HF III (1980) J Chem Phys 72:3837
31. Shepard R, Simons J (1980) Int J Quantum Chem Symp 14:211
32. Evade RHA, Robb MA (1981) Chem Phys Lett 83:362
33. Li X, Paldus J (1994) J Chem Phys 101:8812
34. Jeziorski B, Paldus J, Jankowski P (1995) Int J Quantum Chem 56:129
35. Jankowski P, Jeziorski B (1999) J Chem Phys 111:1857
36. Paldus J, Li X (1999) Adv Chem Phys 110:1
37. Li X and Paldus J (1997) In: Bartlett RJ (ed) Recent advances in CC methods. World Scientific Singapore, pp 183
38. Szalay PG, Gauss J (1997) J Chem Phys 107:9028
39. Heckert M, Heun O, Gauss J, Szalay PG (2006) J Chem Phys 124:124105
40. Berente I, Szalay PG, Gauss J (2002) J Chem Phys 117:7872
41. Nooijen M, Bartlett RJ (1996) J Chem Phys 104:2652
42. Nakatsuji H, Hirao KJ (1978) J Chem Phys 68:2053
43. Nakatsuji H (1979) Chem Phys Lett 67:329
44. Maitra R, Sinha D, Sen S, Shee A and Mukherjee D (2012) In: Theory and applications in computational chemistry: the first decade of the second millennium: AIP Conf. Proc. International Congress TACC-2012 1456:81
45. Maitra R, Sinha D and Mukherjee D (2012) J Chem Phys 137:024105 and erratum 139:229903
46. Sen S, Shee A, Mukherjee D (2012) J Chem Phys 137:074104
47. Shee A, Sen S, Mukherjee D (2013) J Chem Theory Comput 9:2573
48. Sen S, Shee A, Mukherjee D (2013) Mol Phys 111:2477
49. Kutzelnigg W (1982) J Chem Phys 77:3081
50. Kutzelnigg W (1985) J Chem Phys 82:4166
51. Jeziorski B, Monkhorst H (1981) J Phys Rev A 24:1668
52. Mukherjee D (1986) Chem Phys Lett 125:207
53. Mukherjee D (1986) Int J Quantum Chem S20:409
54. Mukhopadhyay D, Mukherjee D (1989) Chem Phys Lett 163:171
55. Meissner L, Kucharski S, Bartlett RJ (1989) J Chem Phys 91:6187
56. Meissner L, Bartlett RJ (1990) J Chem Phys 92:561
57. Mahapatra US, Datta B, Bandyopadhyay B, Mukherjee D (1998) Adv Quantum Chem 30:163
58. Mahapatra US, Datta B, Mukherjee (1998) Mol Phys 94:157
59. Mahapatra US, Datta B, Mukherjee (1999) J Chem Phys 110:6171
60. Evangelista FA, Allen WD, Schaefer HF (2005) J Chem Phys 125:154113
61. Evangelista FA, Simmonett AC, Allen WD, Schaefer HF, Gauss J (2010) J Chem Phys 132:074107
62. Das S, Kállay M, Mukherjee D (2010) J Chem Phys 132:074103
63. Evangelista FA, Allen WD, Schaefer HF (2007) J Chem Phys 127:024102
64. Das S, Kállay M, Mukherjee D (2012) Chem Phys 392:83
65. Bhaskaran-Nair K, Demel O, Pitnner J (2008) J Chem Phys 129:184105
66. Prochnow E, Harding ME, Gauss J (2010) J Chem Theory Comput 6:2339
67. Brabec J, Pittner J, Hubertus JJD, Apra A, Kowalski K (2012) J Chem Theory Comput 8:487
68. Prochnow E, Evangelista FA, Schaefer HF, Allen WD, Gauss J (2009) J Chem Phys 131:064109
69. Nooijen M, Lotrich V (2000) J Chem Phys 113:494
70. Kong L, Shamasundar KR, Demel O, Nooijen M (2009) J Chem Phys 130:114101
71. Datta D, Kong L, Nooijen M (2011) J Chem Phys 134:214116
72. Datta D, Nooijen M (2012) J Chem Phys 137:204107
73. Nooijen M, Demel O, Datta D, Kong L, Shamasundar KR, Lotrich V, Huntington LM, Neese F (2014) J Chem Phys 140:081102
74. Mukherjee D (1995) Recent progress in many-body theories, vol 4. Springer, Berlin, p 127
75. Mukherjee D (1997) Chem Phys Lett 274:561
76. Evangelista FA, Gauss J (2011) J Chem Phys 134:114102
77. Evangelista FA, Hanauer M, Köhn A, Gauss J (2012) J Chem Phys 136:204108
78. Hanauer M, Köhn A (2011) J Chem Phys 132:204111
79. Hanauer M, Köhn A (2012) J Chem Phys 136:204107
80. Kutzelnigg W, Mukherjee D (1997) J Chem Phys 107:432
81. Kutzelnigg W, Shamasundar KR, Mukherjee D (2010) Mol Phys 108:433
82. Shamasundar KR (2009) J Chem Phys 131:174109
83. Jeziorski B, Paldus J (1988) J Chem Phys 88:5673
84. Datta D, Mukherjee D (2009) J Chem Phys 131:044124
85. Datta D, Mukherjee D (2011) J Chem Phys 134:054122
86. Lindgren I (1985) Physica Scripta 32:611
87. Wick GC (1950) Phys Rev 80:268
88. Mukherjee D (1995) In: Schashinger E et al (eds) Recent progress in many-body theories, vol. 4. Plenum Press, New York, pp 127
89. Mahapatra US, Datta B, Bandyopadhyay B, Mukherjee D (1998) Adv Quantum Chem 30:163
90. Kubo R (1962) J Phys Soc (Japan) 17:1100
91. Kong L, Nooijen M, Mukherjee D (2010) J Chem Phys 132:234107
92. Sinha D, Maitra R, Mukherjee D (2013) Comput Theor Chem 1003:62
93. Nooijen M, Shamasundara KR, Mukherjee D (2005) Mol Phys 103:2277

Theor Chem Acc (2014) 133:1467
DOI 10.1007/s00214-014-1467-8

REGULAR ARTICLE

Unitary group approach to the many-electron correlation problem: spin-dependent operators

Xiangzhu Li · Josef Paldus

Received: 13 November 2013 / Accepted: 14 February 2014 / Published online: 4 April 2014

Abstract Following a brief overview of the unitary group approach (UGA) to the many-electron correlation problem, focusing in particular on Shavitt's contribution via his graphical unitary group approach, we present a short review of our earlier results for the evaluation of matrix elements (MEs) of unitary group generators or products of generators in the electronic Gel'fand–Tsetlin basis with the help of spin-adapted second-quantization-like creation and annihilation vector operators at the unitary group level. This formalism is then extended to a spin-dependent case that is required when accounting for relativistic effects by developing explicit expressions for MEs of spin-orbital creation and annihilation operators in terms of the standard spin-adapted UGA basis. This leads naturally to a segmentation of these MEs and enables the evaluation of spin-dependent one-body operators while relying largely on the segment values of the standard spin-independent UGA.

Keywords Unitary group approach (UGA) · Graphical unitary group approach (GUGA) · Correlation problem · Spin-dependent UGA

1 Introduction

This article is a tribute to the late Professor Isaiah Shavitt who belongs to leading scientists of the second half of the twentieth century in the field of atomic and molecular electronic structure calculations. His seminal contributions to the exploitation of Gaussian functions in both statistical and quantum mechanics, to the evaluation of multicenter integrals over the Slater functions, to the many-body perturbation theory (MBPT) and coupled-cluster (CC) methods, and most importantly, to the configuration interaction (CI) methodology became a part of the quantum chemical "folklore." He also carried out a number of important applications providing accurate potential energy surfaces for several small molecular species as well as a characterization of excited states of various molecules, including butadiene and benzene.

The senior author was fortunate to be counted among his friends and to collaborate with him on a number of occasions, benefiting from his vast experience in the design of a sophisticated computer software and his intimate knowledge of quantum chemical methodology and numerical analysis. Thanks primarily to his CI expertise we have collaborated, together with Jiří Čížek, on the first ab initio implementation of the CC method addressing, in particular, the role of triexcited clusters [1]. The latter was the reason for choosing the BH_3 molecule for this purpose, since from among the molecules for which the full CI (FCI) results were available at the time its triexcited configurations were most prominent, at least when described via the Slater-type atomic orbitals. Professor Shavitt kindly provided not only required integrals (via a small packet of punched cards), but also special CI results involving selected classes of configurations for a more detailed comparison with the corresponding CC results.

In handling of triexcited clusters, we employed a truncated version of the general CC formalism that was based on the importance of various cluster components as implied by MBPT rather than involving all the terms of a given

Dedicated to the memory of Professor Isaiah Shavitt and published as part of the special collection of articles celebrating his many contributions.

X. Li · J. Paldus (✉)
Department of Applied Mathematics, University of Waterloo, Waterloo, ON N2L 3G1, Canada
e-mail: paldus@uwaterloo.ca

excitation order as it is done in the CCSDT method. In this way, the diagrams whose contribution was a priori smaller than a fraction of a μH (such as, e.g., powers of T_1) could be safely neglected. Our results clearly indicated the potential of CC methodology as well as the importance of connected triples versus disconnected ones, contrary to the case for quadruples (as, again, implied a priori by MBPT).

In the domain of CI methodology, our close interaction followed the formulation of the unitary group approach (UGA) to the many-electron correlation problem by one of us [2–5]. Professor Shavitt soon fell in love with this formalism and designed an ingenious graphical representation [6–8] of the electronic Gel'fand–Tsetlin (G–T) basis as described in terms of the so-called ABC or Paldus tableaux (see below). This development ultimately resulted in his graphical UGA (GUGA) version of UGA (for most recent overviews of these approaches see [4, 9–13]).

Let us recall that the essence of UGA goes all the way back to 1935 when Pascual Jordan (of Nazi disrepute) used U(∞) generators to represent one- and two-body Hamiltonians [14]. In the late sixties, it was revived by Moshinsky in the context of the nuclear many-body problem [15]. Here, however, four-column irreducible representations (irreps) are required in view of the presence of both the spin and isospin. Consequently, the resulting rather complex formalism has never been implemented in actual applications for nuclei, as far as we know (see, however, [16]).

In early seventies, one of us realized that the U(n)-based formalism could be beneficial for the many-electron problems, since in this case only two-column irreps are involved. This led to a drastic simplification of the G–T tableaux labeling of states that is based on the

$$U(n) \supset U(n-1) \supset \cdots \supset U(2) \supset U(1) \tag{1}$$

chain of subgroups. In fact, the general subduction rules (later called [17] "betweenness conditions") for $U(m) \supset U(m-1)$ were formulated already by Weyl in the very last chapter of the second edition of his seminal book [18]. It was then a small but important step made by Gel'fand and Tsetlin to realize [19, 20] that the highest weights for the subgroups in the chain (1) will uniquely label the basis vectors of general U(n) irrep spaces since U(1) is abelian (for a brief outline see [13]). Moreover, these authors also more or less "guessed" (as rumored) explicit expressions for the generator matrix elements (MEs) [19, 20]. Later on, Bair and Biedenharn [21] provided a detailed derivation and a general formalism for U(n) calculus (see [22, 23]; for a pedagogical review see [17]).

This rather formidable formalism can be, however, drastically simplified when considering many-electron systems by relying on the $U(2n) \supset U(n) \otimes U(2)$ chain. In this case, we require only two-column U(n) irreps of the highest weight $\langle\lambda\rangle = \langle 2^a 1^b 0^c\rangle \equiv (a, b, c)$ with

$$a = \frac{1}{2}N - S, \quad b = 2S, \quad c = n - a - b = n - \frac{1}{2}N - S, \tag{2}$$

of dimension

$$\dim(a, b, c) = \frac{b+1}{n+1}\binom{n+1}{a}\binom{n+1}{c}, \tag{3}$$

considering the states of multiplicity $(2S + 1)$ involving n orbitals and N electrons [2–5] (see also reviews in [9, 10, 13]). The basis vectors of its carrier space (or configurations) may then be labeled by $n \times 3$ ABC tableaux $[P]$ [2–5] (often called *Paldus or Gel'fand–Paldus or electronic G–T tableaux*; see, e.g., [6, 7, 9, 24–33]); $[P] \equiv \| a_i\, b_i\, c_i \|_{n \times 3}$, whose rows (a_i, b_i, c_i), $a_i + b_i + c_i = i$ label the irreps of U(i) in the canonical subgroup chain (1) or, alternatively, by the corresponding two-column Weyl tableaux [3] or, in fact, by the two-column $\Delta a \overline{\Delta c}$ tableaux, where $\Delta x_i = x_i - x_{i-1}$ and $\overline{\Delta x_i} = 1 - \Delta x_i$ $(i = 0, 1, \ldots, n)$, with $x_n \equiv x, x_0 \equiv 0$, $x = a$, b, and c. Clearly, $\Delta a_i + \Delta b_i + \Delta c_i = 1$. Another convenient labeling uses the ternary *step numbers* d_i, $0 \leq d_i \leq 3$ (see, e.g., [2, 3, 6, 7, 10, 11, 34])

$$d_i = 1 - 2\Delta a_i - \Delta c_i = 2\Delta a_i + \overline{\Delta c_i}. \tag{4}$$

The general $U(i) \supset U(i-1)$ step thus involves at most four subreps characterized by these step numbers. Thus, a G–T basis can also be labeled by n-component d-vectors $d = \{d_i\}$.

The entire basis can then be compactly represented by a *Shavit graph* [6–8, 11]. This is a two-rooted graph involving distinct rows of the ABC tableaux for each subgroup U(i) $(i = 1, \ldots, n)$ with one root representing the highest weight $(a, b, c) \equiv (a_n, b_n, c_n)$ for U(n) and the other one the trivial row $(0, 0, 0) \equiv (a_0, b_0, c_0)$. The entire graph is placed on a grid so that the slope of its edges represents the relevant step numbers. Thus, each basis vector or configuration is represented by a path interconnecting both roots. When evaluating MEs of generators or products of generators, the bra and ket state paths coincide outside the range indicated by the generator indices and form a loop within that range. Shavitt then suitably split the available expressions [2, 4] for the generator MEs into the so-called *segment values* associated with different types of bra and ket edges of his graph. This factorization is not unique and can be done in several ways (see, e.g., [34–36]); we must point out here that an often made statement that the MEs of various U(n) generators "can be derived entirely graphically" [37] within GUGA is misleading since the key starting point here (see [11]) is simple expressions for elementary generators as given in [2], which cannot be derived within GUGA and follows most naturally from the SU(2)-based approach representing the corresponding Yamanouchi-Kotani states via Jucys-type diagrams (see, e.g., [38]; see also [39] for a corresponding approach based on the S_N group). Indeed, this is particularly the case for an evaluation of MEs of a two-electron part of the Hamiltonian [34] (see below).

Shavitt graphs thus provide a compact and transparent representation of the entire electronic G–T basis, and the related developments are referred to as the GUGA. It facilitates the design of various strategies not only for the full CI calculations (see, however, [40]), but also for a configuration selection in truncated CI approaches, not to mention the simplest introduction of the so-called distinct row table (DRT) and various coding strategies based on the matrix element-, integral-, or loop-driven algorithms [7, 9, 11]. The GUGA viewpoint thus led to the design of a number of CI codes (see, e.g., [41–56]).

The most recent development that exploits GUGA description of the electronic G–T basis is due to Shepard and collaborators [57–63]. This is the so-called Graphically Contracted Function CI Method that is based on nonlinear variables called *arc factors*. This approach makes it possible to handle extremely large CI expansions exceeding traditional ones by many orders of magnitude. The method enables an efficient computation of energy gradients with respect to the arc factors and the determination of their optimal values. Most recently, the authors developed an efficient recursive algorithm for computation of orbital-level Hamiltonian matrices [62] that characterize the dependence of the energy on nonlinear parameters associated with individual molecular orbitals and enable a computation of gradients with respect to nonlinear arc factors.

At this point, it is worth recalling that the standard UGA-based formalism has also been exploited in numerous other applications other than large-scale CI calculations, namely in MC-SCF approaches [44, 49, 64, 65], perturbation theory [66], propagator method [67], and coupled-cluster methods of both single-reference [68–82] and multi-reference [83, 84] type, as well as in various other investigations, such as quantum dots [85], in handling of composite systems [86–89], valence bond (VB) approaches [90–92], reduced density matrices (RDMs) [93, 94], nuclear magnetic resonance (NMR) [95], or charge migration in fragmentation of peptide ions [96, 97] (see also [10, 98] for other references). We should also mention numerous other innovations and extensions, such as Clifford algebra UGA (CAUGA) [87, 99–103], bonded tableau UGA [104], parafermi algebras [105], and related developments (see also [106]).

Finally, we should also emphasize that while GUGA is extremely useful for providing an insight into the overall structure of the electronic G–T basis and for the ME evaluation in terms of segment values, it cannot provide the actual algebraic expressions for these quantities per se, for these rely on the actual representation theory for U(n). There exist several independent approaches for this purpose that often exploit not only the U(n) representation theory, but also the related symmetric or permutation group S_N, as well as the spin-angular-momentum formalism of SU(2). The U(n)-based approach exploits the tensor operator formalism of Baird and Biedenharn [21] or the Green-Gould formalism [29–31, 33] based on polynomial identities for U(n) generators and the implied projection and tensor operators. Some approaches exploit the S_N representation theory, either per se [26] or in conjunction with the U(n) or SU(2) formalism (see, e.g., [39]). A particularly useful form of the SU(2)-based formalism relies on the graphical methods of spin algebras employing Jucys-type diagrams [38] using the fact that the electronic G–T states correspond to the Yamanouchi-Kotani coupling scheme [15]. This approach was exploited by Paldus and Boyle [34, 107, 108] for the derivation of the explicit expressions for generator MEs, elucidating at the same time various schemes of their partitioning into the segment values and the choice of suitable phase factors. It turned out to be particularly powerful for an optimal handling of the two-electron operators [34] since it avoids a summation over the intermediate states of the original Shavitt formulation [7]. Yet another useful approach, exploited earlier by the authors [35, 36, 89], introduced spin-adapted creation and annihilation operators that in a certain way parallel those of the standard second-quantization formalism (cf., however,[1] [109–112]). This approach will be used in this paper as well since it enables an easy extension to spin-dependent operators.

Spin-dependent operators are no longer expressible as a bilinear form in the orbital U(n) generators but, rather, in terms of the spin-orbital U($2n$) generators. An account of spin-dependent effects that is based on spin-free correlated wave functions thus requires an evaluation of MEs of U($2n$) generators or their products in standard U(n) electronic G–T bases. This problem, which can be generally referred to as a spin-dependent UGA approach, was addressed in several earlier publications [32, 113–117]. In this paper, we apply to this problem our earlier results [35, 36] that introduced spin-free analogues of the standard spin-orbital creation and annihilation operators, labeled as $C^\dagger$ or C, respectively. We shall see that an exploitation of this formalism for spin-dependent UGA offers a great methodological simplification

[1] The concept of spin-adapted creation and annihilation operators, referred to as spin shift or "up" and "down" creation and annihilation operators, was employed in a series of papers by Mukhopadhyay and Pickup [109–112] who attempted to generalize the standard, many-body formalism by constructing spin-adapted one-body propagators [109], configuration state functions [110, 112], hole-particle formalism and Wick's theorem [111], or MBPT [112] rather than aiming for an efficient evaluation of unitary group MEs. In fact, their formulation is not spin-free and their aim has been to avoid "the intricacies of the unitary group" [110]. Consequently, their developments and aims are very different from ours.

by enabling an evaluation of MEs of U(2n) generators in the U(n) electronic G–T bases as a product of segment level expressions. With the exception of a few new top segment values for spin-dependent operators, the same segment level expressions that arise in the standard spin-independent UGA method apply. In this paper, we focus on MEs of spin-dependent one-body operators. Following a brief overview of our formalism for spin-independent operators and of spin-dependent UGA in Sects. 2 and 3, respectively, we address the evaluation of MEs of standard spin-orbital creation and annihilation operators in Sect. 4 and of the MEs of one-body spin-dependent operators in Sect. 5. Section 6 then outlines the relationship of spin-independent and spin-dependent UGA, and brief conclusions are drawn in Sect. 7.

2 Spin-independent operators

We now briefly recall [35, 36] a few basic facts concerning our U(n)-adapted second-quantization-like formalism for the calculation of MEs of U(n) generators E_{ij} satisfying the commutation relations

$$[E_{ij}, E_{kl}] = \delta_{jk}E_{il} - \delta_{il}E_{kj}. \tag{5}$$

A vector operator $\mathbf{\Psi}$ with components Ψ_k then satisfies

$$[E_{ij}, \Psi_k] = \delta_{jk}\Psi_i, \tag{6}$$

and a general tensor operator $\mathbf{T}_\mu$ associated with the U(n) irrep $\langle\mu\rangle \equiv \mu$ is defined by

$$[E_{ij}, T_\mu(W)] = \sum_V \left\langle \begin{matrix}\mu\\V\end{matrix}\middle| E_{ij} \middle| \begin{matrix}\mu\\W\end{matrix}\right\rangle T_\mu(V), \tag{7}$$

where V and W are Paldus or Weyl tableaux labeling components of the tensor $\mathbf{T}_\mu$ and the basis of the carrier space of μ.

The key problem of the U(n) calculus is the evaluation of MEs of tensor operators. Here one relies on the Wigner–Eckart (W–E) theorem

$$\left\langle \begin{matrix}\nu\\W\end{matrix}\middle| T_\mu(V) \middle| \begin{matrix}\lambda\\U\end{matrix}\right\rangle = \langle\nu\|T_\mu\|\lambda\rangle \left\langle \begin{matrix}\nu\\W\end{matrix}\middle| \begin{matrix}\lambda & \mu\\U & V\end{matrix}\right\rangle, \tag{8}$$

where we assume the multiplicity-free case (which holds for two-column irreps) with the first factor on the right-hand side designating the reduced matrix element (RME) and the second factor the U(n) Clebsch–Gordan (C–G) or Wigner coefficient. It is also convenient to define corresponding scaled quantities as

$$\langle\nu\|T_\mu\|\lambda\rangle^{(s)} = \langle\nu\|T_\mu\|\lambda\rangle / c(\lambda, \mu; \nu), \tag{9}$$

$$\left\langle \begin{matrix}\nu\\W\end{matrix}\middle| \begin{matrix}\lambda & \mu\\U & V\end{matrix}\right\rangle^{(s)} = c(\lambda, \mu; \nu)\left\langle \begin{matrix}\nu\\W\end{matrix}\middle| \begin{matrix}\lambda & \mu\\U & V\end{matrix}\right\rangle, \tag{10}$$

so that again

$$\left\langle \begin{matrix}\nu\\W\end{matrix}\middle| T_\mu(V) \middle| \begin{matrix}\lambda\\U\end{matrix}\right\rangle = \langle\nu\|T_\mu\|\lambda\rangle^{(s)} \left\langle \begin{matrix}\nu\\W\end{matrix}\middle| \begin{matrix}\lambda & \mu\\U & V\end{matrix}\right\rangle^{(s)}. \tag{11}$$

Setting $c(\lambda, \mu; \nu) = \langle\nu\|T_\mu\|\lambda\rangle$ we thus have that

$$\left\langle \begin{matrix}\nu\\W\end{matrix}\middle| T_\mu(V) \middle| \begin{matrix}\lambda\\U\end{matrix}\right\rangle = \left\langle \begin{matrix}\nu\\W\end{matrix}\middle| \begin{matrix}\lambda & \mu\\U & V\end{matrix}\right\rangle^{(s)}. \tag{12}$$

Since the U(n) G–T basis relies on the canonical chain (1), a C–G coefficient can be expressed as a product of the U(m) ⊃ U($m-1$) isoscalar factors with m ranging from 1 to n. The scaled C–G coefficients are then given by a product of scaled isoscalar factors.

We can now define a set of fundamental tensor operators $C_i^{\sigma\dagger}$ that transform as a basis for the simplest nontrivial irrep $\langle 1\dot{0}\rangle \equiv \langle 10\cdots0\rangle$ [or $(a, b) = (0, 1)$ using the two-column irrep label for $\langle 2^a1^b0^{n-a-b}\rangle$]

$$[E_{jk}, C_i^{\sigma\dagger}] = \delta_{ik}C_j^{\sigma\dagger}, \tag{13}$$

$$C_i^{\sigma\dagger}\left|\begin{matrix}\lambda\\U\end{matrix}\right\rangle = \sum_W \left\langle \begin{matrix}\lambda+\sigma\\W\end{matrix}\middle| C_i^{\sigma\dagger} \middle| \begin{matrix}\lambda\\U\end{matrix}\right\rangle \left|\begin{matrix}\lambda+\sigma\\W\end{matrix}\right\rangle, \tag{14}$$

$$C_n^{\sigma\dagger}\left|\begin{matrix}\lambda_n\\\lambda_{n-1}\\W_{n-1}\end{matrix}\right\rangle = \left|\begin{matrix}\lambda_n+\sigma\\\lambda_{n-1}\\W_{n-1}\end{matrix}\right\rangle \tag{15}$$

with MEs in (14) given by the W–E theorem, Eq. (8),

$$\left\langle \begin{matrix}\lambda+\sigma\\W\end{matrix}\middle| C_i^{\sigma\dagger} \middle| \begin{matrix}\lambda\\U\end{matrix}\right\rangle = \langle\lambda+\sigma\|C^\dagger\|\lambda\rangle \left\langle \begin{matrix}\lambda+\sigma\\W\end{matrix}\middle| \begin{matrix}\lambda & (0,1)\\U & i\end{matrix}\right\rangle, \tag{16}$$

and $\lambda_n = \lambda_{n\text{-}1}$ in (15) labeling the irreps of U(n) and U($n-1$), respectively. The irrep $\langle\lambda+\sigma\rangle$ is obtained by adding a box in the Weyl tableau W_{n-1} labeled by n assuming that the state n is not occupied in W_{n-1} of λ_n, so that $C_n^{\sigma\dagger}$ adds label n in the σth column of W_{n-1}. Thus, in the case of two-column tableaux, we have only two fundamental vector operators, namely

$$\begin{aligned} C_i^{1\dagger} &: (a,b) \to (a, b+1) \quad \text{and} \\ C_i^{2\dagger} &: (a,b) \to (a+1, b-1). \end{aligned} \tag{17}$$

Thus, the first condition above, Eq. (13), implies that $C_i^{\sigma\dagger}$ is a U(n) vector operator with σ designating the shift component, while the last condition, Eq. (15), determines the RME $\langle\lambda+\sigma\|C^\dagger\|\lambda\rangle$. It is worth noting that the $C^\dagger$ operators are similar to the fundamental Wigner operators of Biedenharn, Louck and others [17, 21]. They both represent unit vector operators but differ in their RMEs, which are equal to unity in the standard approach in which the shift is defined via a row label rather than a column label used here. The latter has the advantage of yielding an n-

independent formalism for RMEs and C–G coefficients. Choosing then a state in which the nth orbital is unoccupied one finds the required RMEs [36]

$$\langle \lambda + \sigma \| C^\dagger \| \lambda \rangle = \begin{pmatrix} \lambda + \sigma & \lambda & (0,1) \\ \lambda & \lambda & (0,0) \end{pmatrix}^{-1} = \left(\frac{N_{\lambda+\sigma} f_\lambda}{f_{\lambda+\sigma}} \right)^{\frac{1}{2}}, \quad (18)$$

where the quantity on the right-hand side of the first equality sign is a $U(n) \supset U(n-1)$ isoscalar factor and the results of [35] for the $U(n)$ isoscalar factors were employed in the last equation. Here, N_μ designates the particle number (i.e., number of boxes) associated with the irrep μ and f_λ the dimension of μ considered as an irrep of the permutation group S_{N_μ} (see [35]). The corresponding shifts $\Delta(\sigma) \equiv \sigma$ for $\sigma = 1$ and $\sigma = 2$ yielding the resulting irreps $(a, b) + \sigma$ are, respectively, (0, 1) and (1, −1), and the corresponding RMEs are

$$\begin{aligned} &\langle (a, b+1) \| C^\dagger \| (a,b) \rangle = [(b+1)(a+b+2)/(b+2)]^{\frac{1}{2}}, \\ &\langle (a+1, b-1) \| C^\dagger \| (a,b) \rangle = [(a+1)(b+1)/b]^{\frac{1}{2}}. \end{aligned} \quad (19)$$

The $U(n)$ C–G coefficients appearing in Eq. (16) may then be factorized into a product of $U(n)$ isoscalar factors (see Eqs. (145) and (154) of [35] and Eq. (24) of [36]) the explicit expressions of which are given in Table 1 of [36]. We also note that the MEs of corresponding conjugate, annihilation-like operators C_i^σ transforming as a contragredient fundamental vector representation, i.e., $[E_{kj}, C_i^\sigma] = -\delta_{ik} C_j^\sigma$, are given by the conjugate of Eq. (14).

Now, the raising generators E_{ij} $(i < j)$ represent the vector operators of $U(j-1)$ since the commutation relations, Eq. (5), imply that $[E_{kl}, E_{ij}] = \delta_{il} E_{kj}$ $(1 \le k, l \le j-1)$, so that $U(j)$ MEs of E_{ij} must be proportional to those of $C_i^{\sigma\dagger}$,

$$\left\langle \begin{matrix} \lambda_j \\ W'_j \end{matrix} \middle| E_{ij} \middle| \begin{matrix} \lambda_j \\ W_j \end{matrix} \right\rangle = \left\langle \begin{matrix} \lambda_j \\ \lambda'_{j-1} \end{matrix} \middle| E \middle| \begin{matrix} \lambda_j \\ \lambda_{j-1} \end{matrix} \right\rangle \left\langle \begin{matrix} \lambda'_{j-1} \\ W'_{j-1} \end{matrix} \middle| C_i^{\sigma\dagger} \middle| \begin{matrix} \lambda_{j-1} \\ W_{j-1} \end{matrix} \right\rangle \quad (20)$$

or to the corresponding scaled quantities, where the relevant proportionality factor can only depend on $U(j)$ and $U(j-1)$ irrep labels. This in turn leads to the expression for E_{ij} MEs in terms of scaled isoscalar factors (note that $x \equiv x_n$, $x = a, b, c$)

$$\begin{aligned} \left\langle \begin{matrix} (a,b) \\ W' \end{matrix} \middle| E_{ij} \middle| \begin{matrix} (a,b) \\ W \end{matrix} \right\rangle &= \left(\prod_{k=j}^{n} \delta_{(a'_k, b'_k),(a_k,b_k)} \right) \\ &\times \left\langle \begin{matrix} (a_j, b_j) \\ (a'_{j-1}, b'_{j-1}) \end{matrix} \middle\| E \middle\| \begin{matrix} (a_j, b_j) \\ (a_{j-1}, b_{j-1}) \end{matrix} \right\rangle^{(s)} \\ &\times \left\{ \prod_{k=i+1}^{j-1} \begin{pmatrix} (a_k,b_k) & (0,1) & (a'_k,b'_k) \\ (a_{k-1},b_{k-1}) & (0,1) & (a'_{k-1},b'_{k-1}) \end{pmatrix}^{(s)} \right\} \\ &\times \begin{pmatrix} (a_i,b_i) & (0,1) & (a'_i,b'_i) \\ (a_{i-1},b_{i-1}) & (0,0) & (a'_{i-1},b'_{i-1}) \end{pmatrix}^{(s)} \prod_{k=1}^{i-1} \delta_{(a'_k,b'_k),(a_k,b_k)}. \end{aligned} \quad (21)$$

The MEs of products of two creation operators $C_i^{\tau\dagger} C_j^{\sigma\dagger}$, and, similarly, of two annihilation operators, can be expressed in terms of the symmetric and antisymmetric tensors (see [36]). On the other hand, products of a creation operator and an annihilation operator form adjoint tensor operators. In view of the $U(n)$ commutation relations (5), the $U(n)$ generators also represent adjoint tensor operators of $U(n)$ belonging to the irrep $\langle 1\dot{0}\bar{1} \rangle = \langle 10 \cdots 0 -1 \rangle$ that are defined as follows

$$[E_{kl}, A_{ij}] = \delta_{il} A_{kj} - \delta_{kj} A_{il} \quad (k, l = 1, 2, \ldots, n). \quad (22)$$

The same relationship holds also for $C_i^{\sigma\dagger} C_j^\tau$ $(\sigma, \tau = 1, 2)$ operator products, so that it must hold that

$$E_{ij} = \sum_{\sigma,\tau} \rho_{\sigma,\tau} C_i^{\sigma\dagger} C_j^\tau = \sum_\tau C_i^{\tau\dagger} C_j^\tau, \quad (23)$$

the last equation holding since it can be shown that $\rho = \|\rho_{\sigma\tau}\|$ must be the 2×2 identity matrix [36]. Thus, the MEs of $U(n)$ generators E_{ij} may be expressed via the MEs of $C_i^{\sigma\dagger}$ and C_j^σ operators, which in turn leads to the expressions for the generator RMEs in terms of isoscalar factors

$$\begin{aligned} &\left\langle \begin{matrix} (a_j, b_j) \\ (a'_{j-1}, b'_{j-1}) \end{matrix} \middle\| E \middle\| \begin{matrix} (a_j, b_j) \\ (a_{j-1}, b_{j-1}) \end{matrix} \right\rangle \\ &= \sum_\sigma \begin{pmatrix} (a_j,b_j) - \sigma & (0,1) & (a_j,b_j) \\ (a_{j-1},b_{j-1}) & (0,1) & (a'_{j-1},b'_{j-1}) \end{pmatrix}^{(s)} \\ &\times \begin{pmatrix} (a_j,b_j) - \sigma & (0,1) & (a_j,b_j) \\ (a_{j-1},b_{j-1}) & (0,0) & (a_{j-1},b_{j-1}) \end{pmatrix}^{(s)}. \end{aligned} \quad (24)$$

At this point, it is useful to emphasize the distinction between the standard creation and annihilation operators $X_{i\sigma}^\dagger, X_{i\sigma}$ and the spin-adapted ones $C_i^{\sigma\dagger}, C_i^\sigma$ in view of the similarity with the well-known relation

$$E_{ij} = \sum_\sigma X_{i\sigma}^\dagger X_{j\sigma} \quad (25)$$

that is exploited in UGA. In fact, this relationship extends to general k-body operators (see Eq. (78) of [36]). In the 2-body case, we thus have

$$e_{ik;jl} \equiv E_{ik}E_{jl} - \delta_{jk}E_{il} = \sum_{\sigma,\tau=\alpha,\beta} X^\dagger_{j\sigma}X^\dagger_{i\tau}X_{k\tau}X_{l\sigma} = \sum_{\sigma,\tau=1,2} C^{\sigma\dagger}_j C^{\tau\dagger}_i C^\tau_k C^\sigma_l. \quad (26)$$

We must emphasize, however, that this correspondence between the vector operators $C^\dagger, C$ and the second-quantization operators $X^\dagger, X$ is only a formal similarity: indeed, the operators $C^\dagger$ and C act on the orbital group U(n) modules while the second-quantization operators $X^\dagger$ and X act on the spin-orbital group U($2n$). The real meaning of the correspondence (26) is that we can either couple the U(n) tensors $C^\dagger$ and C within the U(n) framework to obtain higher rank U(n) tensors, like two-body operators (26) or we can express these U(n) operators as spin contractions of U($2n$) operators.

The adjoint tensor operators are particularly useful in handling of two-electron MEs and, as will be seen below, also when dealing with spin-dependent operators. The action of adjoint tensors A_{ij} on a two-column U(n) irrep (a, b) produces modules that are associated with irreps given by the Littlewood-Richardson rule as a C–G series

$$(a,b) \times \langle 10\cdots 0-1\rangle = (a-1,b+2) + (2-\delta_{b,0})(a,b) + (a+1,b-2) + \cdots \quad (27)$$

the dots indicating more than two-column irreps. Thus, the four adjoint tensors $C^{\sigma\dagger}_i C^\tau_j$ (σ, τ = 1, 2) are associated with three different shifts: $C^{1\dagger}_i C^2_j : (a,b) \mapsto (a-1,b+2)$, $C^{2\dagger}_i C^1_j : (a,b) \mapsto (a+1,b-2)$, and two zero-shifts $C^{\kappa\dagger}_i C^\kappa_j : (a,b) \mapsto (a,b), (\kappa = 1,2)$ (assuming $b \neq 0$).

The multiplicity of zero-shift tensors complicates the evaluation of MEs. We already know that a simple sum of zero-shift operators gives the generator E_{ij}, Eq. (23). We thus have to find the other linear combination of these tensor operators belonging to the zero-shift case. It is shown in [36] that a convenient choice that enables a segmentation of the resulting MEs is given by the operators

$$N^{(+)}_{ij} = \sqrt{(b+3)/(b+2)}\, C^{1\dagger}_i C^2_j, \quad (28)$$

$$N^{(0)}_{ij} = (1-\delta_{b,0})\left(\sqrt{(b+2)/(2b)}\, C^{1\dagger}_i C^1_j - \sqrt{b/2(b+2)}\, C^{2\dagger}_i C^2_j\right), \quad (29)$$

$$N^{(-)}_{ij} = \sqrt{(b+1)/b}\, C^{2\dagger}_i C^1_j, \quad (30)$$

in addition to E_{ij} (note that when $b = 0$ there is only one zero-shift tensor operator).

An actual evaluation of MEs of two-body operators $\langle e_{ik;jl}\rangle = \left\langle \begin{matrix}(a,b)\\ W'\end{matrix} \middle| e_{ik;jl} \middle| \begin{matrix}(a,b)\\ W\end{matrix}\right\rangle$ reduces to a simple product of MEs of the generators involved when their [i, k] and [j, l] regions do not overlap, while for the overlapping case, we have to distinguish different cases as outlined in Table X of [34] (see also Table 8 of [36]). Again, the irrep labels outside the general regions must coincide so that

$$\left\langle \begin{matrix}(a,b)\\ W'\end{matrix} \middle| e_{ik;jl} \middle| \begin{matrix}(a,b)\\ W\end{matrix}\right\rangle = \left(\prod_{m=l}^{n} \delta_{(a'_m,b'_m),(a_m,b_m)}\right) \times \left\langle \begin{matrix}(a_l,b_l)\\ W'_l\end{matrix} \middle| e_{ik;jl} \middle| \begin{matrix}(a_l,b_l)\\ W_l\end{matrix}\right\rangle \left(\prod_{m=1}^{p-1} \delta_{(a'_m,b'_m),(a_m,b_m)}\right), \quad (31)$$

where $m \geq l \equiv \max(i, j, k, l)$ and $m < p \equiv \min(i, j, k, l)$, thus reducing the problem to U(l).

Evaluating the U(l)-based MEs of (31), we then have to distinguish three nontrivial cases depending on the relationship between the indices of $e_{ik;jl}$, namely (1) $i, j, k < l$, (2) $e_{il;jl}$ and $i, j < l$, and (3) $e_{lk;jl}$ and $j, k < l$, requiring different so-called $e - S$, $e - E$, $e - N$, etc. *connecting factors* $\left\langle \begin{matrix}(a_l,b_l)\\ (a'_{l-1},b'_{l-1})\end{matrix} \middle| e \middle| \begin{matrix}(a_l,b_l)\\ (a_{l-1},b_{l-1})\end{matrix}\right\rangle_X$, where $X = A, S, E, N$, and G. For example, in the last case when $j, k < l = i$,

$$\langle e_{lk;jl}\rangle = \left\langle \begin{matrix}(a_l,b_l)\\ (a'_{l-1},b'_{l-1})\end{matrix} \middle| e \middle| \begin{matrix}(a_l,b_l)\\ (a_{l-1},b_{l-1})\end{matrix}\right\rangle_E \times \left\langle \begin{matrix}(a'_{l-1},b'_{l-1})\\ W'_{l-1}\end{matrix} \middle| E_{jk} \middle| \begin{matrix}(a_{l-1},b_{l-1})\\ W_{l-1}\end{matrix}\right\rangle + \left\langle \begin{matrix}(a_l,b_l)\\ (a'_{l-1},b'_{l-1})\end{matrix} \middle| e \middle| \begin{matrix}(a_l,b_l)\\ (a_{l-1},b_{l-1})\end{matrix}\right\rangle_N \times \left\langle \begin{matrix}(a'_{l-1},b'_{l-1})\\ W'_{l-1}\end{matrix} \middle| N^{(\kappa)}_{jk} \middle| \begin{matrix}(a_{l-1},b_{l-1})\\ W_{l-1}\end{matrix}\right\rangle, \quad (32)$$

where κ = +, 0 or − is uniquely fixed by the irreps (a_l, b_l) and (a_{l-1}, b_{l-1}) and involves both $e - E$ and $e - N$ connecting factors. The explicit values of all these connecting factors are given in Eq. (88) and Table 9 of [36]. A general "segmented" form for two-body MEs is then given by Eq. (109) of [36] together with the subsequent points (1)–(6) and Eqs. (110)–(116).

3 Spin-dependent problem

Spin-dependent operators are required when we wish to account for relativistic effects in atoms and molecules [118, 119]. These effects can roughly be classified as strong and weak ones. The relativistic corrections are especially important in heavy atoms where they play a particularly significant role when describing the inner shells. In those cases, they have to be accounted for from the start, usually relying on Dirac-Hartree-Fock method. Fortunately, in most chemical phenomena, only valence electrons play a decisive role and are satisfactorily described by the nonrelativistic Schrödinger equation. Nonetheless, it is occasionally required to consider relativistic effects—even though they may play only a subsidiary role—by accounting for spin-dependent terms [120], such as the spin–orbit, orbit–orbit, spin-other-orbit, and spin–spin interactions. Such terms may be required when exploring the fine structure in high-resolution spectra or in various other important phenomena [121], as the intersystem crossing, phosphorescent lifetimes, molecular pre-dissociation, etc. However, even when spin-dependent forces are present in the Hamiltonian, the nonrelativistic Schrödinger wave functions represent an excellent starting approximation, so that these spin-dependent terms may be taken care of by computing their expectation values in terms of nonrelativistic wave functions (see, e.g., [122]).

The Hamiltonian of a many-electron system has the following general form

$$\begin{aligned}\hat{H} &= \hat{H}_0 + \hat{H}_S = \hat{H}_0 + \hat{H}_S^{(1)} + \hat{H}_S^{(2)} \\ &= \hat{H}_0 + \sum_r \hat{h}_S^{(1)}(r) + \sum_{r<s} \hat{h}_S^{(2)}(rs),\end{aligned} \tag{33}$$

where $\hat{H}_0$ is the standard spin-independent Hamiltonian and $\hat{H}_S$ is a spin-dependent interaction which can be further divided into the one- and two-body parts, $\hat{H}_S^{(1)}$ and $\hat{H}_S^{(2)}$, respectively. In this paper, we shall concentrate on the one-body part $\hat{H}_S^{(1)}$.

We recall that we employ $2n$ spin-orbitals $|i\mu\rangle$ ($i = 1, 2, \ldots, n; \mu = \frac{1}{2}, -\frac{1}{2}$ or α, β) given by products $|i\rangle|\mu\rangle$ of n orbitals $|i\rangle$ and two spin functions $|\mu\rangle$ ($\alpha = \frac{1}{2}$ and $\beta = -\frac{1}{2}$) as well as the associated fermion creation ($X_{i\mu}^{\dagger}$) and annihilation ($X_{i\mu}$) operators. Since $\hat{H}_0$ is spin-independent, we can take a trace over the spin label μ yielding a spin-free formalism that may be described in terms of spatial orbital U(n) generators E_{ij}, Eq. (25),

$$\begin{aligned}\hat{H}_0 &= \sum_{ij} \langle i|z|j\rangle E_{ij} + \frac{1}{2}\sum_{ijkl} \langle ij|v|kl\rangle (E_{ik}E_{jl} - \delta_{jk}E_{il}) \\ &= \sum_{ij} z_{ij}E_{ij} + \frac{1}{2}\sum_{ijkl} v_{ij;kl}e_{ik;jl}.\end{aligned} \tag{34}$$

The one-body spin-dependent terms may then be expressed in terms of the spin-orbital U($2n$) generators

$$\Xi_{i\mu,j\nu} = X_{i\mu}^{\dagger}X_{j\nu} \tag{35}$$

as

$$\hat{H}_S^{(1)} = \sum_{i,j} \sum_{\mu,\nu=-1/2}^{1/2} \langle i\mu|\hat{h}_S^{(1)}|j\nu\rangle \Xi_{i\mu,j\nu}. \tag{36}$$

If the relativistic effects are weak relative to the kinetic and Coulomb potential energies, we can start with a spin-independent problem $\hat{H}_0\Psi = E\Psi$ and handle $\hat{H}_S$ via MBPT. The eigenstates of $\hat{H}_0$ are pure spin states with a good total spin quantum number S that are often generated via CI approaches based on UGA or GUGA. It is thus convenient to extend efficient spin-independent UGA codes to those that can also account for spin-dependent operators.

Thus, in dealing with spin-dependent problems, the U($2n$) $\supset$ U(n) $\otimes$ SU(2) adapted basis provides a natural starting point. We shall designate its basis vectors that belong to the totally antisymmetric representation $[1^N]$ of U($2n$) by

$$\left|[1^N]\begin{matrix}\lambda & S \\ W & M_S\end{matrix}\right\rangle \equiv \left|\begin{matrix}\lambda \\ W\,M_S\end{matrix}\right\rangle, \tag{37}$$

representing simultaneously a standard G–T basis $\left|\begin{matrix}\lambda \\ W\end{matrix}\right\rangle$ of a two-column irrep $\lambda \equiv (a,b) \equiv \langle 2^a, 1^b, 0^{n-a-b}\rangle$ of the orbital group U(n) and a basis $|SM_S\rangle$ of the spin group SU(2). We can employ a shorthand notation on the right-hand side of Eq. (37) since $[1^N]$ and S are fixed by relations $N = 2a + b$ and $S = b/2$, Eq. (2). Moreover, since the subduction of the irrep $[1^N]$ of U($2n$) to the irreps of U(n) and SU(2) is multiplicity-free, the notation of Eq. (37) is unambiguous.

In view of the fact that the MEs of the spin-independent Hamiltonian $\hat{H}_0$ in the basis (37) are diagonal with respect to the spin labels, i.e.,

$$\left\langle\begin{matrix}\lambda' \\ W'\,M_S'\end{matrix}\right|\hat{H}_0\left|\begin{matrix}\lambda \\ W\,M_S\end{matrix}\right\rangle = \delta_{\lambda',\lambda}\delta_{S',S}\delta_{M_S',M_S}\left\langle\begin{matrix}\lambda \\ W'\end{matrix}\right|\hat{H}_0\left|\begin{matrix}\lambda \\ W\end{matrix}\right\rangle, \tag{38}$$

the unperturbed, pure spin eigenstates of $\hat{H}_0$ are linear combinations of the basis (37)

$$\Psi^{\lambda} = \sum_W C_W^{\lambda}\left|\begin{matrix}\lambda \\ W\,M_S\end{matrix}\right\rangle \equiv \sum_W C_W^{\lambda}\left|\begin{matrix}\lambda \\ W\end{matrix}\right\rangle. \tag{39}$$

The coefficients C_W^{λ} being independent of the spin component M_S are completely determined by the spin-independent UGA so that the key to the evaluation of relativistic effects is the evaluation of MEs of spin-dependent operators in the basis (37).

As we have seen above, the well-known Wigner-Eckart theorem represents a powerful tool for the evaluation of MEs. Thus, the MEs of any U($2n$) tensor that may be decomposed into the irreducible tensors of U(n) and SU(2) can be expressed as a product of three factors: (1) the RME that depends on the relevant tensors and irreps of U($2n$), U(n), and SU(2), (2) the U(n) C–G coefficient, and (3) the SU(2) C–G coefficient. In a multiplicity-free case for U(n) irreps, the U(n) C–G coefficients can be further factorized into simple products of isoscalar factors, yielding the ME segmentation formalism for spin-dependent operators. We shall see that this is exactly the case for one-body operators (36).

Finally, let us emphasize that in contrast to a spin-independent case, the U(n) irreps in the U($2n$) $\supset$ U(n)$\otimes$ SU(2) bases are generally changed or shifted by U($2n$) operators. Thus, one-body spin-dependent operators can change the total spin [or U(n) irrep label] by $\Delta S = 0, \pm 1$ (or $\Delta b = 0, \pm 2$) and two-body ones by $\Delta S = 0, \pm 1, \pm 2$ (or $\Delta b = 0, \pm 2, \pm 4$) and thus take us out of the U(n) framework. Nonetheless, as we have indicated above, the U(n)-adapted creation ($C^\dagger$) and annihilation (C) type operators—that represent very useful tensors serving as fundamental building blocks for various U(n) tensors—are also useful in the spin-dependent U($2n$) case. Indeed, since $X^\dagger_{i\mu}$ ($X_{i\mu}$) are vector (contragredient vector) operators when acting on the irrep modules of U(n), their MEs in the U(n) basis are clearly related to those of $C_i^{\sigma\dagger}$ (C_i^σ) operators. In view of this fact, the MEs of one-body operators must be related to those of $C_i^{\sigma\dagger} C_j^\tau$. The latter were carefully examined in [36] and briefly reviewed above.

4 Matrix elements of spin-orbital creation and annihilation operators

We first consider MEs of $X^\dagger_{i\mu}$ and $X_{i\mu}$ operators in the U($2n$) $\supset$ U(n)$\otimes$ SU(2) adapted basis, Eq. (37). These operators represent, respectively, vector and contragredient vector operators of U(n) and SU(2), since

$$[E_{k\ell}, X^\dagger_{i\mu}] = \delta_{i\ell} X^\dagger_{k\mu}, [E_{k\ell}, X_{i\mu}] = -\delta_{ik} X_{\ell\mu} \quad (i, k, \ell = 1, 2, \ldots, n), \tag{40}$$

$$[\mathcal{E}_{\nu\omega}, X^\dagger_{i\mu}] = \delta_{\omega\mu} X^\dagger_{i\nu}, [\mathcal{E}_{\nu\omega}, X_{i\mu}] = -\delta_{\nu\mu} X_{i\omega} \quad (\mu, \nu, \omega = -1/2, 1/2), \tag{41}$$

where $\mathcal{E}_{\nu\omega}$ is a SU(2) generator, $\mathcal{E}_{\nu\omega} = \sum_i X^\dagger_{i\nu} X_{i\omega}$. In view of the fact that the MEs of $X_{i\mu}$ operators are Hermitian conjugates of those for $X^\dagger_{i\mu}$, we shall examine only the latter. Using the Wigner–Eckart theorem, the MEs of $X^\dagger_{i\mu}$ in the basis of Eq. (37) can be expressed as

$$\left\langle \begin{matrix} \lambda' \\ W' M'_S \end{matrix} \middle| X^\dagger_{i\mu} \middle| \begin{matrix} \lambda \\ W\ M_S \end{matrix} \right\rangle = \langle \lambda' \| X^\dagger \| \lambda \rangle \left\langle \begin{matrix} \lambda & (0,1) \\ W & i \end{matrix} \middle| \begin{matrix} \lambda' \\ W' \end{matrix} \right\rangle^{(s)} \times \left\langle S \tfrac{1}{2} M_S \mu | S' M'_S \right\rangle, \tag{42}$$

where the first factor on the right-hand side is a RME of $X^\dagger$ operator in the U($2n$) $\supset$ U(n)$\otimes$ SU(2) basis, the second one is a scaled U(n) C–G coefficient, and the third one is a standard SU(2) C–G coefficient (see, e.g., [123–125]). In general, the RME of $X^\dagger$ should depend on all the U($2n$), U(n), and SU(2) irreps that are involved in the bra and ket vectors. However, as mentioned earlier, since the U($2n$) irrep $[1^N]$ is totally antisymmetric and the SU(2) irrep is conjugate to that of U(n), we can label the RME only by the U(n) irreps λ and λ'. The RMEs in Eq. (42) are in fact scaled RMEs as are the U(n) C–G coefficients. Since the latter have been examined previously and the SU(2) C–G coefficients are well known, the only quantities that remain to be determined are the RMEs.

As we have seen, the coupling $\lambda \times (0, 1) \rightarrow \lambda'$ is straightforward for the two-column irreps. Choosing λ' as (a, b), there are only two possible couplings corresponding to $\lambda = (a, b - 1)$ and $(a - 1, b + 1)$ and, thus, only two nonvanishing RMEs which can be evaluated provided that some specific MEs of $X^\dagger$ are known. To determine such MEs, we express the U($2n$) $\supset$ U(n)$\otimes$ SU(2) basis as a linear combination of the U($2n$) basis (i.e., Slater determinants), employing U($2n$) $\supset$ U(n)$\otimes$ SU(2) subduction coefficients

$$\left| \begin{matrix} (a,b) \\ W\ M_S \end{matrix} \right\rangle = \sum_D \left| \begin{matrix} [1^N] \\ D \end{matrix} \right\rangle \left\langle \begin{matrix} [1^N] \\ D \end{matrix} \middle| \begin{matrix} (a,b) \\ W\ M_S \end{matrix} \right\rangle. \tag{43}$$

The U($2n$) basis vectors, labeled by D, are then fully determined by the spin-orbitals involved. Thus, in order to find the desired RMEs, we next consider the two relevant subduction coefficients.

The U($2n$) basis vectors D that can appear on the right-hand side of Eq. (43) are given by products of N spin-orbitals constituting corresponding U(n) and SU(2) basis vectors, labeled by W and SM_S, respectively, subject to a condition that no spin-orbital appears more than once. For example, suppose that the U(n) basis vector W contains orbitals 1, 2, and 3 and the SU(2) basis element SM_S involves spin functions α, α, and β. Then the sum in Eq. (43) contains three terms associated with spin-orbital determinants $|12\bar{3}|$, $|1\bar{2}3|$, and $|\bar{1}23|$. Clearly, each doubly occupied orbital must be associated with different spins. Thus, the more doubly occupied orbitals are present in the U(n) basis vector and the higher the spin component M_S, the simpler will be the expansion in Eq. (43). Moreover, the values of subduction coefficients are determined by singly occupied orbitals in the U(n) basis vectors.

Let us choose the U($2n$) $\supset$ U(n)$\otimes$ SU(2) basis vector in Eq. (43) in such a way that all two-box rows of W are filled with doubly occupied orbitals labeled $1, 2, \ldots, a$ and all one-box rows with singly occupied orbitals having labels $a+1, a+2, \ldots, a+b$. Further, let $M_S = S$, so that the SU(2) basis contains ($a + b$) spin functions α and a spin functions β. With this choice, only one Slater determinant that involves $N(= 2a + b)$ spin-orbitals $1\alpha, 1\beta, \ldots, a\alpha, a\beta, (a+1)\alpha, \ldots, (a+b)\alpha$ will be present in the sum on the right-hand side of Eq. (43). This immediately implies the following simple subduction coefficient

$$\left\langle \begin{array}{|c|c|}\hline 1 & 1 \\ \hline \vdots & \vdots \\ \hline a & a \\ \hline a_1 & \multicolumn{1}{c}{} \\ \cline{1-1} \vdots & \multicolumn{1}{c}{} \\ \cline{1-1} a_b & \multicolumn{1}{c}{} \\ \cline{1-1} \end{array} \; \begin{array}{|c|c|c|c|c|c|}\hline \alpha & \cdots & \alpha & \alpha & \cdots & \alpha \\ \hline \beta & \cdots & \beta & \multicolumn{3}{c}{} \\ \cline{1-3} \end{array} \right| D \right\rangle = \left\langle \begin{array}{|c|}\hline a_1 \\ \hline \vdots \\ \hline a_b \\ \hline \end{array} \; \begin{array}{|c|c|c|}\hline \alpha & \cdots & \alpha \\ \hline \end{array} \right| D_1 \right\rangle = 1, \tag{44}$$

where $a_1 = a + 1, \ldots, a_b = a + b, D$ is the determinant $|1\bar{1} \cdots a\bar{a}a_1 \cdots a_b|$, and D_1 the determinant $|a_1 \cdots a_b|$. (Note that for typographical reasons—here and in the following text—we represent the U(2n) single-column Weyl tableaux that label totally antisymmetric ket vectors by the corresponding Slater determinants. We also drop the irrep labels since they are implied by the explicit form of the Weyl tableau involved.)

The second useful case is slightly more involved. For the SU(2) basis element, we again choose $M_S = S$, but in the U(n) basis vector, only the first ($a - 1$) rows of W will be doubly occupied with orbitals labeled by 1 to ($a - 1$), while the last two-box row and all the one-box rows will be filled with singly occupied orbitals labeled a through ($a + b + 1$) in such a way that the largest label appears in the last box of the second column. We will show that the subduction coefficient for the above choice of the U(n) and SU(2) bases and the Slater determinant associated with the spin-orbital set $\{1, \bar{1}, 2, \bar{2}, \ldots, (a-1), \overline{(a-1)}, a, (a+1), \ldots, (a+b), \overline{(a+b+1)}\}$ is given by

$$\left\langle \begin{array}{|c|c|}\hline 1 & b_2 \\ \hline 2 & \multicolumn{1}{c}{} \\ \cline{1-1} \vdots & \multicolumn{1}{c}{} \\ \cline{1-1} b_1 & \multicolumn{1}{c}{} \\ \cline{1-1} \end{array} \; \begin{array}{|c|c|c|c|}\hline \alpha & \alpha & \cdots & \alpha \\ \hline \beta & \multicolumn{3}{c}{} \\ \cline{1-1} \end{array} \right| D_2 \right\rangle = (-1)^b \sqrt{(b+1)/(b+2)}, \tag{45}$$

where $b_1 = b + 1$, $b_2 = b + 2$, and D_2 is the determinant $|1 2 \cdots b_1 \overline{b_2}|$. Here, we have used the fact that the subduction coefficients are independent of the ($a - 1$) doubly occupied orbitals. Hence, the corresponding parts in the U(n) and SU(2) Weyl tableaux are not shown, and all the singly occupied labels are shifted by ($a - 1$).

Equation (45) can be proved by induction. We consider first the simplest case of $b = 1$ when

$$\left| \begin{matrix} 1\,2 & \alpha\,\alpha \\ 3 & \beta \end{matrix} \right\rangle = c_1 |1\,2\,\bar{3}| + c_2 |1\,\bar{2}\,3| + c_3 |\bar{1}\,2\,3|. \tag{46}$$

(Here, to simplify the notation, we omit boxes in the Weyl tableaux of the irreps involved.) The unknown coefficients c_1, c_2, and c_3 can be easily determined by requiring that the state is symmetric with respect to a transposition of the orbital labels (1, 2), i.e.,

$$(1,2) \left| \begin{matrix} 1\,2 & \alpha\,\alpha \\ 3 & \beta \end{matrix} \right\rangle = \left| \begin{matrix} 1\,2 & \alpha\,\alpha \\ 3 & \beta \end{matrix} \right\rangle. \tag{47}$$

This implies that $c_1 = 0$ and $c_2 = -c_3$. We thus get that

$$\left| \begin{matrix} 1\,2 & \alpha\,\alpha \\ 3 & \beta \end{matrix} \right\rangle = 2^{-1/2} \{ |1\,\bar{2}\,3| - |\bar{1}\,2\,3| \}. \tag{48}$$

In fact, Eq. (48) holds for a general irrep (1, b), namely

$$\left| \begin{array}{|c|c|}\hline 1 & 2 \\ \hline 3 & \multicolumn{1}{c}{} \\ \cline{1-1} \vdots & \multicolumn{1}{c}{} \\ \cline{1-1} b_2 & \multicolumn{1}{c}{} \\ \cline{1-1} \end{array} \; \begin{array}{|c|c|c|c|}\hline \alpha & \alpha & \cdots & \alpha \\ \hline \beta & \multicolumn{3}{c}{} \\ \cline{1-1} \end{array} \right\rangle = 2^{-1/2} \{ |1\,\bar{2}\,3 \cdots b_2| - |\bar{1}\,2\,3 \cdots b_2| \}, \tag{49}$$

where $b_2 = b + 2$. Considering, next, the action of (2,3) on Eq. (48), we get (see, e.g., Eq. (3–91a) or Table 4.4–2 of [126])

$$\begin{aligned} (2,3) \left| \begin{matrix} 1\,2 & \alpha\,\alpha \\ 3 & \beta \end{matrix} \right\rangle &= -\frac{1}{2} \left| \begin{matrix} 1\,2 & \alpha\,\alpha \\ 3 & \beta \end{matrix} \right\rangle + \frac{\sqrt{3}}{2} \left| \begin{matrix} 1\,3 & \alpha\,\alpha \\ 2 & \beta \end{matrix} \right\rangle \\ &= 2^{-1/2} \{ -|12\bar{3}| + |\bar{1}23| \}. \end{aligned} \tag{50}$$

This implies

$$\left| \begin{matrix} 1\,3 & \alpha\,\alpha \\ 2 & \beta \end{matrix} \right\rangle = -\sqrt{\frac{2}{3}} |1\,2\,\bar{3}| + \frac{1}{\sqrt{6}} |1\,\bar{2}\,3| + \frac{1}{\sqrt{6}} |\bar{1}\,2\,3| \tag{51}$$

or, equivalently,

$$\left\langle \begin{matrix} 1\,3 & \alpha\,\alpha \\ 2 & \beta \end{matrix} \right| |1\,2\,\bar{3}| \right\rangle = -\sqrt{2/3}. \tag{52}$$

In a completely similar way, we can obtain [again omitting boxes for (a, b) = (1, 2)]

$$\left\langle \begin{matrix} 1\,4 & \alpha\,\alpha\,\alpha \\ 2 & \beta \\ 3 & \end{matrix} \right| |1\,2\,3\,\bar{4}| \right\rangle = \sqrt{3/4}. \tag{53}$$

Consequently, Eq. (45) can be proved by induction. It is noticed that although Eq. (45) is derived for the irrep (1,b) (i.e., $a = 1$), it holds for a general state of the irrep (a, b) in which the first ($a - 1$) two-box rows are doubly occupied.

Using the above determined two subduction coefficients, Eqs. (44) and (45), we can calculate the two required RMEs of $X^\dagger$ operators. To obtain $\langle (a,b) \| X^\dagger \| (a, b-1) \rangle$, we consider the matrix element of $X^\dagger_{a_b \alpha}(a_b = a + b)$. For the bra $\left\langle \begin{matrix} \lambda' \\ W' \, M_S' \end{matrix} \right|$ in Eq. (42), we choose the state of (a, b) appearing in Eq. (44) and for the ket state $\left| \begin{matrix} \lambda \\ W \, M_S \end{matrix} \right\rangle$ the state of irrep (a, $b - 1$) which results from the bra by removing the box containing the largest label a_b in the U(n) tableau and the box with α in the SU(2) tableau. In view of Eq. (44), the sought ME equals to 1 and since in this case both the corresponding scaled U(n) C–G coefficient (which equals to a scaled isoscalar factor in our case) and the SU(2) C–G coefficient also equal to 1, we find that

$$\langle (a,b)\| X^{\dagger} \| (a,b-1)\rangle = 1. \tag{54}$$

To evaluate the second possible RME, $\langle (a,b)\| X^{\dagger} \| (a-1,b+1)\rangle$, we carry out the following calculation

$$\left\langle \begin{array}{|c|c|} \hline 1 & 1 \\ \hline \vdots & \vdots \\ \hline a' & a' \\ \hline a & K \\ \hline a_1 \\ \cline{1-1} \vdots \\ \cline{1-1} a_b \\ \cline{1-1} \end{array} \; \begin{array}{|c|c|c|c|c|c|c|} \hline \alpha & \cdots & \alpha & \alpha & \alpha & \cdots & \alpha \\ \hline \beta & \cdots & \beta & \beta \\ \cline{1-4} \end{array} \;\middle|\; X^{\dagger}_{K\beta} \;\middle|\; \begin{array}{|c|c|} \hline 1 & 1 \\ \hline \vdots & \vdots \\ \hline a' & a' \\ \hline a \\ \cline{1-1} a_1 \\ \cline{1-1} \vdots \\ \cline{1-1} a_b \\ \cline{1-1} \end{array} \; \begin{array}{|c|c|c|c|c|c|c|} \hline \alpha & \cdots & \alpha & \alpha & \alpha & \cdots & \alpha \\ \hline \beta & \cdots & \beta \\ \cline{1-3} \end{array} \right\rangle \tag{55}$$

$$= \langle (a,b)\| X^{\dagger} \| (a-1,b+1)\rangle \left(\begin{array}{cc|c} (a-1,b+1) & (0,1) & (a,b) \\ (a-1,b+1) & (0,0) & (a-1,b+1) \end{array} \right)^{(s)} \left\langle \tfrac{b+1}{2}\,\tfrac{1}{2}\,\tfrac{b+1}{2}\,\tfrac{\bar{1}}{2} \middle| \tfrac{b}{2}\,\tfrac{b}{2} \right\rangle$$
$$= (-1)^{b}\sqrt{(b+1)/(b+2)}\,,$$

where $a' = a-1, a_1 = a+1, \ldots, a_b = a+b, K = a_{b+1} = a+b+1$, etc. The first equality results from Eq. (42), where in this case the corresponding scaled U(n) C–G coefficient is equal to the scaled isoscalar factor. Note that the ket in Eq. (55) is given by a single determinant [cf., Eq. (44)], while the bra is a linear combination of a number of determinants. However, only one determinant, namely $|1\bar{1}2\bar{2}\cdots a'\,\overline{a'}\,a\,a_1\cdots a_b\bar{K}|$, contributes to the matrix element, so that the corresponding U($2n$) $\supset$ U(n) $\otimes$ SU(2) subduction coefficient is given by Eq. (45). This fact implies the second equality given by Eq. (55). In this equation, the U(n) scaled isoscalar factor equals 1 (see [36]) and the coupling coefficient of SU(2) is $\sqrt{(2S+1)/(2S+2)}$ (where $2S = b$). This in turn implies that

$$\langle (a,b)\| X^{\dagger} \| (a-1,b+1)\rangle = (-1)^{b}. \tag{56}$$

This concludes derivation of the two required RMEs for the $X^{\dagger}$ operators. The desired MEs of standard second-quantization creation and annihilation operators in the U($2n$) $\supset$ U(n) $\otimes$ SU(2) basis can now be obtained using Eqs. (42), (54), and (56).

At this point, we recall that the scaled U(n) C–G coefficients in Eq. (42) equal the MEs of the $C^{\dagger}$ (or C) operators (see Eq. (25) in [36])

$$\left\langle \begin{array}{cc} \lambda & (0,1) \\ W & i \end{array} \middle| \begin{array}{c} \lambda' \\ W' \end{array} \right\rangle^{(s)} = \left\langle \begin{array}{c} \lambda' \\ W' \end{array} \middle| C_i^{\sigma\dagger} \middle| \begin{array}{c} \lambda \\ W \end{array} \right\rangle, \tag{57}$$

where σ is given by the condition $\lambda + \sigma = \lambda'$. In other words, Eq. (42) expresses the relationship between the MEs of $X^{\dagger}$ (or X) operators in the U($2n$) $\supset$ U(n) $\otimes$ SU(2) basis and the MEs of $C^{\dagger}$ (or C) operators in the U(n) G–T basis.3333

5 Matrix elements of one-body spin-dependent operators

The MEs of U($2n$) generators $\Xi_{i\mu,j\nu}$, Eq. (35), in the U($2n$) $\supset$ U(n) $\otimes$ SU(2) basis may be expresed in terms of MEs of the X operators as follows

$$\left\langle \begin{array}{c} \lambda' \\ W'M'_S \end{array} \middle| \Xi_{i\mu,j\nu} \middle| \begin{array}{c} \lambda \\ WM_S \end{array} \right\rangle = \sum_{\lambda'' W'' M''_S} \left\langle \begin{array}{c} \lambda' \\ W'M'_S \end{array} \middle| X^{\dagger}_{i\mu} \middle| \begin{array}{c} \lambda'' \\ W''M''_S \end{array} \right\rangle \times \left\langle \begin{array}{c} \lambda'' \\ W''M''_S \end{array} \middle| X_{j\nu} \middle| \begin{array}{c} \lambda \\ WM_S \end{array} \right\rangle. \tag{58}$$

Applying Eq. (42) and the fact that the scaled U(n) C–G coefficients are equal to the MEs of the C operators, Eq. (57), we get

$$\left\langle \begin{array}{c} \lambda' \\ W'M'_S \end{array} \middle| \Xi_{i\mu,j\nu} \middle| \begin{array}{c} \lambda \\ WM_S \end{array} \right\rangle = \sum_{\tau=1}^{2} \langle \lambda' \| X^{\dagger} \| \lambda - \tau\rangle \langle \lambda \| X^{\dagger} \| \lambda - \tau\rangle$$
$$\times \left\langle \begin{array}{c} \lambda' \\ W' \end{array} \middle| C_i^{\sigma\dagger} C_j^{\tau} \middle| \begin{array}{c} \lambda \\ W \end{array} \right\rangle \langle (S-\tau)\tfrac{1}{2}(M_S-\nu)\mu | S'M'_S\rangle$$
$$\times \langle (S-\tau)\tfrac{1}{2}(M_S-\nu)\nu | SM_S\rangle, \tag{59}$$

where for σ and τ we require that $\lambda' - \Delta(\sigma) = \lambda - \Delta(\tau)$ with the shifts $\Delta(1) = (0,1)$ and $\Delta(2) = (1,-1)$, while S and S' designate spins associated with the irreps λ and λ', respectively. By an abuse of notation, the shifted spin label $S - \tau$ in the SU(2) coefficients designates $S - \Delta_S(\tau)$ where $\Delta_S(1) = \frac{1}{2}$ and $\Delta_S(2) = -\frac{1}{2}$. Moreover, we must require that $M_S - \nu = M'_S - \mu$ lest the ME vanishes. Equation (59) may be rewritten in a more compact form by employing the known RMEs of the $X^\dagger$ operators and the SU(2) C–G coefficients.

In the following, we always assume that $\lambda = (a, b)$ so that λ' can equal $(a-1, b+2)$, $(a+1, b-2)$, or (a, b). Let us first consider the case $\lambda' = (a-1, b+2)$, i.e., $\Delta S = 1$. Since $\lambda' \equiv (a', b') = (a, b) + \Delta(\sigma) - \Delta(\tau)$ we must have that $\sigma = 1$ and $\tau = 2$ so that $(a, b) + \Delta(1) - \Delta(2) = (a, b) + (0, 1) - (1, -1) = (a-1, b+2)$. The resulting operator $C_i^{1\dagger} C_j^2$ is related to the positive-spin-shift adjoint tensor $N_{ij}^{(+)}$, Eq. (28), when acting on the (a, b) irrep module (see also Eqs. (61a) or (73a) of [36]). We thus get

$$\left\langle \begin{matrix} (a-1,b+2) \\ W'M'_S \end{matrix} \middle| \Xi_{i\mu,j\nu} \middle| \begin{matrix} (a,b) \\ WM_S \end{matrix} \right\rangle = \langle (a-1,b+2) \| X^\dagger \| (a-1,b+1) \rangle \times \langle (a,b) \| X^\dagger \| (a-1,b+1) \rangle \sqrt{\frac{b+2}{b+3}} \times \langle (S+\tfrac{1}{2}) \tfrac{1}{2} (M_S - \nu) \mu | (S+1) M'_S \rangle \times \langle (S+\tfrac{1}{2}) \tfrac{1}{2} (M_S - \nu) \mu | S M_S \rangle \times \left\langle \begin{matrix} (a-1,b+2) \\ W' \end{matrix} \middle| N_{ij}^{(+)} \middle| \begin{matrix} (a,b) \\ W \end{matrix} \right\rangle. \tag{60}$$

We can now evaluate the product of the two RMEs, two SU(2) C–G coefficients, and account for the factor in Eq. (60) for the four possible cases of $(\mu\nu) = (\alpha\alpha)$, $(\beta\beta)$, $(\alpha\beta)$, $(\beta\alpha)$ and define the resulting quantities as $N^+(\mu\nu)_{SM_S}$, so that

$$\left\langle \begin{matrix} (a-1,b+2) \\ W'M'_S \end{matrix} \middle| \Xi_{i\mu,j\nu} \middle| \begin{matrix} (a,b) \\ W\,M_S \end{matrix} \right\rangle = N^+(\mu\nu)_{SM_S} \left\langle \begin{matrix} (a-1,b+2) \\ W' \end{matrix} \middle| N_{ij}^{(+)} \middle| \begin{matrix} (a,b) \\ W \end{matrix} \right\rangle. \tag{61}$$

These new quantities $N^+(\mu\nu)_{SM_S}$, given in Table 1, are functions of the spin labels (SM_S) of the ket and $(\mu\nu)$ labels of the operator $\Xi_{i\mu,j\nu}$. The conditions $S' = S + 1$ [which is implied by the irreps $(a-1, b+2)$ and (a, b)] and $M'_S = M_S - \nu + \mu$ must be satisfied lest the matrix elements vanish.

Next, when $\lambda' = (a+1, b-2)$, the only possible choice for σ and τ is $\sigma = 2$ and $\tau = 1$, implying the negative-spin-shift adjoint tensor $N_{ij}^{(-)}$, Eq. (29), when acting on the (a, b) irrep module (see also Eqs. (61c) or (73b) of [36]). We then get an expression which is similar to Eq. (60). The final result is

$$\left\langle \begin{matrix} (a+1,b-2) \\ W'M'_S \end{matrix} \middle| \Xi_{i\mu,j\nu} \middle| \begin{matrix} (a,b) \\ W\,M_S \end{matrix} \right\rangle = N^-(\mu\nu)_{SM_S} \left\langle \begin{matrix} (a+1,b-2) \\ W' \end{matrix} \middle| N_{ij}^{(-)} \middle| \begin{matrix} (a,b) \\ W \end{matrix} \right\rangle, \tag{62}$$

where the $N^-(\mu\nu)_{SM_S}$ coefficients are given in Table 1 for the four possible cases of $(\mu\nu)$. Again, $M'_S = M_S - \nu + \mu$.

Finally, when $\lambda' = (a, b)$, both terms with $\sigma = \tau = 1$ and $\sigma = \tau = 2$ on the right-hand side of Eq. (59) survive and the MEs of both operators $C_i^{1\dagger} C_j^1$ and $C_i^{2\dagger} C_j^2$ are required. As has been shown in [36], these operators can be expressed in terms of the two zero-shift adjoint tensors, namely E_{ij} and $N_{ij}^{(0)}$, as implied by Eqs. (23) and (29). Inverting these relations, we get (see also Eqs. (73c) and (73d) of [36])

$$C_i^{1\dagger} C_j^1 = \frac{b}{2(b+1)} E_{ij} + \frac{1}{b+1} \sqrt{\frac{b(b+2)}{2}} N_{ij}^{(0)} \tag{63}$$

$$C_i^{2\dagger} C_j^2 = \frac{b+2}{2(b+1)} E_{ij} - \frac{1}{b+1} \sqrt{\frac{b(b+2)}{2}} N_{ij}^{(0)} \tag{64}$$

when acting on the (a, b) irrep module. Via a simple manipulation, we can find expression for the required MEs for the zero-spin-shift case in terms of those for E_{ij} and $N_{ij}^{(0)}$, i.e.,

$$\left\langle \begin{matrix} (a,b) \\ W'M'_S \end{matrix} \middle| \Xi_{i\mu,j\nu} \middle| \begin{matrix} (a,b) \\ WM_S \end{matrix} \right\rangle = \frac{\delta_{\mu,\nu}}{2} \left\langle \begin{matrix} (a,b) \\ W' \end{matrix} \middle| E_{ij} \middle| \begin{matrix} (a,b) \\ W \end{matrix} \right\rangle + N^0(\mu\nu)_{SM_S} \left\langle \begin{matrix} (a,b) \\ W' \end{matrix} \middle| N_{ij}^{(0)} \middle| \begin{matrix} (a,b) \\ W \end{matrix} \right\rangle, \tag{65}$$

where the $N^0(\mu\nu)_{SM_S}$ coefficients are given in Table 1. Again, $M'_S = M_S - \nu + \mu$.

We can thus conclude that in the cases of spin-positive and spin-negative shifts, the MEs of one-body spin-dependent operators are given by the MEs of adjoint tensors $N_{ij}^{(\kappa)}$, $\kappa = +$ or $-$, and in the case of a zero-spin shift by the sum of MEs of the generator E_{ij} and the adjoint tensor $N_{ij}^{(0)}$, as implied by Eqs. (61), (62), and (65). The evaluation of MEs of E_{ij} and $N_{ij}^{(\kappa)}$ operators was described in detail in [36]. There we have shown that this may be achieved via a simple segment level products, and all the required segment values were tabulated in [36]. We thus find that a segmentation formalism applies even for MEs of spin-dependent one-body operators. For example, assuming that $i < j$, the MEs in the zero-shift case are given by

$$\left\langle \begin{matrix} (a,b) \\ W'M'_S \end{matrix} \middle| \Xi_{i\mu,j\nu} \middle| \begin{matrix} (a,b) \\ WM_S \end{matrix} \right\rangle = \left\{ \frac{1}{2} \delta_{\mu,\nu} \prod_{k=j}^{n} \delta_{\lambda'_k \lambda_k} \left\langle \begin{matrix} \lambda_j \\ \lambda'_{j-1} \end{matrix} \middle| E \middle| \begin{matrix} \lambda_j \\ \lambda_{j-1} \end{matrix} \right\rangle + N^0(\mu\nu)_{SM_S} \prod_{k=j+1}^{n} \left\langle \begin{matrix} \lambda'_k \\ \lambda'_{k-1} \end{matrix} \middle| N \middle| \begin{matrix} \lambda_k \\ \lambda_{k-1} \end{matrix} \right\rangle \left\langle \begin{matrix} \lambda'_j \\ \lambda'_{j-1} \end{matrix} \middle| N \middle| \begin{matrix} \lambda_j \\ \lambda_{j-1} \end{matrix} \right\rangle_{C^\dagger} \right\} \times \prod_{k=i+1}^{j-1} \begin{pmatrix} \lambda_k & (0,1) & \lambda'_k \\ \lambda_{k-1} & (0,1) & \lambda'_{k-1} \end{pmatrix}^{(s)} \begin{pmatrix} \lambda_i & (0,1) & \lambda'_i \\ \lambda_{i-1} & (0,0) & \lambda_{i-1} \end{pmatrix}^{(s)} \prod_{k=1}^{i-1} \delta_{\lambda'_k \lambda_k}, \tag{66}$$

Table 1 Coefficients required for the evaluation of MEs of spin-dependent, one-body operators, Eqs. (61), (62), and (65), for three possible shifts, as functions of the SU(2) labels

$\mu\nu$	$N^{+}(\mu\nu)_{SM_S}$	$N^{0}(\mu\nu)_{SM_S}$	$N^{-}(\mu\nu)_{SM_S}$
$\alpha\alpha$	$(-1)^{b}\left[\frac{(S_-+1)(S_++1)}{2(S+1)(2S+3)}\right]^{1/2}$	$\tilde{\delta}_{b0}\frac{M_S}{[2S(S+1)]^{1/2}}$	$(-1)^{b}\left[\frac{S_-S_+}{2S(2S+1)}\right]^{1/2}$
$\beta\beta$	$(-1)^{b+1}\left[\frac{(S_-+1)(S_++1)}{2(S+1)(2S+3)}\right]^{1/2}$	$-\tilde{\delta}_{b0}\frac{M_S}{[2S(S+1)]^{1/2}}$	$(-1)^{b+1}\left[\frac{S_-S_+}{2S(2S+1)}\right]^{1/2}$
$\alpha\beta$	$(-1)^{b}\left[\frac{(S_++1)(S_++2)}{2(S+1)(2S+3)}\right]^{1/2}$	$\tilde{\delta}_{b0}\left[\frac{S_-(S_++1)}{2S(S+1)}\right]^{1/2}$	$(-1)^{b+1}\left[\frac{(S_--1)S_-}{2S(2S+1)}\right]^{1/2}$
$\beta\alpha$	$(-1)^{b+1}\left[\frac{(S_-+1)(S_-+2)}{2(S+1)(2S+3)}\right]^{1/2}$	$\tilde{\delta}_{b0}\left[\frac{(S_-+1)S_+}{2S(S+1)}\right]^{1/2}$	$(-1)^{b}\left[\frac{(S_+-1)S_+}{2S(2S+1)}\right]^{1/2}$

The following short-hand notation is used: $S_+ = S + M_S$, $S_- = S - M_S$, and $\tilde{\delta}_{b0} = 1 - \delta_{b0}$, where S and M_S designate the total spin and the spin component in the ket vector

where $\lambda'_n = \lambda_n = (a, b)$, and the required quantities, which include RMEs of the U(n) generator (E), the N factors, the $N - C^{\dagger}$ connecting factors, and the scaled isoscalar factors for vector operators, can be found in Tables 2, 5, 7, and 1 of [36], respectively.

6 A comparison of spin-dependent and spin-independent UGA

The segmentation of MEs of spin-dependent one-body operators is very similar to that for spin-independent two-body operators. This is a very useful fact in view of an actual implementation of this formalism, since it enables an evaluation of MEs of spin-dependent one-body operators by exploiting presently available UGA or GUGA codes. We now compare both formalisms in greater detail.

The key to both ME evaluations are MEs of adjoint tensors, $N_{ij}^{(\kappa)}$. We recall that the same tensors play an important role when dealing with MEs of U(n) generator products in the UGA spin-independent case. However, the U(n) generator products cannot change the spin of U(n) irreps, while the spin-dependent one-body operators do. This implies that MEs of spin-dependent one-body operators are related to MEs of generator products of a larger group, say U($n + 1$), since the U($n + 1$) generators can change the spin at the U(n) level as is the case for spin-dependent, one-body operators. We thus derive below the relationships between the MEs of U(n) $N_{ij}^{(\kappa)}$ operators in the U(n) G–T basis and the MEs of U($n + 1$) generator products

$$\begin{aligned} e_{n+1,j;i,n+1} &= E_{(n+1)j}E_{i(n+1)} - \delta_{ij}E_{(n+1)(n+1)} \\ &= e_{i,n+1;n+1,j} = E_{i(n+1)}E_{(n+1)j} - E_{ij} \end{aligned} \tag{67}$$

in the U($n + 1$) G–T basis.

While in Sects. 4 and 5, the MEs of spin-dependent operators are derived strictly within the framework of the U(n) group, as enabled by the formalism of our spin-adapted creation and annihilation operators, the same MEs can also be derived by relying on the standard Racah-Wigner spin algebra and exploiting a larger U($n + 1$) group. In fact, the use of a larger U($n + 1$) group follows immediately when we employ graphical methods of spin algebras (see [34]) as was done by Drake and Schlesinger [39] (See also later work by these and other authors [37, 115, 116, 117]. Note, however, that in their latter work [39], they have been unaware of our earlier work [34] which employs graphical methods of spin algebras with Yamanouchi-Kotani coupling scheme and Jucys-type diagrams, a formalism that is directly related to a U(n) formalism). The U($n + 1$)-based formalism (or even U($n + 2$) [116]) for spin-dependent operators that was based on Drake and Schlesinger's developments was then exploited by Yabushita et al. [37] and implemented in actual GUGA codes. In contrast to our approach here, the spin-algebra-based approach that relies on Racah-Wigner calculus and that was employed by Yabushita et al. [37] requires six steps in order to conform with the standard UGA-based formalism, including an introduction of an additional phase factor and a transformation to "real spherical" spin functions to ascertain a real form of the relevant matrix elements. The latter are automatically taken care of in our approach since we systematically employ unitary group formalism. On the other hand, Yabushita et al. also implemented in their codes an adaptation to the D_{2h} point group and its subgroups [37].

Consider, first, MEs of $N_{ij}^{(+)}$ appearing on the right-hand side of Eq. (61). To relate these MEs with those of the U($n + 1$) generator products, we must construct the U($n + 1$) bra and ket states in such a way that they have identical irreps at the U($n + 1$) level but subduce to the U(n) bra and ket states $\left\langle \begin{matrix} (a-1, b+2) \\ W' \end{matrix} \right|$ and $\left| \begin{matrix} (a, b) \\ W \end{matrix} \right\rangle$, respectively, at the U(n) level. This may be easily achieved when we add one box with the orbital label ($n + 1$) to the second column of the Weyl tableau W' as well as to the first

Table 2 Coefficients relating MEs of spin-dependent, one-body operators and spin-independent, two-body operators, Eqs. (70), (73), and (76), for three possible shifts, as functions of the SU(2) labels. See footnote to Table 1 for the notation convention used

$\mu\nu$	$\tilde{N}^{+}(\mu\nu)_{SM_S}$	$\tilde{N}^{0}(\mu\nu)_{SM_S}$	$\tilde{N}^{-}(\mu\nu)_{SM_S}$
$\alpha\alpha$	$(-1)^{b}\left[\frac{(S_-+1)(S_++1)}{(2S+1)(2S+3)}\right]^{1/2}$	$-\tilde{\delta}_{b0}\frac{M_S}{S}$	$(-1)^{b}\left[\frac{S_-S_+}{(2S-1)(2S+1)}\right]^{1/2}$
$\beta\beta$	$(-1)^{b+1}\left[\frac{(S_-+1)(S_++1)}{(2S+1)(2S+3)}\right]^{1/2}$	$\tilde{\delta}_{b0}\frac{M_S}{S}$	$(-1)^{b+1}\left[\frac{S_-S_+}{(2S-1)(2S+1)}\right]^{1/2}$
$\alpha\beta$	$(-1)^{b}\left[\frac{(S_++1)(S_++2)}{(2S+1)(2S+3)}\right]^{1/2}$	$-\tilde{\delta}_{b0}\frac{[S_-(S_++1)]^{1/2}}{S}$	$(-1)^{b+1}\left[\frac{(S_--1)S_-}{(2S-1)(2S+1)}\right]^{1/2}$
$\beta\alpha$	$(-1)^{b+1}\left[\frac{(S_-+1)(S_-+2)}{(2S+1)(2S+3)}\right]^{1/2}$	$-\tilde{\delta}_{b0}\frac{[(S_-+1)S_-]^{1/2}}{S}$	$(-1)^{b}\left[\frac{(S_+-1)S_+}{(2S-1)(2S+1)}\right]^{1/2}$

column of the Weyl tableau W, yielding new bra $\widetilde{W'}$ and ket $\widetilde{W}$ states that are associated with the irrep $(a, b+1)$ of $\mathrm{U}(n+1)$. Expressed in terms of the d-vector notation[2], Eq. (4), we expand the n-component $d'(\equiv W')$ and $d\,(\equiv W)$ vectors (corresponding to W' and W, respectively) to the $(n+1)$-component vectors $\tilde{d}'$ and $\tilde{d}$ by adding 2 to d' and 1 to d as the $(n+1)$th component, respectively, i.e.,

$$\widetilde{W'} \equiv \{\tilde{d}'\} = \{d'2\}, \quad \widetilde{W} \equiv \{\tilde{d}\} = \{d1\}. \tag{68}$$

Employing Eq. (32) and Table 9 of [36] (see also Eq. (92) of [36]) the $\mathrm{U}(n+1)$ ME of $e_{n+1,j;i,n+1}$ involving G–T states of Eq. (68) can be expressed as a product of the $e-N$ connecting factor and the $\mathrm{U}(n)$ ME of $N_{ij}^{(+)}$, namely

$$\begin{aligned}
&\left\langle \begin{matrix}(a,b+1)\\ \tilde{d}'\end{matrix} \middle| e_{n+1,j;i,n+1} \middle| \begin{matrix}(a,b+1)\\ \tilde{d}\end{matrix} \right\rangle \\
&= \left\langle \begin{matrix}(a,b+1)\\ (a-1,b+2)\end{matrix} \middle| e \middle| \begin{matrix}(a,b+1)\\ (a,b)\end{matrix} \right\rangle_N \\
&\quad \times \left\langle \begin{matrix}(a-1,b+2)\\ d'\end{matrix} \middle| N_{ij}^{(+)} \middle| \begin{matrix}(a,b)\\ d\end{matrix} \right\rangle \\
&= \sqrt{\frac{b+1}{b+2}} \left\langle \begin{matrix}(a-1,b+2)\\ d'\end{matrix} \middle| N_{ij}^{(+)} \middle| \begin{matrix}(a,b)\\ d\end{matrix} \right\rangle.
\end{aligned} \tag{69}$$

This implies that

$$\begin{aligned}
&\left\langle \begin{matrix}(a-1,b+2)\\ d'\ M'_S\end{matrix} \middle| \Xi_{i\mu,j\nu} \middle| \begin{matrix}(a,b)\\ d\ M_S\end{matrix} \right\rangle = \tilde{N}^{+}(\mu\nu)_{SM_S} \\
&\quad \times \left\langle \begin{matrix}(a,b+1)\\ \{d'2\}\end{matrix} \middle| e_{n+1,j;i,n+1} \middle| \begin{matrix}(a,b+1)\\ \{d1\}\end{matrix} \right\rangle,
\end{aligned} \tag{70}$$

with $\tilde{N}^{+}(\mu\nu)_{SM_S} = \sqrt{(b+2)/(b+1)}N^{+}(\mu\nu)_{SM_S}$ values listed in Table 2.

Similarly for the MEs of $N_{ij}^{(-)}$ that appear on the right-hand side of Eq. (62), we add one box with the label $(n+1)$ to the first column of W' as well as to the second column of W, obtaining new states $\widetilde{W}'$ and $\widetilde{W}$ of the $\mathrm{U}(n+1)$ irrep $(a+1, b-1)$. Alternatively, in the d-vector notation, we expand the d' and d vectors to $(n+1)$-component vectors $\tilde{d}'$ and $\tilde{d}$ by adding 1 to d' and 2 to d, respectively, i.e.,

$$\{\tilde{d}'\} = \{d'1\}, \{\tilde{d}\} = \{d2\}. \tag{71}$$

The corresponding $\mathrm{U}(n+1)$ ME of $e_{n+1,j;i,n+1}$ is then given by a product of the $e-N$ connecting factor and a ME of $N_{ij}^{(-)}$ in the $\mathrm{U}(n)$ basis. Relying again on Eq. (32) and table 9 of [36] (see also Eq. (92) of [36]), we get

$$\begin{aligned}
&\left\langle \begin{matrix}(a+1,b-2)\\ d'\end{matrix} \middle| N_{ij}^{(-)} \middle| \begin{matrix}(a,b)\\ d\end{matrix} \right\rangle = \sqrt{\frac{b}{b-1}} \\
&\quad \times \left\langle \begin{matrix}(a+1,b-1)\\ \tilde{d}'\end{matrix} \middle| e_{n+1,j;i,n+1} \middle| \begin{matrix}(a+1,b-1)\\ \tilde{d}\end{matrix} \right\rangle
\end{aligned} \tag{72}$$

and

$$\begin{aligned}
&\left\langle \begin{matrix}(a+1,b-2)\\ d'\ M'_S\end{matrix} \middle| \Xi_{i\mu,j\nu} \middle| \begin{matrix}(a,b)\\ d\ M_S\end{matrix} \right\rangle = \tilde{N}^{-}(\mu\nu)_{SM_S} \\
&\quad \times \left\langle \begin{matrix}(a+1,b-1)\\ \{d'1\}\end{matrix} \middle| e_{n+1,j;i,n+1} \middle| \begin{matrix}(a+1,b-1)\\ \{d2\}\end{matrix} \right\rangle,
\end{aligned} \tag{73}$$

where $\tilde{N}^{-}(\mu\nu)_{SM_S} = \sqrt{b/(b-1)}N^{-}(\mu\nu)_{SM_S}$, the values of which are given in Table 2.

The bra and ket states in the MEs of $N_{ij}^{(0)}$, appearing on the right-hand side of Eq. (65), belong to the same irrep. In principle, there exist several possibilities how to construct the desired $\mathrm{U}(n+1)$ basis vectors. An obvious choice that makes the orbital $(n+1)$ either unoccupied or doubly occupied in the bra and ket states does not lead, however, to useful results. A better choice is to add one box labeled with $(n+1)$ to the first column of both bra and ket states, which is always possible (note that it is impossible to add an additional box to the second column when $b = 0$), i.e., to expand the d vectors as follows

$$\{\tilde{d}'\} = \{d'1\}, \quad \{\tilde{d}\} = \{d1\}, \tag{74}$$

obtaining basis states for the $\mathrm{U}(n+1)$ irrep $(a, b+1)$. Employing, again, Eq. (32) and Table 9 of [36], and noting that in this case both the $e-E$ and $e-N$ connecting factors are not zero, we get

[2] Recall that for the $\mathrm{U}(n)$ electronic G–T bases, the G–T tableau, the Weyl tableau, the Paldus tableau or the d-vector notation are all equivalent and there is a one-to-one correspondence between them.

$$\left\langle \begin{matrix}(a,b+1)\\ \tilde{d}'\end{matrix} \middle| e_{n+1,j;i,n+1} \middle| \begin{matrix}(a,b+1)\\ \tilde{d}\end{matrix} \right\rangle$$

$$= -\frac{1}{2}\left\langle \begin{matrix}(a,b)\\ d'\end{matrix} \middle| E_{ij} \middle| \begin{matrix}(a,b)\\ d\end{matrix} \right\rangle$$

$$- \sqrt{\frac{b}{2(b+2)}}\left\langle \begin{matrix}(a,b)\\ d'\end{matrix} \middle| N_{ij}^{(0)} \middle| \begin{matrix}(a,b)\\ d\end{matrix} \right\rangle. \tag{75}$$

Substituting, finally, Eq. (75) into Eq. (65), we find that

$$\left\langle \begin{matrix}(a,b)\\ d'M_S'\end{matrix} \middle| \Xi_{i\mu,j\nu} \middle| \begin{matrix}(a,b)\\ dM_S\end{matrix} \right\rangle = \frac{1}{2}[\delta_{\mu\nu} + \tilde{N}^0(\mu\nu)_{SM_S}]$$

$$\times \left\langle \begin{matrix}(a,b)\\ d'\end{matrix} \middle| E_{ij} \middle| \begin{matrix}(a,b)\\ d\end{matrix} \right\rangle$$

$$+ \tilde{N}^0(\mu\nu)_{SM_S}\left\langle \begin{matrix}(a,b+1)\\ \{d'1\}\end{matrix} \middle| e_{n+1,j;i,n+1} \middle| \begin{matrix}(a,b+1)\\ \{d1\}\end{matrix} \right\rangle, \tag{76}$$

where $\tilde{N}^0(\mu\nu)_{SM_S} = -\sqrt{2(S+1)/S}\, N^0(\mu\nu)_{SM_S}$, whose values are given in Table 2.

7 Conclusions

We have shown that one-electron spin-dependent terms, Eq. (36), in the electronic Hamiltonian, Eq. (33), may be efficiently handled in much the same way as the standard spin-independent two-electron (i.e., Coulomb) terms. Indeed, as clearly implied by Eqs. (70), (73), and (76), the MEs of spin-dependent one-body operators in the $U(2n) \supset U(n) \otimes SU(2)$ basis may be evaluated as MEs of spin-independent two-body operators in a standard $U(n+1)$ electronic G–T basis (see also [37, 39]). Since the MEs of generator products within the spin-independent UGA approach are well known, the above presented development should facilitate the implementation of the spin-dependent UGA formalism. This opens a possible avenue enabling us to handle spin-dependent MBPT terms via a simple modification of the existing UGA and GUGA codes, similarly as done by Yabushita et al. [37]. We emphasize that all the required segment values may be found in [36], and a few additional ones that are specific to a spin-dependent case are given in Tables 1 and 2.

Acknowledgments One of us (X. L.) would like to express his sincere gratitude to his co-author for his hospitality and numerous helpful discussions during his stay in Waterloo. The past support by NSERC (J. P.) is also gratefully acknowledged.

References

1. Paldus J, Čížek J, Shavitt I (1972) Phys Rev A 5:50
2. Paldus J (1974) J Chem Phys 61:5321
3. Paldus J (1976) Phys Rev A 14:1620
4. Paldus J (1976) Many-electron correlation problem: a group theoretical approach. In: Eyring H, Henderson DJ (eds) Theoretical chemistry: advances and perspectives, vol 2. Academic, New York, pp 131–290
5. Paldus J (1981) Unitary group approach to many-electron correlation problem. In: Hinze J (ed) The unitary group for the evaluation of electronic energy matrix elements. Lecture notes in chemistry, no. 22. Springer, Berlin, pp 1–50
6. Shavitt I (1977) Int J Quantum Chem Symp 11:131
7. Shavitt I (1978) Int J Quantum Chem Symp 12:5
8. Shavitt I (1981) The graphical unitary group approach and its application to direct configuration interaction calculations. In: Hinze J (ed) The unitary group for the evaluation of electronic energy matrix elements. Lecture notes in chemistry, no. 22. Springer, Berlin, pp 51–99
9. Robb MA, Niazi U (1984) Comp Phys Rep 1:127
10. Paldus J (1988) Lie algebraic approach to the many-electron correlation problem. In: Truhlar DG (ed) Mathematical frontiers in computational chemical physics. IMA series, vol 15. Springer, Berlin, pp 262–299
11. Shavitt I (1988) Unitary group approach to configuration interaction calculations of the electronic structure of atoms and molecules. In: Truhlar DG (ed) Mathematical frontiers in computational chemical physics. IMA series, vol 15. Springer, Berlin, pp 300–349
12. Robb MA (2000) Theor Chem Acc 103:317
13. Paldus J (2006) Dynamical groups. In: Drake GWF (ed) Handbook of atomic, molecular, and optical physics, chap. 4, 2nd edn. Springer, Berlin, pp 101–114
14. Jordan P (1935) Z Phys 94:531
15. Moshinsky M (1968) Group theory and the many-body problem. Gordon and Breach, New York
16. Moshinsky M, Seligman TH (1971) Ann Phys 66:311
17. Louck JD (1970) Am J Phys 38:3
18. Weyl H (1931) Gruppentheorie und Quantenmechanik, 2nd ed. Hirzel, Leipzig (English translation: The theory of groups and quantum mechanics. Dover, New York, 1950, repr. 1964)
19. Gel'fand IM, Tsetlin ML (1950) Dokl Akad Nauk SSSR 71:825
20. Gel'fand IM, Tsetlin ML (1950) Dokl Akad Nauk SSSR 71:1017
21. Baird GE, Biedenharn LC (1963) J Math Phys 4:1449
22. Biedenharn LC, Louck JD (1981) Angular momentum in quantum physics: theory and application. Addison-Wesley, Reading
23. Biedenharn LC, Louck JD (1981) The Racah-Wigner algebra in quantum theory. Addison-Wesley, Reading
24. Matsen FA, Pauncz R (1986) The unitary group in quantum chemistry. Elsevier, Amsterdam
25. Hinze J (ed) (1981) The unitary group for the evaluation of electronic energy matrix elements. Lecture notes in chemistry, no. 22. Springer, Berlin
26. Pauncz R (1979) Spin eigenfunctions: construction and use. Plenum, New York
27. Wilson S (1984) Electron correlation in molecules, chap. 5. Clarendon, Oxford
28. McWeeny R (1989) Methods of molecular quantum mechanics, 2nd edn, chap. 10. Academic, New York
29. Gould MD, Chandler GS (1984) Int J Quantum Chem 25:553
30. Gould MD, Chandler GS (1984) Int J Quantum Chem 25:603
31. Gould MD, Chandler GS (1984) Int J Quantum Chem 25:1089
32. Gould MD, Chandler GS (1984) Int J Quantum Chem 26:441
33. Gould MD, Chandler GS (1984) Int J Quantum Chem 27:878 (Erratum)
34. Paldus J, Boyle MJ (1980) Phys Scr 21:295
35. Li X, Paldus J (1990) J Math Chem 4:295

36. Li X, Paldus J (1993) J Math Chem 13:273
37. Yabushita S, Zhang Z, Pitzer RM (1999) J Phys Chem A 103:5791
38. Wormer PES, Paldus J (2006) Adv Quantum Chem 51:59
39. Drake GWF, Schlesinger M (1977) Phys Rev A 15:1990
40. Zarrabian S, Sarma CR, Paldus J (1989) Chem Phys Lett 155:183
41. Downward MJ, Robb MA (1977) Theor Chim Acta 46:129
42. Hegarty D, Robb MA (1979) Mol Phys 38:1795
43. Brooks BR, Schaefer III HF (1979) J Chem Phys 70:5092
44. Brooks BR, Laidig WD, Saxe P, Schaefer III HF (1980) J Chem Phys 72:3837
45. Siegbahn PEM (1979) J Chem Phys 70:5391
46. Siegbahn PEM (1980) J Chem Phys 72:1647
47. Siegbahn PEM (1980) Int J Quantum Chem 18:1229
48. Siegbahn PEM (1981) Factorization of the diredt CI coupling coefficients into internal and external parts. In: Hinze J (ed) The unitary group for the evaluation of electronic energy matrix elements. Lecture notes in chemistry, no. 22. Springer, Berlin, pp 119–135
49. Lischka H, Shepard R, Brown FB, Shavitt I (1981) Int J Quantum Chem Symp 15:91
50. Saxe P, Fox DJ, Schaefer III HF, Handy NC (1982) J Chem Phys 77:5584
51. Saunders VR, van Lenthe JH (1983) Mol Phys 48:923
52. Shepard R, Shavitt I, Pitzer RM, Comeau DC, Pepper M, Lischka H, Szalay PG, Ahlrichs R, Brown FB, Zhao J (1988) Int J Quantum Chem Symp 22:149
53. Lischka H, Shepard R, Pitzer RM, Shavitt I, Dallos M, Müller Th, Szalay PG, Seth M, Kedziora GS, Yabushita S, Zhang Z (2001) Phys Chem Chem Phys 3:664
54. Shepard R, Shavitt I, Lischka H (2001) J Comp Chem 23:1121
55. Lischka H, Müller T, Szalay PG, Shavitt I, Pitzer RM, Shepard R (2011) COLUMBUS—a program system for advanced multireference theory calculations. In: Wiley Interdisciplinary Reviews: Computational Molecular Science, vol 1, p 191
56. Lischka H, Shepard R, Shavitt I, Pitzer RM, Dallos M, Müller Th, Szalay PG, Brown FB, Ahlrichs R, Böhm HJ, Chang A, Comeau DC, Gdanitz R, Dachsel H, Ehrhardt C, Ernzerhof M, Höchtl P, Irle S, Kedziora G, Kovar T, Parasuk V, Pepper MJM, Scharf P, Schiffer H, Schindler M, Schüler M, Seth M, Stahlberg EA, Zhao J-G, Yabushita S, Zhang Z, Barbatti M, Matsika S, Schuurmann M, Yarkony DR, Brozell SR, Beck EV, Blaudeau J-P, Ruckenbauer M, Sellner B, Plasser F, Szymczak JJ (2012) COLUMBUS, an *ab initio* electronic structure program, release 7.0
57. Shepard R (2005) J Phys Chem A 109:11629
58. Shepard R (2006) J Phys Chem A 110:8880
59. Shepard R, Minkoff M (2006) Int J Quantum Chem 106:3190
60. Shepard R, Minkoff M, Brozell SR (2007) Int J Quantum Chem 107:3203
61. Gidofalvi G, Shepard R (2009) J Comput Chem 30:2414
62. Shepard R, Gidofalvi G, Hovland PD (2010) Int J Quantum Chem 110:2938
63. G. Gidofalvi and R. Shepard (2010) Mol Phys 108:2717
64. Shepard R, Simons J (1980) Int J Quantum Chem Symp 14:211
65. Evade RHA, Robb MA (1981) Chem Phys Lett 83:362
66. Shavitt I (1979) Report to the national aeronautics and space administration. Batelle Columbus Laboratories, Columbus
67. Born G, Shavitt I (1982) J Chem Phys 76:558
68. Baker H, Robb MA (1983) Mol Phys 50:1077
69. Paldus J, Jeziorski B (1988) Theor Chim Acta 73:81
70. Li X, Paldus J (1994) J Chem Phys 101:8812
71. Li X, Paldus J (1994) Chem Phys Lett 231:1
72. Jeziorski B, Paldus J, Jankowski P (1995) Int J Quantum Chem 56:129
73. Li X, Paldus J (1995) J Chem Phys 102:8059
74. Li X, Paldus J (1995) J Chem Phys 102:8897
75. Li X, Paldus J (1995) J Chem Phys 103:6536
76. Li X, Paldus J (1996) J Chem Phys 104:9555
77. Paldus J, Li X (1996) Can J Chem 74:918
78. Li X, Paldus J (1997) Unitary group based coupled cluster methods and calculation of molecular properties. In: Bartlett RJ (ed) Recent advances in coupled-cluster methods, series on: recent advances in computational chemistry, vol 3. World Scientific, Singapore, pp 183-219
79. Li X, Paldus J (1998) Int J Quantum Chem 70:65
80. Li X, Paldus J (1998) Mol Phys 94:41
81. Li X, Paldus J (2000) J Mol Struct (THEOCHEM) 527:165
82. Jankowski P, Jeziorski B (1999) J Chem Phys 111:1857
83. Maitra R, Sinha D, Mukherjee D (2012) J Chem Phys 137:024105; (2013) 139:229903 (E)
84. Sen S, Shee A, Mukherjee D (2012) J Chem Phys 137:074104
85. Remacle F, Levin RD (2001) Chem Phys Chem 2:20
86. Gould MD, Paldus J (1986) Int J Quantum Chem 30:327
87. Paldus J, Gao M-J, Chen J-Q (1987) Phys Rev A 35:3197
88. Kent RD, Schlesinger M (1993) Phys Rev A 48:4156
89. Li X, Paldus J (1993) J Math Chem 14:325
90. Li X, Paldus J (1992) Int J Quantum Chem 41:117
91. Paldus J, Li X (1992) Lie algebraic approach to valence bond theory of π-electron systems: a preliminary study of excited states. In: Frank A, Seligman TH, Wolf KB (eds) Group theory in physics, AIP conference proceedings, no. 266. American Institute of Physics, New York, pp 159–178
92. Paldus J, Li X (1993) Unitary group approach to valence bond and coupled cluster methods. In: Gruber B (ed) Symmetries in science VI: from the rotation group to quantum algebras. Plenum, New York, pp 573–592
93. Gould MD, Paldus J, Chandler GS (1990) J Chem Phys 93:4142
94. Paldus J, Gould MD (1992) Theor Chim Acta 86:83
95. Kent RD, Schlesinger M, Ponnapalli PS (1985) Phys Rev B 31:1264
96. Remacle F, Levin RD (1999) J Chem Phys 110:5089
97. Remacle F, Levin RD (2000) J Phys Chem A 104:2341
98. Paldus J (1994) Many-electron correlation problem and Lie algebras. In: Kamran N, Olver PJ (eds) Contemporary mathematics, vol 160. American Mathematical Society, Providence, pp 209–236
99. Paldus J, Sarma CR (1985) J Chem Phys 83:5135
100. Zhang Q, Li X (1987) J Phys A Math Gen 20:6185
101. Paldus J, Rettrup S, Sarma CR (1989) J Mol Struct (THEOCHEM) 199:85
102. Li X, Paldus J (1989) Int J Quantum Chem 36:127
103. Gould MD, Paldus J (1987) J Math Phys 28:2304
104. Li X, Zhang Q (1989) Int J Quantum Chem 36:599
105. Gould MD, Paldus J (1986) Phys Rev A 34:804
106. Harter WG, Patterson CW (1976) A unitary calculus for atomic orbitals. Lecture notes in physics, vol 49. Springer, Berlin
107. Paldus J, Boyle MJ (1980) Phys Rev A 22:2299
108. Boyle MJ, Paldus J (1980) Phys Rev A 22:2316
109. Pickup BT, Mukhopadhyay A (1981) Chem Phys Lett 79:109
110. Mukhopadhyay A, Pickup BT (1982) Chem Phys Lett 93:415
111. Pickup BT, Mukhopadhyay A (1984) Int J Quantum Chem 26:101
112. Mukhopadhyay A, Pickup BT (1984) Int J Quantum Chem 26:125
113. Gould MD, Paldus J (1990) J Chem Phys 92:7294
114. Gould MD, Battle JS (1993) J Chem Phys 99:5961
115. Kent RD, Schlesinger M (1990) Phys Rev A 42:1155
116. Kent RD, Schlesinger M, Shavitt I (1990) Int J Quantum Chem 41:89

117. Lin H-L, Cao Y-F (1989) J Phys A Math Gen 22:1509
118. Malli GL (ed) (1983) Relativistic effects in atoms, molecules and solids, NATO Advanced Study Institute, series B: physics, vol 87. Plenum, New York
119. Malli GL (ed) (1994) Relativistic and electron correlation effects in molecules and solids, NATO Advanced Study Institute, series B: physics, vol 318. Plenum, New York
120. Richards WG, Trivedi HP, Cooper DL (1981) Spin-orbit coupling in molecules. Clarendon, Oxford
121. Langhoff SR, Kern CW (1977) Molecular fine structure. In: Schaefer HF III (ed) Applications of electronic structure theory. Plenum, New York, pp 381–437
122. Detrich JH, Roothaan CCJ (1983) Calculation of relativistic effects in atoms and molecules from the Schrödinger wave function. In: Malli GL (ed) Relativistic effects in atoms, molecules and solids, NATO Advanced Study Institute, series B: physics, vol 87. Plenum, New York, pp 169–182
123. Edmonds AE (1957) Angular momentum in quantum mechanics. Princeton University Press, New Jersey
124. Rose ME (1957) Elementary theory of angular momentum. Wiley, New York
125. Brink DM, Satchler GR (1957) Angular momentum. Clarendon, Oxford
126. Chen J-Q, Ping J, Wang F (2002) Group representation theory for physicists. World Scientific, Singapore

Theor Chem Acc (2014) 133:1466
DOI 10.1007/s00214-014-1466-9

REGULAR ARTICLE

Efficient evaluation of exchange integrals by means of Fourier transform of the 1/r operator and its numerical quadrature

Petr Čársky · Roman Čurík

Received: 12 December 2013 / Accepted: 13 February 2014 / Published online: 26 February 2014

Abstract In the present work, we propose an approach to factorize the non-local exchange kernel into a sum of separable terms. We exploit a discretized Fourier transform of the $1/r$ operator, and we devise a method that allows us to employ a manageable number of plane waves in the Fourier expansion while still keeping necessary accuracy. Resulting formulas are amenable for efficient evaluation on graphics processing units (GPU) devices. We discuss the GPU implementation for two-electron repulsion integrals of the $(gk|gk)$ type in the hybrid Gaussian and plane-wave basis. Accuracy and speedups are demonstrated for several practical calculations of electron scattering by cyclopropane, benzene, and adamantane molecules. By that, we want to show that evaluation of $(gk|gk)$ integrals may cease to be a bottleneck in electron scattering calculations. A message to quantum chemists is that the combination of the integral fragmentation and the use of GPU units is a general tool which may improve performance of computational methods of different types.

Keywords Electron scattering · Exchange energy · Plane-wave basis · Efficient evaluation of integrals · GPU computing

Dedicated to the memory of Professor Isaiah Shavitt and published as part of the special collection of articles celebrating his many contributions.

P. Čársky · R. Čurík (✉)
J. Heyrovský Institute of Physical Chemistry, v.v.i.,
Academy of Sciences of the Czech Republic, Dolejškova 3,
18223 Prague 8, Czech Republic
e-mail: roman.curik@jh-inst.cas.cz

1 Introduction

The stimulus for undertaking this study arose from our urgent need to calculate efficiently exchange integrals in electron-molecule scattering calculations. These integrals are of the $[g(1)k(1)|g(2)k(2)]$ type, where g and k symbols, respectively, refer to Gaussians and plane-wave functions. Lengthy evaluation of two-electron integrals in a mixed Gaussian and plane-wave basis has been a bottleneck in the scattering theory [1], and it hampers ab initio applications to larger polyatomic molecules up to now. Incidentally, this problem may also be topical for the mainstream quantum chemistry. In 2002, Fűsti-Molnár and Pulay [2] came up with the idea that "modern computers have overcome the severe memory limitations of their predecessors, and thus, alternative basis sets such as plane waves are becoming attractive alternatives." Up to now, this statement applies to Coulomb integrals only, whereas the exchange energy has to be treated in a traditional way by using a Gaussian basis set or by using a density functional theory (DFT)-type functional. However, even for Coulomb integrals, a Gaussian basis set cannot be substituted fully by a plane-wave basis because compact molecular orbitals (core orbitals) are not suited to be expanded in a plane-wave basis. Therefore, Fűsti-Molnar and Pulay [2] assumed for molecular orbitals the expansion in a mixed Gaussian and plane-wave basis

$$\phi_i(\mathbf{r}) = \sum_{\mu} c_{i\mu}\mu(\mathbf{r}) + \sum_{p} c_{ip}k_p(\mathbf{r}), \tag{1}$$

where the first sum refers to Gaussians and the second one to plane-wave functions

$$k(\mathbf{r}) = \frac{1}{(2\pi)^{3/2}} e^{i\mathbf{k}\cdot\mathbf{r}}. \tag{2}$$

This leads to the expression for the Hartree–Fock energy which contains also integrals in a mixed Gaussian and plane-wave basis. Coulomb Hartree–Fock energy can be now calculated in this way efficiently in an elegant manner [2], but efficient evaluation of the exchange Hartree–Fock energy remains still a problem. Although formulas for the exchange integrals $(gk|gk)$ have been available for a long time [3–8], their use for Hartree–Fock calculations is not profitable.

Since the dark age of ab initio calculations, efficient evaluation of two-electron integrals has not cease to be a hot topic. Many authors contributed to the progress in this field by various approaches, but as stated explicitly in a recent paper [9], $1/r_{12}$ in electronic Hamiltonians is still the ultimate source of all technical and computational difficulties in quantum chemistry. A tempting solution has been to calculate the exchange energy by means of the DFT. Actually, a practical DFT exchange potential was formulated for use in scattering theory [10] prior to its widespread use in quantum chemistry. This potential, called in the literature Hara free-electron–gas exchange approximation (HFEGE), was later reexamined by Riley and Truhlar [11]. They corrected some misprints in Hara's paper and suggested a modification of the Hara formula to correct the large r behavior of the potential. They call their formula asymptotically adjusted free-electron–gas exchange approximation (AAFEGE). It has been believed that the only manageable way of calculating electron scattering in polyatomic molecules is by using local exchange potentials. In spite of its utility and performance [12], we show in the next section that local exchange models result in an average approximation and neglect anisotropy in electron density of asymmetric polyatomic molecules. Hence, for applications to molecules, we recommend to use more rigorous approaches. In Sect. 3, we show that with the actual computational facilities, it is possible now to calculate exchange energy rigorously, accurately, and efficiently. A few applications showing performance and accuracy of our approach are presented in Sect. 4.

As our primary interest is electron scattering, our primary objective was to find an efficient way of evaluation of exchange integrals of the following special form

$$N_{k_1}N_{k_2}(\mathbf{k}_1|V_{\text{ex}}|\mathbf{k}_2) = \sum_{\mu}\sum_{\nu}(\mu\mathbf{k}_1|\nu\mathbf{k}_2)P_{\mu\nu}, \tag{3}$$

as it is encountered in the scattering theory with a static-exchange potential V_{sx}. The symbols $\mathbf{k}_1$ and $\mathbf{k}_2$ in Eq. (3) stand for unnormalized plane-wave functions (2), N_{k_1} and N_{k_2} are their normalization constants, μ and ν are symbols for Gaussian basis functions, and $P_{\mu\nu}$ is a density matrix element. However, our method is general and it can be applied to any kind of two-electron integrals. Hence, in the last section, we discuss possibility of how our approach presented in Sect. 3 could be applied to Hartree–Fock and post-Hartres–Fock calculations in a mixed Gaussian and plane-wave basis as suggested by Fűsti-Molnar and Pulay [2] or in a pure Gaussian basis.

The past decade has seen a tremendous increase in the use of graphics processing units (GPUs) in the field of quantum chemistry. A number of quantum chemistry methods have been ported recently to GPUs, including Gaussian integral generation [13], Hartree–Fock [14], DFT methods [15], perturbation theory [16], and coupled-cluster methods [17, 18]. Existing scientific code must often be entirely rewritten in Cuda or OpenCL programming frameworks to run efficiently on GPU devices. In Sect. 3, we demonstrate that present approach to computation of two-electron repulsion integrals is amenable to GPU evaluation without extensive code modifications. Instead, we show that resulting two-electron integral formulas reduce essentially to a simple Hermitian rank-k update that can be conveniently computed on GPU cards by use of the *Cublas* library. Furthermore, we also propose an algorithm that allows to perform these calculations on arrays that do not fit into limited GPU's onboard memory.

2 Note on the use of density functional theory in scattering calculations

Passage from the rigorous treatment to the DFT approach can be symbolically expressed as

$$(\mathbf{k}_1(1)|K(1,2)|\mathbf{k}_2(2)) \rightarrow (\mathbf{k}_1(1)|V_{\text{lex}}(1)|\mathbf{k}_2(1))\delta(1-2), \tag{4}$$

where the nonlocal K operator

$$K(1,2) = \sum_{i}^{\text{occ}} \phi_i(1)\frac{1}{r_{12}}\phi_i(2) \tag{5}$$

is substituted by a local exchange V_{lex} operator. For a local exchange potential V_{lex}, it holds that $(\mathbf{k}_1|V_{\text{lex}}|\mathbf{k}_2)$ integrals depend only on the difference of the wave vectors $\mathbf{K} = \mathbf{k}_2 - \mathbf{k}_1$ because

$$(\mathbf{k}_1|V_{\text{lex}}|\mathbf{k}_2) = \int \mathrm{e}^{-\mathrm{i}\mathbf{k}_1\cdot\mathbf{r}}V_{\text{lex}}(\mathbf{r})\mathrm{e}^{\mathrm{i}\mathbf{k}_2\cdot\mathbf{r}}\mathrm{d}\mathbf{r} = \int V_{\text{lex}}(\mathbf{r})\mathrm{e}^{\mathrm{i}\mathbf{K}\cdot\mathbf{r}}\mathrm{d}\mathbf{r}. \tag{6}$$

Hence for two different pairs of $\mathbf{k}_1$ and $\mathbf{k}_2$ vectors giving the same $\mathbf{K}$ vector, we should get the same value of $(\mathbf{k}_1|V_{\text{lex}}|\mathbf{k}_2)$. In Fig. 1, we keep the $\mathbf{K}$ vector fixed but vary the $\mathbf{k}_1$ and $\mathbf{k}_2$ pair. HFEGE and AAFEGE (as examples of V_{lex}) give straight lines, as they depend on fixed $\mathbf{K}$ vector, but use of the Hartree–Fock exchange potential V_{ex} (5)

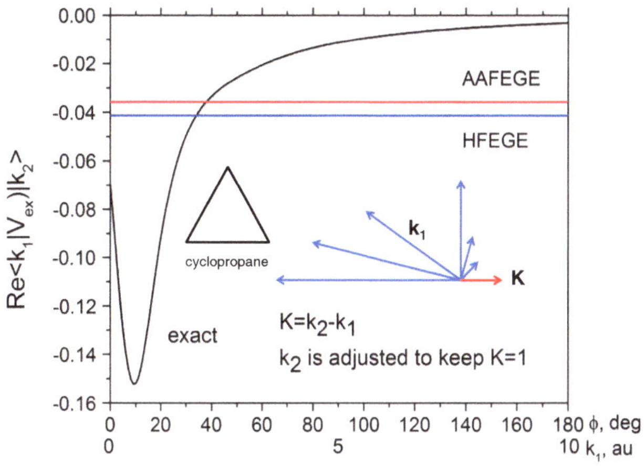

Fig. 1 Exchange integrals $(\mathbf{k}_1|V_{\mathrm{lex}}|\mathbf{k}_2)$ calculated with the AAFEGE, HFEGE, and $(\mathbf{k}_1|V_{\mathrm{ex}}|\mathbf{k}_2)$ of exact exchange potentials for cyclopropane as a function of a simultaneous increase in the norm and angle of the $\mathbf{k}_1$ vector. Both vectors are constrained to lie in the CCC plane. The $\mathbf{k}_1$ vector is stepwise increased and rotated as shown in the figure. The $\mathbf{k}_2$ vector is adjusted to keep the $\mathbf{K}$ vector fixed at the value of 1 a.u. and oriented as shown in the figure

we carried out calculations on elastic collisions of electrons with a helium atom and the cyclopropane molecule as a representative of asymmetric polyatomic molecules. For each target, the calculations were performed both with the AAFEGE local potential and the potential evaluated rigorously. The results are presented in Fig. 2. The left panel shows scattering results for electron scattering by cyclopropane at collision energy of 2.6 eV. The choice of collision energy is based on the presence of weak A_2' shape resonance [22] that promotes an importance of short-range exchange interaction. After comparison with experimental data [22] (shown as circles), we conclude that the local exchange model AAFEGE (green line) fails to reproduce properly the experiment and exact exchange results (red line). The situation is quite different for elastic electron scattering by helium atom shown in the right panel of Fig. 2. Results obtained by application of the AAFEGE exchange potential follow closely both the experimental data [23] and calculations with exact exchange.

gives a different result. Figure 1 shows that local potentials may give excellent results, but that situations are conceivable where it need not be so. When scanning different orientations and lengths of momentum vectors pair $\mathbf{k}_1$ and $\mathbf{k}_2$ shown in Fig. 1, the exact exchange elements $(\mathbf{k}_1|V_{\mathrm{ex}}|\mathbf{k}_2)$ exhibit a nontrivial dependence but the local exchange approximations (HFEGE and AAFEGE) are constants. The two local exchange potentials tend to underestimate the exchange interaction in absolute value for k-values smaller than 2 a.u. and overestimate it for higher k-values. In a way, their constant values can be viewed as average values. Hence, it may be assumed that their successful applications to atoms [19, 20] and diatomics [21] are due to spherical and near-spherical electron distributions, respectively. To substantiate this assumption,

3 Theory

We found inspiration for our work in the series of papers by Gill et al. [9], in particular, their attempts to factorize the Coulomb operator. This is done by approximations in which the two-electron interaction is, in one sense or another, "factorized." In practice, it means to find a suitable basis set that would simulate satisfactorily the action of the $1/r_{12}$ operator by means of the following expansion

$$\frac{1}{r_{12}} = \sum_i \phi_i^*(\mathbf{r}_1)\phi_i(\mathbf{r}_2). \tag{7}$$

We adopted a somewhat different strategy. We decided to use the old idea of using the Fourier or Gaussian

Fig. 2 Elastic differential cross section for scattering of electrons by cyclopropane (*left panel*) and helium atom (*right panel*). Collision energies are 2.6 eV (*left panel*) and 10 eV (*right panel*). Present calculations with exact exchange interaction are displayed by *red curves* while the local AAFEGE exchange model is shown by *green curves*. The results are compared to experimental data for cyclopropane [22] and helium [23], both displayed as *circles* in respective *panels*

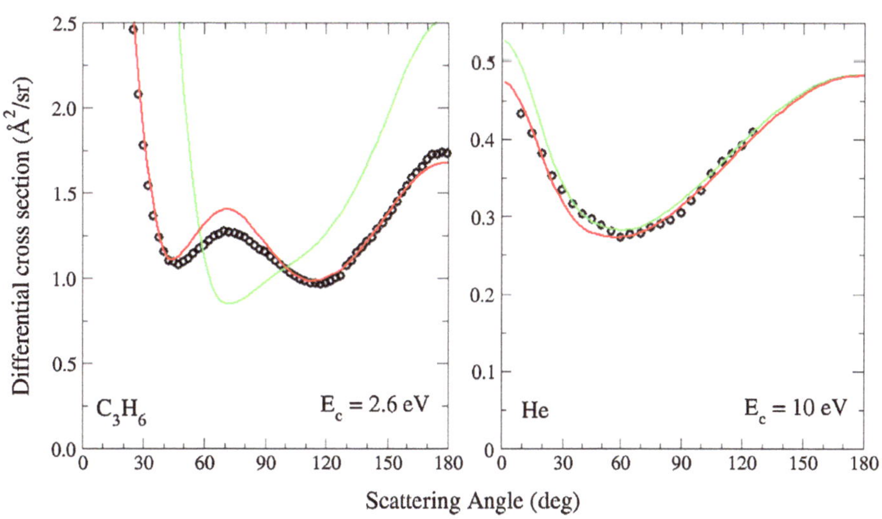

transformation of the $1/r$ operator to obtain analytical formulas for two-electron integrals [4, 5]. The same technique can be also applied to $(gk|gk)$ integrals to obtain the respective formulas in complex arithmetics. Although calculation of $(gk|gk)$ integrals in this way is feasible, it is more time-consuming than calculation of two-electron integrals in a pure Gaussian basis. Hence to avoid this lengthy calculation, we decided to use the Fourier transform

$$\frac{1}{r_{12}} = \frac{1}{2\pi^2}\int \frac{1}{k^2} e^{-i\mathbf{k}\cdot\mathbf{r}_1} e^{i\mathbf{k}\cdot\mathbf{r}_2} d\mathbf{k}, \tag{8}$$

which permits us to use the strategy of factorization. If the integral in Eq. (8) is expressed by means of numerical quadrature

$$\frac{1}{r_{12}} = \frac{1}{2\pi^2}\sum_{pj} \omega_p \omega_j e^{-i\mathbf{k}_{pj}\cdot\mathbf{r}_1} e^{i\mathbf{k}_{pj}\cdot\mathbf{r}_2}, \tag{9}$$

we obtain

$$N_{k_1}N_{k_2}(\mathbf{k}_1|V_{\text{ex}}|\mathbf{k}_2) = N_{k_1}N_{k_2}\frac{1}{2\pi^2} \sum_{\mu\nu}\sum_{pj} \omega_p\omega_j (\mu\mathbf{k}_1\mathbf{k}_{pj})^* (\nu\mathbf{k}_2\mathbf{k}_{pj}) P_{\mu\nu}, \tag{10}$$

where indices p and j, respectively, stand for radial and angular quadrature, and ω_p and ω_j are weights of the grid points. The notation $(\mu\mathbf{k}_1\mathbf{k}_{pj})$ denotes the overlap integral between the gaussian μ and a plane wave function $\exp[i(\mathbf{k}_1 + \mathbf{k}_{pj})]$. We rewrite Eq. (10) in a more compact form as

$$N_{k_1}N_{k_2}(\mathbf{k}_1|V_{\text{ex}}|\mathbf{k}_2) = \sum_i^{\text{occ}}\sum_{\mu\nu}\sum_q \left(a_{\mu q}^{k_1}\right)^* \left(a_{\nu q}^{k_2}\right) c_{i\mu} c_{i\nu}, \tag{11}$$

where q is used as a collective index for grid points in the numerical quadrature

$$q \equiv pj \tag{12}$$

and where by collecting terms in Eq. (10), we obtained

$$a_{\mu q}^{k_1} = N_{k_1}\sqrt{\frac{\omega_p\omega_j}{2\pi^2}}(\mu\mathbf{k}_1\mathbf{k}_{pj}). \tag{13}$$

Summation over μ and ν in Eq. (11) gives us the exchange integral in a compact form

$$N_{k_1}N_{k_2}(\mathbf{k}_1|V_{\text{ex}}|\mathbf{k}_2) = \sum_i^{\text{occ}}\sum_q \left(A_{iq}^{k_1}\right)^* \left(A_{iq}^{k_2}\right), \tag{14}$$

where

$$A_{iq}^{k_1} = \sum_\mu a_{\mu q}^{k_1} c_{i\mu}. \tag{15}$$

It might be expected that the numerical quadrature would run into difficulties for small values of r, unless an excessively large numerical quadrature would be used. However, it turned out that even a moderately large quadrature can give results with accuracy which is more than sufficient for comparison with experimental scattering data.

Choice of the numerical quadrature and its optimization have received a great deal of attention (see for example Refs. [24, 25] and references therein for older literature). However, our literature search did not result in an unequivocal decision what could be the best choice for our purpose. After some experimentation, we found that a scaled adaptive Laguerre quadrature with 15 grid points can be used as an universal radial quadrature for low-energy electron-molecule scattering calculations. The radial points were obtained as scaled Laguerre grid points

$$k_i = x_{15,i}R \tag{16}$$

where $x_{15,i}$ is the ith grid point in the 15-point Laguerre quadrature and the scaling factor R was obtained as

$$R = R_{\max}/x_{15,15} \tag{17}$$

with the fixed $R_{\max} = 11$ a.u. The weights were also scaled as

$$\omega_i = w_i e^{x_i} R, \tag{18}$$

where w_is are weights of Laguerre grid points. The scaled radial points k_i were ordered in six ranges (0, 0.2), (0.2, 0.4), (0.4, 0.6), (0.6, 0.8), (0.8, 1.0), and (1.0, 11.0) and assigned to Lebedev angular quadratures [26, 27] with 26, 50, 86, 110, 194, and 302 grid points, respectively. In the x^y notation of Chien and Gill [25], the quadrature may be denoted as $26^2\ 50^1\ 86^1\ 194^1\ 302^{10}$, indicating that x point Lebedev grid was used for y successive radial points. Test electron-molecule scattering calculations for electron energies <20 eV showed that this numerical quadrature with the total number of 3,402 grid points can be used as a standard quadrature. It provides results for differential and integral cross sections with the accuracy of four to five significant digits. While it is sufficient for scattering calculations, for most quantum chemical calculations, a higher accuracy would be required and therefore, a more systematic optimization of the numerical quadrature should be performed.

In Sect. 4, we present some tests on the performance and accuracy achieved with our standard numerical quadrature. When compared to calculations done with $(gk|gk)$ integrals evaluated rigorously, the time saving is considerable. Yet the merits of the fragmentation approach were not fully exploited. It can be noticed in Eq. (14) that the evaluation of $(gk|gk)$ integrals is reduced to multiplication of two long vectors. This is an ideal task for general-purpose computation on graphics processing units (GPGPU). In Sect. 4, we show that the computer time owing to the evaluation of

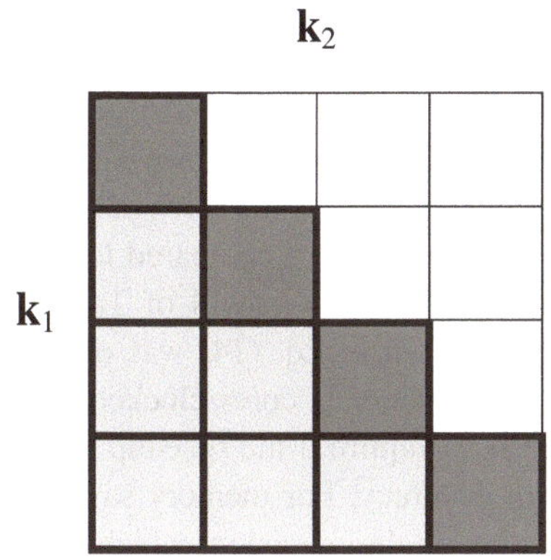

Fig. 3 Splitting of the Hermitian matrix $(\mathbf{k}_1|V_{\mathrm{ex}}|\mathbf{k}_2)$ into 4×4 blocks for serialized evaluation on a GPU device

Table 1 CPU time and speedup in calculations of the preselected sets of exchange integrals $(\mathbf{k}_1|V_{\mathrm{ex}}|\mathbf{k}_2)$ for cyclopropane C_3H_6, benzene C_6H_6, and adamantane $C_{10}H_{14}$

	C_3H_6	C_6H_6	$C_{10}H_{14}$
Rigorous calculation, t_0	1,906	65,651	378,705
A terms, Eq. (15), t_A	7	41	85
$A^*{\cdot}A$, Eq. (14), t_{AA}	317	4,059	10,865
$A^*{\cdot}A$ on GPU, Eq. (14), $t_{AA,GPU}$	3.4	62	102
Speedup on CPU[a]	6	16	35
Speedup on GPU[b]	183	637	2,025

Execution times are given in seconds

[a] This is the ratio $t_0/(t_{AA} + t_A)$

[b] This is the ratio $t_0/(t_{AA,GPU} + t_A)$

A_{iq}^k is marginal and hence, the computer time needed for the evaluation of the exchange energy (3) is almost exclusively due to the multiplication in Eq. (14). With modern computers, we need not be anxious to watch memory limitations. Nowadays, large A_{iq}^k arrays can be kept in the computer memory. However, GPU devices available so far provide less memory, typically 1–6 GB of RAM. Therefore, we propose to distinguish the following two cases:

1. Arrays A_{iq}^k and the resulting Hermitian matrix $(\mathbf{k}_1|V_{\mathrm{ex}}|\mathbf{k}_2)$ of Eq. (14) fit into GPU's onboard memory. In this case, the computational design is reduced to a simple call to a blas3 function that performs the Hermitian rank-k update, for example function *cublasCherk* in Cublas implementation.
2. Arrays A_{iq}^k and $(\mathbf{k}_1|V_{\mathrm{ex}}|\mathbf{k}_2)$ are too large to be stored simultaneously in the GPU memory. For this case, we devise to split the Hermitian matrix $(\mathbf{k}_1|V_{\mathrm{ex}}|\mathbf{k}_2)$ into $N_b \times N_b$ blocks as shown in Fig. 3. Then, the computation on GPU device is serialized to $N_b\,(N_b + 1)/2$ steps going over shaded blocks of Fig. 3. The diagonal blocks (dark shade) are calculated by the Hermitian rank-*k* (update as in the case 1) and they require N_b times smaller GPU memory allocation for A_{iq}^k array and N_b^2 times smaller allocation for the resulting submatrix of $(\mathbf{k}_1|V_{\mathrm{ex}}|\mathbf{k}_2)$. The off-diagonal blocks (light shaded in Fig. 3) are just a result of a general matrix multiplication implemented as a call to blas3 function *cublasCgemm* in Cublas library. In case of the off-diagonal blocks, the GPU device needs to keep two segments of the A_{iq}^k array simultaneously, together with the resulting block of $(\mathbf{k}_1|V_{\mathrm{ex}}|\mathbf{k}_2)$ matrix. In the first segment, the *k* index of A_{iq}^k array spans the rows $\mathbf{k}_1$ of the computed off-diagonal block, while in the second segment, the *k* index spans the columns $\mathbf{k}_2$ of the computed block. Therefore, the GPU memory requirements for computation of the off-diagonal blocks establish the necessary formula that defines the number of blocks $N_b\,(N_b + 1)/2$.

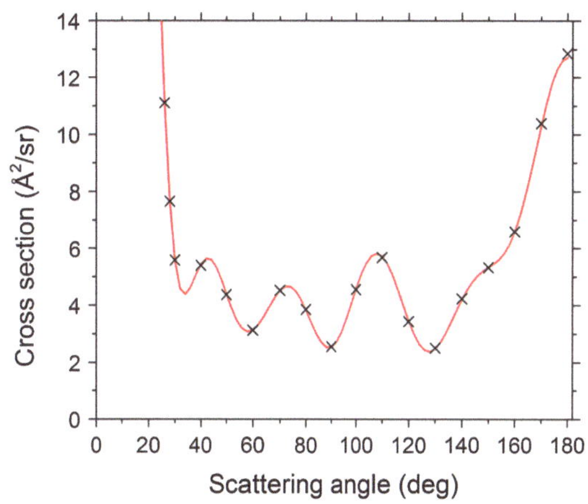

Fig. 4 Angular dependence of the elastic differential cross section for adamantane and incident electron energy of 10 eV. The *red curve* is the result of calculation with factorized exchange integrals, whereas crosses are data points for rigorous calculation

4 Performance and accuracy

Test calculations were performed for three molecules, cyclopropane, benzene, and adamantane. First, we performed geometry optimization of the three molecules by Hartree–Fock calculations with a valence double-zeta basis set with a single set of polarization functions [28]. This was done with the purpose to save expansion coefficients of occupied molecular orbitals for a subsequent evaluation of A_{iq}^k terms [Eq. (15)] and for calculation of differential cross sections. Details of our scattering calculations have been described elsewhere [29]. Here, we only note that the plane-wave basis sets were determined by Legendre–Gauss and Lebedev quadrature as a result of discretization of Lippmann–Schwinger equation [29, 30]. Incident electron energy is set to 10 eV. The numbers of grid points was stepwise increased until convergence was reached in the calculated differential and integral cross sections. In absolute value, the **k** vectors range from 0 to 12 a.u. and the total numbers of k functions in the plane-wave basis sets are 3,878 for cyclopropane, 12,456 for benzene, and 11,400 for adamantane.

The entries in Table 1 are timings for test calculations done in three different ways. In the first one, the exchange integrals were evaluated rigorously by means of complex Shavitt functions $F_n(z)$ [29]. The values of the $F_n(z)$ functions were not calculated explicitly but interpolated from a precalculated table with a 2-D grid. In the other two, the same way of fragmentation, represented by Eqs. (9)–(15), was used, but they differ in a way of how the $A^{*}{\cdot}A$ multiplication in Eq. (14) is performed, viz. if it is done on GPU or not. It may be noticed in Table I that the time for the evaluation of A terms is marginal compared to the time for the $A^{*}.A$ multiplication. The size of the A array is (n, N_q, N_k), where n is the number of occupied molecular orbitals, N_q is the number of grid points in the expansion of $1/r_{12}$, and N_k is the size of the plane-wave basis for the scattering calculation. The traditional calculation with rigorous evaluation of integrals scales as $\sim N^2$, where N is the number of atomic basis set functions, whereas the step $A^{*}{\cdot}A$ scales as $\sim n$ for the same N_k and a fixed N_q. Hence, the speedup increases with the size of the molecule and the size of its atomic basis set. The respective speedup is considerable, but the main merit of fragmentation is that it leads to a simple $A^{*}{\cdot}A$ form amenable to GPU treatment. All the CPU execution times shown in Table 1 (first 3 rows) were obtained by using 12 cores of Intel Xeon 3 GHz workstation. These times are compared to execution times of a single CPU core code (row 4 of Table 1) where the evaluation step shown in Eq. (14) was offloaded to Tesla card M2090 with 512 cuda cores clocked at 1.3 GHz.

We should now inquire if the speedup achieved is not at the expense of accuracy. For memory saving, we evaluate the A terms in single precision which means that the $A^{*}{\cdot}A$ multiplication on GPU brings some loss in precision. We found that this loss in precision was tolerable and that the calculated cross sections were still accurate to 4–5 significant digits. As an illustration, we present in Fig. 4, the angular dependence of the differential cross section for elastic scattering of electrons by adamantane molecule $C_{10}H_{14}$. We note that even for such a large multicenter system results obtained by factorizing, the exchange kernel (red line) are practically indistinguishable from rigorous results obtained via Eq. (3).

It might be assumed that for the purpose of Hartree–Fock and post-Hartree–Fock calculations, the exchange integrals obtained by fragmentation should be calculated with a higher accuracy than that one which was found sufficient for scattering calculations. We decided therefore to perform a test, the result of which is presented in Fig. 5. In the left panel of Fig. 5, it is seen that the maximum error is of about 20 μHartree. The right panel, however, shows the maximum error as high as high 1 mHartree.

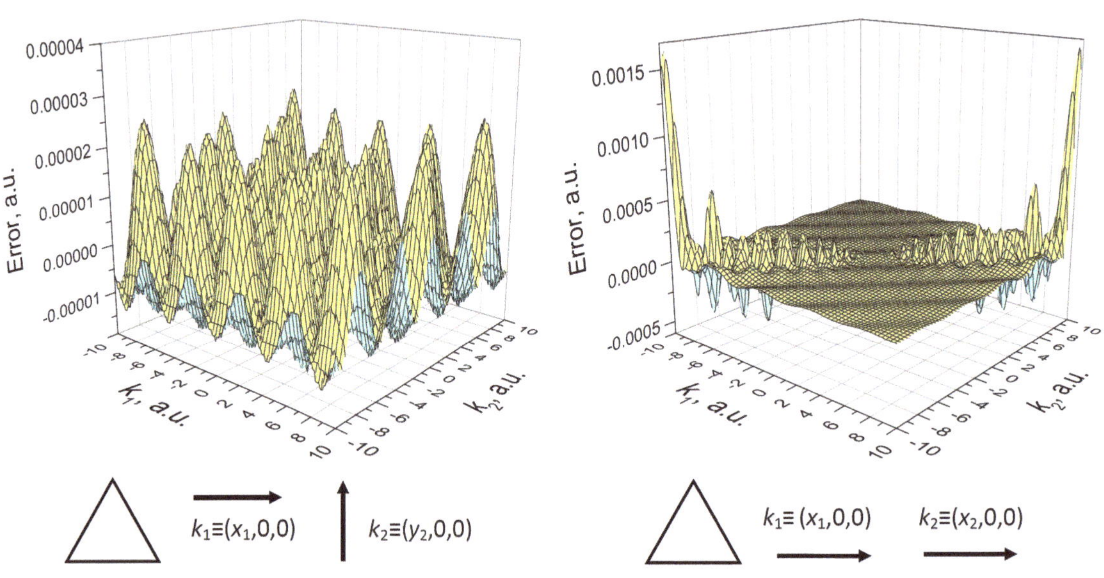

Fig. 5 Test on accuracy of computed $(\mathbf{k}_1|V_{\mathrm{ex}}|\mathbf{k}_2)$ integrals. The errors are plotted as differences in real parts of approximate (evaluated by fragmentation) and exact integrals calculated for cyclopropane. The $\mathbf{k}_1$ and $\mathbf{k}_2$ vectors are constrained to lie in the CCC plane. In the *left panel*, $\mathbf{k}_1$ and $\mathbf{k}_2$ vectors are varied in the perpendicular orientation; in the *right panel*, they are collinear

Apparently, the quadrature assumed in this paper is not large enough for high k-values with a small difference vector $\mathbf{K} = \mathbf{k}_1 - \mathbf{k}_2$. Hence, this small fraction of integrals requires a special treatment.

5 Conclusions

We have shown that Fourier transform of the $1/r_{12}$ operator and its numerical quadrature provides a tool for efficient ab initio evaluation of exchange integrals of the $(gk|gk)$ type. It is every reason to believe that this new approach eliminates the bottleneck in electron-molecule scattering calculations and makes them so feasible for treatments of considerably larger molecules than it was possible so far. Calculation of exchange integrals for adamantane, which lasted previously about 4 days and half, can now be accomplished in 2 min on the same computer. The technique presented in this paper is general and can be also applied to integrals of the $(gg|gk)$ type which is the principal computational task in the Schwinger multichannel approach[1]. Presumably it may initiate a revival of Fūsti–Molnar and Pulay's idea[2] because it eliminates their problem with the evaluation of exchange integrals in a mixed plane-wave and Gaussian basis. Perhaps it is also worth considering to apply the fragmentation technique presented in this paper to evaluation of two-electron integrals in a pure Gaussian basis set. The principle computation task in Hartree–Fock and post-Hartree–Fock methods is the evaluation of two-electron integrals over molecular orbitals. By applying Eq. (9), they can be factorized as

$$(\varphi_r\varphi_s|\varphi_t\varphi_u) = \frac{1}{2\pi^2}\sum_q (\varphi_r\varphi_s\mathbf{k}_q)\sqrt{\omega_q}\sqrt{\omega_q}(\mathbf{k}_q\varphi_t\varphi_u), \tag{19}$$

where the overlap integrals $(\varphi_r\varphi_s\mathbf{k})$ are easy to evaluate. The message of this paper to quantum chemists is that the combination of the integral fragmentation and the use of GPU hardware is a general tool applicable to computational methods of various types. Modern computers allow to keep large blocks of fragmented one-electron integrals in memory and hence efficient use of GPU for their multiplication. The only problem is to find a numerical quadrature that would be superior to ours in securing accuracy.

Acknowledgments We would like to express our sincere gratitude, as well as pay tribute, to Shai Shavitt for his pioneering work on integral evaluation methodology used frequently in quantum chemistry. Moreover, we are extremely thankful for being allowed to produce a related contribution for this Festschrift. We hope that, in some small way, our paper may be seen as a humble continuation of Shavitt's work. We gratefully acknowledge the financial support of the Grant Agency of the Czech Republic for support (Grants P203/12/0665, P208/11/0452, and P208/11/2222). This work was conducted within the framework of the COST Action CM1301 (CELINA), supported also by the Czech Ministry of Education (Grant LD14088).

References

1. Winstead C, McKoy V (1995) In: Yarkony DR (ed) Modern electronic structure theory part II. World Scientific, Singapore
2. Fűsti-Molnar L, Pulay P (2002) J Chem Phys 116:7795
3. Ostlund NS (1975) Chem Phys Lett 34:419
4. Rescigno TN, McCurdy CW Jr, McKoy V (1975) Phys Rev A 11:825
5. Watson DK, McKoy V (1979) Phys Rev A 20:1474
6. Čársky P, Polášek M (1998) J Comp Phys 143:266
7. Čársky P, Reschel T (1998) Coll Czech Chem Commun 63:1264
8. Čársky P (2010) J Phys B 43:175204
9. Limpanuparg T, Hollett JW, Gill PMW (2012) J Chem Phys 136:104102 (and references therein)
10. Hara S (1967) J Phys Soc Jpn 22:710
11. Riley ME, Truhlar DG (1975) J Chem Phys 63:2182
12. Gianturco FA, Sebastianelli F, Lucchese RR, Baccarelli I, Sanna N (2008) J Chem Phys 128:174302
13. Ufimtsev IS, Martínez TJ (2008) J Chem Theory Comput 4:222
14. Ufimtsev IS, Martínez TJ (2009) J Chem Theory Comput 5:1004
15. Genovese L, Ospici M, Deutsch T, Mehaut J-F, Neelov A, Goedecker S (2009) J Chem Phys 131:034103
16. Vogt L, Olivares-Amaya R, Kermes S, Shao Y, Amador-Bedolla C, Aspuru-Guzik A (2008) J Phys Chem 112:2049
17. DePrince AE III, Hammond JR (2011) J Chem Theory Comput 7:1287
18. Ma W, Krishnamoorthy S, Villa O, Kowalski K (2001) J Chem Theory Comput 7:1316
19. Fritsche L, Noffke J, Gollisch H (1984) J Phys B 17:1637
20. Bransden BH, McDowell MRC, Noble CJ, Scott T (1976) J Phys B 9:1301
21. Rumble JR Jr, Truhlar DG (1980) J Chem Phys 72:5223
22. Allan M, Andric L (1996) J Chem Phys 105:3559
23. Brunger MJ, Buckman SJ, Allen LJ, McCarthy IE, Ratnavelu K (1992) J Phys B 25:1823
24. Gill PMW, Chien SH (2003) J Comp Chem 24:732
25. Chien SH, Gill PMW (2006) J Comp Chem 27:730 (and references therein)
26. Lebedev VI (1975) Zh Vychisl Mat Mat Fiz 15:48
27. Lebedev VI (1976) Zh Vychisl Mat Mat Fiz 16:293
28. Dunning TH Jr, Hay PJ (1977) In: Schaefer HF III (ed) Modern theoretical chemistry, vol 3. Plenum Press, New York
29. Čársky P, Čurík R (2012) In: Čársky P, Čurík R (eds) Low-energy electron scattering from molecules, biomolecules and surfaces. CRC Press, Boca Raton
30. Čurík R, Čársky P, Allan M (2008) J Phys B 41:115203

Theor Chem Acc (2014) 133:1468
DOI 10.1007/s00214-014-1468-7

REGULAR ARTICLE

The Wuppertal multireference configuration interaction (MRD-CI) program system

Robert J. Buenker · Robin A. Phillips · Stefan Krebs · Heinz-Peter Liebermann · Aleksey B. Alekseyev · Peter Funke

Received: 1 December 2013 / Accepted: 14 February 2014 / Published online: 6 March 2014

Abstract The goal of a full configuration interaction (CI) calculation in a flexible atomic orbital basis can be approached rather closely for small molecular systems by designating a set of reference configurations from which only single- and double-orbital substitutions are allowed to generate the many-electron basis. The present review discusses how configuration-driven algorithms can be applied to form the corresponding Hamiltonian matrix representation from which approximate energy eigenvalues and eigenvectors can be computed. Tables are constructed which simplify the analysis of configuration relationships that determine how a given interaction is calculated. This Table-CI procedure eliminates the need for carrying out lineup permutations of spin orbitals, which was one of the major bottlenecks in early implementations of the multireference CI method. Although this approach is fundamentally variational in character, it can be easily combined with perturbative techniques that allow one to approximate the corresponding full CI energies to a high degree of accuracy. The speed with which Hamiltonian matrix elements can be calculated makes it feasible to use the Direct CI approach which requires that these values must be recalculated in each iteration of the diagonalization procedure, thereby achieving a considerable increase in the overall efficiency of the computations. The effects of spin–orbit coupling and other terms in the relativistic Hamiltonian have been incorporated in a straightforward manner through the use of effective core potentials.

Keywords Configuration interaction (CI) · Full CI · Multireference single- and double-excitation CI (MRD-CI) · Condon rules · Slater determinants · Table CI · Direct CI · Spin–orbit coupling

1 Introduction

Configuration interaction (CI) is a conceptually simple and eminently practical method of obtaining approximate solutions of the Schrödinger equation for atomic and molecular systems. It has a high degree of applicability for molecules when used in conjunction with the Born–Oppenheimer clamped-nuclei approximation. In mathematical terms, the CI method consists of forming a matrix representation of the electronic Hamiltonian and then determining its characteristic vectors. In the limit of a complete set of basis functions (i.e., nearly complete set of primitives), the result is equivalent to an exact solution of the corresponding Schrödinger equation. From a computational point of view, the main problem is that the dimension of the matrix representation is enormous for even relatively small systems. Not surprisingly, this state of affairs has led to a great deal of research with the twin goals of achieving maximum accuracy with a minimum outlay of computational expense. In addition, in order to obtain maximum benefit from a given CI treatment, it is important to also have the capacity of applying its results to the calculation of numerous physical properties other than energy such as transition probabilities and collision cross sections that are often essential for a suitably quantitative description of associated physical processes. The present

Dedicated to the memory of Professor Isaiah Shavitt and published as part of the special collection of articles celebrating his many contributions.

R. J. Buenker (✉) · R. A. Phillips · S. Krebs · H.-P. Liebermann · A. B. Alekseyev · P. Funke
Fachbereich C-Mathematik und Naturwissenschaften, Bergische Universität Wuppertal, Gaussstr. 20, 42119 Wuppertal, Germany
e-mail: buenker@uni-wuppertal.de; bobwtal@yahoo.de

work describes one such series of computer programs that has been developed at the University of Wuppertal over the past 35 years.

2 Multireference single- and double-excitation CI

The forerunner of CI is the self-consistent field (SCF) method [1, 2]. A version that properly accounts for the antisymmetry of the electronic wave function was developed independently by Fock [3] and Slater [4] shortly after Schrödinger's papers. It is characterized by an approximate wave function that is a single determinant whose elements are one-electron functions (spin orbitals). The latter "orbitals" are optimized under two conditions: minimization of the energy expectation value and mutual orthonormality. The method produces both the occupied orbitals appearing in the determinant but also a potentially infinite number of unoccupied functions that prove to be the basis for the CI method. One can look upon a "Slater" determinant formed by substituting unoccupied for occupied one-electron functions as a representation of an excited state of the molecular system. The possible applications to spectroscopy were obvious.

The choice of orbitals to be included in the Slater determinants was a subject of great interest. Heitler and London [5] showed that one could obtain a relatively compact description of the H_2 molecule by employing the valence bond method of electronic structure in which the individual orbitals could have a high degree of overlap with one another. The advantages of this method were particularly evident in its description of dissociation processes. A single-determinant calculation employing orthonormal basis functions of the type produced by the Hartree–Fock method was notably less effective for large H–H separations [6]. In particular, the molecular ground state function did not correlate with the corresponding atomic limit in this theoretical treatment. However, both basis sets consist of four spin orbitals (two spatial functions multiplied with α- and β-spins), which serve as equivalent basis sets for the same (one electron) linear space. By making all possible substitutions, it is possible to construct six two-electron Slater determinants in each basis, which is referred to as the full CI space. The corresponding CI treatments lead to exactly the same energies and wave functions, as is also evident from the fact that the two sets of many-electron functions are related by a linear transformation and thus must lead to the same set of characteristic vectors/eigenvectors as well as energy eigenvalues.

The above experience certainly demonstrates the advisability of carrying out full CI calculations instead of using only single determinants. The problem is that the number of configurations grows factorially with both the number of electrons in the system and the number of one-electron functions, which are available for population in the determinantal wave functions. The obvious compromise is thus to include only a manageable subset of the possible configurations in a given atomic orbital basis for the purposes of constructing the Hamiltonian matrix. A critical observation in this respect is that the use of an orthonormal basis greatly limits the number of configurations that have nonzero matrix elements with a specific configuration such as that of the Hartree–Fock ground state. Only configurations that differ by at most a double excitation (substitution) from a given reference configuration can interact directly with it (Condon rules) [7]. This fact makes it sound as if one can just ignore triple and higher excitations in the CI calculations and suffer no significant loss of accuracy relative to the corresponding full CI. It is easy to find cases where this conclusion fails, however. For example, in many dissociative processes, triply and quadruply excited configurations tend to gain ever greater importance as the bond distances are stretched beyond their equilibrium values. There are many more triply excited configurations than singly and doubly excited counterparts, so just expanding the CI space to include all of the former also becomes quite impractical for all but the smallest systems and basis sets.

Buenker and Peyerimhoff [8, 9] analyzed this situation and found that the problem was not with the Condon rules for matrix elements, but rather with the basic fact that there are often secondary configurations that make large contributions to the electronic wave function. They concluded that a much more practical means of expanding the CI space than increasing the maximum number of excitations is simply to increase the number of leading configurations from which singly and doubly excitations are allowed. This meant that exploratory calculations had to first be carried out with single-reference single- and double-excitation spaces to identify such "reference" configurations on the basis of their contribution to the final CI wave function. It was also clear that one could further expand the treatment to deal with more than one electronic state in the same calculation. Indeed, it was found that excited configurations generated on the basis of one state's results often had an unexpected impact on other states even though these configurations differed formally by more than a double excitation from the latter's leading terms. A procedure was thus developed by which a "common denominator" of reference configurations was chosen on the basis of a series of test calculations carried out for a representative series of molecular conformations required for a comprehensive treatment of a number of electronic states over the entire potential surface of interest. This meant that reference configurations also were included at a given molecular geometry that had very little influence on the CI solutions for any of the key states. Nonetheless, the procedure has the distinct advantage that it leads to formally the same CI space at each geometry, thereby insuring that the computed potential surfaces and wave functions are suitably continuous over the geometric region of interest. It should be noted, however, that the level of continuity is also affected by the choice of orthonormal one-electron functions employed to construct the configurations, so perfectly continuous results are actually only possible if the full CI space is treated in its entirety.

A word about notation is appropriate. The recommendation in the 1974 study [8] was to employ a CI space that consists of all single and double excitations with respect to an unspecified set of main or reference configurations $\{\phi_m\}$. Quoting from the paper [8] on p. 35: "in this way the class of configurations given explicit consideration can more justifiably be restricted to all single- and double-excitation species with respect to just the members of the $\{\phi_m\}$ subset, since the set so generated includes most of the important triple and higher excitation species (with respect to the leading configuration) which judging from experience cannot safely be neglected". As the method came into wider use, Shavitt [10] suggested to the author that one should use the term "multireference" without hyphen to designate the set of configurations from which the excitation classes are generated. The designation "multireference single- and double-excitation CI" came into common use after this time, but as with other computational electronic structure methods, there came a point where it was advisable to label it with an acronym. It was decided that in the interest of compactness, there was no need to include "single" in the title of the method since it was obvious that it would be counterproductive to exclude this excitation class when all doubly substituted configurations were to be added to the CI space. This concept led to the formulation "multireference through double-excitation configuration interaction" or simply MRD-CI that was first used in print in 1978 [11]. The abstract of that paper gives the definition as "multi-reference double-excitation CI," but the Introduction makes clear that "a suitably flexible CI space would appear to consist of all singly and doubly excited species relative to each of these dominant terms." In succeeding years, other authors have used the acronym "MRSD-CI" to designate the *same* CI space, i.e., making explicit the inclusion of single-excitation configurations that is also implied in the term "MRD-CI." In summary, MRSD-CI is nothing more than a different designation for the CI space first introduced in 1974 by Buenker and Peyerimhoff [8] and should not be construed to be anything different than what is meant by MRD-CI.

One of the more subtle computational difficulties associated with MRD-CI is how to avoid duplications among the configurations that are generated from a given reference space. The problem is that a given configuration can be in the single- and double-excitation classes of more than one reference. The solution is to generate the excitation classes of one reference at a time. Each newly generated configuration is then first checked to see whether it is at least a triple excitation relative to all previously considered reference species. Only in that case is the new configuration included in the final MRD-CI space since otherwise it must already be present there.

The main difficulty in dealing with the MRD-CI space in general is naturally its large dimension for systems of even moderate size and suitably flexible basis sets. It is clear from the outset that a high percentage of the (test) configurations generated from reference species of lesser importance will have a minimal effect on the results. A straightforward and effective way of dealing with this situation is to employ configuration selection based on the Hamiltonian matrix elements of each generated configuration with all of the reference configurations. The main idea is that considerable time can be saved in this process by ignoring the much larger group of matrix elements that occur between test configurations. Two distinct approaches to configuration selection were pursued: group selection and individualized selection. The former possibility is illustrated by Gershgorn and Shavitt's B_k approximation [12] in which a secular matrix is formed with the full dimension of all reference and singly and doubly excited configurations. Which configurations to include in the final selected MRD-CI space is then based on the magnitudes of eigenvector coefficients resulting from the diagonalization of the latter matrix. Buenker and Peyerimhoff [8, 9] advocated an individualized approach by which each test configuration is evaluated on the basis of only its matrix elements with the reference configurations. The advantage of the individualized method is that it avoids diagonalizing a large CI matrix to obtain its results, whereas the group technique has the advantage of considering secondary interactions that result because of a remixing of the reference configurations relative to the case when only the latter are considered. The individualized approach had an additional advantage, however. As advocated in Ref. [9], it is easily possible to use the information about test–reference Hamiltonian matrix elements in a perturbation treatment to obtain a reliable estimate for the unselected MRD-CI space energies (energy extrapolation). It was found that the latter method works best when more roots are considered in the calculations than are actually sought from the treatment. This method has been developed over the succeeding years and is still used in the most recent versions of the Wuppertal MRD-CI programs. The energy results are also used in conjunction with the Davidson correction [13] for higher excitations not considered in the MRD-CI space itself.

Comparison [14, 15] of such extrapolated energies with their explicitly computed full CI counterparts [16] has indicated a high degree of accuracy (within 1.0 kcal/mol) with this combination of variational and perturbative methods. One disadvantage of this approach is that no comparably accurate means of extrapolating the properties of the associated truncated MRD-CI wave functions has ever been found. The fact that most interesting properties such as dipole moments involve one-electron quantum mechanical operators helps to minimize the negative consequences of this state of affairs.

3 Configuration-driven Hamiltonian matrix formation: Table CI

The definition of the MRD-CI space brings with it some special problems that do not exist for single-reference CI calculations. How these problems are handled computationally depends to a large extent on what method is used to solve the corresponding secular equations through diagonalization of the corresponding Hamiltonian matrix. There are two well-defined approaches for accomplishing the latter goal: configuration-driven and integral-driven. In the first case, one loops over pairs of configurations and obtains the corresponding Hamiltonian matrix elements by determining which electron repulsion and other one-electron integrals are required. By contrast, in integral-driven methods, one loops over the aforementioned integrals and determines the configuration pairs to which they contribute. The latter approach makes use of methods developed in nuclear physics. It is traditionally referred to as the graphical unitary group approach (GUGA) and was introduced in quantum chemistry by Paldus [17, 18] and Shavitt [19]. GUGA is easiest to apply when the configuration space is well defined, as is the case for a SD-CI calculation. It becomes more challenging when several reference configurations are used to generate an MRD-CI space, especially if they are chosen solely on the basis of their capacity to make significant contributions to the electronic states of interest. It was recognized early on that configuration selection procedures, whether using individualized [8, 9] or group [12] selection techniques, could not be handled efficiently by GUGA. It was therefore decided that integral-driven MRD-CI algorithms are optimally applied by solving secular problems for an un-truncated MRD-CI space.

The situation is quite different for the configuration-driven approach. In that case, there is no particular advantage to having a simple relationship between the various configurations in the many-electron basis. Such algorithms merely compare configurations pairwise and simply use Condon rules [7] to compute the corresponding Hamiltonian matrix elements. It is important to understand that a configuration as defined above is actually a linear space, which is spanned by a basis of Slater determinants. The size of the basis can be significantly reduced by transforming to a basis of S^2 eigenfunctions since the Hamiltonian employed in a given theoretical treatment is spin-independent. For the same reason, it is advantageous to work exclusively with $M_S = S$ determinants. In principle, it is necessary to compute matrix elements between each pair of Slater determinants on the way to completing the matrix-element calculation for a given pair of configurations. This is the way that the first configuration-driven CI programs actually were written. A large part of the work was involved with what are called "lineup permutations." The two determinants in a given comparison can only have two unlike pairs of spin orbitals in order to have a nonzero Hamiltonian matrix element. This means that the identical spin orbitals need to be lined up in the two products, leaving their unlike counterparts isolated in their respective determinants. The permutation can lead to a change in sign in the matrix element, and this has to be duly noted.

In carrying out the CI calculations, it is important to develop a symbolic representation of the many-electron configurations in terms of spatial orbitals and their occupation. The actual basis functions for a given configuration are symmetry-adapted linear combinations (SAFs) of Slater determinants. The latter are simply assignments of α- and β-spins to open shells [7]. The key point in such a procedure is that for each open-shell type (supercategory) [20], the same linear combinations can be employed for all such configurations by constructing an isomorphic relationship between the determinants in each case. The Hamiltonian matrix calculation involves the above lineup permutations followed by computation of the corresponding nonzero interactions in terms of a relatively small number of electron repulsion integrals (g_{12}). The resulting set of determinantal matrix elements is then transformed to the SAF basis. For this purpose, a double representation [21, 22] of the SAFs in terms of p determinants and q spin-projected determinants, respectively, is convenient (p is the number of $M_S = S$ determinants and q is the corresponding number of SAFs with the correct multiplicity). In this way, the desired set of matrix elements over SAFs can be obtained by evaluating $p_i\ q_j$ determinantal matrix elements for the ith and jth configurations rather than the generally larger number ($p_i\ p_j$) where all determinants are considered without projection.

It is clear that most of the lineup permutations can be avoided by using the same spin assignments within determinants for each configuration with the same number of open shells. The matrix elements for different configuration pairs therefore differ only by the g_{12} integrals needed for a specific case. The main work in the calculations then is

Table 1 Possible relationships between one-electron functions with unequal occupation in configurations differing by single and double excitations, respectively

	Occupation numbers				ΔK	P	No. of R subcases
Single excitations	1	0			0	3	1
	0	1					
	2	0			1	3	2
	1	1					
	2	1			0	3	1
	1	2					
	1	1			−1	–	–
	2	0					
Double excitations	2	0			0	4	1
	0	2					
	2	0	0		1	2	1
	0	1	1				
	2	1	1		−1	–	–
	0	2	2				
	2	1	0		0	2	1
	0	2	1				
	2	2	0		1	2	1
	1	1	2				
	2	2	1	1	0	1	1
	1	1	2	2			
	2	2	0	1	1	1	3
	1	1	1	2			
	2	2	0	0	2	1	6
	1	1	1	1			
	2	1	0		0	2	1
	1	0	2				
	2	1	1	1	−1	–	–
	1	0	2	2			
	2	1	0	0	1	1	3
	1	0	1	1			
	2	1	0	1	0	1	4
	1	0	1	2			
	1	1	0		−1	–	–
	0	0	2				
	1	1	0	0	0	1	1
	0	0	1	1			
	1	1	1	1	−2	–	–
	0	0	2	2			
	1	1	1	0	−1	–	–
	0	0	2	1			

The first row of occupation numbers refers to ϕ_l and the second to ϕ_r (see text for the definitions of ΔK, P, and R)

concentrated in deciding how the spatial orbitals in the two configurations match up. A simple notation for illustrating these relationships consists of placing the initial and final occupation number for distinct orbitals in the same column, but without making an entry where no such difference exists, as shown in Table 1. There are obviously many more cases for still higher-order excitations, but these can be ignored because the corresponding Hamiltonian matrix elements are zero in all cases by virtue of the Condon rules [7] for orthonormal basis sets.

It is important to recognize that the relationships in Table 1 are not dependent on the manner in which the configurations are generated, i.e., by taking a specific excitation class relative to some reference configuration. There are 20 non-trivial cases, four for single excitations and $4 \times 4 = 16$ for doubles. In addition, there is the diagonal case where no orbital excitation is allowed. Of the resulting 21 non-trivial cases in Table 1, more than half can be combined effectively because they involve the same matrix-element operations or are simply eliminated because of the hermitian nature of the Hamiltonian. To show this, it is useful to define a parameter K which numbers the open-shell types for a given multiplicity; specifically, $K = \frac{k}{2} - S + 1$, where k is the number of open shells in the configuration. Each of the 21 cases is thus characterized by a ΔK value, defined as the difference in the supercategory numbers for the bottom (or right-hand) configuration ϕ_r and the top (or left-hand) species ϕ_l in each diagram (or simply half the difference in their respective numbers of open shells). Because of the hermitian nature of the Hamiltonian, it is unnecessary to give further consideration to examples in which $\Delta K < 0$.

Before going further with this analysis, it helps to consider a specific example to illustrate how the above case structure can be used to significantly reduce the size of the computations. The case in question is labeled with $P = 4$ in Table 1. There are two closed-shell orbitals differentially occupied ($\Delta K = 0$). Because determinants in such a configuration pair already differ by a double excitation at the level of spatial orbitals, it is clear that there must be a perfect matching between open-shell spins in order to have a nonzero interaction. The latter condition is only met once for a given determinant D_r in ϕ_r, and when a standard enumeration is adopted, it is clear that the number of the specific determinant D_l in ϕ_l with which it interacts will be the same as for D_r. This relationship obviously holds for any $P = 4$ case, regardless of which orbitals a and b are differentially occupied in the two configurations. The value of the matrix element is easily shown to be the exchange integral K_{ab} between these two closed-shell orbitals. In short, once this simple relationship between two configurations is recognized and the appropriate exchange integral has been retrieved from core storage, the identity and magnitude of all nonzero Hamiltonian matrix elements between ϕ_r and ϕ_l is completely specified. The requisite

SAF transformation can therefore proceed immediately without ever making a lineup permutation among the open-shell spin orbitals.

The situation is more complicated for the other excitation relationships in Table 1. Nonetheless, there are many simplifications in the computational procedure that can be achieved by making use of the standard orders of determinants described above. The only $\Delta K = 2$ case serves as a good illustration [23]. In this instance, ϕ_r possesses four distinct open shells (denoted *a*, *b*, *c*, *d*, with the orbital numbers increasing from left to right). By definition of this case, ϕ_l has two closed shells to balance the four open-shell species. They are chosen among the four open-shell orbitals, so there are six distinct possibilities (labeled as $R = 1$–6). All other open and closed shells must be identical in the two configurations in order to have a nonzero interaction between pairs of determinants. Regardless of which pair of the *a*, *b*, *c*, *d* orbitals is doubly occupied in ϕ_l, it is clear that (a) only determinants D_l and D_r can interact for which all common open-shell spins are identical and (b) when the latter condition is fulfilled, exactly two spin orbitals must be different. If the latter spin orbitals are written schematically as

$$D_l : a\alpha\, a\beta\, b\alpha\, b\beta$$
$$D_r : a\gamma\, b\delta\, c\varepsilon\, d\eta,$$

it is seen that after permuting like spin orbitals to positions in the same column, the Condon relationship can be expressed as:

$$(\pm)\begin{pmatrix} a\gamma' & b\delta' \\ c\varepsilon & d\eta \end{pmatrix}$$

where γ' and δ' are the *opposite* spins of γ and δ, respectively (since the latter were "cancelled out" as common spin orbitals in D_l and D_r). According to the Condon rules [7], the result corresponds to an interaction of

$$(\pm)\left[\begin{pmatrix} a\gamma' & b\delta' \\ c\varepsilon & d\omega \end{pmatrix} - \begin{pmatrix} a\gamma' & b\delta' \\ d\omega & c\varepsilon \end{pmatrix}\right],$$

where the bracketed quantities denote electron repulsion (g_{12}) integrals.

After integration over the spin coordinates, the interaction between D_l and D_r involves at most two spatial integrals $\begin{pmatrix} a & b \\ c & d \end{pmatrix}$ and $\begin{pmatrix} a & b \\ d & c \end{pmatrix}$, whereby the only allowed linear combination coefficients are 0 and ± 1 in each case. In principle, one can expect 16 different spin combinations γ, δ, ε, η, but because of the restriction of $\Delta M_S = 0$, the number of possibilities reduces to only six (two α- and two β-spins). For a given D_r, only one D_l species is capable of an interaction, namely that with identical spin assignments for all (if any) common open shells. All of the above statements hold with only minor modifications if another choice for the distinct closed shells in D_l is taken (a^2c^2, etc.), with only the values of the linear coefficients and the identity of the two basis integrals themselves changing thereby. Six such cases can be distinguished and, as already discussed, they are labeled with the index R, which defines the relationship between non-identical closed and open shells in ϕ_l and ϕ_r, respectively. Likewise, the six possible spin distributions in a given D_r are labeled with the index *Z*. By taking the determinants in standard order, the value of *Z* depends on where the four distinct open shells (*a*, *b*, *c*, *d*) are located in the string of all such singly occupied orbitals.

Table 2 Basis electron repulsion integrals and corresponding pairs of linear coefficients for various combinations of *R* cases and γ, δ, ε, η spin assignments *Z* for the *a*, *b*, *c*, *d* open-shell orbitals

R case		Basic integrals	Linear coefficients for $\gamma\ \delta\ \varepsilon\ \eta$ Spin assignments		
$\Delta K = 2$, $P = 1$			$Z = 1$ $\alpha\,\alpha\,\beta\,\beta$	$Z = 2$ $\alpha\,\beta\,\alpha\,\beta$	$Z = 3$ $\alpha\,\beta\,\beta\,\alpha$
1	a^2b^2 *abcd*	$\begin{pmatrix} a & b \\ c & d \end{pmatrix}, \begin{pmatrix} a & b \\ d & c \end{pmatrix}$	(−1, 1)	(0, −1)	(1, 0)
2	a^2c^2 *abcd*	$\begin{pmatrix} a & c \\ b & d \end{pmatrix}, \begin{pmatrix} a & c \\ d & b \end{pmatrix}$	(0, 1)	(1, −1)	(−1, 0)
3	a^2d^2 *abcd*	$\begin{pmatrix} a & d \\ b & c \end{pmatrix}, \begin{pmatrix} a & d \\ c & b \end{pmatrix}$	(0, −1)	(1, 0)	(−1, 1)

Note that *R* cases with opposite closed shells (e.g., a^2b^2 vs. c^2d^2) have identical interactions for the same spin assignment Z; hence, cases for $R = 4$–6 are not given explicitly. In addition, hole inversion of the γ, δ, ε, η spins for a given *R* case also causes no change in the interaction; hence, only three spin assignments are shown

In this connection, it is important to adopt a convention in which the one-electron functions are numbered consecutively from 1 and the open and closed shells are separately placed in increasing order. This ordering is then maintained within each of the determinantal functions, which in turn are also numbered consecutively from 1. If there are k_r open shells in ϕ_r, the number of distinct combinations for the four non-identical species *a*, *b*, *c*, *d* is the combinatorial number $\begin{pmatrix} k_r \\ 4 \end{pmatrix}$. It is also convenient to define a parameter Q_r, which labels these different possibilities consecutively from 1 as well. Schematically, one can think of performing a permutation of the open shells in each D_r so that *a*, *b*, *c*, *d* are placed immediately below the four positions assigned to the non-identical closed shells in each D_l, while still maintaining increasing order (separately) for both the common and distinct open shells, respectively; hereafter, the parity of this permutation will be referred to as *W*.

To show that the above definitions are purposeful, it is helpful to consider the following. Once the permutation Q_r

is specified for the excitation case under discussion ($\Delta K = 2$), the spin assignment Z for each D_r becomes unambiguously characterized as well as the number of the D_l species (in standard order) with which it interacts. The magnitude of each nonzero Hamiltonian matrix element can then be calculated by combining R with Z and W because this information suffices to evaluate the linear combination coefficients for the two basis g_{12} integrals. The specific values for these coefficients and the identity of the associated integrals are given in Table 2 for all 36 combinations of R and Z. For $R = 2$ and $Z = 3$, for example, the following relationship between D_r and the D_l (with identical common open-shell spins) with which it interacts is found to be

$$D_l : a^2c^2 \qquad \alpha\alpha\, a\beta\, c\alpha\, c\beta$$
$$D_r : a\gamma\, b\delta\, c\varepsilon\, d\eta \qquad a\alpha\, b\beta\, c\beta\, d\alpha.$$

Eliminating the two additional common spin orbitals after permuting them to equivalent positions gives

$$-\begin{pmatrix} a\beta & c\alpha \\ b\beta & d\alpha \end{pmatrix},$$

where the negative sign indicates a net parity change as a result of the spin–orbital permutations. Integration over the spin coordinates then translates into the interaction:

$$-1\begin{pmatrix} a & c \\ b & d \end{pmatrix} + 0\begin{pmatrix} a & c \\ d & b \end{pmatrix},$$

as indicated in Table 2 by the entry (−1, 0) for $R = 2$ and $Z = 3$. The matrix element between D_l and D_r is thus specified completely if the parity W is also known for Q_r.

There is also a $P - 1$ case for both $\Delta K = 0$ and 1, as indicated in Table 1. Note that for $\Delta K = 1$, there are two cases with three R subcases, making the same number of six subcases discussed for $\Delta K = 2$. Similarly for $\Delta K = 0$, there are three $P = 1$ cases with a total of six R subcases. The corresponding matrix-element calculation for the $P = 1$ cases for both $\Delta K = 0$ and 1 can therefore be carried out in a completely analogous manner as for $\Delta K = 2$. A complete discussion of all the $\Delta K = 0$ and 1 cases is given elsewhere [24, 25], including the computation of all diagonal Hamiltonian matrix elements [24].

Because of the above considerations, it is possible to completely avoid explicit comparisons between determinants in evaluating Hamiltonian matrix elements in the MRD-CI calculations. To begin with, it is convenient to order the various configurations so that species with the same number of open shells (same K) appear contiguously in core storage. From the loop structure of the program, the key ΔK value for each configuration pair is then known prior to comparing their occupation numbers. In the $\Delta K = 2$ case, there are only two possibilities, namely either the configurations differ by a double excitation at the orbital level or they do not. This information is obtained quite easily by simply comparing orbital occupancies. Most importantly, *it requires no additional information concerning spin assignments* in the corresponding determinants spanning the two configurations. A label P is again defined to distinguish between the two possibilities; it has a value of $P = 1$ for a double excitation and $P = 0$ when a triple or higher excitation relationship exists (see Table 1 for P labels in all other non-trivial cases). In the event of a double excitation, the foregoing analysis shows that ϕ_l must possess two non-identical closed shells that are balanced by four non-identical open shells in ϕ_r, and hence on this basis, the values of two other labels, R and Q_r, to be defined below can be determined and then recorded on what will hereafter be referred to as the configuration label file A. For higher excitations, a single value of zero is stored on this file, which distinguishes this case from the non-trivial one ($P = 1$) because both R and Q_r can only have nonzero values. In addition, knowledge of the R value for the $P = 1$ case coupled with the orbital numbers a, b, c, d open shells (see Table 2) allows one to determine the identity of the two basis electron repulsion integrals needed for this interaction; the location of these integrals is then recorded on a separate (integral) file in the form of block numbers and addresses within a block [11]. Having made these entries in the two files, the program then proceeds to the consideration of the next pair of configurations and the process is repeated. When all pairs of configurations are dealt with in this manner, the next step is to convert the integral labels into an ordered file B of electron repulsion integrals by means of a sort procedure.

At the completion of the latter step, the final evaluation of Hamiltonian matrix elements can begin by reading into core storage a table containing information about the structure of the determinantal basis functions in a given pair of supercategories K_l and K_r. A separate table is required for each value of K_r, ΔK, and S, but no more than one of these needs be located in core storage at any one time by virtue of the fact that all configurations with the same number of open shells are stored continuously and the program loop structure is arranged accordingly. For $\Delta K = 2$, such a table consists of $\begin{pmatrix} k_r \\ 4 \end{pmatrix}$ parts, each of which is associated with a single value of Q_r. If ϕ_r is spanned by q_r SAFs, each of these subtables consists of $2q_r + 1$ entries, namely the parity W associated with a given Q_r permutation, and a value of Z (labeling one of the six spin–orbital distributions) for each of the q_r projected

Table 3 Storage requirements (in bytes) for the various tables as a function of the number of open shells k and number of SAFs q_r for a given supercategory (open-shell type) K_r (S is the total spin quantum number)

Characterization of case		Table size (bytes)
$\Delta K = 2$	$p = 1$	$\binom{k_r}{4} \cdot (2q_r + 1)$
$\Delta K = 1$	$p = 1$	$(k_r - 2) \cdot \binom{k_r}{3}(2q_r + 1)$
	$p = 2$	$\binom{k_r}{2} \cdot q_r$
	$p = 3$	$\binom{k_r}{2}[1 + q_r(k_r + 2S + 1)]$
$\Delta K = 0$	$p = 1$	$\left[\binom{k}{2} + 1\right] \cdot \binom{k}{2} \cdot \frac{(3q+1)}{2}$
	$p = 2$	$\binom{k+1}{2} \cdot (q + 1)$
	$p = 3$	$\binom{k+1}{2} \cdot (1 + 2k \cdot q)$
	$p = 4$	None required
	$p = 5$	$q\left[(k+1) + 3\left(\frac{k^2}{4} - S^2\right)\right]$

See Table 1 for definitions

determinants PD_r that form a basis for the associated linear space as well as the number of the (only) D_l determinant with which it makes a nonzero interaction. Combining Z and W with the value of R stored in the configuration label file A allows for a complete specification of the interaction, as discussed above. Thus, by using the corresponding two basis electron repulsion integrals on file B, it is possible to compute the desired matrix elements for each of the q_r projected determinants PD_r and its D_l partner. The transformation to matrix elements in the SAF basis can then proceed, and the results can be saved for later use in the diagonalization step. The size of the $\Delta K = 2$ table varies with K_r and S and is seen to have $\binom{k_r}{4}$ $(2q_r + 1)$ entries (see Table 3). The specific values contained therein depend only on the choice of the standard order of determinants (i.e., a series of spin–orbital permutations). Since these positive integer values never exceed the number of D_l determinants p_l or 6 (Z), this information can be stored quite compactly. Clearly, the effect of the tabulated information is to bypass explicit comparisons of the individual determinantal functions.

Very similar procedures have been worked out for all the other excitation cases in Table 1, with only a slightly modified structure for the corresponding tables for the respective matrix-element evaluations. For $\Delta K = 0$ and 1, there are several non-trivial cases (in addition to $P = 1$), however, so in comparing the various configurations, more distinctions must be made than for the $\Delta K = 2$ case considered explicitly above. Procedures for achieving this objective are outlined in the following, while more detailed discussion can be found in earlier papers [24, 25]. For $\Delta K = 1$, the five cases listed in Table 1 are grouped into two double-excitation interactions and one of single-excitation type. The first of these is also characterized by two basis integrals, just as for $\Delta K = 2$, and hence is also referred to as $P = 1$; a total of six distinct relationships among the non-identical open and closed shells in ϕ_r and ϕ_l are possible, as specified by a corresponding R label. Two different open-shell permutations need to be considered in this instance, however, designated by Q_r and Q_l, and together characterized by a single parity value W. The second $\Delta K = 1$ case also corresponds to a double excitation at the spatial orbital level ($P = 2$) but it requires only a single basis integral of the type $\begin{pmatrix} a & b \\ c & c \end{pmatrix}$ and is notably less complicated than $P = 1$. Only two distinct R values are needed, depending on whether there are two or one non-identical closed shells in φ_l, and only Q_r (not Q_l) needs to be specified on the configuration file A.

The remaining case ($P = 3$) is similar in many ways to $P = 2$ (also two R subcases), with the important exception that it involves only a single excitation at the orbital level (Table 1). As a result, in addition to P, Q_r, and R, it is necessary to store the numbers of specific orbitals a and b involved in the single excitation in each instance. Since symmetry is considered explicitly in the program, the irreducible representation label associated with these two orbitals is also included on file A, making six parameters in all; this number can be reduced to five by storing R in the form of a sign for Q_r (since there are two such values in this instance and similarly for $P = 2$). On the integral label, file addresses for the $\begin{pmatrix} a & a \\ a & b \end{pmatrix}$ or $\begin{pmatrix} b & b \\ b & a \end{pmatrix}$ species need to be recorded as well as $\begin{pmatrix} c & a \\ c & b \end{pmatrix}$ and $\begin{pmatrix} c & a \\ b & c \end{pmatrix}$ for each common open and closed shell in the two configurations. The tabulated information is also somewhat more extensive for $P = 3$, since both single and double excitations are possible at the determinantal level; for particular spin assignments in D_r, one D_l is related by a double spin–orbital distribution for each common open shell, while for other Z values only a single D_l can be found which is related by a single excitation. More details may be found elsewhere [25].

For $\Delta K = 0$, the same three P cases can be distinguished in Table 1, in addition to the case first discussed involving a closed-shell to empty-shell double excitation (labeled $P = 4$). Counting the strictly diagonal case ($P = 5$), which can be treated separately from the other four non-diagonal ones, the total number of distinct ΔK and P non-trivial excitation relationships is nine; in the $P = 5$ case at the determinantal level, both diagonal and double-excitation pairings (but not single excitations) occur, whereby in the latter case the interaction is always the negative of some exchange integral [24].

4 Direct MRD-CI

The calculation of eigenvectors of the Hamiltonian matrix is carried out within the context of the Davidson diagonalization method [26]. For this purpose, one needs a series of starting vectors for each root to be determined. The Hamiltonian matrix itself has dimension $N_D \times N$, where N is the number of SAFs in the many-electron basis and N_D is the corresponding number of Slater determinants which span the MRD-CI configuration space ($N_D > N$). In the matrix-element generation scheme described in the previous section, the basis functions are different for columns and rows in this representation, with N PD_β projected determinants in the first case and N_D D_α Slater determinants in the other. Accordingly, it is necessary to describe the above starting vectors ψ^i as two different linear combinations of the basis functions, namely as

$$\psi^i = \Sigma(\alpha)c_{\alpha i}D_\alpha = \Sigma(\beta)b_{\beta i}PD_\beta \tag{1}$$

The corresponding Hamiltonian matrix element between ψ^i and ψ^j is thus

$$\begin{aligned} H_{ij} &= <(\Sigma(\alpha)c_{\alpha i}D_\alpha)H\big(\Sigma(\beta)b_{\beta j}PD_\beta\big)> \\ &= \Sigma(\alpha)c_{i\alpha}<D_\alpha H\Sigma(\beta)b_{\beta j}PD_\beta> \\ &= \Sigma(\alpha)c_{i\alpha}\Sigma(\beta)b_{\beta j}<D_\alpha HD_\beta> \\ &= \Sigma(\alpha)c_{i\alpha}\sigma_{\alpha j} \end{aligned} \tag{2}$$

In the original version of the MRD-CI programs [11], the $H_{\alpha\beta}$ matrix was first stored on an external file and the individual matrix elements were read into core storage and multiplied with the appropriate $b_{\beta j}$ vector coefficients in Eq. (2). The results were then accumulated in the corresponding $\sigma_{\alpha j}$ matrix elements until the entire $H_{\alpha\beta}$ matrix had been considered. The final H matrix on the ψ^i basis was then formed by scalar multiplication of the σ_j and the c_i vectors. The H_{ij} matrix was then diagonalized, and the results were used to generate more starting vectors to be used in subsequent iterations of the Davidson procedure until sufficient convergence for the eigenvectors and eigenvalues was obtained.

The above procedure has the disadvantage of requiring a high percentage of computational time devoted to the operations involved in reading the $H_{\alpha\beta}$ matrix from external storage in each iteration of the diagonalization. For large matrices with dimensions exceeding 200,000, the increase in I/O times begins to outweigh the advantage of only calculating the matrix elements once. The increased efficiency which is gained by the Table-CI technique described in the previous section therefore makes it computationally more effective to compute the matrix elements in each iteration and thereby to avoid storing them on an external file and reading them into core storage for the computation of the σ matrices. This approach is closely related to the Direct CI technique [27] used in integral-driven CI methods. It was implemented by Krebs and Buenker in 1995 [28], and the comparison of computational times may be found in that reference. This general method has also been employed in the configuration selection scheme to improve the performance of the MRD-CI programs in this application. The implementation of the Direct CI procedure increased the range of the programs to as many 10^6 SAFs. The standard version allows for the use of 400 randomly chosen reference configurations to generate the MRD-CI space and up to 40 roots to be extracted from the resulting secular equation.

To give an idea as to the practicality of such calculations, timing experiments have been carried out for the CH_3I system in an AO basis containing 135 primitive Gaussians. Results have been obtained for all four C_s symmetries for 18 roots in each case (14 active electrons in the active space and 350 reference configurations). The corresponding $T = 0$ MRD-CI spaces contain 200 million SAFs on average, whereas the corresponding selected sets have dimensions in the 500,000–600,000 range. An Intel i5-750 computer has been employed with 2.67 GHz and 16 GB memory. Two passes are made at different threshold values in the Davidson diagonalization procedure. The calculations occupy the CPU with an efficiency of 99.9 %, emphasizing the advantages of the Direct CI approach. Typical CPU times are 50–130 min in the first pass to obtain all 18 roots and 90–140 min in the larger second pass.

For applications with a limited set of reference species, another version has been constructed with the internal contraction scheme [29]. The key restriction for this set of CI programs is that a maximum number of 38 (referred to as "internal") orbitals are allowed to be populated in these configurations [30]. Because of the MRD-CI definition, it is clear that no more than two other ("external") orbitals can be occupied in the corresponding configurations that are generated by at most a double excitation relative to any one of the reference species. It is important to reorder the configurations, so that all those with the same occupation of internal orbitals appear consecutively in the list. This can be done using a conventional sort procedure by assigning each configuration a unique number in a ternary system (each orbital has either 0, 1, or 2 occupation), the length of which is equal to the number of internal orbitals.

All configurations with the same internal occupation are referred to collectively as a "lane" in this procedure. If there are two external electrons, such a lane will normally be divided into two sublanes depending on whether two open shells or one closed shell is occupied.

In a given Davidson iteration, one loops over pairs of lanes and compares their respective internal occupations, taking note of those orbitals with distinctive occupations. There are only six possibilities for relative occupations: 1–0, 2–0 and 2–1, and the opposite cases for higher right-hand occupation. Two arrays are defined which contain, respectively, the number of the internal orbital and which of the above six cases describes the occupation relationship. One can find a maximum of four orbital mismatches, the details of which are recorded in the above two arrays. The key point is that the desired improvement in efficiency relative to the previous Direct MRD-CI approach only occurs when there is a minimum of internal orbitals. Each time, a configuration is added to the reference set that adds one or more new internal orbitals, and the result is inevitably a decrease in the sizes of generated lanes, which in turn leads to a decrease in the efficiency of the sorting procedure.

Once all the internal mismatches have been identified, many of the P, R subcases in the Table-CI approach are eliminated. As a result, inspection of the external orbitals of the two configurations can be much more restrictive and less time-consuming than in the original program with randomly ordered configurations. The next step is to loop over configuration pairs within the two sublanes under consideration. At most two orbitals in each configuration need to be considered in this case, so the process is notably less time-consuming than when all internal orbitals need to be scanned. After the external orbital comparisons, the final step is to calculate the interaction if it is nonzero. From the number and type of mismatches, it is easy to assign the values of P and R, the Q_l and Q_r permutations, and the corresponding basis integrals for each case. For single excitations ($P = 3$), it is also necessary to compute the addresses of the required four-index electron repulsion integrals involving common open shells. Diagonal ($P = 5$) cases only occur within the same sublane and can be treated in a straightforward manner. Information about timing comparisons for this method is available in a companion publication [30].

5 Spin–orbit interaction

The treatment of molecules containing heavy atoms poses special problems for CI methods because of the need to consider relativistic effects in the electronic Hamiltonian. If attention is restricted to low-lying states, it is quite useful to employ an accurate representation of the inner-shell electrons in the form of relativistic effective core potentials (RECPs). The key requirement is that enough electrons are maintained in the valence space such that the remaining electrons are truly characterized as atomic core. The multireference CI approach is especially effective in this case because of the high density of states usually found for such systems. There is no particular difficulty in adding scalar relativistic terms to the electrostatic Hamiltonian since they are simply included with the nuclear attraction in the RECP approach. The spin–orbit interaction requires special attention, however, because it mixes states of different spatial and spin symmetry. The Table-CI method can still be applied for the calculations of matrix elements between states of the same Λ–S symmetry, but one must revert to the traditional approach in which explicit comparisons of spin–orbital occupations are made in order to evaluate the corresponding spin–orbit quantities. The use of RECPs has distinct advantages in this respect as well. This is because only one-electron interactions are included, i.e., between valence electrons and the screened nuclei.

As a result, only configurations that differ by at most a single excitation can have a nonzero interaction, thereby greatly reducing the number of configuration pairs that need to be considered explicitly.

In practice, only Abelian groups are employed in conventional CI programs. The inclusion of spin–orbit coupling forces the use of double groups, and this makes it necessary to design distinctive treatments for systems with even and odd numbers of electrons. The general treatment of symmetry in the spin–orbit MRD-CI programs is discussed in an earlier Ref. [31]. A major distinction is that it is no longer feasible to restrict consideration to $M_S = S$ determinants. In a typical case, one needs $M_S = S$ determinants for one configuration and $M_S = S - 1$ species for the other. Use of the Wigner–Eckart theorem [32] helps to minimize the number of matrix elements that need to be computed in each case.

The most straightforward means of applying the above techniques is to form a matrix representation of the full spin–orbit Hamiltonian in a basis of linear combinations of Slater determinants. As discussed in Ref. [31], the dimension of the corresponding secular equation is generally much larger than in conventional MRD-CI calculations because of the fact that functions of typically 10–15 different Λ–S symmetries must be included in order to satisfactorily account for the effects of spin–orbit coupling in a given application. Configuration selection is quite effective in minimizing this problem by restricting the number of many-electron functions to those that interact at least moderately strongly (as determined through the use of an appropriate threshold value).

One can avoid treating such large secular matrices in many cases by employing basis functions that result from the diagonalization of pure Λ–S Hamiltonian matrices. The diagonal elements of the full spin–orbit Hamiltonian matrix are simply the Λ–S eigenvalues. They can be augmented with perturbative corrections to better account for electron correlation at this stage of the theoretical treatment. The corresponding off-diagonal matrix elements are then obtained by forming the spin–orbit matrix on the basis of "truncated" Λ–S eigenfunctions. Diagonalization then leads to the desired mixing of the various Λ–S symmetries and the corresponding approximate eigenvalues and eigenvectors of the full relativistic Hamiltonian including spin–orbit coupling. The resulting functions are then used to compute dipole transition moments and other properties of the system. One can look upon this approach as a two-step procedure to include both electron correlation and relativistic effects into the theoretical description. It uses a relatively small number of highly correlated functions as basis as opposed to the other technique that employs quite large numbers of relatively simple (determinantal) functions. Both methods are variational in nature since their results are obtained in each case by matrix diagonalization. The Λ–S eigenfunctions can be looked upon as a *contracted* many-electron basis analogous to the linear combinations of primitive Gaussian functions typically used as atomic orbitals in forming the one-electron basis for both SCF and CI calculations. Accordingly, the two-step procedure has been designated as the Λ–S contracted spin–orbit CI (LSC-SO-CI) method. It has been used extensively to describe the spectra and electronic structure of diatomic molecules containing such main group elements as Bi, Pb, Sb, Te, Ba, and I [31].

In the other method (MR-SO-CI), one uses what amounts to a straightforward multireference description of the electronic states except that (a) the spin–orbit operator is included in the full Hamiltonian and (b) determinantal functions of more than one Λ–S symmetry are included in the many-electron basis. Its goal is to treat electron correlation and relativistic effects on the same level. It is implemented by forming σ vectors directly as the $H_{\alpha\beta}$ matrix elements are computed, in complete analogy to the Direct MRD-CI approach outlined in the previous section. The resulting wave functions are then used to calculate one-electron properties for use in analyzing the details of the electronic spectra of the molecular systems. Altogether, experience has shown that the LSC-SO-CI and MR-SO-CI methods perform at nearly equivalent levels of accuracies for systems with atoms not heavier than lead. The two-step procedure is actually somewhat easier to optimize because this merely involves the inclusion of more Λ–S eigenfunctions in the contracted basis. A detailed review of calculated results of both methods is given in Ref. [31].

6 Conclusion

Since its introduction by Buenker and Peyerimhoff in 1974 [8, 9], the multireference single- and double-excitation configuration interaction (MRD-CI) method has proven to be an effective means of incorporating electron correlation effects into theoretical treatments of molecular systems. The present review has concentrated on the configuration-driven approach for carrying out this type of calculation. Its main advantage is the freedom it allows in both the choice of reference configurations and the truncation of the resulting CI space to be considered in the associated secular equations. A key advance for configuration-driven CI algorithms is the Table-CI method. Its main purpose is to totally avoid lineup permutations of spin orbitals that are otherwise essential in determining excitation relationships between the Slater determinants that form the many-electron basis for the CI calculations. A preliminary scan of spatial orbital relationships for pairs of configuration suffices to generate labels that are used to identify both the appropriate table to be used as well as the four-index electron repulsion integrals that are required to compute the Hamiltonian matrix elements.

The role of Direct CI procedures to minimize the necessary I/O operations in such calculations has also been discussed and shown to have a significant impact in the overall efficiency of the MRD-CI programs. Further improvements are possible in certain cases by storing the configurations in a way which emphasizes their occupation of so-called internal orbitals, thereby greatly restricting the scope of lineup permutations among spatial orbitals as well. The use of relativistic effective core potentials (RECPs) allows for the inclusion of spin–orbit effects in the theoretical treatment as well as other scalar terms in the Breit–Pauli Hamiltonian without relinquishing the essential variational character of the MRD-CI method. The above changes are especially important for the description of molecules containing heavy atoms such as lead and bismuth.

References

1. Hartree DR (1928) Proc Camb Phil Soc 24:89
2. Hartree DR (1928) Proc Camb Phil Soc 24:111
3. Fock V (1930) Z Physik 61:126
4. Slater JC (1929) Phys Rev 34:1293
5. Heitler W, London F (1927) Z Phys 44:455
6. Slater JC (1963) Quantum theory of molecules and solids, vol 1. McGraw-Hill, New York, pp 22–59
7. Tinkham M (1964) Group theory and quantum mechanics. McGraw-Hill, New York, pp 162–167
8. Buenker RJ, Peyerimhoff SD (1974) Theor Chim Acta 35:33
9. Buenker RJ, Peyerimhoff SD (1975) Theor Chim Acta 39:217
10. Shavitt I (1976) Private communication. The Ohio State University
11. Buenker RJ, Peyerimhoff SD, Butscher W (1978) Mol Phys 35:771
12. Gershgorn Z, Shavitt I (1968) Int J Quantum Chem 2:751

13. Hirsch G, Bruna PJ, Peyerimhoff SD, Buenker RJ (1977) Chem Phys Lett 52:442
14. Buenker RJ, Knowles DB, Rai SN, Hirsch G, Bhanuprakash K, Alvarez-Collado JR (1989) In: Carbó R (ed) Studies in physical and theoretical chemistry. Quantum chemistry-basic aspects, actual trends, vol 62. Elsevier, Amsterdam, p 181
15. Knowles DB, Alvarez-Collado JR, Hirsch G, Buenker RJ (1990) J Chem Phys 92:585
16. Bauschlicher CW Jr, Langhoff SR, Taylor PR, Partridge H (1986) Chem Phys Lett 126:436
17. Paldus J (1974) J Chem Phys 61:5321
18. Paldus J (1975) Int J Quantum Chem Symp 9:165
19. Shavitt I (1977) Int J Quantum Chem Symp 11:131
20. Buenker RJ, Peyerimhoff SD (1968) Theor Chim Acta 12:183
21. Bunge A (1970) J Chem Phys 53:20
22. Davidson ER (1974) Int J Quantum Chem 8:61
23. Buenker RJ (1980) In: Burton P (ed) Proceedings of workshop on quantum chemistry and molecular physics in Wollongong, Australia. University Press, Wollongong, pp 1.5.1–1.5.37
24. Buenker RJ, Phillips RA (1985) J Mol Struct Theochem 123:291
25. Buenker RJ (1981) In: Carbó R (ed) Studies in physical and theoretical chemistry. Current aspects of quantum chemistry, vol 21. Elsevier, Amsterdam, p 17
26. Davidson ER (1975) J Comput Phys 17:87
27. Roos BO (1972) Chem Phys Lett 15:153
28. Krebs S, Buenker RJ (1995) J Chem Phys 103:5613
29. Meyer W (1977) In: Schaefer HF III (ed) Modern theoretical chemistry. Methods of electronic structure, vol 3. Plenum, New York
30. Buenker RJ, Krebs S (1999) In: Hirao K (ed) Recent advances in multireference methods. World Scientific, Singapore, pp 1–29
31. Alekseyev AB, Liebermann H-P, Buenker RJ (2003) In: Hirao K, Ishikawa M (eds) Spin–orbit multireference configuration interaction method and applications to systems containing heavy atoms. World Scientific, Singapore, pp 65–105
32. Tinkham M (1964) Group theory and quantum mechanics. McGraw-Hill, New York, pp 117–123

Theor Chem Acc (2014) 133:1487
DOI 10.1007/s00214-014-1487-4

REGULAR ARTICLE

Finite-temperature full configuration interaction

Zhuangfei Kou · So Hirata

Received: 2 January 2014 / Accepted: 27 March 2014 / Published online: 22 April 2014

Abstract The exact basis-set values of various thermodynamic potentials of a molecule are evaluated by the finite-temperature full configuration-interaction (FCI) method using ab initio molecular integrals over Gaussian-type orbitals. The thermodynamic potentials considered are the grand partition function, grand potential, internal energy, entropy, and chemical potential in the grand canonical ensemble as well as the partition function, Helmholtz energy, internal energy, and entropy in canonical ensemble. Approximations to FCI that are accurate at low and high temperatures are proposed, implemented, and tested. The results of finite-temperature FCI and its approximations are compared with one another as well as with the results of finite-temperature zeroth-order many-body perturbation theory, in which the Fermi–Dirac statistics is exact. Analytical asymptotic properties in the low- or high-temperature limits of some of these thermodynamic potentials are also given.

Keywords Configuration interaction · Thermodynamics · Partition function · Temperature · Canonical ensemble · Grand canonical ensemble · Fermi–Dirac statistics

Dedicated to the memory of Professor Isaiah Shavitt and published as part of the special collection of articles celebrating his many contributions.

Z. Kou · S. Hirata (✉)
Department of Chemistry, University of Illinois
at Urbana-Champaign, 600 South Mathews Avenue,
Urbana, IL 61801, USA
e-mail: sohirata@illinois.edu

S. Hirata
CREST, Japan Science and Technology Agency,
4-1-8 Honcho, Kawaguchi, Saitama 332-0012, Japan

1 Introduction

Late Dr. Isaiah Shavitt has been the champion of configuration-interaction (CI) theory [1–4], which is the first systematically accurate, ab initio electron-correlated molecular orbital (MO) theory. Although today its usage may be diminished with the advent of coupled-cluster (CC) theory [5], its predictions have overturned some experimental conclusions [6, 7] and helped computational quantum chemistry be accepted by a broader community of chemists as a vital engine of chemical research.

The full configuration-interaction (FCI) method [8], in particular, occupies a uniquely important position in ab initio MO theory to this day, as it provides the exact basis-set solutions of the Schrödinger equations of general polyatomic molecules [9–26]. These solutions serve as essential benchmark against which approximate ab initio methods are calibrated and tested [27–38].

In no other fields are such benchmarks more needed than in finite-temperature many-electron theories [39–45]. Although there is a well-established finite-temperature extension to many-body perturbation (MP) theory described in a number of textbooks [46–48], some authors [42–44, 49, 50] have questioned its correctness and proposed alternative formalisms. It is desirable to establish systematic and converging finite-temperature approximations based on CI, CC, and MP theories and, to achieve this, it is imperative to first know the exact converged limit at finite temperatures.

In this article, we present the formalism and algorithms of finite-temperature FCI in either canonical or grand canonical ensemble and implement them in the string-based determinant algorithm with a view to helping establish general finite-temperature many-electron theories. Unlike zero-temperature FCI, the target quantity of finite-temperature FCI is the (grand) partition function (rather than energy), from which various thermodynamic potentials such as internal energy, entropy, Helmholtz energy, and chemical potential can be derived. The different target quantity (such as the partition function, which is a trace, as opposed to the ground-state energy, which is an eigenvalue) suggests different approximations and two such approximations (low- and high-temperature approximations) are proposed and implemented. Furthermore, finite-temperature zeroth-order MP (MP0) method, in which Fermi–Dirac statistics is exact, is implemented and compared with finite-temperature FCI in grand canonical ensemble. In this way, we establish the exact (FCI) and most approximate (MP0) extremes of finite-temperature many-electron theories in a Gaussian-type orbital, leaving in-between to be elucidated by future studies.

2 Theory

2.1 Canonical ensemble

The partition function Q of a system of m non-interacting, identical and thus indistinguishable molecules is given by

$$Q = \frac{(q_{\rm elec} q_{\rm vib} q_{\rm rot} q_{\rm trans})^m}{m!}, \tag{1}$$

where $q_{\rm elec}$, $q_{\rm vib}$, $q_{\rm rot}$, and $q_{\rm trans}$ are molecular partition functions in the electronic, vibrational, rotational, and translational degrees of freedom, respectively. We shall be concentrating on $q_{\rm elec}$ in this study.

In the canonical ensemble, that is, assuming the electrical neutrality of each molecule, the molecular electronic canonical partition function (now renamed Z) is given by

$$Z = {\rm Tr}[\exp(-\beta\hat{H})], \tag{2}$$

where $\beta = 1/k_{\rm B}T$, T is temperature and $\hat{H}$ is the molecular electronic Hamiltonian. One atomic unit of temperature is equal to 3.1577464×10^5 K. The trace is taken over the whole basis-set Hilbert space, that is, the space spanned by all N-electron Slater determinants with all possible spin magnetic quantum numbers (S_z) of a given basis set. There are ${}_{2n}C_N$ such determinants in total. The trace is invariant to a unitary transformation of the many-electron basis. This defines canonical finite-temperature FCI.

From Z, various thermodynamic potentials can be derived, e.g., the Helmholtz energy F by

$$F = -\frac{1}{\beta}\ln Z, \tag{3}$$

and the internal energy U by

$$U = -\frac{\partial}{\partial\beta}\ln Z \tag{4}$$

$$= \frac{{\rm Tr}[\hat{H}\exp(-\beta\hat{H})]}{Z}. \tag{5}$$

These two are related to each other by

$$F = U - TS, \tag{6}$$

where S is the entropy, which is also calculable as

$$S = -k_{\rm B}{\rm Tr}[\hat{W}\ln\hat{W}], \tag{7}$$

with

$$\hat{W} = \frac{\exp(-\beta\hat{H})}{Z}. \tag{8}$$

2.2 Grand canonical ensemble

Alternatively, in the grand canonical ensemble that allows the number of electrons in each molecule to fluctuate, the molecular electronic grand partition function (called Ξ) becomes

$$\Xi = {\rm Tr}[\exp\{-\beta(\hat{H} - \mu\hat{N})\}], \tag{9}$$

where μ is the chemical potential and $\hat{N}$ is the number operator. The trace is taken over the whole basis-set Fock space spanned by all zero- through $2n$-electron Slater determinants with all possible values of S_z, where n is the number of the basis functions. There are 2^{2n} such determinants in total. This defines grand canonical finite-temperature FCI. This is a rather artificial construct for a small molecule, but becomes meaningful for a condensed-phase system viewed as a giant molecule.

The thermodynamic potentials derivable from Ξ are the grand potential Ω,

$$\Omega = -\frac{1}{\beta}\ln\Xi, \tag{10}$$

and the internal energy U,

$$U = -\frac{\partial}{\partial\beta}\ln\Xi + \mu\bar{N} \tag{11}$$

$$= \frac{{\rm Tr}[(\hat{H} - \mu\hat{N})\exp\{-\beta(\hat{H} - \mu\hat{N})\}]}{\Xi} + \mu\bar{N}, \tag{12}$$

where $\bar{N}$ is the average number of electrons, which is also derivable from Ξ by

$$\bar{N} = \frac{1}{\beta}\frac{\partial}{\partial\mu}\ln\Xi \tag{13}$$

$$= \frac{{\rm Tr}[\hat{N}\exp\{-\beta(\hat{H} - \mu\hat{N})\}]}{\Xi}. \tag{14}$$

The grand potential and internal energy (when defined in the same grand canonical ensemble) are related by

$$\Omega = U - TS - \mu\bar{N}. \tag{15}$$

The entropy in the above equation is also written by Eq. (7), but with the following $\hat{W}'$ instead of $\hat{W}$:

$$\hat{W}' = \frac{\exp\{-\beta(\hat{H} - \mu\hat{N})\}}{\Xi}. \tag{16}$$

2.3 Fermi–Dirac statistics

When $\hat{H}$ is additively separable as in an independent-particle model, the grand partition function Ξ and its subordinate functions become simplified, as is well-known [47]. The zeroth-order Hamiltonian of the many-body perturbation theory using the Møller–Plesset partitioning is one such case:

$$\hat{H}_0 = \sum_p \epsilon_p \hat{p}^\dagger \hat{p}, \tag{17}$$

where ϵ_p is the energy of the pth Hartree–Fock (HF) spinorbital and $\hat{p}^\dagger$ and $\hat{p}$ are creation and annihilation operators, respectively, of an electron in the pth spinorbital. Clearly, the Slater determinants composed of these HF spinorbitals are the eigenstates of $\hat{H}_0$.

In the zeroth-order many-body perturbation (MP0) theory using this Hamiltonian in the determinant basis ($\{\Phi_i\}$), we find

$$\Xi_0 = \mathrm{Tr}[\exp\{-\beta(\hat{H}_0 - \mu\hat{N})\}] \tag{18}$$

$$= \sum_i \left\langle \Phi_i \middle| \prod_p \exp\{-\beta(\epsilon_p - \mu)\hat{p}^\dagger\hat{p}\} \middle| \Phi_i \right\rangle \tag{19}$$

$$= \prod_p [1 + \exp\{-\beta(\epsilon_p - \mu)\}], \tag{20}$$

where p runs over all HF spinorbitals. Similarly, we can simplify $\bar{N}_0$ (subscript "0" corresponds to MP0) as

$$\bar{N}_0 = \sum_q f_q, \tag{21}$$

where

$$f_q = \frac{\mathrm{Tr}\left[\hat{q}^\dagger\hat{q}\exp\{-\beta(\hat{H}_0 - \mu\hat{N})\}\right]}{\Xi} \tag{22}$$

$$= \frac{\sum_i \left\langle \Phi_i \middle| \hat{q}^\dagger\hat{q}\prod_p \exp\{-\beta(\epsilon_p - \mu)\hat{p}^\dagger\hat{p}\} \middle| \Phi_i \right\rangle}{\prod_p\left[1 + \exp\{-\beta(\epsilon_p - \mu)\}\right]} \tag{23}$$

$$= \frac{\exp\{-\beta(\epsilon_q - \mu)\}\prod_{p\neq q}[1 + \exp\{-\beta(\epsilon_p - \mu)\}]}{\prod_p[1 + \exp\{-\beta(\epsilon_p - \mu)\}]} \tag{24}$$

$$= [1 + \exp\{\beta(\epsilon_q - \mu)\}]^{-1}, \tag{25}$$

which is identified as the Fermi–Dirac distribution function [47].

The internal energy U_0, entropy S_0, and grand potential Ω_0 of MP0 can also be expressed in terms of f_p:

$$U_0 = \sum_p \epsilon_p f_p, \tag{26}$$

$$S_0 = -k_\mathrm{B}\sum_p \{f_p \ln f_p + (1 - f_p)\ln(1 - f_p)\}, \tag{27}$$

$$\Omega_0 = U_0 - TS_0 - \mu_0\bar{N}_0 \tag{28}$$

$$= \frac{1}{\beta}\sum_p \ln(1 - f_p), \tag{29}$$

where the chemical potential μ_0 in MP0 is determined by Eq. (21). In the last equality, we have used Eq. (21) as well as the identity $\epsilon_p - \mu_0 = \beta^{-1}\ln(f_p^{-1} - 1)$ derivable from Eq. (25). This defines finite-temperature MP0, where the grand canonical ensemble is always implicit.

3 Computer algorithms

3.1 Exact method

The canonical and grand canonical finite-temperature FCI and finite-temperature MP0 have been implemented in our in-house determinant-based FCI code [51]. In it, a determinant is computationally represented by two (α and β) strings of bits, each bit specifying the occupancy of an α or β spinorbital [15]. A wave function is expressed as a matrix in the basis of these two strings, an element of which contains the value of the corresponding CI coefficient. The code has the computational ability to act various operators (such as $\hat{H}$ and $\hat{H}_0$) on any wave function that is a linear combination of determinants with an arbitrary number of electrons.

To obtain Z, Ξ, etc. within a given basis set, i.e., to perform finite-temperature FCI, one in principle needs the whole set of eigenvalues of $\hat{H}$ (see below, however, for various approximations to avoid this). We have obtained it for each of the molecules studied by diagonalization of the $\hat{H}$ matrix either in the basis of all N-electron determinants (canonical FCI) or in the basis of zero- through $2n$-electron determinants (grand canonical FCI) where n is the number of basis functions:

$$Z = \sum_{S_z}\sum_i \left\langle \Psi_i^{(N,S_z)} \middle| \exp(-\beta\hat{H}) \middle| \Psi_i^{(N,S_z)} \right\rangle \tag{30}$$

$$= \sum_{S_z}\sum_i \exp\left(-\beta E_i^{(N,S_z)}\right), \tag{31}$$

$$\Xi = \sum_{N=0}^{2n}\sum_{S_z}\sum_i \left\langle \Psi_i^{(N,S_z)} \middle| \exp\{-\beta(\hat{H} - \mu\hat{N})\} \middle| \Psi_i^{(N,S_z)} \right\rangle \tag{32}$$

$$= \sum_{N=0}^{2n} \sum_{S_z} \sum_{i} \exp\{-\beta(E_i^{(N,S_z)} - \mu N)\}, \tag{33}$$

where $\Psi_i^{(N,S_z)}$ is the ith FCI wave function with N electrons, spin magnetic quantum number S_z, and energy $E_i^{(N,S_z)}$ (in the ascending order with $i = 0$ being the ground state). The value of chemical potential μ is determined by Eq. (14), which is solved computationally by a bisection method. The data referred to as "exact" in the subsequent section are obtained in this way.

The finite-temperature MP0 can be carried out in two ways. One is to run the same computational procedure as the grand canonical finite-temperature FCI but using $\hat{H}_0$ instead of $\hat{H}$. The other is to directly evaluate semi-analytical formulas based on the Fermi–Dirac statistics. We have verified that these two procedures yield the same results.

The codes have also been checked for correctness in the $T = 0$ and $T = \infty$ limits, where asymptotic values of thermodynamic potentials are sometimes known analytically. The values of S obtained by Eqs. (6) and (7) have been verified to agree with each other.

3.2 Low-temperature approximation

In the low-temperature (large-β) limit, the partition functions are dominated by contributions from low-lying states, which can be determined selectively and inexpensively by the Davidson algorithm [11]. Hence, we propose approximating them by

$$Z \approx \sum_{S_z} \sum_{i=0}^{M-1} \left\langle \Psi_i^{(N,S_z)} \right| \exp(-\beta\hat{H}) \left| \Psi_i^{(N,S_z)} \right\rangle \tag{34}$$

$$= \sum_{S_z} \sum_{i=0}^{M-1} \exp\left(-\beta E_i^{(N,S_z)}\right), \tag{35}$$

$$\Xi \approx \sum_{N=0}^{2n} \sum_{S_z} \sum_{i=0}^{M-1} \left\langle \Psi_i^{(N,S_z)} \right| \exp\{-\beta(\hat{H} - \mu\hat{N})\} \left| \Psi_i^{(N,S_z)} \right\rangle \tag{36}$$

$$= \sum_{N=0}^{2n} \sum_{S_z} \sum_{i=0}^{M-1} \exp\left\{-\beta\left(E_i^{(N,S_z)} - \mu N\right)\right\}, \tag{37}$$

where M is the number of the lowest-lying N-electron states with spin magnetic quantum number S_z that we include in the approximate calculations. They are obtained by the Davidson algorithm.

3.3 High-temperature approximation

In the high-temperature (small-β) limit, the summation over states should not be truncated because all states can contribute significantly. This, however, allows us to avoid having to know all eigenstates when evaluating Z or Ξ because trace is invariant to a unitary transformation:

$$Z = \sum_{S_z} \sum_{i} \left\langle \Phi_i^{(N,S_z)} \right| \exp(-\beta\hat{H}) \left| \Phi_i^{(N,S_z)} \right\rangle \tag{38}$$

$$\Xi = \sum_{N=0}^{2n} \sum_{S_z} \sum_{i} \left\langle \Phi_i^{(N,S_z)} \right| \exp\{-\beta(\hat{H} - \mu\hat{N})\} \left| \Phi_i^{(N,S_z)} \right\rangle, \tag{39}$$

where $\Phi_i^{(N,S_z)}$ is the ith determinant (not a FCI wave function) with N electrons and spin magnetic quantum number S_z.

Furthermore, smallness of β suggests the following approximation for Z:

$$Z \approx \exp(-\beta E_{\text{shift}}) \times \sum_{l=0}^{L} \frac{1}{l!} \sum_{S_z} \sum_{i} \left\langle \Phi_i^{(N,S_z)} \right| (-\beta\hat{H}_{\text{shift}})^l \left| \Phi_i^{(N,S_z)} \right\rangle, \tag{40}$$

where the summation over l is truncated after $l = L$, $\hat{H}_{\text{shift}} = \hat{H} - E_{\text{shift}}$, and E_{shift} is an energy shift. The value of E_{shift} is chosen so that the Taylor series of the exponential is rapidly convergent. Computationally, the summand can be obtained by acting $-\beta\hat{H}_{\text{shift}}$ recursively on each determinant at a cost per summand comparable to one cycle of the Davidson algorithm with l trial vectors. In this work, we set E_{shift} at the HF energy. A high-temperature approximation to Ξ is defined analogously.

When this approximation is applied to lower temperatures (larger β's), not only does the error in Eq. (40) grow for a given truncation order L, but also the Taylor series oscillates so widely that its numerical evaluation becomes unstable (due to, e.g., floating-point overflows). In addition to using E_{shift}, we, therefore, invoke the imaginary-time splitting scheme to stabilize the calculation. Using the identity,

$$\exp(-\beta\hat{H}_{\text{shift}}) = \{\exp(-\Delta\beta\hat{H}_{\text{shift}})\}^{N_{\text{split}}}, \tag{41}$$

where $\beta = N_{\text{split}}\Delta\beta$, we can rewrite Eq. (40) as

$$Z \approx \exp(-\beta E_{\text{shift}}) \mathrm{Tr}\, \mathbf{G}^{N_{\text{split}}}, \tag{42}$$

$$(\mathbf{G})_{ij} = \sum_{l=0}^{L} \frac{1}{l!} \left\langle \Phi_i^{(N)} \right| (-\Delta\beta\hat{H}_{\text{shift}})^l \left| \Phi_j^{(N)} \right\rangle. \tag{43}$$

The smaller value of $\Delta\beta$ as compared with β makes $\mathbf{G}$ well behaved and more rapidly convergent with L. However, this does not lead to a speedup in the overall calculation because $\mathbf{G}$ needs to be raised to the N_{split}th power and N_{split} is inversely proportional to $\Delta\beta$. In this study, we use the small fixed value of 0.01 a.u. for $\Delta\beta$, which means that N_{split} and thus the computational cost increases with β. For

Table 1 Finite-temperature canonical FCI results for the internal energy U, entropy S, and Helmholtz energy F of FH

T/K	U/E_h	S/k_B	F/E_h
0	−98.596587	0	−98.596587
10^3	−98.596587	0.000000	−98.596587
10^4	−98.596583	0.000107	−98.596587
10^5	−98.178360	2.659028	−99.020425
10^6	−97.372775	3.782186	−109.35026
10^7	−92.851593	4.130703	−223.66334
10^8	−89.265025	4.188853	−1415.7974
10^9	−88.805385	4.189646	−13356.643

Table 2 Finite-temperature grand canonical FCI results for the chemical potential μ, internal energy U, entropy S, and grand potential Ω of FH

T/K	μ/E_h	U/E_h	S/k_B	Ω/E_h
0	–[a]	−98.596587	0	–[a]
10^3	0.124804	−98.596587	0.000000	−99.844630
10^4	0.134719	−98.596583	0.000113	−99.943772
10^5	0.295683	−98.049383	3.474720	−102.10659
10^6	3.859897	−96.945339	4.957688	−151.24440
10^7	46.86892	−92.055572	5.347656	−730.09520
10^8	504.6548	−88.487404	5.405959	−6847.0025
10^9	5091.666	−88.043292	5.406727	−68126.811

[a] See text for the $T = 0$ limit

Table 3 Finite-temperature canonical FCI results for the internal energy U, entropy S, and Helmholtz energy F of N_2

T/K	U/E_h	S/k_B	F/E_h
0	−107.65253	0	−107.65253
10^3	−107.65253	0.000000	−107.65253
10^4	−107.65231	0.007905	−107.65256
10^5	−106.80732	6.187150	−108.76668
10^6	−104.83040	8.924848	−133.09375
10^7	−104.46698	8.987544	−389.08591
10^8	−104.42979	8.988190	−2950.8236
10^9	−104.42606	8.988196	−28568.385

Table 4 Finite-temperature grand canonical FCI results for the chemical potential μ, internal energy U, entropy S, and grand potential Ω of N_2

T/K	μ/E_h	U/E_h	S/k_B	Ω/E_h
0	–[a]	−107.65253	0	–[a]
10^3	−0.077766	−107.65253	0.000000	−106.56381
10^4	−0.086530	−107.65230	0.008092	−106.44114
10^5	0.057735	−106.66945	7.189548	−109.75454
10^6	1.493091	−104.19035	10.47093	−158.25311
10^7	16.02414	−103.50558	10.58350	−663.00342
10^8	161.6121	−103.41861	10.58500	−5718.0610
10^9	1617.533	−103.40964	10.58501	−56269.651

[a] See text for the $T = 0$ limit

a large β, the cost can even exceed that of the exact method involving diagonalization of the Hamiltonian matrix.

4 Results

We applied the finite-temperature FCI and MP0 to the FH, N_2, and F_2 molecules with the STO-3G basis set using the determinant-based implementations described above. Additionally, the MP0 calculations were performed using the semi-analytical expressions based on the Fermi–Dirac statistics. In the N_2 and F_2 calculations, two 1s core orbitals were excluded from FCI or MP0. The bond lengths were 0.9168 Å (FH), 1.0977 Å (N_2), and 1.41193 Å (F_2). The reference wave functions were obtained by the HF calculations at 0 K in all cases.

The results for FCI calculations are given in Tables 1, 2, 3, 4, 5 and 6. The internal energy U, entropy S, and chemical potential μ of FH obtained with grand canonical FCI calculations and its high- and low-temperature approximations, canonical FCI calculations (when applicable), and MP0 calculations (when applicable) are plotted as a function of temperature in Figs. 1, 2 and 3. In the next section, we discuss only the results of FH as a representative example. The numerical results for N_2 and F_2 are recorded as benchmark with the hope of assisting the future developments of finite-temperature many-electron theories.

The $T = 0$ ($\beta = \infty$) limits of μ and thus Ξ and Ω can be inferred mathematically (see below), but not directly obtained numerically; any value of μ satisfies Eq. (14) insofar as it falls between the lowest energy of the eigenstate with more than $\bar{N}$ electrons and the lowest energy of the eigenstate with less than $\bar{N}$ electrons.

5 Discussion

5.1 Internal energy

The internal energy U of FH is plotted as a function of temperature T in Fig. 1. The grand canonical and canonical FCI U agree with the zero-temperature FCI energy at $T = 0$ and increase monotonically with T toward two separate high-temperature limits. That U has such upper bounds is an artifact caused by the finite number (n) of basis functions. The high-temperature limit of U in the canonical

Table 5 Finite-temperature canonical FCI results for the internal energy U, entropy S, and Helmholtz energy F of F_2

T/K	U/E_h	S/k_B	F/E_h
0	−196.04960	0	−196.04960
10^3	−196.04960	0.000000	−196.04960
10^4	−196.04283	0.254178	−196.05088
10^5	−195.69811	3.953301	−196.95004
10^6	−195.09961	4.766222	−210.19335
10^7	−194.97673	4.787265	−346.58056
10^8	−194.96376	4.787489	−1711.0733
10^9	−194.96246	4.787492	−15356.065

Table 6 Finite-temperature grand canonical FCI results for the chemical potential μ, internal energy U, entropy S, and grand potential Ω of F_2

T/K	μ/E_h	U/E_h	S/k_B	Ω/E_h
0	–[a]	−196.04960	0	–[a]
10^3	0.019696	−196.04960	0.000000	−196.40413
10^4	0.031682	−196.04282	0.254383	−196.62115
10^5	0.415872	−195.56021	4.841924	−204.57925
10^6	5.653602	−194.69582	5.989723	−315.42900
10^7	61.03663	−194.46728	6.027850	−1484.0175
10^8	615.6378	−194.44024	6.028318	−13184.977
10^9	6161.741	−194.43748	6.028323	−130196.37

[a] See text for the $T = 0$ limit

ensemble is the average of all N-electron eigenvalues of the Hamiltonian. The corresponding limit in the grand canonical ensemble is not so simple because of the nonzero chemical potential in that limit, which will be discussed below. A visible increase in U occurs at around $0.1\,E_\mathrm{h}$ or 30,000 K. This may be compared with the lowest excitation energy of $0.43\,E_\mathrm{h}$ and the lowest ionization energy of $0.41\,E_\mathrm{h}$ in this molecule.

The low-temperature approximations to U are systematically more accurate at a given temperature or remain accurate up to higher temperature with increasing the truncation parameter M. Note that $M = 1$ means that the calculation includes the single lowest state in each category with the same N and S_z and hence more than one states in Eqs. (35) or (37); this is why U in this approximation varies with temperature.

The high-temperature approximations are also systematically more accurate with the truncation parameter L, but being truncated Taylor series, they tend to become unstable abruptly at one point. The imaginary-time splitting scheme renders stability to this approximation and gives accurate estimates of U in the wide range of temperature from $10^4\,\mathrm{K}$ to $10^7\,\mathrm{K}$ with just

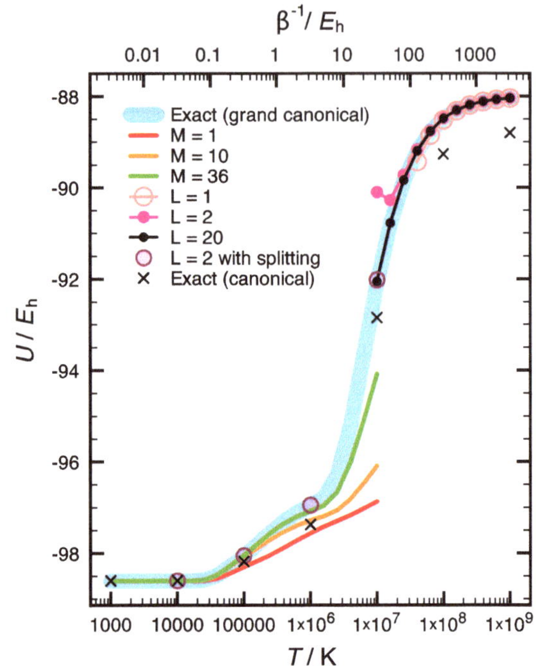

Fig. 1 The internal energy (U) of hydrogen fluoride calculated with grand canonical FCI/STO-3G and its various approximations as a function of temperature (T). M is the number of states (in each group of determinants with the same number of electrons and magnetic spin quantum number) included in the low-temperature approximation. L is the number of terms in the Taylor series expansion in the high-temperature approximation, where the imaginary-time splitting is not used unless otherwise noted. The plot of U from canonical FCI/STO-3G is also superimposed

$L = 2$. However, for this tiny problem, this scheme is in fact more expensive at lower temperatures than FCI itself and its practical utility is still questionable.

5.2 Entropy

The entropy S shows similar temperature dependence as U. It increases monotonically from zero at $T = 0$ (Nernst's law) toward two separate high-temperature limits in the grand canonical and canonical ensembles. Again, the upper boundedness of S is an artifact of the finite-basis-set calculations. The onset ($0.1\,E_\mathrm{h}$) of the rapid increase in S coincides (in the order of magnitude) with the lowest excitation energy. The value of S in the grand canonical ensemble at a given temperature is always greater (much greater at higher temperatures) than the corresponding value in the canonical ensemble.

The high-temperature limit of S in the canonical ensemble is simply $k_\mathrm{B} \ln {}_{2n}C_N$ where ${}_{2n}C_N$ is the total number of N-electron states. This formula can be derived rigorously from Eqs. (7) and (8):

$$\lim_{\beta \to 0} \hat{W} = \frac{1}{Z}, \qquad (44)$$

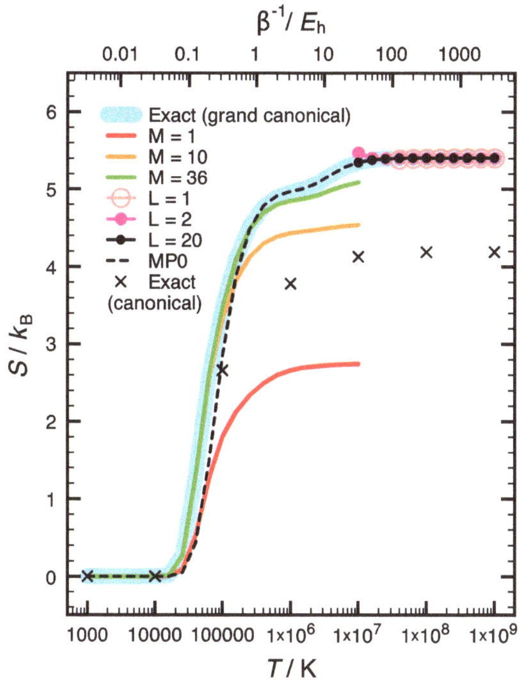

Fig. 2 The entropy (S) of hydrogen fluoride calculated with grand canonical FCI/STO-3G and its various approximations including MP0/STO-3G as a function of temperature (T). M is the number of states (in each group of determinants with the same number of electrons and magnetic spin quantum number) included in the low-temperature approximation. L is the number of terms in the Taylor series expansion in the high-temperature approximation, where the imaginary-time splitting is not used. The plot of S from canonical FCI/STO-3G is also superimposed

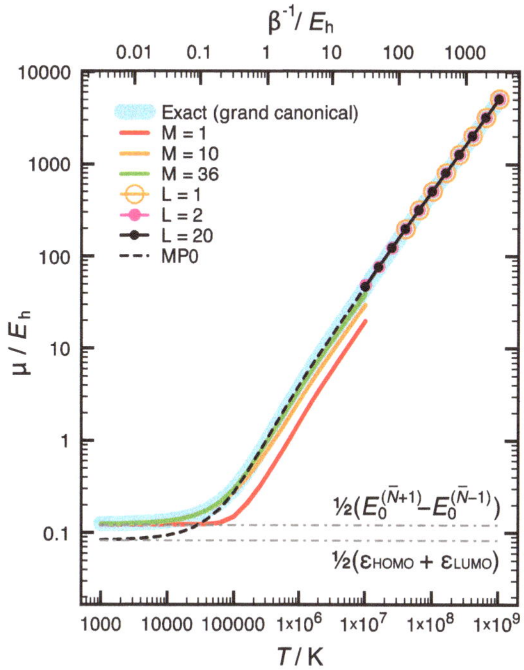

Fig. 3 The chemical potential (μ) of hydrogen fluoride calculated with grand canonical FCI/STO-3G and its various approximations including MP0/STO-3G as a function of temperature (T). M is the number of states (in each group of determinants with the same number of electrons and magnetic spin quantum number) included in the low-temperature approximation. L is the number of terms in the Taylor series expansion in the high-temperature approximation, where the imaginary-time splitting is not used

where Z, in this limit, reduces to the number of N-electron states, which is ${}_{2n}C_N$.

$$\lim_{\beta\to 0} S = -k_B \mathrm{Tr}\left[\frac{1}{Z}\ln\frac{1}{Z}\right] = k_B \ln Z = k_B \ln {}_{2n}C_N, \quad (45)$$

where we have used $\mathrm{Tr}[1] = Z$. For FH, substituting $N = 10$ and $n = 6$, we obtain $S = k_B \ln 66 = 4.18965 k_B$ at $T = \infty$.

The high-temperature limit of S in the grand canonical ensemble is not the simple logarithm of the total number of states. To evaluate it, let us first write the high-temperature limit of Ξ as

$$\lim_{\beta\to 0} \Xi = \sum_{N=0}^{2n} {}_{2n}C_N \exp(\beta\mu N) \quad (46)$$

$$= \{1 + \exp(\beta\mu)\}^{2n} \equiv \Xi_{\beta=0}, \quad (47)$$

where ${}_{2n}C_N$ is the number of states with N electrons (the number density). In the above, we have been able to get rid of $\hat{H}$ in the definition of Ξ (Eq. 9) since it is multiplied by β, which is zero in the limit. We do not, however, eliminate $\beta\mu N$ because μ behaves like β^{-1} in this limit, rendering the product $(\beta\mu)$ nonzero even at $\beta \to 0$. In the last equality, we have used the well-known identity for a binomial distribution:

$$(a+b)^m = \sum_{i=0}^{m} {}_mC_i a^i b^{m-i}. \quad (48)$$

Equation (47) agrees with the grand partition function of the Fermi–Dirac statistics in the $\beta = 0$ limit (Eq. 20). This is expected because the energies of the states no longer matter in this limit and Ξ is determined completely by the numbers of N-electron states, which are the same between FCI and MP0.

We can then analytically evaluate the average number of electrons $\bar{N}$ in the high-temperature limit as

$$\lim_{\beta\to 0} \bar{N} = \frac{1}{\beta}\frac{\partial}{\partial\mu}\ln \Xi_{\beta=0} \quad (49)$$

$$= \frac{2n\exp(\beta\mu)}{1+\exp(\beta\mu)}. \quad (50)$$

This equation should be viewed as the condition by which μ in the high-temperature limit is determined, which yields

$$\lim_{\beta\to 0} \mu = \frac{1}{\beta}\ln\frac{\bar{N}}{2n - \bar{N}}. \quad (51)$$

This justifies our earlier premise that μ is asymptotically inversely proportional to β in that limit.

Substituting these into Eqs. (7) and (16), we find

$$\lim_{\beta\to 0} W' = \frac{\exp(\beta\mu\hat{N})}{\Xi_{\beta=0}}, \tag{52}$$

and

$$\lim_{\beta\to 0} S = -k_B \sum_{N=0}^{2n} {}_{2n}C_N \frac{\exp(\beta\mu N)}{\Xi_{\beta=0}} \ln \frac{\exp(\beta\mu N)}{\Xi_{\beta=0}} \tag{53}$$

$$= -k_B\beta\mu\bar{N} + k_B \ln \Xi_{\beta=0} \tag{54}$$

$$= k_B\{2n\ln(2n) - \bar{N}\ln\bar{N} - (2n-\bar{N})\ln(2n-\bar{N})\}. \tag{55}$$

It is also straightforward to show that it is equal to

$$\lim_{\beta\to 0} S = -k_B 2n\{f\ln f + (1-f)\ln(1-f)\}, \tag{56}$$

where $f = \{1+\exp(-\beta\mu)\}^{-1}$ is the high-temperature limit of the Fermi–Dirac distribution function f_p (Eq. 25). For FH, with $\bar{N} = 10$ and $n = 6$, we obtain the $T = \infty$ limit of S as 5.40673 k_B (see Table 2).

The low-temperature approximations of S with a small number (M) of states are rather poor, which is understandable given that S is the measure of the number of available states. The approximations, however, rapidly improve with increasing S and, with $M = 36$, the agreement is accurate up to 1 E_h or 300,000 K. The high-temperature approximations work well where S is nearly constant (above 10^7 K), but break down at the first sign of variation in S.

Equation (47) explains why MP0 S_0 reproduces grand canonical FCI S at both $T = 0$ and $T = \infty$ limits. In the $T = 0$ limit, $S = S_0 = 0$ according to Nernst's theorem. In the $T = \infty$ limit, according to Eq. (47), the grand partition function and thus S and μ depend solely on the number density ${}_{2n}C_N$ and not on state energies. Further, the number density is common between grand canonical FCI and MP0, which explains their same $T = \infty$ limits. In the intermediate temperatures, they differ from each other.

5.3 Chemical potential

The temperature dependence of μ becomes large at the same order of magnitude as the lowest excitation energy (0.1 E_h or 30,000 K) and is linear with temperature at high temperatures. This behavior has already been explained by Eq. (51). This equation also indicates that whether μ increases or decreases with temperature is determined by the numbers of basis functions (n) and electrons ($\bar{N}$) and is, therefore, another computational artifact. When $\bar{N} > n$, μ increases with temperature; otherwise, μ decreases. In physical reality or in a realistic calculation where $n \gg \bar{N}$, μ is expected to display the $-\ln(2n)/\beta$ behavior.

MP0 μ_0 having the same high-temperature asymptote as grand canonical FCI μ is also expected from the equivalence of independent-particle and non-independent-particle models in the high-temperature limit.

The zero-temperature limits of MP0 μ_0 and grand canonical FCI μ differ from each other. The values of these quantities exactly at $T = 0$ are undefined. This can be understood easily by inspecting the functional form of the Fermi–Dirac distribution function f_p in the case of MP0. It is a Heaviside step function with the step occurring at μ_0. Insofar as μ_0 falls between the energies of the highest-occupied (ϵ_{HOMO}) and lowest-unoccupied (ϵ_{LUMO}) orbitals, the physical description of the MP0 wave function at $T = 0$ is the same. Therefore, μ_0 is indeterminate within that range. A similar situation applies to μ; it is indeterminate in the range between the lowest $(\bar{N}+1)$-electron-state energy ($E_0^{(\bar{N}+1)}$) and the lowest $(\bar{N}-1)$-electron-state energy ($E_0^{(\bar{N}-1)}$) if they are the lowest-energy states with more than $\bar{N}$ electrons and less than $\bar{N}$ electrons.

This is not to say that the limits of μ_0 and μ as we approach $T = 0$ from above are not known. They are indeed the midpoints of the aforementioned respective ranges:

$$\lim_{\beta\to\infty} \mu = \frac{E_0^{(\bar{N}+1)} - E_0^{(\bar{N}-1)}}{2}, \tag{57}$$

$$\lim_{\beta\to\infty} \mu_0 = \frac{\epsilon_{\mathrm{HOMO}} + \epsilon_{\mathrm{LUMO}}}{2}, \tag{58}$$

assuming that there is no degeneracy in the states involved and the lowest-energy state with more than $\bar{N}$ electrons has $\bar{N}+1$ electrons and that with less than $\bar{N}$ electrons has $\bar{N}-1$ electrons. This can be understood from the following argument. In the $\beta = \infty$ limit, only these lowest-energy states need to be considered in evaluating the average number of electrons and determining μ:

$$\lim_{\beta\to\infty} \bar{N} = \frac{(\bar{N}+1)\exp[-\beta\{E_0^{(\bar{N}+1)} - \mu(\bar{N}+1)\}] + (\bar{N}-1)\exp[-\beta\{E_0^{(\bar{N}-1)} - \mu(\bar{N}-1)\}]}{\exp[-\beta\{E_0^{(\bar{N}+1)} - \mu(\bar{N}+1)\}] + \exp[-\beta\{E_0^{(\bar{N}-1)} - \mu(\bar{N}-1)\}]}. \tag{59}$$

We do not need to consider $\bar{N}$-electron states because they cannot shift the average number of electrons. It is straightforward to confirm that this equation is satisfied by Eq. (57). Equation (58) then follows from

$$E_0^{(\bar{N}+1)} = E_0^{(\bar{N})} + \epsilon_{\mathrm{LUMO}}, \tag{60}$$

$$E_0^{(\bar{N}-1)} = E_0^{(\bar{N})} - \epsilon_{\mathrm{HOMO}}, \tag{61}$$

in MP0.

Both low- and high-temperature approximations are rapidly convergent at the exact results for μ for a wide range of temperatures. The high accuracy of the low-temperature approximations may be at least partly due to the fact that only a few lowest-energy states dictate the value of μ.

Acknowledgments We thank U.S. Department of Energy Scientific Discovery through Advanced Computing (SciDAC) program (DE-FG02-12ER46875) for financial support. S. H. is a Camille Dreyfus Teacher-Scholar and a Scialog Fellow of Research Corporation for Science Advancement.

References

1. Shavitt I (1977) Methods of electronic structure theory. Plenum, New York, p 189
2. Lischka H, Shepard R, Brown FB, Shavitt I (1981) Int J Quantum Chem S15:91
3. Shavitt I (1998) Mol. Phys. 94:3
4. Lischka H, Müller T, Szalay PG, Shavitt I, Pitzer RM, Shepard R (2011) Wiley Interdiscip Rev Comput Mol Sci 1:191
5. Shavitt I, Bartlett RJ (2009) Many-body methods in chemistry and physics. Cambridge University Press, Cambridge
6. Shavitt I (1985) Tetrahedron 41:1531
7. Schaefer HF (1986) Science 231:1100
8. Sherrill CD, Schaefer HF (1999) Adv Quantum Chem 34:143
9. Roos B (1972) Chem Phys Lett 15:153
10. Shavitt I, Bender CF, Pipano A, Hosteny RP (1973) J Comput Phys 11:90
11. Davidson ER (1975) J Comput Phys 17:87
12. Siegbahn PEM (1979) J Chem Phys 70:5391
13. Siegbahn PEM (1980) J Chem Phys 72:1647
14. Handy NC (1980) Chem Phys Lett 74:280
15. Knowles PJ, Handy NC (1984) Chem Phys Lett 111:315
16. Olsen J, Roos BO, Jørgensen P, Jensen HJA (1988) J Chem Phys 89:2185
17. Knowles PJ (1989) Chem Phys Lett 155:513
18. Olsen J, Jørgensen P, Simons J (1990) Chem Phys Lett 169:463
19. Harrison RJ (1991) J Chem Phys 94:5021
20. Martinez TJ, Mehta A, Carter EA (1992) J Chem Phys 97:1876
21. Gan ZT, Alexeev Y, Gordon MS, Kendall RA (2003) J Chem Phys 119:47
22. Bytautas L, Ruedenberg K (2004) J Chem Phys 121:10905
23. Shepard R (2005) J Phys Chem A 109:11629
24. Ohtsuka Y, Nagase S (2008) Chem Phys Lett 463:431
25. Booth GH, Thom AJW, Alavi A (2009) J Chem Phys 131:054106
26. Bytautas L, Henderson TM, Jiménez-Hoyos CA, Ellis JK, Scuseria GE (2011) J Chem Phys 135:044119
27. Bauschlicher CW, Langhoff SR, Taylor PR, Handy NC, Knowles PJ (1986) J Chem Phys 85:1469
28. Bauschlicher CW, Taylor PR (1986) J Chem Phys 85:6510
29. Bauschlicher CW, Langhoff SR (1988) J Chem Phys 89:4246
30. Evangelisti S, Bendazzoli GL, Gagliardi L (1994) Chem Phys 185:47
31. Olsen J, Jørgensen P, Koch H, Balkova A, Bartlett RJ (1996) J Chem Phys 104:8007
32. Christiansen O, Koch H, Jørgensen P, Olsen J (1996) Chem Phys Lett 256:185
33. Dutta A, Sherrill CD (2003) J Chem Phys 118:1610
34. Thøgersen L, Olsen J (2004) Chem Phys Lett 393:36
35. Sherrill CD, Piecuch P (2005) J Chem Phys 122:124104
36. Verdicchio M, Bendazzoli GL, Evangelisti S, Leininger T (2013) J Phys Chem A 117:192
37. Giner E, Bendazzoli GL, Evangelisti S, Monari A (2013) J Chem Phys 138:074315
38. Šimová L, Řezáč J, Hobza P (2013) J Chem Theor Comput 9:3420
39. Matsubara T (1955) Prog Theor Phys 14:351
40. Thouless DJ (1957) Phys Rev 107:1162
41. Bloch C, de Dominicus C (1958) Nucl Phys 7:459
42. Kohn W, Luttinger JM (1960) Phys Rev 118:41
43. Luttinger JM, Ward JC (1960) Phys Rev 118:1417
44. Balian R, Bloch C, de Dominicus C (1961) Nucl Phys 25:529
45. Sanyal G, Mandal SH, Guha S, Mukherjee D (1993) Phys Rev E 48:3373
46. March NH, Young WH, Sampanthar S (1967) The many-body problem in quantum mechanics. Cambridge University Press, London
47. Fetter AL, Walecka JD (1971) Quantum theory of many-particle systems. McGraw Hill, New York
48. Mattuck RD (1992) A guide to Feynman diagrams in the many-body problem. Dover, New York
49. Hirata S, He X (2013) J Chem Phys 138:204112
50. He X, Ryu S, Hirata S (2014) J Chem Phys 140:024702
51. Hirata S, Bartlett RJ (2000) Chem Phys Lett 321:216

Theor Chem Acc (2014) 133:1564
DOI 10.1007/s00214-014-1564-8

REGULAR ARTICLE

Massively parallel spin–orbit configuration interaction

Jeffrey L. Tilson · Walter C. Ermler

Received: 31 January 2014 / Accepted: 7 August 2014 / Published online: 22 August 2014

Abstract An implementation of a massively parallel spin–orbit configuration interaction (PSOCI) method is described. This is an extension of a conventional CI method that explicitly includes one-electron spin–orbit operators and certain scalar relativistic effects extracted from relativistic effective core potentials. The performance of the PSOCI code is analyzed on several large-scale computing platforms.

Keywords Spin–orbit coupling · Configuration interaction (CI) · Parallel spin orbit CI (SOCI) · Parallel computing · Eigenanalysis · Global arrays

1 Introduction

Having application codes available that can be used to calculate spectroscopic and structural properties of a range of electronic states of heavy-element-containing molecules and that treat electron correlation, spin–orbit coupling and other relativistic effects in a manner that allows for systematic improvements is invaluable to their detailed understanding. Accurate theoretical models of such systems can be challenging for many reasons. For example, these systems generally have many low-lying valance electronic states in the vicinity of the ground state in many of the formal oxidation states of interest [1–3]. Furthermore, the valence molecular orbital space of such molecules becomes large and diffuse, making it challenging to assign orbitals to valence or core regions. Also, spin–orbit coupling effects can have a large impact on the nature of bonding and electronic excitations in these systems.

The ability to treat these molecular systems at a high level of accuracy requires large-scale computation. This problem is addressed in ab initio spin–orbit configuration interaction (SOCI) calculations. The massively parallel SOCI (PSOCI) code described here uses the so-called *one-step* procedure, a double-group symmetry-driven method where an effective spin–orbit operator is explicitly included in the Hamiltonian operator and evaluated in the context of the complete configuration interaction expansion (see Sect. 2). It has been argued that a determinant-based SOCI procedure is not necessarily more computationally expensive than a configuration state function (CSF)-based approach in contrast to non-relativistic methods [4]. However, notwithstanding the possibility that the determinant method can be somewhat more expensive, this method maintains the ability to utilize highly flexible lists of configurations and will continue to be applicable in addition to more efficient methods such as the graphical unitary group approach (GUGA). The PSOCI method differs from the *two-step* procedure of Dolg et al. [5, 6] and the more compact approach of Balasubramanian [7] in which the spin–orbit contributions are acquired through a limited CI expansion and incorporated into the results of a larger non-spin–orbit CI.

The present implementation of the SOCI method, an extension of the PSOCI code of Tilson et al. [8], has as its goal the processing of *H*-matrices of upwards of

Dedicated to the memory of Professor Isaiah Shavitt and published as part of the special collection of articles celebrating his many contributions.

J. L. Tilson
Renaissance Computing Institute, University of North Carolina at Chapel Hill, Chapel Hill, NC 27517, USA
e-mail: jtilson@renci.org

W. C. Ermler (✉)
University of Texas at San Antonio, San Antonio, TX 78249, USA
e-mail: walter.ermler@utsa.edu

50,000,000–500,000,000 double-group symmetry-adapted configurations on state-of-the-art parallel supercomputers. Since it is based on the use of an electron configurations list, the ability to carry out SOCI calculations that include specific electron configurations permits the detailed study of the importance of electron correlation in such situations as curve crossings, Rydberg states, mixed V states, ion pair states, etc. The authors and others have published several studies on these cases, several of which are referred to below.

Alekseyev et al. [9] have also developed code based on the SOCI method according to Pitzer and Winter [10] for which calculations of up to 1,000,000 double-group symmetry-adapted configurations have been reported. These authors concluded that the SOCI method is capable of accurate predictions of structures and spectra of heavy-element-containing molecules.

2 Background

Procedures based on the use of relativistic effective core potentials (RECPs) that have been derived from all-electron ab initio numerical Dirac–Fock (DF) atomic wave functions can be used to reduce the number of electrons that must be treated explicitly. As a result, the number of electrons can be small enough that, in many instances, near-complete basis set, multi-reference singles plus doubles excitations (MRSD), SOCI calculations incorporating spin–orbit coupling operators and other relativistic effects become feasible [7, 11–13]. RECPs and Gaussian basis sets for representing valence electrons have been reported for all of the elements from Li through Element 118 by Christiansen, Ermler et al. [14–20]. The RECP method is grounded in the fundamental tenet that atomic core and valence electrons can be formally distinguished and, in fact, treated rigorously from first principles as identifiable sets. In addition to distinguishing between atomic core and valence electrons, the effects of relativity should also be incorporated into ECPs, especially for heavy elements. The high nuclear charge in heavy elements leads to tightly bound inner shell electrons whose confinement increases their average kinetic energy such that relativistic effects are large. Relativistic effects on the valence electrons include (1) radial contraction and energetic stabilization of s and $j = l-1/2$, $l > 0$, atomic orbitals, often leading to bond shortening; (2) a concomitant radial expansion and energetic destabilization of $j = l+1/2$ atomic orbitals, usually with somewhat smaller chemical effects, and (3) spin–orbit coupling of the p, d, f and higher angular momentum orbitals into sublevels. As first shown by Lee et al. [21], the use of DF-based RECPs permits the incorporation of all of these effects into molecular electronic structure calculations through a one-electron operator that is added to the Hamiltonian operator for the valence electrons, and at the same time, reduces the computational problem such that only the valence electrons require an explicit theoretical treatment in the context of valence-subshell-only basis sets. The RECPs utilized in this report are due to Lee, Ermler, Christiansen and Pitzer (LECP) [21, 22].

The LECP RECPs are defined using shape-consistent, nodeless two-component pseudospinors [22] extracted from numerical two-component DF atomic spinors [23]. These pseudospinors are used to define RECPs that are expressed as radially local one-electron operators in the context of atomic jj-coupling,

$$U^{\mathrm{RECP}} = U_{LJ}^{\mathrm{RECP}}(r) + \sum_{l=0}^{L-1} \sum_{j=|l-1/2|}^{l+1/2} \sum_{m=-j}^{j} \left[U_{lj}^{\mathrm{RECP}}(r) - U_{LJ}^{\mathrm{RECP}}(r)\right] |ljm><ljm|. \quad (1)$$

In Eq. (1), the projection operators $|ljm><ljm|$ dictate that U^{RECP} be used in the context of a basis set of two-component Dirac spinors. U_{LJ}^{RECP} refers to the so-called residual RECP term, where L and J are taken as one larger than the largest angular momentum quantum numbers of the core electrons. These RECPs are calculated in numerical form, but are re-expressed as expansions in Gaussian-type orbitals (GTOs) to facilitate their use in molecular electronic structure codes that employ GTO basis sets for representing the valence electrons.

The RECPs are recast into the form of angular momentum-averaged relativistic effective core potentials (ARECPs) that may be used in standard non-relativistic electronic structure procedures based on atomic LS-coupling.

$$U^{\mathrm{ARECP}} = U_{L}^{\mathrm{ARECP}}(r) + \sum_{l=0}^{L-1} \sum_{m=-l}^{l} \left[U_{l}^{\mathrm{ARECP}}(r) - U_{L}^{\mathrm{ARECP}}(r)\right] |lm><lm| \quad (2)$$

with

$$U_{l}^{\mathrm{ARECP}} = (2l+1)^{-1}\left[lU_{l,l-1/2}^{\mathrm{RECP}} + (l+1)U_{l,l+1/2}^{\mathrm{RECP}}\right]. \quad (3)$$

An important consequence of this averaging is the outgrowth of a spin–orbit operator [24], which was recast into the following form [10]

$$H^{\mathrm{SO}} = \boldsymbol{s} \cdot \sum_{\mathrm{l=1}}^{\mathrm{L}} H_{l}^{\mathrm{SO}} \sum_{m'=-l}^{l} \sum_{m=-l}^{l} |lm'><lm'|\boldsymbol{l}|lm><lm| \quad (4)$$

where $\boldsymbol{s}$ and $\boldsymbol{l}$ are the spin and orbital angular momentum vector operators and

$$H_{l}^{\mathrm{SO}} = [2/(2l+1)]\Delta U_{l}(r) \quad (5)$$

with

$$\Delta U_l(r) = U^{\text{RECP}}_{l,l+\frac{1}{2}}(r) - U^{\text{RECP}}_{l,l-\frac{1}{2}}(r). \tag{6}$$

The Hamiltonian matrix used in the SOCI method adopted here is constructed in terms of matrix elements involving the U^{ARECP} and H^{SO} operators and determinantal wave functions. These wave functions are comprised of linear combinations of atomic orbitals (viz. contracted GTOs) molecular orbitals (LCAO–MOs) for valence electrons and are defined through standard *LS*-coupling-based multi-configuration-consistent field (MCSCF) procedures.

The Hamiltonian operator for valence electrons i and atoms c with net charge on the nucleus $Z_c' = Z_c - Z_{c,\text{core}}$ that is used to drive the MCSCF step of a calculation is (in a.u.),

$$H^{\text{val}} = \sum_i \left(-\frac{1}{2}\nabla_i^2 - \sum_c Z_c'/r_{ci} \right) + \sum_{i>j} \frac{1}{r_{ij}} + \sum_c U_c^{\text{ARECP}} + \sum_{c>d} Z_c' Z_d'/\text{r}_{cd}. \tag{7}$$

The resulting wave functions are eigenfunctions that transform as irreducible representations of the molecular point group. Thus, existing ab initio molecular structure codes can be used to calculate these LCAO–MO determinantal wave functions.

Generation of the PSOCI solutions requires atomic orbitals (AOs) and molecular orbitals (MOs) from which the configuration space may be generated and within which the solutions are obtained. The AOs are used for subsequent analyses such as a Mulliken population analysis [25]. For the purposes of PSOCI, we typically orthogonalize the AOs using coefficients from an MCSCF procedure. These orbitals are all one-component orbitals generated without any spin–orbit coupling effects, although the ARECPs are included in the MCSCF procedure.

Pitzer and Winter [10] defined a SOCI procedure in the context of double-group theory that leads to molecular wave functions that exhibit the proper intermediate angular momentum coupling for an electronic state. In this method, matrix elements with respect to H^{SO} are added to those for H^{val} to form a total Hamiltonian matrix from which SOCI wave functions and energies are extracted as eigensolutions. The SOCI wave function for state K representing the valence electrons is

$$\Psi_K = \sum C_I D_I \tag{8}$$

where the sum is over all Slater determinants D_I that contribute to the double-group symmetry of the electronic state of interest. Each D_I is composed of LCAO–MOs. The intermediate angular momentum coupling inherent in the molecular SOCI wave function is coded into the CI coefficients of Eq. (8) by virtue of the presence of H^{SO} matrix elements in each element of the Hamiltonian matrix. That is, the final SOCI wave function is an eigenfunction of the total electron angular momentum operator and transforms according to one of the irreducible representations of the molecular double group.

3 Implementation

The PSOCI code is a highly scalable version of the algorithm that implements the previously described mathematical framework as suggested by Pitzer and Winter [10] and developed for parallel computers [8]. PSOCI is written in C++ using a single program multiple data (SPMD) approach where the program utilizes a data-parallel like, loosely coupled, procedure. It employs Global Arrays [26] (GAs) for managing the distributed objects such as the Hamiltonian matrix and SOCI eigenvectors. The GA disk resident array (DRA) subsystem, based on ChemIO [27], is leveraged for the efficient (optional) parallel movement of large data objects to disk. A parallel Davidson diagonalization procedure [28–30] is implemented in the base code with several layers of approximations incorporated, such as multiple (weaker) checks on convergence [30], and multistep thresholds. In particular, all vectors are first converged to a weaker set of convergence criteria. This is followed by an iterative restart and subsequent convergence to the final acceptance criteria based on energy changes and step sizes. The number of restarts may be selected by the user. In each cycle, the assumption is made that once a vector is converged, no retesting is performed. Improvement of the next vector is then begun. Of course, root swapping and other effects will occur that potentially degrade the quality of prior estimates. However, this is a problem primarily limited to the first (lower quality) iterative cycle. Prior to beginning a subsequent cycle with its concomitant tighter convergence criteria, all residuals are recomputed and every vector is re-inspected and improved if necessary. A final complete check is generated upon completion of the diagonalization. Adopting this approach allows the algorithm to defer many potential orthogonalization and residual computations, providing better overall runtime. For difficult to converge cases, the user may select an algorithm that performs more rigorous checking at the expense of greater runtime.

The SOCI procedure incorporates the $U^{\text{ARECP}}s$ and $H^{\text{SO}}s$ of Eqs. (2) and (4). This permits the incorporation of core-electron scalar relativistic effects and valence SO integrals prior to the SOCI procedure using standard non-relativistic codes. The valence electrons that remain are represented by atom-centered, contracted Gaussian-type basis functions (CGTOs). Of course, the use of these operators and basis sets entails certain approximations. First, the use of an ARECP precludes relaxation of the outermost electrons incorporated into the ARECP core with changes in the valence region. The use of a polarizing potential technique to account for this interaction has been proposed and will be incorporated into future releases of our codes [31]. Alternatively, so-called small-core RECPs can be used to account for such effects in a direct manner. Second, certain contributions to the relativistic corrections are not present. These include the Breit and Gaunt interactions, QED effects, and direct valence–valence electron relativistic effects. Most of these excluded terms are quite small and are comparable with bounds considered to be within chemical accuracy (~1 kcal/mol). The *one-step* SOCI technique used here is one of the few techniques where an appreciable amount of scalar relativistic and spin–orbit coupling effects can be incorporated into atomic and molecular calculations while also including large amounts of electron correlation.

The initial step of the SOCI procedure constructs a list of electron configurations using the COLUMBUS program *cgdbg* [32]. This list includes all molecular orbital electron configurations (occupations) consistent with the chosen double-group representation. These spatial configurations are processed as double-group symmetry-adapted functions as part of the SOCI procedure. Only the explicitly treated valence electrons are included in this list. The next step is the construction of MO integrals. These are generally derived using standard (non-SO coupled) multiconfiguration self-consistent field (MCSCF) methods, with the core-valance electron relativistic effects accounted for by the ARECP operator. Generating the MO integrals is a multistep process. In the following, specific programs that are part of the COLUMBUS Suite of Quantum Chemistry Programs [33–36] are indicated parenthetically. It is noted, however, that the PSOCI code is not strongly tethered to any one particular MO generating system. The process begins by computing the basis function integrals with respect to the valence Hamiltonian operator that includes the U^{ARECP} and the H^{SO} operators (*argos*), followed by the determination of optimized molecular orbitals from an MCSCF procedure (*mcscf*) to calculate the MO coefficients. Using these MO coefficients, the AO integrals are transformed into integrals with respect to the MOs (*tran*). These transformed integrals (from *tran*) and the list of configurations (from *cgdbg*) are supplied to our PSOCI code for use in the calculation of SOCI wave functions and energies for a desired number of double-group symmetry electronic states. The resultant SOCI wave functions and density matrices and natural orbitals calculated therefrom are subsequently used to calculate one- electron properties (*prop, exptvl*) and electric dipole transition moments (*tranmom*) [37].

A PSOCI calculation consists of two broad and distinct phases. Phase I is the computation of the Hamiltonian matrix. H (and associated indexing objects), which is stored into memory as large distributed arrays. The data are stored in a customized compressed sparse matrix data format that is distributed across processors. Only nonzero H-matrix elements and associated indexing information are retained in these distributed objects. The construction of a matrix element H_{ij} is a simple configuration-driven process with each core managing complete rows of H. A (spatial) configuration specifies a unique electron occupation pattern; that is, all double-group symmetry-adapted functions arising from a configuration are treated as a block of work. Individual blocks indexed by pairs of spatial configurations are referenced in order to generate the H-matrix. All valid spin projections are processed subject to differences in the number of excitations between configurations and the number of permutations to align configurations. MO integral magnitudes are used for screening when applicable. Checks for interactions that are identically zero are performed. Once an entire row is computed, the rows are uploaded to distributed storage using GA put operations. Computational load is balanced either using a simple shared counter style approach based on the GA read and increment method or using a static work method when using massive numbers of processors. The user may choose to pre-calculate all double-group-adapted functions, store them into a GA object, and communicate to cores as needed or simply process the much smaller configuration data as needed. Once constructed, the entire H-matrix (and associated objects) are (optionally) stored to disk for restart purposes. Generally, data from all cores on a node are copied to one file, though this granularity can be modified. Phase II involves calculating multiple eigenvectors of the H-matrix for the desired number of roots (electronic states). A parallel Davidson method is currently employed [30]. Phases I and II are both computationally demanding (see below).

The H-matrix construction proceeds by each node generating disjoint sub-blocks of the double-group-indexed matrix and storing this into a GA data space. Transfer of data can be fast to local memory and slower to non-local memory requiring appropriate data locality considerations in the algorithm. The algorithm begins by creating as large a GA object as possible given the amount of available distributed memory. At the end of the H-matrix construction phase a complete, packed sparse matrix is stored in distributed memory using GAs on disk for restarting or archiving. The H-matrix construction is a procedure that uses blocks over spatial (occupation) configurations in the context of the symmetries identified in the SOCI method [10]. This results in a reasonably fast construction time, but renders the matrix in some arbitrary order that may not be optimum for the subsequent diagonalization step.

The Phase II iterative diagonalization run time depends both on code performance and on the number of iterations required to achieve convergence to a desired tolerance level. We are actively exploring better ways to load balance work within this phase when using massive numbers of processors as imbalances and sparse matrix bandwidth effects become important. Static work agglomeration schemes analogous to so-called supertasking procedures developed for parallel Fock matrix construction [38] are being explored. As a consequence parallel performance is poorer than in Phase I.

The inter-process communication bandwidth for the matrix–vector product is the fundamental problem for Phase II. For cases where the CI coefficient vector may be replicated, the performance is sufficient to support scientific research. However, when the CI coefficient vector must be distributed across multiple processors, the sparse matrix bandwidth, exacerbated by the spatial-configuration generation scheme of H, results in poorer performance in the diagonalization step. We are currently seeking to reduce the H bandwidth using (reverse) CutHill–McKee techniques [39, 40] implemented for massively parallel distributed data based on the Global Arrays library. The process for reordering the H-matrix, is expected to take about the same amount of time as to actually generate it. However, this additional time is expected to result in significantly decreased diagonalization computation times. The functionality for computing blocks of the H-matrix in line and retaining only matrix–vector products as part of the iterative diagonalization has been incorporated into the code providing an option for not storing the H-matrix in memory. However, for this implementation within PSOCI, the sparse matrix bandwidth is exposed for small problems, rendering the process fairly slow relative to the H-in-memory methods.

As in a conventional CI method, PSOCI must address the storage of the entire Hamiltonian matrix. The H-matrix is sparse but still large enough that distributed memory is required. Specifically, for storage of the H-matrix information, a sparse matrix format is adopted that stores three objects. Symmetry allows storage of the lower triangle of all matrix objects. The store objects consist of the matrix of H elements, a matrix of column entries per row, and a vector of the number of row elements. The GA distributed memory space has a challenging characteristic; that is, one cannot dynamically allocate and de-allocate memory from individual rows. Thus, once allocated, the global H-matrix is fixed in size. For problems of interest to our research group, we have explored estimated sparsity predictors to minimize global storage.

In Fig. 1 the number of nonzero interactions for each double group is displayed for the H-matrix for RuO^{2+} for a valence full CI plus singles (VFCI-S) MRS SOCI

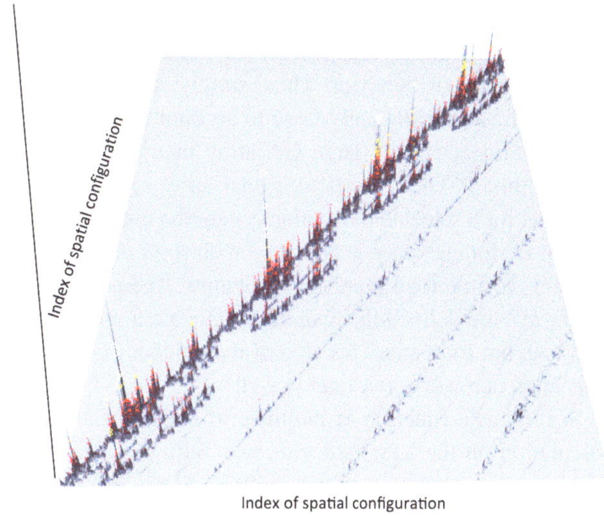

Fig. 1 Depiction of the SOCI Hamiltonian matrix for the RuO^{2+} MRS calculation. This *plot* indicates the number of nonzero double-group configuration interaction terms (height) versus the specified configuration pair (base). The total number of spatial configuration is 607,965, and the total double-group symmetry-adapted function is 6,078,210. The four J symmetries (1, 3, 5, 7) that mix for this problem are evident in the block nature of the block nature

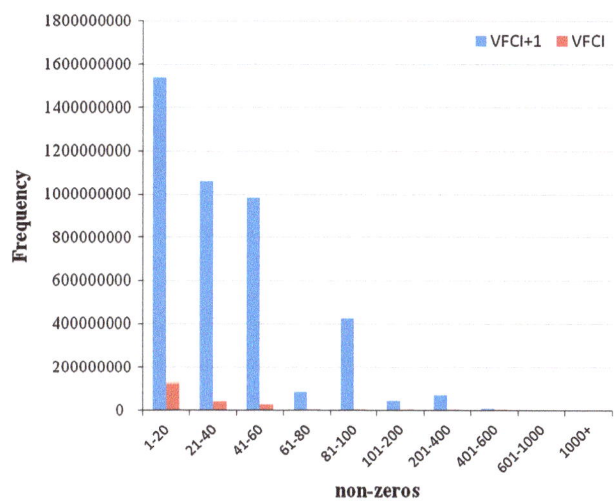

Fig. 2 Histogram of nonzero H-matrix elements for RuO^{2+}

calculation. Values below 30 are not shown. The height of a point depicts the total number of nonzero values for the specified spatial-configuration pair. The diagonal blocks loosely correspond to J projections of 7, 5, 3 and 1. Only the unique elements of H are computed (lower triangle). The number of nonzero elements varies far more slowly than the matrix rank. A plateau is observed at 1,000–2,000 elements. Figure 2 displays a histogram of nonzero values for the entire H-matrix. As the number of nonzero values for a particular configuration pair generally fall far below 1,000, this would correspond to (1,000/607,245) or 0.16 %. A minor problem is evident. A small number of rows display a significantly greater number of interactions; a pseudo-scale-free type of behavior. Thus, simply assigning a GA space with rows sufficiently large to account for these cases results in unnecessarily large GA array memory usage. To address this, PSOCI resorts to a dual array storage. As an example, for a selected molecular system the user may specify that *H* storage space is allocated with rows of size 1,500 elements for the total number of columns. Then, additional GA arrays are defined that can accommodate a much larger row size, but for a small fraction of the number of columns. This approach works reasonably well.

Scaling as a function of numbers of nodes for a PSOCI calculation on the LiSr molecule was followed. This is an 11-electron system for which MRSD SOCI calculations were carried out. The rank of the *H*-matrix is 7,197,320, which was built from 1,891,760 spatial MO configurations. The slopes of time to solution versus numbers of nodes show nearly linear scaling at the largest number of cores. For example, the runtime on 1,000 nodes of the Texas Advanced Computing Center's (TACC) *Lonestar* system was 1,477 s, which may be compared with a very rough estimate of around 18 days if a single core was used.

4 Calculations

Several test problems have been selected for characterization of the PSOCI performance. All calculations were performed using ARECPs defined in *LS*-coupling (Eq. 2) and incorporate one-electron spin–orbit operators derived from the corresponding (*jj*-coupled) RECPs (Eq. 4). The remaining valence electrons are represented by a conventional CGTO basis set in the context of the associated ARECP. Results showing times for building Hamiltonian matrices and time per iteration for eigenanalyses thereof are shown for a series of test cases in Table 1.

The Am^{3+} PSOCI calculation used the LECP so-called Christiansen, Ross, Ermler, Nash, Bursten (CRENB) ARECPs and GTO basis set downloaded from the EMSL Basis Set Exchange [41]. The calculation was at the MRSD level of approximation.

The RuO^{2+} and OsO^{2+} calculations were carried out in the context of the CRENB ARECPs and basis sets which were augmented by adding the aug-cc-pVDZ-PP basis, all obtained from the EMSL Basis Set Exchange [41] including the published diffuse GTFs and two f-type GTFs from the aug-cc-pVDZ-PP as polarization functions [42, 43]. A correlation-consistent valence basis set was used for O and was generally contracted to triple-zeta. Two d-type functions were added as polarization functions [42, 43]. The final set may be designated [Ru,Os:7*s*, 6*p*, 5*d*, 2*f*/O:3*s*, 3*p*, 2*d*] and contains a total of 86 CGTFs representing the 20 valence electrons of RuO^{2+} and OsO^{2+}.

In these calculations the molecular orbital (MO) basis sets were obtained through multi-configuration self-consistent field (MCSCF) calculations that were constructed in the context of the 20 valence electrons. The MOs representing the eight 4*s* and 4*p* electrons of Ru^{2+} and eight 5*s* and 5*p* electrons of Os^{2+} were maintained as doubly occupied in all configurations, while a complete active space self-consistent field (CASSCF) MCSCF procedure was carried out for the remaining Ru^{2+} 4*d*, 5*s*, Os^{2+} 5*d*, 6*s* and O 2*s* and 2*p* subshells. Also, valence/valence contributions to the total wave function that come from the 5p subshell of Ru^{2+} and 6*p* of Os^{2+} are included via dynamic correlation in the subsequent CI calculations. All subsequent CI wave functions maintain the size consistency built into the MCSCF MOs, insuring that their asymptotic behavior is correct in the context of the molecular symmetry that was imposed.

These MCSCF MOs were used to form the basis for defining the SOCI calculations. The calculations correspond to single (S) and double (D) electron excitations into all of the virtual (external) MOs that remained from the

Table 1 PSOCI execution data on four architectures

Machine	Calculation	*H*-matrix rank	Time (s) Phase I	Time (s/iter) Phase II	No. cores
Hopper[a]	Am^{3+}	3,644,641	89	0.5	600
Hopper[a]	RuO^{2+} VFCI	6,078,210	342	1.9	1,200
Hopper[a]	RuO^{2+} MRSD	53,870,274	1,058	5.8	7,200
Hopper[a]	RuO^{2+} MRS	2,047,170	63	0.5	1,200
Hopper[a]	RuO^{2+} MRSD	479,036,418	6,217	N/A	65,536
Edison[b]	OsO^{2+} MRS	2,047,170	189	1.0	2,400
Edison[b]	OsO^{2+} MRSD	53,870,274	605	5.9	7,200
Hopper[a]	OsO^{2+} MRSD	479,036,418	4,117	N/A	57,334
Lonestar[c]	OsO^{2+} MRS	6,078,210	622	N/A	1,536
Lonestar[c]	InH MRS	1,761,306	358	1.3	256
Hatteras[d]	InH MRS	1,761,306	404	1.1	256

[a] National Energy Research Scientific Computing Center (NERSC) Cray XE6

[b] NERSC Cray XC30

[c] Texas Advanced Computing Center (TACC), Dell Linux Cluster

[d] Renaissance Computing Institute (RENCI) Dell Linux Cluster

full basis set (MRS and MRSD). The SOCI wave function was constructed under the C_{2v}^2 double group and included spatial configurations (occupations) constructed from a valence full SOCI (VFCI) representation of the Ru^{2+} 4*d* electrons and O valence electrons, maintaining the Ru^{2+} 4*s* and 4*p* subshells as fully occupied. Results for RuO^{2+} and for analogous calculations for OsO^{2+} using the CRENB ARECP and H^{SO} for Os are given in Table 1. The configuration lists define the wave functions, which are eigenfunctions of one of the irreducible representation of the C_{2v}^2 double group. The numbers of double-group-adapted configurations, which define the rank of the respective Hamiltonian matrices of each symmetry type, are shown in Table 1. The time to construct the largest *H*-matrix (rank 479,036,418) due to MRSD calculations shown in Table 1 was accomplished for RuO^{2+}, taking only about 1.7 h on 65,536 cores of the Hopper system. This matrix proved to be difficult to diagonalize. It is noteworthy that the same rank *H*-matrix for OsO^{2+} took only 1.1 h on the same number of cores. This was due to the higher sparsity of the *H*-matrix because of more diffuse nature of the basis set, the compactness of the MCSCF MOs and the longer bond length. Methods to facilitate such a large diagonalization are being examined. It was found that *H*-matrices of rank 100-200 million could be treated using the current Davidson method, however. The smaller MRSD calculation for RuO^{2+}, H-matrix rank 53,870,274, took 0.3 h to build and was solved for 20 eigenfunctions in a few hundred iterations at 5.8 s/iteration on. Calculations at this level of approximation are relatively fast on 7,200 cores of a parallel supercomputer using the PSOCI code.

Results for the lowest states of Ru^+ and Os^+ are shown in Table 2. The table shows that, in keeping with the well-known assessment of the method, when a large basis set is employed, reasonable accuracy for excitation energies is realized at the MRSD level. As shown in the table, this accuracy carries over to the spin–orbit split electronic states derived from MRSD SOCI calculations.

Table 2 SOCI excitation energies relative to $J = 9/2$ for $Ru^+(^4F)$, $Os^+(^6D)$ in cm^{-1}

	7/2	5/2	3/2	1/2	E (a.u.)
Ru^+ CAS	1,765	2,973	3,839	NA	−93.532
Ru^+ MRS	1,626	2,700	3,443	NA	−93.536
Ru^+ MRSD	1,623	2,678	3,403	NA	−93.662
Ru^+ Exp.[a]	1,523	2,949	3,108	NA	NA
Os^+ CAS	3,819	5,167	6,686	7,671	−90.213
Os^+ MRSD	3,764	4,339	6,023	6,945	−90.329
Os^+ Exp.[a]	3,593	3,928	5,592	6,636	NA

a Moore [45]

The results for InH shown in Table 1 are for MRSD SOCI calculations in the context of a small-core (36 core electrons) ARECP and H^{SO} [43]. The basis set consists of (7*s*, 6*p*, 5*d*, 2*f*/3*s*, 3*p*, 2*d*) CGTOs for In and H, respectively, and applies the CRENB ARECP for a total of 86 basis functions. The AOs, MOs and transformed integrals were calculated and optimized using the *argos*, *mcscf* and *tran* programs of COLUMBUS. A configuration space consisting of 1,761,306 double-group functions of A_1 symmetry was generated. Twenty eigensolutions were calculated. Properties of the ground state of InH calculated using the MRS and MRSD SOCI wave functions that lead to the *H*-matrix listed in Table 1 were calculated using the SOCI wave functions. The results for dipole moment, harmonic frequency and equilibrium bond distance are shown in Table 3. The calculated values are in close agreement with experiment and show clear improvement going from the MRS to MRSD level of approximation. Results from the relativistic CI procedure developed by Balasubramanian [7] mentioned above are shown for comparison.

For the purposes of comparison and consistency, our published results for Am^{3+} [44] were recalculated with the PSOCI code. *H*-matrix size and timing information for this calculation are shown in the first row of Table 1. We extended the study of Am^{3+} to the halide cation $AmCl^+$. Shown in Table 4 are results for the ground state of this cation at four levels of approximation. The MCSCF MOs were used to form the full set of valence configurations (CAS). These CAS configurations were then used as the multi-reference set for the subsequent MRSD SOCI calculations shown in the Table. The last row of Table 4 shows

Table 3 SOCI properties of InH in the context of the CRENB ARECP and SO operators

Prop.	MRS[a]	MRSD[b]	RCI[c]	Exp.[d]
E_{val}^e	−57.3327	−57.4572	–	–
μ^f	0.63	0.53	0.43	–
δ_{In}^g	0.36	0.37	–	–
ω_e^h	1,432	1,440	1,469	1,476
R_e^i	1.860	1.837	1.823	1.838

[a] Multi-reference singles SOCI calculation

[b] Multi-reference singles plus doubles SOCI calculation

[c] Balasubramanian [46]. (The CRENB ARECP and SO operators were used.)

[d] Huber and Herzberg [47]

[e] Total valence electron energy (a.u.) for the experimental bond distance ($R_{exp} = 1.8383$ Å)

[f] Dipole moment (D) at R_{exp}

[g] Mulliken net charge at R_{exp}

[h] Harmonic frequency (cm^{-1})

[i] Equilibrium bond length (Å)

Table 4 Properties of the ground state of $AmCl^+$ [a]

Calculation	D_e (cm^{-1})	R_e (Å)	ω_e (cm^{-1})	$\omega_e x_e$ (cm^{-1})
CAS	26,102	2.565	320.5	0.660
MCSCF	26,883	2.560	321.6	0.659
MRSD	27,482	2.550	326.3	0.695
MRSD + Davidson[b]	27,412	2.553	329.7	0.722

a Tilson et al. [48]

b So-called unlinked clusters correction added (see Ref. 22)

the effect of the addition of the well-known Davidson correction to the MRSD energy. The convergence of the properties of $AmCl^+$ to values of D_e, R_e, ω_e, and $\omega_e x_e$ is seen as a function of the increase in size of the CI space.

5 Conclusions

The code developed using the SOCI method runs in parallel using a distributed memory model and parallel I/O. It is noteworthy that the MRSD SOCI method is able to construct a wave function that has the necessary flexibility to properly model the products and reactants of a bond-breaking process. PSOCI is capable of calculating scores of eigensolutions of high-rank (50–200 million) Hamiltonian matrices for large heavy-element-containing molecular systems. A parallel Cuthill–McKee [39] approach is currently being implemented to facilitate eigenanalysis for very large rank H-matrices (~500 M) executing on thousands of processors. The PSOCI code performs efficiently on state-of-the-art parallel supercomputers. Finally, the PSOCI code has been incorporated into distributed workflows with dynamically allocated computation and networking, rendering it an efficient and dependable platform for calculating accurate wave functions that are eigenfunctions of molecular double-group symmetry irreducible representations.

Run times for the reported test cases are due to parallel computations, ranging from 256 through 65,536 cores. The choices of numbers of cores are functions of both the required distributed memory and desired run time. Computations that include 50–200 million double-group functions are straightforward to perform on the large systems as indicated in Table 1. These larger calculations are of a basis set size and wave function flexibility that has been generally demonstrated to capture the necessary physics of a myriad of molecular problems.

Acknowledgments The Texas Advanced Computing Center (TACC) at the University of Texas at Austin is acknowledged for providing HPC resources that have contributed to the research results reported within this paper. Computer time was also provided by the National Energy Research Scientific Computing Center (NERSC) operated by the U.S. Department of energy and the Renaissance Computing Institute (RENCI) at the University of North Carolina at Chapel Hill. WCE thanks the Computational Biology Initiative (CBI) of the University of Texas at San Antonio for its technological support of his Department of Chemistry Mulliken Linux cluster. JLT was partially supported under the US DOE SciDAC-e award DE-FC02-06ER25764 including software development, benchmarking and computations on Ru and Os systems. The authors thank Dr. S. Brozell for his advice on SOCI electron configurations and Prof. R. M. Pitzer for his counsel on various aspects of the development of the PSOCI code. The authors are grateful to Dr. R. J. Fowler (RENCI) for his support of this work. The reviewers are acknowledged for their comments and suggestions, which resulted in an improved manuscript. Finally, WCE notes his gratitude to his teacher, mentor and collaborator, Isaiah Shavitt, to whom this volume is dedicated.

References

1. Harrison JF (2000) Electronic structure of diatomic molecules composed of a first-row transition metal and main-group element (H-F). Chem Rev 100(2):679–716. doi:10.1021/cr980411m
2. Siegbahn PEM (1993) A comparison of the bonding in the second-row transition-metal oxides and carbenes. Chem Phys Lett 201(1–4):15–23
3. Dolg M, Wedig U, Stoll H, Preuss H (1987) Ab initio pseudopotential study of the first row transition metal monoxides and iron monohydride. J Chem Phys 86(4):2123–2131
4. Sjøvoll M, Gropen O, Olsen J (1997) A determinantal approach to spin-orbit configuration interaction. Theor Chem Acc 97(1):301–312. doi:10.1007/s002140050265
5. Liu W, Dolg M (1998) Benchmark calculations for lanthanide atoms: calibration of ab initio and density-functional methods. Phys Rev A 57(3):1721–1728
6. Liu W, Küchle W, Dolg M (1998) Ab initio pseudopotential and density-functional all-electron study of ionization and excitation energies of actinide atoms. Phys Rev A 58(2):1103–1110
7. Balasubramanian K (1997) Relativistic effects in chemistry: part A. Wiley, New York
8. Tilson JL, Ermler WC, Pitzer RM (2000) Parallel spin-orbit coupled configuration interaction. Comput Phys Commun 128(1–2):128–138
9. Alekseyev AB, Liebermann H-P, Buenker RJ (2004) Spin-Orbit multireference configuration interaction method and applications to systems containing heavy atoms. In: Ishikawa Y (ed) Hirao K. Recent Advances in Relativistic Molecular Theory World Scientific, Singapore, pp 65–106
10. Pitzer RM, Winter NW (1988) Electronic-structure methods for heavy-atom molecules. J Phys Chem 92(11):3061–3063. doi:10.1021/j100322a011
11. Christiansen PA, Ermler WC, Pitzer KS (1985) Relativistic effects in chemical-systems. Annu Rev Phys Chem 36:407–432. doi:10.1146/annurev.physchem.36.1.407
12. Wadt WR, Hay PJ (1985) Ab initio effective core potentials for molecular calculations. Potentials for main group elements Na to Bi. J Chem Phys 82(1):284–298
13. Dolg M, Stoll H (1996) Chapter 152 electronic structure calculations for molecules containing lanthanide atoms. In: Jr Gschneidner Karl A, LeRoy E (eds) Handbook on the physics and chemistry of rare earths, vol 22. Elsevier, Amsterdam, pp 607–729
14. Fernandez Pacios L, Christiansen P (1985) Ab initio relativistic effective potentials with spin-orbit operators. I. Li through Ar. J Chem Phys 82(6):2664
15. Hurley MM, Pacios LF, Christiansen PA, Ross RB, Ermler WC (1986) Ab initio relativistic effective potentials with spin-orbit operators. II. K through Kr. J Chem Phys 84(12):6840–6853

16. LaJohn LA, Christiansen PA, Ross RB, Atashroo T, Ermler WC (1987) Ab initio relativistic effective potentials with spin–orbit operators. III. Rb through Xe. J Chem Phys 87(5):2812–2824
17. Ross RB, Powers JM, Atashroo T, Ermler WC, LaJohn LA, Christiansen PA (1990) Ab initio relativistic effective potentials with spin–orbit operators. IV. Cs through Rn. J Chem Phys 93(9):6654–6670
18. Ross RB, Gayen S, Ermler WC (1994) Ab initio relativistic effective potentials with spin–orbit operators. V. Ce through Lu. J Chem Phys 100(11):8145–8155
19. Nash CS, Bursten BE, Ermler WC (1997) Ab initio relativistic effective potentials with spin-orbit operators. VII. Am through element 118. J Chem Phys 106(12):5133–5142
20. Ermler WC, Ross RB, Christiansen PA (1991) Ab initio relativistic effective potentials with spin-orbit operators. VI. Fr through Pu. Int J Quantum Chem 40(6):829–846. doi:10.1002/qua.560400611
21. Lee YS, Ermler WC, Pitzer KS (1977) Ab initio effective core potentials including relativistic effects. I. Formalism and applications to the Xe and Au atoms. J Chem Phys 67(12):5861–5876
22. Christiansen PA, Lee YS, Pitzer KS (1979) Improved ab initio effective core potentials for molecular calculations. J Chem Phys 71(11):4445–4450
23. Desclaux JP (1975) A multiconfiguration relativistic DIRAC-FOCK program. Comput Phys Commun 9(1):31–45
24. Ermler WC, Lee YS, Christiansen PA, Pitzer KS (1981) AB initio effective core potentials including relativistic effects. A procedure for the inclusion of spin-orbit coupling in molecular wavefunctions. Chem Phys Lett 81(1):70–74
25. Mulliken R (1955) Electronic population analysis on LCAO–MO molecular wave functions. I. J Chem Phys 23(10):1833
26. Nieplocha J, Palmer B, Tipparaju V, Krishnan M, Trease H, Aprà E (2006) Advances, applications and performance of the global arrays shared memory programming toolkit. Int J High Perform C 20(2):203–231. doi:10.1177/1094342006064503
27. Nieplocha J, Foster I, Kendall RA (1998) ChemIo: high performance parallel I/o for computational chemistry applications. Int J High Perform C 12(3):345–363. doi:10.1177/109434209801200304
28. Davidson ER (1989) Super-matrix methods. Comput Phys Commun 53(1–3):49–60
29. Crouzeix M, Philippe B, Sadkane M (1994) The Davidson method. SIAM J Sci Comput 15(1):62
30. Zhou Y, Shepard R, Minkoff M (2005) Computing eigenvalue bounds for iterative subspace matrix methods. Comput Phys Commun 167(2):90–102
31. Marino MM, Ermler WC (1993) Ab initio rep-based relativistic core/valence polarization operator. Chem Phys Lett 206(1–4):271–277
32. Brozell SR (1999) Spin–orbit configuration interaction calculations of actinide and lanthanide systems. Ph.D. dissertation. The Ohio State University
33. Lischka H, Shepard R, Brown FB, Shavitt I (1981) New implementation of the graphical unitary group approach for multireference direct configuration interaction calculations. Int J Quantum Chem 20(S15):91–100. doi:10.1002/qua.560200810
34. Shepard R, Shavitt I, Pitzer RM, Comeau DC, Pepper M, Lischka H, Szalay PG, Ahlrichs R, Brown FB, Zhao J-G (1988) A progress report on the status of the COLUMBUS MRCI program system. Int J Quantum Chem 34(S22):149–165. doi:10.1002/qua.560340819
35. Lischka H, Shepard R, Pitzer RM, Shavitt I, Dallos M, Muller T, Szalay PG, Seth M, Kedziora GS, Yabushita S, Zhang Z (2001) High-level multireference methods in the quantum-chemistry program system COLUMBUS: analytic MR-CISD and MR-AQCC gradients and MR-AQCC-LRT for excited states, GUGA spin-orbit CI and parallel CI density. Phys Chem Chem Phys 3(5):664–673
36. Lischka H, Shepard R, Shavitt I, Pitzer RM, Dallos M, Müller T, Szalay PG, Brown FB, Ahlrichs R, Böhm HJ, Chang A, Comeau DC, Gdanitz R, Dachsel H, Ehrhardt C, Ernzerhof M, Höchtl P, Irle S, Kedziora G, Kovar T, Parasuk V, Pepper MJM, Scharf P, Schiffer H, Schindler M, Schüler M, Seth M, Stahlberg EA, Zhao J-G, Yabushita S, Zhang Z, Barbatti M, Matsika S, Schuurmann M, Yarkony DR, Brozell SR, Beck EV, Blaudeau JP, Ruckenbauer M, Sellner B, Plasser F, Szymczak JJ (2008) COLUMBUS, an ab initio electronic structure program. Release 5.9.2
37. Roostaei B, Ermler WC (2012) Electric dipole transition moments and permanent dipole moments for spin–orbit configuration interaction wave functions. Comput Phys Commun 183(3):594–599
38. Foster IT, Tilson JL, Wagner AF, Shepard RL, Harrison RJ, Kendall RA, Littlefield RJ (1996) Toward high-performance computational chemistry: I. Scalable Fock matrix construction algorithms. J Comput Chem 17(1):109–123. doi:10.1002/(sici)1096-987x(19960115)17:1<109:aid-jcc9>3.0.co;2-v
39. Liu W, Sherman A (1976) Comparative analysis of the Cuthill–McKee and the reverse Cuthill–McKee ordering algorithms for sparse matrices. SIAM J Numer Anal 13(2):198–213. doi:10.1137/0713020
40. Cuthill E, McKee J (1969) Reducing the bandwidth of sparse symmetric matrices. Paper presented at the proceedings of the 1969 24th national conference
41. Schuchardt KL, Didier BT, Elsethagen T, Sun L, Gurumoorthi V, Chase J, Li J, Windus TL (2007) Basis set exchange: a community database for computational sciences. J Chem Inf Model 47(3):1045–1052. doi:10.1021/ci600510j
42. Tilson JL, Ermler WC, Fowler RJ (2011) CI potential energy curves for three states of RuO2+. Chem Phys Lett 516(4–6):131–136
43. Ermler WC, Tilson JL (2012) Generally contracted valence–core/valence basis sets for use with relativistic effective core potentials and spin–orbit coupling operators. Comp Theor Chem 1002:24–30. doi:10.1016/j.comptc.2012.09.020
44. Tilson J, Naleway C, Seth M, Shepard R, Wagner A, Ermler W (2002) An ab initio study of the f–f spectroscopy of americium +3. J Chem Phys 116(13):5494
45. Moore CE (1971) Atomic Energy Levels, Natl. Bur. Stand. (US) Circ. No. 467 (U.S. GP0, Washington, DC)
46. Balasubramanian K (1990) Spectroscopic constants and potential energy curves for indium hydrides (InH and InH^+). J Phys Chem 94:6582–6588
47. Huber KP, Herzberg G (1979) Molecular spectra and molecular structure IV: constants of diatomic molecules. Van Nostrand Reinhold, New York
48. Tilson JL, Naleway C, Seth M, Shepard R, Wagner AF, Ermler WC (2004) An ab initio study of $AmCl^{+1}$: f–f spectroscopy and chemical binding. J Chem Phys 121:5661–5675

Theor Chem Acc (2014) 133:1588
DOI 10.1007/s00214-014-1588-0

REGULAR ARTICLE

Spin–orbit DFT with analytic gradients and applications to heavy element compounds

Zhiyong Zhang

Received: 1 February 2014 / Accepted: 15 October 2014 / Published online: 25 October 2014

Abstract We have implemented the unrestricted DFT approach with one-electron spin–orbit operators in the massively parallel NWChem program. Also implemented is the analytic gradient in the DFT approach with spin–orbit interactions. The current capabilities include single-point calculations and geometry optimization. Vibrational frequencies can be calculated numerically from the analytically calculated gradients. The implementation is based on the spin–orbit interaction operator derived from the effective core potential approach. The exchange functionals used in the implementation are functionals derived for non-spin–orbit calculations, including GGA as well as hybrid functionals. Spin–orbit Hartree–Fock calculations can also be carried out. We have applied the spin–orbit DFT methods to the Uranyl aqua complexes. We have optimized the structures and calculated the vibrational frequencies of both $(UO_2^{2+})_{aq}$ and $(UO_2^{+})_{aq}$ with and without spin–orbit effects. The effects of the spin–orbit interaction on the structures and frequencies of these two complexes are discussed. We also carried out calculations for Th_2, and several low-lying electronic states are calculated. Our results indicate that, for open-shell systems, there are significant effects due to the spin–orbit effects and the electronic configurations with and without spin–orbit interactions could change due to the occupation of orbitals of larger spin–orbit interactions.

Dedicated to the memory of Professor Isaiah Shavitt and published as part of the special collection of articles celebrating his many contributions.

Z. Zhang (✉)
Stanford Research Computing Center, Stanford University, Stanford, CA 94305, USA
e-mail: zyzhang@stanford.edu; zyzhang2006@gmail.com

Keywords Spin–orbit interaction · Density functional theory · Spin–orbit DFT · Effective core potential approach

1 Introduction

With the rapid increase in computing power, first-principles simulations have become indispensable tools for understanding the molecular mechanisms that underlie macroscopic phenomena, especially for systems containing radioactive actinide and lanthanide elements for which experimental characterization may be quite expensive and hazardous. For systems containing heavy elements, it is necessary to take into account the large relativistic effects, both scalar and spin–orbit. The importance of spin–orbit interactions in chemical systems has been firmly established [1–3]. The coupling of spin and orbital angular momentum in general will cause the splitting of spin multiplets, and this effect on the spectra is easily recognizable even at low Z values [4]. Radiationless transitions induced by spin–orbit interaction are one of the most common intramolecular interaction mechanisms. Such transitions frequently lead to predissociation in excited states [5–7]. For systems containing transition metals and even heavier lanthanide and actinide elements, the spin–orbit effects become particularly important, especially in situations where several closely spaced electronic states can be coupled together by the strong spin–orbit interaction [8–12].

The scalar relativistic effects can be adequately taken into account through various approximations that can easily be implemented in modern computational packages. The remaining difficulties are the adequate and efficient account of (1) the spin–orbit effects and (2) quite often the multireference character of the wave function, especially for systems with heavier elements like *f*-electron systems, and the associated multiplicity problem. When the spin–orbit effect is small compared with the spacing of the electronic states, the spin–orbit effect can often be treated with perturbative methods. For systems with strong spin–orbit interactions, the spin–orbit effect needs to be treated on the same footing as correlation effects variationally. The conventional way of treating the spin–orbit effect is spin–orbit configuration interaction (SOCI), where the spin–orbit interaction is treated in the CI step, but such calculations are very expensive [13, 14]. An alternative way is to include the spin–orbit effects within a density functional scheme, employing the common exchange–correlation functionals.

Relativistic effects can be put into molecular electronic calculations in a variety of ways, ranging from full four-component wave function methods to semiempirical methods, as reviewed by Pyykkö [15] and a number of subsequent reviews since then [16–23]. Several all-electron versions of fully relativistic four-component spinor-based DFT have been developed [24–26] and applied to systems containing heavy elements [27–29]. The relativistic effective core potential (RECP) method has been the most extensively used method, though, for two main reasons: (1) the effects of the inner-shell electrons, (which are the fastest moving electrons) whose major relativistic changes in orbital sizes and energies are propagated out to the valence region, are included in the core potentials, so only the valence electrons need to be treated explicitly; (2) many existing nonrelativistic algorithms can be adapted to relativistic calculations using RECPs. The scalar relativistic effects can easily be included in the major computational chemistry packages, and the major additional complication is the need to include the spin–orbit interaction for the valence electrons, which can be quite large even if the electron speeds are not large. In this work, we describe our implementation of spin–orbit interactions in the framework of DFT methods in the heavily parallel computational chemistry package NWChem. Our implementation includes spin–orbit interactions in the total energy calculations as well as analytical gradients with respect to nuclear coordinates for geometry optimization.

We have optimized the structures and calculated the vibrational frequencies of both $(UO_2^{2+})_{aq}$ and $(UO_2^{+})_{aq}$ with and without spin–orbit effects. We also abstained the electron affinity of $(UO_2^{2+})_{aq}$ with and without spin–orbit interactions. The effects of the spin–orbit interaction on the structures and frequencies of these two complexes are discussed. The effect of the spin–orbit effects on the calculated energetics of molecular orbitals, in general good qualitative indicators of chemical reactivities and excited state properties, is discussed. We also carried out calculations for Th_2, and the effects of spin–orbit interactions on the low-lying electronic states, bond lengths, vibrational frequencies, dissociation energies and electronic configurations will be discussed in detail.

2 Spin–orbit operator in relativistic core potential methods

The RECP method is an extension of the nonrelativistic effective core potential approach [30], which has been reviewed by Krauss and Stevens [31]. RECPs are obtained by several algorithms, particularly from wave functions from relativistic atomic calculations [32–34] and from fitting the energy results from all-electron atomic calculations [20]. Corresponding spin–orbit operators are obtained as part of the same process [22, 33] or by a separate process [35].

In the procedure of Christiansen and co-workers [33], a Dirac–Fock atomic wave function is used as the starting point. For each pair of l, j indices, the large-component radial function is used [32] to determine the pseudoorbital by the shape-consistent method [36]; it is defined to be equal to the valence radial function in the valence region and to decrease smoothly and nodelessly through the core region to the value of zero at the nucleus. The pseudoorbital is then used to define the potential for that pair of l, j indices. The electron repulsion interaction among valence electrons is removed from these potentials.

There is no need to explicitly include terms for direct relativistic effects, such as the dependence of mass on velocity, which are important only in the core region, in the valence-electron Hamiltonian. These terms are included as a consequence of using the Dirac–Fock wave functions. Thus, the Hamiltonian for the valence electrons is composed of the nonrelativistic Hamiltonian for the valence electrons plus the RECPs, which include the effects of the core electrons as well as the relativistic effects on the valence electrons in the core region [37]. The RECPs thus represent, for the valence electrons, the dynamical effects of relativity from the core region, the repulsion of the core electrons, the spin–orbit interaction with the nucleus, the spin–orbit interaction with the core electrons, and an approximation to the spin–orbit interaction between the valence electrons [38], which has usually been found to be quite small, especially for heavier element systems [39–41]. The REP operators can be written as a summation of spin-independent potential and the spin–orbit operator, as written below, and the readers are referred to reference [39] for details.

$$U^{\mathrm{AREP}} = U_L^{\mathrm{AREP}}(r) + \sum_{l=0}^{L-1} \left(U_l^{\mathrm{AREP}}(r) - U_L^{\mathrm{AREP}}(r) \right) \hat{O}_l \quad (2.1)$$

$$H^{\mathrm{so}} = \vec{s} \cdot \sum_{l=1}^{L-1} \frac{2}{2l+1} \Delta U_l^{\mathrm{REP}} \vec{l} \hat{O}_l \quad (2.2)$$

$$\Delta U_l^{\mathrm{REP}} = U_{l+1/2}^{\mathrm{REP}}(r) - U_{l-1/2}^{\mathrm{REP}}(r) \quad (2.3)$$

With the core potentials and spin–orbit operators given in the forms of Eqs. (2.1) and (2.2), existing programs for nonrelativistic calculations can be adapted to include relativistic effects. In the literature, the coefficients of the spin–orbit potentials are *not* always defined in the same manner. In our implementation of the spin–orbit interaction in NWChem, the spin–orbit potential is defined as the $\Delta U_l^{\mathrm{REP}}$ in (2.3) scaled by $\frac{2}{2l+1}$:

$$U^{\mathrm{so}} = \frac{2}{2l+1}\Delta U_l^{\mathrm{REP}} \tag{2.4}$$

In the literature, the Stuttgart potentials [20] are defined as $\Delta U_l^{\mathrm{REP}}$ in (2.3) and, hence, have to be multiplied by $\frac{2}{2l+1}$ to use in NWChem. On the other hand, the so-called CRENBL potentials [42] in the published papers are defined as $\frac{l}{2l+1}\Delta U_l^{\mathrm{REP}}$ and, hence, have to be multiplied by $\frac{2}{l}$ (note, on the CRENBL Web site, the spin–orbit potentials already have been corrected with the $\frac{2}{l}$ factor).

3 Spin–orbit DFT with analytic gradients

With spin–orbit interaction, a one-electron spin–orbit interaction operator derived from relativistic ECP procedure is simply added to the Kohn–Sham Hamiltonian:

$$F\phi = (h_0 + h_{\mathrm{so}} + v_c + v_{xc})\phi = \varepsilon\phi \tag{3.1}$$

h_0 is the one-electron Hamiltonian including kinetic, nuclear attraction and RECP operators. The spin–orbit interaction operator is given by

$$h_{\mathrm{so}} = i\vec{v}\cdot\vec{s} = iv_x s_x + iv_y s_y + iv_z s_z. \tag{3.2}$$

v_c and v_{xc} are the coulomb and correlation–exchange potentials usually used in the spin-free case.

To solve the Kohn–Sham equation, a spinor basis that is composed of the direct product of the atomic basis functions (usually Gaussian-type basis functions) and the one-electron spin functions, $|a\sigma\rangle$, where $|a\rangle$ is the atomic basis function and $|\sigma\rangle = |\alpha\rangle$ and $|\beta\rangle$ are the one-electron spin functions. The molecular orbitals can be expanded as linear combinations of the spinor basis functions,

$$|\phi\rangle = \sum_{i,\sigma} C_i^\sigma |i\sigma\rangle, \tag{3.3}$$

and satisfy the Hartree–Fock like equation,

$$F\vec{C} = S\vec{C}\varepsilon \tag{3.4}$$

The column matrix

$$C = \begin{pmatrix} C^\alpha \\ C^\beta \end{pmatrix}, \tag{3.5}$$

is molecular orbital expansion coefficients with C^α and C^β the α and β blocks corresponding to the α and β set of spinors. S is the overlap matrix in the spinor basis $|a\sigma\rangle$ and is block diagonal in α and β, $S = \begin{pmatrix} S^{\alpha\alpha} & 0 \\ 0 & S^{\beta\beta} \end{pmatrix}$, with $S_{ij}^{\alpha\alpha} = S_{ij}^{\beta\beta} = \langle i|j\rangle$. ε is the diagonal matrix of the molecular orbital eigenvalues. C and S satisfy the orthonormality conditions $C^+SC = 1$, where C^+ is the Hermitian conjugate of C.

The density matrix

$$D = \begin{pmatrix} D^{\alpha\alpha} & D^{\alpha\beta} \\ D^{\beta\alpha} & D^{\beta\beta} \end{pmatrix} \tag{3.6}$$

with

$$D_{ab}^{\alpha\beta} = \sum_i \left(C_{a,i}^\alpha\right)^* C_{b,i}^\beta \tag{3.7}$$

where the summation over i is over the molecular orbitals and the density matrix elements satisfy

$$D_{ab}^{\alpha\alpha} = \left(D_{ba}^{\alpha\alpha}\right)^*, \quad D_{ab}^{\beta\beta} = \left(D_{ba}^{\beta\beta}\right)^*, \quad \textit{and} \quad D_{ab}^{\alpha\beta} = \left(D_{ba}^{\beta\alpha}\right)^* \tag{3.8}$$

The Hartree–Fock operator is given by

$$F = \begin{pmatrix} V_0 + \frac{i}{2}V_z + V_c + V_{xc} & \frac{1}{2}V_y + \frac{i}{2}V_x \\ -\frac{1}{2}V_y + \frac{i}{2}V_x & V_0 - \frac{i}{2}V_z + V_c + V_{xc} \end{pmatrix} \tag{3.9}$$

The spin-free matrix elements

$$(V_0)_{ab} = \left\langle a\middle|\hat{h}_0\middle|b\right\rangle, \tag{3.10}$$

and

$$\left(V_{c,xc}\right)_{ab} = \langle a|\hat{v}_c, \hat{v}_{xc}|b\rangle \tag{3.11}$$

are symmetric, and the spin–orbit matrix elements

$$\left(V_{x,y,z}\right)_{ab} = \left\langle a\middle|\hat{V}_{x,y,z}\middle|b\right\rangle, \tag{3.12}$$

are anti-symmetric. The total energy is given by

$$\begin{aligned} E = FD = {} & V_0\left(D^{\alpha\alpha} + D^{\beta\beta}\right) + \tfrac{i}{2}V^x\left(D^{\alpha\beta} + D^{\beta\alpha}\right) \\ & + \tfrac{1}{2}V^y\left(D^{\alpha\beta} - D^{\beta\alpha}\right) + \tfrac{i}{2}V^z\left(D^{\alpha\alpha} - D^{\beta\beta}\right) + E_c + E_{xc} \end{aligned} \tag{3.13}$$

where

$$E_c = \frac{1}{2}\sum_{abcd} D_{ab}D_{cd}(ab|cd) \tag{3.14}$$

is the coulomb energy, and

Table 1 Calculated structural properties of $UO_2^{2+}(H_2O)_5$ and $UO_2^{+}(H_2O)_5$ and vibrational frequencies

	$UO_2^{2+}(H_2O)_5$		$UO_2^{+}(H_2O)_5$	
	Without spin–orbit	With spin–orbit	Without spin–orbit	With spin–orbit
U–O distances (Å)	1.748	1.749	1.806	1.808
U-water distances (Å)	2.497–2.498	2.492–2.495	2.591–2.592	2.587–2.589
O–U–O symmetric(cm^{-1})	942	945	843	845
O–U–O asymmetric (cm^{-1})	1,028	1,032	898	903

Table 2 Calculated properties of Th_2 dimer

	Th–Th distance (Å)	Vibrational frequency (cm^{-1})	Dissociation energy (kcal/mol)
No S.O.			
$s\sigma^2 d\pi^4 d\sigma^{*2}, {}^1\Sigma_g^+$	2.907	138	68.3
$s\sigma^2 d\pi^3 d\sigma^1 d\delta^1 s\sigma^{*1}, {}^5\Phi_g$	2.831	145	72.1
$s\sigma^2 d\pi^4 d\sigma^1 s\sigma^{*1}, {}^3\Sigma_u^+$	2.755	171	81.6
With S.O.			
$s\sigma^2 d\pi^4 d\sigma^{*2}, \Omega = 0$	2.9076	143	67.8
$s\sigma^2 d\pi^4 d\sigma^1 d\delta^1, \Omega = 3$	2.669	196	76.8
$s\sigma^2 d\pi^4 d\sigma^1 s\sigma^{*1}, \Omega = 1$	2.758	183	80.8

$$E_{xc} = \sum_{ab} D_{ab} \langle a | \hat{v}_{xc} | b \rangle \tag{3.15}$$

is the correlation–exchange energy that depends on the form of the exchange–correlation functional used. If the exact Hartree–Fock exchange is used in place of the exchange–correlation functional, we recover the Hartree–Fock theory with spin–orbit interaction.

For a Hamiltonian that depends on a parameter X, in this case the nuclear coordinates, the analytic gradient of the energy with respect to the parameter can simply be obtained by differentiating the energy with respect to the parameter X, based on the Hellman–Feynman theorem

$$\begin{aligned} E^{(X)} = F^{(X)} D &= (V_0)^{(X)} \left(D^{\alpha\alpha} + D^{\beta\beta}\right) + \tfrac{i}{2}(V^x)^{(X)} \left(D^{\alpha\beta} + D^{\beta\alpha}\right) \\ &+ \tfrac{1}{2}(V^y)^{(X)} \left(D^{\alpha\beta} - D^{\beta\alpha}\right) + \tfrac{i}{2}(V^z)^{(X)} \left(D^{\alpha\alpha} - D^{\beta\beta}\right) \\ &+ (E_c)^{(X)} + (E_{xc})^{(X)} - WS^{(X)} \end{aligned} \tag{3.16}$$

Here, the superscript, $^{(X)}$, denotes differentiation with respect to the parameter X. W is the energy-weighted density matrix resulting from the derivative of the overlap matrix with respect to the parameters; W is given by 3.17, and σ are the spin indexes:

$$W_{ab}^{\sigma\sigma} = \sum_i \varepsilon_i (C_{ai}^{\sigma})^* C_{bi}^{\sigma}. \tag{3.17}$$

4 Spin–orbit DFT study of heavy element compounds

We have calculated the structural and electronic properties of $UO_2^{2+}(H_2O)_5$ and $UO_2^{+}(H_2O)_5$ using the Stuttgart small core pseudopotentials and associated spin–orbit interaction operators [43–45]. The calculated U–O distances and U-water distances in the equatorial plane, with and without spin–orbit effects, for both $UO_2^{2+}(H_2O)_5$ and $UO_2^{+}(H_2O)_5$ are listed in Table 1. As seen from the table, for both $UO_2^{2+}(H_2O)_5$ and $UO_2^{+}(H_2O)_5$, the U–O axial bonds are slightly longer by 0.001 and 0.002 Å, respectively, and the U-water distances are shorter by 0.004–0.006 Å, when spin–orbit effects are taken into account. Generally speaking, shorter bond lengths are associated with higher vibrational frequencies, indicating stronger bonding interactions. However, in the case of $UO_2^{2+}(H_2O)_5$ and $UO_2^{+}(H_2O)_5$, the symmetric O–U–O vibrational frequencies turned out to be a few wave numbers higher even though the calculated bond lengths are slightly longer when spin–orbit interactions are considered. The slightly longer U–O bonds can be attributed to the synergistic effects of the contraction and expansion of the $f_{5/2}$ and $f_{7/2}$ atomic spin–orbitals, which both participate in bonding interactions with O. The slight contraction of the equatorial water shell can be attributed to the reduced electrostatic interactions between the equatorial and axial oxygen.

As an example for an open-shell system, we carried out calculations for Th_2. Th clusters have been attracting increasing attention due to potential nuclear energy applications. The calculated structural, electronic structure and energetic properties are summarized in Table 2. We have identified three low-lying electronic states without spin–orbit interactions, the $s\sigma^2 d\pi^3 d\sigma^1 d\delta^1 s\sigma^{*1}\ {}^5\Phi_g$, $s\sigma^2 d\pi^4 d\sigma^1 s\sigma^{*1}\ {}^3\Sigma_u^+$ and $s\sigma^2 d\pi^4 s\sigma^{*2}\ {}^1\Sigma_g^+$ states, respectively. The lowest energy electronic state without spin–orbit interaction is found to be the $^3\Sigma_u^+$ state from the $s\sigma^2 d\pi^4 d\sigma^1 s\sigma^{*1}$ electronic configuration, followed by the $^5\Phi_g$ state of the $s\sigma^2 d\pi^3 d\sigma^1 d\delta^1 s\sigma^{*1}$ configuration and the $^1\Sigma_g^+$ state of the $s\sigma^2 d\pi^4 s\sigma^{*2}$ configuration, 0.411 and 0.574 eV above in energy, respectively. The three low-lying electronic states show progressively longer bond lengths and associated lower vibrational frequencies, as shown in Table 2, indicating progressively weaker bonding interaction, in going from states of lower energy to higher energy. Starting from the three low-lying electronic configurations obtained from non-spin–orbit calculations, we obtained three low-lying electronic states when spin–orbit interactions are included. The lowest energy state is an $\Omega = 1$ state from the $^3\Sigma_u^+$ state of the $s\sigma^2 d\pi^4 d\sigma^1 s\sigma^{*1}$ configuration. The calculated bond length for this state does not show appreciable changes as shown in Table 2, as the electronic configuration remains the same with and without the inclusion of spin–orbit interactions. The highest energy state without spin–orbit, the $^1\Sigma_g^+$ state of the $s\sigma^2 d\pi^4 d\sigma^{*2}$ configuration, gives rise to an $\Omega = 0$ state of this configuration. Again, the bond length for this state remains almost the same as in the case of without spin–orbit interaction for the same reason. For these two cases, even though the calculated bond lengths do not show appreciable changes, the calculated vibrational frequencies show appreciable increase when the spin–orbit interactions are included, indicating that inclusion of spin–orbit effects tends to make the potential energy curve slightly deeper. However, the case of $s\sigma^2 d\pi^3 d\sigma^1 d\delta^1 s\sigma^{*1 5}\Phi_g$ $^5\Phi_g$ state is quite different. With the inclusion of the spin–orbit interactions, due to the much larger spin–orbit effect for the electrons in a $d\delta$ orbital than in a $s\sigma^*$ orbital, the dominant electronic configuration becomes $s\sigma^2 d\pi^4 d\sigma^1 d\delta^1$, giving rise to an $\Omega = 3$ state with considerably shorter bond length and higher vibrational frequency of 2.669 Å and 196 cm^{-1}, compared with 2.831 Å and 145 cm^{-1} without spin–orbit, because a bonding $d\delta$ orbital is occupied instead of an antibonding $s\sigma^*$ orbital. For all the three states considered here, we also calculated the dissociation energy to the lowest energy atomic asymptote and the results are listed in Table 2. As can be seen, the dissociation energy with and without spin–orbit for the first two cases discussed above does not change much, but for the $s\sigma^2 d\pi^4 d\sigma^1 d\delta^1$ $\Omega = 1$ state, there is appreciable increase in the calculated dissociation energy, due to higher occupations of molecular orbitals with (negative) spin–orbit energies.

Our calculations identified the $s\sigma^2 d\pi^4 d\sigma^1 s\sigma^{*1}$ $^3\Sigma_u^+$ state as the ground state. The work of Roos et al. [46] identified the $s\sigma^2 d\pi^4 d\sigma^1 d\delta^{1\,3}\Delta_g$ state as the ground state, with the $s\sigma^2 d\pi^4 d\sigma^1 s\sigma^{*1}$ $^3\Sigma_u^+$ as the third state higher in energy. As they also indicated, the first state higher in energy, the $s\sigma^2 d\pi^4 d\sigma^2$, $^1\Sigma_u^+$ state, is only 400 cm^{-1} higher in energy than the ground state identified in their work. Of the three states considered here, the electronic configurations differ in the occupation of the $7s\sigma^*$, $6d\sigma$ and $6d\delta$ orbitals. Our calculations show that electronic states with occupation of the antibonding $7s\sigma^*$orbital is lower in energy, instead of states with occupation of either the bonding $6d\sigma$ or $6d\delta$ orbital. Even though our results might be counter intuitive, we provide two arguments that may be in favor of our results. First, Roos' results are obtained with CASPT2 calculations and the orbitals used in the perturbation calculations are obtained as the average orbitals that equally describe all possible states of the 8 electrons in 12 orbitals. The orbitals used may not be optimal for the states considered and thus may result in incorrect description of the orders of the correlated states. Secondly, in the atomic case, two electrons occupy the $7s$ atomic orbital, indicating that d orbital correlations are strong and electrons may prefer to occupy the orbitals with 7s character. Nonetheless, we will rather not claim that our calculated results are necessarily correct in comparison with Roos' work. Instead, we only want to point out that for Th_2, due to the very sensitive interplay of the strong correlation of the d electrons and the subtle bonding interactions of the $7s$ and $6d$ orbitals, it is very challenging to determine the order of the low-lying electronic states.

5 Summary

We have implemented the spin–orbit DFT approach in the massively parallel NWChem computational chemistry code. The implementation makes use of the Global Array Tools, allowing for an efficient parallel implementation. Our implementation is based on the RECP description of the scalar relativistic and the effective one-electron spin–orbit interaction operators. However, the implementation of the SODFT approach does not depend explicitly on the spin–orbit integrals and can be applied to one-electron spin–orbit interactions derived from all-electron schemes. The exchange functionals used in the implementation are functionals derived for non-spin–orbit calculations, including GGA as well hybrid functionals. Spin–orbit Hartree–Fock calculations can also be carried out. The effect of using functionals developed for non-spin–orbit interactions in the case of spin–orbit interactions is difficult to quantify but is justified by the good agreement with experimental results and all relativistic calculations, for example, in some of the works cited later in this paper. With this implementation, efficient calculations of single-point energies, analytical gradients and geometry optimizations can be carried out in the DFT framework, taking into account the effective one-electron spin–orbit interactions. Since the implementation is based on the efficient parallelization schemes of

NWChem, scaling properties are similar to the calculations without the spin–orbit interactions.

In addition to computational efficiency, spin-scalar and spin-dependent terms of the effective one-electron equations are treated on an equal footing with both the spin–orbit couplings and correlation effects treated simultaneously and variationally in the SODFT technique. In our SODFT procedure, the one-electron molecular spinors are determined in a fully unrestricted manner and the symmetry-broken solutions ensure that proper dissociation limits can be reached. The SODFT algorithm implemented in NWChem has been applied to a number of interesting problems, including both systems with heavy [12, 47, 48] and lighter elements [49, 50]. In their study of the effects of spin–orbit interaction on the spectroscopic constants (bond lengths, frequencies and dissociation energies) the heavy metal (sixth and seventh row elements) hydrides employing the spin–orbit DFT implemented in NWChem, Lee et al. [49, 50] found that results obtained with the ECP-based SODFT methods are similar to that from the all-electron relativistic density functional approaches and that the spin–orbit effects on the spectroscopic constants are in good agreement with two-component coupled-cluster singles and doubles with perturbative triples results calculated with the same RECPs and basis sets. Quite interestingly, strong effects from spin–orbit interactions on the vibrational frequencies of molecules with relatively light elements were revealed using the SODFT we have implemented in NWChem [49, 50]. In their study of $(CH_2)_nXI$ (n = 1,2 and X = Br, Cl) and similar systems, they have found that the nearly degenerate nonbonding orbitals of iodine can be split due to the inter-halogen interaction and that the spin–orbit interactions further mix these states. The effects of the spin–orbit interactions can be quite significant for open-shell cations even though the properties of the closed-shell neutrals may not be affected much, as shown here for the two systems considered. The implementation has been further extended to include ZORA and Douglas–Kroll implementation of spin–orbit effects in NWChem [51], further extending the range of applications possible. Other implementations of spin–orbit DFT include the ECP SODFT recently implemented in DIRACR [52] and the commercially available packages of all-electron ZORA SODFT in ADF and Turbomole [53, 54],

Acknowledgments This work was done partly using the computational resources at the Molecular Science Computing Facility in the William R. Wiley Environmental Molecular Sciences Laboratory at the Pacific Northwest National Laboratory. The MSCF and EMSL are funded by the Office of Biological and Environmental Research in the US Department of Energy. Part of the work is performed using the computational resources of NNIN, National Nanostructure Infrastructure Network, funded by the National Science Foundation.

References

1. Danovich D, Marian CM, Neuheuser T, Peyerimhoff SD, Shaik S (1998) J Phys Chem A 102:5923
2. Koseki S, Fedorov DG, Schmidt MW, Gordon MS (2001) J Phys Chem A 105:8262
3. Fedorov DG, Koseki S, Schmidt MW, Gordon MS (2003) Int Rev Phys Chem 22:551
4. Agren H, Vahtas O, Minaev B (1996) Adv Quantum Chem 27:71
5. Salem L, Rowland C (1972) Angew Chem Int Ed 11:92
6. Robb MA, Garavelli M, Olivucci M, Bernardi F (2000) Rev Comput Chem 15:87. Lipowitz KB, Boyd DB (eds), Wiley-VCH: New York
7. Yarkony DR (1990) J Phys Chem 94:5572
8. Pyykko P (1978) Adv Quantum Chem 11:353
9. Pitzer RM, Winter NW (1991) Int J Quantum Chem 40:773
10. Balasubramanian K (1997) Relativistic effects in chemistry, part A and B. Wiley, New York
11. Marian CM (2001) Rev Comput Chem 17:99. Lipowitz KB, Boyd DB (ed), Wiley-VCH: New York
12. Zaitsevskii A, Rykova E, Mosyagin NS, Titov AV (2006) Cent Euro J Phys 4:448
13. Yabushita S, Zhang Z, Pitzer RM (1999) J Phys Chem A 103:5791–5800
14. Vallet V, Maron L, Teichteil C, Flament J-P (2000) J Chem Phys 113:1391–1402
15. Pyykkö P (1988) Chem Rev 88:563–594
16. Ermler WC, Ross RB, Christiansen PA (1988) Adv Quantum Chem 19:139–182
17. Wilson S (ed) (1988) Methods in computational chemistry: Vol. 2, relativistic effects in atoms and molecules. Plenum, New York
18. Balasubramanian K (1989) J Phys Chem 93:6585–6596
19. Wilson S, Grant IP, Gyorffy BL (eds) (1991) The effects of relativity in atoms, molecules, and the solid state. Plenum, New York
20. Balasubramanian K (1994) Relativistic effects and electronic structure of lanthanide and actinide molecules. In: Gschneider KA, Eyring L, Choppin GR, Lander GH (eds) Handbook on the physics and chemistry of rare earths, vol 18. Elsevier, Amsterdam, pp 29–158
21. Malli GL (ed) (1994) Relativistic and electron correlation effects in molecules and solids. Plenum, New York
22. Dolg M, Stoll H (1995) Electronic structure calculations for molecules containing lanthanide atoms. In: Gschneider KA, Eyring L (eds) Handbook on the physics and chemistry of rare earths, vol 22. Elsevier, Amsterdam, pp 607–729
23. Balasubramanian K (1997) Relativistic effects in chemistry: part A, theory and techniques. Wiley, New York
24. Liu W-J, Hong G-Y, Dai D-D, Li L-M, Dolg M (1997) The Beijing four component density functional program package (BDF) and its applications to EuO, EuS, YbO and YbS. Theor Chim Acta 96:75–83
25. Varga S, Fricke B, Nakamatsu H, Mukoyama T, Anton J, Geschke D, Heitmann A, Engel E, Bastug T (2000) J Chem Phys 112:3499–3506
26. Anton J, Fricke B, Schwerdtfeger P (2005) Chem Phys 311:97–103
27. Pershina V (2002) J Nucl Radiochem Sci 3:137–141
28. Pershina V, Bastug T, Sarpe-Tudoran C, Anton J, Fricke B (2004) Nucl Phys A 734:200–203
29. Sarpe-Tudoran C, Pershina V, Fricke B, Anton J, Sepp W-D, Jacob T (2003) Eur Phys J D 24:65–67
30. Kahn L, Baybutt P, Truhlar DG (1976) J Chem Phys 10:3826–3853
31. Krauss M, Stevens WJ (1984) Annu Rev Phys Chem 35:357–385
32. Lee YS, Ermler WC, Pitzer KS (1977) J Chem Phys 67:5861–5876

33. See www.clarkson.edu/_pac/reps.html for complete references and a library of potentials
34. Hay PJ, Wadt WR (1985) J Chem Phys 82:270–283
35. Wadt WR (1989) Chem Phys Lett 82:285–288
36. Christiansen PA, Lee YS, Pitzer KS (1979) J Chem Phys 71:4445–4450
37. Christiansen PA, Ermler WC, Pitzer KS (1985) Annu Rev Phys Chem 36:407–432
38. Ross RB, Ermler WC, Christiansen PA (1986) J Chem Phys 84:3297–3300
39. Langhoff SR (1974) J Chem Phys 61:1708–1716
40. Blume M, Watson RE (1963) Proc R Soc Lond A271:565–578
41. Blume M, Freeman AJ, Watson RE (1964) Phys Rev 134:A320–A327
42. Ross RB, Gayen S, Ermler WC (1994) J Chem Phys 100:8145
43. Kuechle W, Dolg M, Stoll H, Preuss H (1994) J Chem Phys 100:7535
44. Cao X, Dolg M, Stoll H (2003) J Chem Phys 118:487
45. Cao X, Dolg M (2004) J Molec Struct (Theochem) 673:203
46. Roos BO, Malmqvist PM, Gagliardi L (2006) J Am Chem Soc 128:17000
47. http://diggy.ruc.dk/bitstream/1800/1595/1/rapport.pdf
48. Choi YJ, Lee YS (2003) J Chem Phys 119:2014
49. Lee M, Kim H, Lee YS, Kim MS (2005) Angew Chem Int Ed 44:2929
50. Lee M, Kim MS (2006) Chem Phys Chem 7:2064
51. Nichols P, Govind N, Bylaska EJ, de Jong WA (2009) J Chem Theory Comput 5:491
52. Park YC, Lim IS, Lee YS (2011) Bull Korean Chem Soc 33:803
53. Velde G, Bickelhaupt FM, Baerends EJ, Guerra CF, van Gisbergen SJA, Snijders JG, Ziegler T (2001) J Comput Chem 22:931
54. Furche F, Ahlrichs R, Hättig C, Klopper W, Sierka W, Weigend F (2013) WIREs Comput Mol Sci. doi:10.1002/wcms.1162

Theor Chem Acc (2014) 133:1475
DOI 10.1007/s00214-014-1475-8

REGULAR ARTICLE

Atomic three- and four-body recurrence formulas and related summations

Frank E. Harris

Received: 31 January 2014 / Accepted: 25 February 2014 / Published online: 25 March 2014

Abstract A new recursive procedure is reported for the evaluation of certain three-body integrals involving exponentially correlated atomic orbitals. The procedure is more rapidly convergent than those reported earlier. The formulas are relevant to ab initio electronic-structure computations on three- and four-body systems. They also illustrate techniques that are useful in the evaluation of summations involving binomial coefficients.

Keywords Three-body integrals · Binomial summations · Exponentially correlated orbitals

1 Introduction

For electronic-structure computations involving exponentially correlated orbitals in atomic systems, it is convenient to generate the necessary integrals using recurrence formulas. For three-body systems, the integrals in question have the generic form

$$\Gamma_{n_1,n_2,n_{12}}(\alpha,\beta,\gamma) = \frac{1}{16\pi^2} \times \int r_1^{n_1-1} r_2^{n_2-1} r_{12}^{n_{12}-1} e^{-\alpha r_1-\beta r_2-\gamma r_{12}} d^3\mathbf{r}_1 d^3\mathbf{r}_2, \tag{1}$$

where $\mathbf{r}_1$ and $\mathbf{r}_2$ (with respective magnitudes r_1 and r_2) are measured from a common origin (ordinarily the position of one of the three particles), $r_{12} = |\mathbf{r}_1 - \mathbf{r}_2|$, and the integration is over all values of $\mathbf{r}_1$ and $\mathbf{r}_2$. Though it may not at first be obvious, the integral in Eq. (1) is symmetric under all simultaneous permutations of its arguments and indices; such permutations merely correspond to renumberings of the particles, including the choice of the particle defining the coordinate origin.

Conventional three-body energy computations require the integrals $\Gamma_{n_1,n_2,n_{12}}$ for a set of nonnegative integer values of n_1, n_2, and n_{12}. Even for $n_1 = n_2 = n_{12} = 0$ these integrals are nonsingular, as can be seen by writing them in terms of the relative coordinates r_1, r_2, and r_{12}, and noting that the volume element (after integrating out the angular coordinates) is proportional to $r_1 r_2 r_{12}\, dr_1 dr_2 dr_{12}$. A general discussion of these three-body integrals can be found in Ref. [1].

The $\Gamma_{n_1,n_2,n_{12}}$ can be generated recursively, starting from

$$\Gamma_{0,0,0}(\alpha,\beta,\gamma) = \frac{1}{(\alpha+\beta)(\alpha+\gamma)(\beta+\gamma)} \tag{2}$$

and using a procedure developed by Sack et al. [2]. That procedure involves the following formulas, in which the Γ, B, and A are assumed to have arguments α, β, γ,

---

Dedicated to the memory of Professor Isaiah Shavitt and published as part of the special collection of articles celebrating his many contributions.

The work of Isaiah Shavitt on ab initio atomic and molecular structure theory was characterized by careful attention to the details of the underlying mathematics and to its development into forms that permitted accurate digital computation. It is in that spirit that this contribution is dedicated to his memory.

F. E. Harris
Department of Physics, University of Utah, Salt Lake City, UT 84112, USA

F. E. Harris (✉)
Quantum Theory Project, University of Florida, Gainesville, FL 32611, USA
e-mail: harris@qtp.ufl.edu

$$\Gamma_{n_1,n_2,n_{12}} = \frac{1}{\alpha+\beta}\left[n_1\Gamma_{n_1-1,n_2,n_{12}} + n_2\Gamma_{n_1,n_2-1,n_{12}} + B_{n_1,n_2,n_{12}}\right], \quad (3)$$

$$B_{n_1,n_2,n_{12}} = \frac{1}{\alpha+\gamma}\left[n_1 B_{n_1-1,n_2,n_{12}} + n_{12}B_{n_1,n_2,n_{12}-1} + A_{n_1,n_2,n_{12}}\right], \quad (4)$$

$$A_{n_1,n_2,n_{12}} = \frac{\delta_{n_1}(n_2+n_{12})!}{(\beta+\gamma)^{n_2+n_{12}+1}}, \quad (5)$$

where $\delta_n = 1$ if $n = 0$ and zero otherwise. It is computationally stable to compute first the array A, then B, and finally Γ.

For some atomic properties, and also in connection with four-body recurrence schemes (vide infra) the Γ are needed with one index equal to -1 but with the others nonnegative, e.g., $\Gamma_{-1,n_2,n_{12}}$. Integrals of this type are convergent, but the recurrence scheme using Eqs. (3)–(5) cannot be used to increase an index of Γ from -1. One method for recursive evaluation of these $\Gamma_{-1,n_2,n_{12}}$ has been presented both by the present author's research group [1] and by Korobov [3]; another method with more rapid convergence is developed in the present communication.

Recursive methods have also been reported for exponentially correlated four-body atomic systems, where the integrals have the generic form

$$\begin{aligned} &I_{n_1,n_2,n_3,m_1,m_2,m_3}(u_1,u_2,u_3,w_1,w_2,w_3) \\ &= \frac{1}{64\pi^3}\int r_1^{m_1-1} r_2^{m_2-1} r_3^{m_3-1} r_{23}^{n_1-1} r_{13}^{n_2-1} r_{12}^{n_3-1} \\ &\quad \times e^{-w_1r_1-w_2r_2-w_3r_3-u_1r_{23}-u_2r_{13}-u_3r_{12}}\, d^3\mathbf{r}_1 d^3\mathbf{r}_2 d^3\mathbf{r}_3. \end{aligned} \quad (6)$$

The integrals described by Eq. (6) are invariant under particle permutations, which include not only permutations of the indices 1, 2, 3 but also changes in the coordinate origin, which correspond to permutations of the type ($w_2 \leftrightarrow u_2$, $w_3 \leftrightarrow u_3$, $n_2 \leftrightarrow m_2$, $n_3 \leftrightarrow m_3$). The symmetry group, isomorphic with that of the 6-j symbols, is the direct product of the six-member group of permutations of (1, 2, 3) and the four-member group of origin changes. The net result is that any one of the six indices of I can be brought to the first index position.

Recurrence formulas for the so-called Hylleraas basis (in which the parameters u_i are zero) were published in 2004 by Pachucki et al. [4]; that work was extended by the present author in 2009 [5] to handle full exponential correlation (general values of all the u_i and w_i). Both these sets of four-body recurrence formulas require an initial integral $I_{0,0,0,0,0,0}$ and various "boundary terms" of the form

$$\begin{aligned} I_{*,n_2,n_3,m_1,m_2,m_3} &= \frac{1}{64\pi^3}\int 4\pi\delta(\mathbf{r}_{23}) r_1^{m_1-1} r_2^{m_2-1} r_3^{m_3-1} r_{13}^{n_2-1} r_{12}^{n_3-1} \\ &\quad \times e^{-w+_1r_1-w_2r_2-w_3r_3-u_2r_{13}-u_3r_{12}}\, d^3\mathbf{r}_1 d^3\mathbf{r}_2 d^3\mathbf{r}_3. \end{aligned} \quad (7)$$

The asterisk, introduced for this purpose in Ref. [4], indicates that $r_{23}^{n_1-1}\exp(-u_1 r_{23})$ is to be replaced by $4\pi\delta(r_{23})$. Insertion of this Dirac delta function enables the integral of Eq. (7) to be reduced to a three-body integral of the type defined in Eq. (1):

$$\begin{aligned} I_{*,n_2,n_3,m_1,m_2,m_3}(,u_2,u_3,w_1,w_2,w_3) &= \Gamma_{m_1,m_2+m_3-1,n_2+n_3-1} \\ &\quad (w_1, w_2+w_3, u_2+u_3). \end{aligned} \quad (8)$$

The vacant first argument of the above I reflects the fact that this integral does not depend on the parameter u_1. When used for four-body recursion, the integrals of Eq. (8) appear only under conditions such that at least one of $m_2 + m_3$ and $n_2 + n_3$ is positive, so the integrals $\Gamma_{\sigma,\mu,\nu}$ will have indices that are nonnegative, except for at most one index of value -1.

The initial integral, $I_{0,0,0,0,0,0}$, can be evaluated in closed form; a formula for it was first presented by Fromm and Hill [6]. Improvements in the Fromm–Hill formula to illuminate its singularity structure and facilitate its computation were subsequently provided by the present author [7].

The recursive four-body formulas increased the importance of having good recursive methods for the three-body exponentially correlated integrals with one index equal to -1. An additional method for dealing with these integrals was briefly sketched by the present author [5], but the material there presented gave neither a full description of the formula nor its method of derivation. The present communication provides the missing derivation and discusses a class of finite summations that are relevant thereto.

2 Recurrence formulas for $\Gamma(-1, n_2, n_{12})$

A starting point for evaluation of $\Gamma(-1, n_2, n_{12})$ is the formula [1] for $\Gamma(-1, 0, 0)$:

$$\Gamma_{-1,0,0}(\alpha,\beta,\gamma) = \frac{1}{\beta^2-\gamma^2}\left[\ln(\alpha+\beta) - \ln(\alpha+\gamma)\right]. \quad (9)$$

We cannot use the procedure of Sack et al. to increase the index -1, but we can use that procedure on the other indices:

$$\Gamma_{-1,n_2,n_{12}} = \frac{1}{\beta+\gamma}\left[n_2\Gamma_{-1,n_2-1,n_{12}} + n_{12}\Gamma_{-1,n_2,n_{12}-1} + G_{n_2,n_{12}}\right] \tag{10}$$

Here $G_{n_2,n_{12}}$ has definition

$$G_{n_2,n_{12}} = \left(-\frac{\partial}{\partial\beta}\right)^{n_2}\left(-\frac{\partial}{\partial\gamma}\right)^{n_{12}} G_{0,0}, \tag{11}$$

with

$$G_{0,0} = \frac{\ln(\alpha+\beta) - \ln(\alpha+\gamma)}{\beta-\gamma}. \tag{12}$$

We have examined several alternative possibilities for the evaluation of $G_{n_2,n_{12}}$. If $|\beta-\gamma|$ is not too small, one can use a variant of the procedure of Sack *et al*, corresponding to the recurrence formula

$$G_{n_2,n_{12}} = \frac{1}{\beta-\gamma}\left[n_2 G_{n_2-1,n_{12}} - n_{12}G_{n_2,n_{12}-1} + K_{n_2,n_{12}}\right], \tag{13}$$

with

$$K_{n_2,n_{12}} = \delta_{n_2}\delta_{n_{12}}[\ln(\alpha+\beta) - \ln(\alpha+\gamma)] - \frac{\delta_{12}(1-\delta_{n_2})(n_2-1)!}{(\alpha+\beta)^{n_2}} + \frac{\delta_2(1-\delta_{n_{12}})(n_{12}-1)!}{(\alpha+\gamma)^{n_{12}}}. \tag{14}$$

Here $\delta_n = 1$ if $n = 0$ and zero otherwise. The use of Eq. (14) is, however, limited by the fact that the formula for $G_{n_2,n_{12}}$ becomes numerically unstable as $\beta-\gamma$ approaches zero. One way of overcoming this difficulty, noted by Korobov [3], is to use downward recursion in n_2 or n_{12}, starting from a $G_{n_2,n_{12}}$ that is deemed negligible.

The alternative approach of Ref. [1] starts by writing $G_{0,0}$ as the following expansion:

$$\begin{aligned} G_{0,0} &= \frac{1}{\beta \quad \gamma}[\ln(\alpha+\beta) - \ln(\alpha+\gamma)] \\ &= -\frac{1}{\beta-\gamma}\ln\left[\frac{(\alpha+\beta)-(\beta-\gamma)}{\alpha+\beta}\right] \\ &= \sum_{\mu=1}^{\infty}\frac{1}{\mu}\frac{(\beta-\gamma)^{\mu-1}}{(\alpha+\beta)^{\mu}}. \end{aligned} \tag{15}$$

Applying the operator $(-\partial/\partial\gamma)^{n_{12}}$ to Eq. (15), we get

$$G_{0,n_{12}} = \sum_{\mu>n_{12}} \frac{(\mu-1)!}{\mu(\mu-n_{12}-1)!}\frac{(\beta-\gamma)^{\mu-n_{12}-1}}{(\alpha+\beta)^{\mu}}. \tag{16}$$

We rewrite this equation in a form that causes the summation to start from zero:

$$G_{0,n_{12}} = \sum_{\mu=0}^{\infty}\frac{(\mu+n_{12})!}{(\mu+n_{12}+1)\mu!}\frac{(\beta-\gamma)^{\mu}}{(\alpha+\beta)^{\mu+n_{12}+1}}. \tag{17}$$

Then, applying $(-\partial/\partial\beta)^{n_2}$ and using Leibniz' formula for repeated differentiation of a product, we first obtain

$$G_{n_2,n_{12}} = \sum_{j=0}^{n_2}(-1)^j\binom{n_2}{j}\sum_{\mu\geq j}\frac{(\mu+n_{12}+n_2-j)!}{(\mu+n_{12}+1)(\mu-j)!} \times \frac{(\beta-\gamma)^{\mu-j}}{(\alpha+\beta)^{\mu+n_{12}+n_2-j+1}}. \tag{18}$$

We next replace μ by $j+k$ and note that the range of k is $(0,\infty)$. Also interchanging the order of the summations, Eq. (18) becomes

$$G_{n_2,n_{12}} = \sum_{k=0}^{\infty}\frac{(n_2+n_{12}+k)!}{k!}\frac{(\beta-\gamma)^k}{(\alpha+\beta)^{n_2+n_{12}+k+1}} \times \sum_{j=0}^{n_2}(-1)^j\binom{n_2}{j}\frac{1}{n_{12}+j+k+1}. \tag{19}$$

The j summation is addressed in Appendix 1; its value, from Eq. (40), is

$$\sum_{j=0}^{n_2}(-1)^j\binom{n_2}{j}\frac{1}{n_{12}+j+k+1} = \frac{n_2!(n_{12}+k)!}{(n_2+n_{12}+k+1)!}. \tag{20}$$

Inserting this result, we reach

$$G_{n_2,n_{12}} = \sum_{k=0}^{\infty}\frac{n_2!(n_{12}+k)!}{k!(n_2+n_{12}+k+1)}\frac{(\beta-\gamma)^k}{(\alpha+\beta)^{n_2+n_{12}+k+1}}. \tag{21}$$

It is evident that the variable involved in the expansion is the dimensionless quantity $\tau = (\beta-\gamma)/(\alpha+\beta)$. This approach is functionally equivalent to that of Korobov and will therefore have the same convergence characteristics. However, if Korobov's formulas are to be used, it should be noted that many are in error by a factor of 2.

For some purposes, it is desirable to have a more symmetric expansion, which we can achieve by defining $x = (\beta+\gamma)/2$, $y = (\beta-\gamma)/2$, and arranging for the expansion variable to be $y/(\alpha+x)$. With that set of variables, we have

$$G_{0,0} = \frac{\ln(\alpha+x+y) - \ln(\alpha+x-y)}{2y}. \tag{22}$$

By a procedure similar to that used in Eq. (15), we can bring $G_{0,0}$ to the form

$$G_{0,0} = \sum_{k=0}^{\infty} \frac{y^{2k}}{(2k+1)(\alpha+x)^{2k+1}}. \tag{23}$$

We now seek to construct the $G_{n_2,n_{12}}$ by applying Eq. (11). To do so, we note that

$$-\frac{\partial}{\partial\beta} = -\frac{1}{2}\left(\frac{\partial}{\partial x} + \frac{\partial}{\partial y}\right), \tag{24}$$

$$-\frac{\partial}{\partial\gamma} = -\frac{1}{2}\left(\frac{\partial}{\partial x} - \frac{\partial}{\partial y}\right). \tag{25}$$

When these formulas are inserted in Eq. (11), we have, for the case $n_2 \geq n_{12}$,

$$G_{n_2,n_{12}} = \frac{(-1)^{n_2+n_{12}}}{2^{n_2+n_{12}}}\left(\frac{\partial^2}{\partial x^2} - \frac{\partial^2}{\partial y^2}\right)^{n_{12}}\left(\frac{\partial}{\partial x} + \frac{\partial}{\partial y}\right)^{n_2-n_{12}} G_{0,0}. \tag{26}$$

Applying binomial expansions to the compound factors in Eq. (26),

$$G_{n_2,n_{12}} = \frac{(-1)^{n_2+n_{12}}}{2^{n_2+n_{12}}}\sum_{\mu=0}^{n_{12}}(-1)^{\mu}\binom{n_{12}}{\mu}\left(\frac{\partial}{\partial x}\right)^{2n_{12}-2\mu}\left(\frac{\partial}{\partial y}\right)^{2\mu} \times \sum_{\nu=0}^{n_2-n_{12}}\left(\frac{\partial}{\partial x}\right)^{n_2-n_{12}-\nu}\left(\frac{\partial}{\partial y}\right)^{\nu}\sum_{k=0}^{\infty}\frac{y^{2k}}{(2k+1)(\alpha+x)^{2k+1}}. \tag{27}$$

When we carry out the indicated differentiations, we note that nonzero contributions only result when $k \geq \mu$, so we change the summation variable k to $j+\mu$, with $j \geq 0$. Evaluation of Eq. (27) then takes the form

$$G_{n_2,n_{12}} = \sum_{j=0}^{\infty}\sum_{\nu=0}^{n_2-n_{12}}\sum_{\mu=0}^{n_{12}}\frac{(-1)^{\mu+\nu}}{2^{n_2+n_{12}}}\binom{n_{12}}{\mu}\binom{n_2-n_{12}}{\nu} \times \frac{(2j+n_2+n_{12}-\nu)!}{(2j-\nu)!(2j+2\mu+1)}\frac{y^{2j-\nu}}{(\alpha+x)^{n_2+n_{12}+2j-\nu+1}}. \tag{28}$$

The summation over μ can now be evaluated. As shown in Appendix 1 at Eq. (41), we have

$$\sum_{\mu=0}^{n_{12}}(-1)^{\mu}\binom{n_{12}}{\mu}\frac{1}{2j+2\mu+1} = \frac{n_{12}!}{2}\frac{1}{(j+\frac{1}{2})_{n_{12}+1}}. \tag{29}$$

The notation $(a)_n$ denotes the Pochhammer symbol, with definition $(a)_0 = 1$, $(a)_1 = a$, $(a)_2 = a(a+1)$, $(a)_n = a(a+1)\cdots(a+n-1)$ for integers $n > 2$. Alternatively,

$$(a)_p = \frac{\Gamma(a+p)}{\Gamma(a)}. \tag{30}$$

The use of Eq. (29) enables us to rewrite the formula for $G_{n_2,n_{12}}$ as

$$G_{n_2,n_{12}} = \frac{n_{12}!}{2^{n_2+n_{12}+1}}\sum_{j=0}^{\infty}\sum_{\nu=0}^{n_2-n_{12}}(-1)^{\nu}\binom{n_2-n_{12}}{\nu} \times \frac{(2j-\nu+n_2+n_{12})!}{(2j-\nu)!(j+\frac{1}{2})_{n_{12}+1}}\frac{y^{2j-\nu}}{(\alpha+x)^{n_2+n_{12}+2j-\nu+1}}. \tag{31}$$

The summation in Eq. (31) can now be reorganized to a form that exhibits it as a power series in $y/(\alpha+x)$. To do so, set $2j-\nu=\sigma$, with $\sigma = 0, 1, 2, \ldots$. We must then restrict ν to nonnegative integers of the same parity as σ, and can write

$$G_{n_2,n_{12}} = \frac{n_{12}!}{2^{n_2}(\alpha+x)^{n_2+n_{12}+1}} \times \sum_{\sigma=0}^{\infty}(-1)^{\sigma}\frac{(n_2+n_{12}+\sigma)!}{\sigma!}S(n_2,n_{12},\sigma)\left(\frac{y}{\alpha+x}\right)^{\sigma}, \tag{32}$$

with

$$S(n_2,n_{12},\sigma) = \sum_{\nu_\sigma}\binom{n_2-n_{12}}{\nu} \times \frac{1}{(\sigma+\nu+1)(\sigma+\nu+3)\cdots(\sigma+\nu+2n_{12}+1)}. \tag{33}$$

The notation ν_σ indicates that ν must be restricted to integers of the same parity as σ.

To proceed further, we need to evaluate the summation S. The evaluation requires a significant number of steps. The result, developed in Appendix 2, takes the form

$$S(n_2,n_{12},\sigma) = \frac{(-1)^{\sigma}2^{n_2}n_2!}{(n_2+n_{12}+1)!}F(n_2+1,-\sigma;n_2+n_{12}+2;2). \tag{34}$$

The quantity $F(a,b;c;x)$ is a hypergeometric function, sometimes written ${}_2F_1(a,b;c;x)$, with definition

$${}_2F_1(a,b;\,c;\,x) = \sum_{j=0}^{\infty}\frac{(a)_j\,(b)_j}{(c)_j\,j!}\,x^j. \tag{35}$$

The quantities $(p)_j$ are Pochhammer symbols, defined after Eq. (29). For a general discussion of the functions ${}_2F_1$, see

Table 1 Coefficients in the expansion of $G_{n_2,n_{12}}$, Eq. (37)

$C_0 = 1$
$C_1 = (\Delta n)$
$C_2 = (\Delta n)^2 + (\Sigma n) + 2$
$C_3 = (\Delta n)^3 + [3(\Sigma n) + 8](\Delta n)$
$C_4 = (\Delta n)^4 + [6(\Sigma n) + 20](\Delta n)^2 + 3(\Sigma n)^2 + 18(\Sigma n) + 24$
$C_5 = \Delta n)^5 + [10(\Sigma n) + 40](\Delta n)^3 + [15(\Sigma n)^2 + 110(\Sigma n) + 184](\Delta n)$
$C_6 = (\Delta n)^6 + [15(\Sigma n) + 70](\Delta n)^4 + [45(\Sigma n)^2 + 390(\Sigma n) + 784](\Delta n)^2 + 15(\Sigma n)^3 + 180(\Sigma n)^2 + 660(\Sigma n) + 720$
$C_7 = (\Delta n)^7 + [21(\Sigma n) + 112](\Delta n)^5 + [105(\Sigma n)^2 + 1050(\Sigma n) + 2464](\Delta n)^3 + [105(\Sigma n)^3 + 1470(\Sigma n)^2 + 6384(\Sigma n) + 8448](\Delta n)$

Here $(\Sigma n) = n_2 + n_{12}$ and $(\Delta n) = n_2 - n_{12}$

Ref. [8]. Despite the appearance of Eq. (34), $S(n_2, n_{12}, \sigma)$ is not really transcendental; with the parameter values given in that equation S reduces to a rational fractional form, so the notation of that equation simply provides a convenient and compact way of specifying the coefficients in the expansion in Eq. (32). With this formula for S, the expansion for $G_{n_2,n_{12}}$ becomes

$$G_{n_2,n_{12}} = \frac{n_2!\, n_{12}!}{(n_2+n_{12}+1)!} \frac{1}{(\alpha+x)^{n_2+n_{12}+1}} \sum_{\sigma=0}^{\infty} \frac{(n_2+n_{12}+\sigma)!}{\sigma!} \times F(n_2+1, -\sigma; n_2+n_{12}+2; 2) \left(\frac{y}{\alpha+x}\right)^{\sigma}. \tag{36}$$

We repeat the definitions: $x = (\beta+\gamma)/2$, $y = (\beta-\gamma)/2$. The expansion given by Eq. (36) should reflect the symmetry of the expansion variable; if we interchange $n_2 \leftrightarrow n_{12}$ and simultaneously interchange $\beta \leftrightarrow \gamma$, the value of G should not change. This invariance can be demonstrated using properties of the hypergeometric function; it can also be seen from the explicit forms of the expansion coefficients. Writing

$$G_{n_2,n_{12}} = \frac{n_2!\, n_{12}!}{(\alpha+x)^{n_2+n_{12}+1}} \sum_{\sigma=0}^{\infty} \frac{(-1)^{\sigma} C_{\sigma}}{\sigma!\,(n_2+n_{12}+\sigma+1)} \left(\frac{y}{\alpha+x}\right)^{\sigma}, \tag{37}$$

the first eight C_σ are given in Table 1. The C_σ have, under the interchange $n_2 \leftrightarrow n_{12}$, the parity of σ. Since y also has this parity under $\beta \leftrightarrow \gamma$, the individual terms in the expansion of $G_{n_2,n_{12}}$ according to Eq. (37) also exhibit its overall symmetry.

When $\beta - \gamma$ is small, the expansion of Eq. (37) converges more rapidly than that of Eq. (21) due to the fact that the expansion parameter in Eq. (37) is half as large as that in Eq. (21). Moreover, as already pointed out, truncated forms of Eq. (37) yield identical values under symmetry interchange, but the same is not true of Eq. (21). We present in Table 2 one numerical example that illustrates these observations.

Table 2 Computations of $G_{n_2,n_{12}}$ for $\alpha = 8.0$, $\beta = 2.0$, $\gamma = 3.0$, using Eqs. (21) or (37) at various truncations.

	Eq. (21)	Eq. (37)
4 terms		
$G_{2,3}(\alpha,\beta,\gamma)$	$1.4418\,19048 \times 10^{-6}$	$1.4412\,84145 \times 10^{-6}$
$G_{3,2}(\alpha,\gamma,\beta)$	1.4411 92027	1.4412 84145
6 terms		
$G_{2,3}(\alpha,\beta,\gamma)$	1.4412 92139	1.4412 82326
$G_{3,2}(\alpha,\gamma,\beta)$	1.4412 81243	1.4412 82326
8 terms		
$G_{2,3}(\alpha,\beta,\gamma)$	1.4412 82476	1.4412 82319
$G_{3,2}(\alpha,\gamma,\beta)$	1.4412 82307	1.4412 82319
Exact result	1.4412 82319	1.4412 82319

Acknowledgments This research has been supported by US Department of Energy Grant DE-SC0002139.

Appendix 1: Some binomial sums

Starting from the equation

$$F(m,n) = \int_0^1 x^m (1-x^2)^n \, dx = \sum_{\mu=0}^{n} \binom{n}{\mu} (-1)^{\mu} \int_0^1 x^{2\mu+m} \, dx = \sum_{\mu=0}^{n} \binom{n}{\mu} \frac{(-1)^{\mu}}{m+1+2\mu}, \tag{38}$$

we evaluate $F(m,n)$ by identifying it as a beta function:

$$F(m,n) = \frac{1}{2}\,B\left(\frac{m+1}{2},\,n+1\right) = \frac{\Gamma(\frac{1}{2}[m+1])\Gamma(n+1)}{2\Gamma(\frac{1}{2}[2n+m+3])} = \frac{n!}{2(\frac{1}{2}[m+1])_{n+1}}. \tag{39}$$

For definition of the beta function and a derivation of Eq. (39), see Ref. [9]. Note also that the notation $(a)_p$ denotes a Pochhammer symbol, defined immediately after Eq. (29).

Expressions of the form $F(m,n)$ are used twice in the main text. Setting $n = n_2$ and $m = 2n_{12} + 2k + 1$, Eqs. (38) and (39) correspond to

$$F(m,n) = \sum_{\mu=0}^{n_2}\binom{n_2}{\mu}\frac{(-1)^\mu}{2(n_{12}+\mu+k+1)} = \frac{\Gamma(n_{12}+k+1)\,n_2!}{2\Gamma(n_2+n_{12}+k+2)} = \frac{(n_{12}+k)!\,n_2!}{2(n_2+n_{12}+k+1)!}, \tag{40}$$

Equivalent to Eq. (20).

Setting $n = n_{12}$ and $m = 2j$,

$$F(m,n) = \sum_{\mu=0}^{n_{12}}\binom{n_{12}}{\mu}\frac{(-1)^\mu}{2j+1+2\mu} = \frac{n_{12}!}{2(j+\frac{1}{2})_{n_{12}+1}}, \tag{41}$$

Equivalent to Eq. (29).

Appendix 2: Evaluation of $S(n_2, n_{12}, \sigma)$

Our starting point for the evaluation of $S(n_2, n_{12}, \sigma)$, defined in Eq. (32), is to write it as an iterated integral. To avoid unnecessary notational complexity, we make the temporary definitions $n = n_2 - n_{12}$, $m = n_{12}$. Thus,

$$S(n_2,n_{12},\sigma) = \sum_{\nu_\sigma}\binom{n}{\nu} \times \frac{1}{(\sigma+\nu+1)(\sigma+\nu+3)\cdots(\sigma+\nu+2m+1)} = \underbrace{\int_0^1 \mathrm{d}z_m\, z_m \int_0^{z_m} \mathrm{d}z_{m-1}\, z_{m-1} \int_0^{z_{m-1}} \cdots \int_0^{z_2} \mathrm{d}z_1\, z_1 \int_0^{z_1} \mathrm{d}z}_{m+1 * \iota\nu\tau\varepsilon\gamma\rho\alpha\lambda\sigma} \times \sum_{\nu_\sigma}\binom{n}{\nu} z^{\sigma+\nu}. \tag{42}$$

Remembering that the index ν_σ is to take only values of the same parity as σ, we evaluate the summation in Eq. (42), obtaining

$$g(z) = \sum_{\nu_\sigma}\binom{n}{\nu} z^{\sigma+\nu} = \frac{z^\sigma}{2}\left[(1+z)^n + (-1)^\sigma(1-z)^n\right]. \tag{43}$$

We now insert the right-hand side of Eq. (43) into Eq. (42), also reversing the integration order, reaching

$$S(n_2,n_{12},\sigma) = \int_0^1 \mathrm{d}z\, g(z) \int_z^1 \mathrm{d}z_1\, z_1 \int_{z_1}^1 \cdots \int_{z_{m-1}}^1 \mathrm{d}z_m\, z_m. \tag{44}$$

We now integrate (from right to left) over the z_i. The z_m integration yields $(1 - z_{m-1}^2)/2$; that over z_{m-1} produces $(1 - z_{m-2}^2)^2/(2\cdot 2^2)$; further integrations over z_{m-2} through z_1 give the overall result $(1 - z^2)^m/2^m m!$. Equation (44) is thereby reduced to

$$\begin{aligned} S(n_2,n_{12},\sigma) &= \frac{1}{2^{m+1}m!}\int_0^1 \mathrm{d}z\, z^\sigma (1-z^2)^m \times \left[(1+z)^n + (-1)^\sigma (1-z)^n\right] \\ &= \frac{1}{2^{m+1}m!}\int_0^1 \mathrm{d}z\, z^\sigma \big[(1+z)^{n+m}(1-z)^m + (-1)^\sigma (1+z)^m (1-z)^{n+m}\big]. \end{aligned} \tag{45}$$

We continue by writing z^σ as its expansion in powers of either $(1+z)$ or $(1-z)$, i.e., as one of

$$z^\sigma = \sum_{j=0}^{\sigma}(-1)^{\sigma-j}\binom{\sigma}{j}(1+z)^j = \sum_{j=0}^{\sigma}(-1)^j\binom{\sigma}{j}(1-z)^j. \tag{46}$$

We insert these expressions into Eq. (45) in a way that leads to

$$S(n_2,n_{12},\sigma) = \frac{(-1)^\sigma}{2^{m+1}m!}\sum_{j=0}^{\sigma}(-1)^j\binom{\sigma}{j} \times \int_0^1 \mathrm{d}z\big[(1+z)^{n+m+j}(1-z)^m + (1-z)^{n+m+j}(1+z)^m\big]. \tag{47}$$

We next process Eq. (47) by carrying out $n+m+j$ integrations by parts, repeatedly differentiating the factors that were originally at powers $n+m+j$ and integrating the other factors. At each step the boundary (integrated) terms vanish. The differentiations produce (for each term) a factor $(n+m+j)!$, while the integrations generate (in the denominator) the product $(m+1)(m+2)\cdots(2m+n+j)$. At each step the (-1) from the integration by parts cancels against a similar quantity from the integration or differentiation of the $(1-z)$ factor. The overall result is

$$S(n_2,n_{12},\sigma)=\frac{(-1)^\sigma}{2^{m+1}m!}\sum_{j=0}^{\sigma}(-1)^j\binom{\sigma}{j}\frac{(n+m+j)!\,m!}{(n+2m+j)!}\times\int_0^1 dz\left[(1-z)^{n+2m+j}+(1+z)^{n+2m+j}\right]. \tag{48}$$

The integral in Eq. (48) has the value $2^{n+2m+j+1}/(n+2m+j+1)$. Inserting that value, canceling $m!$, expanding the binomial coefficient, and replacing n and m by the quantities they represent, we have

$$S(n_2,n_{12},\sigma)=(-1)^\sigma 2^{n_2}\sum_{j=0}^{\sigma}\frac{(-1)^j\sigma^j}{(\sigma-j)!}\frac{(n_2+j)!}{(n_2+n_{12}+j+1)!}\frac{2^j}{j!}. \tag{49}$$

Converting to Pochhammer symbols, Eq. (49) becomes

$$S(n_2,n_{12},\sigma)=\frac{(-1)^\sigma 2^{n_2}n_2!}{(n_2+n_{12}+1)!}\sum_{j=0}^{\sigma}(-\sigma)_j\frac{(n_2+1)_j}{(n_2+n_{12}+2)_j}\frac{2^j}{j!}. \tag{50}$$

Because $(-\sigma)_j$ vanishes for $j>\sigma$, we can extend the summation in Eq. (50) to infinity, thereby causing the sum to correspond to the definition of a hypergeometric function; compare with Eq. (35). The result then reduces to Eq. (34).

References

1. Harris FE, Frolov AM, Smith VH (2004) J Chem Phys 121:6323
2. Sack RA, Roothaan CCJ, Kołos W (1967) J Math Phys 8:1093
3. Korobov VI (2002) J Phys B 35:1959
4. Pachucki K, Puchalski M, Remiddi E (2004) Phys Rev A 70:032502
5. Harris FE (2009) Phys Rev A 79:032517
6. Fromm DM, Hill RN (1987) Phys Rev A 36:1013
7. Harris FE (1997) Phys Rev A 55:1820
8. Abramowitz M, Stegun IA (eds) (1972) Handbook of mathematical functions with formulas, graphs, and mathematical tables (AMS-55). National Bureau of Standards, Washington, DC (reprinted, Dover, 1974)
9. Arfken GB, Weber HJ, Harris FE (2013) Mathematical methods for physicists, 7th edn. Academic Press, New York

Theor Chem Acc (2014) 133:1494
DOI 10.1007/s00214-014-1494-5

REGULAR ARTICLE

Biconfluent Heun equation in quantum chemistry: Harmonium and related systems

Jacek Karwowski · Henryk A. Witek

Received: 5 February 2014 / Accepted: 25 April 2014 / Published online: 11 May 2014

Abstract Schrödinger equation for harmonium and related models may be transformed to the biconfluent Heun equation. The solubility of this equation and its applications in quantum chemistry are briefly discussed.

Keywords Schrödinger equation · Biconfluent Heun equation · Harmonium · Exactly solvable models

1 Introduction

The properties of exact solutions of Schrödinger equation for the hydrogen atom and harmonic oscillator were of fundamental importance for the formulation of theoretical models of atoms, molecules and solids. The search for other exactly solvable potentials resulted in the development of several very general methods including the factorization method of Hull and Infeld [1], the supersymmetric formulation of quantum mechanics [2], and closely linked to it concept of the shape-invariant potentials [3]. In general, the exact wavefunctions corresponding to these solutions were expressed in terms of some orthogonal polynomials multiplied by factors deduced from the form of the potential and from the asymptotic behavior of the underlying differential equation at various limits (0 and ∞ in the case of spherical systems). It was also shown (see, e.g., [2]) that for all known shape-invariant potentials and, consequently, for most of exactly solvable equations, the orthogonal polynomials are special cases of either confluent hypergeometric function ${}_1F_1(a;c;x)$ [4] or hypergeometric function ${}_2F_1(a,b;c;x)$ [5, 6].

Endeavors to analytical solving quantum-mechanical three-body problem have been taken since the earliest years of quantum theory [7–9] and continue until now [10–16]. Bethe and Salpeter in their "Quantum Mechanics of One- and Two-electron Atoms" state that *The differential equation for the two-electron system is not separable. Unlike the solutions for the hydrogen atom, the solutions for the eigenfunctions and energy eigenvalues cannot be expressed in closed analytic form* [17]. This opinion seems to be not a statement of a rigorous mathematical fact but rather an expression of frustrations associated with numerous futile attempts to achieve this task. Though in many aspects the analytical theory of the helium atom reached significant success, it is commonly recognized as a rather unrealistic direction of development. Only a marginal minority of quantum chemists, including the present authors, believes that the analytical approach is promising and may lead to a robust and accurate quantum chemical calculation protocol for many-electron atoms and molecules.

In parallel to the central problem of the helium atom—a system of three quantum particles interacting via Coulomb forces—a search for exactly solvable three-particle problems led to several very interesting discoveries. Probably the most important finding was the so-called *Hooke atom* also referred to as *harmonium*. It was recognized in 1962 by Kestner and Sinanoğlu that the Schrödinger equation describing two electrons interacting by the Coulomb forces

Dedicated to the memory of Professor Isaiah Shavitt and published as part of the special collection of articles celebrating his many contributions.

J. Karwowski (✉)
Institute of Physics, Nicolaus Copernicus University, 87-100 Toruń, Poland
e-mail: jka@fizyka.umk.pl

H. A. Witek
Department of Applied Chemistry and Institute of Molecular Science, National Chiao Tung University, Hsinchu, Taiwan

and confined in a central harmonic (i.e., quadratic or parabolic) potential is separable [18]. The problem was shown quasi-solvable analytically six years later by Santos [19]. In quasi-exactly solvable systems, a single solution expressible in terms of a polynomial multiplied by asymptotic factors may be obtained only for some specific values of constants defining the potential. The results of Santos remained unnoticed for several decades, and the same system was rediscovered a quarter of a century later by Taut [20], becoming a subject of numerous studies (see, e.g., [19–24]). Some related systems, with modified potentials, e.g., containing an additional linear term, were also investigated and applied to the description of a variety of phenomena [25–28].

Independently of the developments motivated by problems in quantum mechanics, mathematical studies of second-order linear ordinary differential equations resulted [29, 30] in formulations going far beyond the classical equations belonging to the hypergeometric class [4–6]. To the most interesting and hardly known in quantum chemistry equations belongs the class of the Heun equations [31] known and studied since 125 years. A relatively recent collection of works on this subject appeared in the proceedings of the Centennial Workshop on Heun's Equation [32]. Important monographs on this and related subjects have been published by Slavyanov and Lay [29] and by Ronveaux [30]. Among equations of the Heun class, the most relevant in the context of harmonium is the biconfluent Heun equation (BHE). It describes harmonium and its more general form with an additional linear term. A discussion of the relations between the BHE and the Schrödinger equation for harmonium is the main subject of this paper.

Hereafter, we use the following notation conventions: A sans serif symbol (e.g., $\mathsf{n}, \mathsf{l}, \mathsf{k}$) always corresponds to a quantum number while the standard one j, k, l, m, n to an integer index; $\mathbb{T}$ stands for a vector with elements t_j;

$$(s)_m = s(s+1)(s+2)\cdots(s+m-1) \tag{1}$$

is the Pochhammer symbol;

$$(s)_{m,a} = s(s+a)(s+2a)\cdots(s+(m-1)a) \tag{2}$$

is the Pochhammer a-symbol (usually referred to in the literature [33] as the Pochhammer k-symbol); $\lfloor a \rfloor$ is the floor function, i.e., the largest integer less than or equal to a.

2 Harmonium

Harmonium may be defined as a quantum three-body problem described by the Schrödinger equation with harmonic interactions between particles $1-3$ and $2-3$ and the Coulombic interaction between particles $1-2$. The problem is separable to three 3D equations also if the $1-2$ interaction is described by an arbitrary potential $\mathsf{V}(r_{12})$ which depends on the distance between the two particles only [28]. The first equation corresponds to the free motion of the center of mass. The second one describes the oscillations of particle 3 and the center of mass of particles 1 and 2 (the spherical harmonic oscillator equation). The third equation reads

$$\mathsf{h}(\mathbf{r})\Phi_{\mathsf{nlm_l}}(\mathbf{r}) = \mathbf{E}_{\mathsf{nl}}\Phi_{\mathsf{nlm_l}}(\mathbf{r}) \tag{3}$$

where $\mathbf{r} = \mathbf{r}_1 - \mathbf{r}_2$,

$$\mathsf{h}(\mathbf{r}) = \frac{\mathsf{p}(\mathbf{r})^2}{2\mu} + \mathsf{V}(\mathbf{r}) + \frac{\mu\omega^2}{2}\mathbf{r}^2, \tag{4}$$

$r = r_{12} = |\mathbf{r}_1 - \mathbf{r}_2|$, μ is the reduced mass of particles 1 and 2 and ω is a constant which depends on the parameters characterizing the harmonic interactions between the particles (see, e.g., [28] for details). Eq. (3) is spherically symmetric, and its solutions can be written as

$$\Phi_{\mathsf{nlm_l}}(\mathbf{r}) = \frac{\phi_{\mathsf{nl}}(r)}{r} Y_{\mathsf{lm_l}}(\hat{\mathbf{r}}), \tag{5}$$

where $\phi_{\mathsf{nl}}(r)$ is the radial part of the wavefunction and $Y_{\mathsf{lm_l}}$ denote the usual spherical harmonics in *3D*.

Assuming

$$\mathsf{V}(r) = \frac{\zeta}{r} + b\,r + c, \tag{6}$$

where ζ, b and c are constants, and setting $b = -r_e\mu\omega^2$, $c = r_e^2\mu\omega^2/2$, where r_e is a new constant introduced to allow for an easy physical interpretation of the potential, we may write the radial Schrödinger equation as

$$\left[-\frac{1}{2\mu}\frac{d^2}{dr^2} + \frac{\mathsf{l}(\mathsf{l}+1)}{2\mu r^2} + \frac{\zeta}{r} + \frac{\mu\omega^2}{2}(r-r_e)^2 - E_{\mathsf{nl}}\right]\phi_{\mathsf{nl}}(r) = 0. \tag{7}$$

Here, we see this equation as a result of exact separation of a 3-particle Schrödinger equation. Alternatively it can be derived for a system of two particles interacting by the Coulomb force and confined in a parabolic external potential.

If $r_e = 0$, then the last equation describes harmonium [19, 20, 24]. For $r_e \neq 0$, it corresponds to a harmonium-like system in which the minimum of the parabolic potential is shifted from $r = 0$ to $r = r_e$. This system, discussed in detail by Ghosh and Samanta [26, 27], is referred to as *shifted harmonium*. If $\omega = 0$ and $\zeta < 0$, the equation describes two interacting particles with opposite electric charges (e.g., positronium or hydrogen atom) [24]. For $\omega = 0$ and $\zeta > 0$, it describes scattering of two particles with the same sign of the charge. Finally, if $\zeta = 0$, we have a spherical harmonic oscillator. For $\omega > 0$, independently of the values of the remaining parameters, all energies derived from Eq. (7) are discrete. On the other hand, if $\omega = 0$ and $\zeta > 0$, then there are no discrete energy levels.

After the substitutions

$$\mathrm{r} = \sqrt{2\mu\,\omega}\,r, \qquad \mathrm{r_e} = \sqrt{2\mu\,\omega}\,r_e, \tag{8}$$

Eq. (7) transforms to

$$\left[-\frac{\mathrm{d}^2}{\mathrm{dr}^2} + \frac{\mathsf{k}(\mathsf{k}-2)}{4\mathrm{r}^2} + \sqrt{\frac{2\mu}{\omega}}\frac{\zeta}{\mathrm{r}} + \frac{(\mathrm{r}-\mathrm{r_e})^2}{4} - \frac{E_{\mathsf{nl}}}{\omega}\right]\phi_{\mathsf{nl}}(\mathrm{r}) = 0. \tag{9}$$

where

$$\mathsf{k} = 2\mathsf{l} + 2 \tag{10}$$

is a quantum number used hereafter in parallel with $\mathsf{l} = \mathsf{k}/2 - 1$ in order to simplify the notation [34]. Square-integrable solutions of Eq. (9) may be expressed as

$$\phi_{\mathsf{nl}}(\mathrm{r}) \sim \mathrm{r}^{\mathsf{k}/2}\,\mathrm{e}^{-(\mathrm{r}-\mathrm{r_e})^2/4}\,P^{\mathsf{k}}_{\mathsf{n}}(\mathrm{r}). \tag{11}$$

The functions $P^{\mathsf{k}}_{\mathsf{n}}(\mathrm{r})$ are square integrable and orthonormal with respect to the weight function (measure)

$$w(\mathrm{r}) = \mathrm{r}^{\mathsf{k}}\mathrm{e}^{-(\mathrm{r}-\mathrm{r_e})^2/2}. \tag{12}$$

This means that

$$\int_0^\infty P^{\mathsf{k}}_{\mathsf{n}}(\mathrm{r})\,P^{\mathsf{k}}_{\mathsf{n'}}(\mathrm{r})\,w(\mathrm{r})\,d\mathrm{r} = C\,\delta_{\mathsf{n},\mathsf{n'}}, \tag{13}$$

where C is a positive constant. Since $\phi_{\mathsf{nl}}(\mathrm{r}) \sim \mathrm{r}^{\mathsf{k}/2}$ for $\mathrm{r} \to 0, P^{\mathsf{k}}_{\mathsf{n}}(0)$ is finite. Therefore, without any loss of generality, we set $P^{\mathsf{k}}_{\mathsf{n}}(0) = 1$ as the normalization condition.

The equation for $P^{\mathsf{k}}_{\mathsf{n}}$ directly results from Eqs. (9) and (11) and reads

$$\left[\frac{\mathrm{d}^2}{\mathrm{dr}^2} + \left(\frac{\mathsf{k}}{\mathrm{r}} + \mathrm{r_e} - \mathrm{r}\right)\frac{d}{d\mathrm{r}} + \left(\mathcal{E}^{\mathsf{k}}_{\mathsf{n}} + \frac{\mathsf{k}\mathrm{r_e} - 2s}{2\mathrm{r}}\right)\right]P^{\mathsf{k}}_{\mathsf{n}} = 0, \tag{14}$$

where

$$s = \sqrt{\frac{2\mu}{\omega}}\,\zeta \tag{15}$$

and the eigenvalue $\mathcal{E}^{\mathsf{k}}_{\mathsf{n}}(\mathrm{r_e}, s)$ is related to the energy of the system as

$$E_{\mathsf{nl}} = \omega\left(\mathcal{E}^{\mathsf{k}}_{\mathsf{n}} + \mathsf{l} + \frac{3}{2}\right). \tag{16}$$

The eigenfunctions

$$P^{\mathsf{k}}_{\mathsf{n}}(\mathrm{r}) = P^{\mathsf{k}}_{\mathsf{n}}\left(\mathrm{r_e}, s, \mathcal{E}^{\mathsf{k}}_{\mathsf{n}}(\mathrm{r_e}, s); \mathrm{r}\right) \tag{17}$$

are square integrable in the sense of Eq. (13). The quantum number $\mathsf{n} = 0, 1, 2, \ldots$ labels all consecutive eigenvalues and eigenfunctions corresponding to fixed $\mathsf{k}, \mathrm{r_e}$ and s.

Equation (14) has been extensively studied by two independent communities. On one hand, the existence of simple polynomial solutions of this equation, discovered in connection with studies on electron correlation, motivated numerous works in the community of quantum chemists [18–28]. However, this equation is known in mathematics since more than a century as the *biconfluent Heun equation* and its properties were studied from both purely mathematical perspective [29, 30, 35–38] and in the context of its applications in different areas of physics [39–41]. Very recently a brief review on its physical applications has been published by Hortaçsu [42]. An analysis of its application to modeling the behavior of two interacting electrons in a uniform magnetic field and a parabolic confinement was published several years ago by Kandemir [34].

3 The traditional approach

In the traditional approach, motivated by the studies on electron correlation problems, one looks for the polynomial solutions of Eq. (14). To this aim, one expresses a solution of Eq. (14) as a power series of r [19, 20]

$$P^{\mathsf{k}}_{\mathsf{n}}(\mathrm{r_e}, s, \mathcal{E}; \mathrm{r}) = \sum_{m=0}^{\infty} a^{\mathsf{k}}_m(\mathrm{r_e}, s, \mathcal{E})\,\mathrm{r}^m, \tag{18}$$

and then formulates conditions under which the expansion terminates, i.e., $P^{\mathsf{k}}_{\mathsf{n}}$ is a polynomial. The normalization $P^{\mathsf{k}}_{\mathsf{n}}(0) = 1$ implies $a^{\mathsf{k}}_0 = 1$.

The substitution of the expansion (18) to Eq. (14) leads to the following three-term recurrence relation

$$\begin{aligned} &B_0 a^{\mathsf{k}}_0 + C_1 a^{\mathsf{k}}_1 = 0,\\ &A_m a^{\mathsf{k}}_m + B_{m+1} a^{\mathsf{k}}_{m+1} + C_{m+2} a^{\mathsf{k}}_{m+2} = 0,\\ &\quad m = 0, 1, 2, \ldots \end{aligned} \tag{19}$$

with

$$A_m = \mathcal{E} - m,\ B_m = (m + \mathsf{k}/2)\,\mathrm{r_e} - s,\ C_m = m(m + \mathsf{k} - 1). \tag{20}$$

The recurrence relation (19) generates a p-th order polynomial if it terminates at a^{k}_p, i.e., if $a^{\mathsf{k}}_p \neq 0$ but $a^{\mathsf{k}}_{p+1} = a^{\mathsf{k}}_{p+2} = \cdots = 0$. This condition may by satisfied if $A_p = 0$ (i.e. $\mathcal{E} = p$) and $a^{\mathsf{k}}_m, m = 0, 1, \ldots, p$ fulfill the set of homogeneous equations (19) for $m = 0, 1, 2, \ldots, p$. This implies that [Eq. (16)]

$$E_{p\mathsf{l}} = \omega\left(p + \mathsf{l} + \frac{3}{2}\right) \tag{21}$$

and

$$W_{p+1}(\mathbb{A}^p,\mathbb{B},\mathbb{C}) = \begin{vmatrix} B_0 & C_1 & 0 & \cdots & 0 & 0 & 0 \\ A_0^p & B_1 & C_2 & \cdots & 0 & 0 & 0 \\ 0 & A_1^p & B_2 & \cdots & 0 & 0 & 0 \\ \vdots & \vdots & \ddots & \ddots & & \vdots & \vdots \\ \vdots & \vdots & & \ddots & \ddots & \vdots & \vdots \\ 0 & 0 & 0 & \cdots & A_{p-2}^p & B_{p-1} & C_p \\ 0 & 0 & 0 & \cdots & 0 & A_{p-1}^p & B_p \end{vmatrix} = 0, \tag{22}$$

where $\mathbb{A}^p = \{A_0^p, A_1^p, \ldots, A_{p-1}^p\}, \mathbb{B} = \{B_0, B_1, \ldots, B_p\}, \mathbb{C} = \{C_1, C_2, \ldots, C_p\}, A_m^p = p - m$.

The determinant depends on s, $\mathrm{r_e}$ and k. Since k is fixed by the selection of a specific angular momentum, Eq. (22) imposes a relation $s_j^p = \varphi_j^p(\mathrm{r_e}), j = 1, 2, \ldots, p+1$, between s and $\mathrm{r_e}$. Thus, for a given degree p of the polynomial and for given $\mathrm{r_e}$, we have a discrete set of $p+1$ values of s for which the polynomial solutions exist. The wavefunctions derived from this procedure are probably the only ones known in the literature in which the r_{12} dependence resulting from the Coulomb interaction between electrons may be expressed exactly in a closed form [43].

Let us note that for given $\{s, \mathrm{r_e}, \mathsf{k}\}$, there exists a complete orthonormal set of solutions of Eq. (14). If there exists p for which the parameters fulfill Eq. (22), then for one of these solutions $\mathcal{E} = p$ and this solution is a polynomial of degree p.

4 Biconfluent Heun equation

In the mathematical literature, the BHE in its canonical form is usually expressed as [30, 35, 36]

$$x y'' + \left(1 + \alpha - \beta x - 2x^2\right) y' + \left((\gamma - \alpha - 2) x - \frac{1}{2}[\delta + \beta(1+\alpha)] \right) y = 0. \tag{23}$$

It is a homogeneous, linear, second-order, differential equation defined in the complex plane. In the two-dimensional space of its particular solutions, one can choose a solution which is finite at $x = 0$. Then the second linearly independent solution behaves at $x = 0$ as $x^{-\alpha}$. The solution finite at $x = 0$ is usually denoted $N(\alpha, \beta, \gamma, \delta; x)$ and referred to as the *biconfluent Heun function*. It is usually expressed as [30]

$$N(\alpha, \beta, \gamma, \delta; x) = \sum_{m=0}^{\infty} \frac{\mathcal{A}_m(\alpha, \beta, \gamma, \delta)}{(1+\alpha)_m} \frac{x^m}{m!}, \tag{24}$$

where

$$\begin{aligned} \mathcal{A}_0 &= 1, \\ \mathcal{A}_1 &= \frac{1}{2}(\delta + \beta(1+\alpha)), \\ \mathcal{A}_{m+2} &= \left((m+1)\beta + \frac{1}{2}[\delta + \beta(1+\alpha)] \right) \mathcal{A}_{m+1} \\ &\quad - (m+1)(m+1+\alpha)(\gamma - 2 - \alpha - 2m)\mathcal{A}_m. \end{aligned} \tag{25}$$

The substitutions

$$\begin{aligned} x &= \mathrm{r}/\sqrt{2} \in \langle 0, \infty \rangle, \\ \alpha &= \mathsf{k} - 1 > 0, \\ \beta &= -\sqrt{2}\,\mathrm{r_e}, \\ \gamma &= 2\mathcal{E} + \mathsf{k} + 1, \\ \delta &= 2\sqrt{2}s, \end{aligned} \tag{26}$$

with all parameters real and $s \geq 0, \mathrm{r_e} \geq 0, \mathcal{E} \geq 0$, restrict the domain of the equation to the real semiaxis and transform Eq. (23) to Eq. (14). In the real semiaxis, we can set

$$N(\alpha, \beta, \gamma, \delta; x) = P(\mathsf{k}, \mathrm{r_e}, s, \mathcal{E}; \mathrm{r}) \tag{27}$$

where $P(\mathsf{k}, \mathrm{r_e}, s, \mathcal{E}; \mathrm{r})$ is a formal solutions of Eq. (14) which may be expressed as in Eqs. (18) and (24). By construction, the normalization $P(\mathsf{k}, \mathrm{r_e}, s, \mathcal{E}; 0) = 1$ is retained. However, $\mathcal{E}$ is an independent parameter, and no boundary conditions have been imposed for $\mathrm{r} \to \infty$. Using the mapping between $\{\alpha, \beta, \gamma, \delta\}$ and $\{\mathsf{k}, \mathrm{r_e}, s, \mathcal{E}\}$ defined by Eqs. (26), one can easily show the equivalence of the recurrence relations (25) and (19). In particular, comparing Eqs. (19), (24) and (26), we get

$$a_m^{\mathsf{k}}(\mathrm{r_e}, s, \mathcal{E}) = \frac{\mathcal{A}_m\left[\mathsf{k} - 1, -\sqrt{2}\,\mathrm{r_e}, 2\mathcal{E} + \mathsf{k} + 1, 2\sqrt{2}s\right]}{2^{m/2}\, m!\, (\mathsf{k})_m}. \tag{28}$$

The asymptotic behavior of $N(\alpha, \beta, \gamma, \delta; x)$ for $x \to \infty$ is given by [35, 36]

$$N(\alpha, \beta, \gamma, \delta; x) \sim K(\alpha, \beta, \gamma, \delta)\, x^{-(\gamma+2+\alpha)/2} e^{\beta x + x^2}, \tag{29}$$

where $K(\alpha, \beta, \gamma, \delta)$ is a constant. Using (26) and (27), we can rewrite Eq. (29) in the limit $\mathrm{r} \to \infty$ as

$$P(\mathsf{k}, \mathrm{r_e}, s, \mathcal{E}; \mathrm{r}) \sim \mathcal{K}(\mathsf{k}, \mathrm{r_e}, s, \mathcal{E})\mathrm{r}^{-(\mathcal{E}+\mathsf{k}+1)} e^{(\mathrm{r}-\mathrm{r_e})^2/2}. \tag{30}$$

By an inspection of Eqs. (12) and (13), we can see that $P(\mathsf{k}, \mathrm{r_e}, s, \mathcal{E}; \mathrm{r})$ is not square integrable, unless

$$\mathcal{K}(\mathsf{k}, \mathrm{r_e}, s, \mathcal{E}) = 0. \tag{31}$$

The last equation plays the role of the quantization condition which determines the eigenvalues. If $\mathcal{E} = \mathcal{E}_{\mathsf{n}}^{\mathsf{k}}(\mathrm{r_e}, s)$ is a root of Eq. (31), then

$$E_{\mathsf{nl}} = \omega\left(\mathcal{E}_{\mathsf{n}}^{\mathsf{k}}(\mathrm{r_e}, s) + \mathsf{l} + \frac{3}{2} \right) \tag{32}$$

and

$$P(\mathsf{k}, \mathrm{r_e}, s, \mathcal{E}^{\mathsf{k}}_{\mathsf{n}}(\mathrm{r_e}, s); \mathrm{r}) = P^{\mathsf{k}}_{\mathsf{n}}(\mathrm{r}); \tag{33}$$

are the square-integrable solutions of Eq. (14).

The structure of expansion (18) for two electrons in a uniform magnetic field was studied by Kandemir [34]. In particular, he reduced the recurrence relation to a closed-form expression. Here, we present a more general analysis. In this derivation, a graphical approach, based on the ideas originally developed by Isaiah Shavitt in his graphical unitary group approach (GUGA) [44, 45], proved to be very useful.

4.1 Expansion coefficients

The recurrence relation (19) may be rewritten as

$$a^{\mathsf{k}}_m = p_{m-1}\, a^{\mathsf{k}}_{m-1} + t_{m-2}\, a^{\mathsf{k}}_{m-2}, \tag{34}$$

where

$$p_m = -\frac{B_m}{C_{m+1}}, \quad t_m = -\frac{A_m}{C_{m+2}}. \tag{35}$$

Equation (34) may be represented by a diagram which facilitates an easy derivation of a closed-form formula for the coefficients of the expansion (18). It also helps to better understand the structure of this expansion and shows interrelations between specific coefficients. The diagram, corresponding to $m = 6$, is presented in Fig. 1. In order to express a specific coefficient, say a^{k}_m, in terms of $p_j, j = 0, 1, \ldots, m-1$ and $t_j, j = 0, 1, \ldots, m-2$, one has to start from the uppermost node corresponding to a^{k}_0 and move to the lower levels of the graph along arcs taking all paths connecting to the node a^{k}_m, visiting each level only once. The contribution from a given path is equal to the product of quantities assigned to the arcs. In the right panel of Fig. 1, these contributions are framed by the square boxes. The value of a^{k}_m is equal to the sum of contributions from all paths. The left panel shows the directions according to which we move along the diagram and, in the nodes, gives the numbers of paths linking this specific node with the one corresponding to a^{k}_0 (i.e., the number of terms in the expression for a given coefficient a^{k}_m). The structure of the graph is self-explanatory, and its extension to larger values of m is obvious.[1]

Using the graph, one can easily write explicit expression for the coefficients a^{k}_m, with $a^{\mathsf{k}}_0 = 1$ set by the normalization condition. Then, we have

$$\begin{aligned} a^{\mathsf{k}}_1 &= p_0, \\ a^{\mathsf{k}}_2 &= p_1 p_0 + t_0, \\ a^{\mathsf{k}}_3 &= p_2 p_1 p_0 + p_2 t_0 + p_0 t_1, \\ a^{\mathsf{k}}_4 &= p_3 p_2 p_1 p_0 + p_3 p_2 t_0 + p_3 p_0 t_1 + p_1 p_0 t_2 + t_2 t_0, \quad \text{etc.} \end{aligned} \tag{36}$$

By introducing

$$Q_m = \prod_{j=0}^{m-1} p_j, \quad R_j = \frac{t_j}{p_j p_{j+1}}, \quad j = 0, 1, 2, \ldots, m-2., \tag{37}$$

we can rewrite Eqs. (36) as

$$a^{\mathsf{k}}_m(\mathrm{r_e}, s, \mathcal{E}) = Q_m \sum_{n=0}^{\lfloor m/2 \rfloor} S^m_n, \tag{38}$$

where

$$\begin{aligned} S^m_0 &= 1, \\ S^m_n &= \sum_{j_1=0}^{m-2n} R_{j_1} \sum_{j_2=j_1}^{m-2n} R_{j_2+2} \cdots \sum_{j_n=j_{n-1}}^{m-2n} R_{j_n+2(n-1)}, \; n \le \lfloor m/2 \rfloor \end{aligned} \tag{39}$$

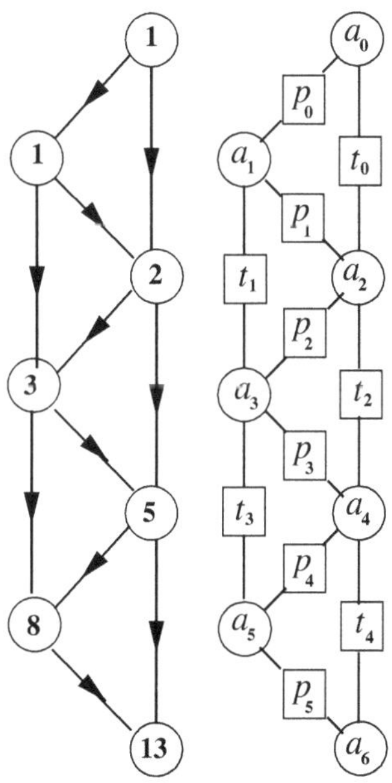

Fig. 1 Graphical representation of three-step recurrence relations exemplified by Eq. (19). In the *left panel*, the ways of constructing paths in the graph and the number of terms in each $a^{\mathsf{k}}_m, m = 0, 1, \ldots, 6$ are shown. In the *right panel*, the arc values are given. See text for details

[1] One can easily see that the recurrence defined by Eq. (34) is structurally identical to the recurrence defining the Fibonacci numbers. Indeed, it is enough to set $p_m = t_m = 1$ for all m and initialize this sequence as $a^{\mathsf{k}}_0 = 1$. The sequence of the Fibonacci numbers may be produced by the diagram shown in the left panel of Fig. 1 and the number of summands needed to compute the coefficient a^{k}_m is equal to the Fibonacci number F_m.

The last equations may be expressed in a compact form as

$$S_0^m = 1,$$
$$S_n^m = \prod_{l=1}^{n} \sum_{j_l=j_{l-1}}^{m-2n} R_{j_l+2(l-1)}, \quad j_0 = 0, \quad n \le \lfloor m/2 \rfloor. \tag{40}$$

Let us note that the product represents n nested sums and is non-commutative.

By the substitution of the explicit expressions (35), (37) and (20), we get

$$Q_m = (-1)^m \prod_{j=0}^{m-1} \frac{B_j}{C_{j+1}} = \frac{(\mathsf{k}\mathfrak{r}_e/2 - s)_{m,\mathfrak{r}_e}}{m!\,(\mathsf{k})_m}. \tag{41}$$

Similarly,

$$R_j = -\frac{A_j\,C_{j+1}}{B_j\,B_{j+1}} = -\frac{(\mathcal{E}-j)(j+1)(j+\mathsf{k})}{(s-(j+\mathsf{k}/2)\,\mathfrak{r}_e)(s-(j+\mathsf{k}/2+1)\,\mathfrak{r}_e)} \tag{42}$$

Note that Eqs. (34)–(40) are valid for an arbitrary three-term recurrence relation, i.e., for all forms of the Heun equation.

Three term recurrences are fulfilled by many quantities related to BHE. In particular, S_m fulfill the following recurrence:

$$S_0^m = 1, \quad S_n^m = S_{n-1}^{m-2} R_{m-2} + S_n^{m-1}. \tag{43}$$

Iterating this relation, we arrive at

$$S_n^m = \sum_{j=\lfloor (n-1)/2 \rfloor}^{m-2} S_{n-1}^j R_j. \tag{44}$$

The iteration terminates at $j = \lfloor (n-1)/2 \rfloor$ because $S_n^{2n-1} = 0$. A three-term recurrence applies also to the determinant W_p. If we set $W_0 = 1$ then, using the Laplace formula, we get:

$$\begin{aligned} W_1 &= B_0 W_0, \\ W_2 &= B_1 W_1 - A_0 C_1 W_0, \\ W_m &= B_{m-1} W_{m-1} - A_{m-2} C_{m-1} W_{m-2}. \end{aligned} \tag{45}$$

The last equation becomes identical with Eq. (34) if we substitute

$$a_m^{\mathsf{k}}(\mathfrak{r}_e, s, \mathcal{E}) = (-1)^m \frac{W_m(\mathbb{A},\mathbb{B},\mathbb{C})}{C_1 C_2 \cdots C_m} = (-1)^m \frac{W_m(\mathbb{A},\mathbb{B},\mathbb{C})}{m!(\mathsf{k})_m}, \tag{46}$$

where $\mathbb{A} = \{A_0, A_1, \ldots, A_{m-2}\}$. As one can easily see,

$$(-1)^m \frac{W_m(\mathbb{A},\mathbb{B},\mathbb{C})}{C_1 C_2 \cdots C_m} = V_m(-\mathbb{T},\mathbb{P}), \tag{47}$$

where

$$V_{m+1}(-\mathbb{T},\mathbb{P}) = \begin{vmatrix} p_0 & 1 & 0 & \cdots & 0 & 0 & 0 \\ -t_0 & p_1 & 1 & \cdots & 0 & 0 & 0 \\ 0 & -t_1 & p_2 & \cdots & 0 & 0 & 0 \\ \vdots & \vdots & \ddots & \ddots & & \vdots & \vdots \\ \vdots & \vdots & & \ddots & \ddots & \vdots & \vdots \\ 0 & 0 & 0 & \cdots & -t_{m-2} & p_{m-1} & 1 \\ 0 & 0 & 0 & \cdots & 0 & -t_{m-1} & p_m \end{vmatrix} = 0, \tag{48}$$

and $\mathbb{T} = \{t_0, t_1, \ldots, t_{m-1}\}, \mathbb{P} = \{p_0, p_1, \ldots, p_m\}$. Then, the expansion coefficients are given by

$$a_m^{\mathsf{k}}(\mathfrak{r}_e, s, \mathcal{E}) = V_m(-\mathbb{T},\mathbb{P}), \quad m = 0, 1, 2, \ldots, \tag{49}$$

with $V_0 = 1$, and the recurrence relations (34) may be rewritten as[2]

$$V_m = p_{m-1} V_{m-1} + t_{m-2} V_{m-2}. \tag{50}$$

Let us define

$$Z_l = \sum_{m=0}^{l} V_m(-\mathbb{T},\mathbb{P})\,\mathfrak{r}^m. \tag{51}$$

Since

$$V_{l+1}\,\mathfrak{r}^{l+1} = Z_{l+1} - Z_l, \tag{52}$$

using Eq. (50), we get

$$Z_{l+1} = Z_l(1 + p_l\,\mathfrak{r}) - Z_{l-1}(p_l - t_{l-1}\,\mathfrak{r})\mathfrak{r} - Z_{l-2}\,t_{l-1}\,\mathfrak{r}^2, \tag{53}$$

where $Z_0 = 1$ and $Z_q = 0$ if $q < 0$.

4.2 Harmonium: the special case of $\mathfrak{r}_e = 0$

In the case of harmonium $\mathfrak{r}_e = 0$, i.e., the minimum of the parabolic potential is located at $\mathfrak{r} = 0$. In this case, the s-dependence of the equations is much simpler since $B_m = -s$ for all values of m. As a consequence,

$$R_j = -\frac{\rho_j}{s^2}, \; j = 0, 1, 2, \ldots, m-2, \tag{54}$$

and

$$S_n^m = (-1)^n \frac{\sigma_n^m}{s^{2n}}, \; n \le \lfloor m/2 \rfloor, \tag{55}$$

where

$$\rho_j = A_j C_{j+1} = (\mathcal{E} - j)(j+1)(j+\mathsf{k}), \; j = 0, 1, 2, \ldots, m-2, \tag{56}$$

and

[2] A large number of relations fulfilled by determinants, including the ones used in this work, may be found in Refs. [46, 47].

$$\sigma_0^m = 1,$$
$$\sigma_n^m = \prod_{l=1}^{n} \sum_{j_l=j_{l-1}}^{m-2n} \rho_{j_l+2(l-1)}, \quad j_0 = 0, \quad n \leq \lfloor m/2 \rfloor. \tag{57}$$

Now, Eq. (38) may be rewritten as

$$a_m^{\mathsf{k}}(0, s, \mathcal{E}) = \frac{s^m}{m!(\mathsf{k})_m} \sum_{n=0}^{\lfloor m/2 \rfloor} (-1)^n \frac{\sigma_n^m}{s^{2n}} \tag{58}$$

with the dependence on s explicitly shown. In the special case of $s = 0$, corresponding to the spherical harmonic oscillator, the only nonzero contribution to Eq. (58) is given by the term with $n = m/2$. Thus, non-vanishing coefficients correspond to $m = 2j$, $j = 0, 1, 2, \ldots$ and

$$a_{2j}^{\mathsf{k}}(0, 0, \mathcal{E}) = \frac{\sigma_j^{2j}}{(2j)!(\mathsf{k})_{(2j)}}, \tag{59}$$

where, according to Eqs. (56) and (57),

$$\sigma_j^{2j} = \prod_{n=0}^{j-1} \rho_{2n} = \prod_{n=0}^{j-1} (\mathcal{E} - 2n)(2n+1)(2n+\mathsf{k}). \tag{60}$$

Combining Eqs. (59) and (60), we get

$$a_{2j}^{\mathsf{k}}(0, 0, \mathcal{E}) = \frac{(-1)^j (-\mathcal{E})_j}{j!\, 2^j \left((\mathsf{k}+1)/2\right)_j}. \tag{61}$$

Equation (55) and the recurrence relations (43) and (44) imply similar recurrences for σ_n^m:

$$\sigma_0^m = 1, \quad \sigma_n^m = \rho_{m-2}\, \sigma_{n-1}^{m-2} + \sigma_n^{m-1}, \tag{62}$$

and

$$\sigma_n^m = \sum_{j=\lfloor (n-1)/2 \rfloor}^{m-2} \sigma_{n-1}^j\, \rho_j. \tag{63}$$

Equation (46) with $\mathfrak{r}_e = 0$ is equivalent to Eq. (58) and, depending on the circumstances using one or another may be more convenient. In particular, comparing Eqs. (46) and (58), we get the expansion of W_m, with $\mathfrak{r}_e = 0$, in terms of powers of s:

$$W_m(\mathbb{A}, \mathbb{B}, \mathbb{C}) = \sum_{n=0}^{\lfloor m/2 \rfloor} (-1)^{n+m}\, \sigma_n^m\, s^{m-2n}. \tag{64}$$

The recurrence relation (62) substituted to the last equation results in Eqs. (45).

5 Odds and ends

5.1 Expansion at $+\infty$

One can also expand solutions of BHE at $+\infty$ and obtain the so-called *recessive Thomé solutions* [29]. The recessive Thomé solution (Eq. (3.1.14) of [29]) is given by

$$P^{[\infty]}(\mathfrak{r}) = \mathfrak{r}^{\mathcal{E}} \sum_{m=0}^{\infty} b_m \mathfrak{r}^{-m}. \tag{65}$$

with $b_0 = 1$ and

$$\tilde{B}_0 b_0 + \tilde{C}_1 b_1 = 0,$$
$$\tilde{A}_m b_m + \tilde{B}_{m+1} b_{m+1} + \tilde{C}_{m+2} b_{m+2} = 0, \tag{66}$$
$$m = 0, 1, 2, \ldots$$

where

$$\tilde{A}_m = (\mathcal{E} - m)(\mathcal{E} - m + \mathsf{k} - 1),\ \tilde{B}_m = -s + \mathfrak{r}_e(\mathcal{E} + \mathsf{k}/2 - m),$$
$$\times \tilde{C}_m = m. \tag{67}$$

An analysis similar to the one leading to Eq. (46) results in the following closed-form expression for the expansion coefficients:

$$b_m = \frac{(-1)^m}{m!} W_m(\tilde{\mathbb{A}}, \tilde{\mathbb{B}}, \tilde{\mathbb{C}}). \tag{68}$$

5.2 Some special cases

The diagram representing three-term recurrence relations may also illustrate the termination of the pertinent expansions. For example, let $a_n^{\mathsf{k}} \neq 0$ for $n < 3$, but $a_3^{\mathsf{k}} = t_1 a_1^{\mathsf{k}} + p_2 a_2^{\mathsf{k}} = 0$. If we wish to construct a solution for which $a_m^{\mathsf{k}} = 0$ if $m \geq 3$ then also a_4^{k} should vanish. But this is possible only if $t_2 = 0$. If two consecutive coefficients (in this example a_3^{k} and a_4^{k}) vanish, all remaining ones with indices larger than 4 also vanish. The condition $a_3^{\mathsf{k}} = 0$ is referred to as the *closing condition* and is equivalent to Eq. (22). In general, if the expansion terminates at a_p^{k}, i.e., $a_p^{\mathsf{k}} \neq 0$ but $a_{p+1}^{\mathsf{k}} = a_{p+2}^{\mathsf{k}} = \cdots = 0$, then

$$t_p = 0 \tag{69}$$

and

$$a_{p+1}^{\mathsf{k}} = \frac{(-1)^{p+1}}{(p+1)!\,(\mathsf{k})_{p+1}} W_{p+1}(\mathbb{A}, \mathbb{B}, \mathbb{C}) = 0 \tag{70}$$

The first of these conditions implies

$$A_p = \mathcal{E} - p = 0, \tag{71}$$

which is equivalent to Eq. (21). The second condition reduces to $W_{p+1} = 0$. The following examples, corresponding to cases known from the literature, illustrate the procedure. In order to simplify the formulas, we set $\mathfrak{r}_e = 0$ and $\mu = 1/2$.

- $p = 0$: In this case, the closing condition reduces to $B_0 = -s = 0$. Then, the solution corresponds to $p = 0$ states of the spherical harmonic oscillator. $P_0^{\mathsf{k}}(\mathfrak{r}) = 1, E = \omega\,(\mathsf{k}+1)/2$, and ω is an arbitrary positive number.

- $p = 1$: In this case $P_1^{\mathsf{k}}(\mathrm{r}) = 1 + B_0\mathrm{r} = 1 - s\mathrm{r}$. The closing condition

$$B_0B_1 - A_0C_1 = s^2 - \mathsf{k} = 0 \tag{72}$$

leads to

$$\omega = \frac{\zeta^2}{\mathsf{k}}. \tag{73}$$

and, consequently,

$$E = \zeta^2 \frac{\mathsf{k}+3}{2\mathsf{k}}. \tag{74}$$

Parameter ζ is arbitrary but ω is defined by Eq. (73). In the case of electron-electron interaction $\zeta = 1$ and, thus, $\omega = 1/\mathsf{k}$ and $E = (\mathsf{k}+3)/(2\mathsf{k})$.

5.3 Harmonic oscillator

The BHE transforms to the spherical harmonic oscillator equation if $s = 0$ and $\mathrm{r_e} = 0$. In this case $p_m = 0$ and the three-term recurrence (34) simplifies to $a_m^{\mathsf{k}} = t_{m-2}a_{m-2}^{\mathsf{k}}$. An inspection of Fig. 1 shows that the only nonzero coefficients are $a_0^{\mathsf{k}} = 1, a_2^{\mathsf{k}} = t_0, a_4^{\mathsf{k}} = t_0t_2, \ldots, a_{2n}^{\mathsf{k}} = t_0t_2 \ldots t_{2n-2}$. The condition for the termination of the recurrence is the quantization condition and if a_{2n}^{k} is the last term of the expansion then $n = \mathsf{n}$, where n is the principal quantum number, and $\mathcal{E}_{\mathsf{n}}^{\mathsf{k}} = 2n$, i.e., $E_{\mathsf{nl}} = \omega(2\mathsf{n} + (\mathsf{k}+1)/2)$. The explicit form of the expansion coefficients may be easily obtained using Eqs. (35). We get

$$a_{2j}^{\mathsf{k}} = \frac{2^{-j}(-\mathsf{n})_j}{j!\,((\mathsf{k}+1)/2)_j}. \tag{75}$$

One can see that this expression may also be obtained from Eq. (61) upon the substitution $\mathcal{E} = 2n$. Thus,

$$P_{\mathsf{n}}^{\mathsf{k}}(\mathrm{r}) = {}_1F_1\left(-\mathsf{n}; \frac{\mathsf{k}+1}{2}; \frac{\mathrm{r}^2}{2}\right), \tag{76}$$

as it should be for the spherical harmonic oscillator (see, e.g., [48]).

6 Several remarks on the non-polynomial solutions

As it was already mentioned, the family of the Heun equations, including the BHE, was studied by the mathematicians since more than a century [29, 30, 35–38]. A very rich bibliography of the texts published on the Heun functions throughout the years has been collected in the framework of *The Heun Project: Heun functions, their generalizations and applications* created at the University of Sofia [49]. In this section, some general results derived in the mathematical literature are briefly discussed and transformed to a form suitable for quantum chemical implementations.

6.1 Some general properties BHE

The constant $K(\alpha,\beta,\gamma,\delta)$ in Eq. (29) which determines the asymptotic behavior of $N(\alpha,\beta,\gamma,\delta;x)$ for $x \to \infty$ is given by [35, 36]

$$K(\alpha,\beta,\gamma,\delta) = C(\alpha,\gamma)J_\lambda\left(\frac{\alpha+\gamma}{2}, \beta, \frac{3\alpha-\gamma}{2}, \delta + \beta\frac{\gamma-\alpha}{2}\right), \tag{77}$$

where

$$C(\alpha,\gamma) = \frac{\Gamma(1+\alpha)}{\Gamma\left(\frac{\alpha-\gamma}{2}\right)\Gamma\left(\frac{\alpha+\gamma}{2}+1\right)}, \tag{78}$$

$$J_\lambda(a,b,c,d) = \int_0^\infty x^{\lambda-1}\mathrm{e}^{-bx-x^2}N(a,b,c,d;x)\mathrm{d}x \tag{79}$$

and $\lambda = (\gamma+\alpha)/2 + 1$. The integral is absolutely convergent in a rather narrow range of the parameters: $0 < \lambda < 1 + (\alpha+\gamma)/2$ [35]. However, for our aims, the absolute convergence is not necessary. Besides, in some cases, the singularities which appear when the integral is divergent determine the energy eigenvalues.

Using Eqs. (26), (27) and (77), we can express $\mathcal{K}(\mathsf{k}, \mathrm{r_e}, s, \mathcal{E})$ from Eq. (30) as

$$\mathcal{K}(\mathsf{k}, \mathrm{r_e}, s, \mathcal{E}) = c_k(\lambda)\sum_{m=0}^{\infty} a_m^\lambda(\mathrm{r}'_\mathrm{e}, s', \mathcal{E}')\, G_{\mathrm{r_e}}(\lambda+m), \tag{80}$$

where

$$c_{\mathsf{k}}(\lambda) = \frac{\Gamma(\mathsf{k})\,2^{\lambda/2}}{\Gamma(\mathsf{k}-\lambda)\Gamma(\lambda)}\mathrm{e}^{-\mathrm{r_e^2}/2}, \tag{81}$$

$$\begin{aligned} &\lambda = k + \mathcal{E} + 1,\\ &\mathrm{r}'_\mathrm{e} = \mathrm{r_e},\\ &s' = s - \mathrm{r_e}\,(\mathcal{E} + \mathsf{k}/2 + 1),\\ &\mathcal{E}' = -\mathcal{E} - 2, \end{aligned} \tag{82}$$

and, according to Eqs. (3.462) and (3.478) of Ref. [50],

$$G_{\mathrm{r_e}}(\nu) = \int_0^\infty \mathrm{r}^{\nu-1}\mathrm{e}^{-(\mathrm{r}-\mathrm{r_e})^2/2}\mathrm{d}\mathrm{r} = \begin{cases} \Gamma(\nu)D_{-\nu}(-\mathrm{r_e})\mathrm{e}^{-\mathrm{r_e^2}/4}, & \text{if } \mathrm{r_e} \neq 0,\\ 2^{\nu/2-1}\,\Gamma(\nu/2), & \text{if } \mathrm{r_e} = 0. \end{cases} \tag{83}$$

where $D_{-\nu}(-\mathrm{r_e})$ is the parabolic cylinder function.

The coefficients a_m^λ in Eq. (80) can be expressed in a closed form using Eq. (49). By the substitution of the parameters defined in Eqs. (82), we get

$$a_m^{\lambda}(\mathfrak{r}_e', s', \mathcal{E}') = V_m(-\mathbb{T}', \mathbb{P}'), \tag{84}$$

where $\mathbb{T}' = \{t_0', t_1', \ldots, t_{m-1}'\}, \mathbb{P}' = \{p_0', p_1', \ldots, p_m'\}$ and

$$\begin{aligned} t_n' &= \frac{\mathcal{E}+n+2}{(\mathcal{E}+\mathsf{k}+n+2)(n+2)}, \quad n = 0, 1, \ldots, m-1, \\ p_n' &= \frac{s - \mathfrak{r}_e(\mathcal{E}+\mathsf{k}/2+n+1)}{(\mathcal{E}+\mathsf{k}+n+1)(n+1)}, \quad n = 0, 1, \ldots, m. \end{aligned} \tag{85}$$

6.2 Harmonium

The formulas simplify if we set $\mathfrak{r}_e = 0$, i.e., if we consider harmonium. Then

$$G_0(\lambda+m) = 2^{(\lambda+m)/2-1}\Gamma\left(\frac{\lambda+m}{2}\right). \tag{86}$$

Thus,

$$G_0(\lambda+m) = 2^{(\lambda+m)/2-1} \begin{cases} \Gamma(\frac{\lambda}{2})(\frac{\lambda}{2})_l, & \text{if } m = 2l,\ l = 0, 1, 2, \ldots, \\ \Gamma(\frac{\lambda+1}{2})(\frac{\lambda+1}{2})_l, & \text{if } m = 2l+1. \end{cases} \tag{87}$$

Let us denote $d_m^{\mathsf{k}}(\lambda) = c_{\mathsf{k}}(\lambda)G_0(\lambda+m)$. Then, using the duplication formula for the gamma function

$$\Gamma(\lambda) = \frac{2^{\lambda-1}}{\sqrt{\pi}}\,\Gamma\left(\frac{\lambda}{2}\right)\Gamma\left(\frac{\lambda+1}{2}\right), \tag{88}$$

we get

$$d_m^{\mathsf{k}}(\lambda) = \frac{\sqrt{\pi}\Gamma(\mathsf{k})}{\Gamma(\mathsf{k}-\lambda)} \times \begin{cases} 2^j\,\Gamma\left(\frac{\lambda+1}{2}\right)^{-1}\left(\frac{\lambda}{2}\right)_j, & \text{if } m = 2j,\ j = 0, 1, 2, \ldots, \\ 2^{j+1/2}\,\Gamma\left(\frac{\lambda}{2}\right)^{-1}\left(\frac{\lambda+1}{2}\right)_j, & \text{if } m = 2j+1, \end{cases} \tag{89}$$

According to Eq. (58),

$$a_m^{\lambda}(0, s, -\mathcal{E}-2) = \sum_{n=0}^{\lfloor m/2 \rfloor}(-1)^n\,\tilde{\sigma}_n^m\,\frac{s^{m-2n}}{m!(\lambda)_m}, \tag{90}$$

where $\tilde{\sigma}_n^m$ is determined in terms of

$$\tilde{\rho}_j = -(\mathcal{E}+2+j)(j+1)(j+\lambda) \tag{91}$$

rather than in terms of ρ_j as defined in Eq. (56). Alternatively we can express a_m in terms of determinants V_m using Eq. (84).

Integrals $J_\lambda(a, b, c, d)$ fulfill the following recurrence relation [35]:

$$\begin{aligned} &(c+a-2-2\lambda)J_{\lambda+2} - \frac{1}{2}(d+b(2\lambda+1-a))J_{\lambda+1} \\ &+ \lambda(\lambda-a)J_\lambda = 0. \end{aligned} \tag{92}$$

In some analyses, this recurrence relation may be very useful. In terms of the parameters describing harmonium, it reads

$$\begin{aligned} &2(\mathcal{E}+3)J_{\mathcal{E}+\mathsf{k}+3} + \sqrt{2}(s - \mathfrak{r}_e(\mathcal{E}+2))J_{\mathcal{E}+\mathsf{k}+2} - (\mathcal{E}+\mathsf{k}+1) \\ &\times J_{\mathcal{E}+\mathsf{k}+1} = 0 \end{aligned} \tag{93}$$

As an example of interesting consequences of this relation let us note that for $\mathsf{k} = 2$ and for arbitrary s and $\mathfrak{r}_e$, the values of $\mathcal{E}$ for which condition $J_{\mathcal{E}+4} = 0$ is fulfilled may also be derived from $J_{\mathcal{E}+3} = 2J_{\mathcal{E}+5}$.

6.3 Spherical harmonic oscillator: the special case of $\mathfrak{r}_e = 0$ and $s = 0$

If $\mathfrak{r}_e = 0$ and $s = 0$ then, according to Eqs. (59) and (90) for $m = 2j, j = 0, 1, 2, \ldots$

$$a_{2j}^{\lambda}(0, 0, -\mathcal{E}-2) = \frac{(-1)^j\tilde{\sigma}_j^{2j}}{(2j)!(\lambda)_{(2j)}} \tag{94}$$

and $a_{2j+1} = 0$. According to Eqs. (60) and (91)

$$\tilde{\sigma}_j^{2j} = (-1)^j\prod_{n=0}^{j-1}(\mathcal{E}+2+2n)(2n+1)(2n+\lambda). \tag{95}$$

Combining Eqs. (89), (94), (95) and using several times the duplication formula (88), we get

$$\mathcal{K}(\mathsf{k}, 0, 0, \mathcal{E})) = \frac{\sqrt{\pi}}{2^{\lambda}}\frac{\Gamma(2c-2a)\,{}_2F_1(a, b; c; 1)}{\Gamma(1-2a)\Gamma(c)}, \tag{96}$$

where $a = \mathcal{E}/2+1, b = \lambda/2, c = (\lambda+1)/2$. The hypergeometric function may be expressed by the gamma functions using equation ([50], 9.122)

$${}_2F_1(a, b; c; 1) = \frac{\Gamma(c)\Gamma(c-a-b)}{\Gamma(c-a)\Gamma(c-b)}. \tag{97}$$

After the substitution to Eq. (96) and some simple transformations, we get a surprisingly simple expression

$$\mathcal{K}(\mathsf{k}, 0, 0, \mathcal{E}) = \Gamma\left(\frac{\mathsf{k}+1}{2}\right)\Gamma\left(-\frac{\mathcal{E}}{2}\right)^{-1}. \tag{98}$$

The same result may be obtained by the straightforward evaluation of J_λ according to Eq. (79) with the Heun function given by Eq. (76). Roots of equation $\mathcal{K}(\mathsf{k}, 0, 0, \mathcal{E}) = 0$ are equal to $\mathcal{E} = 2\mathrm{n}, \mathrm{n} = 0, 1, 2, \ldots$, and the substitution of these roots to the appropriate expansion gives the radial wavefunctions (76).

7 Final remarks

The quest of square-integrable analytical solutions of BHE, apart of the mathematical interest, has been motivated by numerous applications in theory of atomic and molecular structure. The applications range from the analysis of the behavior of the wavefunction in the vicinity of the Coulomb singularity [18, 20, 43] and the construction of the exact density functionals ([51] and references therein) to the studies on the dependence of the charge density distribution in a molecule on the masses of the constituent particles [52, 53]. Until now, these applications are restricted to the well-known polynomial solutions. An extension to the non-polynomial ones is an interesting and important challenge. Possibly, some further studies on the properties of the solutions of BHE may result in tractable algorithms for deriving the eigenvalues and deriving the analytic forms of square-integrable non-polynomial solutions. Maybe, by expressing the formal solution of BHE in terms of the Sturm functions or the hypergeometric functions rather than in powers of r, one could easily reduce the set of solutions to the space of square-integrable functions. Certainly, by using the Heun equation, we can see the problem of harmonium from a wider perspective and get new, powerful technical tools to study its properties.

Acknowledgments J.K. gratefully acknowledges a research Grant from the National Chiao Tung University. National Science Council of Taiwan (Grant NSC 102-2113-M-009-015-MY3) and Ministry of Education (MOE-ATU project) are also acknowledged for financial support.

References

1. Infeld L, Hull TE (1951) The factorization method. Rev Mod Phys 23:21–68
2. Cooper F, Khare A, Sukhatme U (1995) Supersymmetry and quantum mechanics. Phys Rep 251:267–285
3. Gendenshtein LE (1983) JETP Lett. (Pis'ma Zh. Eksp. Teor. Fiz.) Derivation of exact spectra of the Schrödinger equation by means of supersymmetry **38**, 356–359
4. Slater LJ (1960) Confluent hypergeometric functions. Cambridge University Press, Cambridge
5. Bailey WN (1935) Generalized hypergeometric series. Cambridge University Press, Cambridge
6. Slater LJ (1966) Generalized hypergeometric functions. Cambridge University Press, Cambridge
7. Hylleraas EA (1930) Über den Grundterm der Zweielektronenprobleme von H, He, Li^+, Be^{++} usw. Z Phys 65:209–225
8. Kato T (1951) On the existence of solutions of the helium wave equation. Trans Am Math Soc 70:212–218
9. Fock VA (1954) Izv. Akad. Nauk Ser. Fiz. 18:161–172. English translation (1958) On the Schrödinger equation of the helium atom, K. Norske Vidensk. Selsk. Forh. 31:138–152
10. Morgan JD III (1986) Convergence properties of Fock's expansion for S-state eigenfunctions of the helium atom. Theor Chim Acta 69:181–223
11. Abbott PC, Maslen EN (1987) Coordinate systems and analytic expansions for three-body atomic wavefunctions: I. Partial summation for the Fock expansion in hyperspherical coordinates. J Phys A: Math Gen 20:2043–2075
12. Gottschalk JE, Abbott PC, Maslen EN (1987) Coordinate systems and analytic expansions for three-body atomic wavefunctions: II. Closed form wavefunction to second order in r. J Phys A Math Gen 20:2077–2104
13. Gottschalk JE, Maslen EN (1987) Coordinate systems and analytic expansions for three-body atomic wavefunctions: III. Derivative continuity via solutions to Laplace's equation. J Phys A Math Gen 20:2781–2803
14. Pluvinage Ph (1982) Premiers termes du développement de Fock pour les états S de HeI et sa séquence isoélectronique. J Physique 43:439–458
15. Myers CR, Umrigar CJ, Sethna JP, Morgan JD III (1991) Fock's expansion, Kato's cusp conditions, and the exponential ansatz. Phys Rev A 44:5537–5546
16. Witek HA (2013) Toward analytical wave function of helium atom, 8th Congress of the International Society for Theoretical Chemical Physics, Budapest
17. Bethe HA, Salpeter EE (1957) Quantum mechanics of one- and two-electron atoms, Springer, Berlin (See also the Dover edition, New York (2008))
18. Kestner NR, Sinanoğlu O (1962) Study of electron correlation in helium-like systems using an exactly solvable model. Phys Rev 128:2687–2692
19. Santos E (1968) Calculo aproximado de la energia de correlacion entre dos electrones. Anal R Soc Esp Fis Quim 64:177–193
20. Taut M (1993) Two electrons in an external oscillator potential: particular analytic solutions of a Coulomb correlation problem. Phys Rev A 48:3561–3566
21. Cioslowski J, Pernal K (2000) The ground state of harmonium. J Chem Phys 113:8434–8443
22. Matito E, Cioslowski J, Vyboishchikov SF (2010) Properties of harmonium atoms from FCI calculations: calibration and benchmarks for the ground state of the two-electron species. Phys Chem Chem Phys 12:6712–6716
23. Maksym PA, Chakraborty T (1990) Quantum dots in a magnetic field. Phys Rev Lett 65:108–111
24. Karwowski J, Cyrnek L (2004) Harmonium. Ann Phys (Leipzig) 13:181–193
25. Bose SK, Gupta N (1998) Exact solution of nonrelativistic Schrödinger equation for certain central physical potentials. Nuovo Cimento B 113:299–328
26. Samanta A, Ghosh SK (1990) Correlation in exactly solvable two-particle quantum system. Phys Rev A 42:1178–1183
27. Ghosh SK, Samanta A (1991) Study of correlation effects in an exactly solvable model two-electron system. J Chem Phys 94:517–522
28. Karwowski J, Szewc K (2010) Separable N-particle Hookean systems. J Phys Conf Series 213:012016
29. Slavyanov SY, Lay W, Seeger A (2000) Special functions. A unified theory based on singularities. Oxford University Press, New York
30. Ronveaux A (1995) Heun's differential equations. Oxford University Press, Oxford
31. Heun K (1889) Zur Theorie der Riemann'schen Functionen Zweiter Ordnung mit vier Verzweigungspunkten. Math Annalen 33:161–179
32. Seeger A, Lay W (eds) (1990) Centennial workshop on Heun's equations—theory and applications. Max-Planck-Institut für Metallforschung, Institut für Physik, Stuttgart
33. Díaz R, Pariguan E (2007) On hypergeometric functions and Pochhammer k-symbol. Divulgaciones Matemáticas 15:179–192

34. Kandemir BS (2005) Two interacting electrons in a uniform magnetic field and a parabolic potential: the general closed-form solution. J Math Phys 46:032110
35. Batola F (1982) Quelques relations fondamentales entre solutions de l'équation biconfluente de Heun. Arch Ration Mech Anal 78:275–291
36. Batola F (1982) Une généralisation d'une formule d'Erdelyi-Tricomi. Ark Mat 20:87–99
37. Ariola ER, Zarzo A, Dehesa JS (1991) Spectral properties of the biconfluent Heun differential equation. J Comput Appl Math 37:161–169
38. Exton H (1996) A non-trivial special case of the biconfluent Heun equation $[0, 1, 1_3]$: Orthogonality of its solutions. Le Matematiche 51:105–111
39. Leaute B, Marcilhacy G (1986) On the Schrödinger equation of rotating harmonic, three-dimensional and doubly anharmonic oscillators and a class of confinement potentials in connection with the biconfluent Heun differential equation. J Phys A Math Gen 19:3527–3533
40. Exton H (1995) The exact solution of two new types of Schrödinger equation. J Phys A Math Gen 28:6739–6741
41. Caruso F, Martins J, Oguri V (2013) Solving a two-electron quantum dot model in terms of polynomial solutions of a biconfluent Heun equation, arXiv:1308.0815v1 [quant-ph]
42. Hortaçsu M (2013) Heun functions and their uses in physics. In: U Camci and I Semiz (eds) Proc 13th Reg Conf Math Phys, Antalya, Turkey, pp. 23–39, World Scientific, arXiv:1101.0471 [math-ph]
43. Piela L (2007) Ideas of quantum chemistry. Elsevier, Amsterdam
44. Shavitt I (1977) Graph theoretical concepts for the unitary group approach to the many-electron correlation problem. Int J Quantum Chem 12(S11):131–148
45. Shavitt I (1978) Matrix element evaluation in the unitary group approach to the electron correlation problem. Int J Quantum Chem 14(S12):5–32
46. Muir T (1882) A treatise on the theory of determinants. MacMillan and Co., London
47. Vein R, Dale P (1999) Determinants and their applications in mathematical physics. Springer, New York
48. Greiner W (2001) Quantum mechanics, An introduction. Spinger, Berlin
49. Sofia University, The Heun Project: Heun functions, their generalizations and applications, http://theheunproject.org/
50. Gradshteyn IS, Ryzhik IM (2007) Table of integrals, series, and products. Elsevier, Amsterdam
51. Zhu W, Trickey SB (2006) Exact density functionals for two-electron systems in an external magnetic field. J Chem Phys 125:094317
52. Müller-Herold U (2006) On the emergence of molecular structure from atomic shape in the $1/r^2$ harmonium model. J Chem Phys 124:014105
53. Karwowski J (2013) Some remarks on the mass density distribution. Croat Chem Acta 86:531–539

Theor Chem Acc (2014) 133:1521
DOI 10.1007/s00214-014-1521-6

REGULAR ARTICLE

Construction of complex STO-*N*G basis sets by the method of least squares and their applications

Rei Matsuzaki · Shigeko Asai · C. William McCurdy · Satoshi Yabushita

Received: 1 February 2014 / Accepted: 14 June 2014 / Published online: 11 July 2014

Abstract Electronic resonance state energies and photoionization cross sections of atoms and molecules are calculated with the complex basis function method by using mixture of appropriate complex basis functions representing one-electron continuum orbitals and the usual real basis functions for the remaining bound state orbitals. The choice of complex basis functions has long been a central difficulty in such calculations. To address this challenge, we constructed complex Slater-type orbital represented by *N*-term Gaussian-type orbitals (cSTO-*N*G) basis sets using the method of least squares. Three expansion schemes are tested: (1) expansion in complex Gaussian-type orbitals, (2) expansion in real Gaussian-type orbitals, and (3) expansion in even-tempered real Gaussian-type orbitals. By extending the Shavitt–Karplus integral transform expression to cSTO functions, we have established a mathematical foundation for these expansions. To demonstrate the efficacy of this approach, we have applied these basis sets to the calculation of the lowest Feshbach resonance of H_2 and the photoionization cross section of the He atom including autoionization features due to doubly excited states. These calculations produce acceptably accurate results compared with past calculations and experimental data in all cases examined here.

Keywords Complex basis function method · Slater-type orbital · Gaussian-type orbital · Least squares fitting · Feshbach resonance · Autoionization

Dedicated to the memory of Professor Isaiah Shavitt and published as part of the special collection of articles celebrating his many contributions.

R. Matsuzaki · S. Asai · S. Yabushita (✉)
Department of Chemistry, Faculty of Science and Technology, Keio University, 3-14-1 Hiyoshi, Kohoku-ku, Yokohama 223-8522, Japan
e-mail: yabusita@chem.keio.ac.jp

C. W. McCurdy
Department of Chemistry, University of California, Davis, CA 95616, USA

C. W. McCurdy
Chemical Sciences Division, Lawrence Berkeley National Laboratory, Berkeley, CA 94720, USA

1 Introduction

With the advent of new experimental techniques for spectroscopic measurements in the extreme ultraviolet and X-ray [1] and new techniques for observing the dissociative attachment of electrons to molecules [2, 3], the description of metastable electronic states of anions and autoionizing states of neutral molecules embedded in the electronic continuum is becoming an increasingly important aspect of the theory of excited molecular dynamics.

In contrast to the highly developed array of methods for bound state electronic structure calculations in quantum chemistry, generally applicable techniques for treating metastable or pure continuum states of molecules are still in a comparatively early stage of development. Most of the available theoretical methods for describing such states are adaptations of theoretical methods applicable only to small molecules. The ones which promise to allow the extension of the techniques of bound state quantum chemistry to these problems are based on the so-called L^2 basis expansion methods. Among them, the complex scaling method [4–7] and the complex basis function (CBF) method [8–12] enable us to compute directly resonance energies (position and widths) and also the analytic continued frequency-dependent polarizability for the calculation of photoionization cross sections. The Stieltjes moment method [13, 14] also employs the frequency-dependent polarizability with L^2 basis functions to extract bound-free transition moments. The stabilization method [15–17] is the simplest L^2 basis function method to compute resonance positions and sometimes even widths.

There is another class of theoretical methods for dealing with continuum phenomena based on scattering theory such as R-matrix theory [18], K-matrix theory [19, 20], the Schwinger variational method [21], and the complex Kohn variational method [22]. These techniques solve the Schrödinger equation with continuum boundary conditions, and although they are more general, they are limited to small molecules, particularly if they also include ab initio treatments of electron correlation. One of the advantages of the L^2 basis expansion methods is that they allow the adaptation of the existing techniques of bound state theory to continuum problems and can therefore potentially treat larger systems of chemical interest. However, the choice of the appropriate basis functions has been a persistent difficulty in these L^2 methods. For example, Ruberti et al. [14] have discussed a difficulty in using Gaussian-type orbitals (GTOs) with the Stieltjes moment theory for describing photoionization cross sections above certain energy.

Studies of basis functions for bound electronic states have a long history, and many kinds of basis sets are now available with various characteristics. However, there have been only a limited number of studies of basis sets for continuum states. In the R-matrix method [23], Nestmann et al. [24] constructed a GTO set by fitting GTOs for a neutral target and later Faure et al. [25] generalized the fitting method to charged and neutral targets. Fiori et al. [26] fitted GTOs to a distortion factor instead of wave function itself. Kaufmann et al. [27] optimized GTOs to represent Rydberg states and continuum states by maximizing overlap with target wave functions. The target of the CBF method is not only just to represent continuum wave functions, but also to analytically continue the Hamiltonian and allow the calculation of complex-valued energies corresponding to metastable states, or to calculate the continuum contribution to the frequency-dependent polarizability. Real-valued GTOs cannot be used directly for these purposes. In the CBF method, our previous calculations have used large numbers of basis functions, but without choosing their parameters carefully to describe continuum states. The difficulty of basis set choice has prevented us from applying the method to a wealth of interesting many electron problems. One of the motivations of this study is to change the selection of complex orbital exponents "from art to science" by using more physical complex Slater-type orbitals (cSTOs).

One of the early studies of basis functions for bound electronic structure was construction of the STO-*N*G basis functions [28, 29], which are the Slater-type orbitals represented by the *N*-term Gaussian-type orbitals. In the present paper, we construct basis functions for the CBF method by extending such a basis set construction to a complex version. The essential idea is to represent Slater-type orbitals (STOs) with complex orbital exponents as a linear combination of *N*-term complex GTOs (cGTOs), called complex Slater-type orbital represented by *N*-term Gaussian-type orbitals (cSTO-*N*G).

Among several methods previously developed to construct STO-*N*G basis sets, we choose the method of least squares for accuracy and convenience. There are two possibilities for constructing the cSTO-*N*G basis set: expansion in terms of GTOs with complex orbital exponents or with real orbital exponents. We call these two expansions cSTO-*N*cG and cSTO-*N*rG, respectively. The latter basis function is complex valued only because the contraction coefficients are complex, and a favorable feature is that the time-consuming incomplete error function evaluations can be avoided. However, during the optimization of the real orbital exponents for cSTO-*N*rG, we have encountered a linear dependence problem. To circumvent this problem, we further optimized the real orbital exponents with a constraint that they make a geometric sequence, that is, their initial value and ratio are optimized. Such basis set is called cSTO-*N*reG, which stands for the expansion of *N*-term real even-tempered GTOs [30]. We demonstrate the efficiency and effectiveness of these cSTO-*N*G sets by calculating the well-studied Feshbach resonance energy of $(1\sigma_u)^2$ H_2 and the photoionization cross section of He atom in energy interval containing the doubly excited autoionizing states below the first excited states of the cation.

In Sect. 2, we explain cSTO-*N*G and their optimization procedure. In Sects. 3 and 4, applications to two-electron systems are described. Section 5 summarizes some conclusions.

2 Construction of cSTO-*N*cG and cSTO-*N*rG basis sets

2.1 cSTO and cGTO

The general definitions of STO and GTO functions are as follows,

$$\psi^s(\alpha) = N^s(\alpha) r^{n-1} e^{-\alpha r} Y_{lm}(\hat{\mathbf{r}}), \tag{1}$$

$$\psi^g(\zeta) = N^g(\zeta) r^{n-1} e^{-\zeta r^2} Y_{lm}(\hat{\mathbf{r}}), \tag{2}$$

where α and ζ are the respective orbital exponents. If these are complex valued, $\psi^s(\alpha)$ and $\psi^g(\zeta)$ are denoted as cSTO and cGTO, respectively. Since we intend to use these basis functions within the framework of the CBF methods, the normalization constants $N^s(\alpha)$ and $N^g(\zeta)$ are determined by the conditions,

$$(\psi^s(\alpha), \psi^s(\alpha)) = 1, \tag{3}$$

$$(\psi^g(\zeta), \psi^g(\zeta)) = 1, \tag{4}$$

where the round brackets stand for the so-called c-product [7, 31] in which the integration over $\boldsymbol{r}$ is to be carried out without complex conjugation,

$$(\psi, \phi) = \int \psi(\boldsymbol{r})\phi(\boldsymbol{r})\mathrm{d}\boldsymbol{r}. \tag{5}$$

Two well-recognized advantages of STOs over GTOs in the bound state calculations are for describing the cusp behavior at the nucleus and the damping behavior in the asymptotic region. We emphasize another advantage of using cSTOs over cGTOs in representing continuum one-electron orbitals with the CBF methods. These two types of functions exhibit quite different oscillatory behaviors, as is easily understood by separating their radial parts into the real and imaginary parts as follows,

$$r^{n-1}\mathrm{e}^{-\alpha_r r}\mathrm{e}^{-i\alpha_i r} = r^{n-1}r^{-\alpha_r r}\{\cos(\alpha_i r) - i\sin(\alpha_i r)\}, \tag{6}$$

$$r^{n-1}\mathrm{e}^{-\zeta_r r^2}\mathrm{e}^{-i\zeta_i r^2} = r^{n-1}r^{-\zeta_r r^2}\{\cos(\zeta_i r^2) - i\sin(\zeta_i r^2)\} \tag{7}$$

where $\alpha = \alpha_r + i\alpha_i$, and $\zeta = \zeta_r + i\zeta_i$. From Eq. (6), it is clear that the radial part of cSTO shows an oscillation with the wave number equals to $-\alpha_i = -\mathrm{Im}\alpha$ with the envelope function of real STO, while that of cGTO in Eq. (7) exhibits an oscillation of the wave number of $-\zeta_i r = -\mathrm{Im}\zeta r$, which is not constant, but increases in proportion to r. As we will explain in more detail later, we intend to represent a continuum orbital, that behaves asymptotically as e^{ikr}, where k is a complex wave number for a resonance state calculation, and a real wave number for a photoionization problem. Note that the target charge is not explicitly considered in this simple argument. From the preceding discussion, it is obvious that this behavior can be expressed better with a cSTO, which has a functional behavior of $\mathrm{e}^{-i\alpha_i r}$.

Considering the above three advantages of cSTO, a cSTO-*N*G basis set is expected to show good performance both in accuracy and computational cost. Moreover, their orbital exponents α's can be selected based on the physical situation to be handled. As explained before, we can consider various types of expansion in GTOs. The first two are denoted as cSTO-*N*cG and cSTO-*N*rG. It is easily understandable that cSTO-*N*cG is better than cSTO-*N*rG for accuracy, but cSTO-*N*rG is better for the computational cost.

We can show the formal possibility of constructing cSTO-*N*cG and cSTO-*N*rG basis sets by generalizing the Gaussian-transform method for a 1s-STO of Kikuchi [32] and Shavitt and Karplus [33, 34]

$$\mathrm{e}^{-\alpha r} = \frac{\alpha}{2\sqrt{\pi}}\int_0^\infty s^{-3/2}\mathrm{e}^{-\alpha^2/4s}\mathrm{e}^{-sr^2}\mathrm{d}s. \tag{8}$$

Although they formulated and employed the above integral transformation only for the real orbital exponents α and s in Eq. (8), it is readily shown that α can take a complex value. First, we note that this integral for the real integral variable s converges if and only if $|\arg \alpha| < \pi/4$. This means that cSTO can be expressed as an integral transformation of GTO as long as the above condition is satisfied. Next, we change the integration path for s from the real axis to a path in the complex plane using Cauchy's integral theorem. To see the condition for the appropriate integral path, we consider the behavior of the integrand at the asymptotic region and the origin. The respective convergence conditions at these points are as follows,

$$\lim_{|s|\to\infty} |\arg s| < \pi/2, \tag{9}$$

$$\lim_{|s|\to 0} |2\arg\alpha - \arg s| < \pi/2. \tag{10}$$

By combining the above two conditions and choosing the integration path appropriately, we have obtained the integral relationship

$$\mathrm{e}^{-\alpha r} = \frac{\alpha}{2\sqrt{\pi}}\int_\Gamma s^{-3/2}\mathrm{e}^{-\alpha^2/4s}\mathrm{e}^{-sr^2}\mathrm{d}s \tag{11}$$

for $|\arg \alpha| < \pi/2$. The subscript Γ stands for the integration path extending to infinity and satisfying both Eqs. (9) and (10). Equation (8) describes only 1s-STO, and its generalization to basis functions with higher principle quantum numbers n was given by Wright [35]. Using the same logic and differentiating both sides of Eq. (11) $n - 1$ times with respect to $-\alpha$ and by mathematical induction, the following generalized formula can be obtained,

$$r^{n-1}\mathrm{e}^{-\alpha r} = \frac{1}{2^n\sqrt{\pi}}\int_\Gamma s^{-(n+1)/2}\mathrm{e}^{-\alpha^2/4s}H_n\left[\frac{\alpha}{2\sqrt{s}}\right]\mathrm{e}^{-sr^2}\mathrm{d}s, \tag{12}$$

where $H_n\,[x]$ is the nth Hermite polynomial. Eq. (12) shows that general cSTOs can be expressed as an integral transformation of cGTOs. By changing the integration contour appropriately and performing the integral with a quadrature, we have relations that allow a cSTO to be represented in a discrete set of cSTO-*N*rG and cSTO-*N*cG. Note that, in this paper, we do not intend to use the above integral expressions in actual calculations, but from these relations, we have a mathematical foundation for the construction of both cSTO-*N*rG and cSTO-*N*cG. As is described later, cSTO-*N*rG will be constructed only for cSTO with a small argument of the complex orbital exponents, while cSTO-*N*cG will be optimized even with a large argument close to 90°.

2.2 Method of least squares

The purpose of the calculation is to find the best $\{c_j\}$ and $\{\zeta_j\}$ such that

$$\psi^s(\alpha) \approx \sum_j c_j \psi^g(\zeta_j). \tag{13}$$

Among several possible techniques for this purpose, we use the least squares method [28, 29], and the parameters are determined by searching for the minimum of the square error:

$$I[\alpha,\{c_j\},\{\zeta_j\}] = \langle \psi^s(\alpha) - \sum_i c_i \psi^g(\zeta_i) | \psi^s(\alpha) - \sum_j c_j \psi^g(\zeta_j) \rangle \tag{14}$$

Here and in what follows, the angle brackets denote the ordinary scalar product; thus, the above square error is a real quantity and can be expressed by using vector and matrix notation as follows,

$$I[\alpha,\{c_j\},\{\zeta_j\}] = \langle \psi^s(\alpha)|\psi^s(\alpha)\rangle - \mathbf{c}\cdot\mathbf{m} - \mathbf{c}^*\cdot\mathbf{m}^* + \mathbf{c}^*\cdot\mathbf{Sc} \tag{15}$$

$$S_{ij} = \langle \psi^g(\zeta_i)|\psi^g(\zeta_j)\rangle, \tag{16}$$

$$m_j = \langle \psi^s(\alpha)|\psi^g(\zeta_j)\rangle. \tag{17}$$

The expansion coefficients $\{c_j\}$ are linear parameters and can be determined easily by linear algebra, that is, by setting the first derivative of I with respect to $\{c_j^*\}$ zero,

$$\frac{\partial I}{\partial \mathbf{c}^*} = -\mathbf{m}^* + \mathbf{Sc} = 0 \tag{18}$$

Then, the optimum coefficients are determined as follows,

$$\mathbf{c} = \mathbf{S}^{-1}\mathbf{m}^* \tag{19}$$

and the square error can be expressed as follows,

$$I[\alpha,\{\zeta_j\}] = \langle \psi^s(\alpha)\psi^s(\alpha)\rangle - \mathbf{m}\cdot(\mathbf{S}^*)^{-1}\mathbf{m}^*. \tag{20}$$

The orbital exponents $\{\zeta_j\}$ are nonlinear parameters, and the optimization is not trivial. In general, ζ_j are complex valued and their real and imaginary parts must be individually optimized. For this purpose, the Newton–Raphson iteration method has been used.

One may need to optimize $\{\zeta_j\}$ for each complex value α, but there exists a scaling property, which simplifies the calculation. This scaling property for complex basis functions is an extension of the well-known one between real STOs and real GTOs [28, 29] and can be expressed as follows,

$$I[|\alpha|e^{i\arg\alpha},\{|\alpha|^2\zeta_j\}] = I[e^{i\arg\alpha},\{\zeta_j\}]. \tag{21}$$

Using this relation, one can obtain the optimum $\{\zeta_j'\}$ for any α from the optimum $\{\zeta_j\}$ for $\alpha' = \exp i(\arg\alpha)$ with the same argument by a simple scaling relation $\zeta_j' = |\alpha|^2\zeta_j$. In this study, optimum$\{\zeta_j\}$ for $\alpha = \exp(-i\theta)$ ($\theta = 0°, 1°, \ldots, 88°$) are determined by the least squares method. The determined basis sets will be used in the applications along with the scaling property.

2.3 Calculation procedure

The Newton–Raphson optimization to determine the complex basis exponents is started at $\theta = 0°$, for which $\psi^s(\alpha)$ is just the ordinary STO, whose GTO expansion was studied by several workers in the early days of electronic structure theory [28, 29]. Such a real parameter optimization is easier than the complex-valued optimization needed in this work. Once optimum $\{\zeta_j\}$ at $\theta = 0°$ is determined, the optimization for the next value at $\theta = 1°$ is carried out by using the previously determined values as the initial guess, and this procedure is continued up to $\theta = 88°$. The optimization using many primitive cGTO (ex. cSTO-10cG), in particular, is difficult task and needs some care. We have used the analytic gradient and Hessian for the Newton–Raphson method, with the trust region method for step size evaluation using quadruple precision. The program for these optimizations was developed using Mathematica 9.0 [36].

2.4 Results

We constructed cSTO-*N*cG and cSTO-*N*rG representations for the 1s, 2p, and 3d cSTOs. As examples, the optimized orbital exponents are shown in Fig. 1, where we plot (a) the absolute values and (b) the arguments for 1s-cSTO-4cG, (c) the real exponents for 1s-cSTO-4rG, and (d) the real exponents for 1s-cSTO-4reG. In Fig. 1c, as θ increases, the calculated results for cSTO-4rG show two orbital exponents ζ_1 and ζ_2 approaching each other, and at $\theta = 30°$, they have the same value, implying that the primitive GTOs become linear dependent and the iteration was stopped. One way to overcome the problem is to impose a constraint that $\{\zeta_j\}$ makes a geometric series and leads us to the cSTO-*N*reG set. Figure 1d shows that, although the orbital exponents are obtained without linear dependence from $\theta = 0°$–$60°$, at $\theta = 65°$, all the orbital exponents took almost the same values and the iteration was stopped. Although these linear dependence problems can be overcome by using, for example, GTOs with different principle quantum numbers such as 1s-, 3s-, and 5s-GTOs, in this work, we only apply thusly optimized cSTO-*N*rG and cSTO-*N*reG sets to determine the location of narrow resonances because for such calculations cSTOs with small θ values are sufficient. Figure 1a, b shows that construction of cSTO-*N*cG is successfully carried out without a linear dependence problem up to large angles θ and we use the cSTO-*N*cG set for the calculation of photoionization cross sections, which need cSTOs having a large θ. The orbital

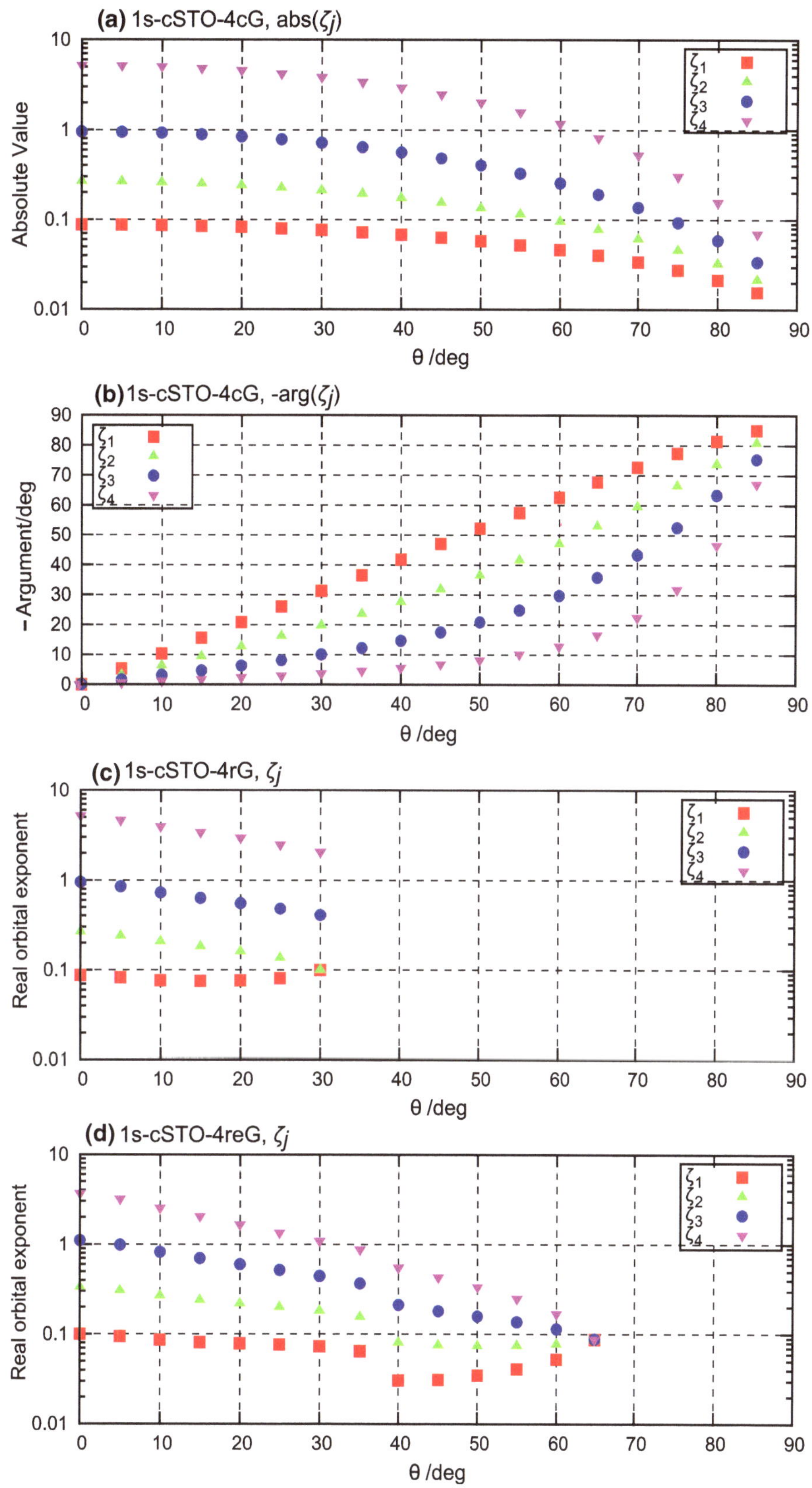

Fig. 1 Orbital exponents of cGTOs optimized for 1s-cSTO-4cG, 1s-cSTO-4rG, and 1s-cSTO-4reG. **a** Absolute value of 1s-cSTO-4cG, **b** argument of 1s-cSTO-4cG, **c** real value of 1s-cSTO-4rG, **d** real value of 1s-cSTO-4reG. θ is negative of the argument of complex orbital exponents of cSTO. In **c**, ζ_1 and ζ_2 approach to the same value and they become linear dependent at $\theta = 30°$. In **d**, all ζ_j take similar values and become linear dependent at $\theta = 65°$

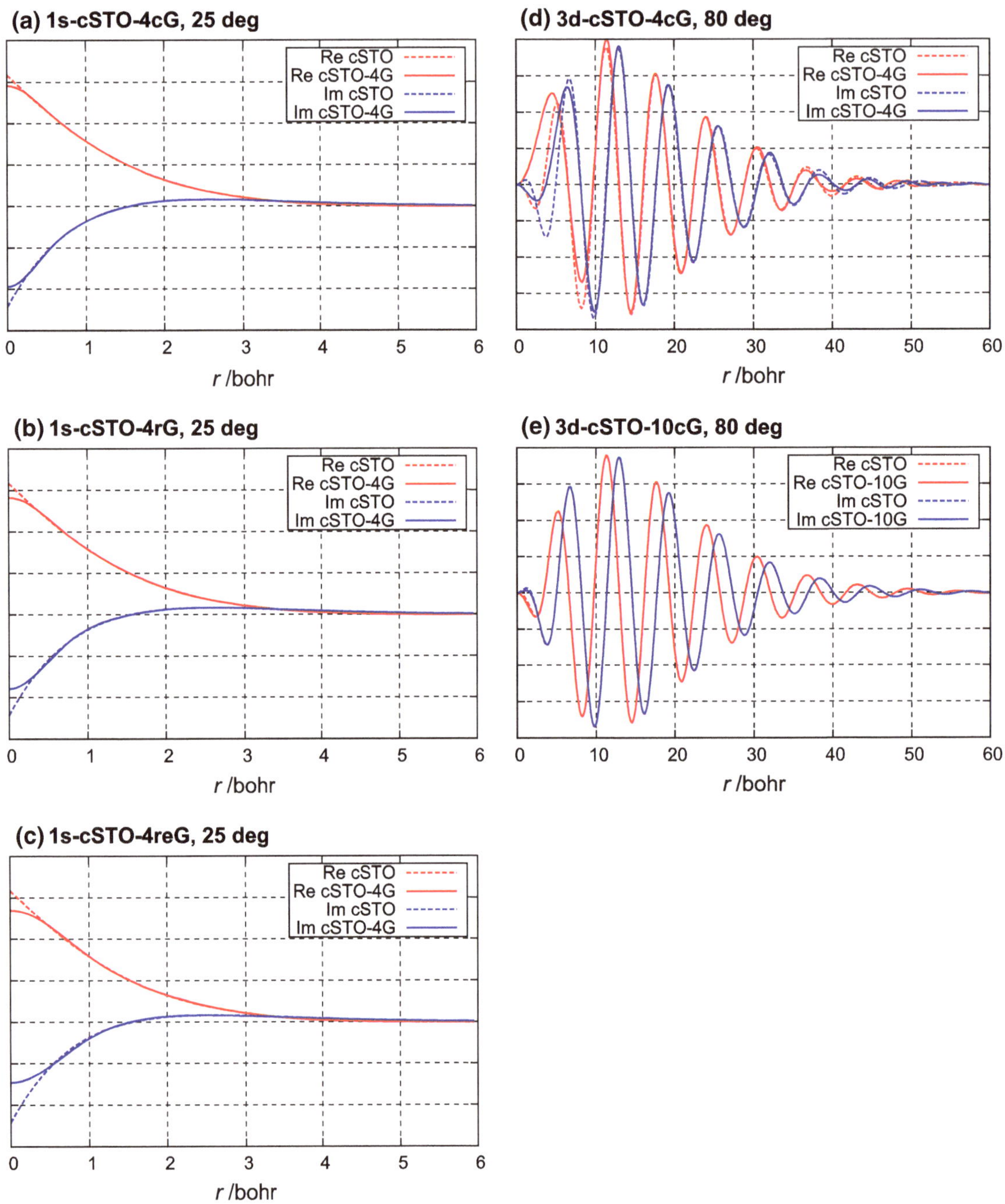

Fig. 2 Comparison of the radial functions of cSTO and cSTO-*N*G constructed by the method of least squares. **a** 1s-cSTO-4cG, **b** 1s-cSTO-4rG, **c** 1s-cSTO-4reG at $\theta = 25°$. **d** 3d-cSTO-4cG and **e** 3d-cSTO-10cG at $\theta = 80°$. *Solid lines* and *dashed lines* describe cSTO-*N*G and cSTO, respectively, and *red lines* and *blue lines* describe the real part and the imaginary part, respectively

exponents of other cSTO-*N*rG and cSTO-*N*cG show a similar behavior.

In Fig. 2a–c, the radial parts of 1s-cSTO-4cG, 1s-cSTO-4rG, and 1s-cSTO-4reG are compared with 1s-cSTO at $\theta = 25°$ and all the expansion schemes represent the behavior of cSTO very well for this small angle. In Fig. 2d, e, the radial parts of 3d-cSTO-4cG and 3d-cSTO-10cG at $\theta = 80°$ are compared with 3d-cSTO. This is a rather surprising result, because as shown in Eqs. 6 and 7, cGTO and cSTO show quite different oscillatory behaviors; cSTO has a behavior with a constant wave number, while cGTO has a behavior reflecting a linearly increasing wave number.

Figure 3 shows that the minimized square errors defined by Eq. (20) for the cSTO-*N*rG, cSTO-*N*cG, and cSTO-*N*reG expansions for 1s-cSTO, from which one can see that

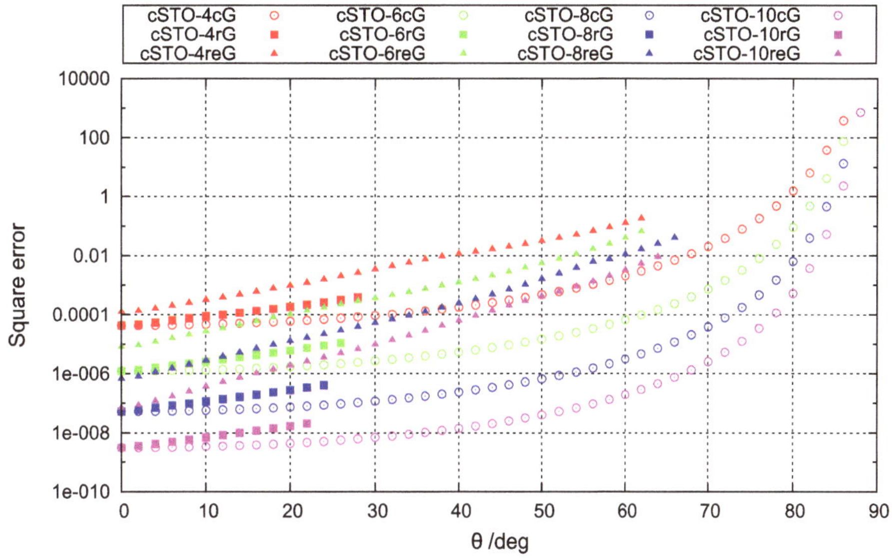

Fig. 3 Square error of 1s-cSTO-*N*cG, 1s-cSTO-*N*rG, and 1s-cSTO-*N*reG as a function of the argument of complex orbital exponent for cSTO. *Red, green, blue,* and *pink marks* represent cSTO-*N*G ($N = 4$, 6, 8 and 10) set, respectively. *Empty circle, filled square,* and *filled triangle* represent cSTO-*N*cG, cSTO-*N*rG, and cSTO-*N*reG, respectively. Only data without the linear dependence problem were shown. See the text for details. The square errors for 2p-cSTO-*N*G and 3d-cSTO-*N*G sets exhibit a similar behavior as that of 1s-cSTO-*N*G

as θ increases, the square errors increase. As one can expect based on the degree of parametric freedom, cSTO-*N*cG is the most accurate and cSTO-*N*reG is the worst. The calculated results for 2p-cSTO and 3d-cSTO behave in a similar manner.

3 Application to autoionizing states

3.1 Method

We have applied the optimized cSTO-*N*cG, cSTO-*N*rG, and cSTO-*N*reG sets to the $^1\Sigma_g^+$ doubly excited autoionizing states of H_2 by the CBF method at the bond length of 1.4 a.u. The electronic wave functions are evaluated by the complex CI method (CCI) [37]. This system is a simple example of Feshbach resonance of electron-molecular ion collision process and has been studied with various theoretical methods including CCI method. In the CCI method, one uses a mixture of real and complex basis function sets, and after determining appropriate one-electron orbitals by solving, for example, the Roothan–Hartree–Fock equation, one diagonalizes the complex symmetric Hamiltonian matrix constructed from configuration functions as is done in the ordinary CI method. The difference is that the matrix element is evaluated by the c-product rule, under which the Hamiltonian matrix is not Hermitian but complex symmetric, and the eigenvalues become complex valued [4]. To determine the resonance energy by the CCI method, one varies variational parameter θ contained in the complex basis functions and calculates the excitation energy $E(\theta)$ as a function of θ. In many cases, the common argument of complex orbital exponents is used as θ. If one can use a complete basis set, the resonance energy would be independent of θ, but in practical calculation, the resonance energy weakly depends on θ because of incompleteness of the basis set. The best variational parameter θ to describe the resonance energy is determined by requiring that $E(\theta)$ is stationary with respect to θ,

$$\frac{dE(\theta)}{d\theta} = 0. \tag{22}$$

The accuracy of the resonance energy critically depends on the basis set; therefore, the basis functions must be chosen carefully. We used 1s Gaussian of DZP from Huzinaga–Dunning for σ_g and σ_u symmetries and two diffuse 1s primitive GTOs with orbital exponents equal to 0.0411 and 0.0137 for only σ_u symmetry. Additionally, 2p Gaussian with exponent 0.8 was used for σ_g, σ_u, π_g, and π_u symmetries and 2p Gaussians with exponents 0.2 and 0.05 are used only for π_g and π_u symmetries. These 2p Gaussians were added to describe angular correlation effects. These are used as a real basis function set centered on the H atom nuclei.

Previous studies [37, 38] of the $^1\Sigma_g^+$ doubly excited autoionizing states of H_2 show that the lowest resonance is dominated by the *s* and *d* partial waves. Therefore, we use three 1s-cSTO-*N*Gs and three 3d-cSTO-*N*Gs. As discussed in Introduction, the selection of complex orbital exponents is not an easy task at all, and in this work, we attempt to propose a systematic way to select the orbital exponents for cSTOs. A hint for the selection can be obtained by relating the CBF method to the complex Kohn variation method [22, 39]. In the latter method, the outgoing continuum wave function is represented as a linear combination of L^2 basis functions and a few non L^2 functions satisfying the outgoing asymptotic behavior. In the CBF method, only L^2 functions are used; thus, additional basis functions need to be included to satisfy the exact outgoing boundary condition as much as possible. As discussed in previous section using Eq. 6, the imaginary part of a complex orbital exponent of cSTO represents the wave number approximately and this feature can be used to guess the appropriate orbital exponents. According to the previous studies [37, 40, 41], the resonance position of the $^1\Sigma_g^+$ doubly excited autoionizing states of H_2 is about 12.6 eV above the $H_2^+(1s\sigma_g)$ state at the bond length of 1.4 a.u. Then, the wave number of the continuum orbital at the asymptotic region is calculated as $k_0 = 0.96$ a.u. If the imaginary part of the exponent α_0 of cSTO is equated to the wave number, a relation $|\alpha_0| \sin\theta = k_0$ is obtained. From this relation alone, the orbital exponents cannot be determined; instead, a common practice has been to have a set of $\{\alpha_j\}$ and treat θ as a variational parameter. Because we have cSTO-*N*rG sets only with small arguments and our preliminary calculations with a larger number of cSTO-*N*cG basis functions have stationary angles close to 20°, we have assumed an argument of $\theta = \pi/10 = 18°$ as a tentatively stationary angle then we have obtained $|\alpha_0| = 3.1066$. Even though the orbital exponent α_0 is matched to the wave number at θ_0, obviously only one cSTO is not enough to accurately obtain the resonance energy, and two more cSTOs are added to represent both of s-wave and d-wave in an even-tempered manner such that $\alpha_n = \alpha_0 R^n$ $(n = -1, 0, 1)$. We set the ratio $R = 1.75$ for 1s-cSTOs and $R = 1.5$ for 3d-cSTOs.

The argument θ for these cSTO-*N*G set has the same value and is used as a complex scaling parameter for θ-trajectory. These cSTO-*N*G sets were located on the middle of the molecule to describe the partial waves of continuum wave function. The evaluations of the matrix elements and SCF and CI calculations were carried out using complex version of the COLUMBUS program. The details were explained in references [10, 37, 42].

We calculated complex excitation energies with CISD, namely Full CI with cSTO-*N*cG, cSTO-*N*rG, and cSTO-*N*reG, and showed the energies in the complex energy plane to make the θ-trajectories. As mentioned before, since the cSTO-*N*rG basis set for large θ values is not available, we have limited the use of cSTO-*N*rG only for small θ values. To determine the stationary point on the complex energy plane, we fit the θ-trajectory to a rational fraction with the data $\{(e^{-i\theta}, E(\theta))\}$ and obtained the stationary point of the fitted rational fraction numerically.

3.2 Results

Figure 4 shows the θ-trajectories and stationary points using cSTO-*N*cG ($N = 4,8,10$) sets, cSTO-*N*rG ($N = 8,10$) sets, and cSTO-10reG. The trajectories of cSTO-8cG, cSTO-10cG, and cSTO-10rG show almost the same behavior implying that these cSTO-*N*G sets represent cSTO basis functions accurately in the calculations of the complex eigenvalue problem of the electronic resonance state. The trajectories of cSTO-8rG and cSTO-10reG exhibit a little difference from that with cSTO-10cG but still have a similar cusp behavior and suggest that cSTO-8rG and cSTO-10reG are also useful basis functions for determining resonance energies. Comparing the three types of cSTO-10G sets, cSTO-10reG is the worst expansion, and this is reflected in the above difference. The θ-trajectory obtained with cSTO-4cG shows a different shape, but the stationary point thus obtained is located in a similar energy region.

The resonance positions E_r and widths Γ obtained are summarized in Table 1 along with previous theoretical results. Here, E_r values were computed by subtracting the energy of $H_2^+(1s\sigma_g)$ calculated with the same basis set at $\theta = 0°$. In spite of a smaller number of CBFs, the results are in good agreement with previous theoretical results. Note that, strictly speaking, the direct comparison of these resonance energies is meaningful only if they are derived from the same theoretical method.

The success of the cSTO-*N*rG and cSTO-*N*reG sets means that the resonance energy can be computed by using only real primitive basis functions and analytically continued to the complex energy plane by the complex contraction coefficients. This is good news for computational efficiency. In the current computation, cSTO-*N*rG can be used to compute the resonance position and width because the Feshbach resonance of H_2 is very narrow; however, it seems difficult to treat more broad resonance with the cSTO-*N*rG set.

4 Application to photoionization of He

4.1 Method

The next application is photoionization of He atom including autoionization features by the CBF method and CCI (CISD) method. In this subsection, the outline of this method is described, but for more details, the reader is referred to references [8–10]. In the dipole approximation, the total photoionization cross section $\sigma(\omega)$ can be related to the imaginary part of the frequency-dependent polarizability,

$$\sigma(\omega) = \frac{4\pi}{c\omega} \operatorname{Im} \alpha(\omega), \tag{23}$$

$$\alpha(\omega) = -\lim_{\epsilon \to +0} \langle \Psi^{(0)} \Big| \mu \frac{1}{E^{(0)} + \omega - H + i\epsilon} \mu \Big| \Psi^{(0)} \rangle, \tag{24}$$

where $\Psi^{(0)}$ and $E^{(0)}$ are the initial bound state and its eigen energy and c, ω, μ are the light speed, photon

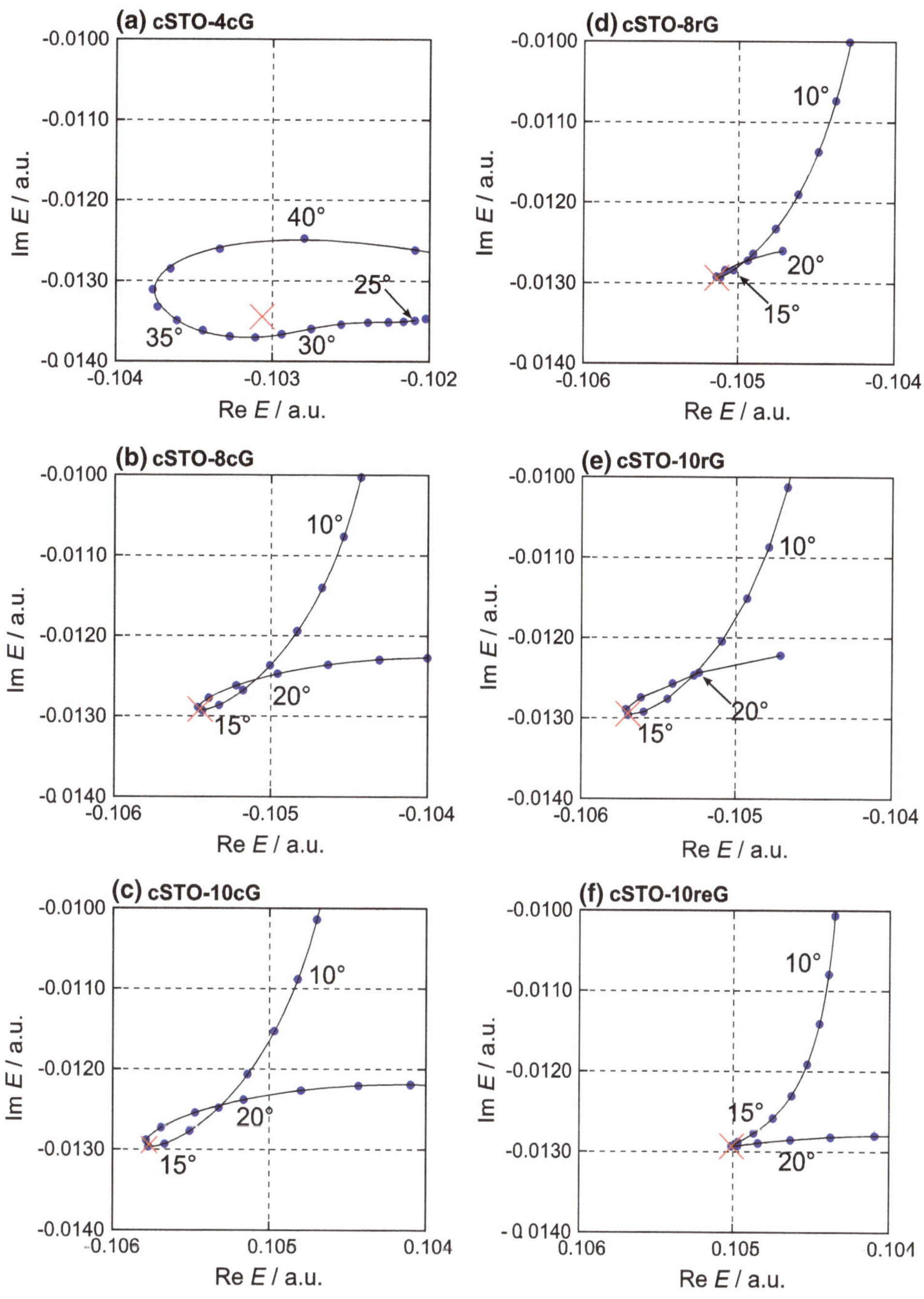

Fig. 4 Calculated results of θ-trajectory for the lowest ${}^{1}\Sigma_{g}^{+}$ resonance of H_2 at bond length of 1.4 bohr. *Circles* represent calculated complex eigenenergies with complex CI using cSTO-*N*cG with **a** $N = 4$, **b** $N = 8$, **c** $N = 10$, cSTO-*N*rG with **d** $N = 8$ **e** $N = 10$, and **f** cSTO-10reG. *Solid line* is the curve of the rational fraction fitted with these calculated results. *Cross* is the stationary point calculated from these fitted rational fractions

energy, and velocity dipole operator, respectively. In this study, we use dipole velocity operator as in our previous work [43–45]. In the CBF method, the matrix element of Green's operator is approximated by eigenvalues $\{E_n\}$ and eigenfunctions $\{\Phi_n\}$ of the Hamiltonian by means of c-product

$$(\Phi_n, H\Phi_m) = E_n\delta_{nm}, \tag{25}$$

$$(\Phi_n, \Phi_m) = \delta_{nm}. \tag{26}$$

Using $\{\Phi_n\}$ and $\{E_n\}$, the frequency-dependent polarizability is approximated in the discrete representation

$$\alpha(\omega) \approx -\sum_n \frac{d_n^2}{E_0 + \omega - E_n}. \tag{27}$$

In this formula, d_n is a dipole transition moment $(\Phi_n, \mu\Psi^{(0)})$. In a many electron system, both $\{E_n\}$ and $\{\Phi_n\}$ are obtained from complex CI calculations. Our target states contain autoionizing states which are described with two-electron excited configurations. We have used the same computer program used for H_2 calculation.

As in the previous calculation of H_2 autoionizing states, basis set must be treated carefully. We employed a large set of even-tempered real GTOs (18s16p4d1f) from our previous work [43] to represent the bound state one-electron orbitals. The continuum configurations with which the autoionizing two-electron excited configurations have configuration interactions are known to be of (1s,*k*p) type; therefore, we intend to represent this *k*p one-electron orbitals by solving the hydrogen atom 1s → *k*p problem. The one-electron orbital *k*p (Coulombic p-wave) of He at $\omega^{\mathrm{He}} \approx 60\,\mathrm{eV}$ region can be approximated by that of H with the photon energy of $\omega^{\mathrm{H}} = I^{\mathrm{H}} - I^{\mathrm{He}} + 60\,\mathrm{eV} \approx 1.3$ Hartree with taking into account the difference in the ionization potentials of H (I^{H}) and He (I^{He}). Then, the five complex orbital exponents $\{\alpha_j\}$ of cSTOs were determined by extending the procedure in Ref. [44], that is, they were independently optimized variationally to stabilize the frequency-dependent polarizability of the hydrogen 1s → *k*p problem with the photon energy $\omega = 1.3$ Hartree. One of the five cSTOs with the orbital exponent closest to the real axis in the complex α plane was essentially a real function and was simply removed to avoid linear dependence with the real GTOs explained above. The optimized complex orbital exponents for the remaining four 2p-cSTO basis functions are shown in Table 2 in a form of the absolute value and the argument (rounded to integer). Each of the four 2p-cSTO basis functions is represented by a 2p-cSTO-*N*cG contracted basis function with the appropriate scaling relation explained before. Since cSTO-*N*rG is not available for higher θ values, we have applied only cSTO-*N*cG basis functions in this study.

4.2 Results

Figure 5 shows $\sigma(\omega)$ calculated using cSTO-4cG and cSTO-10cG functions. These results show that the optimized basis set for hydrogen provides good agreement with experimental results not only at the optimized kinetic energy but also at the neighborhood of this kinetic energy. The autoionizing features are also well described in this calculation. These results show the efficiency of both optimized cSTO set and cSTO-*N*cG set. Comparing the experimental cross sections with the two calculated results using cSTO-4cG and cSTO-10cG basis sets, the differences

Table 1 Position E_r and width Γ of the ${}^1\Sigma_g^+$ autoionizing state of H_2 molecule

References	Method	E_r (eV)	Γ (eV)
a	CCI/cSTO-10cG	12.58	0.7044
a	CCI/cSTO-10rG	12.59	0.7044
a	CCI/cSTO-10reG	12.60	0.7037
a	CCI/cSTO-4cG	12.62	0.8108
a	CCI/cSTO-8cG	12.59	0.7211
a	CCI/cSTO-8reG	12.63	0.7404
Yabushita and McCurdy [37]	CCI/cGTO	12.60	0.7402
Yabushita and McCurdy [37]	CMCSCF/cGTO	12.62	0.7347
Sajeev and Moiseyev [40]	Reflection-free complex absorbing potential	12.56	0.617
Moiseyev and Corcoran [51]	Complex scaling	12.86	1.21
Isaacson [41]	Siegert eigenvalue ($\zeta_{2s} = 0.48$, $\zeta_{3s} = 0.46$)[b]	12.58	0.92
Isaacson [41]	Siegert eigenvalue ($\zeta_{2s} = \zeta_{3s} = 0.70$)[b]	12.59	0.89
Isaacson [41]	Siegert eigenvalue ($\zeta_{2s} = \zeta_{3s} = 1.00$)[b]	12.61	0.83
Morales et al. [52]	Exterior complex scaling B-spline	12.74[c]	0.714[d]
Honigmann et al. [53]	Complex coordinate		0.707[d]
Shimamura et al. [54]	R-matrix	12.80	0.751
Hara and Sato [55]	Projection operator	12.62	0.756
Sánchez [56]	Projection operator B-spline	12.82[d]	0.735[d]
Hazi et al. [57]	Projection operator CI-A[e]	12.70	0.69
Hazi et al. [57]	Projection operator CI-B[e]	12.70	0.65
Collins and Schneider [58]	Linear algebra		0.699[d]

[a] Present work

[b] The Siegert eigenvalue method using different basis sets

[c] The energy was read from figure of Ref. [52], and the energy of $H_2^+(1s\sigma_g)$ was taken from Wind [63]

[d] Value was read from the figure of each reference

[e] Projection operator and the Stieltjes moment method using different projection operators

Table 2 Complex orbital exponents of cSTOs for cSTO-*N*cG set used for the calculation of He autoionization

No.	Absolute value	Argument (°)
1	1.13513	17
2	1.33919	39
3	1.54996	63
4	1.64740	82

These are optimized complex orbital exponents of cSTOs for the 1s → *kp* channel of hydrogen photoionization using velocity form at $\omega = 1.3$ Hartree. Arguments are rounded to the nearest integer for calculation convenience

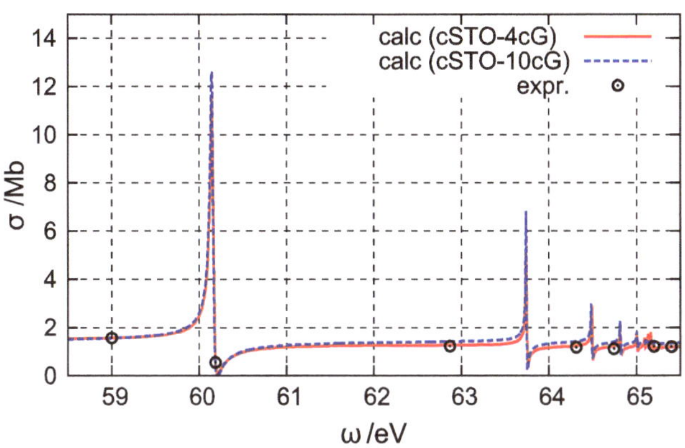

Fig. 5 Calculation result of photoionization total cross section (Mb) of He $^1S(1s)^2 \rightarrow {}^1P(1s)^1(kp)$ as a function of photon energy in eV using cSTO-4cG (*solid line*) and cSTO-10cG (*dashed line*). Experimental data (*circles*) are from Samson [50]. The peaks are assigned to the resonance states (sp, 22+), (sp, 23+), (sp, 24+), and so on

can be seen to be very small. This comparison suggests that exact representation of the cSTO with cGTO is not very important in the calculation of photoionization total cross sections. This result is encouraging from the viewpoint of computational cost because the use of cSTO-*N*cG with smaller *N* can reduce the integral evaluation cost. Since the total photoionization cross section is proportional to the square of overlap of the initial bound state and the final continuum state in the target region, thereby the behavior of continuum orbital in the asymptotic region is not important. Some details of this point were discussed for hydrogen photoionization calculation in a previous study [44].

The asymmetry line profiles of the (sp, 2*n*+) series arising from the autoionizing states appear in the total cross section in Fig. 5. The profiles are very similar to previous theoretical [43, 46] and experimental results [47, 48]. For a concrete comparison with other studies, we extracted the resonance position E_r, width Γ, and the shape parameter q for the lowest autoionizing state (sp,22+) by fitting the total cross section to the well-known Fano profile [49] formula in a least squares sense.

$$\sigma_{\rm fit}(\omega) = \sigma_a \frac{(q+\varepsilon)^2}{1+\varepsilon^2} \quad \text{with} \quad \varepsilon = \frac{\omega - E_r}{\Gamma/2}. \tag{28}$$

The fitted parameters of E_r, Γ, and q are in good agreement with previous experimental and theoretical results, as shown in Table 3.

5 Conclusion

In this paper, we have constructed three types of basis sets contracted with primitive GTOs, namely cSTO-*N*cG, cSTO-*N*rG, and cSTO-*N*reG. Among them, the cSTO-*N*rG set was obtained only for the small arguments of the orbital exponents of cSTOs, and some improvement would be necessary if it is to be applied to the calculation of broad resonances. All of these sets represent cSTO very well especially for small arguments of cSTO exponents. The calculation of the resonance position for H_2 using cSTO-*N*cG and cSTO-*N*reG produced very sharp θ-trajectories, and the resonance positions obtained are in good agreement with previous studies. Calculation of He autoionization represents experimental data very well. We have discussed the selection procedures for the complex orbital exponents for cSTOs based on the physical situations of the outgoing

Table 3 Resonance position E_r, width Γ, and the Fano shape parameter q of the lowest ^{1}P autoionizing state of He atom

References	Method	E_r (eV)	Γ (eV)	q
a	CCI/cSTO-4cG	60.155	0.0408	−2.68
a	CCI/cSTO-10cG	60.152	0.0385	−2.83
Morita and Yabushita [43]	CCI/cGTO	60.141	0.0358	
Domke et al. [59]	Experiment	60.147	0.037	−2.75
Morgan and Ederer [60]	Experiment	60.151	0.038	−2.60
Ho [61]	Complex scaling	60.153	0.0371	
McCurdy and Martín [62]	ECS B-spline	60.122	0.0373	
Hamacher and Hinze [46]	R-matrix	60.19	0.04019	−2.68

[a] Present work

electron. A corresponding physical guide to selecting the exponents of cGTOs is not obvious and that fact complicates the selection of complex Gaussian exponents. Expanding cSTOs in Gaussians is a way to address this difficulty. Future application to larger molecules is expected.

Acknowledgments Work by S.Y. was supported in part by Grants-in-Aid for Scientific Research and by the MEXT-Supported Program for the Strategic Research Foundation at Private Universities, 2009–2013. The computations were partly carried out using the computer facilities at the Research Center for Computational Science, Okazaki National Institutes. Work by CWM was supported through the Scientific Discovery through Advanced Computing (SciDAC) program funded by the US Department of Energy, Office of Science, Advanced Scientific Computing Research, and Basic Energy Sciences.

References

1. Leone SR, McCurdy CW, Burgdörfer J, Cederbaum LS, Chang Z, Dudovich M, Feist J, Greene C, Ivanov M, Kienberger R, Keller U, Kling M, Loh ZH, Pfeiffer T, Pfeiffer AN, Santra R, Schafer K, Stolow A, Thumm W, Vrakking M (2014) Nat Photon 8:162
2. Ram NB, Krishnakumar E (2012) J Chem Phys 136:164308
3. Slaughter DS, Haxton DJ, Adaniya H, Weber T, Rescigno TN, McCurdy CW, Belkacem A (2013) Phys Rev A 87:052711
4. Reinhardt WP (1982) Annu Rev Phys Chem 33:223
5. Aguilar J, Combes JM (1971) Commun Math Phys 22:269
6. Junker BR (1982) Adv At Mol Phys 18:207
7. Moiseyev N (2011) Non-hermitian quantum mechanics. Cambridge University Press, Cambridge
8. Rescigno TN, McCurdy CW (1985) Phys Rev A 31:607
9. Yu CH, Pitzer RM, McCurdy CW (1985) Phys Rev A 32:2134
10. Yabushita S, McCurdy CW, Rescigno TN (1987) Phys Rev A 36:3146
11. Honigmann M, Buenker RJ, Liebermann HP (2006) J Chem Phys 125:234304
12. Honigmann M, Liebermann HP, Buenker RJ (2010) J Chem Phys 133:044305
13. Langhoff PW, Corcoran CT (1974) J Chem Phys 61:146
14. Ruberti M, Yun R, Gokhberg K et al (2013) J Chem Phys 139:144107
15. Hazi AU, Taylor HS (1970) Phys Rev A 1:1109
16. Taylor HS, Hazi AU (1976) Phys Rev A 14:2071
17. Jordan KD, Voora VK, Simons J (2014) Theor Chem Acc 133:1445
18. Zatsarinny O, Bartschat K (2013) J Phys B At Mol Opt Phys 46:112001
19. Cacelli I, Carravetta V, Rizzo A, Moccia R (1991) Phys Rep 6:283
20. Montuoro R, Moccia R (2003) Chem Phys 293:281
21. Lucchese RR, Takatsuka K, Mckoy V (1986) Phys Rep 131:147
22. Rescigno TN, Orel AE, McCurdy CW (1997) Phys Rev A 55:342
23. Tennyson J (2010) Phys Rep 491:29
24. Nestmann BM, Peyerimhoff SD (1990) J Phys B At Mol Phys 773:L773
25. Faure A, Gorfinkiel JD, Morgan LA, Tennyson J (2002) Comput Phys Commun 144:224
26. Fiori M, Miraglia JE (2012) Comput Phys Commun 183:2528
27. Kaufmann K, Baumeistert W, Jungen M (1989) J Phys B At Mol Phys 22:2223
28. O-ohata K, Taketa H, Huzinaga S (1966) J Phys Soc Jpn 21:2306
29. Stewart RF (1970) J Chem Phys 52:431
30. Schmidt MW, Ruedenberg K (1979) J Chem Phys 71:3951
31. Moiseyev N, Certain PR, Weinhold F (1978) Mol Phys 36:1613
32. Kikuchi R (1954) J Chem Phys 22:148
33. Shavitt I, Karplus M (1962) J Chem Phys 36:550
34. Shavitt I, Karplus M (1965) J Chem Phys 43:398
35. Wright JP (1963) Quarterly progress report solid state and molecular theory group. MIT, Cambridge, p 35
36. Wolfram S (2013) Mathematica ver 9.0. Wolfram Research Inc., USA
37. Yabushita S, McCurdy CW (1985) J Chem Phys 83:3547
38. Takagi H, Nakamura H (1983) Phys Rev A 27:691
39. Miller WH, Jansen op de Haar BMDD (1987) J Chem Phys 86:6213
40. Sajeev Y, Moiseyev N (2007) J Chem Phys 127:034105
41. Isaacson AD (1991) J Chem Phys 95:8254
42. Yasuike T, Yabushita S (2000) Chem Phys Lett 316:257
43. Morita M, Yabushita S (2008) Chem Phys 349:126
44. Morita M, Yabushita S (2008) J Comput Chem 29:2317
45. Morita M, Yabushita S (2008) J Comput Chem 29:2471
46. Hamacher P, Hinze J (1989) J Phys B At Mol Opt Phys 22:3397
47. Domke M, Xue C, Puschmann A et al (1991) Phys Rev Lett 66:1306
48. Chan WF, Cooper G, Brion CE (1991) Phys Rev A 44:186
49. Fano U (1961) Phys Rev 124:1866
50. Samson JAR, He ZX, Yin L, Haddad GN (1994) J Phys B At Mol Phys 27:887
51. Moiseyev N, Corcoran C (1979) Phys Rev A 20:814
52. Morales F, McCurdy CW, Martín F (2006) Phys Rev A 73:014702
53. Honigmann M, Hirsch G, Buenker RJ et al (1999) Chem Phys Lett 305:465
54. Shimamura I, Noble CJ, Burke PG (1990) Phys Rev A 41:3545
55. Hara S, Sato H (1984) J Phys B At Mol Opt Phys 17:4301
56. Sánchez I, Martín F (1997) J Chem Phys 106:7720
57. Hazi AU, Derkits C, Barsley JN (1983) Phys Rev A 27:1751
58. Collins LA, Schneider BI (1983) Phys Rev A 27:101
59. Domke M, Schulz K, Remmers G et al (1996) Phys Rev A 53:1424
60. Morgan HD, Ederer DL (1984) Phys Rev A 29:1901
61. Ho YK (1981) Phys Rev A 23:2137
62. McCurdy CW, Martín F (2004) J Phys B At Mol Opt Phys 37:917
63. Wind H (1965) J Chem Phys 42:2371

Theor Chem Acc (2014) 133:1481
DOI 10.1007/s00214-014-1481-x

REGULAR ARTICLE

SDS: the 'static–dynamic–static' framework for strongly correlated electrons

Wenjian Liu · Mark R. Hoffmann

Received: 9 January 2014 / Accepted: 7 March 2014 / Published online: 2 April 2014

Abstract A genetic 'static–dynamic–static' (SDS) framework is proposed for describing strongly correlated electrons. It permits both simple and sophisticated parameterizations of many-electron wave functions. One particularly simple realization amounts to constructing and diagonalizing the Hamiltonian matrix in the same number of many-electron basis functions in the primary (static), external (dynamic) and secondary (static) subspaces of the full Hilbert space. It combines the merits of both internally and externally contracted configuration interaction as well as intermediate Hamiltonian approaches. When the Hamiltonian matrix elements between the contracted external functions, with the coefficients determined by first order perturbation, are approximated as the diagonal elements of the zeroth-order Hamiltonian H_0, we obtain a multi-state multi-reference second-order perturbation theory (denoted as SDS-MS-MRPT2) that scales computationally with the fifth power of the molecular size. Depending on how H_0 is defined, various variants of SDS-MS-MRPT2 can be obtained. For simplicity, we here choose H_0 as a multi-partitioned Møller–Plesset-like diagonal operator. Further combined with the string-based macroconfiguration technique, an efficient implementation of SDS-MS-MRPT2 is realized and tested for prototypical systems of variable near-degeneracies. The results reveal that SDS-MS-MRPT2 can well describe not only standard benchmark systems but also problematic systems. Taking SDS-MS-MRPT2 as a start, the accuracy may steadily be increased by relaxing the contraction of the external functions and/or iterating the diagonalization–perturbation–diagonalization procedure. As such, the SDS framework offers a very powerful scenario for handling strongly correlated systems.

Keywords Strongly correlated electrons · Static–dynamic–static · Minimal MRCI · Multi-state multi-reference perturbation theory

1 Introduction

Given the remarkable progress in many-body theories, accurate descriptions of electron correlation in molecular systems of variable near-degeneracies are still challenging and remain an active area of research. One framework that has provided not only accurate results but also qualitative insight is effective Hamiltonian (EH), based on which various multi-reference (MR) or quasi-degenerate (QD) perturbation theories (PT) have been proposed [1–26] in the past (see Refs. [27] and [28] for careful comparisons). Yet, the premises of MRPTs and QDPTs that electron correlation can be separated into static and dynamic components and then that they can respectively be treated variationally and perturbatively are not always justified. As a consequence, low-order MRPTs and QDPTs usually depend strongly on the

Dedicated to the memory of Professor Isaiah Shavitt and published as part of the special collection of articles celebrating his many contributions.

W. Liu (✉)
Beijing National Laboratory for Molecular Sciences, Institute of Theoretical and Computational Chemistry, State Key Laboratory of Rare Earth Materials Chemistry and Applications, College of Chemistry and Molecular Engineering, Center for Computational Science and Engineering, Peking University, Beijing 100871, People's Republic of China
e-mail: liuwjbdf@gmail.com

M. R. Hoffmann (✉)
Chemistry Department, University of North Dakota, Grand Forks, ND 58202-9024, USA
e-mail: mark.hoffmann@email.und.edu

zeroth-order Hamiltonian as well as the model space. Even spurious results may result occasionally. Going to higher orders in the perturbation expansion may introduce more problems than actually solved.

A more refined separation of static and dynamic correlation invokes a process in which the dynamic correlation informs (or provides feedback to) the static correlation. This produces either explicitly or implicitly an intermediate effective Hamiltonian (IEH) [29–38], whose lowest few eigenpairs seek to reproduce the exact eigenvalues and the projections of the exact wave functions onto the model space, whereas the remaining eigenpairs are of no particular significance. Such IEH approaches are conceptually more sophisticated than the EH-based MRPTs and QDPTs. Nevertheless, they are not free of drawbacks. In particular, the improved description of the static correlation comes at the expense of some of the dynamic correlation. Moreover, the so-called intruder state problem is effectively avoided only at lowest orders but raises its head again in higher-order expansions of the IEHs.

All the above approaches stay within the 'first static then dynamic' scenario, along with the use of 'common orbitals for all configurations.' Since dynamic correlation is system independent, whereas static correlation is system dependent, an inverse scenario, viz., 'first dynamic then static' [39] should also be possible, particularly when combined with the idea of 'different orbitals for different charge/spin distributions.' Compared with the former, orbital relaxations arising from dynamic correlation are here fully accounted for, such that only few nearly degenerated and dynamically correlated many-electron functions are necessary for an adequate description. Yet, due to technical reasons related to the evaluation of the Hamiltonian and metric matrix elements over nonorthogonal determinants, we do not intend to further explore this scenario here. Instead, we would like to suggest a framework in which the static and dynamic contributions to electron correlation are treated more equitably than in the previous approaches that feature 'common orbitals for all configurations.' The framework, characterized by 'static–dynamic–static (SDS) multi-state (MS),' is genetic in the sense that it permits whatsoever treatment of the dynamic correlation as well as successive generations of both dynamic and static multi-reference functions from the preceding generations. For instance, a second-order treatment of the dynamic correlation leads naturally to SDS-MS-MRPT2, which already combines the advantages of both internally and externally contracted configuration interaction (CI) as well as the IEH approaches. Taking SDS-MS-MRPT2 as a start, the accuracy can steadily be increased by relaxing the contraction of the secondary functions and/or iterating the diagonalization–perturbation–diagonalization procedure.

The general features of the SDS framework are first described in Sect. 2.1, where distinctions from existing EH- or IEH-based MRPTs are also elucidated. The SDS-MS-MRPT2 variant is then formulated in Sect. 2.2. While any partitioning of the Hamiltonian can be adopted, only the Møller–Plesset-like partitioning is considered here. The implementation employs a novel hybrid of the macroconfiguration description [40] of electron distributions among groups of orbitals and determinant strings [41–44] (cf. 'Appendix' section). As a first pilot application, several well-studied model problems are investigated: the symmetric bond stretching of H_2O (Sect. 3.1); the lowest singlet-triplet energy gap of CH_2 (Sect. 3.2); the entire ground-state potential energy curve of LiF using a single (multi-determinant) reference (Sect. 3.3); and the energy separation between the lowest two singlet curves of LiF (Sect. 3.4). While clearly not comprehensive, these systems represent prototypical problematic situations for MRPTs. The size extensivity of SDS-MS-MRPT2 is also explored numerically by taking dimers and trimers of H_2 as examples (Sect. 3.5). The account ends with a brief summary of the suggested method in Sect. 4.

2 Theory

2.1 Generality

The full configuration interaction (FCI) solutions of the Schrödinger equation can generally be classified into three sets in an energetic ascending order: N_P low-lying nearly degenerate (primary) states, N_S intermediate (secondary) states and N_Q high-lying (external) states. Such a classification holds for most molecular systems, especially those with a reasonable gap between the lowest unoccupied and the highest occupied molecular orbitals. It is just that the specific values of N_P, N_S and N_Q are case dependent. If N_P is just one and the state is well separated from the rest, the system is only weakly correlated and can well be described by a single-reference prescription. On the other hand, if N_P is exceedingly large, the system would be ultra-strongly correlated and cannot be handled by standard approaches. What is particularly interesting is the intermediate case, i.e., the commonly called strongly correlated systems, with N_p being a limited number. The crucial issues are then: (1) how to construct appropriate many-electron functions to mimic such a general feature of the exact solutions and (2) how to account for the contributions of such functions to the N_P states of primary interest.

To achieve the first goal, we begin with partitioning a given set of spin orbitals $\{\varphi_p : p = 1, \ldots, N\}$ into three disjoint sets: (1) core (**C**), (2) active (**A**) and (3) virtual (**V**) of dimensions $n_{\mathbf{C}}$, $n_{\mathbf{A}}$, and $n_{\mathbf{V}}$, respectively. Given the total number of electrons (n_e), we construct a model space $M =$

$\{\Phi_\mu : \mu = 1, \ldots, d_M\}$ that is spanned by the determinants Φ_μ in which the core spin orbitals are occupied and the remaining $n_a = n_e - n_C$ active electrons are distributed in the active spin orbitals in all possible ways. The determinants can further be spin and spatially symmetrized to form configuration state functions (CSF). This space is just the commonly called 'complete active space' [45] with n_a electrons in $n_\mathbf{A}$ orbitals, CAS(n_a, $n_\mathbf{A}$). Such a construction of the model space is operationally very simple. Yet, such a space usually contains too many functions that are energetically too high to have appreciable weights in the target N_P primary states. For this reason and for more generality, a subset of $\{\Phi_\mu\}$, governed by, e.g., the expectation values $\langle\Phi_\mu|H|\Phi_\mu\rangle$ or generated from some macroconfigurations [40] (i.e., distributions of n_a electrons in prescribed orbital groups), is to be chosen as the reference space, $M_R = \{\Phi_v, v = 1, \ldots, d_R \leq d_M\}$. A SA-MCSCF (state-averaged multi-configuration self-consistent-field) can then be carried out within this space. The lowest N_P solutions $\{\Psi_I^{(0)}, I = 1, \ldots, N_P\}$, which provide either semi-quantitatively or qualitatively correct descriptions of the N_p exact states, are adopted to form the primary subspace projected by the operator P_m,

$$P_m = \sum_I^{N_P} |\Psi_I^{(0)}\rangle\langle\Psi_I^{(0)}|, \quad \Psi_I^{(0)} = \sum_{\mu=1}^{d_R} C_{\mu I}^{(0)} \Phi_\mu. \tag{1}$$

The remaining part (P_s) of M_R is then designated as the secondary space, viz.,

$$P_s = P - P_m, \quad P = \sum_{\mu=1}^{d_R} |\Phi_\mu\rangle\langle\Phi_\mu| = \sum_{K=1}^{d_R} |\Psi_K^{(0)}\rangle\langle\Psi_K^{(0)}|. \tag{2}$$

The second equality of P only arises if the entire M_R space is diagonalized, which is not necessary though. Since the linearly independent singly and doubly excited CSFs (determinants) $\{\Phi_q\}$ relative to $\{\Phi_\mu \in M_R\}$ span exactly the first-order interacting space [46, 47], it appears natural to choose the N_P first-order functions $\{\Psi_I^{(1)}\}$ as the external functions for a minimal dynamic correction to $\{\Psi_I^{(0)}\}$:

$$\begin{aligned} |\Psi_I^{(1)}\rangle &= Q_1 \frac{1}{E_I^{(0)} - H_0} Q_1 H |\Psi_I^{(0)}\rangle \\ &= \sum_{q \in Q_1} |\Phi_q\rangle C_{qI}^{(1)}, \quad I = 1, \ldots, N_P, \end{aligned} \tag{3}$$

$$Q_1 = \sum_q |\Phi_q\rangle\langle\Phi_q|, \tag{4}$$

$$\begin{aligned} |\Phi_q\rangle \in \Big\{ &|\Phi_{i,\mu}^a\rangle, |\Phi_{ij,\mu}^{ab}\rangle, \mu = 1, \ldots d_R; \quad i,j \in \mathbf{C} \oplus \mathbf{A} \cup \\ &\times a, b \in \mathbf{A} \oplus \mathbf{V} \backslash \mathbf{i}, \mathbf{j}, \mathbf{a}, \mathbf{b} \in \mathbf{A} \}. \end{aligned} \tag{5}$$

Note that internal excitations (labeled only by active orbital indices) that map one reference determinant into the (incomplete) model space M_R are excluded in the definition of $|\Phi_q\rangle$ (5). Without loss of generality, the $\{\Psi_I^{(1)}\}$ functions (3) can be Gram–Schmidt orthonormalized, leading to

$$|\Xi_I^{(1)}\rangle = \sum_{q \in Q_1} |\Phi_q\rangle \tilde{C}_{qI}^{(1)}, \quad \langle\Xi_I^{(1)}|\Xi_J^{(1)}\rangle = \delta_{IJ}. \tag{6}$$

As for a minimal set of secondary (buffer) states, there exist two options. One is to take the $\{\Theta_I = \Psi_{N_P+I}^{(0)}, I = 1, \ldots, N_P\}$ functions from partial diagonalization of the projected Hamiltonian *PHP* with *P* given in Eq. (2). The other is to use the $\{\Theta_I\}$ functions arising from the Gram–Schmidt orthonormalization of the following functions

$$|\tilde{\Theta}_I\rangle = \left(1 - P_m - \sum_J^{N_P} |\Xi_J^{(1)}\rangle\langle\Xi_J^{(1)}|\right) V |\Psi_I^{(1)}\rangle, \quad V = H - H_0. \tag{7}$$

That is, the action of the Hamiltonian on the first-order functions $\{\Psi_I^{(1)}, I = 1, \ldots, N_P\}$ generates N_P new functions belonging to P_m, P_s or $Q = 1 - P_m - P_s$. Only those belonging to P_s are retained here. Note that the $\{\Theta_I\}$ functions constructed in either way do not interact directly with the primary functions $\{\Psi_I^{(0)}\}$, i.e., $\langle\Psi_I^{(0)}|H|\Theta_J\rangle = 0$. This is because $\{\Psi_I^{(0)}\}$ are eigenvectors of the projected Hamiltonian *PHP*. In other words, the interactions between the primary and secondary states are treated variationally, since they are, by assumption, energetically close to each other.

The above considerations lead to the following Ansatz for the wave functions of the N_P low-lying states:

$$\begin{aligned} |\Psi_I\rangle &= \sum_J^{N_P} C_{JI} |\Psi_J^{(0)}\rangle + \sum_J^{N_P} C_{(J+N_P)I} |\Xi_J^{(1)}\rangle + \sum_J^{N_P} C_{(J+2N_P)I} |\Theta_J\rangle, \\ J &= 1, \ldots, N_P. \end{aligned} \tag{8}$$

Taking $\{\Psi_I^{(0)}\}$, $\{\Xi_I^{(1)}\}$ and $\{\Theta_I\}$ as independent functions, the remaining task is just to solve the following $3N_P$-dimensional eigenvalue problem

$$HC = CE. \tag{9}$$

The so-proposed procedure has the following distinct features:

1. Equations (8) and (9) define a minimal MRCI approach, with the dimension being only $3N_P$. It is both internally and externally contracted and further augmented with secondary functions. For this reason, the approach can be designated as ixc-MRCISD+s. By replacing the summation over the compound index q (i.e., $\{\mu; i,j \in \mathbf{C} \oplus \mathbf{A} \cup a, b \in \mathbf{A} \oplus \mathbf{V} \backslash i, j, a, b \in \mathbf{A}\}$) in Eq. (3) with $\{i,j \in \mathbf{C} \cup a, b \in \mathbf{V}\}$, Eq. (8) would reduce to the doubly contracted MRCISD [48],

provided that the primary functions $\{\Psi_I^{(0)}\}$ are fully decontracted and the secondary functions $\{\Theta_I\}$ are discarded. By further replacing the summation over $\{i,j \in \mathbf{C} \cup \mathbf{a}, \mathbf{b} \in \mathbf{V}\}$ with $\{a, b \in \mathbf{V}\}$ alone, the so-called externally contracted MRCISD [49, 50] can be obtained. On the other hand, if $\{\Xi_J^{(1)}\}$ are replaced with uncontracted external singles and contracted external doubles (i.e., $\sum_\mu |\Phi_{ij,\mu}^{ab}\rangle C_{\mu J}^{(0)}$, $i,j \in \mathbf{C} \oplus \mathbf{A} \cup$ $a, b \in \mathbf{V}$), Eq. (8) would correspond to the internally contracted MRCISD [51–54].

2. The approach is also related to MS-MRPTs [6–12] in structure. This can be seen by setting $C_{JI} = C_{(J+N_\mathrm{P})I}$ and $C_{(J+2N_\mathrm{P})I} = 0$ in Eq. (8), leading to

$$\Psi_I^{[1]} = \sum_J^{N_\mathrm{P}} C_{JI}\left(\Psi_J^{(0)} + \Xi_J^{(1)}\right). \tag{10}$$

The Hamiltonian matrix (9) correct to second order is then of dimension N_P, viz.,

$$\begin{aligned}\left(H_\mathrm{eff}^{[2]}\right)_{IJ} &= \frac{1}{2}\left\{\left\langle \Psi_I^{(0)} + \Xi_I^{(1)}|H|\Psi_J^{(0)}\right\rangle + \left\langle \Psi_I^{(0)}|H|\Psi_J^{(0)} + \Xi_J^{(1)}\right\rangle\right\} && (11)\\ &= \left\langle \Psi_I^{(0)}|H|\Psi_J^{(0)}\right\rangle + \frac{1}{2}\left\{\left\langle \Xi_I^{(1)}|H|\Psi_J^{(0)}\right\rangle + \left\langle \Psi_I^{(0)}|H|\Xi_J^{(1)}\right\rangle\right\}. && (12)\end{aligned}$$

It is well known that such MS approaches do not have the full flexibility for the revision of the components of the wave function in the primary reference functions and may additionally suffer from the (in)famous intruder state problem. The situation can be improved by further introducing the secondary functions $\{\Theta_I\}$, i.e.,

$$\tilde{\Psi}_I^{[1]} = \sum_J^{N_\mathrm{P}} C_{JI}\left(\Psi_J^{(0)} + \Xi_J^{(1)}\right) + \sum_J^{N_\mathrm{P}} C_{(J+N_\mathrm{P})I}\Theta_J, \tag{13}$$

which leads to an IEH matrix of dimension $2N_\mathrm{P}$:

$$\left(\tilde{H}_\mathrm{eff}^{[2]}\right)_{IJ} = \left(H_\mathrm{eff}^{[2]}\right)_{IJ}, \quad I, J = 1, \ldots, N_\mathrm{P}, \tag{14}$$

$$\begin{aligned}\left(\tilde{H}_\mathrm{eff}^{[2]}\right)_{(I+N_\mathrm{P})J} &= \left\langle \Theta_I|H|\Xi_J^{(1)}\right\rangle,\\ \left(\tilde{H}_\mathrm{eff}^{[2]}\right)_{I(J+N_\mathrm{P})} &= \left\langle \Xi_I^{(1)}|H|\Theta_J\right\rangle,\end{aligned} \tag{15}$$

$$\left(\tilde{H}_\mathrm{eff}^{[2]}\right)_{(I+N_\mathrm{P})(J+N_\mathrm{P})} = \langle \Theta_I|H|\Theta_J\rangle. \tag{16}$$

In view of Eq. (7), the $\tilde{\Theta}_I$ (and hence Θ_I) functions are formally of second order, such that the couplings $\langle \Theta_I|H|\Xi_J^{(1)}\rangle$ (15) are of fourth order. In contrast, if the states $\{\Psi_K^{(0)}, K = N_\mathrm{P}+1, \ldots, 2N_\mathrm{P}\}$ are taken for $\{\Theta_I\}$, the couplings would be of second order. However, such an argument is not really meaningful for both $\{\Theta_I\}$ and $\{\Xi_I^{(1)}\}$ are treated here as independent basis functions. It will be shown later on that the two choices of $\{\Theta_I\}$ do give rise to similar results. Had all the $\{\Psi_K^{(0)}, K = N_\mathrm{P}+1, \ldots, d_\mathrm{R}\}$ states been taken for the secondary states, which is usually the case for IEH approaches such as GVVPT2 [16, 17], several side effects may arise: (a) Such secondary states are usually much more numerous than the target primary states (i.e., $d_\mathrm{R} \gg 2N_\mathrm{P}$), and some of them may actually be higher in energy than the external functions, thereby losing the meaning of static correlation. Moreover, it does not make physical sense to treat high-lying secondary states variationally but those external states of lower energy perturbatively. (b) Probably more seriously, due to the one-to-one fixed combination [i.e., $\Psi_J^{(0)} + \Xi_J^{(1)}$ in $\tilde{\Psi}_I^{[1]}$ (13)], the relaxation of the primary states $\Psi_I^{(0)}$ furnished by the secondary states is accompanied with the reduction of dynamic correlation, as compared with the intermediate normalization. All such problems are avoided by treating the external functions $\{\Xi_I^{(1)}\}$ as independent basis functions, the key feature of the present approach. In sum, ixc-MRCISD+s combines the merits of both internally and externally contracted CI as well as IEH approaches. Numerical results will reveal that the $\{\Psi_I^{(0)}\}$, $\{\Xi_I^{(1)}\}$ and $\{\Theta_I\}$ functions do have decreasing weights in the wave functions Ψ_I, thereby justifying the SDS characterization.

3. The approach can progressively be extended to potentially exact solutions. First of all, the first-order external functions $\{|\Psi_I^{(1)}\rangle\}$ can be partially decontracted. For instance, those $\{\Phi_q\}$ functions of coefficients larger than a preset threshold can be treated as independent functions. Alternatively, the fluctuation potential can be separated into different classes according to the number of electrons excited into or out of the active space [13–15], so as to generate partially contracted first-order functions. The extended set of secondary functions can still be generated according to Eq. (7). Secondly, after solving the eigenvalue problem (9), a new generation of primary functions as linear combinations of the original primary, external and secondary functions can be obtained. The new generation of external functions, which include triples and quadruples relative to the original primary functions, can then be generated by single and double excitations from the new primary functions. The new generation of secondary functions result again from Eq. (7). The procedure can be continued to reach a desired accuracy. Since each new

generation inherits optimal characteristics of the preceding generation, the procedure is genetic. Two other options can further be considered: (a) instead of the first-order determination of the external functions, any other state-specific, internally contracted multi-reference approach (e.g., multi-configuration coupled electron pair approximation (CEPA) [55]) can be employed; (b) an energy functional [56–59] that is approximately size extensive can be used in place of the straight eigenvalue problem (9).

In summary, unlike the existing EH- or IEH-based MRPTs, which can also be described nominally as SDS methods, the present approach is intrinsically a CI, yet with the flexibility of using a variety of expansion vectors as well as approximations for the Hamiltonian.

2.2 SDS-MS-MRPT2

The ixc-MRCISD+s approach defined by Eqs. (8) and (9) is infinite order even though the contraction coefficients of the external functions $\{|\Phi_q\rangle\}$ are determined only to first order. Computationally, it scales as the sixth power of the molecular size N. To achieve an order N^5 scaling, we consider a second-order variant, denoted as SDS-MS-MRPT2. Depending on how the zeroth-order Hamiltonian H_0 in Eq. (3) is defined, various variants of SDS-MS-MRPT2 can be obtained. For simplicity, we consider here a multi-partitioned (state-dependent) Møller–Plesset-like [60] diagonal operator

$$H_0^I = \sum_{J\in P} |\Psi_J^{(0)}\rangle E_J^{(0)}(I)\left\langle \Psi_J^{(0)}| + \sum_{q\in Q} |\Phi_q \right\rangle E_q^{(0)}(I)\langle\Phi_q|, \tag{17}$$

$$E_I^{(0)}(I) = \left\langle \Psi_I^{(0)}|F^I|\Psi_I^{(0)}\right\rangle, \quad E_q^{(0)}(I) = \langle\Phi_q|F^I|\Phi_q\rangle, \tag{18}$$

where

$$\begin{aligned} F^I &= \mathrm{Const}(I) + \sum_{p\sigma} \epsilon_p^I a_{p\sigma}^\dagger a_{p\sigma}, \\ \mathrm{Const}(I) &= \left\langle \Psi_I^{(0)}|H|\Psi_I^{(0)}\right\rangle - \left\langle \Psi_I^{(0)}|H_0^I|\Psi_I^{(0)}\right\rangle, \end{aligned} \tag{19}$$

Here, ϵ_p^I are the diagonal elements of the spin-averaged Fock matrix f_{pq}^I

$$\begin{aligned} f_{pq}^I &= h_{pq} + \sum_{kl}\left\{(pq|kl) - \frac{1}{2}(pl|kq)\right\}\gamma_{lk}^I, \\ \gamma_{lk}^I &= \sum_\sigma \left\langle \Psi_I^{(0)}|a_{k\sigma}^\dagger a_{l\sigma}|\Psi_I^{(0)}\right\rangle \end{aligned} \tag{20}$$

constructed with the quasi-canonical orbitals for state $\Psi_I^{(0)}$. The state-dependent constant $\mathrm{Const}(I)$ introduced to F^I (19) is merely to set a common energy reference for all the CSFs (determinants). With the so-defined H_0^I (17) operator, which leads to a barycentric zeroth-order energy [61], the first-order functions $\{\Psi_I^{(1)}\}$ (3) can straightforwardly be obtained without the need to solve a linear system of equations. To achieve the desired $\mathcal{O}(N^5)$ scaling, the Hamiltonian matrix elements over $\Xi_I^{(1)}$ are further approximated as

$$H_{(I+N_\mathrm{P})(J+N_\mathrm{P})} = \left\langle \Xi_I^{(1)}|H|\Xi_J^{(1)}\right\rangle, \quad I,J = 1,\ldots,N_\mathrm{P} \tag{21}$$

$$\approx \frac{1}{2}\left[\left\langle \Xi_I^{(1)}|H_0^J|\Xi_J^{(1)}\right\rangle + \left\langle \Xi_J^{(1)}|H_0^I|\Xi_I^{(1)}\right\rangle\right] \tag{22}$$

$$= \frac{1}{2}\sum_{q\in Q_1}\left[E_q^{(0)}(I) + E_q^{(0)}(J)\right]\tilde{C}_{qI}^{(1)}\tilde{C}_{qJ}^{(1)}. \tag{23}$$

It is this step that reduces the previous ixc-MRCISD+s (9) down to SDS-MS-MRPT2, which is no longer an upper bound to FCI. However, SDS-MS-MRPT2 is still a CI-like rather than a genuine perturbation approach.

3 Test calculations

3.1 H_2O

The symmetric stretching of the OH bonds in water has long been taken as a model problem for assessing the ability of a method to describe variable near-degeneracies. Although the earlier DZP model problem is still of interest, the more recent FCI calculations [62] using a cc-pVDZ basis set [63] provides a more thorough assessment, and it is this set of calculations with which we compare. The model space is spanned by all the $M_s = 0\,A_1$ irreducible representation determinants that could be generated from distributing the 8 valence electrons among the 6 valence orbitals. The energetically lowest $S = 0$ eigenfunction was chosen as the primary space (N.B. Spin contamination was monitored for this and all the other model problems of the study, and was found to be less than 1×10^{-9} a.u.). The lowest lying (i.e., oxygen 1s-like) molecular orbital was also correlated in the SDS-MS-MRPT2 calculation. The results for various geometries are given in Table 1. The reference geometry has the hydrogens at (0., ±1.515263, −1.049898) a.u., with oxygen at the origin; the other geometries are obtained by elongating simultaneously the OH bonds by 1.5, 2., 2.5 and 3 times. It is seen that the nonparallellity error (NPE, i.e., the difference between the maximum and minimum deviations from FCI for the calculated point energies) of SDS-MS-MRPT2 is rather similar to that of GVVPT2 [16, 17]. This is very appealing since SDS-MS-MRPT2 has a very simple structure,

Table 1 Deviations (in a.u.) of CASSCF, GVVPT2, SDS-MS-MRPT2 and MR-CISD from FCI for symmetrically stretched H_2O

	FCI[a]	CASSCF	GVVPT2	SDS-MS-MRPT2[b]	SDS-MS-MRPT2[c]	MRCISD
$1.0R_e$	−76.241860	0.012157	0.011296	0.015994	0.017069	0.004425
$1.5R_e$	−76.072348	0.007331	0.005926	0.010293	0.012277	0.003937
$2.0R_e$	−75.951665	0.006353	0.005286	0.008499	0.010268	0.003208
$2.5R_e$	−75.917991	0.007336	0.007226	0.010022	0.010568	0.002837
$3.0R_e$	−75.911946	0.007742	0.007933	0.010692	0.010790	0.002753
NPE	0	0.039309	0.006010	0.007495	0.006801	0.001673

NPE nonparallelity error

[a] Ref. [62]

[b] With secondary functions defined by Eq. (7)

[c] Without secondary functions

Table 2 Coefficients of the SDS-MS-MRPT2 wave functions for H_2O as a function of geometry

	C_{Primary}	C_{external}	$C_{\text{secondary}}$
$1.0R_e$	0.984	0.178	−0.020
$1.5R_e$	0.982	0.185	−0.040
$2.0R_e$	0.983	0.175	−0.050
$2.5R_e$	0.986	0.163	−0.029
$3.0R_e$	0.987	0.159	−0.012

whereas GVVPT2 involves a sophisticated nonlinear resolvent. For a better understanding of SDS-MS-MRPT2, the coefficients of the primary, external and secondary functions are further documented in Table 2. It is seen that the coefficient of the primary vector (i.e., the CASSCF wave function) does not vary appreciably with geometry (i.e., from a high of 0.987, when the bonds are essentially broken, to a low of 0.982, when the bonds begin to break), while that of the external vector (i.e., the dynamic correlation contribution) varies significantly (i.e., from a low of 0.159, at large geometry, to a high of 0.185, when the bonds begin to break). The variation in the contribution from the secondary space is comparable to that in the external space. However, the maximum contribution from the secondary (0.05 at $2R_e$) does not coincide with the maximum from the external (0.185 at $1.5R_e$). These results show that the secondary space plays a role even in problems as simple as the symmetric stretching of H_2O, and its contribution varies with geometry as does the contribution from the external space.

To better elucidate the effects of the secondary space, additional tortured calculations were performed on the first excited state of H_2O. At the reference geometry, the orbitals were generated from a nonphysical SA–MCSCF calculation with a relative weighting of 100:1 for the ground and first excited states. While the results for the ground state are unremarkable, examination of the first excited state is interesting. The corresponding MCSCF and single-state (SS) MRPT [i.e., Eq. (10) with $N_P = 1$] energies are higher than the MRCISDTQ value (−75.860 427) by 0.417 and 1.050 a.u., respectively. The (intermediate normalized) correlation vector length of SS-MRPT is as large as 3.198. In contrast, SDS-MS-MRPT2 produces marked improvement over MCSCF, with the energy higher than that by MRCISDTQ by 0.355 a.u.. The SDS-MS-MRPT2 (unity normalized) coefficients are 0.618, 0.124 and −0.776 for the primary, secondary and external states. The situation can certainly be improved by further iterations as discussed before.

3.2 CH_2

The adiabatic excitation of the ground state $^3B_1$ of CH_2 to the first excited state $^1A_1$ is one of the simplest models for assessing the ability of a method to describe equitably electronic states of different spins. The two states are significantly different in nature: While $^3B_1$ is well described by a single determinant, $^1A_1$ is profoundly two configurational. The DZP FCI calculations [64] of Bauschlicher and Taylor remain the standard. As in the FCI paper, we used the modified DZP basis sets of Bauschlicher and Taylor, which applies different polarization functions (triplet: $\zeta_d = 0.74$; singlet: $\zeta_d = 0.51$) to a standard Huzinga–Dunning DZ basis [65, 66]. We also used the same geometries: H = (0., ±1.871093, 0.82525) for $^3B_1$ and H=(0., ±1.644403, 1.32213) for $^1A_1$, with carbon at the origin. The model spaces are spanned by all the $M_s = 1$ and $M_s = 0$ determinants of B_1 and A_1 irreducible representations for the $^3B_1$ and $^1A_1$, respectively. The energetically lowest $S = 1$ and $S = 0$ eigenfunctions were chosen as the corresponding primary spaces. As in the Bauschlicher and Taylor study, the carbon 1*s*-like lowest energy molecular orbital was not correlated in the SDS-MRPT2 calculation. As can be seen from Table 3, SDS-MS-MRPT2, using the

Table 3 Deviations of CASSCF, GVVPT2, SDS-MS-MRPT2 and MRCISD from FCI for the singlet and triplet energies (in a.u.) and excitation energy (ΔE in kcal/mol) of CH_2

	FCI[a]	CASSCF	GVVPT2	SDS-MS-MRPT2	MRCISD
$^3B_1$	−39.046260	0.080306	0.009966	0.015224	0.001332
$^1A_1$	−39.027183	0.081654	0.015408	0.016058	0.001379
ΔE	11.97	0.85	3.41	0.52	0.03

[a] Ref. [64]

spin-dependent analog of Eq. (20), does provide equitable descriptions of the two states, such that the error in the adiabatic excitation energy is very small (i.e., 0.5 kcal/mol) and is even smaller than that (3.4 kcal/mol) by GVVPT2. The underlying reason is that the relative importance of primary, external and secondary contributions in SDS-MS-MRPT2 is very similar for $^3B_1$ and $^1A_1$ (i.e., primary: 0.990 and 0.989; secondary: 0.141 and 0.147; external: −0.017 and −0.019).

3.3 The ground state of LiF

Thanks to the avoided crossing of the qualitatively ionic configuration (at short internuclear distances) with the asymptotic neutral configuration, the ground state $1^1\Sigma^+$ of LiF is difficult not only for single-reference methods, but can even be for a single-state multi-reference approach. The difficulty is compounded by the use of modest active spaces, as found in the pioneering multi-reference studies of Bauschlicher and Langhoff [67]. The difficulty is exhibited in the abrupt change in the optimal molecular orbitals appropriate for an ionic configuration from those appropriate for a neutral configuration. Because of this, a balanced treatment using orbitals from a single-state method (even CASSCF used here) is problematic even for variational approaches like MRCISD. Of course, the difficulty can be addressed with larger active spaces or with state averaging (i.e., a MS approach).

The SDS-MS-MRPT2 calculations were performed using the same basis set and active space as did Bauschlicher and Langhoff [67]. A $(9s4p/3s2p)$ double split valence basis was used for lithium [68] and a $(9s6p1d/4s3p1d)$ augmented polarized double split valence basis was used for fluorine [65–67]. The model space is spanned by all the $M_S = 0\,A_1$ determinants can be obtained by distributing the five $2p$ electrons of fluorine and the $2s$ electron of lithium among the σ (a_1 in C_{2v} symmetry), π (b_1 and b_2), σ^* and π^* orbitals. While the current software cannot make full use of non-Abelian point group symmetry, the MCSCF program ensures that only the orbitals of correct Λ-quantum number were used. As in

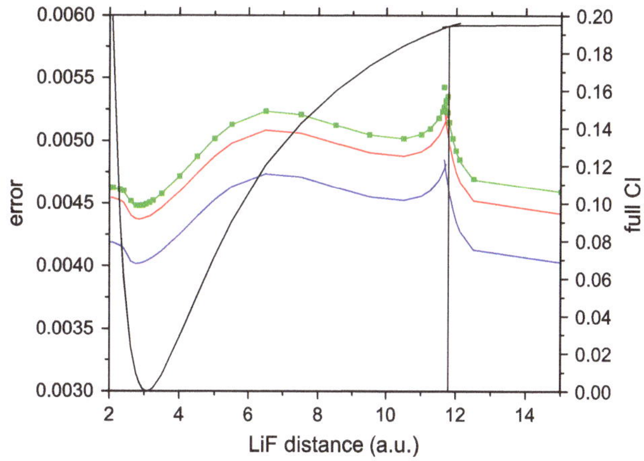

Fig. 1 The FCI potential energy curve of $1^1\Sigma^+$ LiF (*black line*, right ordinate) and energy errors for GVVPT2 (*blue line*, left ordinate), SDS-MS-MRPT2 with secondary states defined by the projection (7) (*red line*, left ordinate), SDS-MS-MRPT2 with secondary states being the $N_P + 1$ to $2N_P$ solutions of the projected Hamiltonian *PHP* (*olive line*, left ordinate) and SDS-MS-MRPT2 without secondary states (*olive squares*, left ordinate). The crossover of the FCI ionic and covalent solutions is indicated by the *vertical line*

the original Bauschlicher and Langhoff study, the energetically lowest three orbitals were kept doubly occupied. Since the FCI energy depends on the precise nature of such core-like orbitals, FCI calculations were performed at all the reported geometries. Indeed, the dependence of solutions on the core orbitals is so severe for LiF that it persists even for FCI. The FCI energies relative to its minimum are plotted in Fig. 1, using the right ordinate. It is seen that the FCI solution changes abruptly (i.e., discontinuously in the first derivative) from one type of solution to the other at slightly less than 11.8 a.u. For comparison, single-state CASSCF predicts this change to occur close to 12.0 a.u. Consequently, post-MCSCF perturbative treatments are required to use the qualitatively wrong set of orbitals in the interval between 11.8 and 12.0 a.u. Nonetheless, the present SDS-MS-MRPT2 does predict a smooth curve with a transition around 11.5 a.u., just like GVVPT2 which guarantees the continuity. Since the agreement of SDS-MS-MRPT2/GVVPT2 with FCI is so good, only the deviations of SDS-MS-MRPT2/GVVPT2 from FCI are plotted in Fig. 1, using the left ordinate. Also, plotted in Fig. 1 are the results of calculations without the secondary state. It can be seen that these calculations (olive squares) essentially coincide with those using the second lowest state as the secondary state, but are not in as good agreement with those from the projected secondary state (7) or GVVPT2. Note in passing that metastable solutions of SDS-MS-MRPT2/GVVPT2 were also found beyond the transition point (i.e., nonionic solutions at less than 11.5 a.u. and ionic solutions at greater than 11.5 a.u) that qualitatively resemble the correct FCI curves.

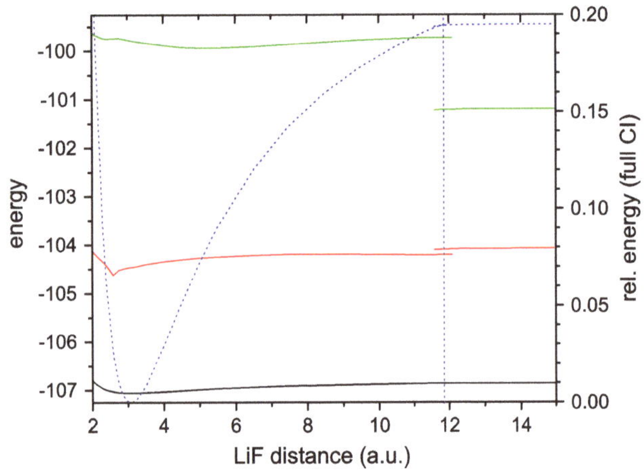

Fig. 2 Expectation values of primary (*black*), secondary (*red*) and external functions (*olive*) as function of LiF bond length. The FCI energy curve (*blue dashed*, right ordinate) and ionic-covalent transition point are included for reference

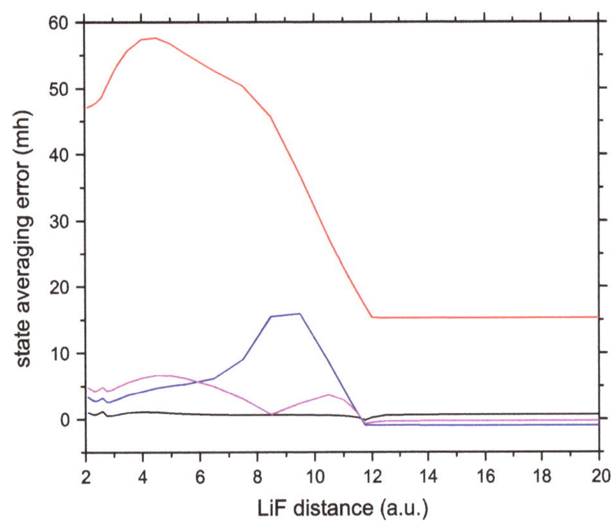

Fig. 3 Difference between calculations using state-averaged and single-state CASSCF orbitals for the energies of $1^1\Sigma^+$ LiF. The FCI, CASSCF, SS-MRPT2 and SDS-MS-MRPT2 results are in *black, red, blue* and *magenta*, respectively

Further insight into the characteristics of the primary, secondary and external functions of SDS-MS-MRPT2 can be obtained by plotting their expectation values as a function of the internuclear distance. It can be seen from Fig. 2 that the values vary smoothly with geometry, except for the ionic-covalent transition point, and corroborate the protocol of treating the primary–external interactions perturbatively and the primary–secondary interactions variationally. Moreover, it is hardly surprising that qualitatively different external (and secondary) functions are needed for describing the ionic and covalent configurations.

We take the opportunity to demonstrate the orbital effects by plotting the energy difference between calculations using the two-state-averaged and single-state CASSCF orbitals, see Fig. 3. It is seen that the changes in the CASSCF energies are profound and highly variable, i.e., going from 15.3 mE_h at the dissociation limit to a maximum of 57.6 mE_h at 4.5 a.u. The situation is markedly improved by SS-MRPT2, whose maximum deviation from FCI is 16.0 mE_h at 9.5 a.u. and minimum deviation is -1.0 mE_h at the dissociation limit. Yet, there still exists a broad peak around the inversion point to be discussed below. In contrast, SDS-MS-MRPT2 performs much better, with the variations in the range only from -0.7 to 6.6 mE_h and without sharp features. It is interesting to note that even FCI has a weak dependence on the orbitals, ranging from -0.1 to 1.2 mE_h.

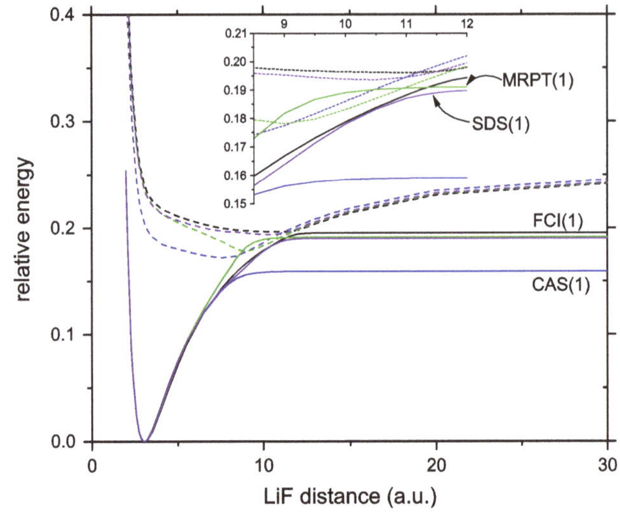

Fig. 4 Potential energy curves of 1,2 $^1\Sigma^+$ LiF relative to the minimum of the lowest curve. The $1^1\Sigma^+$ ($2^1\Sigma^+$) curves of FCI, SA-CASSCF, SS-MRPT2 and SDS-MS-MRPT2 are in *solid* (*dashed*) *black, blue, olive* and *violet*, respectively. *Inset* shows magnified view of the region near avoided crossing

3.4 LiF excitation energies

The excitation energy from $1^1\Sigma^+$ to $2^1\Sigma^+$ of LiF as a function of geometry is difficult to calculate accurately [67]. The difficulty arises from several factors. First, the excitation energies at short distances and near the avoided crossing differ by two orders of magnitude (see Fig. 4). Second, the region near the avoided crossing is problematic because of the significantly different natures of optimal orbitals for the ionic and covalent states (see above). Third, the interplay between static and dynamic correlation is subtle near the avoided crossing, to the extent that a two-state SA-CASSCF treatment predicts a closest approach distance displaced by over 3 a.u. from that of FCI. Specifically, the FCI minimum excitation energy (ca. 3.6 mE_h) occurs near 11.75 a.u., whereas that (ca. 21 mE_h) of SA-CASSCF is reached around 8.5 a.u. Consequently,

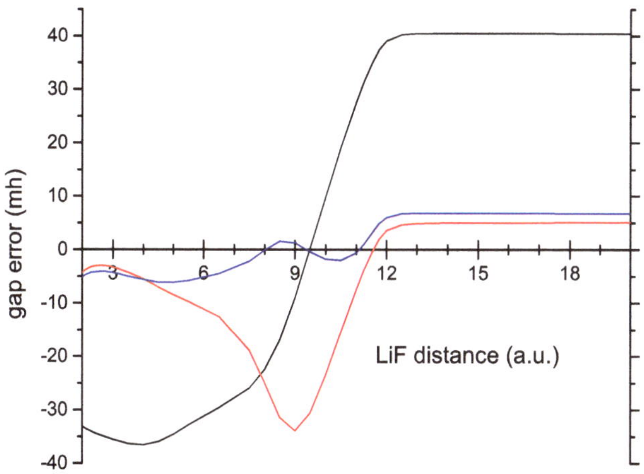

Fig. 5 Deviations of SA-CASSCF (*black*), SS-MRPT2 (*red*) and SDS-MS-MRPT2 (*blue*) from FCI for the energy gap between $2^1\Sigma^+$ and $1^1\Sigma^+$ of LiF

the dynamic correlation must in some sense undo the problems introduced by SA-CASSCF.

SA-CASSCF calculations, with an equal weight for the two states, were performed to generate the orbitals for subsequent SS-MRPT2, SDS-MS-MRPT2 and FCI calculations. It turns out that SS-MRPT2 not only fails to predict the avoided crossing, but also produces an inverted order of states for a range of distances (see Fig. 4, especially the inset). Specifically, from ca. 9.0 to ca. 10.5 a.u., the SS-MRPT2 state based on the second CASSCF wave function is energetically lower than that based on the lowest CASSCF wave function. In particular, the inversion reaches a maximum of ca. 7 mE_h near 9.5 a.u. As can be seen from Fig. 5, SDS-MS-MRPT2 has a NPE of 13.0 mE_h for the gap over the entire range, which amounts to 5 % of the FCI gap variation of 260.2 mE_h. Moreover, there are not any abrupt changes in the gap error. In contrast, the NPE for SS-MRPT2 is three times larger (39.1 mE_h) and has a relatively sharp maximum. Even if one always chose the lowest energy solution, which would result in two geometries where the first derivatives are discontinuous, the NPE would still be 31.8 mE_h. The SA-CASSCF NPE error for the gap is about twice as large (i.e., 76.9 mE_h).

3.5 Numerical test of size extensivity

The so-called size extensivity, viz., the property of the calculated energy to scale linearly with system size, is highly desirable. Unfortunately, achieving this ideal is difficult, especially for multi-reference approaches. A strict enforcement of this property is generally accompanied by additional approximations to the wave function [69]. Incorporating other desirable properties, such as smooth and global dissociation curves, is generally accompanied by some variational character, which in turn often introduces or enhances size inextensivity. A pragmatic approach is to view size inextensivity as yet another source of error and to assess its magnitude and characteristics. It is this approach that is adopted herein.

Consider an arrangement of multiple H_2 molecules, separated by sufficiently large distances such that interactions between the monomers are negligible. For a size-extensive method, the energy of a system composed of n subsystems is exactly n times the energy of a single monomer. To assess the characteristics of SDS-MS-MRPT2, two H_2 molecules were placed in parallel with their centers separated by 40 a.u. Another system of three H_2 molecules was constructed similarly. The experimental equilibrium distance was adopted for each monomer. The model spaces were generated by constructing all the Σ_g^+ determinants from the $M_s = 0$ distributions of electrons among the $1\sigma_g$ and $1\sigma_u$ molecular orbitals. The cc-pVDZ basis set [63] was used. The SDS-MS-MRPT2 results with and without the secondary states are given in Table 4. Several observations can be made here. First, the size-extensivity errors (SEE) of SS-MRPT2 are fairly small (in the sub-milihartree range), which are further reduced by an order of magnitude by SDS-MS-MRPT2. Even in the worst case, the SEEs of SDS-MS-MRPT2 are just dozens of mircohartees. It appears that the final diagonalization step of SDS for relaxing the dynamic correction to the original static correlation tends to correct for SEE. Second, the SEEs of SDS-MS-MRPT2 grow as $n_o(n_o - 1)/2$, following the underlying MRPT2. Such functional dependence is also similar to that of MRCISD, suggesting that coupled-

Table 4 SS-MRPT2, GVVPT2 and SDS-MS-MRPT2 energies (in a.u.) of noninteracting H_2 molecules

	MRPT2		GVVPT2		SDS-MS-MRPT2	
	E	ΔE	E	ΔE	E	ΔE
H_2	−1.15990714	0	−1.15985885	0	−1.15991642[a]	0[a]
					−1.15985632[b]	0[b]
$(H_2)_2$	−2.31991288	−0.09861	−2.31982751	−0.10981	−2.31984155[a]	−0.00870[a]
					−2.31970817[b]	0.00447[b]
$(H_2)_3$	−3.48001970	−0.29829	−3.47989228	−0.31573	−3.47977656[a]	−0.02728[a]
					−3.47955577[b]	0.01319[b]

$\Delta E = E((H_2)n) - nE(H_2)$ (in mE_h)

[a] With secondary functions defined by Eq. (7)

[b] Without secondary functions

cluster-based corrections [56–58] can be employed to reduce the SEE [59]. Third, the SEEs of SDS-MS-MRPT2 with and without secondary functions bracket the size-extensive results. This suggests that it is possible to modify SDS-MS-MRPT2 such that no SEE would arise regardless of the treatment of dynamic correlation. Alternatively, the SEEs could first be eliminated from the underlying dynamic calculation, with better balanced choice of H_0 [70]. However, the already very small SEEs of SDS-MS-MRPT2 raise the question whether such refinements are of practical value at all. By contrast, it is more straightforward to relax the contraction of the external functions and/or iterate the diagonalization–perturbation–diagonalization procedure, so as to improve steadily the accuracy of SDS-MS-MRPT2.

4 Conclusions

A new framework for treating strongly correlated electrons is suggested, which is related to, but more flexible than, intermediate Hamiltonian approaches that follow a sequence of first static, then dynamic and then again static. It is unique in that it neither develops a final answer in a space of the same dimensionality as the original unperturbed (contracted) reference functions (e.g., as do MS-CASPT2 and MS-MCQDPT) nor expands to a final intermediate Hamiltonian that has the same dimensionality as the entire original model space (e.g., as does GVVPT2). Instead, for N_P reference functions, N_P correlation functions are generated, as are another N_P functions in the model space. That is, the final space is of dimension $3N_P$, in which the Hamiltonian (or an approximation to it) is represented and diagonalized. In essence, the external vectors, generated in whatsoever way, are treated as independent functions, such that the approach is best characterized as a CI rather a genuine intermediate Hamiltonian approach. In the SDS-MS-MRPT2 realization, the diagonal (first order) external function block of the Hamiltonian is approximated by a Møller–Plesset-like operator that maintains the $\mathcal{O}(n^5)$ scaling of second-order perturbation theory with system size. Yet, SDS-MS-MRPT2 is still a CI-like rather than a perturbation approach.

Because of the flexibility and the controlled dimensionality of the EH, the wave functions are both resilient to small, unphysical denominators (which often plague MRPTs) and retain the ability for static correlation to respond to the effect of dynamic correlation. Indeed, numerical tests on the acknowledged problematic avoided crossing of the lowest two energy curves of LiF showed that SDS-MS-MRPT2 did not require any level shifts of either orbital or state energies. Of course, level shifts can simply be introduced if a situation so requires. Numerical tests on the symmetrically stretched H_2O, the singlet-triplet excitation energy of CH_2, and the dissociation of LiF showed that the performance of SDS-MS-MRPT2 is very close to that of GVVPT2, which is rigorously continuous and has previously been shown to be accurate. Moreover, SDS-MS-MRPT2 reproduced faithfully the kink in the FCI curve of the ground state of LiF, when the unusual choice of active space inherent in the LiF model problem was used. Although not strictly size extensive, the already very small SEE of SDS-MS-MRPT2 can likely be cured by simple modifications.

Last but not least, the suggested framework is general and applicable to most, if not all, methods for treating dynamic correlation. In particular, relaxing the contraction of the external functions and/or iterating the diagonalization–perturbation–diagonalization procedure would allow for progressive accuracies.

Acknowledgments The research of this work was supported by the NSFC (Project Nos. 21033001, 21273011 and 21290192) and NSF (Grant No. EPS-0814442).

Appendix: String-based macroconfigurations

Since their introduction into quantum chemistry by Knowles and Handy [41], representation of determinants as separate strings of $m_s = \frac{1}{2}$ (alpha) occupancies and $m_s = -\frac{1}{2}$ (beta) occupancies have become well entrenched [42–44]. This representation has several distinct advantages, e.g., the simple coupling coefficients (either -1, 0 or 1) and the compact independent indexing of alpha and beta strings (i.e., pruned binomial trees). However, extension of this concept, originally designed for FCI, to more complicated calculations that have multiple groups of orbitals (e.g., MRCISD with core, valence and virtual orbitals) has generally been accompanied by a reduction in some of the advantages of the original method. In the SDS framework introduced in this work, including its MRPT2 realization, there is a special need to limit excitations between particular groups of orbitals.

The partitioning of orbitals into arbitrary groups together with assigning fixed number of electrons for each group, i.e., a macroconfiguration $\kappa_i = G_1^{N_1^{\kappa_1}} G_2^{N_2^{\kappa_2}} \ldots G_g^{N_g^{\kappa_g}}$, has been shown to lead to both reliable means of generating complete and incomplete model spaces and computational efficiencies [40]. The advantages of articulating parts of the Hilbert space with macroconfigurations and the advantages of representing determinants with strings would seem to be incompatible. However, we demonstrate that this is not the case.

The critical insight that allows full use of both the concepts of macroconfigurations and strings is the notion of

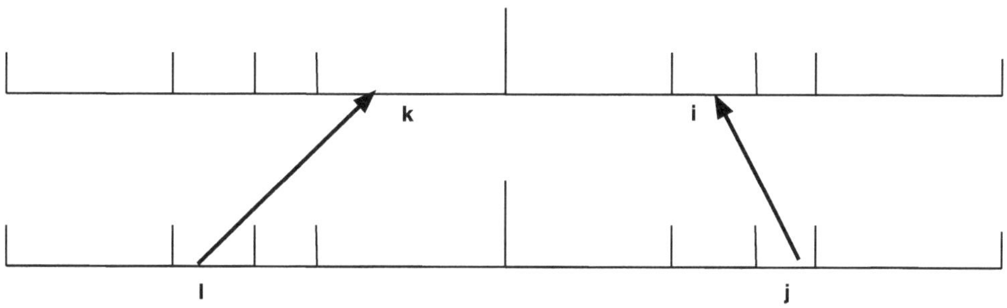

Fig. 6 Representation of a double excitation from a ket determinant involving groups 2 and 7 to a bra determinant involving groups 4 and 6

spectator groups. Suppose that a class of excitations from a set of ket determinants to another set of bra determinants can be characterized as a one-electron excitation from group G_j to G_i and another from G_l to G_k. A concrete example is depicted in Fig. 6. In this example, there are four orbital groups, corresponding to eight spin-orbital groups. The indices of the generated strings within groups G_k and G_i can be calculated straightforwardly as walks on a (universal) pruned binary tree, as in the original work of Knowles and Handy. The indices of the strings within G_l and G_i are even simpler, as they can be generated sequentially. The key to efficiency is that the product of the resulting coupling coefficient multiplied by the two-electron integral [i.e., $(ij|kl)$] is identical for every bra/ket pair that shares identical walks in the spectator groups (i.e., groups 1, 3, 5 and 8 in the considered example). It can be appreciated that the number of bras and kets sharing this matrix element could be very large (e.g., if group 8 is the beta excited orbitals and there are two electrons in this group, then its contribution alone can be on the order of 10,000 for a modest molecule using a triple split valence one-electron basis set).

Sharing the single matrix element by a truly large number of bra/ket pairs can only occur if the indexing of the bra and ket determinants can be done efficiently. This is accomplished by precomputing the base index of each group of the bra and ket macroconfigurations. For example, each walk within group 3 is offset from the previous one by the product of the total number of walks in group 2 multiplied by the total number of walks in group 1. Then, fragments of walks up to and including group 3 can be computed as the contribution from groups 1 and 3 added to the contribution from group 2. Since the contributions from groups 1 and 3 do not change as a result of the class of excitations connecting the ket and bra macroconfigurations, the sequence of contributions to the overall indices from the spectator groups can be precomputed at the macroconfiguration level.

The described algorithm is clearly not dependent on the number of spin-orbital groups. This allows one further efficiency in that Abelian molecular point group symmetry restrictions can be imposed at the macroconfiguration level, provided that each spin-orbital group contain only spin orbitals that transform as the same irreducible representation. In practice, this requires a sequestering of spin orbitals for each group of occupation-restricted orbitals. For example, if the CASSCF active orbitals transform as, say, two of the irreducible representations of the molecule, then four groups of spin orbitals would be generated.

The described algorithm is applicable to any partitioning of the Hilbert space that can be represented as direct sum of determinants generated from macroconfigurations. This includes all finite-order PT, including multi-reference variants, of present interest, as well as (multi-reference) CI and CEPA methods. Consequently, the computational realization was straightforwardly validated against well-established computer programs.

References

1. Hirao K (1992) Chem Phys Lett 190:374
2. Hirao K (1992) Chem Phys Lett 196:397
3. Andersson K, Malmqvist PA, Roos BO, Sadlej AJ, Wolinski K (1990) J Phys Chem 94:5483
4. Andersson K, Malmqvist PA, Roos BO (1992) J Chem Phys 96:1218
5. Celani P, Werner HJ (2000) J Chem Phys 112:5546
6. Finley J, Malmqvist PA, Roos BO, Serrano-Andres L (1998) Chem Phys Lett 288:299
7. Shiozaki T, Győrffy W, Celani P, Werner HJ (2011) J Chem Phys 135:081106
8. Nakano H (1993) J Chem Phys 99:7983
9. Nakano H, Nakayama K, Hirao K, Dupuis M (1997) J Chem Phys 106:4912
10. Granovsky AA (2011) J Chem Phys 134:214113
11. Spiegelmann F, Malrieu JP (1984) J Phys B 17:1235
12. Angeli C, Borini S, Cestari M, Cimiraglia R (2004) J Chem Phys 121:4043
13. Angeli C, Cimiraglia R, Evangelisti S, Leininger T, Malrieu JP (2001) J Chem Phys 114:10252
14. Angeli C, Cimiraglia R, Malrieu JP (2001) Chem Phys Lett 350:297
15. Angeli C, Cimiraglia R, Malrieu JP (2002) J Chem Phys 117:9138
16. Khait YG, Song J, Hoffmann MR (2002) J Chem Phys 117:4133

17. Jiang W, Khait YG, Hoffmann MR (2009) J Phys Chem A 113:4374
18. Mahapatra US, Datta B, Mukherjee D (1999) J Phys Chem A 103:1822
19. Mao S, Cheng L, Liu W, Mukherjee D (2012) J Chem Phys 136:024105
20. Mao S, Cheng L, Liu W, Mukherjee D (2012) J Chem Phys 136:024106
21. Rolik Z, Szabados Á, Surján PR (2003) J Chem Phys 119:1922
22. Szabados Á, Tóth G, Rolik Z, Surján PR (2005) J Chem Phys 122:114104
23. Chen F (2009) J Chem Theory Comput 5:931
24. Chen F, Fan Z (2013) J Comput Chem. doi:10.1002/jcc.23471
25. Lei Y, Wang Y, Han H, Song Q, Suo B, Wen Z (2013) J Chem Phys 137:144102
26. Xu E, Li S (2013) J Chem Phys 139:174111
27. Chaudhuri RK, Freed KF, Hose G, Piecuch P, Kowalski K, Wloch M, Chattopadhyay S, Mukherjee D, Rolik Z, Szabados Á, Tóth G, Surján PR (2005) J Chem Phys 122:134105
28. Hoffmann MR, Datta D, Das S, Mukherjee D, Szabados Á, Rolik Z, Surján PR (2009) J Chem Phys 131:204104
29. Kirtman B (1981) J Chem Phys 75:798
30. Malrieu JP, Durand Ph, Daudey JP (1985) J Phys A 18:809
31. Heully JL, Daudey JP (1988) J Chem Phys 88:1041
32. Mukhopadhyay D, Datta B, Mukherjee D (1992) Chem Phys Lett 197:236
33. Malrieu JP, Heully JL, Zaitsevskii A (1995) Theor Chim Acta 90:167
34. Khait YG, Hoffmann MR (1998) J Chem Phys 108:8317
35. Meissner L (1998) J Chem Phys 108:9227
36. Landau A, Eliav E, Kaldor U (1999) Chem Phys Lett 313:399
37. Nikolic D, Lindroth E (2004) J Phys B 37:L285
38. Eliav E, Borschevsky A, Shamasundar KR, Pal S, Kaldor U (2009) Int J Quantum Chem 109:2909
39. Liu W (2010) Mol Phys 108:1679
40. Khait YG, Song J, Hoffmann MR (2004) Int J Quantum Chem 99:210
41. Knowles PJ, Handy NC (1984) Chem Phys Lett 111:315
42. Knowles PJ, Handy NC (1989) Comp Phys Comm 54:75
43. Olsen J, Roos BO, Jørgensen P, Jensen HJA (1988) J Chem Phys 89:2185
44. Kallay M, Surjan PR (2001) J Chem Phys 115:2945
45. Roos BO, Linse P, Siegbahn PEM, Blomberg MRA (1981) Chem Phys 66:197
46. McLean AD, Liu B (1973) J Chem Phys 58:1066
47. Werner HJ (1987) Adv Chem Phys 49:1
48. Wang Y, Suo B, Zhai G, Wen Z (2004) Chem Phys Lett 389:315
49. Siegbahn PEM (1977) Chem Phys 25:197
50. Siegbahn PEM (1983) Int J Quantum Chem 23:1869
51. Meyer W (1977) In: Schaefer HF III (ed) Modern theoretical chemistry. Plenum, New York
52. Werner HJ, Reinsch EA (1982) J Chem Phys 76:3144
53. Siegbahn PEM (1980) Int J Quantum Chern 18:1229
54. Werner HJ, Knowles PJ (1988) J Chem Phys 89:5803
55. Fink R, Staemmler V (1993) Theor Chim Acta 87:129
56. Gdanitz RJ, Ahlrichs R (1988) Chem Phys Lett 143:413
57. Szalay PG, Bartlett RJ (1993) Chem Phys Lett 214:481
58. Szalay PG (2008) Chem Phys 349:121
59. Khait YG, Jiang W, Hoffmann MR (2010) Chem Phys Lett 493:1
60. Zaitsevskii A, Malrieu JP (1995) Chem Phys Lett 233:597
61. Huron B, Malrieu JP, Rancurel F (1973) J Chem Phys 58:5745
62. Olsen J, Jørgensen P, Koch H, Balkova A, Bartlett RJ (1996) J Chem Phys 104:8007
63. Dunning TH Jr (1989) J Chem Phys 90:1007
64. Bauschlicher CW, Taylor PR (1987) J Chem Phys 86:2844
65. Dunning TH Jr (1970) J Chem Phys 53:2823
66. Dunning TH Jr, Hay PJ (1977) In: Schaefer HF III (ed) Methods of electronic structure theory, vol 2. Plenum, New York
67. Bauschlicher CW, Langhoff SR (1988) J Chem Phys 89:4246
68. Huzinaga S (1965) J Chem Phys 42:1293
69. Nooijen M, Shamasundar KR, Mukherjee D (2005) Mol Phys 103:2277
70. Heully JL, Malrieu JP, Zaitevskii A (1996) J Chem Phys 105:6887

Theor Chem Acc (2014) 133:1514
DOI 10.1007/s00214-014-1514-5

REGULAR ARTICLE

Singlet–triplet separations of di-radicals treated by the DEA/DIP-EOM-CCSD methods

Ajith Perera · Robert W. Molt Jr. · Victor F. Lotrich · Rodney J. Bartlett

Received: 15 February 2014 / Accepted: 30 May 2014 / Published online: 20 June 2014

Abstract The singlet–triplet splittings of the di-radicals methylene, trimethylene–methane, ortha-, meta- and para-benzynes, and cyclobutane-1,2,3,4-tetrone have become test systems for the applications of various multi-reference (MR) coupled-cluster methods. We report results close to the basis set limit computed with double ionization potential (DIP) and double electron attachment (DEA) equation-of-motion coupled-cluster methods. These di-radicals share the characteristics of a 2-hole 2-particle MR problem and are commonly used to assess the performance of MR methods, and yet require more careful study unto themselves as benchmarks. Here, using our CCSD(T)/6-311G(2d,2p) optimized geometries, we report DIP/DEA-CC results and single-reference (SR) CCSD, CCSD(T), ΛCCSD(T) and CCSDT results for comparison.

Keywords Singlet–triplet separations · Methylene · DIP · DEA · Benzyne · Multi-reference · Equation-of-motion coupled-cluster

1 Introduction

The treatment of some classes of multi-reference problems can be found by exploiting single-determinant reference states composed of a *different* number of particles from that of the MR target state, with subsequent addition or removal of excess electrons to define the N-particle states of interest. The idea originates in work by Nooijen and Bartlett [1] on using the double ionization potential similarity transformed equation-of-motion (DIP-STEOM-CC) and its complementary double electron attached variant (DEA-STEOM-CC). This was used to treat the vibrational frequencies of ozone from a formally O_3^{-2} reference and certain states of the NO dimer [2] relative to the closed shell $(NO)_2^{+2}$. An equation-of-motion coupled-cluster method without the second similarity transformation of STEOM, for adding (DEA) or removing (DIP) two electrons from an underlying single-reference system has been proposed [3–5] with planned extension to three (TIP/TEA) [6, 7] or four (QIP/QEA). The method has several attractive features that make the approach worth pursuing. These include the following:

1. The ansatz has a global *extensive* part, $\exp(T^{n\pm2})|n \pm 2\rangle$ based on a closed shell $n \pm 2$ reference vacuum and a local correlation *intensive* part, whose wave function is $|\psi\rangle = R^{n\mp2}\exp(T^{n\pm2})|n \pm 2\rangle$, where $R^{n\mp2}$ is the CI-like right-hand eigenvector in EOM-CC. This has the advantage that instead of asking a fully extensive MR-CC method to account for dynamic and non-dynamic correlation effects, the intensive part allows one to target the usually local multi-reference behavior.
2. As long as the $|n \pm 2\rangle$ reference is closed shell, the target states are automatically spin-eigenfunctions. The guarantee of a spin-eigenfunction is not always achievable in most SR-CC or MR-CC methods.
3. The DIP/DEA-EOM-CC wavefunction is operationally single reference making it as easy to apply as single-reference CC, with no decisions for the user but basis set, level of correlation and a choice of the one spatial active orbital to doubly occupy in the $n + 2$ vacuum

Dedicated to the memory of Professor Isaiah Shavitt and published as part of the special collection of articles celebrating his many contributions.

A. Perera (✉) · R. W. Molt Jr. · V. F. Lotrich · R. J. Bartlett
Quantum Theory Project, University of Florida, Gainesville, FL 32655, USA
e-mail: perera.ajith@gmail.com

(usually the LUMO), or to un-occupy in the $n - 2$ vacuum (usually the HOMO).

4. The DIP/DEA-EOM-CC is invariant to active orbital rotations by virtue of the active orbitals being in either the occupied or the unoccupied space.
5. Multiple states can be obtained from the EOM matrix diagonalization providing excited states as well as the ground state. So in the event that other occupied orbitals in DIP or virtuals in DEA interact strongly with the chosen active orbital, then that solution occurs as well, and will appear as one of the eigenvectors. This helps to confirm the particular orbitals that manifest MR character and if there are more than two, might suggest a subsequent TIP/TEA or QIP/QEA calculation. The price paid for these attractive features is that each stage of the calculation has to be converged: first the SR-CC solution for the $n \pm 2$ closed shell system, then the DEA/DIP-EOM solution itself. Because of orbital dependence in these calculations, the DIP solution in particular can sometimes be difficult to converge.

This contribution addresses single–triplet splitting in di-radicals, which have been considered a proving-ground for multi-reference methodologies. These include generalized SR-CC [8–10], renormalized CC [11–13], ΛCCSD(T) [14–17], spin-flip [18–20] and hybrids like reduced multi-reference [21–24]. There are also more complete state-universal (SU) forms along with the state-specific Brillouin-Wigner (BW) [25–28], Mk version [29–32], or the internally contracted form [33, 34]. The singlet states involved tend to manifest most of the multi-reference character, since they can vary from a two-determinant open-shell singlet, for which there are other methods available (TD-CCSD [35, 36]) to a GVB when a bond starts to form, or to situations where all four determinants, as shown below (see Fig. 1), could have a large role in the wave function. The triplet, on the other hand, can frequently be described well by standard ROHF or UHF SR-CC using the high-spin determinant as the reference, but here, both states are described with DIP/DEA. The following outlines the theory and discusses its applications to singlet–triplet splittings (S-T splitting) in a variety of di-radicals.

2 Theory

The wavefunction ansatz for a potentially MR n-particle state is

$$|\psi_k\rangle = R_k^{n\mp2}\exp\left(T^{n\pm2}\right)|n\pm2\rangle \tag{1}$$

The $n \pm 2$ solution is a single-reference CC one, but for $n \pm 2$ electrons while the $R_k^{n\mp2}$

$$R^{n-2} = \Sigma\left[r_{ij}\{ij\} + r_{ijk}^{a}\left\{a^{\dagger}ijk\right\}\right] = R_{2h}^{n-2} + R_{3h1p}^{n-2} \tag{2}$$

and

$$R^{n+2} = \Sigma\left[r^{ab}\left\{a^{\dagger}b^{\dagger}\right\} + r_i^{abc}\left\{a^{\dagger}b^{\dagger}c^{\dagger}i\right\}\right] = R_{2p}^{n+2} + R_{3p1h}^{n+2} \tag{3}$$

restores the n-particle solutions, by either removing or adding two electrons to the underlying SR-CC solution,

$$\left|\Phi^{n\pm2}\right\rangle = \exp\left(T^{n\pm2}\right)|n\pm2\rangle \tag{4}$$

for the $n \pm 2$ electrons. The operators are indicated in normal order { }, with $i, j, k,\ldots$ indicating occupied orbitals or holes and $a, b, c,\ldots$ particles or unoccupied orbitals.

Then inserting this ansatz into the Schrödinger equation using the usual EOM-CC strategy, we obtain

$$\bar{H}^{n\pm2}R_k^{n\mp2}|n\pm2\rangle = \omega_k R_k^{n\mp2}|n\pm2\rangle \tag{5}$$

which provides the n-particle k-states of interest via the EOM-CC diagonalization. The quantity $\bar{H}^{N\pm2} = \exp(-T^{n\pm2})H\exp(T^{n\pm2})$, is defined from the CC reference-state wavefunction, $\Phi^{(n\pm2)}$,

$$E_{CC}^{n\pm2} = \langle n\pm2|\bar{H}^{n\pm2}|n\pm2\rangle \tag{6}$$

$$Q^{n\pm2}\bar{H}^{n\pm2}|n\pm2\rangle = 0. \tag{7}$$

In two steps, DIP/DEA-EOM-CC [3] provides solutions for different k-states via matrix diagonalization with two-orbital, two-electron MR character. To involve more orbitals and electrons in a MR description, say 3 or 4, requires the TIP/TEA and QIP/QEA extensions. Because $\bar{H}^{(n\pm2)}$ is non-Hermitian, there is also a left-hand eigenvector, $\langle n\pm2|L_k\bar{H}^{n\pm2} = \langle n\pm2|L_k\omega_k$, which, unlike the right-hand eigenvector, is not connected to $\bar{H}^{n\pm2}$. Both are required in the treatment of properties. The treatment of

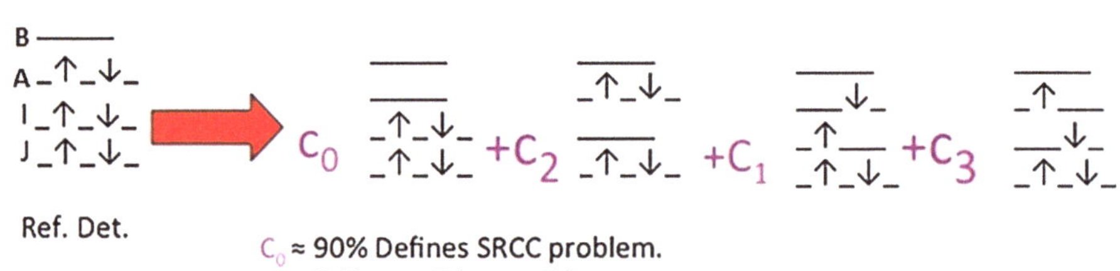

Fig. 1 Reference determinant and achieved static correlation in DIP-CCSD scheme

properties in EOM-CC is well known [37], and the same methods can be used for the DIP/DEA-EOM-CC solutions.

The above procedure is MR for the target states. For a normal n-particle reference double anion state, as shown on the left in Fig. 1, the HOMO is I and the LUMO is A. Starting from the artificial closed shell $|n+2\rangle$ vacuum that doubly occupies A (see Fig. 1), the DIP-EOM operator, $\hat{R}_k^{DIP} = \hat{R}_k^{n-2}$ generates all four of the determinants that a MR space would demand. As each of these determinants will be weighted by the coefficients in the CI-like operator, R_k^{n-2}, their contribution to the final wavefunction is determined by matrix diagonalization allowing any coefficient to be as large as required in the MR description.

Analogously, an $|n-2\rangle$ vacuum state would consist of removing electrons from the HOMO (I) to define a "core" vacuum, and $\hat{R}_k^{DEA} = \hat{R}_k^{n+2}$ produces the same four determinants by adding two electrons in all possible ways into the two quasi-degenerate orbitals. In these methods, the MR space is naturally introduced *after* doing a SR-CC calculation for the $n \pm 2$ problem. The *size-extensive* SR-CC result handles dynamic correlation; the EOM diagonalization properly introduces the *intensive*, or local, part of the correlated wavefunction to account for non-dynamic or static correlation. The former pertains to the left–right correlation for bond breaking subject to an incorrectly separating RHF reference function, while the latter means the kind of correlation required to get the multiplets of complicated open shell systems like transition metal atoms right, of which being a spin-eigenfunction is important.

If we started from a more traditional MR viewpoint, we would add to the four determinants shown in Fig. 1, single and double excitation to build a MR-CI wavefunction. The role of the 3h1p term in Eq. 3 does exactly that for single excitations of the two-hole MR space. So we view the r_{ij} as the coefficients of the four-determinant MR space, c, with the $r_{ijk}^{a}(r)$ corresponding to additional single excitations (s) among these four determinants, such that $R = C{+}S$. In principle, from the CI part of the DIP ansatz, we could add 4h2p, 6h4p…., leading to the full CI limit. The exponential part of the ansatz is SR-CC theory, but for $|n \pm 2\rangle$. Other recent work has numerically considered 4h2p and 4p2 h terms beyond the 3h1p and 3p1h terms considered above subject to an active orbital space [38]. Assuming n_o and n_v are the number of occupied and virtual orbitals, respectively, the 3h1p DIP-EOM and STEOM-CCSD scales as $n_o^4 n_v$ while the 3p1 h DEA-EOM and STEOM-CCSD scales as $n_v^4 n_o$. The TIP, QIP … analogs increase the 3h1p DIP and 3p1h DEA scaling by $(n_o n_v)^p$, where $p = 1$ for TIP, $p = 2$ for QIP… of respectively.

The distinction between the CI part and the CC part is pertinent. A response theory like EOM-CC focuses on the ground state, with all other states being derived from the ground state. These 3h1p terms would arise from the MR-CI viewpoint for the target state. It turns out that these terms are extremely important numerically, much more so than are the reference-state triples contributions [5]. In Ref. [5], we demonstrated that 3h1p terms are sufficient to account for triples in the ground and target state. It also suggests the alternative designation (MR)-EOM-DIP-CC, with the MR space created by the DIP operator, but then single excitations are taken among those functions as they would be in MR-CIS, while the CC part of the problem can be restricted to just CCSD. If 4h2p and 2 h4p terms are important, then this addition would be analogous to MR-CISD.

There is no doubt that the large effects of single excitations are mostly due to the need for orbital relaxation following the introduction of the $|n \pm 2\rangle$ vacuum. Any orbitals can be used to build this vacuum state, so there is no unusual dependence on actually using the "core" orbitals for DEA cases, as other choices can work fine. Similarly, the use of $n+2$ dianion orbitals are possible, though they would be strongly discouraged for DIP calculations on the basis of being unphysical, yet in modest basis sets they are sufficient. The n-particle HF orbitals are generally superior to dianion orbitals since some of the unphysical aspects that can occur in the dianion orbitals are suppressed, but not all. Regardless of orbitals choice, the SR-CC for the $n+2$ vacuum has to offer the full CI description of the dianion in the limit, which would be unphysical for most problems. But for many applications the degree of coupling to the continuum can be controlled in various ways with constraining potentials or with continuum absorbing potentials, e.g., So, the choice of orbitals is the first step. This is our argument in favor of n-particle orbitals. The n-particle reference orbital can be generated by a variety of ways. Our choices have been n-particle system using fractional occupations (to better account for near degeneracy), energetically close triplet state's orbitals, the unrestricted natural orbitals etc. The working assumption for using n-particle orbitals is that if one starts from n-particle orbitals and constructs a $n+2$ state, the subsequent CC step would more easily provide the correct description for DIP calculations. Nevertheless, like other problems where continuum effects would be formally expected (finite field treatment of electric properties [39, 40] and DFT for anions [41]), the fact that the solutions are represented in a L^2 basis also partially ameliorates the problem. Nevertheless, we should be aware of this issue. Prior wok considered several different orbital choices [3], while future work will investigate the orbital dependence of results more thoroughly.

3 Methodology

The geometries are optimized at the RHF-CCSD(T) [42–44] /6-311++G(2d,2p) [45] level of theory for the singlet states and with its UHF counterpart for triplet states. The one exception to this is the singlet state of para-benzyne, whose complex electronic structure warrants special consideration. The harmonic vibrational frequencies are computed to verify that the optimized structures are minima and to obtain the zero-point energies. In the case of para-benzyne, the symmetry is lowered from D_{2h} to the C_{2v} subgroup to obtain a symmetry broken reference state that has partial diradical character (unpaired electrons localized on carbon atoms at para-positions). This will be further discussed in the section on benzynes. The single-point energy calculations employ cc-pVDZ, cc-pVTZ and cc-pVQZ basis sets [46]. Both the dication and neutral reference orbitals are used for DEA calculations, while for DIP either neutral, dianion or Brueckner are used on a case by case basis. The basis set limit results are obtained with the two point extrapolation scheme [47, 48]. The optimized geometries, the vibrational frequencies, CCSD, CCSD(T), ΛCCSD(T) energies are obtained with the ACES II [49] program while the DEA results are obtained with ACES III [50]. Spherical basis functions are used.

4 Results and discussion

4.1 Methylene

The singlet–triplet splitting of the prototype diradical methylene has been the subject of numerous studies [19, 51–61] and spurred a historically significant narrative of the interplay between theory and experiment [51, 53]. Isaiah Shavitt's role in this saga deserves special mention [52]. The methylene radical has a triplet $^3B_1$ ground state, which is single-reference in nature. The electronic configuration of the $M_s = 0$ component of the $^3B_1$ state is $(3a_1^1\alpha b_1^1\beta) - (b_1^1\alpha 3a_1^1\beta)$, but can be well described using the $M_s = 1$ component $(3a_1^1\alpha b_1^1\alpha)$ as a reference for SR-CC. The corresponding singlet configuration $(3a_1^1\alpha b_1^1\beta) + (b_1^1\alpha 3a_1^1\beta)$ gives a $^1B_1$ state that is higher in energy than the $^1A_1$ state (the state labeling corresponds to the C_{2v} irreducible representations (irrep.) with the x direction belonging to the B_1 irrep.). The singlet $^1A_1$ state is partially multi-reference in nature, represented in zeroth order by two reference determinants, $c_1(3a_1^2) + c_2(2a_1^2b_1^2)$. The experimentally measured singlet–triplet (S-T) splitting is between the $^1A_1$ and $^3B_1$ states. All of this spectroscopic information may be found in the aforementioned literature.

The most accurate experiments [62] measure the S-T splitting of methylene with zero-point energy (ZPE) to be 8.998 ± 0.014 kcal/mol (T_0) and the S-T splitting without zero-point energies to be 9.215 kcal/mol (T_e) (also see Refs. [57, 63]). The computed ZPE correction at the CCSD(T)/6-311++G(2d,2p) level is +0.44 kcal/mol (triplet having the larger ZPE correction). The computed Born–Oppenheimer and relativistic corrections are −0.11 [64], −0.04 [56], respectively. Though there is little but historical reason for freezing the core, when doing so there is a core polarization correction of +0.37 kcal/mol (triplet has the larger core polarization effect) which we obtain at the CCSD(T)/cc-pCV5Z level, also confirmed in previous estimates [60]. Subtracting these corrections from the experimental value, we obtain a frozen core value of 8.99 kcal/mol to offer an experimental analog that can be compared with valence-only computed single-point energies. The often-quoted experimental value, 9.37 kcal/mol is the all electron Te value subtracting only the Born–Oppenheimer and relativistic corrections.

Unlike some prior MR-CC calculations on CH_2 singlet–triplet splitting calculations that relied on CISD/DZP [51] geometries, we choose to obtain geometries for each state as above. It has been shown that in general CCSD(T) geometries on single-reference cases are accurate to within 0.01 Å [47]. Obviously this estimated error does not directly translate into the estimated errors for systems that have MR character, but the general insensitivity to small geometry changes should not affect the results much—nonetheless we consider the effect. The 6-311++G(2d,2p) basis is approximately of the same quality as cc-pVTZ in practice, but has some diffuse functions, should they be necessary. There is indeed some variation in the geometries of the singlet; the previously used value of 1.118 Å is slightly longer than our 1.109 Å value; the previous angle of 102.4° is slightly wider than our 101.7°. For the triplet, our bond length is 1.077 Å, compared to the previously reported 1.083 Å; the previous angle of 132.4° is very slightly tighter than our 133.3°.

A collection of previously published S-T splitting using various forms of MR-CC methods are presented in Table 1. Our DEA-CCSD results along with CCSD, CCSD(T), ΛCCSD(T) and CCSDT results obtained at the CCSD(T)/6-311++G(2d,2p) geometry in various basis sets, are shown in Table 2.

The S-T splitting is obtained as a difference of the two separate individual energy calculations for the singlet and triplet. Those individual singlet and triplet calculations do not necessarily have to be by the same method, and in fact, the majority of the results in Table 1 use composite methods in which the singlet and triplet are treated by two different methods (primarily the singlet as a MR state while the triplet is SR). RMR indicates the reduced multi-reference approach of Li and Paldus [58]. The TD-CCSD of Balkova and Bartlett [54], an early SU-CCSD application, provides two roots simultaneously at each geometry. This SU-CCSD calculation introduced GVB CCSD as a two-determinant reference. The MR-BW is a state-specific MR

Table 1 Methylene singlet–triplet splitting obtained with MR methods (in kcal/mol[a])

$^3B_1$	$^1A_1$	Basis			
		cc-pVDZ	cc-pVTZ	cc-pVQZ	cc-pV5Z
CCSD[b]	RMR	12.04	10.42	9.84	9.74
SU-CCSD[b]	RMR	12.20	10.71	10.17	10.08
SU-CCSD[b]	SU-CCSD	12.04	10.52	9.92	9.76
$CCSD(T)_f^2$	RMR(T)	11.89	10.19	9.55	9.40
$SU\text{-}CCSD(T)_f^b$	RMR(T)	12.20	10.76	10.22	10.10
$SU\text{-}CCSD(T)_{se}^b$	RMR(T)	12.05	10.55	10.00	9.88
MR BWCCSD[c]			10.52	9.90	9.72
MR BWCCSDT[c]			10.18	9.51	
TD-CCSD[d]	9.08[d]				
FCI[e]	11.14[e]				
Experiment[f]					8.99

[a] If only one method is listed for the two states, the same method is used twice

[b] Reference [58]; geometries come from Ref. [51], a CISD/DZP estimate. The subscript se (size extensivity) indicate that disconnected off-diagonal Fock matrix elements are ignored, while f means they are included. All our CCSD(T) results for open shells use the procedure of Ref. [65] that is formulated in spin-orbitals and thus does not require any further approximation

[c] Reference [55]

[d] The TD-CCSD calculations are done with a DZP basis set (for comparison with full CI) and ANO basis set of Widmark et al. See Ref. [54] for details (the cited result in the table is for ANO basis)

[e] This FCI result comes from Ref. [61], using the TZ2P basis set, freezing one core orbital and the corresponding virtual

[f] This is based on the experimental result of Ref. [62], with computational estimates of the nonadiabaticity and relativistic effects subtracted based on refs. [56, 64], respectively. The zero-point energy differences are accounted for in the experimental protocol. Our estimate of the core polarization comes from CCSD(T)/cc-pCV5Z calculations

Table 2 Methylene singlet–triplet splitting at the CCSD(T)/6-311++G(2d,2p) geometry (in kcal/mol[a])

	cc-pVDZ	cc-pVTZ	cc-pVQZ	3–4 Extrap.
CCSD	12.8	11.2	10.6	10.2
CCSD(T)	12.1	10.3	9.7	9.19
ΛCCSD(T)	12.1	10.4	9.8	9.28
DIP-EOM-CCSD	10.3	7.4	5.5	4.1
DEA-EOM-CCSD	11.6	10.4	9.6	8.94
DEA-STEOM-CCSD	14.0	12.6	12.2	11.8
CCSDT	11.8	10.1	9.5	9.00
Experiment-core-ZPE-NDB				8.99

[a] Performed by the authors using CCSD(T)/6-311++G(2d,2p) for geometries and frequencies. Core orbitals are kept frozen for energy calculations. The largest T_2 amplitudes of DIP and DEA calculations of the singlet state are −0.0812 and −0.0710, respectively, and both of them are HOMO–LUMO excitations

description, as is MR-Mk-CCSD. The latter is size-extensive, but no invariant to active orbital rotations, while MR-BW requires an a *posterior* extensivity correction.

The predicted energy differences differ from the experimental results by up to 0.2 kcal/mol, despite basis saturation. In the sense of agreement to experiment, the two-determinant CCSD would appear to perform very well. The TD-CCSD (GVB) SU-CCSD calculation done used a large ANO basis. It appears to have the right mix of MR character for the $^1A_1$ and $^3B_{1u}$ states.

In Table 2, we show the new results as a function of basis set. The DIP and DEA-EOM are direct methods in the sense that they provide multiple roots at a given geometry, but the differences shown reflect the two geometries involved. Whereas CCSD is clearly off by >1 kcal/mol, the basis set limit CCSD(T) is quite accurate. Though it is expected that ΛCCSD(T) should be better than CCSD(T) for MR problems [16, 17], we see little difference here. CCSDT is excellent regardless of the MR character (a previously reported aug-cc-pVTZ CCSDT result is 0.52 kcal/mol larger than the current extrapolated value [66]). DEA-EOM-CCSD performs very well. DEA-STEOM seems not to do as well as its EOM counterpart. The most notable result is for DIP-EOM. Methylene is sufficiently small that its dianion state, which at least has to converge as a SR-CC ($n + 2$) to start the procedure, appears to fail as a function of basis, despite using n-particle HF orbitals instead of di-anion orbitals. For larger molecules and smaller basis sets, the continuum effects encountered in the description of the di-anion are ameliorated enough to successfully use the method, but apparently not for methylene near the basis set limit, contrary to some other results in the literature [38]. When we use their TZ2P basis and the di-anion orbitals, the S-T gap is 6.0 eV. Using UHF triplet orbitals (and the TZ2P basis), the DIP result is 10.8 eV consistent with their active space restricted value (we note that the convergence of DIP equations with such choices of orbitals is extremely slow, and in some cases, it is difficult to achieve the desired strict levels of convergence).

4.1.1 Trimethylenemethane (TMM)

The singlet–triplet splitting of TMM has been the subject of numerous studies [19, 55, 58, 59, 67–73]. TMM has five relevant quasi-degenerate spectroscopic states, $^1A_1$, $^1B_2$, $^3A'(^3B_2)$, $^1B_1$ and $^3B_1$. The three lowest energy states are $^3A'(^3B_2)$, $^1A_1$ and $^1B_1$. The experimentally measured singlet–triplet splitting for TMM is between the $^1A_1$ singlet state and the $^3A_2'$ triplet state. The $^1A_1$ state is C_{2v} and planar, and is best described by two-determinant electronic

Fig. 2 **a** $^1A_1$ C_{2v} structure **b** $^3A_2'D_{3h}$ structure. **c** Relationship among the TMM states. The primary axis is always the *z* axis, perpendicular to the plane of the molecule. The *x* axis, in C_{2v}, is the B_1 irreducible representation

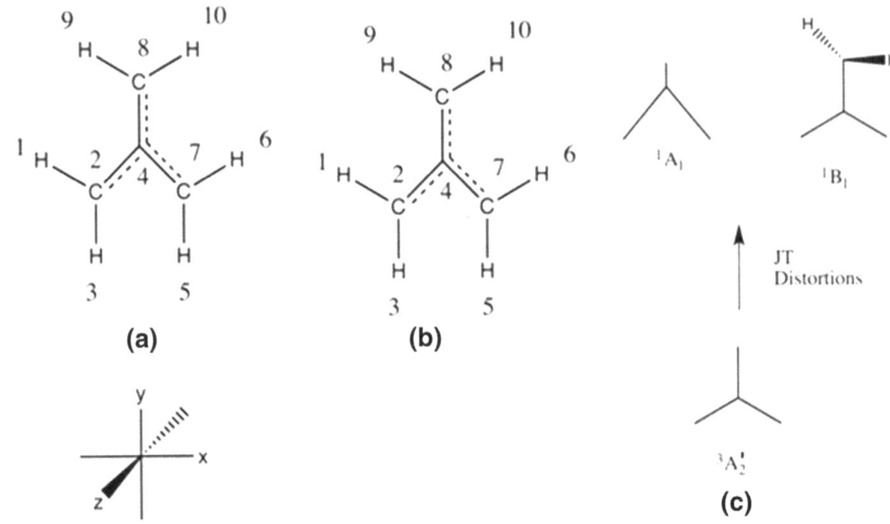

Table 3 The structure of $^1A_1$ C_{2v} state (bond lengths are in Angstroms, and angles are in degrees)

	CCSD(T)/ 6-311++G(2d,2p)	MC-SCF(10,10)/ cc-pVDZ	SF-DFT/ 6-31G(d)
Lengths (Å)			
12	1.081	1.080	1.074
24	1.491	1.496	1.453
48	1.342	1.370	1.338
89	1.082	1.080	1.077
Angles			
248	121.9	121.1	120.9
476	120.3	120.2	120.4
489	121.2	121.2	121.5
475	120.7	120.9	121.2

[a] The CCSD(T)/6-311++G(2d,2p) result is from the current study, while the MC-SCF(10,10)/cc-pVDZ and SF-DFT/6-31G(d) are from Refs. [19, 66], respectively. All lengths and angles are specified in terms of the atom numbering scheme in Fig. 1. All dihedrals are zero by the symmetry of the system in question

Table 4 The structure of $^3B_2$ (C_{2v})/$^3A_2'$(D_{3h}) state (bond lengths are in Angstroms and angles are in degrees)[a]

Lengths (Å)	CCSD(T)/ 6311++G(2d,2p)	MC-SCF(10,10)/ cc-pVDZ	SF-DFT/ 6-31G(d)
C2C4	1.418	1.438	1.402
H8H9	1.081	1.081	1.076

[a] All lengths and angles are specified in terms of the atom numbering scheme in Fig. 3. All dihedrals are 90° by the symmetry of the system in question; all angles are 120° by symmetry (we note that some authors have allowed a small relaxation from the symmetry-governed angle, which never relaxed by more than 1°). References are the same as given in Table 3 for the singlet geometry

configurations, $a_2^2 + 2b_1^2$. The $^3A_2'$ state has D_{3h} symmetry and is commonly referred to as $^3B_2$ in its Abelian subgroup designation. In Fig. 2, we show our axes for the $^1A_1$ structure and $^3A_2'(D_{3h})$/$^3B_2$ (C_{2v}) structures, as these are the relevant states for the experimentally observed singlet–triplet splitting. In Tables 3 and 4, we compare our CCSD(T)/6-311++G(2d,2p) geometries to other work.

It is not clear, a priori, which geometric estimate is best. Clearly, the (10,10) active space of the CAS calculation encapsulates the non-dynamic correlation in question, but with a double-zeta basis set and no dynamic correlation. Our CCSD(T) methodology is clearly superior in dynamic correlation, and our basis set is superior, but it is unclear how well it fares in ultimate accuracy. There have also been prior studies using CCSD/cc-pVDZ geometries [58], but these would be inferior to both current estimates. The SF-DFT/6-31G(d) [19] geometries differ most widely from the other two methodologies. The differences can be quite significant; the lengths differ by 0.04 Å in SF-DFT compared to MC-SCF(10,10) and CCSD(T). We note that there is no consistent agreement between even any pair of methods in describing the carbon–carbon bond lengths. It would be of interest to see a carefully done MR-CC geometry optimization to settle this question. As long as such variation exists, doubt persists as to how much effect the geometry will have on the electronic energies obtained with multi-reference wavefunctions.

The experimental value for the singlet/triplet energy difference is (T_0) 16.1 ± 0.1 kcal/mol [68]. However, in the literature, it is common to use a "derived experimental" value instead. There is some subtlety to "subtracting out" the effect of zero-point energies (ZPE) on the electronic energy of the system. Some authors choose to subtract off a calculated ZPE from the above experimental value. In doing so, one assumes that the calculated geometry and harmonic frequencies are the same as those in the experiment. A commonly used estimate for the harmonic

ZPE difference is Krylov's SF-DFT/6-31G(d) estimate of −2.03 kcal/mol, with the singlet ZPE being lower in energy [19]. The harmonic CCSD(T)/6-311++G(2d,2p) ZPE value is −4.00 kcal/mol (triplet state having larger ZPE correction). The core polarization as calculated by CCSD(T)/cc-pCVQZ is −0.7 kcal/mol (triplet state having larger core polarization effect); we note how very large both this and the methylene polarizations are! Nevertheless, our current best estimate of the "derived experimental" difference of purely electronic energies subject to frozen core (T_e) is 19.4 kcal/mol.

The $^1A_1$ singlet state is qualitatively two-determinantal, $c_1(a_2^{(2)}) + c_2(2b_1^{(2)})$. There are some conflicting reports on the nature of this state. Using CCSD(T), this is a second-order saddle-point, in agreement with DFT and MC-SCF calculations [67]. However, Krylov and coworkers find that this is a minimum using SF-DFT [59]. Experimentally, this state is observed to be a metastable intermediate to the stable cyclic form [68]. We would write the electronic configuration of the D_{3h} $^3A_2$ state in terms of the Abelian C_{2v} subgroup as $(2b_1^1\alpha a_2^1\alpha)$ and assign the $^3B_2$ designation. Also, note that in D_{3h} symmetry, the $^1A_1(c_1(a_2^2) + c_2(2b_1^2))$ and $^1B_2$ $(2b_1^1\alpha a_2^1\beta - 2b_1^1\beta a_2^1\alpha)$ states are degenerate. The $^{1/3}B_1$ pair is a nonplanar C_{2v} molecule of electronic configuration $(5b_2^1\alpha a_2^1\beta \pm 5b_2^1\beta a_2^1\alpha)$.

Table 5 lists various calculations of the electronic energy difference using a plethora of many-body methods.

Given that basis set extrapolations have not been done all reported energies in Table 5, it might vary within ~2 kcal/mol due to valence basis completeness. This says nothing of the error intrinsic to imperfect geometries. The MC-SCF(10,10) and CASPT2(10,10) calculations are impressively accurate, as is SU-CCSD. The Brillouin-Wigner estimates get disconcertingly worse with the addition of triples. The SR-CCSDT result is close in the small basis used. As shown in Table 6, our CCSDT in a cc-pVTZ basis is 23.1 kcal/mol.

Our results are shown in Table 6. Compared to the CCSDT value of 23.1 kcal/mol, the CCSD(T), ΛCCSD(T) and DEA-CCSD would appear to have larger errors. In this example, DEA-STEOM does well as DIP-EOM, although as in methylene, the latter ceases to converge in the QZ basis.

Since SF-DFT/6-31G(d) finds the $^1A_1$ state to be a minimum in conflict with the MCSCF and the current CCSD(T) results, there are some questions about using this geometry as the starting point for the S-T splitting calculations. Here, we have also studied the DEA, DIP and CCSD(T) S-T splitting at the SF-DFT geometries with the cc-pVTZ basis set to assess the effect of geometry on the S-T splitting. The CCSD(T), DIP and DEA S-T splitting at the SF-DFT geometry using the cc-pVTZ basis set are

Table 5 The singlet–triplet splitting of TMM computed with MR methods (in kcal/mol)

$^3A_2$	$^1A_1$	$\Delta E_{S\text{-}T}$
CCSD[a]	RMR	30.6
CISD[a]	2RCISD	38.3
SU-CCSD[a]	SU-CCSD	19.9
CCSD(T)f[a]	RMR(T)	25.8
MCSCF(10,10)[b]		18.9
SF-CIS(D)[b]		20.6
CASPT(2)		19.1
4R BWCCSD it[c]		17.8
4R BWCCSDT-1α[c]		14.9
4R BWCCSDT-α[c]		15.6
SS-EOM-CCSD(MCSCF)[d]		19.8
Experiment-ZPE-core[e]		19.4

[a] From Ref. [56], CC bond lengths are optimized at CCSD(T)/cc-pVTZ and 2R RMR CCSD(T)/cc-pVTZ (or RMR(T)/cc-pVTZ) for singlet and triplet, respectively. The carbon–hydrogen bond lengths are constrained to those given in Ref. [67]

[b] From Ref. [19], all calculations use cc-pVTZ on carbons, cc-pVTZ on hydrogens with a geometry derived from SF-DFT/6-31G(d)

[c] Reference [55]. All calculations use cc-pVTZ with geometries from CAS-SCF(4,4)/cc-pVDZ

[d] Reference [70]. The geometries are derived from CAS(4,4)/cc-pVDZ reported in Ref. [71]

[e] The experimental value is given in reference 69; the CCSD(T)/6-311++G(2d,2p) ZPE and the CCSD(T)/cc-pCVQZ core polarization estimate from this study are used

Table 6 TMM singlet–triplet splittings at CCSD(T)/6-311++G(2d,2p) geometries (in kcal/mol)[a]

	cc-pVDZ	cc-pVTZ	cc-pVQZ	3–4 Extrap.
CCSD	46	48	48	48
CCSD(T)	22.7	25.5	25.6	25.6
ΛCCSD(T)	28.5	30.6	31.0	31.3
DEA-EOM-CCSD	24.1	24.4	24.2	24.1
DEA-STEOM-CCSD	19.4	19.8	20.1	20.3
DIP-EOM-CCSD[b]	21.5	21.1	20.8	20.6
DIP-EOM-CCSD[c]	19.0	18.9	N/C	
CCSDT	21.8	23.1		
Experiment-ZPE-core				19.4

[a] Performed by the authors using CCSD(T)/6-311++G(2d,2p) for geometries and frequencies. The singlet neutral orbital DIP-EOM-CCSD/cc-pVQZ did not converge, and the CCSDT/cc-pVQZ is not performed. The largest T_2 amplitudes of DIP and DEA calculations of the singlet states are −0.0357 and −0.04386, and both of them correspond to HOMO–LUMO excitations. The nature of the reference orbital (in the DIP case, $n + 2$ or neutral orbitals) has no significant effect on the magnitude of T_2 amplitudes

[b] The reference ($n + 2$) state uses doubly negative anion orbitals

[c] The reference ($n + 2$) state uses singlet neutral orbitals

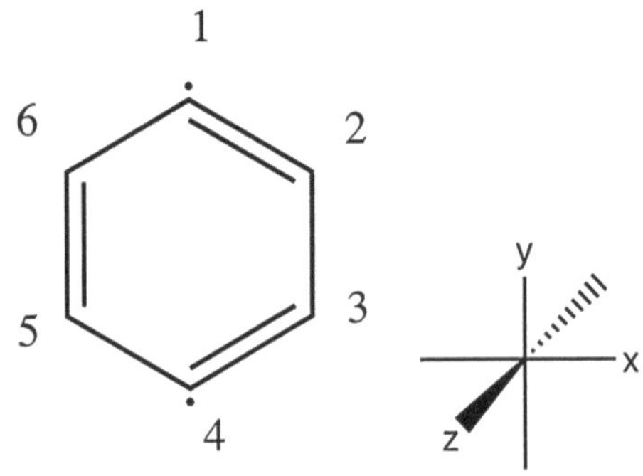

Fig. 3 Para-benzyne in the canonical choice of axes consistent with spectroscopy texts. The primary axis is always the z axis, perpendicular to the plane of the molecule. The x axis, in D_{2h}, is the B_{3u} irreducible representation. We use this same numbering scheme to describe the lengths and angles of *o*-, *m*- and *p*-benzyne

26.3, 18.3, 23.4 kcal/mol, respectively. This shows that the effect of geometry is rather small. However, the DIP results, regardless of the nature of the orbitals that are used to construct the $n+2$ vacuum, or the geometry used, warrant additional comments. As expected, the $^3B_2$ state is well described by the DIP solution (the DIP vector primarily consists of $b_2\alpha a_2\beta + b_2\beta a_2\alpha$ determinants) and well behaved in terms of converging to the desired solution. The DIP solution of the $^1A_1$ state correctly consists of nearly equal weights of $a_2\alpha a_2\beta$ and $b_2\alpha b_2\beta$ determinants, but the near degeneracy of these two orbitals (both occupied in the $n+2$ vaccuum) render poor convergence during the diagonalization step. It is also worth mentioning that these problems are exacerbated as the basis set becomes diffuse.

4.2 Benzynes

First let us consider the more complicated case of para-benzyne. Para-benzyne is indeed infamous for its computational challenges [58, 72–74]. A description of its computational difficulties has been given by Crawford et al. [72]. To make matters worse, experimental research has also come to contradictory conclusions [75, 76]. The molecule is shown in Fig. 3 along with its associated geometric parameters in Table 7. The singlet state is ^1A_g with the dominant configurations $6a_g^2 + 5b_{2u}^2$, while the triplet is $^3B_{2u}$ with configuration $6a_g^{(1)}5b_2u^{(1)}$ in D_{2h} symmetry. This designation follows the axis specification in Fig. 3.

The key features of the electronic structure of *p*-benzyne are well established and only summarized here for completeness. The $6a_g$ orbital and the $5b_{2u}$ orbitals of *p*-benzyne are more or less localized on both carbon radical centers (numbers 1 and 4 in Fig. 3) and must be included in the correct description of its electronic structure. The RHF description, $6a_g^2$ (or $5b_{2u}^2$) leading to a ^1A_g state, does not reflect the true biradical nature since in this description both electrons are localized at either of the two carbon atoms. It is observed that such a reference, even with high levels of correlation, does not yield the correct structure and vibrational spectrum. However, by lowering the symmetry to C_{2v} and switching to UHF, we can allow these two orbitals to mix to obtain a better single-determinant description. The CCSD(T) geometries and the vibrational frequencies reported in this study are obtained in this manner and are consistent with the previous findings.

Table 7 The geometry of p-benzyne, singlet followed by triplet (bond lengths are in Angstroms, and the angles are in degrees)[a]

	TCSCF-Mk-CCSD/cc-pVTZ	CCSD(T)/cc-pVDZ	CCSD(T)/6-311++G(2d,2p)
Lengths Å			
12	1.363	1.390	1.377
23	1.430	1.427	1.418
CH	1.080	1.097	1.082
Angles			
612	124.8	125.1	125.2
123	117.6	117.4	117.4
H23	118.4	120.2	119.7
Lengths Å			
12	1.376	1.395	1.386
23	1.405	1.417	1.412
CH	1.081	1.098	1.083
Angles			
612	127.0	126.3	126.4
123	116.5	116.8	116.8
H23	121.1	121.1	121.0

[a] All lengths and angles are specified in terms of the atom numbering scheme in Fig. 3. The letter H is specified for the hydrogens, given that they are all equivalent in this structure. The CCSD(T)/cc-pVDZ [58] geometry differs significantly from our own. The Mk-CCSD geometry derives from private communication [77]

However, there is a drawback. In addition to the fact that the spatial symmetry is broken, the spin contamination of this spin symmetry broken UHF reference is unusually large, $(\langle 2\bar{S}+1\rangle = 2.83)$. This is because of the presence of two closely lying triplet states (triplets arising from $6b_{2u}$ and $7b_{2u}$ orbitals) that can mix. We note that the $(2\bar{S}+1)$ of CCSD(T) is 2.12 compared to 2.83 of the reference UHF state. This observation is consistent with findings in the literature [72, 78]. However, in this particular case, it is also important to note that the UHF reference (in C_{2v} symmetry) has large T_1 amplitudes (those T_1 amplitudes are strictly zero by symmetry in D_{2h}). The Brueckner reference eliminates large T_1 amplitudes present in the symmetry broken solution. As further discussed below, the

CCSD(T) and ΛCCSD(T) give improved S-T splitting for *p*-benzyne when the Brueckner reference is used instead of the symmetry broken UHF reference.

There are two reported experimental values for the singlet–triplet splitting of para-benzyne. Wenthhold et al. [76] estimate the splitting to be 3.8 ± 0.3 kcal/mol; Leopold et al. [75] estimate the splitting to be 2.1 ± 0.4 kcal/mol. The highest level calculations on the ZPE difference are the current ones. We have performed geometry optimizations and obtained harmonic vibrational frequencies using CCSD(T)/6-311++G(2d,2p), producing a ZPE difference of 0.14 kcal/mol between the two states, with the singlet having larger ZPE. The TC-SCF-Mk-CCSD/cc-pVTZ geometry optimization of Allen et al. [77] is likely to be the most trustworthy geometry yet available on this heavily multi-reference system; yet, the broken-symmetry CCSD(T) geometry produces a geometry in relatively close agreement with it, thus giving confidence to our estimate of the ZPE. Table 5 illustrates this concordance.

Table 8 Geometry of singlet and triplet state of meta-benzyne[a]

	TC-SCF-Mk-CCSD/ cc-pVTZ	CCSD(T)/ 6-311++G(2d,2p)
Singlet		
Lengths Å		
12	1.363	1.378
2H	1.076	1.077
34	1.373	1.381
4H	1.080	1.082
45	1.397	1.406
5H	1.084	1.086
Angles		
123	96.0	99.4
234	138.5	135.7
345	116.6	117.4
456	113.7	114.4
H61	120.8	120.5
Triplet		
Lengths Å		
12	1.379	1.385
2H	1.082	1.083
34	1.377	1.383
4H	1.080	1.081
45	1.397	1.404
5H	1.082	1.084
Angles		
123	115.0	115.2
234	124.9	124.7
345	116.8	116.9
456	121.4	121.5
H61	122.4	122.3

[a] All bond lengths are in Å, and the angles are in degrees. All lengths are described by numbered line segments corresponding to Fig. 2, as are the angles. The letter H indicates the hydrogen atom as a line segment point/angle vertex. All dihedrals are zero by symmetry. Reference [77] gives the Mk-CCSD geometry

We briefly discuss meta-benzyne and ortho-benzyne whose multi-reference quality is not so drastic. Figure 3 shows our numbering system for meta-benzyne, and Table 8 gives the geometries calculated in our work using CCSD(T)/6-311++G(2d,2p) as well as TC-SCF-Mk-CCSD/cc-pVTZ. The agreement between the two methodologies is extremely high, excepting two singlet angles around the diradical centers. This gives considerable confidence in the geometries used. The meta-benzyne singlet is a C_{2v} structure whose wavefunction follows $^1A_1$ symmetry; its triplet is also C_{2v} corresponding to $^3B_1$. The electron configuration of the triplet is $10a_2^1 8b_2^1$ (the ZPE difference corresponding to a CCSD(T)(2)/6-311++G(2d,2p) harmonic vibrational frequency on the CCSD(T)/6-311++G(2d,2p) optimized meta-benzyne structure yields a difference of 0.70 kcal/mol with the triplet having larger ZPE).

Table 9 lists the values for the electronic energies calculated using a variety of methodologies. There are, again, different geometry sources. It is hard to infer whether the errors from the commonly used CCSD(T)/cc-pVDZ geometry influence the quality of these high-level single-point energy calculations. We can see the drastic failures of traditional CCSD in both cases (see the discussion below for further insight into this behavior). RMR acting on the para-benzyne singlet is also clearly insufficient, as it predicts the wrong ordering of states. RMR combined with SU-CCSD underestimates the energy splitting significantly.

Table 9 Singlet–triplet splitting of ortho-, meta- and para-benzyne using MR methods[a]

Singlet	triplet	Ortho	meta	para
RMR[a]	CCSD	33.4	13.8	−2.1
RMR[a]	SU-CCSD	34.5	16.2	0.7
RMR(T)[a]	$CCSD(T)_t$	35.5	17.7	3.2
RMR(T)[a]	$CCSD(T)_{se}$	36.3	18.7	4.1
Expt. 1[c]		37.5 ± 0.3	21.0 ± 0.3	3.8 ± 0.3
Expt. 2[b]		37.7 ± 0.7		2.1 ± 0.4
Expt. 1—ZPE[b]		37.8 ± 0.3	20.4 ± 0.3	3.4 ± 0.3

[a] From Ref. [58], all calculations use a modified cc-pVTZ in which the f functions on carbons and the d functions on hydrogens are omitted. All geometries come from reference 72, which used unrestricted CCSD(T)/cc-pVDZ geometries and harmonic vibrational frequencies for para-benzyne, and restricted CCSD(T)/cc-pVDZ for the meta-benzyne's geometries and harmonic vibrational frequencies. The ZPE differences are −0.3, 0.7 and 0.3 kcal/mol respectively Ref. [79]

[b] From Ref. [76]

[c] From Ref. [75]

All the remaining computations of para-benzyne including CCSD(T) tend to favor the Lineberger experiment, being closer to 3.4 than 1.7 after removing the zero-point value.

The results obtained in the current study are reported in Tables 9 and 10. In agreement with the literature, the CCSD of *p*-benzyne predicts the wrong S-T ordering by nearly 18 kcal/mol. However, the inclusion of perturbative triples vastly improves the results by not only predicting the correct S-T ordering but also predicting a S-T splitting that is in close agreement with experiment. It is important to note that these CCSD and CCSD(T) energy calculations employ the RHF single-determinant reference. We have already discussed the problems of using such a reference for geometry optimization and vibrational frequency calculations. Nevertheless, the RHF reference CC calculations seem to yield correct S-T splitting after inclusion of the perturbative triples.

One could also use the symmetry broken UHF reference. With this partially correct zeroth-order description, the UHF-CCSD seems to give a very small splitting with the correct ordering (0.1 kcal/mol at cc-pVTZ). However, to make matters worse, the UHF-CCSD(T), in contrast to both UHF-CCSD and RHF-CCSD(T), predicts the wrong ordering of energy levels (for example, −0.5 kcal/mol at cc-pVTZ). Another noteworthy anomalous behavior of *p*-benzyne is that both UHF and RHF reference-based ΛCCSD(T) give a very small incorrectly ordered S-T splitting. Two possible reasons for this behavior are unusually large spin contamination (the computed multiplicity $(2\bar{S}+1)$ at the CCSD and CCSD(T) level are 2.21 and 2.12, respectively) and the large T_1 amplitudes. In fact, Brueckner-CCSD(T) with the UHF reference (UB-CCSD(T)) yields the correct S-T ordering, and the UB-CCSD(T) 2.5 kcal/mol splitting at cc-pVTZ is consistent, albeit not as quantitative, with the experimental results and those from MR methods (the average multiplicity of UB-CCSD and UB-CCSD(T) are 2.17 and 2.10, respectively). Since the multiplicity of UB-CCSD and UB-CCSD(T) are nearly the same as the UHF-CCSD, we suspect that large T_1 amplitudes present in the UHF-CCSD are primarily responsible for the anomalous behavior). Furthermore, as shown in Table 3, when Brueckner reference (RB or UB) is used, ΛCCSD(T) follows the observed trend that it is in close agreement with CCSD(T). We also believe that further inclusion of correlation effects by more costly methods such as CCSDT-3 or CCCSDT will also show the correct UHF behavior. The DEA-CCSD of Table 10 below for para-benzyne is conspicuously in agreement with the 3.4 kcal/mol observation.

For meta-benzyne, given that methods in Table 9 have not performed a basis set extrapolation, one may reasonably believe that all of the proposed methods may change by ~ 2 kcal/mol. The DEA-CCSD value is basis set extrapolated in the valence space and is comparable to other CC methods. It would seem that maybe the 2h4p terms are needed to account for the remaining energy difference between DEA-CCSD's 18.5 and experiment, 20.4 kcal/mol. The same logic applies for DIP-CCSD.

Table 10 Ortho-, meta- and para-benzyne singlet–triplet splittings (in kcal/mol)[a]

	cc-pVDZ	cc-pVTZ	cc-pVQZ	3–4 Extrap.
Ortho				
CCSD	28.6	30.6	31.3	31.8
CCSD(T)	33.8	37.0	37.4	38.0
ΛCCSD(T)	33.4	36.6	36.9	37.1
DEA-EOM-CCSD	33.5	36.1	36.9	37.5
DEA-STEOM-CCSD	30.6	34.3	34.2	34.0
DIP-EOM-CCSD	37.5	42.6	46.3	49.0
Experiment-ZPE-core				37.4 ± 0.3 37.5 ± 0.5
Meta				
CCSD	9.9	10.5	10.7	10.8
CCSD(T)	20.4	22.1	22.4	22.7
ΛCCSD(T)	20.9	23.1	20.9	19.3
DEA-EOM-CCSD	17.1	18.3	18.4	18.5
DEA-STEOM-CCSD	14.0	15.3	18.4	20.6
DIP-EOM-CCSD	18.0	19.7	18.3	17.3
Experiment-ZPE-core				20.5 ± 0.3
Para				
CCSD	−17.8	−19.2	−19.5	−19.7
CCSD(T)	4.0	3.6	3.5	3.4
ΛCCSD(T)	1.9	2.3	2.4	2.5
DEA-EOM-CCSD	3.0	3.4	3.4	3.4
DEA-STEOM-CCSD	4.9	4.6	4.3	4.0
DIP-EOM-CCSD	3.9	4.4	N/C	
Experiment-ZPE-core				4.4 ± 0.3 2.7 ± 0.4

[a] Performed by the authors using geometries from CCSD(T)/6-311++G(2d,2p), harmonic frequencies CCSD(T)/6-311++G(2d,2p) for para-benzyne; geometries come from CCSD(T)/6-311++G(2d,2p), harmonic frequencies from MBPT(2)/6-311++G(2d,2p) for meta-benzyne. Core orbitals are dropped for energy calculations. For *o*-, *p*- and *m*-benzynes, the largest T_2 amplitudes of DIP and DEA of the singlet state are −0.0621, −0.0385 and −0.0602, and −0.0833, −0.0766 and −0.0909, respectively. The listed largest excitations are of HOMO–LUMO type

Among the three, *o*-benzyne is the simplest in terms of its electronic structure. Being on adjacent carbon atoms, the orbitals occupied by the lone pair electrons can interact strongly to form well-separated pairs of bonding and antibonding orbitals, a characteristic of strong covalent interactions. As such single-reference methods favor systems that are dominated by dynamic correlation effects, they would be expected to perform well. This is evident from the results in Tables 9 and 10. For example, *o*-benzyne is the only isomer where the CCSD S-T splitting is qualitatively correct. The basis set limit CCSD(T) results are in near-perfect agreement with experiment. A desirable feature of any MR method is that when the SR character dominates, the result must approach the single-reference limit.

4.2.1 Cyclobutane-1,2,3,4-tetrone $(CO)_4$

The $(CO)_n$ ($n = 1, 2, \ldots$) molecules are oligomers of carbon monoxide (CO). Simple molecular orbital theory concepts yield the following qualitative picture of the bonding of $(CO)_4$. The LUMO of CO is a degenerate set of π orbitals (designated as π^*). The interaction of the 4 in plane π^* orbitals leads to a b_{2g} orbital (in D_{2h} symmetry labeling of $(CO)_4$), and the interaction of 4 out-of-plane π^* orbitals leads to an a_{2u} orbital (only the two lowest energy orbitals are of interest). The four HOMO σ orbitals interact to form four orbitals of form a_{2g}, e_{1u}, b_{1g} orbitals. There is no ambiguity about the fact that out of 8 electrons, 6 of them occupy the a_{2g} and e_{1u} orbitals. The remaining 2 electrons have three possible orbital locations: b_{1g}, b_{2g} and a_{2u}. Among these three orbitals, b_{1g} is the most antibonding, and as a result, the 2 electrons prefer the b_{2g} and a_{2u}. However, since both b_{2g} and a_{2u} originate from the degenerate π^* orbitals of CO, they nearly degenerate in $(CO)_4$ and orthogonal. So, Hund's rule applies, and the ground state is a triplet $^3B_{1u}$. These conclusions based on the qualitative MO arguments have been confirmed by recent experiment by Wang et al. [80], who conclude that the triplet is (T) 1.5 kcal/mol lower in energy than the singlet. A comprehensive account of MO theory arguments and theoretical calculations at the DFT and CCSD(T) levels of theory S-T splitting of $(CO)_n$ series has been recently presented by Borden et al. [81, 82].

In this study, we have optimized the geometry of both $^1A_1$ and $^3B_{1u}$ state using the CCSD(T)/6-311G++(2d,2p) level of theory while keeping the core electrons frozen for the post-HF stage. The CC and CO bond lengths of the singlet state are 1.559 and 1.187 Å, and the corresponding CC and CO bond lengths of the triplet $^3B_{1u}$ state are 1.569 and 1.190 Å. The frequency calculation shows that at this level of theory, the singlet state is a transition state and the triplet state is a minimum. This is in contrast to the previous and our own DFT results which find both the singlet and triplet geometries to be minima. Except in this feature, it is interesting to note that the DFT geometries do not significantly differ from the CCSD(T) geometries. For example, MO6/cc-pVTZ CC and CO bond lengths for singlet and triplet states are 1.547 and 1.172 and 1.550 and 1.180 Å, respectively. The CCSD(T)/6-311G++(2d,2p) ZPE difference (triplet having a smaller ZPE) is 0.48 kcal/mol.

The S-T splitting computed with a selected set of methods is shown in Table 5. In this case, the "derived" experimental S-T splitting that can be directly compared with the valence-only computed result is 1.19 kcal/mol. The DIP calculations are repeated: one using the less physically meaningful double anion reference orbitals and the second the recommended n-particle HF orbitals. In an exception to the generally observed trend that all the

Table 11 The S-T splitting of $(CO)_4$ $^1A_1$ and $^3B_{1u}$ states (in kcal/mol)[a]

	cc-pVDZ	cc-pVTZ	cc-pVQZ	3–4 Extrap.
CCSD	19.2	20.1	20.2	20.5
CCSD(T)	5.5	5.9	5.9	5.9
ΛCCSD(T)	6.4	7.0	7.0	7.0
CCSDT-3	4.5	4.9	4.8	4.8
DIP-EOM-CCSD[b]	1.0	−1.6	−2.3	−2.8
DIP-EOM-CCSD[c]	−2.9	0.9	0.3	−0.1
DIP-EOM-CCSD[d]	3.0	1.3	0.7	0.3
DEA-STEOM-CCSD[e]	3.9	2.4	1.6	1.1
DEA-EOM-CCSD[f]	1.9	−0.4	−0.6	−0.8
DEA-EOM-CCSD[g]	3.8	1.9	1.3	0.9
Experiment-ZPE core				1.5–0.48 + 0.17 = 1.19
Experiment-ZPE (for all electron calcs.)				1.5–0.48 = 1.02

[a] The largest T2 amplitudes of DIP and DEA calculations of the singlet state are −0.0630 and −0.0813, respectively, and both of them correspond to HOMO–LUMO excitations

[b] The reference ($n + 2$) state uses doubly negative anion orbitals (core orbitals are frozen)

[c] The reference ($n + 2$) state uses doubly negative anion orbitals (all electrons are correlated)

[d] The reference ($n + 2$) state uses singlet neutral orbitals (core orbitals are frozen)

[e] The reference ($n - 2$) state uses doubly positive cation orbitals (core orbitals are frozen)

[f] The reference ($n - 2$) state uses doubly positive cation orbitals (all electrons are correlated)

[g] The reference ($n - 2$) state uses singlet neutral orbitals (all electrons are correlated)

methods predict the correct S-T ordering, the former shows the incorrect sign. As far as the S-T splitting is concerned, we observe a wide variation among various methods. The CCSD, CCSD(T) and ΛCCSD(T) behave uniformly as the basis set quality is improved. However, both CCSD(T) and ΛCCSD(T) overestimate the S-T splitting by nearly 4–5 kcal/mol (not surprisingly CCSD grossly overestimates the splitting). The DIP with neutral orbitals is uniform as the basis set quality is improved, though they underestimate the splitting by 0.7 kcal/mol. The DEA-STEOM shows uniform convergence with basis set, and its basis set extrapolated value along with that for the DEA-EOM-CCSD using the recommended n-particle orbital is perhaps best when compared to experiment. The DEA results have the wrong sign for the first time in our experience (Table 11).

5 Conclusion

The DIP/DEA framework is proven to be able to correctly treat difficult two-orbital two-electron, multi-reference problems within a computationally effective single-reference framework. In choosing a single-reference charged SCF, and adding/subtracting the appropriate particles to return to the multi-reference system, we circumvent some of the challenges of static or non-dynamic correlation, though there are others as in the residual continuum effects encountered despite the use of L^2 basis in the choice of dianion reference cases. Choosing such pathological molecules such as methylene, TMM and benzynes, and still seeing substantial success, illustrates some of the suitability of the methodology.

Acknowledgments This work was supported by the United States Army Research Office, under Grant No. 54344CH. We are pleased to contribute this paper to this Special Issue in honor of Isaiah Shavitt, who one of us (RJB) had the pleasure of knowing and collaborating with from 1976. Shavitt was a gentleman and a scholar. Our book on Many-Body Methods in Chemistry and Physics: MBPT and Coupled-Cluster Theory, was realized though Shi Shavitt's attention to detail, clarity and persistence. "To the greater craftsman."

References

1. Nooijen M, Bartlett RJ (1997) J Chem Phys 107:6812
2. Tobita M, Perera SA, Musial M, Bartlett RJ, Nooijen M, Lee JS (2013) J Chem Phys 119:10713
3. Bartlett RJ (2010) Mol Phys 108:2905
4. Musial M, Bartlett RJ (2007) Rev Mod Phys 79:291
5. Musial M, Perera A, Bartlett RJ (2011) J Chem Phys 134:114108
6. Musial M, Katarzyna K, Lyakh DL, Bartlett RJ (2013) J Chem Phys 138:194103
7. Musial M, Olszowka M, Lyakh DL, Bartlett RJ (2012) J Chem Phys 137:174102
8. Jeziorski B, Monkhorst HJ (1981) Phys Rev A 24:1668
9. Lindgren I, Mukherjee D (1987) Phys Rep 51:93
10. Lindgren I (1987) Int J Quantum Chem Symp 12:33
11. Kowalski K, Piecuch PJ (2000) Chem Phys 113:8490
12. Piecuch P, Kowalski K, Pimienta ISO, McGuire MJ (2002) Int Rev Phys Chem 21:527
13. Kowalski K, Piecuch P (2002) J Chem Phys 116:7411
14. Stanton JF (1997) Chem Phys Lett 281:130
15. Kucharski SA, Bartlett RJ (1998) J Chem Phys 108:5243
16. Taube AG, Bartlett RJ (2008) J Chem Phys 128:044110
17. Taube AG, Bartlett RJ (2008) J Chem Phys 128:044111
18. Krylov AI (2001) Chem Phys Lett 350:522
19. Slipchenko LV, Krylov AI (2002) J Chem Phys 117:4694
20. Krylov I (2005) J Chem Phys A 109:10638
21. Li X, Paldus JJ (1997) Chem Phys 107:6257
22. Li X, Paldus J (2000) J Mol Phys 98:1185
23. Li X, Paldus JJ (2000) Chem Phys 113:9966
24. Li X, Paldus JJ (2006) Chem Phys 124:174101
25. Hubač I, Neogrády P (1994) Phys Rev A 50:4558
26. Mášik J, Hubač I (1997) Collect Czech Chem Commun 62:829
27. Mášik J, Hubač I, Mach P (1998) J Chem Phys 108:6571
28. Pittner J (2003) J Chem Phys 118:10876
29. Mukhopadhyay D Jr, Mukherjee D (1989) Chem Phys Lett 163:171
30. Mukhopadhyay D Jr, Mukherjee D (1991) Chem Phys Lett 177:441
31. Meissner L, Kucharski SA, Bartlett RJ (1989) J Chem Phys 91:6187
32. Meissner L, Bartlett RJ (1990) J Chem Phys 92:561
33. Evangelista FA, Gauss J (2011) J Chem Phys 134:114102
34. Mahapatra US, Datta B, Bandyopadhyay D, Mukherjee D (1998) Adv Quantum Chem 30:163
35. Balková A, Bartlett RJ (1994) J Chem Phys 101:8972
36. Balková A, Bartlett RJ (1992) Chem Phys Lett 193:364
37. Stanton JF, Bartlett RJ (1993) J Chem Phys 99:5178
38. Shen J, Piecuch P (2013) J Chem Phys 138:194102
39. Sekino H, Bartlett RJ (1985) J Chem Phys 82:4225
40. Pople JA, McIver JW Jr, Ostlund NS (1967) Chem Phys Lett 1:465
41. Rienstra-Kiracofe JC, Tschumper GS, Schaefer HF, Nandi S, Ellison GB (2002) Chem Rev 102:231
42. Noga J, Bartlett RJ (1982) J Chem Phys 76:1910
43. Raghavachari K, Trucks GW, Head-Gordon M, Pople JA (1989) Chem Phys Lett 157:479
44. Urban M, Noga J, Cole S, Bartlett RJ (1985) J Chem Phys 83:4041
45. Krishnan R, Binkley JS, Seeger R, Pople JA (1980) J Chem Phys 72:650
46. Dunning TH (1989) J Chem Phys 90:1007
47. Helgaker T, Gauss J, Jørgensen P, Olsen J (1997) J Chem Phys 106:6430
48. Bak KL, Jørgensen P, Olsen J, Helgaker T, Klooper W (2000) J Chem Phys 112:9229
49. ACES II is a product of the Quantum Theory Project, University of Florida, Authors: Stanton JF, Gauss J, Perera A, Yau A, Watts JD, Nooijen M, Oliphant N, Szalay PG, Lauderdale WJ, Gwaltney SR, Beck S, Balková A, Bernholdt DE, Baeck K-K, Rozyczko P, Sekino H, Huber C, Pittner J, Bartlett RJ, Integral packages included are VMOL (Almölf J, Taylor PR), VPROPS (Taylor PR) and ABACUS (Helgaker T, Jensen HJAa, Jørgensen P, Olsen J, Taylor PR)
50. Lotrich V, Flocke N, Ponton M, Yau A, Perera A, Deumens D, Bartlett RJ (2008) J Chem Phys 128:194104
51. Bauschlicher CW, Shavitt I (1978) J Am Chem Soc 100:739
52. Shavitt I (1985) Tetrahedron 41:1531

53. Schaefer HF (1986) Science 231:1100
54. Balkova A, Bartlett RJ (1995) J Chem Phys 102:7116
55. Demel O, Pittner J (2008) J Chem Phys 128:104108
56. Davidson ER, Feller D, Phillips P (1980) Chem Phys Lett 76:416
57. McKellar ARW, Bunker PR, Sears TJ, Evenson KM, Saykally RJ, Langhoff SR (1983) J Chem Phys 79:5251
58. Li X, Paldus J (2008) J Chem Phys 129:174101
59. Slipchenko LV, Krylov AI (2005) J Chem Phys 123:084107
60. Hirata S, Yanai T, De Jong WA, Nakajima T, Hirao K (2004) J Chem Phys 120:3297
61. Sherill CD, Leininger ML, Van Huis TJ, Schaefer HF (1998) J Chem Phys 108:1040
62. Jensen P, Bunker PR (1988) J Chem Phys 89:1327
63. Bunker PR, Jensen P, Kraemer WP, Beardsworth R (1986) J Chem Phys 85:3724
64. Handy NC, Yamaguchi Y, Schaefer HF (1986) J Chem Phys 84:4481
65. Watts JD, Gauss J, Bartlett RJ (1993) J Chem Phys 98:8718
66. Shen J, Piecuch P (2012) J Chem Theory Comp 8:4968
67. Cramer CJ, Smith BA (1996) J Phys Chem 100:9664
68. Wenthold PG, Hu J, Squires RR, Lineberger WC (1996) J Am Chem Soc 118:475
69. Slipchenko LV, Krylov AI (2003) J Chem Phys 118:6874
70. Demel O, Shamasundar KR, Kong L, Nooijen M (2008) J Phys Chem A 112:11895
71. Brabec J, Pittner J (2006) J Phys Chem A 110:11765
72. Crawford TD, Kraka E, Stanton JF, Cremer D (2011) J Chem Phys 114:10638
73. Wang EB, Parish CA, Lischka H (2008) J Chem Phys 129:044306
74. Marquardt R, Balster A, Sander W, Kraka E, Cremer D, Radziszewski J (1998) Angew Chem Int Ed 37:955
75. Leopold DG, Miller A, Lineberger WC (1986) J Am Chem Soc 86:1379
76. Wenthold PG, Squires RR, Lineberger WC (1998) J Am Chem Soc 120:5279
77. Private communication with Prof. Westly Allen
78. Stanton JF (1994) J Chem Phys 101:371
79. Evangelista FA, Allen WD, Schaefer HF III (2007) J Chem Phys 127:024102
80. Guo J-C, Hou G-L, Li SD, Wang X-B (2012) J Phys Chem Lett 3:304
81. Bao X, Zhou X, Lovitt CF, Venkatraman A, Horavt DA, Gleiter R, Hofmann R, Borden WT (2012) J Am Chem Soc 134:10259
82. Zhou X, Hrovat DA, Bordon WT (2010) J Phys Chem A 114:1304

Theor Chem Acc (2014) 133:1492
DOI 10.1007/s00214-014-1492-7

REGULAR ARTICLE

Comparison of one-dimensional and quasi-one-dimensional Hubbard models from the variational two-electron reduced-density-matrix method

Nicholas C. Rubin · David A. Mazziotti

Received: 20 March 2014 / Accepted: 19 April 2014 / Published online: 21 May 2014

Abstract Minimizing the energy of an N-electron system as a functional of a two-electron reduced density matrix (2-RDM), constrained by necessary N-representability conditions (conditions for the 2-RDM to represent an ensemble N-electron quantum system), yields a rigorous lower bound to the ground-state energy in contrast to variational wave function methods. We characterize the performance of two sets of approximate constraints, (2,2)-positivity (DQG) and approximate (2,3)-positivity (DQGT) conditions, at capturing correlation in one-dimensional and quasi-one-dimensional (ladder) Hubbard models. We find that, while both the DQG and DQGT conditions capture both the weak and strong correlation limits, the more stringent DQGT conditions improve the ground-state energies, the natural occupation numbers, the pair correlation function, the effective hopping, and the connected (cumulant) part of the 2-RDM. We observe that the DQGT conditions are effective at capturing strong electron correlation effects in both one- and quasi-one-dimensional lattices for both half filling and less-than-half filling.

Keywords Two-electron reduced density matrix · N-representability conditions · Strong electron correlation · Hubbard models

Dedicated to the memory of Professor Isaiah Shavitt and published as part of the special collection of articles celebrating his many contributions.

N. C. Rubin · D. A. Mazziotti (✉)
Department of Chemistry and The James Franck Institute, The University of Chicago, Chicago, IL 60637, USA
e-mail: damazz@uchicago.edu

1 Introduction

The authors are honored to dedicate this article in memory of Isaiah Shavitt whose remarkable contributions transformed electronic structure theory. Observables that depend on pairwise interactions can be directly computed with the two-electron reduced density matrix (2-RDM) without the N-electron wave function [10, 36]. Integration of the N-electron density matrix over all electrons save two yields the 2-RDM

$$^2D(12;\bar{1}\bar{2}) = \binom{N}{2} \int {}^N D(12..N;\bar{1}\bar{2}..N)d3..dN. \quad (1)$$

Minimization of the energy with respect to the 2-RDM results in unphysical ground states because the variational space of two-electron density matrices is larger than the set of 2-RDMs that can be contracted from an ensemble N-electron density matrix [52]. The constraints on the 2-RDM to ensure a valid N-electron density matrix preimage are known as *N-representability conditions* [8, 10, 12, 24]. While the set of necessary and sufficient constraints needed to satisfy Eq. (1) was unknown until recently [39], an approximate class of N-representability constraints has been demonstrated to be sufficient for calculating ground-state properties of the metal-to-insulator transition of hydrogen chains [50], ground states and charge distributions of quantum dots [48], quantum phase transition [17, 49] dissociation channels, [33] and quantum lattice systems [2–4, 20, 55–57].

Variational minimization of the energy as a functional of the 2-RDM is expressible as a special convex optimization problem known as a *semidefinite program* [33, 37, 41, 42, 53, 58]. The convexity of the N-representable set of 2-RDMs ensures a rigorous lower bound to the ground-state energy. Because the variational 2-RDM method generates large semidefinite programs, the second-order interior-point methods must be exchanged for first-order methods. We solve the semidefinite program by a boundary-point method developed for RDMs [37]. The boundary-point method is one to two orders of magnitude faster than the first-order algorithm, previously applied to the one-dimensional Hubbard model [20].

In this paper, we examine the electron correlation of one-dimensional and quasi-one-dimensional Hubbard models with two sets of approximate N-representability conditions. While recent RDM calculations have examined linear [20] as well as 4×4 and 6×6 Hubbard lattices [2, 57], there has not been an exploration of RDMs on quasi-one-dimensional Hubbard lattices with a comparison to the one-dimensional Hubbard lattices. How does the electron correlation change as we move from a one-dimensional to a quasi-one-dimensional Hubbard model? How are these changes in correlation reflected in the required N-representability conditions on the 2-RDM? One- and two-particle correlation functions are used to compare the electronic structure of the half-filled states of the 1×10 and 2×10 lattices with periodic boundary conditions. The degree of correlation captured by approximate N-representability conditions is probed by examining the one-particle occupations around the Fermi surfaces of both lattices and measuring the entanglement with a size-extensive correlation metric, the Frobenius norm squared of the cumulant part of the 2-RDM [23].

2 Theory

In this section, we review the salient features of variational reduced density methodology and discuss the approximate N-representability conditions used in this work.

2.1 2-RDM method

In variational 2-RDM theory, the energy functional is minimized with respect to the 2-RDM

$$E = \mathrm{Tr}({}^2\mathrm{K}\,{}^2\mathrm{D}), \tag{2}$$

where

$${}^2K_{kl}^{ij} = \frac{1}{N-1}\left(h_k^i \delta_l^j + h_l^j \delta_k^i\right) + u_{ij}^{kl} \tag{3}$$

and

$${}^2D_{kl}^{ij} = \mathrm{Tr}(\hat{a}_i^\dagger \hat{a}_j^\dagger \hat{a}_l \hat{a}_k, {}^2D). \tag{4}$$

In Eq. (3), ${}^2K_{kl}^{ij}$ are the elements of the reduced Hamiltonian matrix, in Eq. (4), ${}^2D_{kl}^{ij}$ are the elements of the 2-RDM in a spin-orbital basis set, h and u are tensors containing the one- and two-electron integrals, and the $\hat{a}$, $(\hat{a}^\dagger)$ are the fermionic annihilation (creation) operators. The variational space in which the energy is minimized can be constrained by a hierarchical set of N-representability constraints on the 2-RDM called $(2,p)$-positivity conditions. These constraints have recently been shown when $p = r$ to form a complete set of N-representability conditions [39] where r is the rank of the one-electron spin-orbital basis set.

The (p,p)-positivity conditions [31, 35] on the p-RDM restrict the $(p+1)$ metric (or overlap) matrices of the form

$$M = \langle\psi|\hat{C}\hat{C}^\dagger|\psi\rangle \tag{5}$$

to be positive semidefinite where the operators $\hat{C}$ are linear combinations of all possible products of p creation and/or annihilation operators. A matrix M is *positive semidefinite*, denoted by $M \succeq 0$, if and only if all of its eigenvalues are nonnegative. The three distinct (2,2)-positivity conditions [8, 15, 31, 32, 35, 40, 60] are given by

$${}^2D \succeq 0 \tag{6}$$

$${}^2Q \succeq 0 \tag{7}$$

$${}^2G \succeq 0, \tag{8}$$

where

$${}^2D_{ij}^{pq} = \langle\psi|a_p^\dagger a_q^\dagger a_j a_i|\psi\rangle \tag{9}$$

$${}^2Q_{ij}^{pq} = \langle\psi|a_p a_q a_j^\dagger a_i^\dagger|\psi\rangle \tag{10}$$

$${}^2G_{ij}^{pq} = \langle\psi|a_p^\dagger a_q a_j^\dagger a_i|\psi\rangle. \tag{11}$$

Physically, the semidefinite conditions on the ${}^2D, {}^2Q$, and ${}^2G$ matrices restrict the probabilities of finding particle-particle, hole-hole, and particle-hole pairs to be nonnegative, respectively. Even though the nonnegativity constraints in Eqs. (6–8) are non-redundant, these matrices contain equivalent information as each matrix can be expressed in a one-to-one mapping of another by the fermionic anticommutation relations. The (2,2)-positivity conditions are often denoted as DQG. Contraction of the positive semidefinite ${}^2D, {}^2Q$, and ${}^2G$ matrices generates one-particle ${}^1D$ and one-hole ${}^1Q$ matrices that are also positive semidefinite.

The (2,3)-positivity conditions [39] on the 2-RDM may be formed by taking all convex combinations of the (3,3)-positivity conditions that depend only on the 2-RDM. Here, we consider two important (2,3)-positivity conditions, proposed by Erdahl [12, 60]

$$T_1 = {}^3D + {}^3Q \succeq 0 \tag{12}$$

$$T_2 = {}^3E + {}^3F \succeq 0, \tag{13}$$

where

$$^3D^{qrs}_{ijk} = \langle\psi|\hat{a}^\dagger_q\hat{a}^\dagger_r\hat{a}^\dagger_s\hat{a}_k\hat{a}_j\hat{a}_i|\psi\rangle \tag{14}$$

$$^3E^{qrs}_{ijk} = \langle\psi|\hat{a}^\dagger_q\hat{a}^\dagger_r\hat{a}_s\hat{a}^\dagger_k\hat{a}_j\hat{a}_i|\psi\rangle \tag{15}$$

$$^3F^{qrs}_{ijk} = \langle\psi|\hat{a}_q\hat{a}_r\hat{a}^\dagger_s\hat{a}_k\hat{a}^\dagger_j\hat{a}^\dagger_i|\psi\rangle \tag{16}$$

$$^3Q^{qrs}_{ijk} = \langle\psi|\hat{a}_q\hat{a}_r\hat{a}_s\hat{a}^\dagger_k\hat{a}^\dagger_j\hat{a}^\dagger_i|\psi\rangle. \tag{17}$$

Previous investigations indicated that the T_1 condition is less important that T_2 [16, 60], and therefore, it is excluded from the approximate (2,3)-positivity conditions used in this work. The (2,2)-positivity plus the T_2 condition is denoted in this work as DQGT.

2.2 Semidefinite programming

The computational implementation of the energy minimization with respect to the 2-RDM is formulated as a *semidefinite program* (SDP). A SDP is a generalization of a linear program where the objective variable is kept positive semidefinite. The program is constructed by considering the minimization of the linear energy functional in Eq. (2) subject to constraints

$$\min \mathrm{Tr}(\mathrm{K\ X}) \tag{18}$$

$$\text{such that } \mathrm{Tr}(A_i X) = b_i \tag{19}$$

$$X \succeq 0, \tag{20}$$

where K and X are block matrix representations of the reduced Hamiltonian and the reduced density matrices

$$K = \begin{pmatrix} 0 & 0 & 0 & 0 & 0 \\ 0 & 0 & 0 & 0 & 0 \\ 0 & 0 & {}^2K & 0 & 0 \\ 0 & 0 & 0 & 0 & 0 \\ 0 & 0 & 0 & 0 & 0 \end{pmatrix} \tag{21}$$

and

$$X = \begin{pmatrix} {}^1D & 0 & 0 & 0 & 0 \\ 0 & {}^1Q & 0 & 0 & 0 \\ 0 & 0 & {}^2D & 0 & 0 \\ 0 & 0 & 0 & {}^2Q & 0 \\ 0 & 0 & 0 & 0 & {}^2G \end{pmatrix}. \tag{22}$$

The constraint matrices A_i in Eq. (19) contain the mappings among ${}^2D, {}^2Q$, and ${}^2G$, the contractions to ${}^1D$ and ${}^1Q$, and the fixed-trace condition. Semidefinite programs for quantum chemical Hamiltonians have been solved with a variety of algorithms [27, 33, 37, 54, 60]. In this work, we utilize the boundary-point method [26, 37], a type of quadratic regularization method. The floating-point operations and memory scaling for the boundary-point method are r^6 for DQG (DQGT r^9) and r^4 (DQGT r^6), where r is the rank of the one-electron basis set.

3 Model

The single-band ladder extension of the one-dimensional Hubbard model [18, 19, 22] has been utilized as a minimalist model to study spin-liquid behavior [11, 14, 43] and high-temperature superconductors [1, 5, 21, 25, 44]. The ladder model is a quasi-one-dimensional system with a fourfold degenerate Fermi surface and correlations in two-dimensions.

3.1 Hamiltonian

The Hamiltonian of the ladder single-band model in position space is defined as follows:

$$\hat{H} = t\sum_{n,\lambda,\sigma}(\hat{a}^\dagger_{n,\lambda,\sigma}\hat{a}_{n+1,\lambda,\sigma} + \hat{a}^\dagger_{n+1,\lambda,\sigma}\hat{a}_{n,\lambda,\sigma}) \tag{23}$$

$$+ t_\perp\sum_{n,\sigma}(\hat{a}^\dagger_{n,a,\sigma}\hat{a}_{n,b,\sigma} + \hat{a}^\dagger_{n,b,\sigma}\hat{a}_{n,a,\sigma}) \tag{24}$$

$$+ U\sum_{n}\hat{a}^\dagger_{n,\lambda,\sigma}\hat{a}_{n,\lambda,\sigma}\hat{a}^\dagger_{n,\lambda,-\sigma}\hat{a}_{n,\lambda,-\sigma} \tag{25}$$

where t is a parameter controlling transport between rungs of the ladder, $t_\perp$ is a parameter controlling transport between the two sides of a ladder's rung, U is a parameter controlling the one-site repulsion between electrons, the index n is the rung number, the index $\lambda = a(b)$ corresponds to a ladder leg, and the index σ indicates the spin of the electron created at rung n on leg $a(b)$. We impose periodic boundary conditions along the legs of the ladder forming a Hubbard ribbon.

3.2 Spin and spatial symmetry adaptation

One can take advantage of any spin or spatial symmetry in the Hamiltonian by symmetry adapting the metric matrices and thereby reducing the size of the 2-RDM to be optimized [16]. For the ladder model, we transform the RDMs to bonding and antibonding spaces and then Fourier transform to take advantage of the translational symmetry. We consider linear combination of creation and annihilation operators to form two disjoint one-electron subspaces

$$\hat{a}^B_{n,\sigma} = \frac{1}{\sqrt{2}}(\hat{a}_{n,a,\sigma} + \hat{a}_{n,b,\sigma}) \tag{26}$$

$$\hat{a}^A_{n,\sigma} = \frac{1}{\sqrt{2}}(\hat{a}_{n,a,\sigma} - \hat{a}_{n,b,\sigma}), \tag{27}$$

where $\hat{a}^B_{n,\sigma}$ and $\hat{a}^A_{n,\sigma}$ are annihilation operators for the bonding and antibonding orbitals with spin $\sigma \in \{\alpha, \beta\}$. The

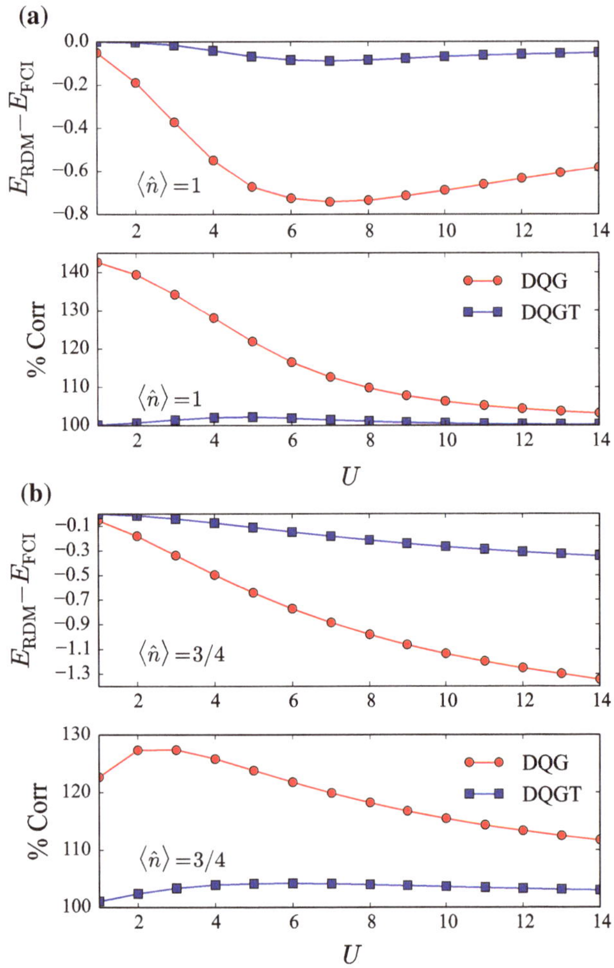

Fig. 1 Absolute deviation of ground-state energy of the variational 2-RDM method with DQG and DQGT constraints and the percent correlation defined as $(E_{RDM} - E_{HF})/(E_{FCI} - E_{HF}) \times 100$ are shown for 2×4 lattices at **(a)** $\langle \hat{n} \rangle = 1$ and **(b)** $\langle \hat{n} \rangle = 3/4$

one-body part of the Hamiltonian divides into antibonding $\hat{H}_A$ and bonding $\hat{H}_B$ parts:

$$\hat{H} = \hat{H}_A + \hat{H}_B + \hat{H}int^{AB}. \tag{28}$$

When expressed in the bonding and antibonding basis, the interaction term decomposes into four two-body operators representing inter- and intra-subspace pair scattering and inter- and intra-subspace pair exchange.

Spatial symmetry is imposed in each one-electron space by the Bloch transformation

$${}^B\hat{a}_{n,\sigma} = \frac{1}{\sqrt{N_x}} \sum_{k_b} e^{-ik_b n} \hat{a}_{k_b,\sigma} \tag{29}$$

$${}^A\hat{a}_{n,\sigma} = \frac{1}{\sqrt{N_x}} \sum_{k_a} e^{-ik_a n} \hat{a}_{k_a,\sigma}, \tag{30}$$

where $\hat{a}_{k_b,\sigma}$ annihilates an electron with momentum k_b in the bonding band and $\hat{a}_{k_a,\sigma}$ annihilates an electron with momentum k_a in the antibonding band. The Hamiltonian and 2-RDM can also be spin adapted. As discussed in Ref. [16], because the three triplet blocks are equivalent in the singlet case, each metric matrix has only two distinct spin blocks defined by the folded operators

$$\hat{C}^{0,0}_{i,j;i\leq j} = \frac{1}{\sqrt{2}} (\hat{a}^\dagger_{i,\alpha} \hat{a}^\dagger_{j,\beta} + \hat{a}^\dagger_{j,\alpha} \hat{a}^\dagger_{i,\beta}) \tag{31}$$

$$\hat{C}^{1,0}_{i,j;i<j} = \frac{1}{\sqrt{2}} (\hat{a}^\dagger_{i,\alpha} \hat{a}^\dagger_{j,\beta} + \hat{a}^\dagger_{j,\alpha} \hat{a}^\dagger_{i,\beta}). \tag{32}$$

These new $\hat{C}$ operators are substituted into Eq. (5) generating symmetric and antisymmetric parts of ${}^2D$ and ${}^2Q$. Spin symmetry adaptation of ${}^2G$ and T_2 can be achieved by the same methodology [34]. The size of the 2-RDM can be further reduced by additional symmetries [13, 56, 57], but they have not been exploited in the present calculations.

4 Results

In Sect. 4.1, we compare the ground-state energies of the 2×4 ladder system at and below half filling from the variational 2-RDM method with those from full configuration interaction (FCI). Section 4.2 contains the analysis of the 2×10 and 1×10 lattices through the α, β-two-point pair correlation function, a measure of one-particle effective hopping, one-electron natural occupation numbers, and the squared Frobenius norm of the cumulant (connected) part of the 2-RDM. Results from 2-RDM calculations with DQG and DQGT conditions are compared.

4.1 Energies of the Hubbard ladder

We report the errors in the ground-state energies from the variational 2-RDM method for the 2×4 ladder Hubbard model for a range of interaction strengths where $t = t_\perp = 1$ a.u. Comparisons are made to the ground-state energies from FCI. The FCI calculation determines the ground-state energies by computing the lowest eigenvalue of the N-electron Hamiltonian matrix in the basis of all possible Slater determinants. In the FCI calculation, the spin orbitals are products of a spin function (α or β) and a "spatial" orbital which can be defined either in the position representation or in the momentum representation. All 2-RDM calculations were optimized until the primal feasibility norm was below 1.0×10^{-5} and the primal-dual gap was below 1.0×10^{-4}. In Fig. 1, the error in the ground-state energy and the percentage of the correlation energy recovered from the 2-RDM method with DQG and DQGT conditions are reported for (a) $\langle \hat{n} \rangle = 1$ and (b) $\langle \hat{n} \rangle = 3/4$ fillings. In the case of half filling, the DQG and the DQGT energies deviate at most from those from FCI by -0.74 and

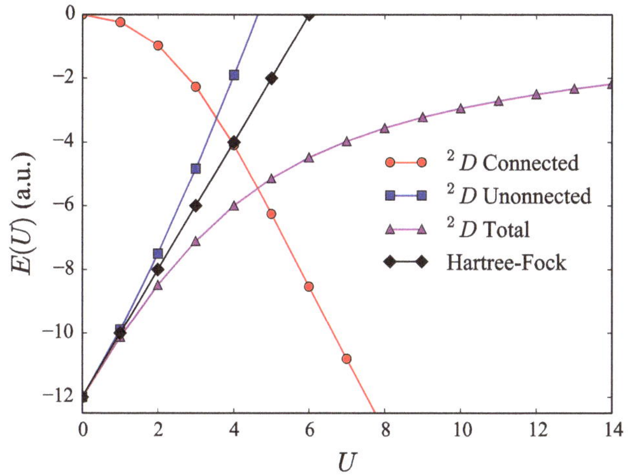

Fig. 2 For the 2×4 lattice at $\langle \hat{n} \rangle = 1$ the connected, unconnected, and total energies from the variational 2-RDM method with DQGT constraints as well as the Hartree-Fock total energies are shown

-0.087 a.u., respectively. The maximum deviations occur in the intermediate interaction regime $U \in [4, 8]$, where there is a large degree of competition between delocalization and localization. For the $\langle \hat{n} \rangle = 3/4$ filling, the DQG and DQGT conditions result in a larger absolute error than all corresponding values at half filling, which is consistent with previous observations in the literature [55] that correlated systems with an imbalance between the number of particles and holes require more stringent N-representability constraints.

The 2-RDM can be expressed as the wedge product of 1-RDMs (unconnected) plus a cumulant (connected) part [28, 29] denoted as $^2\Delta$

$$^2D = ^1D \wedge ^1D + ^2\Delta, \tag{33}$$

where the $\wedge$ is the Grassmann wedge product [9, 30, 51]. The unconnected term captures the statistically independent part of the electron pair probability. The energies from the unconnected and connected components [20] are

$$E_1 = \mathrm{Tr}[^2\mathrm{K}\ (^1\mathrm{D} \wedge ^1\mathrm{D})] \tag{34}$$

$$E_2 = \mathrm{Tr}[^2\mathrm{K}\ ^2\Delta]. \tag{35}$$

These energies as well as the Hartree–Fock mean-field energy are plotted in Fig. 2. The Hartree–Fock energy grows linearly as U is increased which is closely mirrored by the unconnected piece. Consequently, all the correlated information of the 2-RDM that results in localization is contained in its connected part.

4.2 One- and two-particle correlations

The α, β-two-point pair correlation function for the 2×10 and 1×10 lattices is examined at half filling to explore the variational 2-RDM method with DQG and DQGT

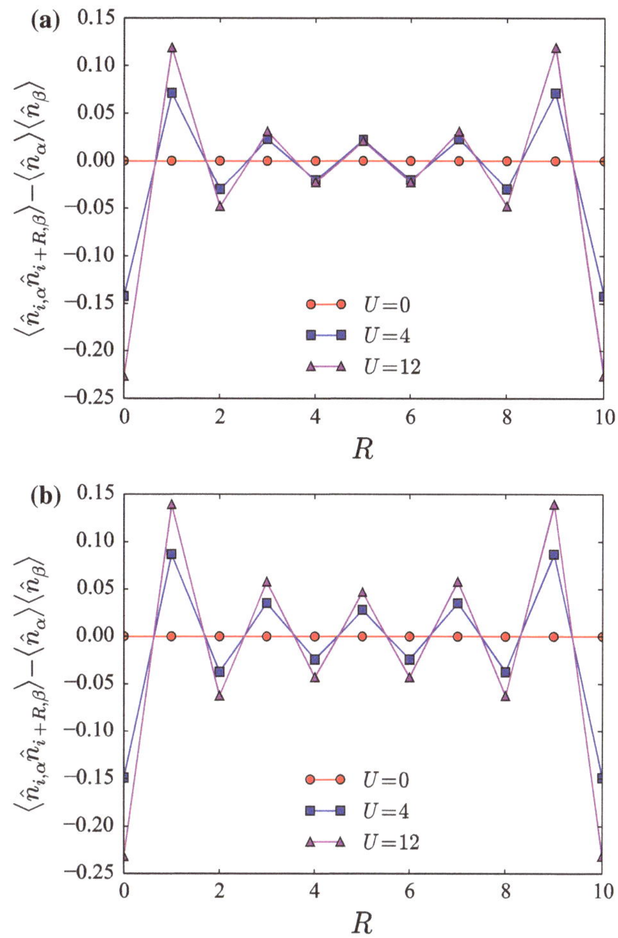

Fig. 3 The α, β-two-point pair correlation functions of the (**a**) 2×10 and (**b**) 1×10 lattices at half filling are computed as a function of R from the variational 2-RDM method with DQGT where R is the distance along each Hubbard strand between the pairs. Because of the periodic boundary conditions, the correlation function values are unique until the lattice inversion center at $R = 5$

conditions. We can express the two-point spin-up, spin-down correlation as

$$\langle \hat{n}_{i\alpha}\hat{n}_{i+R\beta} \rangle - \langle \hat{n}_\alpha \rangle \langle \hat{n}_\beta \rangle, \tag{36}$$

where $\langle \hat{n}_\alpha \rangle$ is the total density of α-electrons in the system

$$\langle \hat{n}_\alpha \rangle = N_\alpha/(2N_L), \langle \hat{n}_\beta \rangle = N_\alpha/(2N_L). \tag{37}$$

This correlation function is an extension of the local double-occupancy, $\langle \hat{n}_{i,\alpha}\hat{n}_{i,\beta} \rangle$, which has been used to examine the Mott transition of the Hubbard model defined on various lattices and temperatures [6, 7, 45, 59].

The pair correlation function probes the expectation of an antiferromagnetic pair R-sites away from each other in the long direction of the lattice. In the $U = 0$ limit where the α and β electrons are delocalized across the lattice, the correlation function is zero for all values of R. As U is increased, antiferromagnetic pairing becomes less favorable and order is induced. In Fig. 3a, b we plot the two-

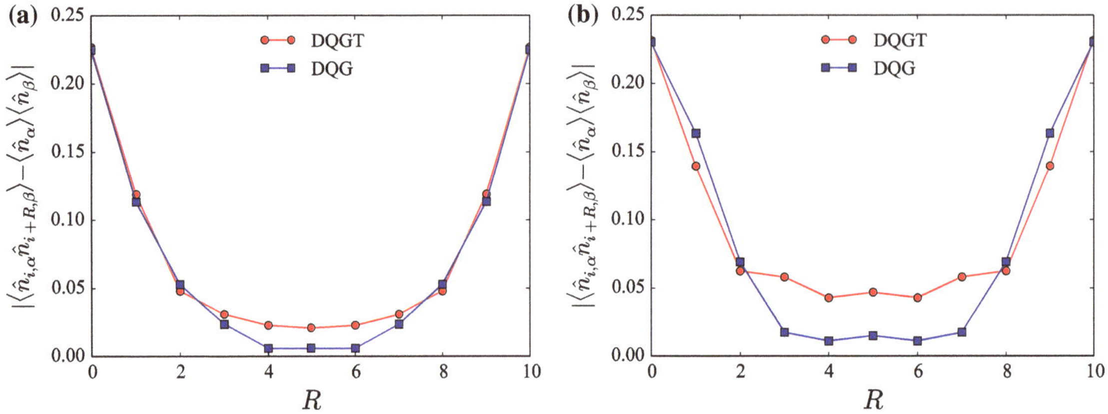

Fig. 4 The α, β-two-point pair correlation functions calculated with DQG and DQGT constraints are given for the (**a**) ladder 2×10 and (**b**) linear 1×10 lattices at half filling $\langle \hat{n} \rangle = 1$

particle correlation function as a function of R for representative U values. The correlation function decays across the lattice until the inversion center is reached. An exponential fitting of the absolute value of the pair correlation function, not shown, indicates that DQGT predicts a similar decay on the one-dimensional and quasi-one-dimensional lattices. In Fig. 4, we plot the absolute value of the pair correlation function for $U = 12$ from DQG and DQGT to study the differences generated in the (a) ladder and (b) linear lattices from approximate N-representability conditions. In the ladder case, the results from DQG and DQGT are in fairly close agreement while in the linear case, DQG deviates significantly from DQGT.

We can use the effective hopping as a one-particle correlation function. Defined in Eq. (38), the effective hopping is the likelihood of transport normalized by the non-interacting limit of the model

$$t_{\text{eff}} = \frac{\langle \hat{a}^{\dagger}_{i\sigma}\hat{a}_{j\sigma} + \hat{a}^{\dagger}_{j\sigma}\hat{a}_{i\sigma}\rangle_U}{\langle \hat{a}^{\dagger}_{i\sigma}\hat{a}_{j\sigma} + \hat{a}^{\dagger}_{j\sigma}\hat{a}_{i\sigma}\rangle_{U=0}}. \tag{38}$$

For low U, where the kinetic energy has the largest energy contribution, approximate N-representability conditions underestimate the effective hopping (kinetic energy), and at high U, where the repulsive interaction term dominates, approximate N-representability conditions underestimate the localization, which causes the effective hopping to be overestimated. In Fig. 5a, we plot t_{eff} for the 2×4 lattice calculated with DQG and DQGT compared against t_{eff} from FCI. As expected, DQG yields a lower t_{eff} than FCI before $U = 4$ and a higher t_{eff} than FCI after $U = 4$. The effective hopping from DQGT shows a similar trend while exhibiting a much smaller deviation from the FCI curve, which reinforces the accuracy of the DQGT conditions for lattices with strongly correlated electrons. We compare t_{eff} from DQG and DQGT for the (b) 2×10 and (c) 1×10 lattices. Both the linear and ladder Hubbard models exhibit the same underestimation of t_{eff} at low U and overestimation of t_{eff} at high U. Figure 5d compares t_{eff} calculated from DQGT for the linear and ladder models. The increase in lattice dimension facilitates transport down the chain.

4.3 Natural occupation numbers and entanglement

We examine the one-electron occupation numbers of the natural orbitals around the Fermi surface. The natural orbitals are the eigenfunctions of the 1-RDM. Select occupation numbers of the α-spin block of the 1-RDM in the quasi-momentum basis are provided in Table 1. For both lattices, the $N_\alpha - 1$ and N_α occupation numbers calculated with DQG are larger than the occupation numbers with DQGT, and the $N_\alpha + 1$ and $N_\alpha + 2$ occupation numbers calculated with DQG are smaller than the occupation numbers with DQGT. The difference between DQG and DQGT occupation numbers is greater for the 2×10 lattice than 1×10 lattice for all U values. Furthermore, the higher degree of multireference character of the occupation numbers when calculated with DQGT constraints indicates that polyradical character induced by a transition to the Mott insulating state is better captured by DQGT conditions.

We measure correlation explicitly by calculating the squared Frobenius norm of the cumulant portion of the 2-RDM $^2\Delta$

$$||^2\Delta||^2 = \text{Tr}(^2\Delta^{\dagger}\,^2\Delta) \tag{39}$$

as a function of system size and coupling. The squared Frobenius norm of the cumulant [23] is size-extensive and contains spin entanglement information not captured by the correlation energy or von Neumann entropy. Using this metric, we compare the degree of correlation in each ladder system. In Fig. 6, we plot the norm-per-lattice site for the (a) 2×10 and (b) 1×10 lattices at half filling. For the

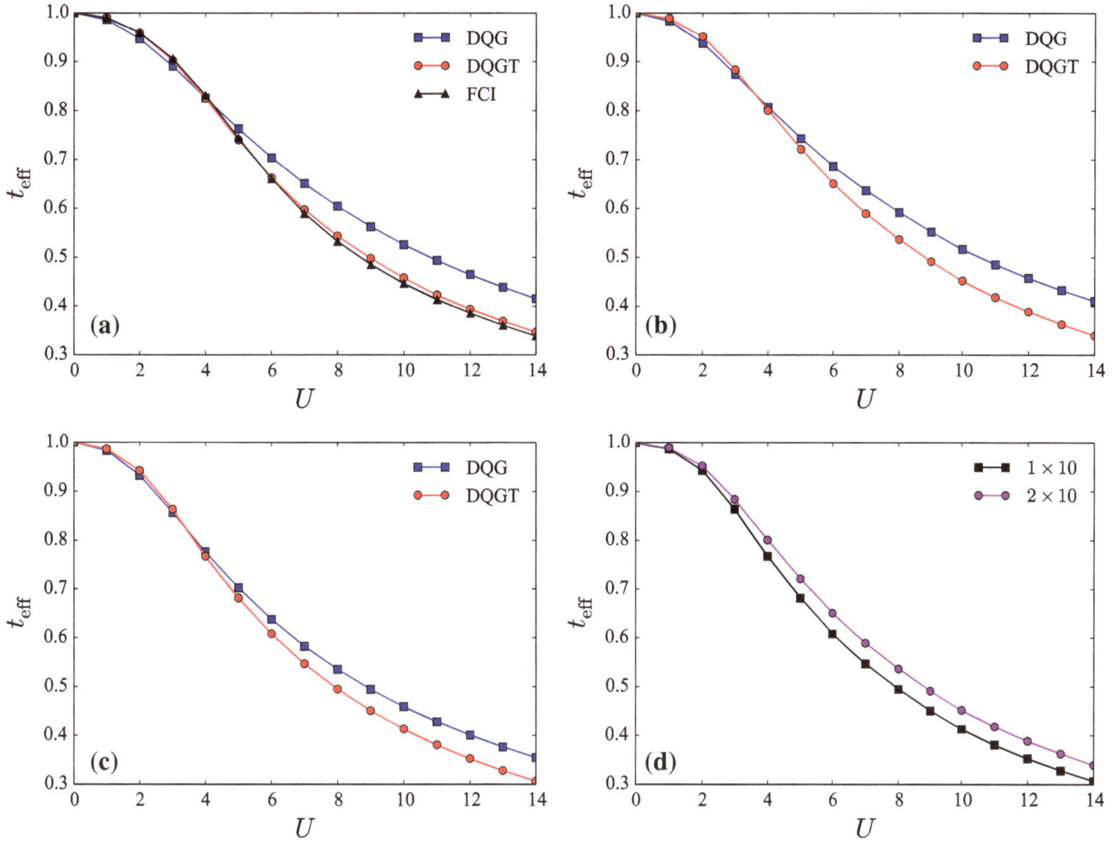

Fig. 5 The effective hopping t_{eff} for the 2×4 lattice is calculated with DQG, DQGT, and FCI. In (**a**), we plot t_{eff} for the 2×4 lattice calculated with DQG and DQGT compared against t_{eff} from FCI. We compare t_{eff} from DQG and DQGT for the (**b**) 2×10 and (**c**) 1×10 lattices. Both the linear and ladder Hubbard models exhibit the same underestimation of t_{eff} at low U and overestimation of t_{eff} at high U. Part (**d**) compares t_{eff} calculated from DQGT for the linear and ladder models. The increase in lattice dimension facilitates transport down the chain

Table 1 Natural-orbital occupation numbers of the α-spin block of the 1-RDM in the quasi-momentum basis

Lattice	Conditions	U	$N_\alpha - 1$	N_α	$N_\alpha + 1$	$N_\alpha + 2$
1 × 10	DQG	4	0.9022	0.8064	0.1936	0.0978
		8	0.7828	0.6566	0.3434	0.2172
		12	0.7114	0.6014	0.3986	0.2886
	DQGT	4	0.8993	0.7895	0.2104	0.1008
		8	0.7634	0.6291	0.3710	0.2366
		12	0.6870	0.5812	0.4188	0.3130
2 × 10	DQG	4	0.8081	0.7667	0.2333	0.1919
		8	0.6629	0.6280	0.3720	0.3371
		12	0.6067	0.5827	0.4173	0.3933
	DQGT	4	0.7863	0.6897	0.3103	0.2138
		8	0.6338	0.5710	0.4290	0.3662
		12	0.5811	0.5462	0.4538	0.4189

2×10 lattice, DQG predicts a significantly larger norm at high U. In the case of the 1×10 lattice, DQG and DQGT agree on the amount of correlation in the lattice.

5 Conclusion

Calculation of ground-state properties of strongly correlated model systems is highly important for understanding a plethora of condensed phase N-body physics. One of the main limitations of wave function methodologies is the exponential scaling of the Hilbert space with system size. In exchange for the exponentially scaling N-particle wave function, we have reviewed how to compute the ground-state energy with respect to the polynomial scaling 2-RDM. The variational 2-RDM has some key benefits: (1) it provides a lower bound to wave function methods, (2) can be numerically implemented as a semidefinite program which is solved with a polynomial scaling algorithm, and (3) and gives easy access to pair correlation functions important for characterizing condensed-matter systems.

We have demonstrated that the variational 2-RDM method with moderate N-representability constraints can be used to calculate the ground-state energies of ladder Hubbard models accurately. In keeping with recent results for 4×4 and 6×6 two-dimensional Hubbard models, we

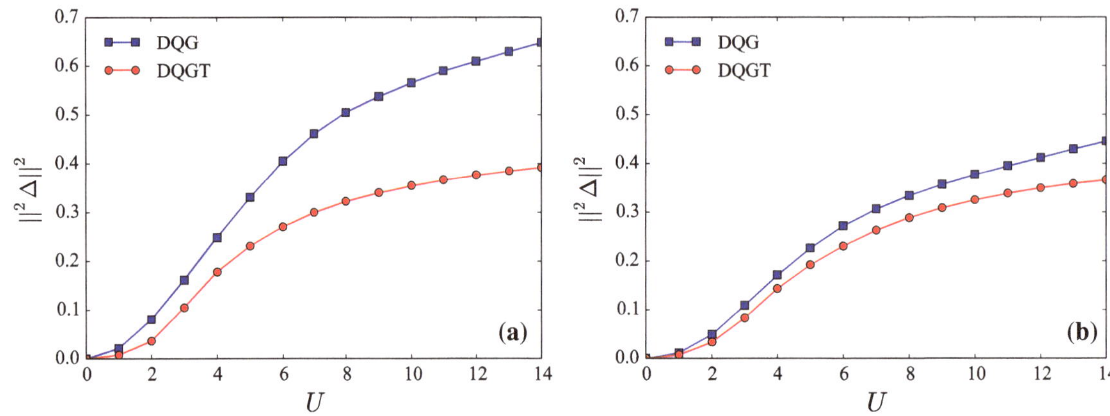

Fig. 6 Frobenius norm squared of the cumulant part of the 2-RDM is shown for the (**a**) 2 × 10 and (**b**) 1 × 10 lattices where the 2-RDMs are computed with variational 2-RDM calculations with DQG and DQGT conditions

observe that partial (2,3)-positivity (DQGT) conditions are effective at capturing strong electron correlation effects in both one- and quasi-one-dimensional lattices for both half filling and less-than-half filling. We have found that certain correlation functions can be accurately predicted with (2,2)-positivity (DQG) conditions. Furthermore, 2-RDM methods offer a way to analyze the correlation per site in a lattice model with a size-extensive metric and give direct access to occupation numbers. The 2-RDM methods complement recently developed wave function-based methods [46], and they may be useful in the context of approximate embedding calculations [47]. In general, the variational RDM method offers a new approach to studying lattice models of varying topology and filling.

Acknowledgments D.A.M. gratefully acknowledges the NSF, the ARO, and the Keck Foundation for their generous support. The authors express their gratitude to Dr. Jonathan J. Foley for assistance with the configuration interaction calculations.

References

1. Anderson PW (1987) Science 235:1196
2. Anderson JS, Nakata M, Igarashi R, Fujisawa K, Yamashita M (2013) Comput Theor Chem 1003:22
3. Barthel T, Hübener R (2012) Phys Rev Lett 108:200404
4. Baumgratz T, Plenio MB (2012) New J Phys 14:023027
5. Cava R, Siegrist T, Hessen B, Krajewski J, P W Jr, Batlogg B, Takagi H, Waszczak J, Schneemeyer L, Zandbergen H (1991) J Solid State Chem 94:170
6. Chen Y-H, Wu W, Tao H-S, Liu W-M (2010) Phys Rev A 82:043625
7. Chen Y-H, Tao H-S, Yao D-X, Liu W-M (2012) Phys Rev Lett 108:246402
8. Coleman AJ (1963) Rev Mod Phys 35:668
9. Coleman AJ, Absar I (1980) Int J Quant Chem 18:1279
10. Coleman AJ (2000) Reduced density matrices: coulson's challenge. Springer, Berlin
11. Dagotto E, Rice TM (1996) Science 271:618
12. Erdahl RM (1978) Int J Quantum Chem 13:697–718
13. Essler FHL, Korepin VE, Schoutens K (1991) Phys Rev Lett 67:3848
14. Fabrizio M, Parola A, Tosatti E (1992) Phys Rev B 46:3159
15. Garrod C, Percus JK (1964) J Math Phys 5:1756
16. Gidofalvi G, Mazziotti DA (2005) Phys Rev A 72:052505
17. Gidofalvi G, Mazziotti DA (2006) Phys Rev A 74:012501
18. Gutzwiller MC (1963) Phys Rev Lett 10:159
19. Gutzwiller MC (1965) Phys Rev 137:A1726
20. Hammond JR, Mazziotti DA (2005) Phys Rev A 71:062503
21. Hiroi Z, Azuma M, Takano M, Bando Y (1991) J Solid State Chem 95:230
22. Hubbard J (1963) Proc R Soc Lond A 276:238
23. Juhász T, Mazziotti DA (2006) J Chem Phys 125:174105
24. Kummer H (1967) J Math Phys 8:2063
25. Landee CP, Turnbull MM, Galeriu C, Giantsidis J, Woodward FM (2001) Phys Rev B 63:100402
26. Malick J, Povh J, Rendl F, Wiegele A (2009) SIAM J Optim 20:336
27. Mazziotti DA (2004) J Chem Phys 121:10957
28. Mazziotti DA (1998) Chem Phys Lett 289:419
29. Mazziotti DA (1998) Int J Quantum Chem 70:557
30. Mazziotti DA (1998) Phys Rev A 57:4219
31. Mazziotti DA, Erdahl RM (2001) Phys Rev A 63:042113
32. Mazziotti DA (2002) Phys Rev A 65:062511
33. Mazziotti DA (2004) Phys Rev Lett 93:213001
34. Mazziotti DA (2005) Phys Rev A 72:032510
35. Mazziotti DA (2006) Phys Rev A 74:032501
36. Mazziotti DA (ed) (2007) Reduced-density-matrix mechanics: with application to many-electron atoms and molecules. Advances in Chemical Physics, vol 134. Wiley, New York
37. Mazziotti DA (2011) Phys Rev Lett 106:083001
38. Mazziotti DA (2012) Chem Rev 112:244
39. Mazziotti DA (2012) Phys Rev Lett 108:263002
40. Nakata M, Nakatsuji H, Ehara M, Fukuda M, Nakata K, Fujisawa K (2001) J Chem Phys 114:8282
41. Nakata M, Braams BJ, Fujisawa K, Fukuda M, Percus JK, Yamashita M, Zhao Z (2008) J Chem Phys 128:164113
42. Nesterov Y, Nemirovskii AS (1993) Interior point polynomial method in convex programming: theory and applications. SIAM, Philadelphia
43. Noack R, White S, Scalapino D (1996) Physica C 270:281
44. Norrestam R, Nygren M, Bovin J-O (1991) Angew Chem Int Ed 30:864
45. Ohashi T, Kawakami N, Tsunetsugu H (2006) Phys Rev Lett 97:066401

46. Petruziel FR, Holmes AA, Changlani HJ, Nightingale MP, Umrigar CJ (2012) Phys Rev Lett 109:230201
47. Rodriguez-Guzman RR, Jimenez-Hoyos CA, Schutski R, Scuseria GE (2013) Phys Rev B 87:235129
48. Rothman AE, Mazziotti DA (2008) Phys Rev A 78:032510
49. Schwerdtfeger CA, Mazziotti DA (2009) J Chem Phys 130:224102
50. Sinitskiy AV, Greenman L, Mazziotti DA (2010) J Chem Phys 133:014104
51. Slebodzinski W (1970) Exterior forms and their application. Polish Scientific Publishers, Warsaw
52. Tredgold RH (1957) Phys Rev 105:1421
53. Vandenberghe L, Boyd S (1996) SIAM Rev. 38:49
54. Verstichel B, van Aggelen H, Van Neck D, Ayers PW, Bultinck P (2009) Phys Rev A 80:032508
55. Verstichel B, van Aggelen H, Poelmans W, Van Neck D (2012) Phys Rev Lett 108:213001
56. Verstichel B, van Aggelen H, Poelmans W, Wouters S, Neck DV (2013) Comput Theor Chem 1003:12
57. Verstichel B, Poelmans W, De Baerdmacker S, Wouters S, and Van Neck D (2013). arXiv:1307.1002
58. Wright S (1997) Primal-dual interior-point methods. SIAM, Philadelphia
59. Wu W, Chen Y-H, Tao H-S, Tong N-H, Liu W-M (2010) Phys Rev B 82:245102
60. Zhao Z, Braams BJ, Fukuda M, Overton ML, and Percus JK, (2004) J Chem Phys 120:2095–2104

Theor Chem Acc (2014) 133:1513
DOI 10.1007/s00214-014-1513-6

REGULAR ARTICLE

Anchoring the potential energy surface for the Br + H_2O → HBr + OH reaction

Meiling Zhang · Yanjun Hao · Yundong Guo · Yaoming Xie · Henry F. Schaefer

Received: 31 March 2014 / Accepted: 24 May 2014 / Published online: 17 June 2014

Abstract The forward and reverse reactions Br + H_2O → HBr + OH are important in atmospheric and environmental chemistry. Five stationary points on the potential energy surface for the Br + H_2O → HBr + OH reaction, including the entrance complex, transition state, and exit complex, have been studied using the CCSD(T) method with correlation-consistent basis sets up to cc-pV5Z-PP. Contrary to the valence isoelectronic F + H_2O system, the Br + H_2O reaction is endothermic (by 31.8 kcal/mol after zero-point vibrational, relativistic, and spin–orbit corrections), consistent with the experimental reaction enthalpy. The CCSD(T)/cc-pV5Z-PP method predicts that the reverse reaction HBr + HO → Br + H_2O has a complex but no *classical* barrier. When zero-point vibrational energies are added, the transition state lies 0.25 kcal/mol above the separated products. This is consistent with the negative temperature dependence for the rate constant observed in experiments. The entrance complex is predicted to lie 2.6 kcal/mol below separated Br + H_2O. The exit complex is predicted to lie 1.8 kcal/mol below separated HBr + OH.

Keywords Atmospheric chemistry · Water reactions · Bromine atoms · OH radicals · Potential energy surfaces · Ab initio computations

Dedicated to the memory of Professor Isaiah Shavitt and published as part of the special collection of articles celebrating his many contributions.

Electronic supplementary material The online version of this article (doi:10.1007/s00214-014-1513-6) contains supplementary material, which is available to authorized users.

M. Zhang
School of Biomedical Engineering, Tianjin Medical University, Tianjin 300070, People's Republic of China

Y. Hao
College of Physical Science and Technology, Sichuan University, Chengdu 610064, Sichuan, People's Republic of China

Y. Guo
School of Engineering and Technology, Neijiang Normal University, Neijiang 641112, Sichuan, People's Republic of China

Y. Xie · H. F. Schaefer (✉)
Center for Computational Quantum Chemistry, University of Georgia, Athens, GA 30602, USA
e-mail: qc@uga.edu

1 Introduction

Following the F + H_2O and Cl + H_2O reactions, the Br + H_2O reaction would appear to come next in the quest to understand the reaction mechanisms of the halogen atom-water systems, X + H_2O → HX + HO (X = F, Cl, Br) reactions. The Br + H_2O reaction has been widely studied because Br atoms destroy ozone in the atmosphere [1, 2] and because some brominated compounds play important roles in combustion chemistry acting as fire retardants [3].

Much research has been directed toward the forward and reverse Br + H_2O → HBr + HO reactions [4–15] because it is a prototype for A + BCD chemical reactions, and this relatively simple chemical system displays interesting rate constant complexity. Sims et al. [4] first reported that the rate of the OH + HBr reaction below 249 K and the rate constants increase monotonically with decreasing temperature from 295 to 23 K. The theoretical studies by Clary et al. [5] in the same year predicted negative temperature dependence in agreement with the experimental results [4]. Jaramillo et al. [7] later observed inverse temperature dependence for the OH + HBr reaction at temperature below 200 K using uniform supersonic flow reactors with laser-induced fluorescence detection. Che et al. [8] used crossed molecular beam techniques and found the reaction cross section to decrease with increasing collision energy, and they suggested no barrier for the OH + HBr reaction.

Theoretically, Nitzamov et al. [6] studied the rate constants and energy partitioning for the OH + HBr reaction via quasi-classical trajectories. Liu and coworkers [9] predicted that the OH + HBr reaction has a small barrier (<1.0 kcal/mol, from CCSD(T)/6-311+G(2df,2p) single-point energies), and their theoretical rate constants are comparable with the experimental results [4, 11–14].

In 2013, Li et al. [15] used the ROHF-UCCSD(T)-F12 method with the aug-cc-pVTZ basis set and a frozen-core (FC) approximation to study the characteristics of the three X···H_2O (X = F, Cl, Br) complexes, but no other stationary points in the potential surface were reported. Since the reaction rate constants and their temperature dependence are sensitive to the accuracy of the potential energy surface, we will, in the present paper, adopt high-level ab initio coupled-cluster methods along with correlation-consistent cc-pV5Z-PP basis sets to study all stationary points on the potential surface for the Br + H_2O → HBr + OH reaction.

2 Computational methods

In this research, the coupled-cluster single and double substitutions method with a perturbative treatment of triple excitations CCSD(T) [16–18] was adopted based on unrestricted Hartree–Fock reference wave functions. For the CCSD(T) computations, the cc-pVDZ, cc-pVTZ, and cc-pVQZ quality basis sets were used, where cc-pVnZ is an abbreviation for the correlation-consistent polarized valence basis sets of Dunning et al. [19–22]. With the coupled-cluster methods used in this research, the core orbitals are frozen. That is, the 1s-like molecular orbital is frozen for O, while the 1s2s2p3s3p3d-like molecular orbitals are frozen for Br.

In parallel computations, we adopted the relativistic pseudopotentials (PPs) and the corresponding correlation-consistent cc-pVnZ-PP basis sets for Br [23]. Therein, the bromine inner core ($1s^22s^22p^6$) is replaced by an energy-consistent pseudopotential, which was adjusted to atomic multiconfigurational Dirac–Hartree–Fock results [23]. With this cc-pVnZ-PP basis sets for Br, in our present CCSD(T) study, the outer core-like molecular orbitals (3s3p3d) are frozen. With the relativistic pseudopotentials, we adopted basis sets as large cc-pV5Z-PP for Br, in conjunction with the corresponding cc-pV5Z basis sets for O and H.

The optimized geometries and resulting energies for the reactants, entrance complex, transition state, exit complex, and products were predicted. Harmonic vibrational frequencies predicted at the same level (up to cc-pVQZ and cc-pVQZ-PP basis sets) were used for the characterization of stationary points and zero-point vibrational energies (ZPVE). The CFOUR program of Stanton, Gauss, Harding, Szalay, and coworkers was used for the coupled-cluster computations [24]. Most DZ results are not shown in text, but they are available in Supplementary Material.

3 Nonrelativistic results

We consider first the nonrelativistic predictions of this research. Shown in Fig. 1 and Table 1 are the geometrical structures and relative energies for the reactants, entrance complex, transition state, exit complex, and products involved in the title reaction, as optimized at the CCSD(T)/cc-pVnZ, (n = D, T, Q) levels of theory. The harmonic vibrational frequencies (cm^{-1}) for all the stationary points of the Br + H_2O → HBr + HO potential energy surface are reported in Table 2. For comparison, the limited available experimental results are also shown in Tables 1 and 2 [25–28]. Our ZPVE value for water (13.52 kcal/mol) is easily tested. Barletta et al. [29] have determined the exact ZPVE of water to be 4,638.31 cm^{-1}, or 13.26 kcal/mol. This means that there is an error of (13.52–13.26 =) 0.26 kcal/mol in our water ZPVE.

Compared with the available experimental geometrical parameters, our theoretical results are in satisfactory agreement. For example, the CCSD(T)/cc-pVQZ predicted O–H distance (0.958 Å) and ∠H–O–H (104.1°) for H_2O are close to the experimental values O–H(r_e) = 0.9575 Å, ∠H–O–H = 104.51° [30]. For the products OH and HBr, the experimental bond distances are O–H(r_e) = 0.9697 Å and H–Br(r_e) = 1.4144 Å [30], compared with our CCSD(T)/cc-pVQZ results O–H(r_e) = 0.970 Å and H–Br(r_e) = 1.420 Å.

It is seen in Fig. 1 that the CCSD(T) method shows the Br + H_2O → HBr + HO reaction to endothermic by 28.0, 31.3, and 31.8 kcal/mol with the cc-pVDZ, cc-pVTZ, and cc-pVQZ basis sets, respectively. The results appear to show convergence approaching with the size of the basis set. Further corrections to the classical endothermicity will be discussed below. In the laboratory, only the reverse reaction has yet been observed, *i.e.*, BrH + OH → Br + H_2O at a range of temperatures, from 48 to 224 K [7] and 230–360 K [13].

The entrance complex Br···H_2O predicted by the CCSD(T)/cc-pVQZ method lies 3.5 kcal/mol (without ZPVE) below the separated reactants (Br and H_2O). This complex (C_s symmetry) involves a weak interaction

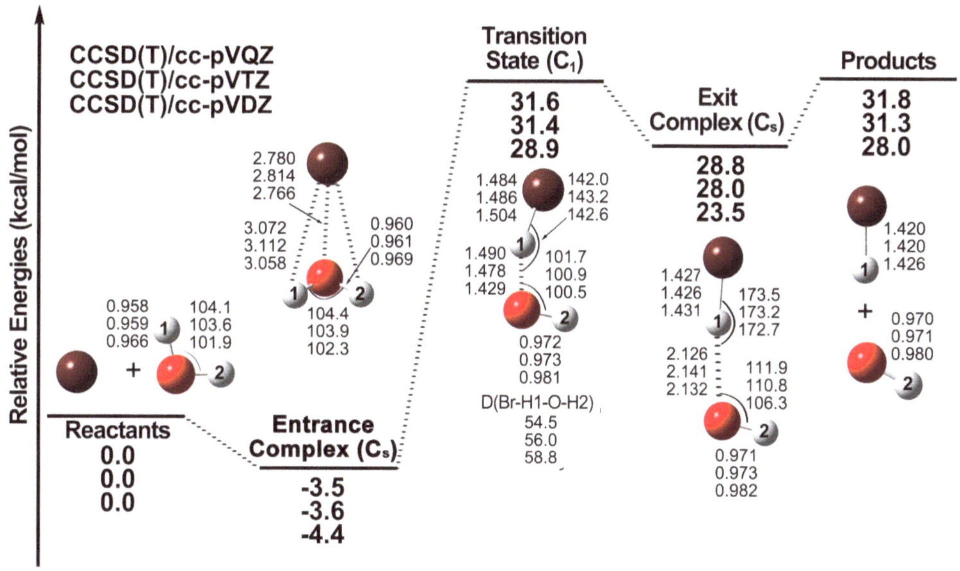

Fig. 1 Stationary points on nonrelativistic the Br + H_2O potential energy surface. Optimized geometrical parameters for the structures of the reactants, entrance complex, transition state, exit complex, and products are from the CCSD(T) method with basis sets cc-pVDZ, cc-pVTZ, and cc-pVQZ, respectively. The bond lengths and angles are in angstroms and degrees, respectively. Energies do not include zero-point vibrational corrections or spin–orbit effects

Table 1 Relative energies (in kcal/mol) for the stationary point of the reaction of Br with H_2O using the CCSD(T) method, with the cc-pVnZ (n = D, T, Q) basis sets, as well as the analogous cc-pVnZ(-PP) basis sets

	cc-pVTZ			cc-pVQZ			Expt.
	ΔE	+ZPVE	+SO	ΔE	+ZPVE	+SO	ΔH_0 (T = 0 K)
CCSD(T)/cc-pVnZ							
Br + H_2O	0.00	0.00	0.00	0.00	0.00	0.00	30.18, 30.97[a]
Entrance complex	−3.57	−2.76	−2.76	−3.45	−2.62	−2.62	
TS	31.41	27.73	31.24	31.64	27.84	31.35	
Exit complex	27.98	24.76	28.27	28.77	25.34	28.85	
HBr + OH	31.27	26.91	30.42	31.80	27.45	30.96	

	cc-pVTZ(-PP)			cc-pVQZ(-PP)			cc-pV5Z(-PP)		
	ΔE	+ZPVE	+SO	ΔE	+ZPVE	+SO	ΔE	+ZPVE	+SO
CCSD(T)/cc-pVnZ(-PP)									
Br + H_2O	0.00	0.00	0.00	0.00	0.00	0.00	0.00	0.00	0.00
Entrance complex	−3.64	−2.82	−2.82	−3.54	−2.71	−2.71	−3.41	−2.59	−2.59
TS	31.83	28.06	31.57	32.07	28.29	31.80	32.36	28.58	32.09
Exit complex	28.63	25.37	28.88	29.39	26.09	29.60	29.85	26.56	30.07
HBr + OH	31.88	27.51	31.02	32.38	28.02	31.53	32.70	28.33	31.84

The cc-pVnZ(-PP) basis sets have the 10 core electrons of bromine replaced by a relativistic effective potential. At each level of theory are reported (a) ΔE based on electronic energies only; (b) ΔE plus zero-point vibrational energy (ZPVE) corrections; and (c) ΔE + ZPVE corrections + spin–orbit (SO) corrections

[a] References [38–41, 43]

between Br and H_2O, similar to its analogues F···H_2O [31] and Cl···H_2O [32]. This complex (Br···H_2O) was not reported in some earlier theoretical studies [6, 9], but it was carefully examined in the recent research by Li et al. [15]. Using the ROHF-UCCSD(T)-F12 method, the latter authors found the entrance complex to lie 3.5 kcal/mol lower than Br + H_2O. It would have been good to see a simple pattern of convergence with respect to basis set from TZ to QZ to 5Z. However, the pseudopotential TZ/QZ difference in the entrance complex well depth is 0.11 kcal/mol, but that for QZ/5Z is a bit larger, 0.12 kcal/mol.

The barrier from the Br + H_2O side is predicted to be 31.6 kcal/mol at the CCSD(T)/cc-pVQZ level of theory. The same theoretical method has been used to predict reliable barriers for the F + H_2O [31] and Cl + H_2O [32]

Table 2 Nonrelativistic harmonic vibrational frequencies (cm^{-1}) and zero-point vibrational energies (kcal/mol) for the stationary points of the Br + H_2O → HBr + OH reaction

	ZPVE (kcal/mol)	Vibrational frequencies (cm^{-1})					
		ω_1	ω_2	ω_3	ω_4	ω_5	ω_6
CCSD(T)/cc-pVTZ							
H_2O	13.52	3,946	3,841	1,669			
Entrance complex	14.33	3,924	3,818	1,662	255	251	113
TS	9.84	3,730	1,637	737	418	272	714i
Exit complex	10.30	3,731	2,597	371	249	156	98
HBr	3.80	2,660					
OH	5.35	3,745					
HBr + OH	9.15						
CCSD(T)/cc-pVQZ							
H_2O	13.52	3,951	3,844	1,659			
Entrance complex	14.35	3,926	3,817	1,651	266	264	115
TS	9.72	3,733	1,692	713	406	255	672i
Exit complex	10.09	3,736	2,600	292	261	135	56
HBr	3.80	2,661					
OH	5.36	3,750					
HBr + OH	9.16						
Experiment[a]							
H_2O	13.47	3,943	3,832	1,649			
HBr	3.79	2,649					
OH	5.34	3,735					

[a] Reactant and product experimental results shown for comparison [25–28]

reactions. The transition state is a first-order saddle point, confirmed by the vibrational analysis with one imaginary frequency, 974i, 714i, and 672i cm^{-1} at the CCSD(T)/cc-pVDZ, CCSD(T)/cc-pVTZ, and CCSD(T)/cc-pVQZ levels, respectively (Table 2). As can be seen from Fig. 1, there are significant changes in the geometry (bond angles and bond distances) from the entrance complex to the transition state. This late transition state is a nonplanar structure with C_1 symmetry, with the departing O–H distance at the TS stretching to 1.490 Å at the CCSD(T)/cc-pVQZ level. This is 0.530 Å longer than the O–H distance in the entrance complex (0.960 Å) or for the reactant (0.958 Å). Compared with the nearly linear O–H–Br angle in the empirical transition state reported in Ref. [6], our predicted value of 142° (Fig. 1) is quite different. However, our result is much closer to those (147°, 149°, or 152°) at the MP2, BHLYP, and MP4SDQ levels, respectively, obtained by Liu et al. [9].

The exit complex Br–H···O–H for the Br + H_2O → HBr + OH reaction is predicted to lie (31.8–28.8) = 3.0 kcal/mol below the products (separated HBr plus OH). The internuclear separations H–Br and O–H for the exit complex are very similar to those for the separated products HBr and OH, indicating this complex is a product-like intermediate. Consequently, the harmonic vibrational frequencies of O–H and H–Br in the complex (Table 2) are close to those for separated HBr plus OH.

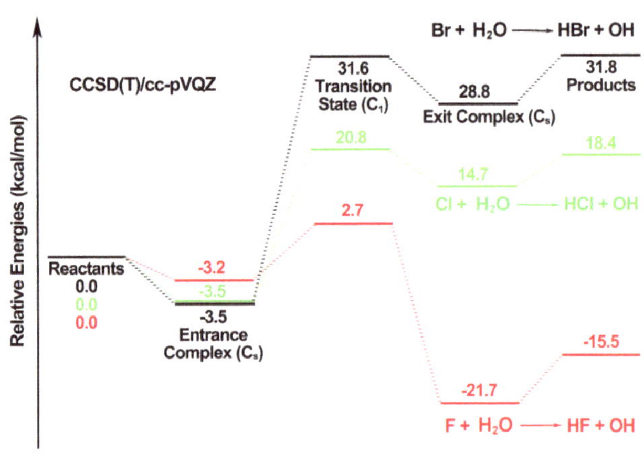

Fig. 2 Energetic comparison of stationary points for the F + H_2O, Cl + H_2O, and Br + H_2O reactions with the CCSD(T)/cc-pVQZ method. Energies do not include zero-point vibrational corrections or relativistic or spin–orbit effects

Figure 2 exhibits the similarities and differences between our Br + H_2O potential energy surface and those for the F + H_2O [31] and Cl + H_2O [32] systems. Like the potential energy surfaces for F + H_2O and Cl + H_2O systems, we see both an entrance complex and an exit complex for Br + H_2O, with their geometries qualitatively similar to those for the F + H_2O and Cl + H_2O systems. For these systems in Fig. 2, all the three X···H_2O entrance complexes (X = F, Cl, Br) lie below the corresponding reactants X + H_2O by 3.2 to 3.5 kcal/mol, while the X–H···O–H exit complexes lie below the X–H + O–H products by 3.0–6.2 kcal/mol at the CCSD(T)/cc-pVQZ level of theory.

However, there are major qualitative differences between the three reactions. The Br + H_2O and Cl + H_2O systems are endothermic reactions, as opposed to the exothermic F + H_2O reaction. Thus, for the F + H_2O system (like the more exhaustively studied F + H_2 reaction [33, 34]), the reaction proceeds more readily in the forward direction, while for the Cl + H_2O and Br + H_2O systems, the reactions are much more likely to occur in the reverse direction under normal conditions. Furthermore, the Br + H_2O system is more endothermic energy than is Cl + H_2O. We can also see that for the three halogen atom-water reactions (X = F, Cl, Br), the relative energies of the transition states are obviously different. The energy barrier from the Br + H_2O side (31.6 kcal/mol) is higher than that for the Cl + H_2O system (20.8 kcal/mol), and much higher than that for the F + H_2O system (2.7 kcal/mol). For the important reverse XH + OH → X + H_2O reactions, the barriers from the XH + OH side are 18.2, 2.4, and -0.2 kcal/mol for X = F, Cl, and Br, respectively, predicted by the CCSD(T)/cc-pVQZ method. For the HBr + OH → Br + H_2O reaction, the transition state lies lower than HBr + OH, leading to a negative temperature dependence for the rate constants [4, 7]. Accordingly, the Br + H_2O reaction has a late transition state with its geometry similar to the product HBr plus OH, while the F + H_2O reaction has an early transition state with its geometry similar to the reactant F plus H_2O. The transition state for the Cl + H_2O reaction lies in between.

4 Scalar relativistic results

Pseudopotentials (PP) were originally proposed to reduce the computational cost for the heavy atoms with the replacement of the core orbitals by an effective potential. Modern pseudopotentials implicitly include relativistic effects by means of adjustment to quasi-relativistic Hartree–Fock or Dirac–Hartree–Fock orbital energies and densities [35]. In the present research, we adopted Peterson's correlation-consistent cc-pVnZ-PP (n = D, T, Q, 5) basis sets [23] with the corresponding relativistic pseudopotential for the Br atom. The corresponding cc-pVnZ (n = D, T, Q, 5) basis sets were used for the O and H atoms. The optimized geometries and relative energies for the stationary points are reported in Table 1 and Fig. 3, and the harmonic vibrational frequencies and zero-point vibrational energies are reported in Table 4.

The relativistic pseudopotential (PP) predictions are close to the all-electron results. Of course, there are no changes for H_2O and OH, since no pseudopotential is used for the oxygen atom. The change for the bond distance for the product HBr is also small (<0.002 Å). For the entrance complex, the Br···H_2O distance is reduced by ~0.01 Å, while the O–H distance shows no change. For the exit complex, no change is found for the O–H distance, a small change (<0.002 Å) for the H–Br distance, and the change of HO···HBr distance is <0.007 Å. For the transition state, there is no change for O–H distance, a change (<0.006 Å) for H–Br distance, and slightly larger changes (0.008–0.012 Å) for Br–H···O–H distance.

The relative energies for the entrance complex predicted by the cc-pVnZ-PP basis sets are almost the same as those predicted by the all-electron cc-pVnZ basis sets, except for cc-pVDZ-PP, which predicts a 0.1 kcal/mol deeper well. The barrier associated with the transition state is predicted a little higher (0.5 kcal/mol) by the PP basis sets. For the exit complex, the relative energy is also slightly higher (by 0.6 kcal/mol). The reaction energy is also predicted higher with the PP basis sets by 0.6 kcal/mol.

Table 3 Dipole moments (D, debye) from the CCSD(T) method with the cc-pVTZ-PP, cc-pVQZ-PP, and cc-pV5Z-PP basis sets for all the stationary points of the Br + H_2O → HBr + OH reaction

	cc-pVTZ-PP	cc-pVQZ-PP	cc-pV5Z-PP	Expt[a]
H_2O	1.92	1.90	1.88	1.8546 ± 0.0040
Entrance complex	2.24	2.21	2.23	
TS	1.77	1.69	2.36	
Exit complex	2.53	2.51	2.51	
HBr	0.87	0.83	0.82	0.8272 ± 0.0003
OH	1.67	1.67	1.67	1.655 ± 0.001

Reactant and product experimental results are also shown for comparison

[a] Ref. [42]

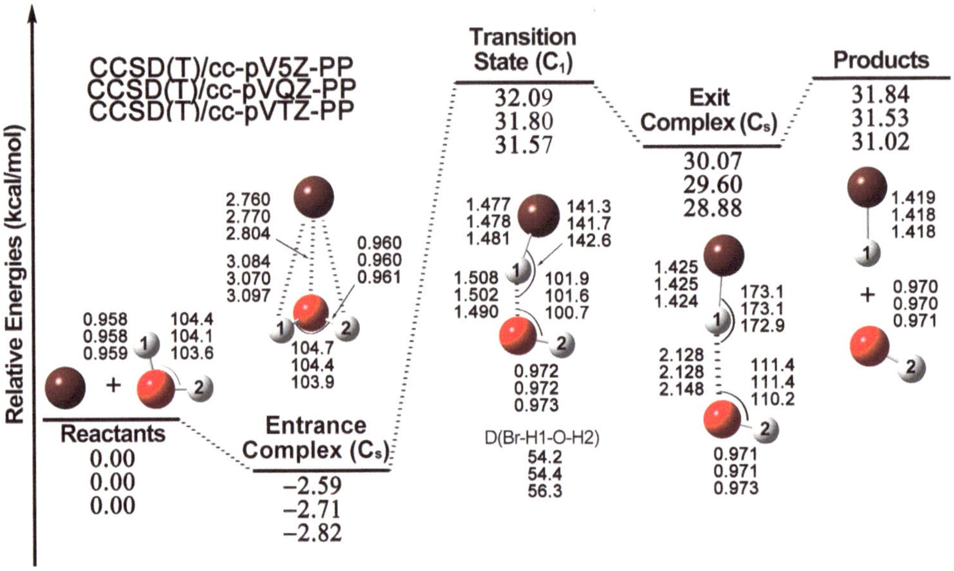

Fig. 3 Stationary points on the Br + H_2O potential energy surface. Optimized geometrical parameters for the structures of the reactants, entrance complex, transition state, exit complex, and products from the CCSD(T) method with the relativistic basis sets cc-pVTZ-PP cc-pVQZ-PP, and cc-pV5Z-PP, respectively. The bond lengths and angles are in angstrom and degree, respectively. Energies include both spin–orbit and zero-point vibrational corrections

Table 4 Relativistic vibrational frequencies (cm^{-1}) and zero-point vibrational energies (kcal/mol) for the stationary points of the Br + H_2O → HBr + OH reaction

	ZPVE (kcal/mol)	ΔZPVE (kcal/mol)	Vibrational frequencies (cm^{-1})					
			ω_1	ω_2	ω_3	ω_4	ω_5	ω_6
CCSD(T)/cc-pVTZ-PP								
H_2O	13.52	0.00	(Same as in Table 2)					
Entrance complex	14.34	0.82	3,924	3,817	1,662	259	254	115
TS	9.75	−3.77	3,730	1,677	728	416	269	678i
Exit complex	10.26	−3.26	3,732	2,590	367	243	150	96
HBr	3.79		2,649					
OH	5.35		(Same as in Table 2)					
HBr + OH	9.14	−4.38						
CCSD(T)/cc-pVQZ-PP								
H_2O	13.52	0.00	(Same as in Table 2)					
Entrance complex	14.34	0.82	3,927	3,818	1,649	262	261	116
TS	9.74	−3.78	3,732	1,727	703	400	251	640i
Exit complex	10.22	−3.30	3,737	2,583	364	226	146	96
HBr	3.79		2,651					
OH	5.36		(Same as in Table 2)					
HBr + OH	9.15	−4.37						

Obtained from the CCSD(T) method with basis sets cc-pVTZ-PP and cc-pVQZ-PP, respectively. The column ΔZPVE reports the difference in ZPVE between reactants (Br + H_2O) and the stationary point

Dipole moments for all the stationary points are shown in Table 3, and these predictions will hopefully also be useful for future experimental research. Good agreement with experiment is found for the dipole moments of H_2O, HBr, and OH [42]. It is seen that the dipole moment is largest (2.51 D) at the exit complex.

Table 4 lists the harmonic vibrational frequencies and zero-point vibrational energies predicted by the PP basis sets. Compared with the results in Table 2, the changes are very small. For the entrance complex, most vibrational frequencies are the same, except for small changes for the two lowest frequencies. For the transition state, the imaginary frequency changes from 974i to 940i cm^{-1} (DZ), from 714i to 678i cm^{-1} (TZ), and from 672i to 640i cm^{-1} (QZ). For the exit complex, there are changes as large as 72 cm^{-1}, but the change for the ZPVE is still as small as 0.1 kcal/mol.

5 Spin–orbit coupling

The scalar relativistic effects considered via pseudopotentials involve only the core electrons. For the valence electrons, spin–orbit (SO) coupling effects arising from the heavy Br atom may be very important. Since we are interested in the stationary points on the Br + H_2O potential surface, the reliable evaluation of the spin–orbit coupling becomes more straightforward. Recently, Czakó has studied the related Br + CH_4 reaction [36]. He notes that the SO correction is essentially quenched at the products (HBr + CH_3), exit complex, and even the late transition state (geometrically similar to the products). The SO splitting between the ground and excited SO states of the Br atom is known from experiment, namely 3,685 cm^{-1} [37]. Since the ground SO state ($^2P_{3/2}$) is fourfold degenerate and the excited SO state ($^2P_{1/2}$) is twofold degenerate, the SO ground state lies below the spin-averaged (non-SO) energy by $3{,}685/3 = 1{,}228$ cm^{-1} = 3.51 kcal/mol. The Br + H_2O reaction in our present study is analogous to the Br + CH_4 reaction, and thus, the SO correction may be readily evaluated, i.e., it actually increases the reaction barrier by 1,228 cm^{-1} (3.51 kcal/mol). Our CCSD(T)/cc-pVQZ(-PP) predicted Br + H_2O energy barrier after the ZVPE and SO corrections will be 31.80 (=28.29 + 3.51) kcal/mol. With the same logic, the endothermic reaction energy also increases to 31.53 (=28.02 + 3.51) kcal/mol (Table 1).

The evaluation of the experimental endothermicity of our reaction requires some care. For the 0 K heats of formation, we use the values of Ruscic et al. [38] for the H_2O (−57.10) and OH (8.85 kcal/mol) molecules. For HBr and Br, the experimental heats of formation seem less secure [39–41]. We chose the NIST values [40] of 26.74 (Br) and −8.67 kcal/mol (HBr). With these assumptions, an experimental endothermicity of 30.18 kcal/mol is deduced. Using slightly different experimental heats of formation, de Oliveira-Filho [43] finds a value of 30.97 kcal/mol. We include both values in Table 1.

6 Conclusions

The present CCSD(T) studies provide reliable stationary point information for the Br + H_2O → HBr + HO potential energy surface, as well as zero-point harmonic vibrational energies, vibrational frequencies, and dipole moments. A classical barrier of 31.6 kcal/mol (with the all-electron cc-pVQZ basis set) or 32.4 kcal/mol (with ECP and cc-pV5Z-PP basis set for Br) from the Br + H_2O side is predicted, and these values reduced to 27.8 (QZ AE) or 28.6 (5Z ECP) kcal/mol after the ZPVE corrections (Table 1). The final energy barrier with SO correction becomes 31.3 (QZ AE) or 32.1 (5Z ECP) kcal/mol. The 5Z ECP relative energy with ZPVE and SO corrections for the entrance complex is −2.6 kcal/mol and that for exit complex is 30.1 kcal/mol. Compared with the analogous systems F + H_2O and Cl + H_2O, the Br + H_2O potential energy surface has some very different features.

7 Addendum: comparison with de Oliveira-Filho, Ornellas, and Bowman

After the completion of the present research, a beautiful paper by de Oliveira-Filho et al. [43] appeared on the same reaction. The methodology used by Oliveira-Filho is somewhat different from ours. First, they used the CCSD(T)-F12a/cc-pVTZ-F12 method for their most reliable structural predictions. Our most reliable structural predictions were made with CCSD(T)/cc-pV5Z(-PP), with the analogous DZ, TZ, and QZ basis set results included test convergence. Thus, our structural predictions include scalar relativistic effects, while those of Bowman do not.

Figure 4 compares our best geometry predictions with the best predictions of de Oliveira-Filho [43]. For the reactants, exit complex, and products, the bond distance differences are negligible, no more than 0.001 Å. For the entrance complex, the interatomic distances are the same to within 0.002 Å. The transition state geometry should be the most challenging. Even there, the largest difference in bond distances is only 1.516–1.508 = 0.008 Å. At the transition state, the dihedral Br–H–O–H angles are 54.2 and 53.3. These two potential energy surfaces, computed completely independently, are remarkably similar in terms of stationary point geometries.

Finally, we compare the 0 K stationary point energies predicted by us and by Bowman and coworkers. In the present research, the CCSD(T)/cc-pV5Z-PP energies are appended with harmonic ZPVE corrections predicted at the CCSD(T)/cc-pVQZ-PP method and with our spin–orbit corrections (see above). In Bowman's final energetics, single-point energies were computed using a many-layered composite scheme (see pages 708 and 709 of their paper) including relativistic effects via the Douglas–Kroll method and spin–orbit contributions using the Breit–Pauli operator.

Note that the energetics in Fig. 4 include both relativistic and zero-point vibrational corrections. The most obvious difference in the energetics occurs for the depth of

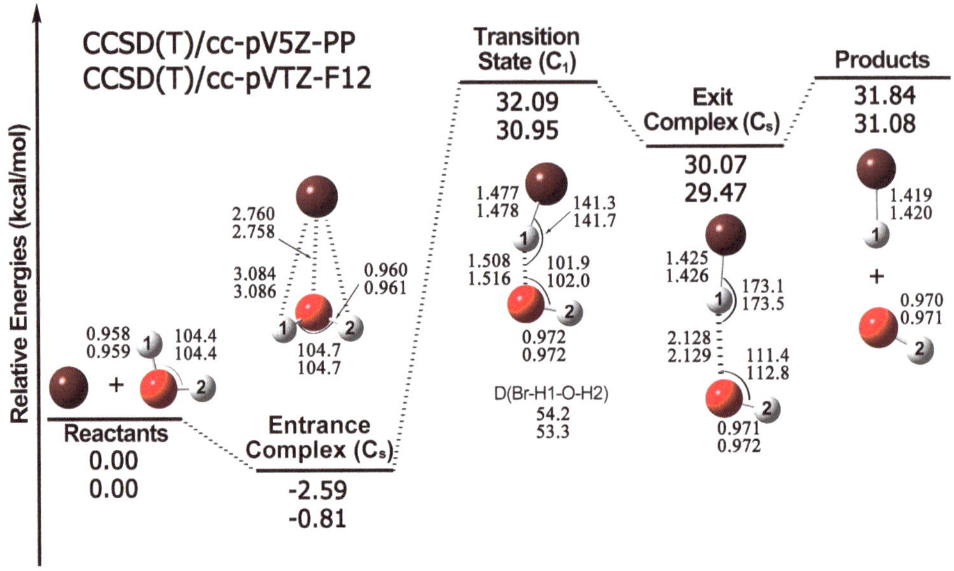

Fig. 4 Comparison between the present structural and energetic predictions (*upper*) and those of Bowman and coworkers (*lower*) [43]. Both methods include zero-point vibrational, scalar relativistic, and spin–orbit effects. The energetics is for absolute zero temperature

the entrance complex well, the binding energy between the Br atom and the H_2O molecule. The classical binding energy is 3.57 and 3.45 kcal/mol (Table 1), predicted by the CCSD(T)/cc-pVTZ and CCSD(T)/cc-pVQZ methods, respectively, in the present study. For comparison, Bowman's analogous predicted binding energy is 3.40 kcal/mol (36.01–32.61) by the CCSD(T)-F12a/cc-pVTZ-F12 method (Table 1 in ref. 43). In our research, after the ZPVE and scalar relativistic corrections, the binding energy becomes 2.59 kcal/mol (Table 1) predicted with the CCSD(T)/cc-pV5Z-PP method. Note our assumption that the spin–orbit correction would not affect this value, since the Br atom in entrance complex is ~3.0 Å far from the H_2O molecule. However, Bowman's final corrected value at the CCSD(T)-F12a/cc-pVTZ-F12 level is as small as 0.81 kcal/mol (based on Tables S1 and S2 in Supporting Information of ref. 43). The difference is due to the fact that the bromine atom spin–orbit coupling predicted in the research of Bowman is significantly quenched at the entrance complex. This seems counterintuitive (in the light of the long Br···H distance of 3.08 Å), but Bowman's treatment of the spin–orbit interaction is far more sophisticated than ours. The present 0 K heats of formation for the transition state, exit complex, and products are higher than those of Bowman by 1.14, 0.60, and 0.76 kcal/mol. We predict the transition state to lie 0.25 kcal above the products, while Bowman predicts the transition state to lie below the products by 0.13 kcal/mol. Given the differences in the theoretical methods, the agreement seems satisfactory.

Acknowledgments We thank Dr. Gabor Czakó for very helpful discussions. Correspondence with Professors Antonio de Oliveira-Filho and Joel Bowman are sincerely appreciated. This research was supported by Tianjin Natural Science Foundation (11JCYBJC14500), the National Natural Science Foundation of China (Grant No. 10904111), and China Postdoctoral Science Foundation (20100470792), as well as by the U.S. Department of Energy, Office of Basic Energy Sciences, Chemical Sciences Division, Fundamental Interactions Branch.

References

1. Sander SP, Friedl RR, Barker JR, Golden DM, Kurylo MJ, Wine PH, Abbatt J, Burkholder JB, Kolb CE, Moortgat GK, Huie RE, Orkin VL (2009) Chemical kinetics and photochemical data for use in stratospheric studies evaluation number 16. NASA, Jet Propulsion Laboratory, California Institute of Technology, Pasadena, CA
2. Holloway AM, Wayne RP (2010) Atmospheric chemistry. RSC Publishing, Cambridge
3. Clark DR, Simmons RF, Smith DA (1970) Trans Faraday Soc 66:1423–1435
4. Sims IR, Smith IWM, Clary DC, Bocherel P, Rowe BR (1994) J Chem Phys 101:1748–1751
5. Clary DC, Nyman G, Hernandez R (1994) J Chem Phys 101:3704–3714
6. Nizamov B, Setser DW, Wang H, Peslherbe GH, Hase WL (1996) J Chem Phys 105:9897–9911
7. Jaramillo VI, Gougeon S, Le Picard SD, Canosa A, Smith MA, Rowe BR (2002) Int J Chem Kinet 34:339–344
8. Che DC, Matsuo T, Yano Y, Bonnet L, Kasai T (2008) Phys Chem Chem Phys 10:1419–1423
9. Liu JY, Li ZS, Dai ZW, Huang XR, Sun CC (2001) J Phys Chem A 105:7707–7712
10. Tsai PY, Che DC, Nakamura M, Lin KC, Kasai T (2011) Phys Chem Chem Phys 13:1419–1423
11. Ravishankara AR, Wine PH, Langford AO (1979) Chem Phys Lett 63:479–484
12. Atkinson DB, Jaramillo VI, Snith MA (1997) J Phys Chem A 101:3356–3359
13. Bedjianian Y, Riffault V, Le Bras G, Poulet G (1999) J Photochem Photobiol A Chem 128:15–25
14. Wilson WE, O'Donovan JT, Fristrom RM (1969) 12th Symposium combustion. The Combustion Institute, Pettsburgh

15. Li J, Li YL, Guo H (2013) J Chem Phys 138:141102–141104
16. Purvis GD, Bartlett RJ (1982) J Chem Phys 76:1910–1918
17. Scuseria GE, Janssen CL, Schaefer HF (1988) J Chem Phys 89:7382–7387
18. Raghavachari K, Trucks GW, Pople JA, Head-Gordon M (1989) Chem Phys Lett 157:479–483
19. Dunning TH (1989) J Chem Phys 90:1007–1023
20. Kendall RA, Dunning TH, Harrison RJ (1992) J Chem Phys 96:6796–6806
21. Woon DE, Dunning TH (1993) J Chem Phys 98:1358–1371
22. Wilson AK, Woon DE, Peterson KA, Dunning TH (1999) J Chem Phys 110:7667–7676
23. Peterson KA, Figgen D, Goll E, Stoll H, Dolg M (2003) J Chem Phys 119:11113–11123
24. CFOUR, a quantum chemical program package written by Stanton JF, Gauss J, Harding ME, Szalay PG, with contributions from Auer AA, Bartlett RJ, Benedikt U, Berger C, Bernholdt DE, Bomble YJ, Cheng L, Christiansen O, Heckert M, Heun O, Huber C, Jagau T-C, Jonsson D, Jusélius J, Klein K, Lauderdale WJ, Matthews DA, Metzroth T, O'Neill DP, Price DR, Prochnow E, Ruud K, Schiffmann F, Schwalbach W, Stopkowicz S, Tajti A, Vázquez J, Wang F, Watts JD, with the integral packages MOLECULE (Almlöf J, Taylor PR), PROPS (Taylor PR), ABACUS (Helgaker T, Jensen HJ Aa, Jorgensen P, Olsen J), and ECP routines by Mitin AV, Wullen C van (2010)
25. Hoy AR, Bunker PR (1979) J Mol Spectrosc 74:1–8
26. Benedict WS, Gailar N, Plyler EK (1956) J Chem Phys 24:1139–1165
27. Rank DH, Rao BS, Wiggins TA (1965) J Mol Spectrosc 17:122–130
28. Huber KP, Herzberg G (1979) Constants of diatomic molecules. Van Nostrand Reinhold Company, New York
29. Barletta P, Shirin SV, Zobov NF, Polyansky OL, Tennyson J, Valeev EF, Császár AG (2006) J Chem Phys 125:204307
30. Lide DR (2014) Structure of free molecules in the gas phase, in Section 9, CRC Handbook of Chemistry and Physics, 94th Edition. In: Haynes WM, (ed), CRC Press/Taylor and Francis Group, Boca Raton, FL
31. Li G, Zhou L, Li Q-S, Xie Y, Schaefer HF (2012) Phys Chem Chem Phys 14:10891–10895
32. Guo Y, Zhang M, Xie Y, Schaefer HF (2013) J Chem Phys 139:041101–041103
33. Schaefer HF (1985) J Chem Phys 89:5336–5343
34. Werner HJ, Kallay M, Gauss J (2008) J Chem Phys 128:034305–034308
35. Metz B, Stoll H, Dolg M (2000) J Chem Phys 113:2563–2569
36. Czakó G (2013) J Chem Phys 138:134301–134305
37. Moore CE (1971) Atomic energy levels, Volume II, page 159, NSRDS-NBS 35, Washington, DC
38. Ruscic B, Wagner AF, Harding LB, Asher RL, Feller D, Dixon DA, Peterson KA, Song Y, Qian X, Ng C-Y, Liu J, Chen W, Schwenke DW (2002) J Phys Chem A 106:2727–2747
39. Ruscic B (2014) Active thermochemical tables, version 1.112, Argonne national laboratory, Argonne, IL. http://atct.anl.gov
40. Linstrom PJ, Mallard WG (eds) (2014) NIST Chemistry WebBook, NIST Standard Reference Database Number 69, National Institute of Standards and Technology, Gaithersburg MD
41. Shenyavskaya EA, Yungman VS (2004) J Phys Chem Ref Data 33:923–957
42. Lide DR (2014) "Dipole Moments" in section 9. In: Haynes WM (ed) CRC handbook of chemistry and physics, 94th edn. CRC Press/Taylor and Francis Group, Boca Raton, FL
43. de Oliveira-Filho AGS, Ornellas FR, Bowman JM (2014) J Phys Chem Lett 5:706–712

Theor Chem Acc (2014) 133:1473
DOI 10.1007/s00214-014-1473-x

REGULAR ARTICLE

Performance of density functionals for computation of core electron binding energies in first-row hydrides and glycine

Iogann Tolbatov · Daniel M. Chipman

Received: 14 February 2014 / Accepted: 21 February 2014 / Published online: 25 March 2014

Abstract A number of density functionals are benchmarked for calculation of 1s core electron binding energies for carbon, nitrogen, and oxygen nuclei in glycine, and for comparison in the first-row hydrides methane, ammonia, and water. The goal is to establish methods having potential to aid the analysis of experimental X-ray photoelectron spectra on compounds such as amino acids, DNA nucleosides, and large polypeptides in various environments. Several promising density functionals are identified that can reproduce experimental results within 0.2 eV on average for the absolute binding energies and also for the intramolecular and intermolecular shifts in the studied molecules.

Keywords Quantum chemistry · Density functional theory · Core electron binding energies · X-ray photoelectron spectroscopy · Glycine

Dedicated to the memory of Professor Isaiah Shavitt and published as part of the special collection of articles celebrating his many contributions.

Electronic supplementary material The online version of this article (doi:10.1007/s00214-014-1473-x) contains supplementary material, which is available to authorized users.

I. Tolbatov · D. M. Chipman (✉)
Notre Dame Radiation Laboratory, University of Notre Dame, Notre Dame, IN 46556, USA
e-mail: chipman.1@nd.edu

I. Tolbatov
Department of Physics, University of Notre Dame, Notre Dame, IN 46556, USA

1 Introduction

Biomolecules on metal surfaces are being widely studied for their practical applicability in areas such as biosensors [1–5], organic semiconductors [6–10], and biocatalysis [11–15]. X-ray photoelectron spectroscopy is a valuable nondestructive tool in this area of research, as recently demonstrated by its use to characterize the adsorption of thymidine [16], histidine [17], and histidine tripeptide [17] on gold and copper surfaces. To facilitate the interpretation of such photoelectron spectra, we have initiated a project to establish accurate and efficient methods for computation of core electron binding energies (CEBEs) in amino acids, DNA nucleotides, and polypeptides. As a first step, the present paper benchmarks the performance of various density functional methods for calculation of carbon, nitrogen, and oxygen CEBEs in the model compound glycine, and for comparison also in the first-row hydrides methane, ammonia, and water.

Glycine is the smallest of the 21 amino acids that form proteins [18]. It plays an essential role in living animals [19, 20] and plants [21, 22] and has even been found in space [23, 24] advocating the idea of panspermia. Its main function in animals is to serve as a building block for protein formation. Previous analyses of glycine in the gas phase include several experimental investigations [25–27]. A computational [28] investigation of gas phase glycine based on density functional theory showed an average absolute deviation from the experimental [25] CEBEs of just 0.2 eV, though the particular method used was later declared obsolete because it depended on a fortuitous cancelation of errors [29]. Shifts in the glycine CEBEs from the gas phase have also been studied in neutral, acidic, and basic solutions by means of X-ray photoelectron spectroscopy and by theoretical methods [30].

High accuracy in CEBE calculations can be reliably achieved with ab initio post-Hartree–Fock methods. For example, it was shown recently that the observed shifts of carbon 1s photoelectron energies in some organic compounds could be reproduced on average to within 0.03 eV by means of Møller–Plesset many-body perturbation theory and coupled cluster approaches [31]. However, such high accuracy methods are generally too expensive to be applied to large molecules.

The most promising approaches for efficient electronic structure calculations on large molecules are generally based on density functional theory with Kohn–Sham orbitals [32–35]. The most efficient such method for CEBEs is based on Koopmans' theorem, but this approach has quite limited accuracy [36–39]. Better accuracy can be obtained from calculations based on an effective core potential [40–45], an equivalent core approximation [46–48], a fractionally occupied transition state [49–52], or with a ΔSCF approach [29, 31, 53–57]. Time-dependent density functional theory is also widely used for CEBE calculation [58–62], wherein the best results are usually given with functionals having a long-range correction [63, 64].

A study using the ΔSCF approach for CEBEs of first-row atoms in many small gas phase molecules found that the quality of the results was very sensitive to the choice of functional [29], with the best performance found corresponding to a mean unsigned error (MUE) from experiment of just 0.16 eV. Similar accuracy was also found for the first-row atoms in gas phase thymine [65, 66].

In the present study, we use the ΔSCF approach with Hartree–Fock and a wide variety of pure and hybrid density functional approaches to study CEBEs in glycine, methane, ammonia, and water. Each approach is evaluated for its accuracy in reproducing experimental values for the absolute CEBEs in all four molecules, as well as for the intramolecular and intermolecular chemical shifts between like nuclei in the same or different molecules. Several promising candidates are found that can be recommended for future testing to establish accurate and efficient methods for calculations of CEBEs and their chemical shifts in large biomolecules.

2 Computational methods

Separate initial state and final core–hole state calculations provided ΔSCF values of the CEBEs. The maximum overlap method [67, 68] was used to prevent variational collapse of the final hole state. This simply replaces the usual aufbau criterion for occupying orbitals in each iteration with a criterion that the occupied orbitals be selected to overlap as much as possible with those of the previous iteration. The Ahlrichs' VTZ basis set [69] was used, based on the very good results it provided in a recent MCSCF-MRPT study of CEBEs in simple hydrides [39]. All calculations were performed with the Q-Chem 4.0 program [70, 71].

Hartree–Fock (HF) and a variety of exchange, correlation, and hybrid functionals were considered in this study. The local spin density approximation is represented by the exchange functional S (Slater and Dirac 1930) [72] together with the correlation functionals VWN (Vosko, Wilk, and Nusair) [73], PZ81 (Perdew and Zunger) [74], and PW92 (Perdew and Wang 1992) [75].

Generalized gradient approximation (GGA) functionals examined in this work include the exchange functionals B (Becke 1988) [76] and PW91 (Perdew and Wang 1991) [77, 78], together with the correlation functionals PW91 (Perdew and Wang 1991) [77, 78], P86 (Perdew 1986) [79], and LYP (Lee, Yang, and Parr) [80]. The correlation part of PBE (Perdew, Burke, and Ernzerhof 1996) [81] was used as well in the schemes PBEOP (PBE exchange with the reparametrized one-parameter progressive correlation functional) [82] and PBE0 (PBE hybrid with 25 % HF exchange) [83]. Other hybrid schemes considered were B97 [84], B97-1 [85], B97-2 [86] (a family of Becke 1997 hybrids), BOP (Becke 1988 with the OP correlation functional) [82], B3PW91 [87], B3LYP [88], and B3LYP5 [73, 76]. In addition, the empirical density functional EDF1 [89] was utilized.

Meta-GGA functionals incorporate the Laplacian of the electron density and generally depend on the electron kinetic energy density. These features allow a systematic improvement of results for many quantum chemical calculations. We have used the exchange functional BR89 (Becke-Roussel 1989 represented in analytic form) [87, 90] and the correlation functional B94 (Becke 1994) [90, 91]. Minnesota functionals tested are M05 [92], M05-2X [93], M06-L [94], M06-HF [95], M06 [96], M06-2X [96], and M11 [97]. Another global hybrid studied is BMK [98]. A hybrid extension of the nonempirical exchange–correlation TPSS (Tao, Perdew, Staroverov, and Scuseria) [99] and functional TPSSh [100] is also examined.

Several range-corrected and asymptotically corrected schemes have been studied as well. The long-range-corrected functionals considered include μBOP [101] (a long-range-corrected version of BOP) and the hybrids ωB97 and ωB97X [102]. The asymptotically corrected exchange–correlation potential considered is LB94, which is a linear combination of an LDA exchange–correlation potential and the LB (van Leeuwen and Baerends) [103] exchange potential. Short-range-corrected functionals that were specifically formulated for the treatment of core-excited states include SRC1 and SRC2 [104], which we have implemented with the BLYP functional and a recommended set of parameters for designing the shape of the long-range and short-range Hartree–Fock components (SRC1: $C_{\mathrm{SHF}} = 0.50$, $\mu_{\mathrm{SR}} = 0.56\ a_0^{-1}$, $C_{\mathrm{LHF}} = 0.17$, $\mu_{\mathrm{LR}} = 2.45\ a_0^{-1}$;

Table 1 Experimental values for CEBEs of C, N, and O nuclei in CH_4, NH_3, H_2O, and glycine

Molecule	Orbital	Binding energy measured (eV)	Experimental references [25, 105–108]
CH_4	C 1s	290.83	Nordfors et al. [105]
			Pireaux et al. [106]
NH_3	N 1s	405.60	Jolly et al. [107]
H_2O	O 1s	539.91	Nordfors et al. [108]
glycine	C (methyl carbon) 1s	292.25	Slaughter et al. [25]
	C (carboxyl carbon) 1s	295.15	
	N 1s	405.58	
	O (keto oxygen) 1s	538.2	
	O (hydroxyl oxygen) 1s	540.0	

Table 2 Experimental values for chemical shifts

Chemical shift	Experimental value (eV)
$\Delta E^C\left(\mathrm{Gly}_{\mathrm{methyl}\,C}, \mathrm{Gly}_{\mathrm{carboxyl}\,C}\right)$	−2.90
$\Delta E^O(\mathrm{Gly}_{\mathrm{keto}\,O}, \mathrm{Gly}_{\mathrm{hydroxyl}\,O})$	−1.75
$\Delta E^C\left(\mathrm{CH_4}, \mathrm{Gly}_{\mathrm{methyl}\,C}\right)$	1.42
$\Delta E^C\left(\mathrm{CH_4}, \mathrm{Gly}_{\mathrm{carboxyl}\,C}\right)$	4.32
$\Delta E^N(\mathrm{NH_3}, \mathrm{Gly}_N)$	−0.02
$\Delta E^O(\mathrm{H_2O}, \mathrm{Gly}_{\mathrm{keto\,O}})$	−1.71
$\Delta E^O\left(\mathrm{H_2O}, \mathrm{Gly}_{\mathrm{hydroxyl}\,O}\right)$	0.09

SRC2: $C_{\mathrm{SHF}} = 0.55$, $\mu_{\mathrm{SR}} = 0.69\ a_0^{-1}$, $C_{\mathrm{LHF}} = 0.08$, $\mu_{\mathrm{LR}} = 1.02\ a_0^{-1}$) [104].

For purposes of discussion, the various functionals considered are separated into seven categories, those being HF, LSDA, GGA, hybrid GGA, meta-GGA, meta-GGA hybrid, and range or asymptotically corrected functionals. Results from a larger collection that further includes some possible combinations that are mixed between the categories, for instance exchange LSDA + correlation GGA, exchange GGA + correlation meta-GGA, etc., were also calculated. Those mixed category results generally turned out to be less promising and so are relegated to the Electronic Supplementary Material, where details of the molecular geometries used are also documented.

3 Results and discussion

Table 1 lists experimentally measured values of the eight absolute CEBEs under consideration, those being the three carbon nuclei in methane and the methyl and carboxyl carbons in glycine, the two nitrogen nuclei in ammonia and glycine, and the three oxygen nuclei in water and the keto and hydroxyl oxygens in glycine. Table 2 shows the seven experimental values for the following intermolecular and intramolecular chemical shifts between like nuclei.

Table 3 MUE values in eV for CEBEs and chemical shifts computed by HF and various density functionals

Type of DFT functional	Density functional	MUE for	
		CEBEs	Chemical shifts
HF	HF	0.37	0.59
LSDA	SPW92	3.84	0.26
	SPZ81	3.76	0.26
	SVWN	3.83	0.26
GGA	BOP	0.12	0.17
	BLYP	0.27	0.16
	BP86	0.17	0.18
	BPBE	0.35	0.19
	BPW91	0.16	0.19
	EDF1	0.32	0.17
	PBEOP	0.12	0.17
	PW91LYP	0.24	0.16
	PW91P86	0.15	0.19
	PW91PBE	0.38	0.19
	PW91PW91	0.19	0.18
Hybrid GGA	B3LYP	0.46	0.26
	B3LYP5	0.35	0.26
	B3PW91	0.15	0.27
	B97	0.51	0.25
	B97-1	0.68	0.27
	B97-2	0.23	0.27
	PBE0	0.20	0.30
Meta-GGA	BR89B94	0.55	0.24
	M06L	0.11	0.19
Meta-GGA hybrid	BMK	2.04	0.35
	M05	0.24	0.30
	M052X	0.88	0.47
	M06	0.26	0.32
	M062X	0.26	0.42
	M06HF	0.45	0.61
	M11	0.92	0.94
	TPSSh	0.31	0.18
Range or asymptotically corrected	BLYP-SRC1	1.32	0.99
	BLYP-SRC2	3.93	0.37
	LB94	20.05	5.65
	μBOP	0.52	0.35
	ωB97	0.60	0.35
	ωB97X	0.73	0.35

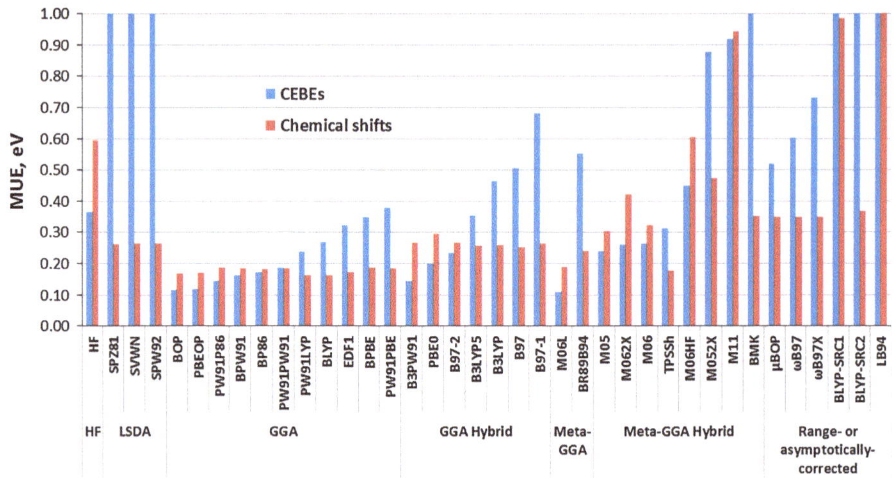

Fig. 1 MUE values for calculated CEBEs and chemical shifts

Intermolecular chemical shifts

$$\Delta E^{C}(\mathrm{CH_4}, \mathrm{Gly}_{\mathrm{methyl}\,C}) = BE(C\,1\mathrm{s\,in\,CH_4}) - BE(\mathrm{methyl}\,C\,1\mathrm{s\,in\,glycine}) \quad (1)$$

$$\Delta E^{C}(\mathrm{CH_4}, \mathrm{Gly}_{\mathrm{carboxyl}\,C}) = BE(C\,1\mathrm{s\,in\,CH_4}) - BE(\mathrm{carboxyl}\,C\,1\mathrm{s\,in\,glycine}) \quad (2)$$

$$\Delta E^{N}(\mathrm{NH_3}, \mathrm{Gly}_{N}) = BE(N\,1\mathrm{s\,in\,NH_3}) - BE(N\,1\mathrm{s\,in\,glycine}) \quad (3)$$

$$\Delta E^{O}(\mathrm{H_2O}, \mathrm{Gly}_{\mathrm{keto}\,O}) = BE(O\,1\mathrm{s\,in\,H_2O}) - BE(\mathrm{keto}\,O\,1\mathrm{s\,in\,glycine}) \quad (4)$$

$$\Delta E^{O}(\mathrm{H_2O}, \mathrm{Gly}_{\mathrm{hydroxyl}\,O}) = BE(O\,1\mathrm{s\,in\,H_2O}) - BE(\mathrm{hydroxyl}\,O\,1\mathrm{s\,in\,glycine}) \quad (5)$$

Intramolecular chemical shifts

$$\Delta E^{C}(\mathrm{Gly}_{\mathrm{methyl}\,C}, \mathrm{Gly}_{\mathrm{carboxyl}\,C}) = BE(\mathrm{methyl}\,C\,1\mathrm{s\,in\,glycine}) - BE(\mathrm{carboxyl}\,C\,1\mathrm{s\,in\,glycine}) \quad (6)$$

$$\Delta E^{O}(\mathrm{Gly}_{\mathrm{keto}\,O}, \mathrm{Gly}_{\mathrm{hydroxyl}\,O}) = BE(\mathrm{keto}\,O\,1\mathrm{s\,in\,glycine}) - BE(\mathrm{hydroxyl}\,O\,1\mathrm{s\,in\,glycine}) \quad (7)$$

Table 3 shows the MUEs from experiment for various computational methods. A full listing of the computed CEBEs and chemical shifts for each computational method is given in the Electronic Supplementary Material. The MUE results are also shown graphically in Fig. 1, which is cut off at 1 eV because methods having errors larger than that are of little interest. The functionals in Fig. 1 are ordered primarily according to the category and within each category according to the accuracy of the CEBE calculation.

HF treatment gives only modest accuracy for the CEBEs, with a MUE of 0.37 eV, and even less accuracy for the chemical shifts, with a MUE of 0.59 eV. The LSDA functionals' performance is not very good either, with very large MUEs of almost 4 eV for the CEBEs, although the chemical shifts are given fairly well with MUE of 0.26 eV.

The GGA category of functionals performs best overall. The BOP and PBEOP hybrids yield very good MUEs of 0.12 eV for CEBEs and 0.17 eV for chemical shifts. The BP86, BPW91, PW91P86, and PW91PW91 functionals also all yield MUEs of less than 0.20 eV for both CEBEs and chemical shifts. While the remaining GGA functionals have MUEs of 0.24 eV or more for the CEBEs, the errors are systematic enough to cancel somewhat in the chemical shifts, giving MUEs below 0.20 eV in all cases.

Among the hybrid GGAs, B3PW91 and PBE0 have MUEs of 0.20 eV or less for CEBEs, but these two along with all the other hybrid GGAs have somewhat higher MUEs of 0.25–0.30 eV for the chemical shifts.

The meta-GGA functional M06L performs very well, with MUEs of 0.11 eV for CEBES and 0.19 eV for chemical shifts, while the BR89B94 functional gives a high MUE of 0.55 eV for the CEBEs.

None of the meta-GGA hybrids perform consistently well, with MUEs for CEBEs all being 0.24 eV or more and MUEs for chemical shifts all being 0.30 eV or more.

Perhaps surprisingly, the range-corrected methods perform poorly, with MUEs of 0.52 eV or more for CEBEs and 0.35 eV or more for chemical shifts, and the asymptotically corrected LB94 functional performs very poorly for both CEBEs and chemical shifts.

Unsurprisingly, we can conclude that a functional's good performance in the calculation of CEBEs almost necessarily means a good description of chemical shifts, but the converse is not true.

4 Conclusion

A variety of pure and hybrid functionals has been studied for ΔSCF calculation of CEBEs and chemical shifts of C, N, and O nuclei in the first-row hydrides and glycine. Seven functionals are found that perform consistently well, giving MUEs of less than 0.20 eV for both CEBEs and chemical shifts. These are the GGA functionals BOP, PBEOP, PW91P86, BPW91, BP86, and PW91PW91 and the meta-GGA functional M06L. These can all be recommended for further study to identify suitable candidates for the interpretation of photoelectron spectra of large biomolecules.

Acknowledgments This material is based upon work supported by the Department of Energy under Award Number DE-SC0002216. This is Contribution No. NDRL-5005 from the Notre Dame Radiation Laboratory.

References

1. Ebersole RC, Miller JA, Moran JR, Ward MD (1990) Spontaneously formed functionally active avidin monolayers on metal surfaces: a strategy for immobilizing biological reagents and design of piezoelectric biosensors. J Am Chem Soc 112:3239–3241
2. Homola J (2003) Present and future of surface plasmon resonance biosensors. Anal Bioanal Chem 377:528–539
3. Du H, Disney MD, Miller BL, Krauss TD (2003) Hybridization-based unquenching of DNA hairpins on Au surfaces: prototypical "molecular beacon" biosensors. J Am Chem Soc 125:4012–4013
4. Haes AJ, Zou S, Schatz GC, Van Duyne RP (2004) Nanoscale optical biosensor: short range distance dependence of the localized surface plasmon resonance of noble metal nanoparticles. J Phys Chem B 108:6961–6968
5. Du H, Strohsahl CM, Camera J, Miller BL, Krauss TD (2005) Sensitivity and specificity of metal surface-immobilized "molecular beacon" biosensors. J Am Chem Soc 127:7932–7940
6. Gerischer H, Heller A (1991) The role of oxygen in photooxidation of organic molecules on semiconductor particles. J Phys Chem 95:5261–5267
7. Hill I, Rajagopal A, Kahn A, Hu Y (1998) Molecular level alignment at organic semiconductor-metal interfaces. Appl Phys Lett 73:662–664
8. Hamers RJ, Coulter SK, Ellison MD, Hovis JS, Padowitz DF, Schwartz MP, Greenlief CM, Russell JN (2000) Cycloaddition chemistry of organic molecules with semiconductor surfaces. Acc Chem Res 33:617–624
9. Gao Y (2010) Surface analytical studies of interfaces in organic semiconductor devices. Mater Sci Eng R-Rep 68:39–87
10. Lee WH, Park J, Sim SH, Lim S, Kim KS, Hong BH, Cho K (2011) Surface—directed molecular assembly of pentacene on monolayer graphene for high-performance organic transistors. J Am Chem Soc 133:4447–4454
11. Zayats M, Pogorelova SP, Kharitonov AB, Lioubashevski O, Katz E, Willner I (2003) Au nanoparticle-enhanced surface plasmon resonance sensing of biocatalytic transformations. Chem Eur J 9:6108–6114
12. Wu L, Payne GF (2004) Biofabrication: using biological materials and biocatalysts to construct nanostructured assemblies trends. Biotechnol 22:593–599
13. Willner I, Baron R, Willner B (2006) Growing metal nanoparticles by enzymes. Adv Mater 18:1109–1120
14. Mahapatro A, Johnson DM, Patel DN, Feldman MD, Ayon AA, Agrawal CM (2006) Surface modification of functional self–assembled monolayers on 316L stainless steel via lipase catalysis. Langmuir 22:901–905
15. Carley A, Davies P, Roberts M (2011) Oxygen transient states in catalytic oxidation at metal surfaces. Catal Today 169:118–124
16. Plekan O, Feyer V, Ptasińska S, Tsud N, Cháb V, Matolín V, Prince KC (2010) Photoemission study of thymidine adsorbed on Au (111) and Cu (110). J Phys Chem C 114:15036–15041
17. Feyer V, Plekan O, Ptasińska S, Iakhnenko M, Tsud N, Prince KC (2012) Adsorption of histidine and a histidine tripeptide on Au (111) and Au (110) from acidic solution. J Phys Chem C 116:22960–22966
18. Wagner I, Musso H (1983) New naturally—occurring amino-acids. Angew Chem Int Ed 22:816–828
19. Zhang Y, Li X, Peng L, Wang G, Ke K, Jiang Z (2012) Novel glycine—dependent inactivation of NMDA receptors in cultured hippocampal neurons. Neurosci Bull 28:550–560
20. Euden J, Mason SA, Viero C, Thomas NL, Williams AJ (2013) Investigations of the contribution of a putative glycine hinge to ryanodine receptor channel gating. J Biol Chem 288:16671–16679
21. McFarland JW, Ruess RW, Kielland K, Pregitzer K, Hendrick R (2010) Glycine mineralization in situ closely correlates with soil carbon availability across six North American forest ecosystems. Biogeochemistry 99:175–191
22. Rothstein DE (2010) Effects of amino-acid chemistry and soil properties on the behavior of free amino acids in acidic forest soils. Soil Biol Biochem 42:1743–1750
23. Belloche A, Menten KM, Comito C, Müller HSP, Schilke P, Ott J, Thorwirth S, Hieret C (2008) Detection of amino acetonitrile in Sgr B2(N). Astron Astrophys 482:179–196
24. Escamilla-Roa E, Moreno F (2013) Adsorption of glycine on cometary dust grains: II-effect of amorphous water ice. Planet Space Sci 75:1–10
25. Slaughter A, Banna M (1988) Core-photoelectron binding-energies of gaseous glycine—correlation with its proton affinity and gas–phase acidity. J Phys Chem 92:2165–2167
26. Plekan O, Feyer V, Richter R, Coreno M, de Simone M, Prince KC, Carravetta V (2007) Investigation of the amino acids glycine, proline, and methionine by photoemission spectroscopy. J Phys Chem A 111:10998–11005
27. Plekan O, Feyer V, Richter R, Coreno M, de Simone M, Prince KC, Carravetta V (2007) Photoemission and the shape of amino acids. Chem Phys Lett 442:429–433
28. Chong D (1996) Density functional calculation of core-electron binding energies of glycine conformers. Can J Chem 74:1005–1007
29. Takahata Y, Chong DP (2003) DFT calculation of core-electron binding energies. J Electron Spectrosc Relat Phenom 133:69–76
30. Ottosson N, Børve KJ, Spångberg D, Bergersen H, Sæthre LJ, Faubel M, Pokapanich W, Öhrwall G, Björneholm O, Winter B (2011) On the origins of core-electron chemical shifts of small biomolecules in aqueous solution: insights from photoemission and ab initio calculations of glycine(aq). J Am Chem Soc 133:3120–3130
31. Holme A, Borve KJ, Saethre LJ, Thomas TD (2011) Accuracy of calculated chemical shifts in carbon 1s ionization energies from single-reference ab initio methods and density functional theory. J Chem Theory Comput 7:4104–4114
32. Chong DP (ed) (1995) Recent advances in density functional methods: Part I. Recent advances in computational chemistry, vol 1. World Scientific Publishing Co Pte Ltd, Singapore
33. Chong DP (ed) (1997) Recent advances in density functional methods: Part II. Recent advances in computational chemistry, vol 1. World Scientific Publishing Co Pte Ltd, Singapore

34. Maruhn J, Reinhard P, Suraud E (2010) Density functional theory. In: Simple models of many-fermion systems. Springer, Berlin, pp 143–161
35. Nalewajski RF (2012) Density functional theory. In: Perspectives in electronic structure theory. Springer, Berlin, pp 255–368
36. Inoue C, Kaneda Y, Aida M, Endo K (1995) Simulation of XPS of poly (vinyl alcohol), poly (acrylic acid), poly (vinyl acetate), and poly (methyl methacrylate) polymers by an ab initio MO method using the model molecules. Polym J 27:300–309
37. Endo K, Maeda S, Aida M (1997) Simulation of C1s spectra of C- and O- containing polymers in XPS by ab initio MO calculations using model oligomers. Polym J 29:171–181
38. Bureau C, Chong DP, Endo K, Delhalle J, Lecayon G, Le Moël A (1997) Recent advances in the practical and accurate calculation of core and valence XPS spectra of polymers: from interpretation to simulation? Nucl Instr Meth Phys Res B 131:1–12
39. Shirai S, Yamamoto S, Hyodo S (2004) Accurate calculation of core-electron binding energies: multireference perturbation treatment. J Chem Phys 121:7586–7594
40. Pettersson LG, Wahlgren U, Gropen O (1983) Effective core potential calculations using frozen orbitals applications to transition metals. Chem Phys 80:7–16
41. Panas I, Siegbahn P, Wahlgren U (1987) Model studies of the chemisorption of hydrogen and oxygen on nickel surfaces. I. The design of a one-electron effective core potential which includes 3d relaxation effects. Chem Phys 112:325–337
42. Mattsson A, Panas I, Siegbahn P, Wahlgren U, Akeby H (1987) Model studies of the chemisorption of hydrogen and oxygen on Cu (100). Phys Rev B 36:7389–7401
43. Nyberg M, Hasselström J, Karis O, Wassdahl N, Weinelt M, Nilsson A, Pettersson LG (2000) The electronic structure and surface chemistry of glycine adsorbed on Cu (110). J Chem Phys 112:5420–5427
44. Oltedal V, Borve K, Saethre L, Thomas T, Bozek J, Kukk E (2004) Carbon 1s photoelectron spectroscopy of six-membered cyclic hydrocarbons. Phys Chem Chem Phys 6:4254–4259
45. Carroll TX, Thomas TD, Sæthre LJ, Børve KJ (2009) Additivity of substituent effects core-ionization energies and substituent effects in fluoromethylbenzenes. J Phys Chem A 113:3481–3490
46. Johansson B, Mårtensson N (1980) Core-level binding-energy shifts for the metallic elements. Phys Rev B 21:4427–4457
47. Grunze M, Brundle C, Tomanek D (1982) Adsorption and decomposition of ammonia on a W (110) surface: photoemission fingerprinting and interpretation of the core level binding energies using the equivalent core approximation. Surf Sci 119:133–149
48. Plashkevych O, Privalov T, Agren H, Carravetta V, Ruud K (2000) On the validity of the equivalent cores approximation for computing X-ray photoemission and photoabsorption spectral bands. Chem Phys 260:11–28
49. Hadjisavvas N, Theophilou A (1985) Rigorous formulation of Slater's transition-state theory for excited states. Phys Rev A 32:720–724
50. Wang S, Schwarz W (1996) Simulation of nondynamical correlation in density functional calculations by the optimized fractional orbital occupation approach: application to the potential energy surfaces of O and SO. J Chem Phys 105:4641–4648
51. Triguero L, Plashkevych O, Pettersson L, Agren H (1999) Separate state vs transition state Kohn-Sham calculations of X-ray photoelectron binding energies and chemical shifts. J Electron Spectrosc Relat Phenom 104:195–207
52. Goddard JD, Orlova G (1999) Density functional theory with fractionally occupied frontier orbitals and the instabilities of the Kohn-Sham solutions for defining diradical transition states: ring-opening reactions. J Chem Phys 111:7705–7712
53. Bagus PS (1965) Self-consistent-field wave functions for hole states of some Ne-like and Ar-like ions. Phys Rev 139:A619–A634
54. Naves de Brito A, Correia N, Svensson S, Ågren HA (1991) Theoretical study of X-ray photoelectron spectra of model molecules for polymethylmethacrylate. J Chem Phys 95:2965–2974
55. Takahashi O, Pettersson LG (2004) Functional dependence of core-excitation energies. J Chem Phys 121:10339–10345
56. Myrseth V, Sæthre LJ, Børve KJ, Thomas TD (2007) The substituent effect of the methyl group carbon 1s ionization energies, proton affinities, and reactivities of the methylbenzenes. J Org Chem 72:5715–5723
57. Saethre LJ, Borve KJ, Thomas TD (2011) Chemical shifts of carbon 1s ionization energies. J Electron Spectrosc Relat Phenom 183:2–9
58. Furche F, Ahlrichs R (2002) Adiabatic time-dependent density functional methods for excited state properties. J Chem Phys 117:7433–7447
59. Stener M, Fronzoni G, De Simone M (2003) Time dependent density functional theory of core electrons excitations. Chem Phys Lett 373:115–123
60. Fronzoni G, Stener M, Decleva P, Wang F, Ziegler T, Van Lenthe E, Baerends E (2005) Spin–orbit relativistic time dependent density functional theory calculations for the description of core electron excitations: $TiCl_4$ case study. Chem Phys Lett 416:56–63
61. Rappoport D, Furche F (2005) Analytical time-dependent density functional derivative methods within the RI-J approximation, an approach to excited states of large molecules. J Chem Phys 122(064105):1–8
62. Chong DP (2005) Density functional calculation of K-shell spectra of small molecules. J Electron Spectrosc Relat Phenom 148:115–121
63. Yanai T, Tew DP, Handy NC (2004) A new hybrid exchange-correlation functional using the Coulomb-attenuating method (CAM-B3LYP). Chem Phys Lett 393:51–57
64. Besley NA, Peach MJG, Tozer DJ (2009) Time-dependent density functional theory calculations of near-edge X-ray absorption fine structure with short-range corrected functionals. Phys Chem Chem Phys 11:10350–10358
65. Takahata Y, Okamoto AK, Chong DP (2006) DFT calculation of core-electron binding energies of pyrimidine and purine bases. Int J Quantum Chem 106(13):2581–2586
66. Takahata Y, Marques ADS (2010) Accurate core-electron binding energy shifts from density functional theory. J Electron Spectrosc Relat Phenom 178:80–87
67. Gilbert ATB, Besley NA, Gill PMW (2008) Self-consistent field calculations of excited states using the maximum overlap method (MOM). J Phys Chem A 112:13164–13171
68. Hanson-Heine MWD, George MW, Besley NA (2013) Calculating excited state properties using Kohn-Sham density functional theory. J Chem Phys 138(064101):1–8
69. Schafer A, Horn H, Ahlrichs R (1992) Fully optimized contracted Gaussian-basis sets for atoms Li to Kr. J Chem Phys 97:2571–2577
70. Shao Y, Molnar LF, Jung Y, Kussmann J, Ochsenfeld C, Brown ST, Gilbert AT, Slipchenko LV, Levchenko SV, O'Neill DP et al (2006) Advances in methods and algorithms in a modern quantum chemistry program package. Phys Chem Chem Phys 8:3172–3191
71. Krylov AI, Gill PMW (2013) Q-Chem: an engine for innovation. WIREs Comput Mol Sci 3:317–326
72. Dirac P (1930) Note on exchange phenomena in the Thomas atom. Proc Cambridge Philos Soc 26:376–385
73. Vosko S, Wilk L, Nusair M (1980) Accurate spin-dependent electron liquid correlation energies for local spin-density calculations—a critical analysis. Can J Phys 58:1200–1211

74. Perdew J, Zunger A (1981) Self-interaction correction to density-functional approximations for many-electron systems. Phys Rev B 23:5048–5079
75. Perdew J, Wang Y (1992) Accurate and simple analytic representation of the electron-gas correlation-energy. Phys Rev B 45:13244–13249
76. Becke A (1988) Density-functional exchange-energy approximation with correct asymptotic-behavior. Phys Rev A 38:3098–3100
77. Perdew JP (1991) Unified theory of exchange and correlation beyond the local density approximation. In: Ziesche P, Eschrig H (eds) Electronic structure of solids' 91. Akademie-Verlag, Berlin, pp 11–20
78. Perdew JP, Wang Y (1992) Accurate and simple analytic representation of the electron-gas correlation energy. Phys Rev B 45:244–249
79. Perdew J (1986) Density-functional approximation for the correlation-energy of the inhomogeneous electron-gas. Phys Rev B 33:8822–8824
80. Lee C, Yang W, Parr R (1988) Development of the Colle-Salvetti correlation-energy formula into a functional of the electron-density. Phys Rev B 37:785–789
81. Perdew J, Burke K, Ernzerhof M (1996) Generalized gradient approximation made simple. Phys Rev Lett 77:3865–3868
82. Tsuneda T, Suzumura T, Hirao K (1999) A new one-parameter progressive Colle–Salvetti-type correlation functional. J Chem Phys 110:10664–10678
83. Adamo C, Scuseria G, Barone V (1999) Accurate excitation energies from time-dependent density functional theory: assessing the PBE0 model. J Chem Phys 111:2889–2899
84. Becke A (1997) Density-functional thermochemistry. V. Systematic optimization of exchange-correlation functionals. J Chem Phys 107:8554–8560
85. Hamprecht F, Cohen A, Tozer D, Handy N (1998) Development and assessment of new exchange-correlation functionals. J Chem Phys 109:6264–6271
86. Wilson PJ, Bradley TJ, Tozer DJ (2001) Hybrid exchange-correlation functional determined from thermochemical data and ab initio potentials. J Chem Phys 115:9233–9242
87. Becke A, Roussel M (1989) Exchange holes in inhomogeneous systems: a coordinate-space model. Phys Rev A 39:3761–3767
88. Stephens P, Devlin F, Chabalowski C, Frisch MJ (1994) Ab initio calculation of vibrational absorption and circular dichroism spectra using density functional force fields. J Phys Chem 98:11623–11627
89. Adamson RD, Gill PM, Pople JA (1998) Empirical density functionals. Chem Phys Lett 284:6–11
90. Proynov E, Gan Z, Kong J (2008) Analytical representation of the Becke-Roussel exchange functional. Chem Phys Lett 455:103–109
91. Becke A (1994) Thermochemical tests of a kinetic-energy dependent exchange-correlation approximation. Int J Quant Chem S28:625–632
92. Zhao Y, Schultz NE, Truhlar D (2005) Exchange-correlation functional with broad accuracy for metallic and nonmetallic compounds, kinetics, and noncovalent Interactions. J Chem Phys 123(161103):1–4
93. Zhao Y, Schultz N, Truhlar D (2006) Design of density functionals by combining the method of constraint satisfaction with parametrization for thermochemistry, thermochemical kinetics, and noncovalent interactions. J Chem Theory Comput 2:364–382
94. Zhao Y, Truhlar DG (2006) A new local density functional for main-group thermochemistry, transition metal bonding, thermochemical kinetics, and noncovalent interactions. J Chem Phys 125(194101):1–18
95. Zhao Y, Truhlar DG (2006) Density functional for spectroscopy: no long-range self-interaction error, good performance for Rydberg and charge-transfer states, and better performance on average than B3LYP for ground states. J Phys Chem A 110:13126–13130
96. Zhao Y, Truhlar DG (2008) The M06 suite of density functionals for main group thermochemistry, thermochemical kinetics, noncovalent interactions, excited states, and transition elements: two new functionals and systematic testing of four M06-class functionals and 12 other functionals. Theor Chem Acc 120:215–241
97. Peverati R, Truhlar DG (2011) Improving the accuracy of hybrid meta-GGA density functionals by range separation. J Phys Chem Lett 2:2810–2817
98. Boese A, Martin J (2004) Development of density functionals for thermochemical kinetics. J Chem Phys 121:3405–3416
99. Tao J, Perdew J, Staroverov V, Scuseria G (2003) Climbing the density functional ladder: nonempirical meta-generalized gradient approximation designed for molecules and solids. Phys Rev Lett 91(146401):1–4
100. Staroverov V, Scuseria G, Tao J, Perdew J (2003) Comparative assessment of a new nonempirical density functional: molecules and hydrogen-bonded complexes. J Chem Phys 119:12129–12137
101. Song J, Hirosawa T, Tsuneda T, Hirao K (2007) Long-range corrected density functional calculations of chemical reactions: redetermination of parameter. J Chem Phys 126(154105):1–7
102. Chai J, Head-Gordon M (2008) Systematic optimization of long-range corrected hybrid density functionals. J Chem Phys 128(084106):1–15
103. Van Leeuwen R, Baerends E (1994) Exchange-correlation potential with correct asymptotic behavior. Phys Rev A 49:2421–2431
104. Besley NA, Peach MJ, Tozer DJ (2009) Time-dependent density functional theory calculations of near-edge X-ray absorption fine structure with short-range corrected functionals. Phys Chem Chem Phys 11:10350–10358
105. Nordfors D, Agren H (1991) Calculation of core electron-binding energies from structural formulas. J Electron Spectrosc Relat Phenom 56:1–11
106. Pireaux J, Svensson S, Basilier E, Malmqvist P, Gelius U, Gaudano R, Siegbahn K (1976) Core-electron relaxation energies and valence-band formation of linear alkanes studied in gas-phase by means of electron-spectroscopy. Phys Rev A 14:2133–2145
107. Jolly W, Bomben K, Eyermann C (1984) Core-electron binding-energies for gaseous atoms and molecules. At Data Nucl Data Tables 31:433–493
108. Nordfors D, Nilsson A, Martensson N, Svensson S, Gelius U, Agren H (1981) X-ray excited photoelectron-spectra of free molecules containing oxygen. J Electron Spectrosc Relat Phenom 56:117–164

Theor Chem Acc (2014) 133:1493
DOI 10.1007/s00214-014-1493-6

REGULAR ARTICLE

Why edge inversion? Theoretical characterization of the bonding in the transition states for inversion in $F_nNH_{(3-n)}$ and $F_nPH_{(3-n)}$ (n = 0–3)

Lu T. Xu · Tyler Y. Takeshita · Thom H. Dunning Jr.

Received: 13 February 2014 / Accepted: 21 April 2014 / Published online: 14 May 2014

Abstract As first noted by Dixon et al. (J Am Chem Soc 108:2461–2462, 1986), heavily fluorinated pyramidal phosphorus compounds, e.g., $F_nPH_{(3-n)}$ with $n > 1$, invert through a T-shaped transition state (edge inversion) rather than the D_{3h}-like transition states (vertex inversion) found in the corresponding nitrogen compounds and less fluorinated phosphorus compounds. Subsequent studies by Dixon and coworkers established that this is a general phenomenon and has important chemical consequences. But what is the reason for the change in the structure of the transition state? Recent theoretical investigations have resulted in the discovery of a new type of chemical bond, the recoupled pair bond. In particular, it was found that recoupled pair bond dyads account for the hypervalency of the elements beyond the first row. In this paper, we show that recoupled pair bond dyads also account for the existence of the edge inversion pathway in heavily fluorinated phosphorus compounds and likely account for the presence of the lower energy inversion pathways in pyramidal compounds of other elements beyond the first row.

Keywords Edge inversion · Vertex inversion · Transition state · Recoupled pair bond · Recoupled pair bond dyad · Generalized valence bond (GVB) theory

Dedicated to the memory of Professor Isaiah Shavitt and published as part of the special collection of articles celebrating his many contributions.

L. T. Xu · T. Y. Takeshita · T. H. Dunning Jr. (✉)
Department of Chemistry, University of Illinois at Urbana-Champaign, 600 S Mathews Avenue, Urbana, IL 61801, USA
e-mail: thdjr@illinois.edu

L. T. Xu
e-mail: luxu1@illinois.edu

1 Introduction

The structures, energetics and properties of molecules formed from elements in the first row of the periodic table, Li to Ne, can be dramatically different from those formed from elements in the subsequent rows—the so-called *first-row anomaly*. The anomaly manifests itself in a number of ways, such as the inability of the first row p-block elements to form hypervalent species. N and P are an example of this anomaly: P is able to form hypervalent molecules such as PF_5 and PCl_5, while N only forms NF_3 and NCl_3. Another manifestation of the difference between N and P has drawn quite a bit of attention. The ground states of NH_3, NF_3, PH_3 and PF_3 are all pyramidal, as expected. However, NH_3, NF_3 and PH_3 invert through a transition state with D_{3h} symmetry, while the transition state for inversion in PF_3 is T-shaped with C_{2v} symmetry.

Investigations of the T-shaped pnictogen transition states for inversion were inspired by the experimental synthesis of the first molecule containing a T-shaped, tricoordinated hypervalent phosphorous structure, 5-aza-2,8-dioxa-3,7-di-tert-butyl-l-phosphabicyclo[3.3.0]octa-3,6-diene (ADPO) [1, 2]. Another hypervalent 10-P-3 compound, an intermediate, was discovered by Lochschmidt and Schmidpeter at approximately the same time [3]. Following up on this discovery, Dixon and co-workers [4–8] explored the structures of the transition states for inversion in phosphorus compounds. They performed calculations on PH_3, FPH_2, F_2PH and PF_3 and found that F_2PH and PF_3 have T-shaped transition states while FPH_2 and PH_3 have D_{3h}–like transition states. They pointed out that when inversion occurs through a T-shaped structure, the lone pair orbital is in the molecular plane, not perpendicular to the plane as in the D_{3h}-like transition states. Inversion through a T-shaped transition state is referred to as *edge inversion*, while inversion through a D_{3h}-like transition state is referred to as *vertex inversion*.

Arduengo, Dixon and Roe experimentally verified the edge inversion mechanism in a bi-cyclo[3.3.0] octane ring system that is a saturated analog of ADPO by measuring the barrier to inversion at the tricoordinated phosphorous atom [5]. Although ADPO contains a T-shaped phosphorus structure in its ground state, its saturated analog contains a pyramidal phosphorus structure in its ground state and a T-shaped phosphorus structure in its transition state. They argued that the conjugated π system of ADPO stabilizes the T-shaped structure while the saturated analog of ADPO is not so stabilized. The electronic structure of all of these molecules can be easily understood once one recognizes the ability of phosphorous to form both covalent and recoupled pair bonds in simple tricoordinated molecules such as PF_3 [9], as will be shown in this paper.

Although edge inversion has been well established for pyramidal molecules with central atoms beyond the first row and with electronegative ligands, the reason for the difference in inversion pathway upon fluorination has been the subject of some debate. Both perturbation molecular orbital arguments involving the HOMO–LUMO gap [10] as well as pseudo-Jahn–Teller effects [11] have been used to rationalize the T-shaped transition state structures. Woon and Dunning, on the other hand, noted that the two axial bonds in the transition state for inversion in PF_3 closely resemble those in the $PF_2(A^2\Pi)$ state, which has a recoupled pair bond dyad, and concluded that the ability of the P atom to form a very stable recoupled pair bond dyad is the source of this anomaly [9]. In fact, the formation of recoupled pair bonds is the basis for the bonding in hypervalent molecules, as shown in studies on SF_n [12], PF_n [9], ClF_n [13] and a number of related compounds. The T-shaped transition state for inversion of PF_3 should, in fact, be considered hypervalent, even though it is tricoordinated, because it possesses one of the hallmarks of hypervalent compounds—a recoupled pair bond dyad.

In the present study, we systematically investigated the ground and transition states for inversion of NH_3, PH_3 and their F substituents, $F_nNH_{(3-n)}$ and $F_nPH_{(3-n)}$ ($n = 1$–3), and show that other than PF_3, F_2PH is the only species with a T-shaped transition state. Accurate predictions of the structures and energies of the $F_nNH_{(3-n)}$ and $F_nPH_{(3-n)}$ ($n = 0$–3) species were obtained by using coupled cluster methods [14–17] with large correlation consistent basis sets [18–21]. We then used generalized valence bond (GVB) theory [22, 23] to obtain insights into the nature of the bonding in these molecules and explain the similarities and differences in the structural and energetic trends of the N and P species.

The layout of the paper is as follows: Sect. 2 describes the computational methods we used for this study, including a brief overview of GVB theory; Sect. 3 presents the optimized geometries and energetics for the ground and transition states of $F_nNH_{(3-n)}$ and $F_nPH_{(3-n)}$, proposes the three major questions we need to answer, analyzes the GVB wave functions of related atoms and molecules, discusses the role recoupled pair bonding plays in the transition states and answers the questions raised; and finally, we summarize our findings in Sect. 4.

2 Computational methods

The calculations presented in this study were performed with the Molpro suite of quantum chemical programs (version 2008.1 and 2010.1) [24]. In order to provide accurate geometries and energetics, the structures, energies and frequencies of the $F_nNH_{(3-n)}$ and $F_nPH_{(3-n)}$ molecules were determined with single-reference restricted singles and doubles coupled cluster theory with perturbative triples [CCSD(T)] [14–17]. For geometry optimizations and energies, augmented correlation consistent basis sets of quadruple zeta quality (aug-cc-pVQZ) were used for the first row atoms (H, N and F), and the corresponding d-function augmented set [aug-cc-pV(Q + *d*)Z] was used for the second row P atom. For frequency calculations, aug-cc-pVTZ basis sets were used for the first row atoms and an aug-cc-pV(T + *d*)Z basis set was used for the P atom [18–21]. The frequency calculations were performed at geometries optimized using the same basis sets. The shorthand notation AVXZ (X = T, Q) will be used to represent the sets of a specific quality (including the extra d-function on the P atom). The frequency calculations enabled us to conclusively identify the ground states (all real frequencies) and transition states (one imaginary frequency) for all of the species.

The inversion barrier for each molecule is calculated as the difference between the electronic energy of the transition state and the ground state, i.e., it does not include the zero point energy correction. We did this to focus on the effect of the changes in the electronic structure of the molecules upon inversion; the reader can easily correct these numbers using the vibrational frequencies given in the tables. Energies are quoted with two significant figures after the decimal place for comparison with other theoretical calculations.

Multireference configuration interaction (MRCI) calculations based on valence complete active space self-consistent field (CASSCF) wave functions with the quadruples corrections (+Q) [25–30] were used to determine the excitation energies of the N and P atoms using the AVQZ basis set.

To characterize the nature of the bonding in the $F_nNH_{(3-n)}$ and $F_nPH_{(3-n)}$ species, we used the generalized valence bond (GVB) method [22, 23]. The GVB wave function is well suited to analysis of the bonding in molecules as it describes bond-breaking processes properly. The GVB wave function is also inherently more accurate than the Hartree–Fock (HF) wave function, including the most important non-dynamical correlation effects represented in a valence CASSCF wave function. At the same time, the GVB wave function is concise, offering a clear physical picture of the electronic structure of a molecule that is readily connected with those of the atoms or fragments of which it is composed.

In the GVB framework, a covalent bond is formed by singlet coupling two electrons in a pair of overlapping, singly occupied GVB orbitals concentrated on the two atoms involved in the bond. A recoupled pair bond, on the other hand, results from a two-center, three-electron interaction. Nevertheless, it is also easily described in GVB theory because each electron has its own orbital [31]. The recoupled pair bond dyad, which is of particular interest here, is simply two singlet-coupled bonding pairs—one from the original recoupled pair bond and the other from a covalent bond formed with the electron left over from forming the recoupled pair bond. The remarkable stability of the recoupled pair bond dyad is a direct result of the ionicity of these two bonds—a dyad is only found when the two ligands are very ionic [32, 33]. The GVB orbitals, orbital overlaps and spin coupling functions provide a concise picture of the electronic structure of the molecule. The fully variational GVB method is equivalent to the spin-coupled VB method [34], and the CASVB [35–37] program implemented in Molpro was used to perform the calculations with the AVQZ basis sets.

The GVB/spin-coupled VB wave function for a molecular system of n_a active electrons with total spin S and projection M is:

$$\Psi_{\mathrm{GVB}} = \hat{a}\phi_{d1}\phi_{d1}\phi_{d2}\phi_{d2}\cdots\phi_{dn_d}\phi_{dn_d}\varphi_{a1}\varphi_{a2}\cdots\varphi_{an_a}\alpha\beta\alpha\beta\ldots\alpha\beta\Theta^{n_a}_{S,M}. \quad (1)$$

In the above equation, $\hat{a}$ is the antisymmetrizer; the set of orbitals, $\{\phi_{di}\}$, are the set of n_d doubly occupied core and valence orbitals, and the set of orbitals, $\{\varphi_{ai}\}$, are the set of n_a singly occupied active valence orbitals. The total number of electrons is $N_e = 2n_d + n_a$. The doubly occupied valence orbitals do not directly participate in bonding, although, as we shall see, they can affect which type of bonds are formed. The active orbitals are distinct, singly occupied and non-orthogonal orbitals. The spatial product of orbitals in Eq. (1) is multiplied by a product of $\alpha\beta$ spin functions associated with the doubly occupied orbitals times a spin function, $\Theta^{n_a}_{S,M}$, for the electrons in the active orbitals. This spin function is a linear combination of spin eigenfunctions, also known as spin basis functions, which represent the unique ways in which the spins of the n_a electrons in the active orbitals can be coupled to give a total spin of S. Kotani spin functions [38] were used in our study. The Kotani functions are orthogonal to each other, and, therefore, the contribution of each spin function to the total GVB wave function, the weight w_k, is simply the square of its coefficient.

We found that one of the spin eigenfunctions was dominant for all of the molecular systems that we studied here, and it is the perfect pairing (PP) function:

$$\Theta^{n_a}_{S,M,\mathrm{PP}} = \frac{(\alpha\beta-\beta\alpha)}{\sqrt{2}}\frac{(\alpha\beta-\beta\alpha)}{\sqrt{2}}\cdots\frac{(\alpha\beta-\beta\alpha)}{\sqrt{2}} \quad (2)$$

As shown in Eq. (2), the perfect pairing spin function singlet couples all of the active electrons into electron pairs. A pair of singlet-coupled orbitals can describe a lone pair, in which case the orbitals are highly overlapping and concentrated on one atom. Or the orbital pair can describe a bond if the orbitals are overlapping and concentrated on two atoms. The GVB calculations for the states reported in this paper are all 6-in-6 calculations, i.e., there are six active electrons in six GVB orbitals. All spin functions were included in the calculations, although, as noted, the PP spin function was always dominant ($w_{\mathrm{PP}} = 0.91$–$0.99+$). This means that there are three pairs of singlet-coupled orbitals, or, since we kept the lone pair orbitals doubly occupied in all calculations, three bonds in both the ground and transition states of the $F_nNH_{(3-n)}$ and $F_nPH_{(3-n)}$ molecules.

3 Results and discussion

3.1 Similarities and differences in the $F_nNH_{(n-3)}$ and $F_nPH_{(3-n)}$ ($n = 0$–3) ground states and transition states for inversion

3.1.1 Molecular structures

The optimized structures and geometrical parameters of the ground state (GS, X^1A_1) and transition state (TS, $^1A_1$) of all of the $F_nNH_{(3-n)}$ and $F_nPH_{(3-n)}$ ($n = 0$–3) molecules are tabulated in Tables 1, 2, 3 and 4 and illustrated in Fig. 1. The ground states of $F_nNH_{(3-n)}$ and $F_nPH_{(3-n)}$ are pyramidal with the singly occupied 1s orbitals of $H(^2S)$ and the singly occupied 2p orbitals of $F(^2P)$ forming normal 2-electron, 2-center covalent bonds with the three singly occupied 2p and 3p orbitals on $N(^4S)$ and $P(^4S)$, respectively. However, the inversion transition states fall into two different structural categories. NH_3, NF_3 and PH_3 have planar transition states with D_{3h} symmetry ($\theta_{\mathrm{XYX}} = 120°$). The transition states of FNH_2 and FPH_2 are planar with C_{2v} symmetry and similar bond angles: $\theta_{\mathrm{HNF}} = 114.0°$ and

Table 1 The geometries, frequencies and total energies for the ground states (GS) of the $F_nNH_{(3-n)}$ ($n = 0$–3) molecules from CCSD(T) calculations with the indicated basis sets

		R_{NH}	R_{NF}	θ_{HNH}	θ_{FNH}	θ_{FNF}	ω_1	ω_2	ω_3	ω_4	ω_5	ω_6	E
NH_3	AVTZ	1.015		106.4			1,063	1,672	1,672	3,463	3,592	3,592	−56.480527
	AVQZ	1.013		106.5			1,059	1,674	1,674	3,476	3,606	3,607	−56.495733
FNH_2	AVTZ	1.021	1.433	104.8	101.1		927	1,270	1,341	1,623	3,410	3,507	−155.541406
	AVQZ	1.019	1.426	105.0	101.3								−155.582143
F_2NH	AVTZ	1.027	1.400		99.8	103.0	504	910	992	1,343	1,467	3,367	−254.621481
	AVQZ	1.025	1.394		100.0	103.1							−254.687975
NF_3	AVTZ	1.375				101.7	498	498	654	924	924	1,045	−353.714384
	AVQZ	1.369				101.8							−353.806789

Distances are in Å, angles in degrees, frequencies in cm^{-1} and energies in Hartrees

Table 2 The geometries, frequencies and total energies for the inversion transition states (TS) of the $F_nNH_{(3-n)}$ ($n = 0$–3) molecules from CCSD(T) calculations with the indicated basis sets

		R_{NH}	R_{NF}	θ_{HNH}	θ_{FNH}	θ_{FNF}	ω_1	ω_2	ω_3	ω_4	ω_5	ω_i	E
NH_3	AVTZ	0.998		120.0			1,582	1,582	3,626	3,837	3,837	868	−54.471739
	AVQZ	0.996		120.0			1,582	1,583	3,633	3,846	3,847	850	−56.487226
FNH_2	AVTZ	0.996	1.385	132.0	114.0		1,078	1,189	1,552	3,674	3,898	1,148	−155.515766
	AVQZ	0.995	1.381	132.0	114.0								−155.557208
F_2NH	AVTZ	0.998	1.352		123.7	112.6	493	1,092	1,171	1,336	3,786	1,373	−254.559030
	AVQZ	0.997	1.348		123.7	112.6							−254.626575
NF_3	AVTZ		1.343			120.0	420	420	802	1,313	1,313	1,192	−353.581416
	AVQZ		1.338			120.0							−353.674851

Distances are in Å, angles in degrees, frequencies in cm^{-1} and energies in Hartrees. ω_i denotes the imaginary frequency associated with transition state motion

Table 3 The geometries, frequencies and total energies for the ground states (GS) of the $F_nPH_{(3-n)}$ ($n = 0$–3) molecules from CCSD(T) calculations with the indicated basis sets

		R_{PH}	R_{PF}	θ_{HPH}	θ_{FPH}	θ_{FPF}	ω_1	ω_2	ω_3	ω_4	ω_5	ω_6	E
PH_3	AVTZ	1.417		93.5			1,013	1,143	1,143	2,409	2,416	2,416	−342.699014
	AVQZ	1.415		93.6			1,012	1,143	1,143	2,421	2,429	2,429	−342.710887
FPH_2	AVTZ	1.420	1.613	92.1	97.7		809	927	975	1,138	2,378	2,383	−441.866170
	AVQZ	1.419	1.607	92.1	97.8								−441.905727
F_2PH	AVTZ	1.424	1.591		95.6	98.8	354	841	848	971	1,023	2,344	−541.056075
	AVQZ	1.423	1.585		95.7	98.9							−541.123513
PF_3	AVTZ		1.572			97.4	344	344	484	862	862	894	−640.266595
	AVQZ		1.567			97.5							−640.362060

Distances are in Å, angles in degrees, frequencies in cm^{-1} and energies in Hartrees

$\theta_{HPF} = 114.9°$. The transition state of F_2NH is also of C_{2v} symmetry with θ_{HNF} at 123.7°. These latter three C_{2v} transition states are, in fact, D_{3h}-like and fall into the same structural category as NH_3, NF_3 and PH_3.

The transition states for F_2PH and PF_3, on the other hand, are outliers. In the transition state for inversion of F_2PH, the central P atom and two of the F atoms, the axial F atoms, lie almost on a straight line with $\theta_{FPF} = 169.8°$ and $\theta_{HPF} = 84.9°$. This quasi-linear structure is also present in PF_3, where the FPF axial angle is 176.0° and the angle between the axial and equatorial F atom is 88.0°. These two transition states are referred to as T-shaped transition states as first characterized by Dixon and coworkers [4–8] and are referred to as *edge* transition states in contrast to the transition states for the other six $F_nNH_{(3-n)}$ and $F_nPH_{(3-n)}$ molecules, which are referred to as *vertex* transition states.

Table 4 The geometries, frequencies and total energies for the inversion transition states (TS) of the $F_nPH_{(3-n)}$ (n = 0–3) molecules from CCSD(T) calculations with the indicated basis sets

		R_{PH}	R_{PF}	θ_{HPH}	θ_{FPH}	θ_{FPF}	ω_1	ω_2	ω_3	ω_4	ω_5	ω_i	E
PH_3	AVTZ	1.380		120.0			1,017	1,017	2,597	2,676	2,676	1,098	−342.645628
	AVQZ	1.379		120.0			1,023	1,023	2,605	2,684	2,684	1,096	−342.657904
FPH_2	AVTZ	1.376	1.590	130.2	114.9		719	871	1,001	2,626	2,712	1,255	−441.785348
	AVQZ	1.375	1.585	130.2	114.9								−441.825182
F_2PH	AVTZ	1.414	1.655(ax)		84.9	169.8	422	582	740	1,319	2,404	348	−540.975049
	AVQZ	1.413	1.649(ax)		84.9	169.8							−541.040581
PF_3	AVTZ		1.566(eq)			88.0	389	503	578	769	888	312	−640.182433
			1.635(ax)										
	AVQZ		1.561(eq)			88.0							−640.276004
			1.630(ax)										

Distances are in Å, angles in degrees, frequencies in cm^{-1} and energies in Hartrees. ω_i denotes the imaginary frequency associated with transition state motion. The labels, "eq" and "ax," refer to the equatorial and axial bonds in F_2PH and PF_3

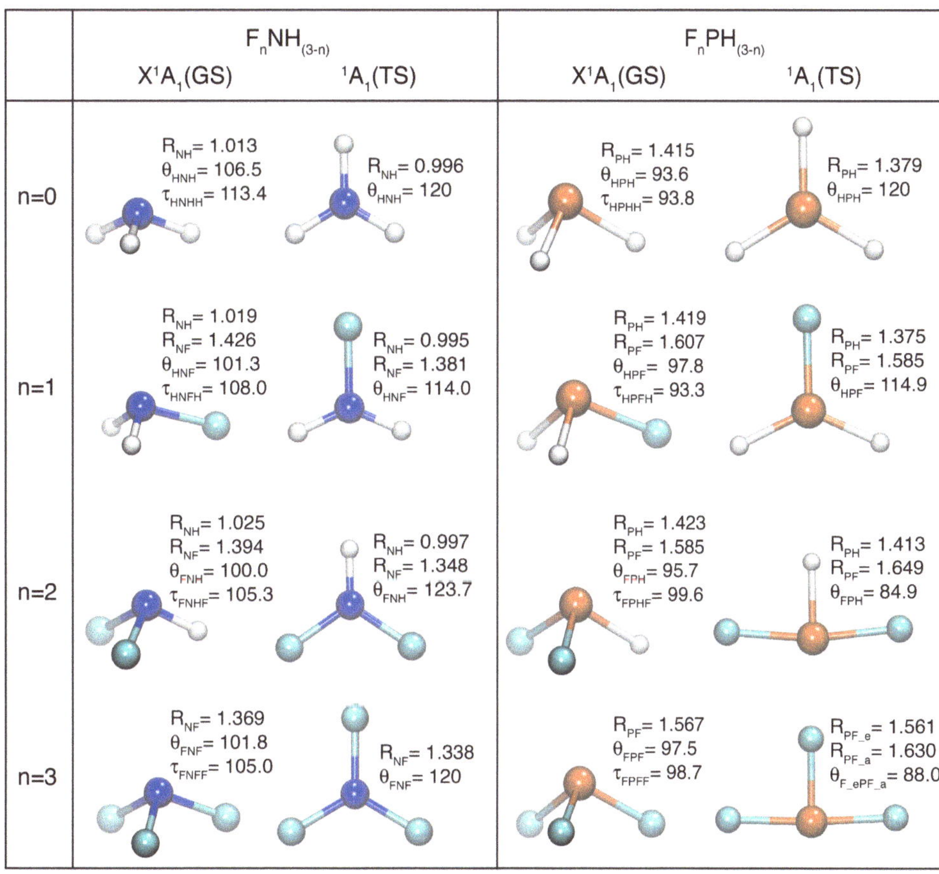

Fig. 1 Optimized geometries of the $F_nNH_{(3-n)}$ and $F_nPH_{(3-n)}$ (n = 0–3) ground states (GS, X^1A_1) and the transition states for inversion (TS, $^1A_1$) obtained from CCSD(T)/AVQZ calculations. Bond distances are in Å and bond angles in degree. θ corresponds to bond angle and τ corresponds to dihedral angle. N is color coded in *blue*, P in rusty *orange*, F in cyan and H in *gray*

The above observations lead to **Question #1**: What is the cause of the differences in the structures of the transition states of (F_2NH, F_2PH) and (NF_3, PF_3), as well as those of (PH_3, FPH_2) and (F_2PH, PF_3)?

Besides the obvious structural differences, a closer examination of the bond distances shows that for all of the molecules that have D_{3h} or D_{3h}-like transition states, i.e., NH_3, FNH_2, F_2NH_3, NF_3, PH_3 and FPH_2, the bond lengths in the transition states are all shorter than the ground state bond lengths. However, the differences in the bond lengths in F_2PH and PF_3 do not follow the same simple pattern. For F_2PH, the PH bond distance is slightly shorter (Δ = −0.010 Å) and the PF bond distances are significantly longer (Δ = +0.064 Å) in the transition state than in the

Table 5 The changes in the bond lengths, ΔR, in $F_nNH_{(3-n)}$ and $F_nPH_{(3-n)}$ (n = 0–3), between the ground states (GS) and the inversion transition states (TS): $\Delta R = R(TS) - R(GS)$, in Å

n	ΔR_{NH}(Å)	ΔR_{NF}(Å)	ΔR_{PH}(Å)	ΔR_{PF}(Å)	ΔR_{PF}(Å)	
					Axial	Equatorial
0	−0.017 (−1.7)		−0.036 (−2.5)			
1	−0.024 (−2.4)	−0.045 (−3.2)	−0.044 (−3.1)	−0.022 (−1.4)		
2	−0.028 (−2.7)	−0.046 (−3.3)	−0.010 (−0.7)		0.064 (4.0)	
3		−0.031 (−2.3)			0.063 (4.0)	−0.006 (−0.4)

The percentage change with respect to the bond distance in the ground state is given in parentheses

ground state. For PF_3, the transition state has two types of PF bonds. The two axial PF bonds are much longer ($\Delta = +0.063$ Å) than the ground state PF bond, and the equatorial PF bond is slightly shorter ($\Delta = -0.006$ Å) than the ground state PF bond.

Table 5 summarizes the bond distance changes between the ground state and transition state of all eight molecules. The percentage changes with respect to the bond lengths in the ground state are also listed (in parentheses). The percentage change for molecules in the D_{3h}-like category ranges from −1.4 % to −3.1 %. On the other hand, the percentage change for the PH bond in F_2PH is only −0.7 %, which is very close to the −0.4 % change for the equatorial PF bond in F_2PF. The percentage change for the axial PF bonds when $n = 2$ and $n = 3$ is 4.0 %, which is larger than any of the differences in the other molecules. These data lead to **Question #2**: Why do the bond distances of the D_{3h}-like transition states decrease and those of the T-shaped transition states behave very differently?

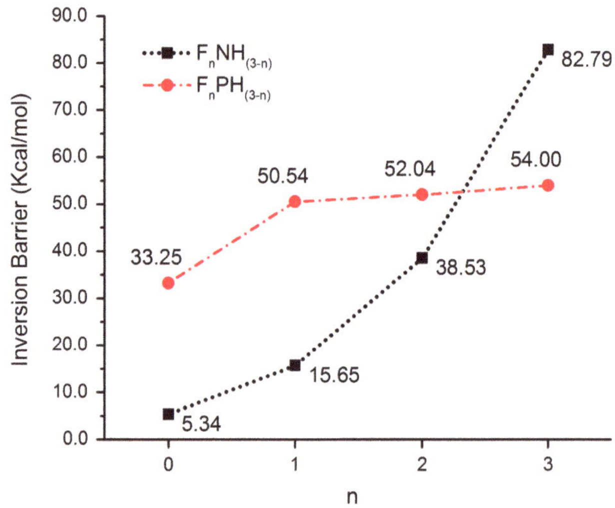

Fig. 2 Computed inversion barriers for $F_nNH_{(3-n)}$ and $F_nPH_{(3-n)}$ (n = 0–3) from CCSD(T)/AVQZ calculations. The inversion barrier is the electronic energy difference between the transition state and the ground state for each molecule without zero-point energy correction

3.1.2 Inversion barrier

Here, as noted in Sect. 2, the inversion barrier is defined as the difference in the electronic energies of the transition state and the ground state, i.e., we ignore differences in the zero-point energies even though vibrational frequencies are reported, because we want to focus on the variations in the barriers caused by the changes in the electronic structure of the molecules. Figure 2 shows the plot of the inversion barrier with respect to n, the number of F atoms in the molecule. When N is the central atom, the barrier height increases monotonically from 5.34 to 82.79 kcal/mol as n increases from 0 to 3. The rate of increase increases with n as well, essentially doubling with each additional F atom: 10.31 kcal/mol from $n = 0$–1, 22.88 kcal/mol from $n = 1$–2 and 44.26 kcal/mol from $n = 2$–3. When P is the central atom, the barrier height increases from 33.25 kcal/mol to just 54.00 kcal/mol as n increases from 0 to 3. The rate of increase is 17.29 kcal/mol from $n = 0$–1, but only 1.50 kcal/mol from $n = 1$–2 and 1.96 kcal/mol from $n = 2$–3. Thus, from $n = 1$–3, the change in the height of the inversion barrier is vastly different in the N series than in the P series: The total increase in the barrier height for $F_nNH_{(3-n)}$ is 77.45 kcal/mol while the barrier height increase for $F_nPH_{(3-n)}$ is less than a third of that, 20.75 kcal/mol. So **Question #3** is: Why does the barrier change very little from $n = 1$–3 in the $F_nPH_{(3-n)}$ series, instead of increasing dramatically like the $F_nNH_{(3-n)}$ series does?

As an aside, we did locate higher lying "transition states" in both F_2PH and PF_3. The geometry of the D_{3h}-like transition state in F_2PH is $R_{PH} = 1.375$ Å, $R_{PF} = 1.571$ Å and $\theta_{HPF} = 124.8°$. This structure lies 32.72 kcal/mol above the lower transition state, at 84.76 kcal/mol. It is a true transition state with only one imaginary frequency: 1,158.2i cm^{-1} (because of its high energy, we did not follow the reaction path from this transition state to see where it led). When PF_3 is constrained to have a D_{3h} structure, $R_{PF} = 1.633$ Å. The D_{3h} structure lies 33.12 kcal/mol higher than the T-shaped transition state

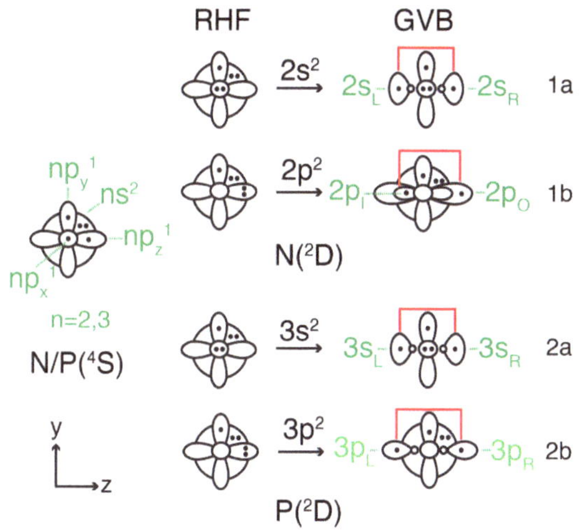

Fig. 3 HF and GVB diagrams for N and P atoms; $n = 2$ for N and $n = 3$ for P atom. The subscripts "L", "R", "I" and "O" represent left, right, inner and outer. The electrons in orbital pairs connected with a red line are singlet coupled

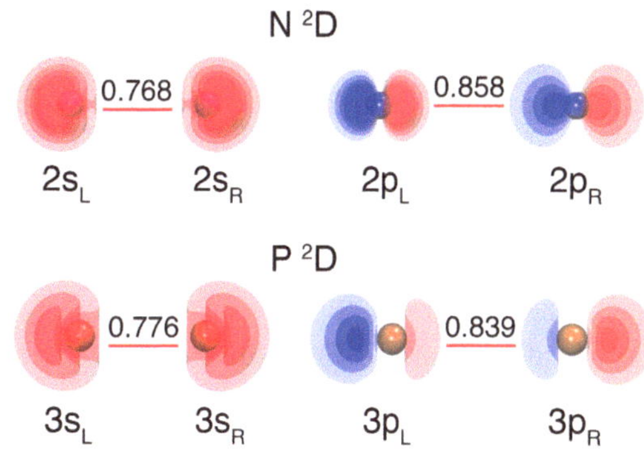

Fig. 4 GVB orbitals and overlaps for the N(^{2}D) and P(^{2}D) states. A *red line* means that the electrons in these singly occupied orbitals are singlet coupled; the values of the orbital overlaps are given above the line. Contours are ±0.10, ±0.15, ±0.20 and ±0.25. *Red contour* represents positive, and *blue contour* represents negative orbital phase

and 87.12 kcal/mol above the ground state PF_3. However, this configuration has three imaginary frequencies ($2 \times 214i$, $596i$ cm^{-1}) and therefore is not a true transition state.

3.2 p-Recoupled pair bonding in F_2PH and PF_3 and s-recoupled pair bonding in the other molecules

In the last section, we posed three questions. In this section, we show that the answers to all of these questions center on the ability of the P atom to form recoupled pair bonds with F and, more specifically, p-recoupled pair bond dyads with two F atoms. In molecules other than F_2PH and PF_3, the inversion transition states involve formation of s-recoupled pair bond dyads.

3.2.1 GVB description of the N and P atoms

In order to understand the differences between $F_nNH_{(3-n)}$ and $F_nPH_{(3-n)}$ ($n = 0$–3), we need to understand the differences between the central atoms in these molecules. Figure 3 presents the GVB orbital diagrams for the ground and low-lying excited states of the N and P atoms. The diagrams represent the valence electrons and orbitals. The ground states of the N and P atoms are ^{4}S states with three singly occupied valence p orbitals in each atom. In the N/P(^{4}S) diagram, the big circle represents the valence s orbital, the small circle represents the out-of-plane p_x orbital, and the two dumbbell shapes represent the two in-plane (p_y, p_z) orbitals. The dots represent electron occupations. The ground states of $F_nNH_{(3-n)}$ and $F_nPH_{(3-n)}$ ($n = 0$–3) have three normal covalent bonds formed with the electrons in these three singly occupied p orbitals. The pyramidal structures of these molecules are a natural result of the orientations of the three p orbitals in the atom.

Although the connection of the ground states of the N and P atoms with the ground states of the $F_nNH_{(3-n)}$ and $F_nPH_{(3-n)}$ molecules is straightforward, the transition states do not correlate with the ground state atoms. Rather, they correlate with the first excited (^{2}D) states of the atoms. In this state, one electron from a p orbital is excited into one of the other p orbitals with the original p orbital no longer occupied in the configuration (this is schematically represented by the configuration $s^2p_x^2p_y^1$ or $s^2p_z^2p_y^1$ in Fig. 3). Now, the atoms can form both covalent bonds (with the singly occupied p_y orbital) as well as recoupled pair bonds and recoupled pair bond dyads (with the s^2 or p^2 lone pairs). The calculated excitation energy from the ^{4}S state to the ^{2}D state is 55.33 kcal/mol for the N atom and 32.23 kcal/mol for the P atom (MRCI + Q/AVQZ); these numbers are to be compared to 54.97 and 32.48 kcal/mol from the NIST Atomic Spectra Tables [39]. Because of this energetic difference, recoupled pair bonds and recoupled pair bond dyads in P will lie at much lower energies than in N.

As noted above, there are two doubly occupied orbitals in the ^{2}D state, i.e., $2s^2$ and $2p^2$ for N and $3s^2$ and $3p^2$ for P atom. Although these orbitals are doubly occupied in the HF wave function, in the GVB wave function s^2 and p^2 lone pairs are each described by two singly occupied, non-orthogonal lobe orbitals. Figure 4 shows the s and p GVB lobe orbitals for the N(^{2}D) and P(^{2}D) states. The s lobe orbitals of both atoms are very similar in shape and orientation, although the P orbitals are more diffuse and span a larger spatial region. The s lobe orbitals are well separated spatially, residing on opposite sides of the central atom, and have overlaps of only 0.768 (N) and 0.776 (P); we refer to them as the s_L and s_R lobe orbitals. In the GVB orbital diagrams 1a and 2a in Fig. 3, the s_L and s_R lobe orbitals are connected with a red line, representing the fact that they are singlet coupled.

On the other hand, the 2p lobe orbitals of $N(^2D)$ and the 3p lobe orbitals of $P(^2D)$ are very different. Both 2p lobe orbitals on the $N(^2D)$ state look like p orbitals, one tighter than the original $2p_z$ orbital and the other one more diffuse. They lie in the same spatial region, and we refer to them as inner and outer lobe orbitals, $2p_I$ and $2p_O$. In the $N(^2D)$ GVB orbital diagram 1b in Fig. 3, these orbitals are represented with two dumbbell shapes, one inside of the other with a red line connecting them. The 3p lobe orbitals of $P(^2D)$, on the other hand, are distorted $3p_z$ orbitals, with one more concentrated on the left side of the P atom and the other more concentrated on the right side; they have an overlap of 0.839 and are referred to as $3p_L$ and $3p_R$ lobe orbitals. In diagram 2b in Fig. 3, they are represented with two half-dumbbell shapes connected with a red line.

Both of the lone pairs on the N and P atoms are potentially available for recoupling to form bonds. However, the ease with which a lone pair can be recoupled is dependent on two factors: (1) the spatial orientation of the lobe orbitals and (2) the overlap of the orbitals. To form a strong recoupled pair bond or recoupled pair bond dyad, the lobe orbitals must be localized in different spatial regions and the overlap of the lone pair orbitals must be significantly less than one (the smaller, the better). It also helps if the ligand is very electronegative, because this will reduce the Pauli exchange-repulsion between the bonds (or between the bond and the electron in the left over orbital) formed by recoupling the lone pair [31].

As noted above, the N and P s_L and s_R orbitals are spatially well separated and their overlaps, 0.768 (N) and 0.776 (P), are similar. So, s-recoupled pair bonds can be formed in both N and P. The overlap between the N $2p_I$ and $2p_O$ orbitals (0.858) is slightly larger than that between the P $3p_L$ and $3p_R$ orbitals (0.839), and both overlaps are much larger than the overlaps between the ($2s_L$, $2s_R$) and ($3s_L$, $3s_R$) lobe orbitals. But, more importantly, the $2p_I$ and $2p_O$ orbitals occupy essentially the same spatial region, while the $3p_L$ and $3p_R$ orbitals are spatially separated. Therefore, it is far more favorable for P to participate in p-recoupled pair bonding than N. However, the ($3p_L$, $3p_R$) orbitals of the P atom are not as spatially separated as the ($3s_L$, $3s_R$) orbitals, so it will be more difficult to form recoupled bonds with the 3p lone pair than with the 3s lone pair.

From the GVB orbital diagram of the $P(^2D)$ state in Fig. 3, one can easily see that three strong bonds can be formed: with the singly occupied ($3s_L$, $3s_R$, $3p_y$) orbitals in 2a and the singly occupied ($3p_L$, $3p_R$, $3p_y$) orbitals in 2b. In 2a, the remaining lone pair is an out-of-plane $3p_x$ lone pair, and in 2b, it is an in-plane 3s pair. As will be demonstrated and discussed in Sects. 3.2.2 and 3.2.3, with 2a, the resulting structures of the transition states correspond to that of the D_{3h}-like structures, which have an out-of-plane lone pair. With 2b, the resulting structures will be those of the T-shaped transition states for inversion with an in-plane lone pair. Since the lobe orbitals in 1a, 2a and 2b in the $N(^2D)$ and $P(^2D)$ states have different properties than the 2p and 3p orbitals in the $N(^4S)$ and $P(^4S)$ states, the bonds formed with these orbitals in the transition states will have different lengths, strengths and spatial orientations than those in the ground states.

Which lobe orbitals will be used to form bonds depends on the strength of the resulting bonds as well as the resulting interactions with the electrons in the other orbitals in the molecule. In particular, we note that the formation of bonds with the ($3s_L$, $3s_R$, $3p_y$) orbitals has a lone pair perpendicular to the yz plane that will have repulsive interactions with the lone pair orbitals in F that are also perpendicular to the yz plane. This is not the case when bonds are formed with the ($3p_L$, $3p_R$, $3p_y$) orbitals where the lone pair lies in the molecular plane.

3.2.2 *GVB orbitals of the transition states*

As discussed above, the ground states of the eight molecules in our study have three normal covalent bonds formed by singlet coupling the singly occupied ligand orbitals with the three singly occupied p orbitals of $N(^4S)$ and $P(^4S)$. Figure 5 shows the GVB bonding orbitals of the eight transition states, along with the doubly occupied lone pair orbital. When there are equivalent bonds, the GVB orbitals of only one bond are shown. The overlap between the two orbitals that form a bond is also shown. The GVB wave functions of all of the eight transition state molecules are predominantly PP spin coupled with w_{PP} ranging from 0.91 to more than 0.99.

NH_3 and PH_3 each have three equivalent bonds. Upon bond formation, the orbitals on the two atoms polarize, hybridize, expand or contract, delocalize, etc. in response to the presence of the other atoms. However, it is clear that, for instance, one of the orbitals participating in the NH bond closely resembles the 2s lobe orbital of N shown in Fig. 4, although with somewhat more 2p character than in the N atom, and the other orbital resembles the 1 s orbital of the H atom. The lone pair orbital is essentially the out-of-plane $2p_x$ orbital in the $N(^2D)$ state. This is consistent with the orbital diagram 1a in Fig. 3. The three equivalent NH bonds result from resonance between the 2s-recoupled pair bond dyad and the normal covalent bond with the $2p_y$ orbital. The resulting structure has D_{3h} symmetry because this arrangement reduces the Pauli repulsion between the three bonds, and so, the GVB orbitals on the N atom in the NH_3 transition state are resonance averages (hybrids) of the two 2s lobe orbitals and the $2p_y$ orbital. The situation is similar in PH_3. The reason for the shorter bond distances in

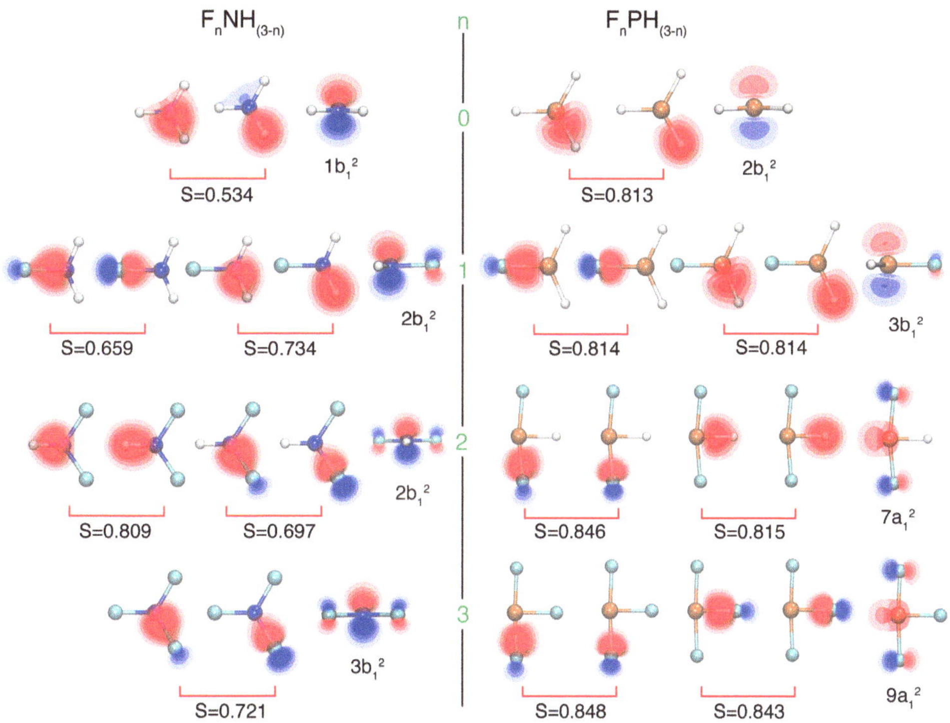

Fig. 5 Selected GVB orbitals and overlaps for the inversion transition states of $F_nNH_{(3-n)}$ and $F_nPH_{(3-n)}$ (n = 0–3). The GVB calculations are 6-in-6 calculations. When two or three bonds are equivalent, the two GVB orbitals constituting one bond are shown. The doubly occupied orbital on the central atom for each molecule is shown following the GVB orbitals. Contours are ±0.10, ±0.15, ±0.20 and ±0.25

the transition states is that the orbitals on N and P have large s orbital components in the transition states, and s orbitals are closer to the nucleus than p orbitals, especially for the P atom. Therefore, the bonds formed with these orbitals are shorter than those formed with p orbitals in the ground states (Question #2).

The N atom is more electronegative than the P atom, so NH bonds are more polarized toward the N atom than PH bonds. A result of this is that the three orbitals centered on N have much higher overlaps (0.782) with each other than do the three orbitals centered on P (0.442). This leads to larger Pauli repulsion between the bond pairs in NH_3. One impact of this repulsive interaction is that the H 1s-like bond orbital develops a node in the region of the N atom. As a result, the overlap of the bond pair is smaller in NH_3 (0.534) than in PH_3 (0.813).

Upon a single substitution of H by F, the geometries do not change much. But there are significant changes in some of the orbitals. The N or P orbital involved in the bond to the F atom has delocalized onto the F atom as would be expected for a polar covalent bond. This delocalization builds $N^{\delta+}F^{\delta-}$ ($P^{\delta+}F^{\delta-}$) character into the GVB wave function. The orbitals for the other two bond pairs resemble those in NH_3 and PH_3. Due to the interaction with the out-of-plane, doubly occupied $2p_x$ orbitals on F, the doubly occupied, out-of-plane $2p_x$ and $3p_x$ lone pair orbitals on N and P acquire antibonding character. Also, since F is very electronegative, the overlaps between the orbitals on the central atoms are reduced and the node in the H 1s-like orbital in NH_3 disappears upon F substitution. Regardless of these small changes, the nature of the bonding orbitals on the central atoms in FNH_2 and FPH_2 is similar to that of NH_3 and PH_3: hybrids of an s-recoupled pair bond dyad and a p-covalent bond.

Once another H atom is substituted with an F atom, there are marked differences in the geometrical and orbital structure of F_2NH and F_2PH. F_2NH is very similar to FNH_2, although two F atoms pull more electron density away from the central N atom than one F atom and the orbitals are less dense around the N atom. The overlaps between the bonding orbitals increase for both the NH and NF bonds. The out-of-plane, doubly occupied $2p_x$ orbital on N has acquired antibonding character on both F atoms, evidence of increased repulsive interactions between the electrons in the N lone pair and those in the doubly occupied $2p_x$ orbitals on F. On the other hand, F_2PH does not resemble FPH_2. The geometry of the transition state is now T-shaped, and the lone pair orbital is no longer a $3p_x$-like orbital perpendicular to the molecular plane but a 3s lobe-like orbital in the molecular plane. This latter orbital has acquired a measure of antibonding character because of the interaction of the electrons in this orbital with the doubly occupied in-plane 2p orbitals on the F atom. In addition, the orbitals participating in the axial FPF bond in the F_2PH transition state are 3p lobe orbitals instead of 3s lobe orbitals.

The two axial PF bonds in F_2PH are, in fact, a p-recoupled pair bond dyad (Question #1). The P–F bond distance is very similar to that found in the $A^2\Pi$ state of PF_2, 1.649 Å (F_2PH) versus 1.639 Å (PF_2), a molecule known to have a p-recoupled pair bond dyad [9]. This is the reason for the unusual increase in the length of the PF axial bonds between the ground and transition state: p-recoupled pair bond dyads have bond distances that are significantly longer than the corresponding covalent bonds [9, 12, 13, 31, 40]. In contrast, the P orbital participating in the equatorial PH bond is the singly occupied 3p orbital (See 2b in Fig. 3). This bond is a normal polar covalent bond, which leads to only a slight decrease in the bond length (−0.7 % in Table 3) going from the ground state to the transition state (Question #2). Note that the p-recoupled pair bond dyad in F_2PH is almost linear, $\theta_{FPF} = 169.2°$, as suggested in the orbital diagram 2b in Fig. 3, and as the geometry of $PF_2(A^2\Pi)$ suggests ($\theta_{FPF} = 180°$). This is due to the dominant 3p character in the 3p lobe orbitals and the repulsive forces between the two PF bonds, the PH bond and the doubly occupied distorted 3s-like orbital. The quasi-linear structure is consistent with our finding that p-recoupled pair bond dyads are structurally rigid and prefer nearly collinear arrangements [9, 12, 13, 31, 40].

When all of the H atoms are substituted with F atoms, the NF_3 transition state has three equivalent NF bonds and D_{3h} symmetry. The doubly occupied orbital is an out-of-plane N $2p_x$-like orbital with antibonding character on all three F atoms. The overlap between the two bonding orbitals constituting a NF bond increases from 0.697 in F_2NH to 0.721 in NF_3 as more F atoms pull more electron density into the bonding region. PF_3 is similar to F_2PH, with the only difference being that the PH covalent bond in F_2PH becomes a PF polar covalent bond. The two axial bonds are, again, a p-recoupled pair bond dyad formed with the P 3p lobe orbitals. The axial PF bonds in PF_3 are similar in length to those in F_2PH, 1.630 Å (PF_3) versus 1.649 Å (F_2PH) and are much longer than the PF bond in the PF_3 ground state (1.567 Å). For the equatorial bond, on the other hand, the percentage change in the bond length is just −0.4 % (0.006 Å), consistent with the equatorial PF bond being a covalent bond formed with a 3p orbital. Therefore, the seemingly random bond length changes in F_2PH and PF_3 (Question #2) are, in fact, not random: The p-recoupled pair bond dyads of the transition states have much longer bond distances than the covalent bonds, and the equatorial covalent bonds of the transition states are only slightly shorter than the covalent bonds of the ground states. The doubly occupied P orbital in PF_3 is similar to that in F_2PH, a distorted 3 orbital with some antibonding character associated with the F in-plane lone pairs.

3.2.3 Further comparison of s- and p-recoupled pair bond dyads

The analysis of the GVB orbitals of (PH_3, FPH_2, F_2PH, PF_3) shows that the dramatic change in the structures of the

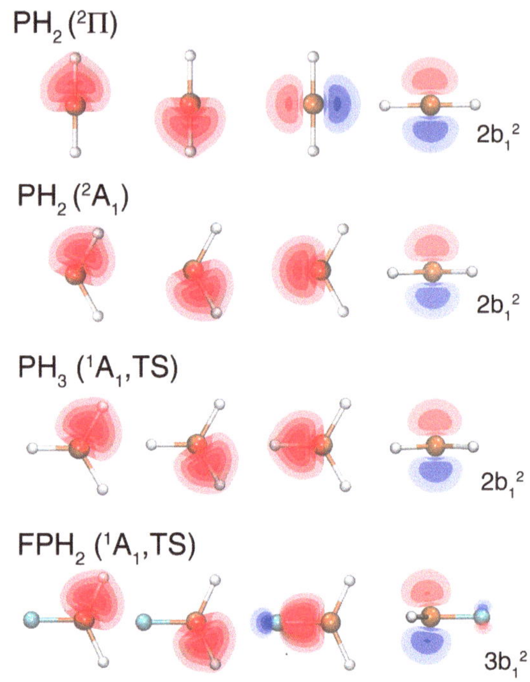

Fig. 6 GVB orbitals centered on the P atom for the linear $PH_2(^2\Pi)$ configuration, the bent $PH_2(^2A_1)$ excited state, and for the transition states for inversion in $FPH_2(^1A_1)$ and $PH_3(^1A_1)$

transition states for F_2PH and PF_3 is a result of the formation of a p-recoupled pair bond dyad in these species (Question #1). In this section, we take a step back to examine the triatomic molecules, because recoupled pair bond dyads also exist in NH_2, NF_2, PH_2 and PF_2, although in their excited states not in their ground states. Here, we focus on PH_2 and PF_2 as two examples of s- and p-recoupled pair bond dyads.

The ground state of PH_2 is a $^2B_1$ state, bound by two normal covalent bonds. The first excited state is a $^2A_1$ state. The $^2A_1$ state derives from a linear $^2\Pi$ state that contains a 3s-recoupled pair bond dyad, and relaxes to a bent geometry upon geometry optimization, incorporating additional 3p character into the P bonding orbitals. So let us compare the recoupled pair bond dyad in the $^2\Pi$ and $^2A_1$ states of PH_2 with the bonds in the PH_3 and FPH_2 transition states to understand how the s-recoupled pair bond dyad evolves in these species. Figure 6 shows the GVB orbitals in PH_2, PH_3 and FPH_2 that are centered on the P atom along with the doubly occupied, out-of-plane 3p-like lone pair orbitals. The $PH_2(^2\Pi)$ state has two 3s-like lobe orbitals as well as singly and doubly occupied 3p(π)-like orbital on the P atom. The 3s lobe orbitals are slightly delocalized onto the H atoms compared to the 3s lobe orbitals in the P atom (Fig. 4). In order to strengthen the bond and reduce the repulsion between the electrons in these orbitals and because mixing between (hybridization of) s and p orbitals is facile, the $^2\Pi$ state rearranges to become a strongly bent

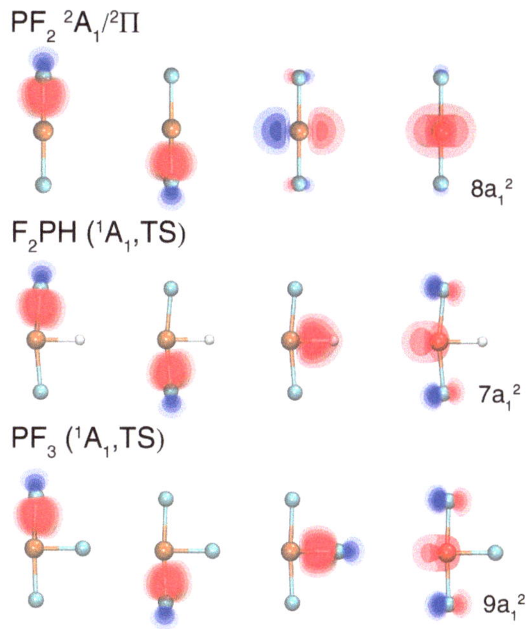

Fig. 7 GVB orbitals centered on the P atom for the linear $PF_2(^2\Pi)$ excited state and the transition states for $F_2PH(^1A_1)$ and $PF_3(^1A_1)$

$^2A_1$ state with a bond angle of 121.8°, reducing the energy by 19.66 kcal/mol. The 3p orbital is pushed away from the two PH bonds, picking up 3s character. So the three singly occupied P orbitals in the $^2A_1$ state of PH_2 are not pure 3s lobe orbitals or 3p orbitals, but a mixture of the two. However, their origins are clear.

The singly occupied P orbitals in the PH_3 transition state look very much like the corresponding orbitals in the $PH_2(^2A_1)$ state. The doubly occupied out-of-plane orbital is largely unchanged. The orbitals of the FPH_2 transition state are similar except that the P bonding orbital delocalizes onto the F atom (consistent with the fact that the F atom is more electronegative than the P atom), and the out-of-plane lone pair orbital acquires antibonding character due to repulsive interactions of the electrons in this orbital with those in the F 2p orbital.

Similar comparisons for PF_2, F_2PH and PF_3 are given in Fig. 7. The ground state of PF_2 is also a $^2B_1$ state, and the lowest-lying excited state is a $^2\Pi$ state. In PF_2, in contrast to PH_2, the first excited state is linear. The $PF_2(^2\Pi)$ state contains a 3p-recoupled pair bond dyad, a singly occupied 3p-like orbital and a doubly occupied polarized 3s lone pair orbital ($8a_1$) [9]. As noted previously, p-recoupled pair bonds prefer linear or quasi-linear geometries [9, 12, 13, 31, 40]. Comparing the three molecules containing the p-recoupled pair bond dyad, one can see that the 3p-like lobe orbitals remain largely unchanged. The biggest changes occur for the 3p-like orbital and the doubly occupied 3s-like orbital as the third bond forms in F_2PH and F_2PF. The 3p-like P orbital localizes in the bonding region as the PH bond forms and then delocalizes onto the more electronegative F atom when the PF bond forms. The 3s-like orbitals gain antibonding character and are pushed away from the covalent bonds. Comparing Figs. 6 and 7, one can see the s-recoupled pair bond dyad is indeed much more structurally flexible than the p-recoupled pair bond dyad.

As shown above, it is possible for the same central atom to form different types of recoupled pair bonds and recoupled pair bond dyads. As shown in Fig. 4, the overlap between the $3s_L$ and $3s_R$ lobe orbitals of the P atom is 0.776, while the overlap between $3p_L$ and $3p_R$ is 0.839. Therefore, it is much easier to form an s-recoupled pair bond than a p-recoupled pair bond. In fact, forming a p-recoupled pair bond dyad requires electronegative ligands, which separates the two highly overlapping p lobe orbitals and reduces the Pauli repulsion between the resulting bond pairs [31]. The H atom is not sufficiently electronegative to recouple the P $3p^2$ lone pair. Therefore, the $P(^2D)$ state forms an s-recoupled pair bond dyad with two H atoms and a p-recoupled pair bond dyad with two F atoms. Although the F atom is also able to recouple a $3s^2$ lone pair, an s-recoupled pair bond dyad does not form in F_2PH and PF_3 because the P lone pair orbital would then be perpendicular to the molecular plane, resulting in strong Pauli repulsive interactions with the electrons in the doubly occupied orbitals on the F atoms. These repulsions are minimized if the lone pair on the P atom is in a 3s-like orbital, which is polarized away from the F atoms as is the case for the p-recoupled pair bond dyad. Therefore, the p-recoupled pair bond dyad is preferred, which gives rise to the low-barrier inversion pathways in F_2PH and PF_3 (Question #3). Even though the lone pair on the P atom is not involved in bond formation, it clearly influences which types of bonds are formed.

3.3 Transition state formation pathways

To summarize the above GVB analysis of the transition states, GVB diagrams of the pathways for forming the inversion transition states of two representative molecules, NF_3 and PF_3, are shown in Fig. 8. The NF_3 transition state goes through a pathway involving the formation of an s-recoupled pair bond dyad. The two F atoms recouple the $2s^2$ electrons of $N(^2D)$ and form the $NF_2(^2\Pi)$ configuration, which rearranges to the bent $^2A_1$ state. The third F atom forms a normal (polar) covalent bond with the singly occupied orbital in the $NF_2(^2A_1)$ state, leading to the NF_3 D_{3h} transition state. The doubly occupied orbital on N is perpendicular to the molecular plane. The three NF bonds are equivalent as a result of the ease with which s and p orbitals hybridize (mix). The very high barrier to inversion in NF_3 is, in large part, due to the strong repulsions between the electrons in the lone pair orbital on the N atom and those in the lone pair orbitals on the F atom.

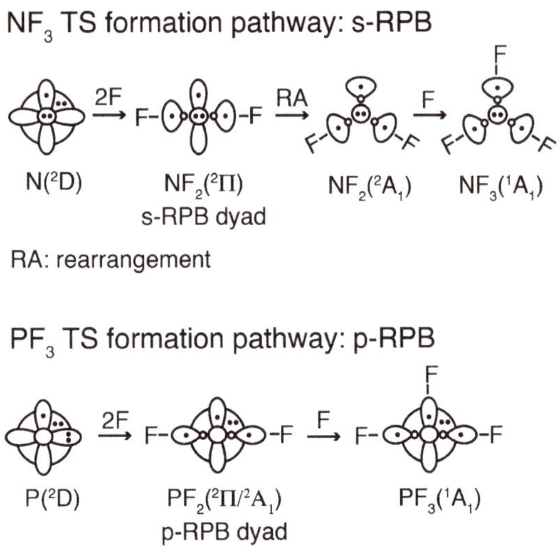

Fig. 8 Formation pathway diagrams of the $^1A_1$ inversion transition states for NF_3 and PF_3

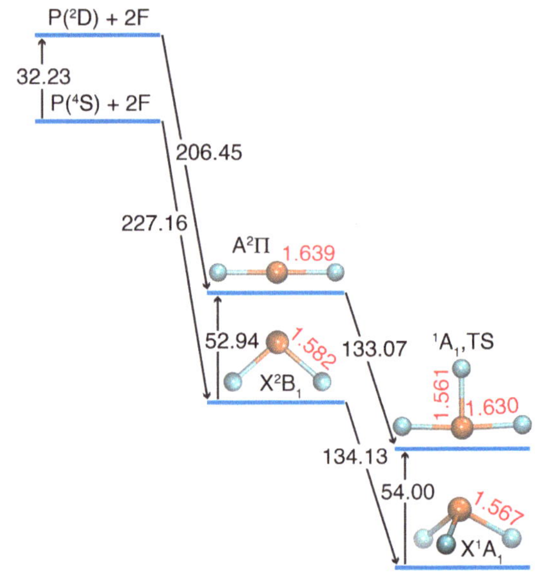

Fig. 9 Energy and bond length changes during the formation of the ground state and inversion transition state of PF_3, beginning with the atoms. The energies are in kcal/mol, and bond distances are in Å

The PF_3 transition state goes through a p-recoupled pair bond dyad pathway. The two F atoms recouple the $3p^2$ electrons of $P(^2D)$ to form the $PF_2(^2\Pi)$ state. The third F atom then forms a normal covalent bond with the remaining singly occupied P 3p orbital, resulting in a T-shape transition state structure. The lone pair orbital on P in PF_3 is a distorted 3s-like orbital polarized away from the F atoms in order to reduce the repulsions between the electrons in this orbital and those in the F lone pair orbitals. The collinear structure of the p-recoupled pair bond dyad remains almost unchanged in the process, with the dyad bonds bending slightly away from the lone pair. Not only are the axial bond lengths nearly the same in $PF_2(A^2\Pi)$ and $PF_3(^1A_1$, TS), but the F_2P–F ground and transition state bond energies are nearly the same, which means that the magnitude of the inversion barrier is nearly the same as the energy difference between the $PF_2(X^2B_1)$ and $PF_2(A^2\Pi)$ states, 54.00 versus 52.94 kcal/mol. This process is illustrated in Fig. 9. The bond energy of the equatorial PF bond in the transition state is 133.07 kcal/mol (the bond energy reported in ref. [9] is incorrect), similar to the 134.13 kcal/mol bond energy of the covalent PF bond in the ground state. Also, the energy lowering for forming a p-recoupled pair bond dyad from the atoms in $PF_2(A^2\Pi)$ is 206.45 kcal/mol, which is only 20.71 kcal/mol less stable than forming the two covalent bonds in $PF_2(X^2B_1)$, 227.16 kcal/mol. In fact, as noted earlier, the strength of recoupled pair bond dyads is the reason for the existence of hypervalent species such as PF_5 [9] and SF_6 [12].

4 Conclusions

In this article, we report accurate CCSD(T) calculations on the ground and transition states for inversion of the $F_nNH_{(3-n)}$ and $F_nPH_{(3-n)}$ ($n = 0$–3) molecules along with a detailed analysis of the GVB wave functions for these molecules at the calculated stationary points. All of the molecules go through D_{3h}-like transition states, except for PF_3 and F_2PH, which have T-shaped transition states, a fact first reported by Dixon and coworkers [4–8]. For all of the molecules with D_{3h}-like transition states, the bond distances are shorter than those in their ground states, a result of the increased s character in the N and P bond orbitals in the transition states. However, for F_2PH and F_2PF, the transition state is T-shaped and the axial PF bonds are much longer than those in the ground state, while the equatorial PH and PF bonds are slightly shorter than those in the F_2PH and PF_3 ground states.

Likewise, there is a dramatic difference in the dependence of the barrier height for inversion as the number of F atoms (n) increases. The height of the inversion barrier height increases dramatically for the $F_nNH_{(3-n)}$ series as n increases, from 5.34 (NH_3) to 82.79 (NF_3) kcal/mol. However, for the $F_nPH_{(3-n)}$ series, the barrier height increases substantially from $n = 0$ to $n = 1$ (33.25–50.54 kcal/mol), but thereafter it increases only modestly (from 50.54 to 54.00 kcal/mol).

The explanation for the anomalous behavior of the F_2PH and PF_3 molecules is simple. In the transition states of both F_2PH and PF_3, the nearly collinear PF bonds are a result of the formation of p-recoupled pair bond dyads. A hydrogen atom is not sufficiently electronegative to recouple the lone pair in the $P(^2D)$ excited state, and therefore, these bonds are only found in F_2PH and PF_3. Formation of the p-recoupled pair bond dyads in the transition states of these two molecules is favored because it minimizes the repulsion between the electrons in the lone pair on P and those in the lone pairs on the F atoms. Since a p-recoupled pair bond dyad prefers a nearly collinear arrangement of the F–P-F atoms, the transition states in F_2PH and PF_3 are T-shaped. The lengthening of the axial PF bonds is a result of the formation of p-recoupled pair bond dyads in F_2PH and PF_3, while the third bond in each species is a covalent bond similar to that in the ground state.

The ground state of F_2PH and PF_3 arises from the addition of an H or F atom to the ground state of PF_2, the X^2B_1 state, while the T-shaped transition arises from the corresponding additions to the first excited state of PF_2, the $A^2\Pi$ state. The energy of the $PF_2(X^2B_1)$ and $PF_2(A^2\Pi)$ states differs by 52.94 kcal/mol [9]. If the strengths of the PH and PF bonds in the F_2PH and PF_3 transitions states are similar to those in the ground states, as would be expected, one would predict barrier heights close to this value, which, indeed, is the case: 52.04 kcal/mol (F_2PH) and 54.00 kcal/mol (PF_3). In addition, it should be noted that the lengths of the covalent PH and PF bonds are similar in the ground and transition states.

In summary, the transition states for NH_3, FNH_2, F_2NH, NF_3, PH_3 and FPH_2 involve the formation of s-recoupled pair bond dyads. Neither the H or F atoms are able to recouple the $2p^2$ lone pair in the $N(^2D)$ state, so all of the transition states in the $F_nNH_{(3-n)}$ series possess s-recoupled pair bond dyads (or hybrids thereof). H is not able to recouple the $3p^2$ lone pair of P atom, and thus, the PH_3 and FPH_2 transition states also contain s-recoupled pair bond dyads. Extensive mixing between the s-recoupled pair bond dyad and the remaining covalent bond results in the D_{3h}-like transition state structures and vertex inversion pathways. The F atom is sufficiently electronegative to recouple both the $3s^2$ and the $3p^2$ electrons of the P atom, but the p-recoupled pair bond dyad is preferred in F_2PH and PF_3 over the s-recoupled pair bond dyad because the repulsion between the out-of-plane lone pair on P and the lone pairs on F is smaller in the former case. This leads to much lower inversion barriers than expected for F_2PH and PF_3 based on the trends in the $F_nNH_{(3-n)}$ and the first two members of the $F_nPH_{(3-n)}$ series. The p-recoupled pair bond dyad is rigid and stays almost collinear when other bonds are formed, which results in the T-shaped transition states and edge inversion in F_2PH and PF_3.

The T-shaped pnictogen structures in the ground state and transition states of compounds such as ADPO and its saturated analog also result from the formation of p-recoupled pair bond dyads by the pnictogen elements from the second row and beyond. Factors that stabilize the T-shaped structures of these compounds include electronegative ligands (facilitating the formation of the recoupled pair bond dyad) and conjugated π system (allowing delocalization into the space usually occupied by the lone pair). Such compounds are expected to be widespread; their chemistry has been reviewed in detail by Arduengo and Stewart [2].

Acknowledgments This work was supported by funding from the Distinguished Chair for Research Excellence in Chemistry and the National Center for Supercomputing Applications at the University of Illinois at Urbana-Champaign.

References

1. Culley SA, Arduengo AJ III (1984) J Am Chem Soc 106:1164–1165
2. Arduengo AJ III, Stewart CA (1994) Chem Rev 94:1215–1237
3. Lochschmidt S, Schmidtpeter A (1985) Z Naturforsch Teil B 40:765–773
4. Dixon DA, Arduengo AJ III, Fukunaga T (1986) J Am Chem Soc 108:2461–2462
5. Arduengo AJ III, Dixon DA, Roe DC (1986) J Am Chem Scc 108:6821–6823
6. Dixon DA, Arduengo AJ III (1987) J Chem Soc Chem Comm 2:498–500
7. Dixon DA, Arduengo AJ III (1987) J Am Chem Soc 109:338–341
8. Dixon DA, Arduengo AJ III, Lappert M (1991) Heteroatom 2:541–544
9. Woon DE, Dunning TH Jr (2010) J Phys Chem A 114:8845–8851
10. Clotet A, Rubio J, Illas F (1988) J Mol Struc (Theochem) 164:351–361
11. Creve S, Nguyen MT (1998) J Phys Chem A 102:6549–6557
12. Woon DE, Dunning TH Jr (2009) J Phys Chem A 113:7915–7926
13. Chen L, Woon DE, Dunning TH Jr (2009) J Phys Chem A 113:12645–12654
14. Purvis GD III, Bartlett RJ (1982) J Chem Phys 76:1910–1918
15. Raghavachari K, Trucks GW, Pople JA, Head-Gordon M (1989) Chem Phys Lett 157:479–483
16. Knowles PJ, Hampel C, Werner HJ (1993) J Chem Phys 99:5219–5227
17. Watts JD, Gauss J, Bartlett RJ (1993) J Chem Phys 98:8718–8733
18. Dunning TH Jr (1989) J Chem Phys 90:1007–1023
19. Kendall RA, Dunning TH Jr, Harrison RJ (1992) J Chem Phys 96:6796–6806
20. Woon DE, Dunning TH Jr (1993) J Chem Phys 98:1358–1371
21. Dunning TH Jr, Peterson KA, Wilson AK (2001) J Chem Phys 114:9244–9253
22. Goddard WA III, Blint RJ (1972) Chem Phys Lett 14:616–622
23. Goddard WA III, Dunning TH Jr, Hunt WJ, Hay PJ (1973) Acc Chem Res 6:368–376
24. MOLPRO is a package of ab initio programs written by Werner HJ, Knowles PJ, Knizia G, Manby FR, Schütz M, Celani P, Korona T, Lindh R, Mitrushenkov A, Rauhut G, Shamasundar KR, Adler TB, Amos RD, Bernhardsson A, Berning A, Cooper DL, Deegan MJO, Dobbyn AJ, Eckert F, Goll E, Hampel C, Hesselmann A, Hetzer G, Hrenar T, Jansen G, Köppl C, Liu Y, Lloyd AW, Mata RA, May AJ, McNicholas SJ, Meyer W, Mura ME, Nicklaß A, O'Neill DP, Palmieri P, Peng D, Pflüger K, Pitzer R, Reiher M, Shiozaki T, Stoll H, Stone AJ, Tarroni R, Thorsteinsson T, Wang M. See http://www.molpro.net

25. Knowles PJ, Werner HJ (1985) Chem Phys Lett 115:259–267
26. Werner HJ, Knowles PJ (1985) J Chem Phys 82:5053–5063
27. Knowles PJ, Werner HJ (1988) Chem Phys Lett 145:514–522
28. Werner HJ, Knowles PJ (1988) J Chem Phys 89:5803–5814
29. Ragavachari K, Trucks GW, Pople JA, Head-Gordon M (1989) Chem Phys Lett 157:479–483
30. Langhoff SR, Davidson ER (1974) Int J Quant Chem 8:61–72
31. Dunning TH Jr, Woon DE, Leiding J, Chen L (2013) Acc Chem Res 46:359–368
32. Lindquist BA, Woon DE, Dunning TH Jr (2014) J Phys Chem A 118:1267–1275
33. Woon DE, Dunning TH Jr, to be published
34. Gerratt J, Cooper DL, Karadakov PB, Raimondi M (1997) Chem Soc Rev 26:87–100 and references therein
35. Cooper DL, Thorsteinsson T, Gerratt J (1997) Int J Quant Chem 65:439–451
36. Thorsteinsson T, Cooper DL (1998) Int J Quant Chem 70:637–660
37. Cooper DL, Thorsteinsson T, Gerratt J (1998) Adv Quant Chem 32:51–67
38. Cooper DL, Karadakov PB (2009) Int Rev Phys Chem 28:169–206
39. NIST Atomic Spectra Database (ver. 5.1), [Online]. Available at: http://physics.nist.gov/asd [2014, February 2]
40. Leiding J, Woon DE, Dunning TH Jr (2011) J Phys Chem A 115:4757–4764

Theor Chem Acc (2014) 133:1511
DOI 10.1007/s00214-014-1511-8

REGULAR ARTICLE

A comparison of singlet and triplet states for one- and two-dimensional graphene nanoribbons using multireference theory

Shawn Horn · Felix Plasser · Thomas Müller · Florian Libisch · Joachim Burgdörfer · Hans Lischka

Received: 31 January 2014 / Accepted: 22 May 2014 / Published online: 15 June 2014

Abstract This study examines the radical nature and spin symmetry of the ground state of the quasi-linear acene and two-dimensional periacene series. For this purpose, high-level ab initio calculations have been performed using the multireference averaged quadratic coupled cluster theory and the COLUMBUS program package. A reference space consisting of restricted and complete active spaces is taken for the π-conjugated space, correlating 16 electrons with 16 orbitals with the most pronounced open-shell character for the acenes and a complete active-space reference approach with eight electrons in eight orbitals for the periacenes. This reference space is used to construct the total configuration space by means of single and double excitations. By comparison with more extended calculations, it is shown that a focus on the π space with a 6-31G basis set is sufficient to describe the major features of the electronic character of these compounds. The present findings suggest that the ground state is a singlet for the smaller members of these series, but that for the larger ones, singlet and triplet states are quasi-degenerate. Both the acenes and periacenes exhibit significant polyradical character beyond the traditional diradical.

Keywords Singlet–triplet splitting · MR-AQCC · Unpaired electron density · Natural orbitals

Dedicated to the memory of Professor Isaiah Shavitt and published as part of the special collection of articles celebrating his many contributions.

Electronic supplementary material The online version of this article (doi:10.1007/s00214-014-1511-8) contains supplementary material, which is available to authorized users.

S. Horn · H. Lischka (✉)
Department of Chemistry and Biochemistry, Texas Tech University, Lubbock, TX 79409-1061, USA
e-mail: hans.lischka@univie.ac.at

F. Plasser
Interdisciplinary Center for Scientific Computing, Ruprecht-Karls-University, Im Neuenheimer Feld 368, 69120 Heidelberg, Germany

T. Müller
Institute of Advanced Simulation, Jülich Supercomputer Centre, Forschungszentrum Jülich, 52425 Jülich, Germany

F. Libisch · J. Burgdörfer
Institute for Theoretical Physics, Vienna University of Technology, Wiedner Hauptstrasse 8–10, 1040 Vienna, Austria

H. Lischka
Institute for Theoretical Chemistry, University of Vienna, Waehringerstrasse 17, 1090 Vienna, Austria

1 Introduction

Over the past decade, there has been significant rise in graphene research due to its potential application in nanoelectronics [1] and organic semiconductors [2]. This enthusiasm was triggered by the so-called scotch tape isolation [3] performed in 2004 by Geim and Novoselov. After isolating a one-atom-thick sheet of graphite, which they would coin as "graphene," many experimental and theoretical groups began to explore the peculiar electronic properties of this zero-bandgap semiconductor [4]. Quasi-linear acenes (Fig. 1a) and two-dimensional nanoribbons [5] (Fig. 1b) have been frequently used to investigate the fascinating electronic properties of graphene. The synthesis of quasi-linear *n*-acenes is possible up to 9-acene, but beyond 5-acene, measures must be taken to overcome its high reactivity. For hexacene, emersion in silicone oil solution has been used by Ref. [6]. Heptacene has been

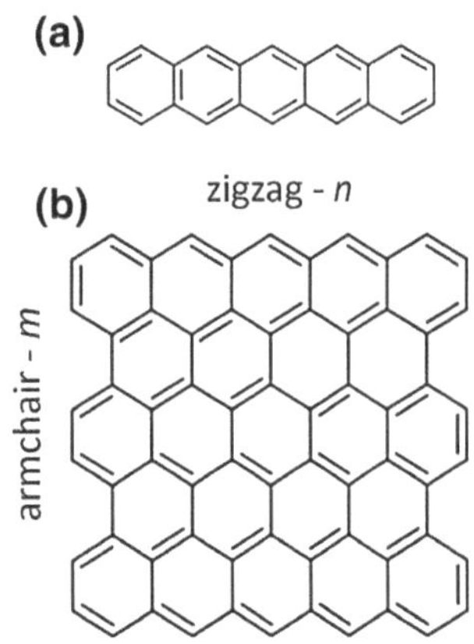

Fig. 1 Structures **a** n-acene and **b** (ma, nz) periacene

photogenerated in a polymethyl methacrylate matrix [7]. Octacene and nonacene were kept at very low temperatures ($\sim$30 K) in a solid argon matrix by Tonshoff and Bettinger [8]. Additionally, although not preserving the true molecules, significant substitution by bulky groups has been used to stabilize the acenes [9].

Extensive theoretical work has been performed on acenes [10–13] and graphene flakes [14–18] using density functional theory (DFT) calculations. However, because of the polyradical character of these systems, an unrestricted approach had to be used [10] with concomitant spin contamination, demonstrating the energetic instability of the restricted DFT procedure [11]. Furthering this, Scuseria et al. [12] used a spin-projected UHF (SUHF) to overcome the deficiencies in a standard single-determinant HF approach. As an alternative, the density matrix renormalization group (DMRG) [19, 20] and the active-space variational two-electron reduced density matrix (2-RDM) methods [21, 22] aim at an exact solution within a full configuration space spanned by a limited basis set (minimal or non-polarized double zeta basis sets) in the π space. The spin-flip configuration interaction method [23] and coupled cluster with singles, doubles, and non-iterative triples CCSD(T) [24, 25] have been applied as well. Recently, multireference averaged quadratic coupled cluster (MR-AQCC) calculations [26], with all σ molecular orbitals (MOs) frozen, have been performed.

The extraordinary spin polarization and half-metallic properties of zigzag graphene nanoribbons (GRN) (Fig. 1) [5] have been demonstrated in many investigations based on general considerations using Clar's sextet theory [27, 28] and by explicit quantum chemical calculations using DFT [14, 16, 18, 29], DMRG [19, 20], 2-RDM [21, 22], and MR-AQCC methods [26]. Singlet–triplet (S-T) splitting gives important information about the radical character of a compound. It was first estimated by Angliker et al. [6] that the ground state of nonacene and beyond was a triplet state. This estimate was based on the extrapolation of the experimentally observed singlet–triplet splitting available between benzene and pentacene to nonacene. But in 2010, Tonshoff and Bettinger [8] concluded from the existence of a finite optical band gap deduced from Vis/NIR spectroscopy that nonacene was in fact a singlet, while, e.g., in [30], EPR spectra for substituted nonacene were observed indicating a non-singlet state. Additionally, several theoretical groups [19, 31–33] report S-T splittings for acenes and S–S excitation energies have been examined as well [34]. For a detailed discussion on these findings, see Sect. 3.

The present study explores the spin symmetry and radical nature of graphene nanoribbons via quasi-linear acenes as well as two-dimensional periacenes. There are two questions this study seeks to answer: (1) What is the spin state of the ground state of acenes and periacenes, and can this knowledge then be used to extrapolate this to a larger graphene nanoribbon? (2) What can be said about the radical nature of graphene?

The first question posed is addressed with calculations using multiconfigurational and multireference ab initio methods. Multireference averaged quadratic coupled cluster (MR-AQCC) [35] calculations are particularly useful for large aromatic systems and radical systems [26, 36, 37], of which the molecules in this study are both. This approach aims at a compact representation of complicated electronic wave functions by constructing a reference space containing the most important quasi-degenerate configurations (non-dynamical electron correlation) and representing the dynamical correlation by means of single and double excitations [38]. Size extensivity contributions are taken care of by the AQCC approach. To answer the second question, an analysis of the radical nature of the acenes is performed by two means: tracking the evolution of both the natural orbital (NO) occupations with increasing chain length and the total number of effectively unpaired electrons (N_U). The method for determining N_U was first proposed by Takatsuka et al. [39] and was further developed by Staroverov and Davidson [40]. The methodology eventually used in this paper comes from Head-Gordon [41]. Additionally, we investigate the influence of freezing the σ orbitals on the NO occupations and the S-T splittings. The effect of polarized basis sets on these quantities is considered as well.

2 Computational methods

For the purposes of this study, the "zigzag" edge of the acenes and periacenes is taken to be along the x-axis in the x–y plane. The acenes were examined from n = 2–13, n being the number of fused benzene rings in the chain.

The periacenes were studied from (5*a*,2*z*) to (5*a*,5*z*), in which *m* and *n* count the number of benzene rings along each direction and *a* and *z* denote armchair and zigzag boundaries, respectively (Fig. 1b). These molecules belong to the D_{2h} symmetry group, with b_{1u}, b_{2g}, b_{3g}, a_u being the irreducible representations that correspond to the π orbitals. The geometries for the acenes and periacenes were taken from Ref. [26], which had been optimized with second-order Møller–Plesset perturbation theory [42] including the resolution of the identity approach (RI-MP2) [43, 44] with an SV(P) basis set [45] under D_{2h} symmetry.

As a first option in constructing the multireference wave function, a complete active-space (CAS) approach was chosen. In this, eight electrons were correlated with eight orbitals (two taken from each π symmetry). This resulting CAS(8,8), although modified at times, was used throughout the calculations of the periacenes. CAS self-consistent field (CASSCF) calculations were performed to determine the MOs. This CAS(8,8) was used as a reference space in constructing a multireference (MR) expansion in configuration state functions (CSF) with all single and double excitations [38]. This MR expansion was used in two ways. The multireference averaged quadratic coupled cluster (MR-AQCC) method [35] is our preferred method in the reliable treatment of static and dynamic electron correlation effects. In calculations of higher triplet states, which were performed to determine the symmetry of the lowest one, the MR-AQCC method suffered from the problem of intruder states. In these cases, and for the purpose of comparison with MR-AQCC calculations on the lowest triplet state, the multireference configuration interaction with singles and doubles (MR-CISD) method was employed, also using the same configuration space as for the MR-AQCC approach. To account for quadruple and higher excitations, the renormalized Davidson correction [38, 46] (denoted + Q) was used as follows:

$$E_Q = \frac{(1 - c_0^2)(E_{CI} - E_{SCF})}{c_0^2} \qquad (1)$$

in which c_0^2 is the sum of the squared coefficients of the reference configurations in the MR-CISD expansion.

We found that the periacenes were well described by the MR-AQCC/CAS(8,8) approach. In the case of the larger members of the *n*-acene series (*n* larger than 8), a significant number of intruder states appeared. This resulted in configurations (mostly singly excited either from the doubly occupied orbitals into the CAS or from the CAS to the virtual orbital space) not included in the references. Configurations exceeding a threshold of 0.01 in their weight in the CSF expansion were considered intruder states. Moreover, in such cases, the reference space was not preserved throughout the calculations. Consequently, the symmetry numbering of the output natural orbitals (NOs) did not correspond to that of the MOs used for the construction of the reference space in the input. We therefore set up an extended scheme as second option, in which the number of internal orbitals was significantly increased. However, including these orbitals as a CAS proved to be too costly. Following the procedures used by Plasser et al. [26], a set of active orbitals was introduced by moving some reference doubly occupied orbitals into the restricted active space (RAS) from which only single excitations were allowed in the process of constructing the references. Similarly, auxiliary orbitals were introduced by moving virtual orbitals into the auxiliary active space (AUX). Only a single electron, at most, was permitted into the AUX. The set of reference CSFs constructed from this RAS/CAS/AUX scheme was then used to construct the entire CSF expansion space by means of single and double excitations as described above for the CAS(8,8) reference space.

The RAS is composed of orbitals that are generally in the 1.84*e*–1.90*e* range for NO occupation. The number of RAS orbitals used for $n = 2$–6 was *n*, and for $n = 7$–13, six orbitals were maintained as in the 6-acene. Their contribution to the reference space is greater than that of the doubly occupied space, but less than that of the CAS. Based on experience with the calculations on the singlet state of the *n*-acenes [26], a CAS(4,4) was taken for all *n*, correlating 4 electrons with 4 orbitals. These orbitals exhibited the most pronounced open-shell character in the CAS(8,8) calculation. They are, therefore, given the largest weight in the reference space. The range of their NO occupation is from 0.25*e* to 1.75*e*. The auxiliary space (AUX) is the excitation space for the reference configurations. The AUX orbitals generally have an NO occupation of 0.07*e*–0.15*e*.

This calculation is referred to as RAS/CAS(4,4)/AUX in the following. A total of 16 electrons/orbitals are used, at most, in this reference space, as opposed to the 8 in the CAS(8,8) reference space. This RAS/CAS(4,4)/AUX representation is used at the MCSCF, MR-CISD, and MR-AQCC levels. For the *n*-acenes, $n = 9$–11, one additional configuration was added to the reference space to accommodate for a significant intruder state in the MR-AQCC wave function.

In terms of molecular size, the most extended set of calculations were performed by freezing all σ orbitals and using the 6-31G basis set [47]. Starting from a closed-shell self-consistent field (SCF) calculation, all occupied and virtual σ orbitals were frozen by transforming the one- and two-electron integrals into a new basis, keeping only the π orbitals. The effect of the frozen σ orbitals was folded into effective one-electron Hamilton matrix elements according to the formalism described by Shavitt [48]. To validate the use of the 6-31G basis set and the freezing of σ orbitals, two additional sets of calculations were performed for the n-acenes up to $n = 9$. The first was identical to the scheme outlined above, except a 6-31G* basis set [47] was used. The second series of calculations also used a 6-31G* basis set, but froze only the σ core orbitals. In these calculations though, a smaller reference space was chosen. For $n = 2$–4, a CAS(4,4) was sufficient as a reference space, which is the same CAS(4,4) as in the RAS/CAS(4,4)/AUX. For $n = 5$, one orbital was added to both the RAS and the AUX space based on their proximity to CAS occupations. For $n = 6$–7, two RAS and AUX orbitals were added, and an additional amount of σ orbitals equal to the number of 1s core orbitals was frozen. This procedure was tested for $n = 7$, and no significant deviation in S-T splitting energies was found. This second calculation was only used up to $n = 7$.

The effective unpaired electron densities and total number of effectively unpaired electrons (N_U) were computed [39–41]. To avoid overemphasizing the contribution of the natural orbitals (NOs) that are nearly occupied or nearly unoccupied, we chose to use the nonlinear model suggested in [41] where N_U is given by

$$N_U = \sum_{i=1}^{M} n_i^2 (2 - n_i)^2 \quad (2)$$

in which n_i is the occupation of the ith NO and M is the number of NOs. In this, the open-shell character is given the largest weight.

The orbital occupation scheme of the doubly occupied orbitals was obtained by performing a DFT calculation with the Becke–Perdew functional [49, 50] and a 6-31G* basis set [47]. Geometry optimizations were performed with the TURBOMOLE [51, 52] package, and all other calculations use COLUMBUS [53–55].

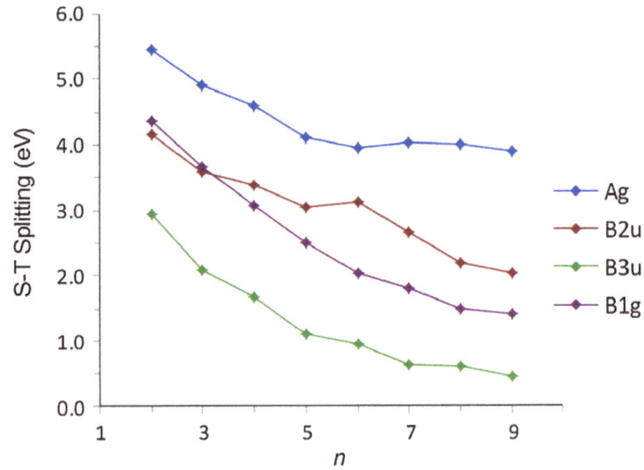

Fig. 2 Singlet–triplet splitting of n-acenes ($n = 2$–9) with respect to the ^1A_g ground state using the π-MR-CISD + Q/CAS(8,8)/6-31G approach

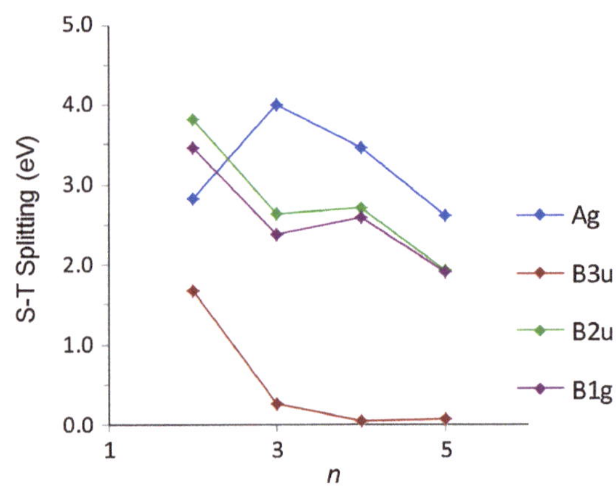

Fig. 3 Singlet–triplet splitting of (5a,nz) periacenes ($n = 2$–5) with respect to the ^1A_g ground state using the π-MR-CISD + Q/CAS(8,8)/6-31G approach

3 Results

3.1 Singlet–triplet splitting

A direct product of all the π symmetries for D_{2h} results in the following symmetries for the triplet states: ^3A_g, $^3B_{3u}$, $^3B_{2u}$, and $^3B_{1g}$. The S-T splitting for the π-MR-CISD + Q/CAS(8,8)/6-31G calculation for each of these symmetries (i.e., the excitation energies from the 1^1A_g state to the respective lowest triplet state in these symmetries) is compared in order to determine the lowest energy state. It is shown in Figs. 2 and 3 (Tables S1–2 in the Supplementary Material contain the data in tabular form; the same procedure has been followed for the other figures showing S-T splittings) that the $^3B_{3u}$ state has the smallest S-T splitting for all molecules. In light of this, only the $^3B_{3u}$ symmetry was considered in the remaining work.

The S-T splitting for the MCSCF, MR-CISD, and MR-AQCC calculations for the acenes (Fig. 4, Table S3) remains positive in all instances, indicating that the system maintains singlet ground state character for these computational levels and all values of n investigated. However, the relative theoretical–experimental error, as shown in Table 1, shows that the π-MR-AQCC/RAS/CAS(4,4)/AUX/6-31G results deviate by 0.27 ± 0.06 eV from the experimental data [31, 56–59]. Correcting our results by this value shows that our calculations predict 11-acene as the first case where the triplet state is lower than the singlet state, as it can be seen from S-T splitting of 0.26 in Table S3. This value of $n = 11$ compares quite well with the value of $n = 9$ deduced by Angliker et al. [6] from

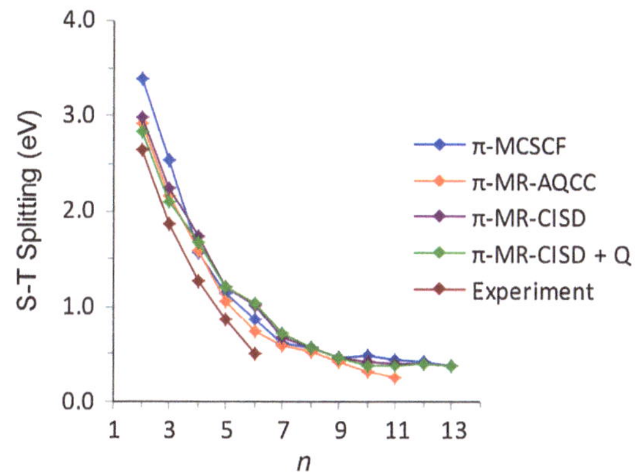

Fig. 4 Singlet–triplet splitting for *n*-acenes ($n = 2$–13) at four different levels of theory using a π-RAS/CAS(4,4)/AUX reference space and the 6-31G basis set in comparison with experimental data

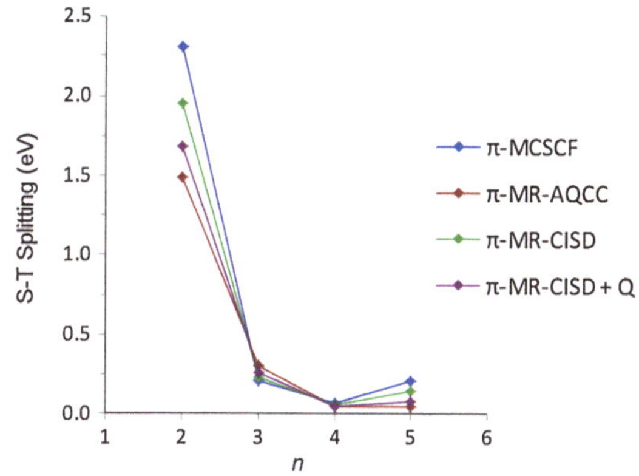

Fig. 5 Singlet–triplet splitting for (5*a*, *nz*) periacenes ($n = 2$–5) at four different levels of theory with the σ system frozen using a CAS(8,8) reference configuration set

Table 1 Average difference and standard deviation of the S-T splitting (eV) between theory and experiment for the linear acenes $n = 2$–6

Method	Average ± standard deviation
π-MCSCF/6-31G	0.48 ± 0.22
π-MCSCF/6-31G*	0.48 ± 0.18
Total-MCSCF/6-31G*	0.70 ± 0.13
π-MR-CISD/6-31G	0.41 ± 0.06
π-MR-CISD + Q/6-31G	0.33 ± 0.13
π-MR-AQCC/6-31G	0.27 ± 0.05
π-MR-AQCC/6-31G*	0.29 ± 0.08
Total-MR-AQCC/6-31G*	0.40 ± 0.19

experimental data as mentioned above. The difference between these two *n* values comes from a slight leveling off of the MR-AQCC results (Fig. 4).

π-MR-CISD/6-31G and π-MR-CISD + Q/6-31G calculations have been performed for $n = 12$ and 13 with the present RAS/CAS/AUX scheme, for which the MR-AQCC calculation showed significant intruder states. It is interesting to note that until $n = 10$, the former two methods as well as MCSCF show very good agreement with the MR-AQCC data, and only starting at $n = 10$, no further reduction of the S-T splitting is observed for these methods. This good agreement with the MR-AQCC reference data is partly due to the flexible reference configuration set, but probably also due to error cancelation in computing the energy difference between singlet and triplet states.

While it is possible that acenes maintain a singlet ground state as *n* approaches infinity, they surely have nearly degenerate singlet and triplet ground states. A similar situation is present in the much larger periacene system, in which the S-T splitting, as shown in Fig. 5 (Table S4), drops to nearly zero (0.05 eV) by the (5*a*,4*z*). From DFT calculations, there is no clear consensus on the spin multiplicity of the acenes beyond octacene. Houk et al. [31] and Rayne et al. [32] show that nonacene is a triplet, while Bendikov et al. [10] report a singlet state, at least through decacene. Going beyond DFT, Casanova and Head-Gordon [23] developed a single-reference ($^5A_1$) spin-flip configuration interaction method that predicts a singlet up to 20-acene, though the S-T splitting is nearly zero (0.09 eV). MR-CISD was used recently with an S-T splitting for decacene (Table 3 in Ref. [33]) consistent with our MR-CISD values (Fig. 4, Table S3). DMRG calculations [19] have been interpreted to result in a small but finite singlet–triplet gap. Hajgató et al. [25] concluded that the S-T splitting in the large *n* limit would be around 0.17 eV and went later on [24] to show with single-reference CCSD(T) that this limit is ~0.06 eV. Their study included up to $n = 11$ in the acene series and gave an S-T splitting for $n = 11$ of 0.31 eV. This value falls within the same near-degeneracy that our adjusted MR-AQCC predicts. To our knowledge, there is no current experimental S-T data for the periacene series. Additionally, we have only found one other group [60] that reported theoretical S-T data for periacenes. With R(U)B3LYP/3-21G, they found a −0.33 eV S-T splitting for the (5*a*,5*z*) periacene, indicating a triplet ground state. For the MR-AQCC/CAS(8,8)/6-31G calculation, we find a splitting of 0.05 eV. Accordingly, all these calculations strongly suggest that an actual graphene nanoribbon has a ground state with nearly degenerate singlet and triplet states.

The comparison between the basis sets (6-31G and 6-31G*) and the two different electronic systems (π-conjugated and total) for the linear acenes, as shown in Fig. 6 (Table S5), leads to the conclusion that the π-conjugated system with a 6-31G basis set is quite adequate to describe the S-T splitting for these molecules. There is virtually no

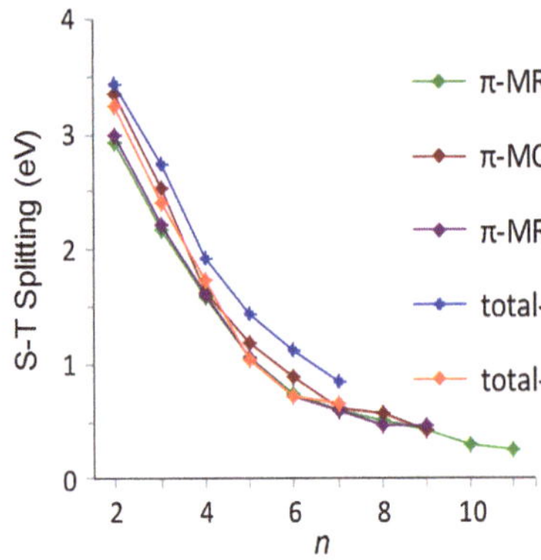

Fig. 6 Singlet–triplet splitting for the *n*-acenes computed at MCSCF and MR-AQCC levels for three different cases: π-6-31G, π-6-31G*, and total-6-31G*

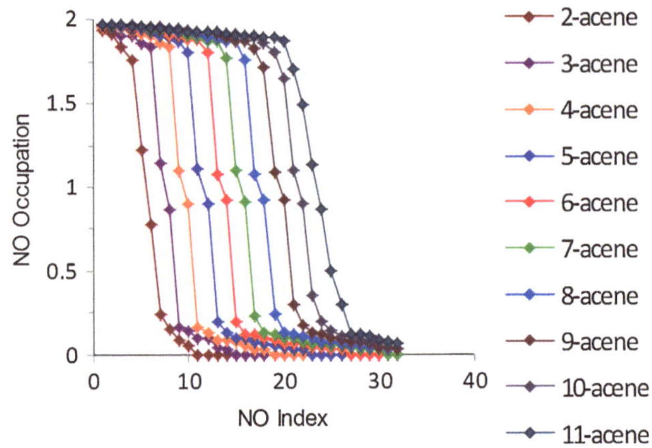

Fig. 7 Natural orbital (NO) occupations of the $^3B_{3u}$ state of *n*-acenes ($n = 2$–11) obtained from π-MR-AQCC/RAS/CAS(4,4)/AUX/6-31G calculations

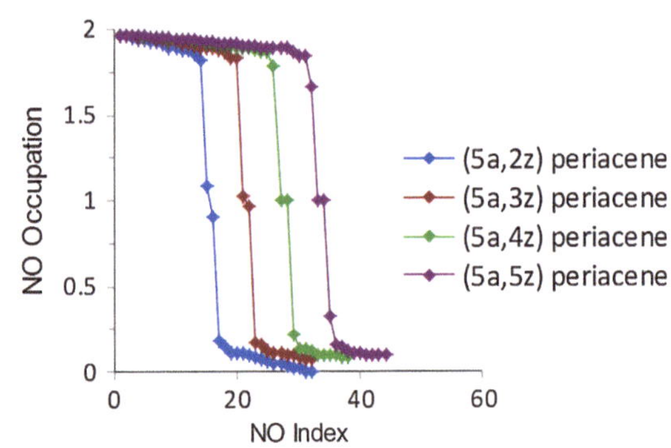

Fig. 8 Natural orbital (NO) occupations of the $^3B_{3u}$ state of (5*a*,*nz*) periacenes ($n = 2$–5) obtained from π-MR-AQCC/CAS(8,8)/6-31G calculations

difference in the relative S-T splittings for the three MR-AQCC variants. This is true in particular for the larger systems beyond pentacene where the three values are always within an interval of 0.1 eV. Only the S-T splittings computed at the MCSCF level deviate a bit from the MR-AQCC results. Also, what can be seen from these graphs is that a systematic error with respect to experiment, as shown in Table 1, still arises in the MR-AQCC, regardless of the basis or level of σ–π correlation. Thus, it is not the freezing of the σ orbitals that leads to the error, and accordingly, the SCF approach seems to be sufficient for the description of the σ system. While the MR-AQCC corrects for the size extensivity issue present in MR-CISD, it is likely the truncated nature of the calculations that still causes this error.

3.2 Radical character

It is primarily the zigzag edge of graphene nanoribbons that contributes to its high radical character. Nakada et al. [61] showed analytically that there is a degenerate flat band near the Fermi level on the zigzag edge, which is not present on the armchair edge. Jiang et al. [15] went on to show that the carbon atoms on the zigzag edge of a graphene nanoribbon are more chemically reactive than those of a graphene sheet, nanotube, and nanoribbon armchair edge, having a bond dissociation energy at least 1.2 eV times higher than any of them when bonded to hydrogen. The radical character of the system is examined via two means in this paper: the NO occupation and effective number of unpaired electrons.

3.2.1 NO occupation

The NO occupations are derived from the spin-averaged one-electron density matrix, thus leading to a spectrum of occupation from zero to two. This can be seen for the $^3B_{3u}$ state for the acenes in Fig. 7 and for the $^3B_{3u}$ state for the periacenes in Fig. 8. The HONO and LUNO for each molecule can be recognized by the occupations closest to one for the whole series. The acene series starts for $n = 2$ with a HONO/LUNO difference of about 0.56, which is reduced to about 0.27. With increasing *n*, additional NO occupations deviating from the limit values of zero and two appear, indicating the evolving polyradical character. This evolution of increasing radical character is analogous to the one found for the lowest singlet state of acenes in the MR-AQCC [26], DMRG [19], 2-RDM [21], and projected Hartree–Fock calculations [12]. DMRG calculations on larger acenes [20] show for the singlet state a HONO/LUNO gap of ~0.5*e*, which is classified by the authors as single radical occupation being, in their conclusions, similar to the just-mentioned other calculations. The singlet and triplet NO occupation plots differ primarily for the initial members of the acene series, which starts with HONO/LUNO occupations for the singlet case close to zero and two, respectively, whereas in the triplet case, the aforementioned open-shell NOs already appear at the very beginning of the series. However, after a certain size of the acene chain, the NO occupation graphs look very similar. The same situation is seen with the periacenes in Fig. 3 of Ref. [26] and Fig. 8. HONO/LUNO plots for the 11-acene and (5*a*,5*z*) periacene for both the ^1A_g and $^3B_{3u}$ states are included in the supplementary material (Figures S1–4).

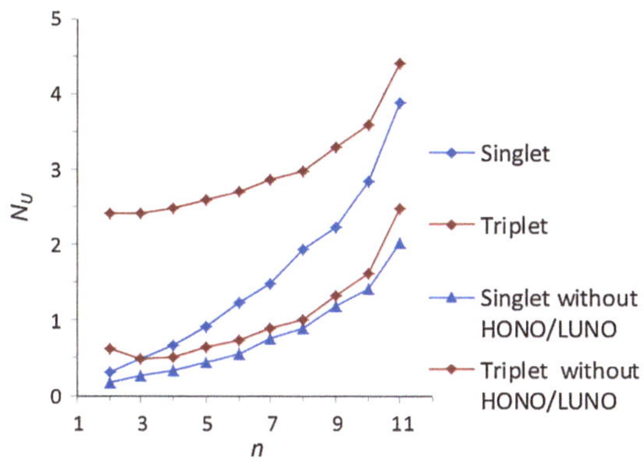

Fig. 9 Effective number of unpaired electrons for the ^1A_g and $^3B_{3u}$ states of the *n*-acenes with and without the HONO and LUNO included using the π-MR-AQCC/RAS/CAS(4,4)/AUX/6-31G method

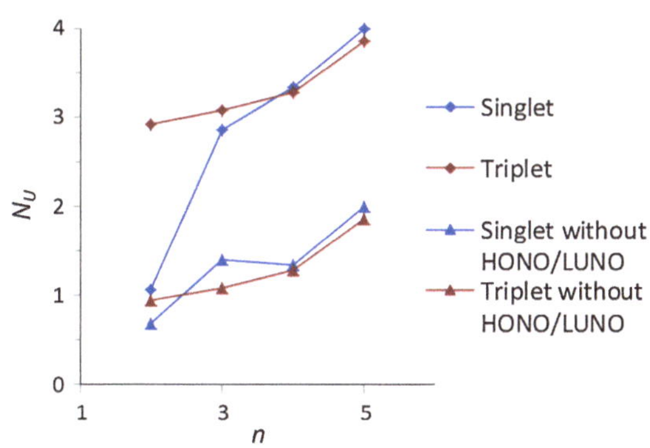

Fig. 10 Effective number of unpaired electrons for the ^1A_g and $^3B_{3u}$ states of the (5*a*,*nz*) periacene series with and without the HONO and LUNO included using the π-MR-AQCC/CAS(8,8)/6-31G method

3.2.2 *Effective number of unpaired electrons*

In Fig. 9, the N_U values (Eq. 2) for the singlet and triplet states of the acene series are presented. The triplet curve starts from a value of 2.4 for $n = 2$, which is consistent with the strong open-shell character shown in Fig. 7. The singlet state starts with closed-shell character, but then, it rapidly catches up with the triplet state. It has been found for a true biradical (the stacked tetracyanoethylene anion dimer complex with two potassium cations as counter ions) [62] that the N_U values could be modeled by the HONO/LUNO contributions alone. The situation is completely different here. There is still a strong increase in N_U even after the HONO/LUNO contribution is deducted (Fig. 9). A similar situation is found for the periacenes (Fig. 10). It is interesting to note that the partial number of unpaired electrons is almost identical for the singlet and triplet states in all cases considered. This suggests similarities between the many-particle wave functions in both cases and that the single major difference lies in the frontier orbitals, which are only in the triplet case necessarily occupied by unpaired electrons. As soon as the singlet acquires radical character its total N_U reaches the value of the triplets. Plots of the unpaired densities including a Mulliken population analysis are given in Figs. 11 and 12. The unpaired density is always located on the zigzag edge. There is a specific concentration on the center, but a delocalization over the whole edge is visible.

Fig. 11 Unpaired electron density for the ^1A_g (top) and $^3B_{3u}$ (bottom) states of the 11-acene (isovalue 0.005 a.u.) of the π-MR-AQCC/RAS(6)/CAS (4,4)/AUX(6)/6-31G calculation with individual atomic populations computed from a Mulliken analysis

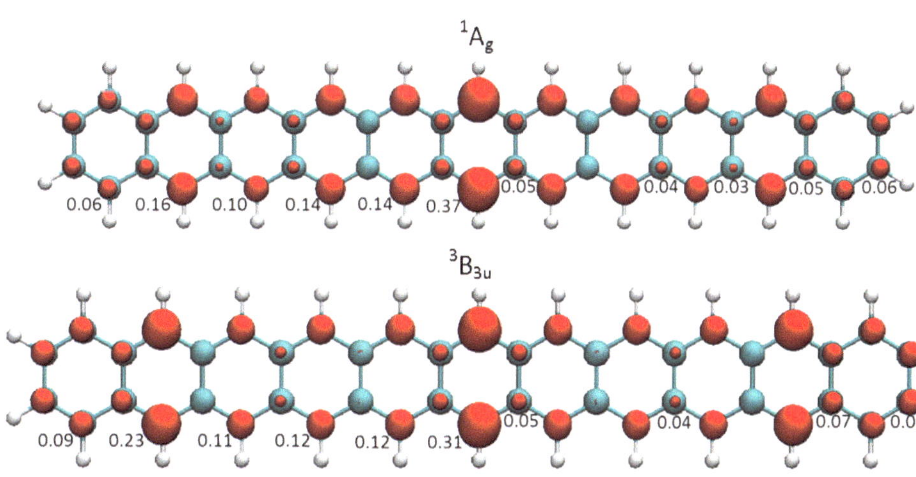

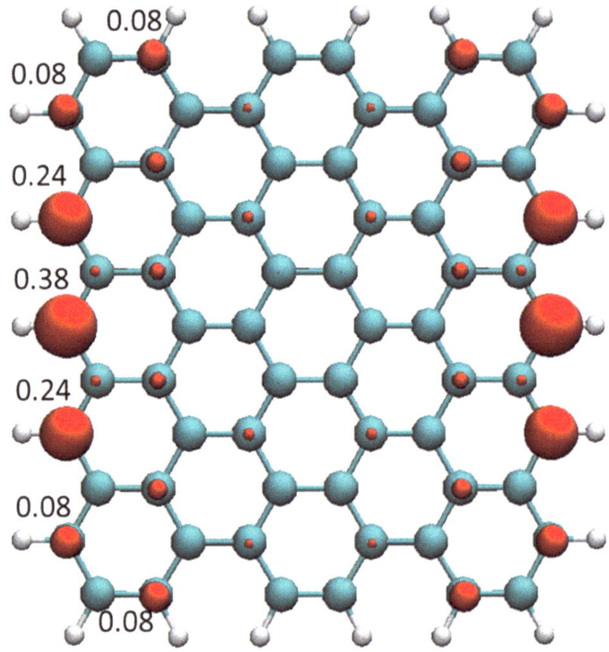

Fig. 12 Unpaired electron density for the $^3B_{3u}$ state of the (5*a*,5*z*) periacene (isovalue 0.005 a.u.) of the π-MR-AQCC/CAS(8,8)/6-31G calculation with individual atomic populations computed from a Mulliken analysis

4 Conclusions

The purpose of this work was twofold: first, to determine whether the ground state of graphene nanoribbons is a singlet or a triplet and second, to qualify/quantify the multiradical nature of these systems. This was accomplished by performing high-level ab initio multiconfigurational and multireference calculations using the COLUMBUS program on quasi-linear acenes and two-dimensional periacenes. It is clearly seen that the validity of the results of these calculations is independent of both the basis set and the amount of correlated σ electrons. For both systems, a near-degeneracy of the singlet and triplet states is found for sufficiently extended systems. For the n-acene series, this happens at around $n = 11$. In particular, in the periacene case, the S-T splitting drops rapidly to nearly zero eV by the (5*a*,4*z*). It is clear, however, from the calculations that graphene is multiradical in nature as the number of unpaired electrons increases with chain length. As in the case of the singlet state, the unpaired densities in the periacenes are concentrated along the zigzag edges with only minor extension into the inner parts of the nanosheet. The multiradical character of the acenes and periacenes leads to very high reactivity, which will be amenable to tuning, either by structural defects or by heteroatoms. Further work will examine various forms of defects in regard to their stability and electronic properties.

Acknowledgments This work was supported by the National Science Foundation under Project No. CHE-1213263, by the Austrian Science Fund (SFB F41, ViCoM, and Project P20893-N19), and the Robert A. Welch Foundation under Grant No. D-0005. Shawn Horn is funded by a research fellowship at Texas Tech University. Computer time at the Vienna Scientific Cluster (Project Nos. 70151 and 70376) and by the Chemistry Computational Cluster of Texas Tech University is gratefully acknowledged.

References

1. Geim A, Novoselov K (2007) Nat Mater 6:183
2. Bendikov M, Wudl F, Perepichka D (2004) Chem Rev 104:4891
3. Novoselov K, Jiang D, Schedin F, Booth T, Khotkevich V, Morozov S, Geim A (2005) Proc Natl Acad Sci 102:10451
4. Lu G, Yu K, Wen Z, Chen J (2013) Nanoscale 5:1353
5. Son Y, Cohen M, Louie S (2006) Nature 444:347
6. Angliker H, Rommel E, Wirz J (1982) Chem Phys Lett 87:208
7. Mondal R, Shah B, Neckers D (2006) J Am Chem Soc 128:9612
8. Tonshoff C, Bettinger H (2010) Angew Chem Int Ed 49:4125
9. Zade SS, Bendikov M (2010) Angew Chem Int Edit 49:4012
10. Bendikov M, Duong H, Starkey K, Houk K, Carter E, Wudl F (2004) J Am Chem Soc 126:7416
11. Jiang D, Dai S (2008) J Phys Chem A 112:332
12. Rivero P, Jimenez-Hoyos C, Scuseria G (2013) J Phys Chem B 117:12750
13. Zimmerman P, Bell F, Casanova D, Head-Gordon M (2011) J Am Chem Soc 133:19944
14. Hod O, Barone V, Scuseria G (2008) Chem Phys Lett 466:72
15. Jiang D, Sumpter B, Dai S (2007) J Chem Phys 126:134701
16. Nagai H, Nakano M, Yoneda K, Kishi R, Takahashi H, Shimizu A, Kubo T, Kamada K, Ohta K, Botek E, Champagne B (2010) Chem Phys Lett 489:212
17. Barone V, Hod O, Peralta J, Scuseria G (2011) Acc Chem Res 44:269
18. Jiang DE, Dai S (2008) Chem Phys Lett 466:72
19. Hachmann J, Dorando J, Aviles M, Chan G (2007) J Chem Phys 127:134309
20. Mizukami W, Kurashige Y, Yanai T (2012) J Chem Theory Comput 9:401–407
21. Gidofalvi G, Mazziotti D (2008) J Chem Phys 129:134108
22. Pelzer K, Greenman L, Gidofalvi G, Mazziotti D (2011) J Phys Chem A 115:5632
23. Casanova D, Head-Gordon M (2009) Phys Chem Chem Phys 11:9779
24. Hajgato B, Huzak M, Deleuze M (2011) J Phys Chem A 115:9282
25. Hajgato B, Szieberth D, Geerlings P, de Proft F, Deleuze M (2009) J Chem Phys 131:22
26. Plasser F, Pasalic H, Gerzabek M, Libisch F, Reiter R, Burgdorfer J, Muller T, Shepard R, Lischka H (2013) Angew Chem Int Ed 52:2581
27. Wassmann T, Seitsonen AP, Saitta AM, Lazzeri M, Mauri F (2010) J Am Chem Soc 132:3440
28. Balaban AT, Klein DJ (2009) J Phys Chem C 113:19123
29. Pisani L, Chan JA, Montanari B, Harrison NM (2007) Phys Rev B 75:064418
30. Purushotharman B, Bruzek M, Parkin S, Miller A, Anthony J (2011) Angew Chem Int Ed 50:7013
31. Houk K, Lee P, Nendel M (2001) J Org Chem 66:5517
32. Rayne S, Forest K (2011) Comput Theor Chem 976:105
33. Chakraborty H, Shukla A (2013) J Phys Chem A 117:14220
34. Knippenberg S, Starcke JH, Wormit M, Dreuw A (2010) Mol Phys 108:2801

35. Szalay P, Bartlett R (1993) Chem Phys Lett 214:481
36. Antol I, Eckert-Maksic M, Lischka H, and Maksic Z (2007) Eur J Org Chem 3173
37. Wang E, Parish C, Lischka H (2008) J Chem Phys 129:044306
38. Szalay P, Muller T, Gidofalvi G, Lischka H, Shepard R (2012) Chem Rev 112:108
39. Takatsuka K, Fueno T, Yamaguchi K (1978) Theor Chim Acta 48:175
40. Staroverov V, Davidson E (2000) Chem Phys Lett 330:161
41. Head-Gordon M (2003) Chem Phys Lett 372:508
42. Moller C, Plesset M (1934) Phys Rev 46:0618
43. Vahtras O, Almlof J, Feyereisen M (1993) Chem Phys Lett 213:514
44. Weigend F, Haser M (1997) Theor Chem Acc 97:331
45. Schafer A, Horn H, Ahlrichs R (1997) J Chem Phys 97:2571
46. Luken W (1978) Chem Phys Lett 58:421
47. Hehre W, Ditchfield R, Pople J (1972) J Chem Phys 56:2257
48. Hosteny R, Dunning T Jr, Gilman R, Pipano A, Shavitt I (1975) J Chem Phys 62:4764
49. Becke A (1988) Phys Rev A 38:3098
50. Perdew J (1986) Phys Rev B 33:8822
51. Haser M, Ahlrichs R (1989) J Comput Chem 10:104
52. Treutler O, Ahlrichs R (1995) J Chem Phys 102:346
53. Lischka H, Shepard R, Pitzer R, Shavitt I, Dallos M, Muller T, Szalay P, Seth M, Kedziora G, Yabushita S, Zhang Z (2001) Phys Chem Chem Phys 3:664
54. Lischka H, Shepard R, Shavitt I, Pitzer R, Dallos M, Müller T, Szalay P, Brown F, Ahlrichs R, Böhm H, Chang A, Comeau D, Gdanitz R, Dachsel H, Ehrhardt C, Ernzerhof M, Höchtl P, Irle S, G K, Kovar T, Parasuk V, Pepper M, Scharf P, Schiffer H, Schindler M, Schüler M, Seth M, Stahlberg E, Zhao J-G, Yabushita S, Z Z, Barbatti M, Matsika S, Schuurmann M, Yarkony D, Brozell S, Beck E, Blaudeau J-P, Ruckenbauer M, Sellner B, Plasser F, Szymczak J (2012) COLUMBUS, an ab initio electronic structure program, release 7.0
55. Lischka H, Muller T, Szalay PG, Shavitt I, Pitzer RM, Shepard R (2011) Wires Comput Mol Sci 1:191
56. Birks J (1970) Photophysics of aromatic molecules. Wiley, London
57. Schiedt J, Weinkauf R (1997) Chem Phys Lett 266:201
58. Sabbatini N, Indelli M, Gandolfi M, Balzani V (1982) J Phys Chem 86:3585
59. Burgos J, Pope M, Swendberg C, Alfano R (1977) Phys Status Solidi B 83:249
60. Wang J, Zubarev D, Philpott M, Vukovic S, Lester W, Cui T, Kawazoe Y (2010) Phys Chem Chem Phys 12:9839–9844
61. Nakada K, Fujita M, Dresselhaus G, Dresselhaus M (1996) Phys Rev B 54:17954
62. Cui Z, Lischka H, Mueller T, Plasser F, Kertesz M (2013) Chem Phys Chem. doi:10.1002/cphc.201300784

Theor Chem Acc (2014) 133:1517
DOI 10.1007/s00214-014-1517-2

REGULAR ARTICLE

Mechanisms of f–f hypersensitive transition intensities of lanthanide trihalide molecules: a spin–orbit configuration interaction study

Miho Hatanaka · Satoshi Yabushita

Received: 25 December 2013 / Accepted: 6 June 2014 / Published online: 21 June 2014

Abstract The optical properties of intra-$4f^N$ transitions (f–f transitions) in lanthanide compounds are usually insensitive to the surrounding environment due to the shielding effect of the outer 5s and 5p electrons. However, there are exceptional transitions, the so-called hypersensitive transitions, whose oscillator strengths change sensitively to a small change of the surrounding environment. The mechanism of the hypersensitive transitions was explained mostly with the dynamic-coupling (DC) model. In this study, the oscillator strengths of hypersensitive transitions in lanthanide trihalides (LnX_3; Ln = Pr, Nd, Pm, Sm, Eu, Tb, Dy, Ho, Er, Tm; X = Cl, Br, I) were calculated by the multi-reference spin–orbit configuration interaction (CI) method, and the origin of the hypersensitive transition intensities was examined. To compare the intensities derived from the DC model and from the ab initio CI computations, we evaluated two Judd–Ofelt intensity parameters: τ_2(dc) by the DC model and τ_2(ab) by the CI computations. Although these two parameters showed similar overall behaviors, their Ln dependences were different, suggesting the involvement of other mechanism(s) in τ_2(ab). Close examination of the spatial distributions of the transition densities and the integrand of the transition dipole moments (TDMs) suggested that the Judd–Ofelt theory contributions were also involved in τ_2(ab) with the opposite sign relative to the TDMs with the DC model in all the hypersensitive transitions of LnX_3. Moreover, the different Ln dependences in τ_2(dc) and τ_2(ab) were related to the different amount of the mixing of ligand-to-metal charge transfer configurations into the dominant $4f^N$ configurations, especially for Eu and Tb.

Keywords f–f transition · Multi-reference spin–orbit configuration interaction (MRSOCI) method · Transition dipole moment · Judd–Ofelt theory · Dynamic-coupling model · Charge transfer

1 Introduction

Lanthanide compounds have attracted attention due to their potential applications as various materials. Their intra-$4f^N$ electronic transition, the so-called f–f transition, is one of the most discriminative properties and used extensively for many optics, such as lasers, fibers, optical displays, and biosensors [1–3]. The probability of photo absorption is represented by oscillator strength, a dimensionless quantity. The oscillator strengths of f–f transitions are small, typically in the order of 10^{-6}, reflecting Laporte forbidden transitions, and their absorption and emission spectra have sharp peaks in visible, near infrared, and near ultraviolet regions. Despite the usefulness of these systems, there are only a few relevant ab initio studies of the f–f transition intensities that explicitly included ligand electrons [4–7] because these calculations are difficult due to the importance of both relativistic and electron correlation effects.

Dedicated to the memory of Professor Isaiah Shavitt and published as part of the special collection of articles celebrating his many contributions.

M. Hatanaka · S. Yabushita (✉)
Department of Chemistry, Faculty of Science and Technology, Keio University, 3-14-1 Hiyoshi, Kohoku-ku, Yokohama 223-8522, Japan
e-mail: yabusita@chem.keio.ac.jp

Present Address:
M. Hatanaka
Fukui Institute of Fundamental Chemistry, Kyoto University, 34-4 Takano-Nishihirakicho, Sakyo-ku, Kyoto 606-8103, Japan

Instead of ab initio calculations for entire molecular systems, the f–f oscillator strengths have long been studied extensively with the semi-empirical Judd–Ofelt (JO) theory [8, 9], in which only the lanthanide electrons were explicitly treated and the effects due to the ligands were handled by the traditional crystal field theory. According to this theory, Laporte forbidden f–f transitions become allowed due to the odd parity components of crystal field generated by the ligands, which induce the mixing between $4f^N$ states and the opposite parity intra-metal excitation configurations, such as $4f^{N-1}5d$ or $4f^{N-1}g$. When such perturbed $4f^N$ states are considered, the f–f oscillator strengths can be evaluated in terms of three parameters, the so-called JO intensity parameters τ_λ ($\lambda = 2, 4, 6$). Other transition mechanisms, such as vibronic transitions due to the coupling between electronic states and vibrational modes, and magnetic dipole transitions cannot be neglected for a general discussion of f–f transition intensities, especially in a centro-symmetric environment, in which all the genuine electric dipole transitions are Laporte forbidden.

Although most of f–f oscillator strengths could be explained with these mechanisms, there were still some problems. The most famous example is "a hypersensitive transition" [10], one of f–f transitions, which shows exceptional behavior among various physical phenomena caused by the 4f electrons. Excitation energies and transition intensities of typical f–f transitions are insensitive to the surrounding environment because 4f electrons are shielded by the closed-shell 5s and 5p electrons from outside. However, the intensities of the hypersensitive transitions change dramatically by a small change of surrounding environment, while their excitation energies are insensitive. Such transitions obey the selection rules of $|\Delta J| \leq 2$, $|\Delta L| \leq 2$, and $|\Delta S| = 0$, which are identical to those of electric quadrupole transitions and included in those of electric dipole transitions caused by the odd parity crystal field ($|\Delta J| \leq 6$, $|\Delta L| \leq 6$, and $|\Delta S| = 0$); however, they are covering the outside of those of magnetic dipole transitions ($|\Delta J| \leq 1$, $|\Delta L| = 1$, and $|\Delta S| = 0$) [11]. Therefore, the origin of hypersensitive transitions has been discussed not from the aspect of magnetic transitions but from electric transitions. The most notable example of the hypersensitive transitions is lanthanide trihalide (LnX_3) in vapor phase [12, 13]. A typical example is the transition $^4I_{9/2} \rightarrow {}^4G_{5/2}$ in Nd^{3+} compounds. This oscillator strength is as much as 530×10^{-6} for NdI_3 in vapor phase compared with the value of 5.6×10^{-6} in the LaF_3 crystal [12, 13].

Hypersensitive transitions are attractive and important phenomena not only from the aspects of fundamental chemistry but also on a practical side. Their characters, such as the sharp peaks, invariance of the excitation energies, and hypersensitivity of only the intensities, have been widely used in optical materials in solids, solutions, and molecules. They are very suitable for use as luminescence probes, such as in vivo luminescence probes to measure pH and the concentration of metal cations, neutral molecules, anions, and so on [1–3, 14]. The luminescent transitions that have large intensities obey the same selection rule as the hypersensitive transitions for the absorption. Therefore, to develop new luminescent materials, it is helpful to understand the detailed mechanism and to develop an appropriate theoretical scheme of the quantitative quantum mechanical calculations.

Historically, hypersensitive transitions have been interpreted with modifying the JO theory [10, 15–37]. One of the semi-empirical models based on the JO theory called dynamic-coupling (DC) model [31–33] succeeded in explaining the large hypersensitive transition intensities of LnX_3 molecules in vapor phase. In this model that had been proposed originally in the theory of optical rotation [38, 39], the crystal field generated by the oscillating induced dipole moments on the ligands was considered explicitly by including the configuration mixing of intra-ligand excitation configurations into $4f^N$ wave functions. One nice feature is that the f–f transition oscillator strengths in this DC model can be written as in the original JO theory formulation only by replacing the JO intensity parameters τ_λ with new ones τ_λ(dc).

We have previously studied the oscillator strengths and the JO intensity parameters of $LnBr_3$ with ab initio method and have clarified that the dominant mechanism of hypersensitive transition intensities could be attributable to the DC model and that the effects of f-d mixing, molecular vibration, polarization shielding, and ligand-to-metal charge transfer (LMCT) have a lesser importance [4, 5]. Through these works, the importance of LMCT effect especially in $EuBr_3$ [5] was pointed out; however, such effect is still open to discussion.

In this study, we evaluate the oscillator strength of hypersensitive transitions in LnX_3 (X = Cl, Br, I) molecules with the multi-reference spin–orbit (MRSO) CI method and discuss the origin of hypersensitive transition intensities, especially focusing on the Ln dependence, halogen dependence, and also the effect of LMCT. To compare our ab initio results and the semi-empirical concepts, such as the JO theory and the DC model, we evaluate the oscillator strengths and two kinds of JO intensity parameters τ_λ(dc) and τ_λ(ab) in Sect. 4.1. These two parameters τ_λ(dc) and τ_λ(ab), the details of which are defined in Sect. 3 and Ref. [5], represent the contributions only from the DC theory and those from all the effects considered in the ab initio calculations, respectively. If the origin of hypersensitive transition intensities can be explained by the DC model alone, these two JO intensity parameters τ_λ(dc) and τ_λ(ab) are expected to show similar behaviors. Though the halogen dependence of τ_λ(ab) is similar to that of τ_λ(dc), the Ln dependence turns out to be different. The different behaviors of these parameters indicate that there exist non-negligible effects other than the DC model. The relations between the JO theory effect and other effects, such as the DC model, were investigated previously based on the JO theory [40–44]. In these earlier studies, each of physically distinct effects was considered by including the corresponding type of perturbing configurations into the dominant $4f^N$ wave functions. On the other hands, our τ_λ(ab) contains all the effects simultaneously, by representing them as one-electron excitations from the dominant $4f^N$ wave functions. Therefore, our τ_λ(ab) is proportional to the square of the transition dipole moments (TDMs) calculated with ab initio method $\mathbf{M}_{FI}$(ab). This TDM can be decomposed as $\mathbf{M}_{FI}(\text{ab}) = \mathbf{M}_{FI}(\text{JO}) + \mathbf{M}_{FI}(\text{DC}) + \mathbf{M}_{FI}(\text{LMCT}) + \ldots$, where $\mathbf{M}_{FI}$(JO) and $\mathbf{M}_{FI}$(DC) are contributions considered in the JO theory and the DC model, respectively, and $\mathbf{M}_{FI}$(LMCT) is that from configuration mixings due to LMCT as discussed in Ref. [5]. Therefore, τ_λ(ab) must contain not only the magnitude of each transition mechanism but also their relative phases.

To analyze these contributions and their relative phases, we take two different theoretical approaches. In Sect. 4.2, as an extension of our previous decomposition analyses in Refs. [4, 5], we focus on the TDMs, especially on the integrands, to see the Ln dependence and ligand dependence. Finally, the effect of LMCT configuration mixing is discussed in Sect. 4.3, and the reason of the different behaviors of τ_λ(dc) and τ_λ(ab), especially in EuX_3, is clarified.

2 Theory

We explain two semi-empirical theories that are used for comparison of our ab initio results. It should be noted that our ab initio calculations do not assume the crystal fields, while most previous semi-empirical models employed specific crystal fields to evaluate the f–f transition intensities.

2.1 Judd–Ofelt theory

Judd [8] and independently Ofelt [9] presented treatises about the calculation method for Laporte forbidden f–f transition intensities of Ln systems with using coupling scheme for angular momenta. The basic idea is that f–f transitions can be allowed because of the mixing of opposite parity configurations, such as $4f^{N-1}5d$ or $4f^{N-1}g$ states to $4f^N$ states due to the presence of odd parity crystal field generated by the ligands. In this theory [8, 9], the oscillator strengths f between two $4f^N$ multiplets ${}^{2S+1}L_J$ and ${}^{2S'+1}L'_{J'}$ are written as

$$f = \frac{\omega}{2J+1} \sum_{\lambda=2,4,6} \tau_\lambda \left| \left\langle {}^{2S+1}L_J \middle\| \mathbf{U}^{(\lambda)} \middle\| {}^{2S'+1}L'_{J'} \right\rangle \right|^2 \quad (1)$$

where ω is the excitation energy from ${}^{2S+1}L_J$ to ${}^{2S'+1}L_J$, $\langle {}^{2S+1}L_J \| \mathbf{U}^{(\lambda)} \| {}^{2S'+1}L'_{J'} \rangle$ is the reduced matrix elements of the λ-th rank reduced tensor operators $\mathbf{U}^{(\lambda)}$. Here, τ_λ ($\lambda = 2, 4, 6$) are called "JO intensity parameters" and can be expressed as

$$\tau_\lambda = \frac{4\pi m_e c \chi_{ed}}{3\hbar} (2\lambda + 1) \sum_k (2k+1)^{-1} \sum_q \left|A_{kq}\right|^2 \Xi^2(k, \lambda) \quad (2)$$

where A_{kq} is the crystal field coefficient; $\Xi^2(k, \lambda)$ are the parameters that contain excitation energies and transition moments between $4f^N$ and the opposite parity states, such as $4f^{N-1}d$, $4f^{N-1}g$, and also continuum states; χ_{ed} is the Lorentz field correction, whose value is unity in vapor; and m_e, c, $\hbar$ are the electron mass, the light speed, and the reduced Planck constant, respectively.

In general, the parameters τ_λ have been treated as adjustable parameters and simply determined experimentally only from Eq. (1), not from Eq. (2) [11, 14]. Three τ_λ parameters can be determined empirically from the experimental data of oscillator strengths and the excitation energies for at least three transitions. Once they are determined, all the remaining oscillator strengths in the Ln system can be evaluated from Eq. (1) by substituting the values of τ_λ, ω, and the reduced matrix elements given in Refs. [45–48]. In other words, the oscillator strength can be divided into two parts, one is τ_λ that is unique for the compound and the other that depends only on the angular momentum couplings for individual transitions. From a viewpoint of analysis, these JO parameters τ_λ are also useful for measuring the hypersensitivity of f–f transition intensities. Because the electronic states (${}^{2S+1}L_J$ terms) of the initial and the final states are generally different for each Ln element, comparing the hypersensitivity for the different Ln systems would be quite difficult, if it were made only through comparing their oscillator strengths. In contrast, the JO intensity parameters do not contain the information of individual electronic structures, but they can reflect only the hypersensitivity of the f–f transition intensities. In this way, oscillator strengths of many Ln systems have been successfully explained by the JO theory. Moreover, the original JO theory has been extended and generalized by many contributors to include electron correlations [16, 23–25], relativistic effects [16, 24–30], and magnetic dipole transitions [16, 49].

In spite of the utility of the JO theory, there were still problems for the hypersensitive transition intensities. As long as the JO intensity parameters were treated as adjustable parameters for experimental data, the JO theory usually gave good agreement between experimental and calculated dipole strengths even for hypersensitive transitions. However, the original JO theory alone could not give a theoretically sound explanation for the hypersensitivity. The most severe problem was that if the JO parameters of LnX_3 molecules in D_{3h} symmetry were evaluated with ab initio methods in terms of Eq. (2), they were underestimated significantly compared to those from experimental data [12, 13]. To reproduce the significantly large oscillator strength of LnX_3 in D_{3h} symmetry within static crystal field model, symmetry reduction by vibration was considered [13, 20]. However, it was criticized [4, 17]. The vibronic effect on the f–f transition intensities is negligibly small because 4f electrons are not affected by surrounding environment; thereby, the potential energy surfaces of $4f^N$ states have similar shapes [11, 14]. Therefore, only the electronic transitions between the same vibrational quantum states are allowed [4], and most f–f transitions including hypersensitive transitions exhibit sharp peaks and a negligibly small vibronic effect. Note that, as discussed in Introduction, if Ln^{3+} is in a centro-symmetric environment, the zero-phonon electric dipole transitions are forbidden and weak vibronic f–f transitions can be exposed [17, 42, 50–54]. However, according to the observation and theoretical calculation of octahedral $ErCl_6{}^{3-}$ system [54], the sum of vibronic transition intensities for the hypersensitive transition of Er^{3+} ($^4I_{15/2} \rightarrow {}^4G_{11/2}$) is as small as 1×10^{-6}. Therefore, the vibronic coupling effect cannot be an origin of hypersensitivity even though it is essential in centro-symmetric systems.

2.2 Dynamic-coupling model

To give an explanation for the notably large oscillator strengths of hypersensitive transitions of LnX_3, Mason and Peacock et al. [31–33] proposed the "dynamic-coupling model," which was extended later for general cases by Reid and coworkers [34–37, 43, 44]. In what follows, a simple explanation of this model is provided by taking LnX_3 as examples. Let put a Ln^{3+} cation on the origin and consider an f electron whose position vector is $\mathbf{r} = (x, y, z)$. The electron is under the influence of the static crystal field due to the three ligands X^-, whose effects are considered in the JO theory. In the case of LnX_3 with D_{3h} structure, the lowest order crystal field generated by the point charges q on the ligands is third order and can be expressed as follows,

$$V_{\mathrm{CF}}^{q}(\mathbf{r}) = C\frac{q}{R_L^4}(x^3 - 3xy^2), \tag{3}$$

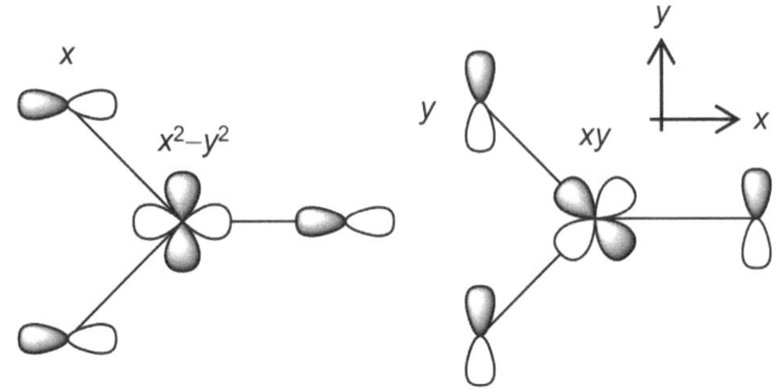

Fig. 1 Nonzero Coulombic correlation between the metal ion quadrupole moments and the ligand dipole moments in D_{3h} LnX_3 molecules

where C is a proportional constant and R_L is the distance between Ln and X.

Here, if the molecule is put in an oscillating electric field due to light, the f electron receives additional perturbation fields that contain not only the external electric field in Eq. (3) but also that newly created by the oscillating induced dipoles on the ligands. The latter field can act as a strong time-dependent crystal field, in modern terminology, an inhomogeneous near field [55]. The crystal field generated by the induced dipole moments $\boldsymbol{\mu}_{\mathrm{ind}} = \alpha\mathbf{E}$ on the ligands due to the oscillating external electric field $\mathbf{E}$ can be written as follows,

$$\begin{aligned} V_{CF}^{\mu}(\mathbf{r}) &= -\frac{\alpha\mathbf{E}}{q}\cdot\nabla V_{CF}^{q}(\mathbf{r}) \\ &= -\frac{3\alpha C}{R_L^4}\{E_x\cdot(x^2-y^2) - E_y\cdot 2xy\}, \end{aligned} \tag{4}$$

where α is the frequency-dependent polarizability of the ligand, and E_x and E_y are the components of the oscillating electric field $\mathbf{E}$. Since the external electric field $\mathbf{E}$ is oscillating, the crystal field generated by the induced dipole moments is also oscillating. The two terms in Eq. (4) can be regarded respectively as the Coulombic interaction between the quadrupole component of x^2–y^2 on Ln and the field gradient due to the x component of the induced dipoles on the ligands, and the interaction between their components of xy and y, as schematically shown in Fig. 1. In the oscillating electric field, all the induced dipole moments are oriented in the same direction, and their directions are altered after a half cycle of the oscillation. Because the transition quadrupole moments between 4f orbitals of Ln are affected by the time-dependent crystal field in Eq. (4), f–f transitions can be induced resonantly. This coupling between the transition quadrupole moment on Ln and the oscillating induced dipole moments on the ligands is the physical picture of the dynamic-coupling (DC). Note that this coupling has a nonzero contribution only when the direct product of the irreducible representations for the transition quadrupole moments on Ln^{3+} and the induced dipole moments on the ligands contain the totally symmetric representation in the point group of the complex [31, 32]. Therefore, the transition caused by the DC model is forbidden in the cases with the inversion symmetry.

To include the effect of the induced dipole moments on ligands in the framework of the JO theory, Mason et al. [31–33] considered the mixing of intra-ligand excitation configurations into $4f^N$ states. The initial and final $4f^N$ states in a Ln complex, neglecting overlap, are expanded to the first order of perturbation as follows,

$$
\begin{aligned}
\left|M_a L_g\right\rangle &= |M_a L_g) + \sum_{c,e} |M_c L_e) \frac{(M_c L_e|V|M_a L_g)}{E_a + E_g - E_c - E_e} \\
\left|M_b L_g\right\rangle &= |M_b L_g) + \sum_{c',e'} |M_{c'} L_{e'}) \frac{(M_{c'} L_{e'}|V|M_b L_g)}{E_b + E_g - E_{c'} - E_{e'}},
\end{aligned}
\tag{5}
$$

where V is the electrostatic potential between the charge distribution of the Ln ion and that of each ligand; M_c and L_e are zero-order orthonormal eigenstates of the Ln ion and single-ligand subsystems, respectively; and $|M_cL_e)$ is the simple product function of M_c and L_e [38]. Here, $c = a, b$ and $e = g$ means the ground states of Ln and the ligand sub-systems, respectively. The TDM between these perturbed $4f^N$ states is represented by

$$
\begin{aligned}
\left\langle M_a L_g|\mathbf{r}|M_b L_g\right\rangle &= \left\langle M_a L_g|\mathbf{r}(M) + \mathbf{r}(L)|M_b L_g\right\rangle \\
&= \sum_{c' \neq a} (M_a|\mathbf{r}(M)|M_{c'}) \frac{(M_{c'} L_g|V|M_b L_g)}{E_b - E_{c'}} \\
&+ \sum_{c \neq b} (M_c|\mathbf{r}(M)|M_b) \frac{(M_a L_g|V|M_c L_g)}{E_a - E_c} \\
&+ \sum_{e' \neq g} (L_g|\mathbf{r}(L)|L_{e'}) \frac{(M_a L_{e'}|V|M_b L_g)}{E_b + E_g - E_a - E_{e'}} \\
&+ \sum_{e \neq g} (L_e|\mathbf{r}(L)|L_g) \frac{(M_a L_g|V|M_b L_e)}{E_a + E_g - E_b - E_e}.
\end{aligned}
\tag{6}
$$

Here, the dipole operator $\mathbf{r}$ is divided into two contributions $\mathbf{r}(M)$ and $\mathbf{r}(L)$ which operate only within the M_c and L_e states, respectively, excluding the possibility of charge transfer effect. Focusing on the matrix elements of dipole operators in Eq. (6), the first and second terms, which contain $(M_a|\mathbf{r}(M)|M_{c'})$ and $(M_c|\mathbf{r}(M)|M_b)$, express intra-metal excitations derived from the JO theory and the third and fourth terms, which contain $(L_g|\mathbf{r}(L)|L_{e'})$ and $(L_e|\mathbf{r}(L)|L_g)$, express intra-ligand excitations derived from the DC model. When the Coulombic potential between non-overlapping transition charge distributions of $M_a \leftrightarrow M_b$ and $L_g \leftrightarrow L_e$ in $(M_aL_{e'}|V|M_bL_g)$ and $(M_aL_g|V|M_bL_e)$ is expanded in terms of multipole expansion, the leading term is Coulombic correlation of electric dipole moments in the ligands and quadrupole moments in the Ln ion [31–33].

From these third and fourth terms in Eq. (6), the f–f transition oscillator strengths can finally be represented in terms of the JO type equation (Eq. 1). Only the formula of the JO intensity parameter τ_λ is changed as follows,

$$
\begin{aligned}
\tau_\lambda(dc) &= \frac{4\pi m_e c}{3\hbar} \chi_{\mathrm{ed}} (\lambda + 1) \left\langle f \| C^{(\lambda)} \| f \right\rangle^2 \left\langle 4f|r^2|4f\right\rangle^2 \\
&\sum_{m=0}^{\lambda+1} (2 - \delta_0^m) \left| \sum_L \alpha(L) R_L^{-\lambda-2} C_{-m}^{\lambda+1}(L) \right|^2
\end{aligned}
\tag{7}
$$

where $\langle f \| C^{(\lambda)} \| f \rangle$ is the reduced matrix element of the λ-th rank Racah tensor connecting the f-orbital functions, $\langle 4f|r^2|4f\rangle$ is the squared radial expectation values of 4f orbital of Ln, $\alpha(L)$ is the mean polarizability of the ligand L, and $C_{-m}^{\lambda+1}(L)$ is the $(\lambda + 1)$-th rank Racah tensor depending on the ligand structure.

Here, as was already pointed out, it is noticed from the values for matrix elements of irreducible tensor operators [45–48] that the matrix elements of $\mathbf{U}^{(2)}$ have notably large values only in the hypersensitive transitions among possible f–f transitions, whereas those of $\mathbf{U}^{(4)}$ and $\mathbf{U}^{(6)}$ are less directly related to the hypersensitive transitions [10]. Therefore, it can be interpreted that the characteristics of hypersensitivity are involved in the term of $\lambda = 2$ in Eq. (7), namely τ_2. In fact, the magnitude relation $\tau_2 >> \tau_4, \tau_6$ holds in LnX_3 molecules in vapor phase [31, 32], and these oscillator strengths can be evaluated only by the term of $\lambda = 2$ in Eq. (7). Comparison of the τ_2(dc) term for LnX_3,

$$
\begin{aligned}
\tau_2(dc) &= \frac{112\pi m_e c}{15\hbar} \chi_{\mathrm{ed}} \left\langle 4f|r^2|4f\right\rangle^2 \\
&\times \sum_{m=0}^{3} (2 - \delta_0^m) \left| \sum_L \alpha(L) R_L^{-4} C_{-m}^3(L) \right|^2.
\end{aligned}
\tag{8}
$$

and Eq. (4) implies that the effect of the crystal field due to the oscillating induced dipole moments on the ligands is represented in Eq. (8) as the sum of the squared term of $\alpha(L)$ times R_L^{-4}. Note, however, that this original DC model neglects shielding effect due to the 5s and 5p closed subshells of Ln ion and the mixing of LMCT configurations into $4f^N$ states. Later, the shielding effect was studied and genuine τ_λ value in Eq. (8) turned out to be typically overestimated if the shielding factor was neglected [55, 56]. However, the latter LMCT mixings have not been studied very well.

3 Calculation methods

Ab initio calculations are performed for LnX_3 (Ln = Pr, Nd, Pm, Sm, Eu, Tb, Dy, Ho, Er, Tm; X = Cl, Br, I) by the multi-reference spin–orbit configuration interaction (MRSOCI) method using the COLUMBUS program package [57, 58]. We use the model core potentials (MCPs) by Sakai et al. The valence shells of MCP [59, 60] for Ln are 4f5s5p, and those of MCP-DZPs [61, 62] are 3s3p for Cl, 3d4s4p for Br, and 4d5s5p for I. We use (10s,7p,7d,6f)/[5s,3p,4d,4f] basis sets for Ln [59, 60], (5s,5p,4d)/[3s,3p,2d] basis set for Cl, and (9s,8p,8d)/[4s,4p,3d] basis sets for Br and I [61, 62]. The geometries of LnX_3 were D_{3h} with the experimental bond lengths [63]. The semi-empirical one-body spin–orbit (SO) Hamiltonian is used [64]. The value of effective nuclear charge $Z_{eff}(A)$ of each atom A is determined to reproduce the experimental SO splittings among the lower $4f^N$ states of Ln^{3+} and those between $^2P_{3/2}$ and $^2P_{1/2}$ of neutral halogens. The one-electron orbitals used for the MRSOCI calculations were obtained by the state-averaged open-shell SCF method in which the energy expression to be optimized is the ensemble average energy derived from all the $4f^N$ configurations with an equal weight as described in Ref. [65]. With this approach, symmetry-adapted one-electron orbitals for LnX_3 molecule in D_{3h} structure are readily obtained.

For the MRSOCI calculations, we generated all the singly excited configuration state functions (CSFs) obtained from all the possible $4f^N$ reference configurations of the complete-active-space (CAS) type. This means all the singly excited configurations from $4f^N5s^25p^6$ of Ln^{3+} and also from valence $(ns^2np^6)_3$ of $(X^-)_3$ to all the virtual orbitals are included. For the diagonalization in the MRSOCI calculation step, we use two schemes, one-step and two-step diagonalization schemes, which were called "standard SOCI" and "contracted SOCI" methods [66] in our previous study [5]. The one-step diagonalization scheme, where the sum of spin-free and spin–orbit Hamiltonian is diagonalized in the basis of all the CSFs with all possible spin multiplicities, is used for PrX_3, EuX_3, TbX_3, and TmX_3. For the other LnX_3, we apply the two-step scheme, where spin–orbit Hamiltonian is diagonalized in the basis of the limited numbers of spin-free electronic states obtained by the diagonalization of spin-free Hamiltonian. We have confirmed [5] that the excitation energies of hypersensitive transitions obtained by these calculation schemes were in good agreement with the experimental data and were almost independent of X^-.

To evaluate the two JO intensity parameters τ_2(ab) and τ_2(dc), we apply different calculation schemes. For τ_2(ab), we take the major $\lambda = 2$ term of Eq. (1) and the smaller terms of $\lambda = 4$ and 6 are neglected by the reason explained before. The ab initio intensity parameters τ_2(ab) are evaluated without relying on any crystal field treatments for X^- by substituting to Eq. (1) the values of MRSOCI excitation energies, oscillator strengths, and the matrix element of $\mathbf{U}^{(2)}$ [45–48]. The DC model intensity parameters τ_2(dc) are obtained by substituting to Eq. (8) experimental bond length [63], squared radial expectation values of 4f orbital of Ln^{3+} determined by the state-average SCF method, and the polarizability of X^- obtained by the first-order SOCI method.

In Sect. 4.2, the transition density and the integrand of TDM are calculated for a pair of eigenstates which carry the largest TDM as in the previous study [5]. The values of the planar distribution functions of the transition densities were calculated at 0.1 bohr intervals and then integrated numerically over the perpendicular (z) direction. The values of the radial distribution functions of integrands of TDMs were calculated at 0.1 bohr intervals for radial part and $\pi/180$ radian intervals for angular part, and then integrated numerically over the angular part. All the integrations were performed with the *Mathematica* program [67].

In Sect. 4.3, to express only valence spaces, we removed some diffuse basis functions from the above basis sets and kept only (8s,7p,7d,6f)/[3s,3p,4d,4f] basis sets for Ln, (4s,4p)/[2s,2p] for Cl, and (8s,7p,7d)/[3s,3p,2d] for Br and I as in Ref. [5].

4 Results and discussions

4.1 Oscillator strengths and Judd–Ofelt intensity parameters of LnX_3

The calculated oscillator strengths of hypersensitive transitions of LnX_3 (X = Cl and I) are shown in Table 1 along with those of $LnBr_3$ obtained in Ref. [5]. They are in reasonable agreement with the experimental ones [13] even though the values are as small as from 10^{-6} to 10^{-4}. The JO parameters τ_2(dc) and τ_2(ab) were evaluated from these results and shown in Fig. 2. Most parameters in τ_2(dc) are common to all the LnX_3 except for the squared radial expectation values of 4f orbital of Ln $\langle 4f|r^2|4f\rangle$ and the polarizability of the ligand $\alpha(L)$. The ligand dependence of τ_2(dc) is reflected only by the frequency-dependent polarizability. However, in the case of LnX_3, the frequency-dependent polarizabilities can be approximated by the frequency-independent ones, since the formal charge of LnX_3 is $Ln^{3+}(X^-)_3$, and the excitation energies of X^- with the rare gas configurations are much higher than those of the hypersensitive transitions of Ln^{3+}. The values of frequency-independent ones for halogen anions calculated with the MCP method were $\alpha(Cl^-) = 27$, $\alpha(Br^-) = 37$ [5], and $\alpha(I^-) = 59$ a.u, in reasonable agreement with the experimental ones [68], $\alpha(Cl^-) = 20$, $\alpha(Br^-) = 30$, and $\alpha(I^-) = 47$ a.u. As shown in Fig. 2, the τ_2(dc) value

Table 1 Oscillator strengths $f \times 10^6$ (dimensionless) of LnX_3

Ln	$4f^N$	Transitions[a]	f ($LnCl_3$)		f ($LnBr_3$)		f (LnI_3)	
			Calc.	Expl.[b]	Calc.[c]	Expl.[b]	Calc.	Expl.[b]
Pr	$4f^2$	$^3H_4 \rightarrow {}^3F_2$	27.8	15	36.1	20	74.9	40.0
Nd	$4f^3$	$^4I_{9/2} \rightarrow {}^4G_{5/2}$	135.0	120	195.4	330	444.0	530.0
Pm	$4f^4$	$^5I_4 \rightarrow {}^5G_2$	104.7	–	125.3	–	260.0	–
Sm	$4f^5$	$^6H_{5/2} \rightarrow {}^6F_{1/2}$	10.7	–	15.8	–	15.7	–
Eu	$4f^6$	$^7F_0 \rightarrow {}^7F_2$	7.7	–	9.2	–	10.6	–
		$^7F_0 \rightarrow {}^5D_2$	0.4	–	0.5	–	0.90	–
Tb	$4f^8$	$^7F_6 \rightarrow {}^7F_5$	4.8	–	6.0	–	10.0	–
Dy	$4f^9$	$^6H_{15/2} \rightarrow {}^6F_{11/2}$	34.6	32	43.9	–	86.3	–
Ho	$4f^{10}$	$^5I_8 \rightarrow {}^5G_6$	138.7	178	174.6	–	358.5	500.0
Er	$4f^{11}$	$^4I_{15/2} \rightarrow {}^2H_{11/2}$	18.7	34	28.3	58	67.9	95.5
		$^4I_{15/2} \rightarrow {}^4G_{11/2}$	78.2	85	99.3	99	237.9	–
Tm	$4f^{12}$	$^3H_6 \rightarrow {}^3H_4$	8.2	–	9.6	12	14.7	10.7
		$^3H_6 \rightarrow {}^3F_4$	13.5	12	15.6	15.3	32.6	25.3

[a] List of hypersensitive transitions is taken from Ref. [13]

[b] Experimental values are from Ref. [13]

[c] These data are the same as those in Ref. [5]

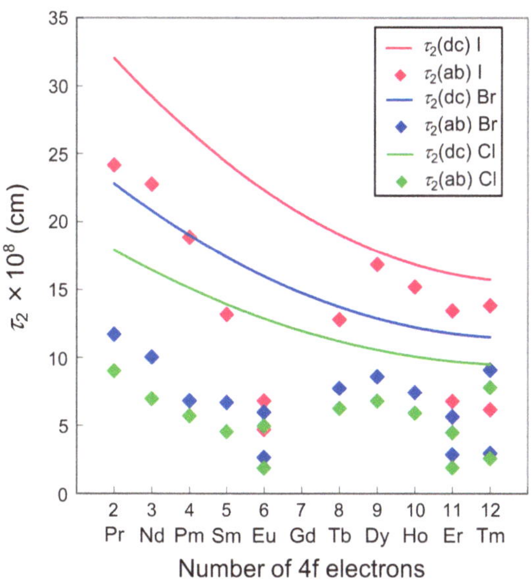

Fig. 2 JO intensity parameters τ_2(dc) and τ_2(ab) of LnX_3 (X = Cl, Br, I) as functions of the number of 4f electrons

increases monotonically from X = Cl to I, because the polarizability of halogen increases monotonically. A similar ligand dependence is also seen in τ_2(ab) values, that is, both τ_2(dc) and τ_2(ab) increase as halogen changes from Cl to Br and to I.

Unlike the ligand dependence, the Ln dependences of τ_2(dc) and τ_2(ab) are different especially between Eu and Tb for all LnX_3 series. The value of τ_2(ab) decreases within the early Ln series and also within the late Ln series, but increases between Eu and Tb, while τ_2(dc) decreases monotonically. The origin of Ln dependence of τ_2(dc) is explained only from the squared radial expectation values of 4f orbital of Ln, whose value decreases monotonically with the atomic number of Ln. As clarified in the previous study, one of the reasons why τ_2(ab) values are smaller than τ_2(dc) can be explained with the polarization shielding effect by 5 s and 5p electrons of Ln [5]. However, the different Ln dependences, especially between Eu and Tb, cannot be explained only by the shielding effect. As mentioned in Introduction, τ_λ(ab) is proportional to $(\mathbf{M}_{FI}\text{(JO)} + \mathbf{M}_{FI}\text{(DC)} + \mathbf{M}_{FI}\text{(LMCT)} + \cdots)^2$ and can be affected by these individual magnitudes and their relative phases. Therefore, in the following section, the relation between the effect of the DC model and other effects, such as the effect of the JO theory and LMCT, will be studied to examine the cause of different behavior of τ_2(dc) and τ_2(ab). It should be noted that the easiest way to examine the relation between these effects is to divide the TDM based on the atomic orbital (AO) centers. In our previous studies [4, 5], we investigated the weight of each contribution and clarified that the DC model had dominant contributions. However, these weights show basis set dependence, especially for those matrix elements between Ln and X_3, and some ambiguity remains in the division of the transition populations. Therefore, other theoretical tools, which depend on the basis sets as little as possible, are devised and applied in the following sections.

4.2 Spatial distribution of transition density and integrand of transition dipole moment

4.2.1 Physical view of the electronic excitation of hypersensitive transition

We next try to describe the hypersensitive transitions and their intensities visually by using the transition densities and the integrands of TDMs, because their values are less sensitive to the basis sets and helpful for understanding the physical view of hypersensitive transitions and the origin of their intensities. Firstly, we describe the physical view of the electronic excitation of the hypersensitive transition by using the one-electron transition density function $\rho^{IF}(\mathbf{r})$ between the initial state Ψ_I and the final state Ψ_F of an N-electron system which is defined as follows,

$$\rho^{IF}(\mathbf{r}_1) = N\int\cdots\int \Psi_I(\{\mathbf{r}_k\sigma_k\})\Psi_F^*(\{\mathbf{r}_k\sigma_k\})\,d\mathbf{r}_2\cdots d\mathbf{r}_N d\sigma_1\cdots d\sigma_N. \quad (9)$$

Here, $\{\mathbf{r}_k\sigma_k\}$ stand for the electron spatial and spin coordinates collectively; thus, the above transition density function is given as a function of the non-integrated spatial variable $\mathbf{r}_1$. The initial and final electronic wave functions Ψ_I and Ψ_F are calculated with the MRSOCI method. Though the wave functions contain various spin multiplicities, the transition density function $\rho^{IF}(\mathbf{r})$ is independent of spin coordinates. Note that our spin–orbit CI implementation uses the so-called real spherical form of spin functions [58]; therefore, in the actual computation, we can avoid handling complex CI coefficients, and all the quantities appearing in Eqs. (9–13) above and below are of real values. To project the transition density function onto the LnX_3 molecular plane, Eq. (9) is integrated in a direction perpendicular to the LnX_3 molecular plane and a two-dimensional transition density function $D(x, y)$ is obtained as follows,

$$D(x,y) = \int \rho^{IF}(x,y,z)dz. \quad (10)$$

As an example, the planar distribution functions of the transition densities $D(x, y)$ of $PrBr_3$ are shown in Fig. 3. Because the irreducible representation of the transition that carries the largest TDM is always degenerate E′, both components denoted as $D^{(x)}(x, y)$ and $D^{(y)}(x, y)$ for the two kinds of polarizations, called Type (x) and Type (y), respectively, of D_{3h} are shown, where their directions of the TDMs are x and y, respectively. Both of the transition densities have large values only around the position of Pr, which is located on the origin, because the initial and final state wave functions in most f–f transitions have reference ($4f^N$) weights typically higher than 0.95. The shapes of $D^{(x)}(x, y)$ and $D^{(y)}(x, y)$ suggest that they consists of transition quadrupole moments with components of x^2–y^2 and xy, respectively.

To observe the distribution of $D(x, y)$ around the halogen atoms more closely, those in the three Br regions (1), (2), and (3) are extended in Fig. 4. Here, the transition density in regions (2) and (3) contains the components of the TDMs in the direction to the origin, but in the case of Type (x), the sum of these two y components cancels each other, while that of the remaining x components remains and is shown as

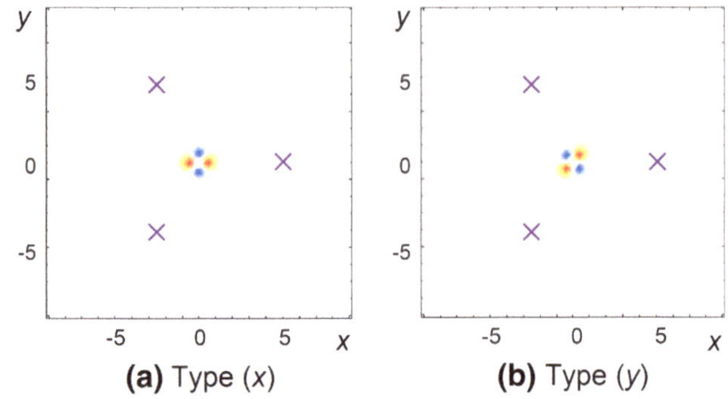

Fig. 3 Planar distribution functions of transition densities $D^{(x)}(x, y)$ and $D^{(y)}(x, y)$ of $PrBr_3$. Those of hypersensitive transitions (E′) whose x and y components of TDMs have large values are called as Type (x) and shown in **a**, and Type (y) in **b**, respectively. *Red* and *blue* areas represent positive and negative values, respectively. The *x-marks* represent the positions of Br atoms

arrows in (2) and (3) in this Fig. 4a. Similar cancellation occurs in Type (y) component as well as shown in Fig. 4b. Therefore, the transition moments on Br generate the crystal fields in the directions of x and y for Type (x) and Type (y) transitions, respectively. Recalling the coupling scheme in the DC model mentioned in Sect. 2.2, the structures of the transition density around the Pr and Br atoms in Fig. 4 are quite similar to the structures shown in Fig. 1 for the DC scheme caused by the periodically oscillating induced dipole moments on the ligands. This similarity of structures of the transition densities in Figs. 4 and 1 is not a coincidence, but shows just two expressions for the same physical phenomenon using the time-independent and time-dependent languages, respectively.

4.2.2 *Physical view of the intensity of hypersensitive transition*

Next, to observe the origin of the hypersensitive transition intensities, we focus on the integrands of the TDMs. Once these transition density functions are available, the x component TDM, for example, can be obtained by the following integrations,

$$M_x^{FI} = \iiint x\rho^{IF}(\mathbf{r})d\mathbf{r} = \iint xD^{(x)}(x,y)dxdy. \quad (11)$$

The x times two-dimensional distribution function $xD^{(x)}(x, y)$ has large values around the localized region of Ln as shown in Fig. 5. However, from the shape of $xD^{(x)}(x, y)$, the contribution from this region to the TDM becomes canceled out after the angular integration, and the contributions only from the regions (1), (2), and (3) remain and become relatively large.

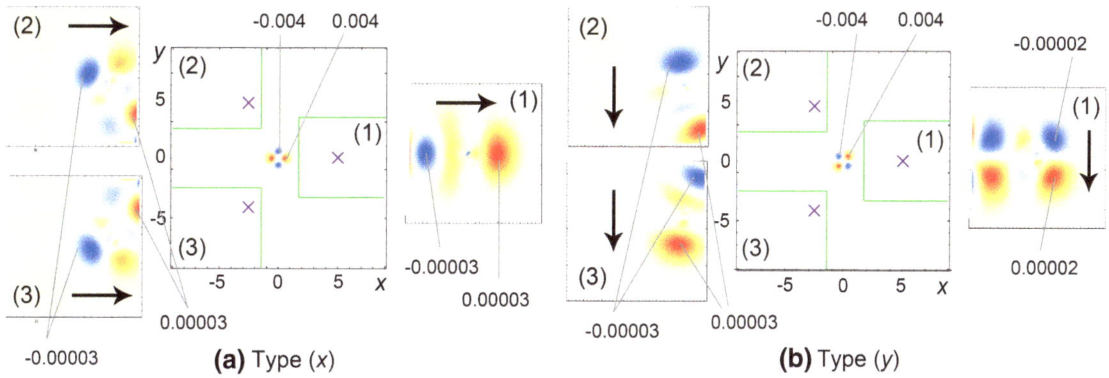

Fig. 4 Extended views of **a** $D^{(x)}(x, y)$ for Type (x) transition and **b** $D^{(y)}(x, y)$ for Type (y) transition of $PrBr_3$. The *x-marks* represent the positions of Br atoms. The directions of *arrows* represent those of transition moments on Br atoms. Because the values in the regions (1–3) are too small, the scaled up views are shown in the surrounding three *squares*. The values of distribution functions are given in a.u

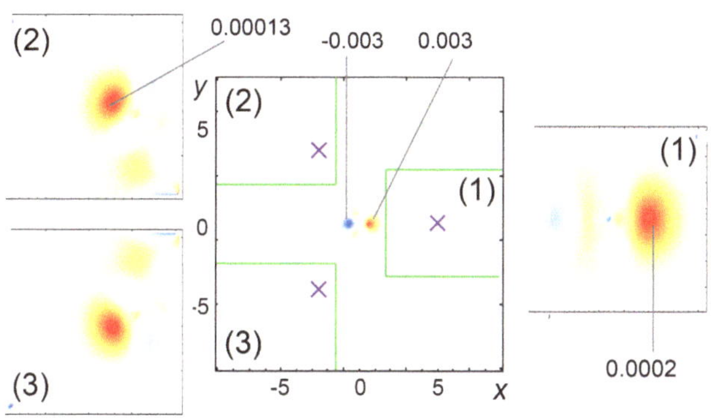

Fig. 5 Planar distributions of $xD^{(x)}(x, y)$ of $PrBr_3$ for Type (x) transition. Because the values in the regions (1–3) are too small, the scaled up views are shown in the surrounding three *squares*. The values of distribution functions are given in a.u

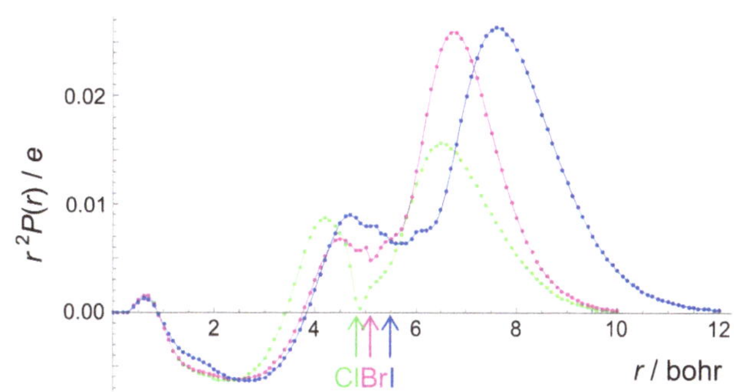

Fig. 6 Plots of the radial functions $r^2P(r)$ versus r (in a.u.). *Green*, *pink*, and *blue* lines are those of $PrCl_3$, $PrBr_3$, and PrI_3, respectively. Three *arrows* represent the positions of halogen atoms

From the above reason, we focus on the radial distribution functions of the integrand of TDM, which can be obtained by multiplying transition density and the dipole operator x or y and integrating it over the angular part as follows,

$$M_x^{FI} = \iiint (r \sin\theta \cos\phi)\rho^{IF}(r, \theta, \phi) r^2 \sin\theta dr d\theta d\phi = \int_0^{\infty} r^2 P(r)\, dr. \quad (12)$$

It should be noted that the shape of the radial distribution $r^2P(r)$ of Type (x) transition is exactly the same as that of Type (y) transition. The radial distribution functions $r^2P(r)$ of PrX_3 (X = Cl, Br, I) are shown in Fig. 6. The nuclei are located at the origin for Pr, $r = 4.8$ bohr for Cl, $r = 5.1$ bohr for Br, and $r = 5.5$ bohr for I, respectively. Here, the curves of $r^2P(r)$ are not smooth around the position of the three halogen nuclei because the wave functions vary sharply near the nuclei. Comparing these functions suggests the general feature is independent of the halogens, that is, the values of $r^2P(r)$ are negative in the region of $0.8 < r < 3.5$ bohr and become positive and larger around in the regions of $6 < r < 8$ bohr.

To examine these radial distribution functions more closely, their integrand of TDM was decomposed into three kinds of components based on the AO centers, which are denoted as L for Pr and X for halogen, as follows,

$$r^2P(r) = \sum_{s,t}^{AO} r^2 P_{s,t}(r) = r^2P_{LL}(r) + r^2P_{LX}(r) + r^2P_{XX}(r)\ . \quad (13)$$

and shown in Fig. 7. The area between the $r^2P(r)$ curve and the horizontal axis expresses the value of TDM because it can be obtained by the integration of $r^2P(r)$ over r as Eq. (12) stands for, the region where $r^2P(r)$ has a positive value has a positive contribution to the TDM, and vice versa. This interpretation also applies to each of the decomposed ones $r^2P_{s,t}(r)$. Therefore, the areas of $r^2P_{LL}(r)$,

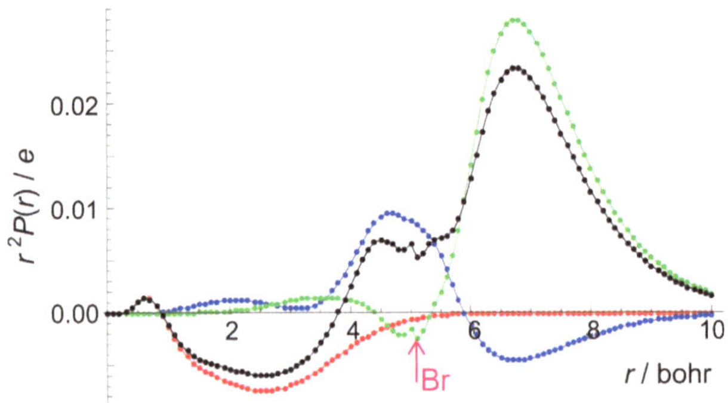

Fig. 7 Plots of the radial functions $r^2P(r)$ and its components versus r (in bohr) for $PrBr_3$. *Red, blue, green,* and *black* lines are those of $r^2P_{LL}(r)$, $r^2P_{LX}(r)$, $r^2P_{XX}(r)$ and their sum of $r^2P(r)$, respectively. The *arrow* represents the position of Br^-

$r^2P_{LX}(r)$, and $r^2P_{XX}(r)$ represent the values of the sum of the matrix elements of the transition dipole moment within Ln, those between Ln and X, and those within X, respectively. We have previously decomposed TDM based on the AO centers and shown the results in Table 2 and Fig. 2 of Ref. [5]. From that work, it is clear that the three areas above mainly represent the sum of matrix elements of the transition dipole moment between $4f^N$ and intra-Ln excitation CSFs, those between $4f^N$ and ligand-to-metal charge transfer (LMCT) excitation CSFs, and those between $4f^N$ and intra-ligand excitation CSFs, respectively. In other words, the areas of $r^2P_{LL}(r)$ and $r^2P_{XX}(r)$ represent the effects of the JO theory and the DC model, respectively.

From the above decomposition analysis of the TDM, it is clear that the intra-ligand excitations have the largest contribution to the TDM because the large values of $r^2P(r)$ in the region of $6 < r < 8$ bohr come from those of $r^2P_{XX}(r)$. Additionally, its contribution has the opposite sign relative to the intra-Ln excitations because the areas of $r^2P_{LL}(r)$ and $r^2P_{XX}(r)$ have opposite signs. It suggests that the effect considered in the JO theory has an opposite sign contribution against that considered in the DC model. This sign relation has been pointed out in other studies based on experiments [40] and semi-empirical calculations using the crystal field theory [69, 70]. This relation is also consistent with our previous analysis shown in Table 2 in Ref. [5].

The values of $r^2P_{LX}(r)$ are very small in the region between Ln and Br. They are positive near the ligand positions ($4 < r < 6$ bohr) and become negative in the region outside the ligands ($r > 6$ bohr). In the current CI wave function, the basis functions on Ln have nonzero population tail in the position of the ligands in a minute measure. Therefore, the area of $r^2P_{LX}(r)$ in the region of $r > 4$ bohr can be regarded as a part of the effect considered in the DC model. This result is consistent with the fundamental assumption in the JO theory and the DC model that the effect of orbital overlap between metal and ligand, caused by back donation from X_3 to Ln, namely LMCT, was negligible. Note that at this point, we cannot say unambiguously whether the effect of LMCT is responsible for the difference in the Ln dependence between τ_2(dc) and τ_2(ab), because what Fig. 7 shows is just the contributions of nonzero matrix elements to the total value of the TDM. The effect of LMCT configurations will be discussed in Sect. 4.3.

Next, to examine the origin of the Ln dependence in the TDM of the hypersensitive transition, $r^2P(r)$ for other $LnBr_3$ are shown in Fig. 8. Amazingly, the shapes of $r^2P(r)$ are very similar among all the $LnBr_3$, even though the scale size of y axis of each figure is very much different reflecting the different magnitude of their oscillator strengths. The similarity of the shape of $r^2P(r)$ suggests that the contribution and relative phase of each effect, such as the JO theory and the DC model, are quite similar among all the LnX_3 series. To sum up, all the hypersensitive transition intensities for LnX_3 series have the same origins which are mainly explained by the DC model. Additionally, both the ratio of the magnitudes and the relative phase between the DC model and the lesser important JO theory mechanism are also quite similar among all the LnX_3, even though the characters of their initial and final states, such as the numbers of f electrons, spin, and orbital angular momenta, are totally different.

4.3 Effect of ligand-to-metal charge transfer

In Sect. 4.2, the different behavior of τ_2(dc) and τ_2(ab) especially between Eu and Tb could not be explained from the shape of integrands of TDMs. One of the remaining possibilities for the reason for the difference between these two τ_2's is the LMCT configuration mixing whose TDM matrix elements have almost zero values. Although the contributions of LMCT configuration mixing, orbital overlap, and covalency in the metal–ligand bonding to hypersensitive transition intensities have been studied [20–22, 71–76], their importance seems to be still open to discussion.

Therefore, we discuss the effect of LMCT in a different perspective. There are some difficulties for extracting LMCT effects from the CI wave functions or the TDMs. Firstly, the evaluation of the weights of LMCT mixing in CI wave functions is difficult because most virtual MOs contained in CSFs are delocalized between Ln and X. Secondly, the amount of mixing of LMCT CSFs is too small to analyze because the weight of the reference $4f^N$ CSFs exceeds 95 percent. Therefore, to remove the dominant 4f components, we focus on the $N_{occ} \times (N_{act} + N_{vir})$ rectangular block of transition density matrix elements, where N_{occ}, N_{act}, and N_{vir} are the numbers of the doubly occupied MOs, active MOs (4f), and virtual MOs, respectively. This rectangular block

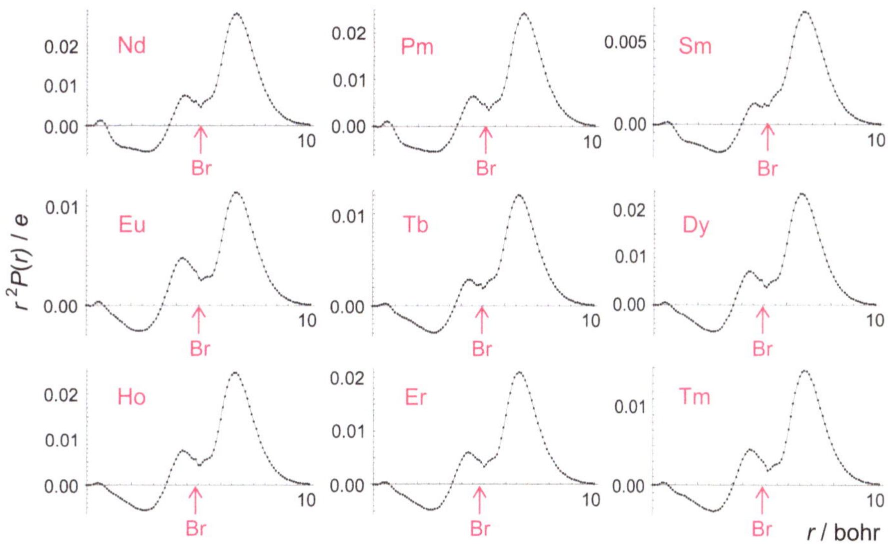

Fig. 8 Plots of the radial functions $r^2P(r)$ versus r (in bohr) for $LnBr_3$. (Ln = Nd, Pm, Sm, Eu, Tb, Dy, Ho, Er, Tm). The *arrows* represent the positions of Br^-

contains the relevant quantity associated with products of the CI coefficients for the pairs of CSFs which are the reference CSFs in Ψ_I and the one-electron excitation CSFs from the occupied MOs to the (active + virtual) MOs in Ψ_F. To extract the amount of charge transfer effect from the rectangular block transition density matrix, "corresponding orbital" [77] or "natural transition orbital" [78, 79] transformation was carried out based on a singular value decomposition concept. By using the corresponding orbitals, we define a quantity, called γ, the charge population change on Ln averaged over individual hole-particle corresponding orbital pairs. See Appendix for details. This γ is expected to have a larger value for a transition whose final state has a larger LMCT weight.

The calculated values of γ along with τ_2(ab) for LnX_3 are shown in Table 2. We can confirm our idea that γ has a larger value for a transition whose final state has a larger LMCT weight from its ligand dependence and Ln dependence. Focusing on the ligand dependence of γ for each Ln, the γ value increases with the atomic number of halogen atom X. Generally, as halogen atom X becomes heavier, the LMCT excited states appear in the lower energy region, because their energy levels decrease as the electronegativity of the ligand decreases [11, 80–83]. Therefore, the amount of the LMCT character in the final states increases with the atomic number of X, since the energy gap between the LMCT states and the still lower-lying final state of the hypersensitive transition becomes smaller. On the other hand, focusing on the Ln dependence of γ, it increases with the atomic number of Ln except between Eu and Tb. It can be understood by the energy levels of the LMCT excited states. It is generally said that the energy levels of the LMCT excited states in Ln systems become lowered from Pr to Eu and once increased at Gd and then become lowered again from Tb to Tm [11, 80–83]. The reason of the lowest energy levels of LMCT states of Eu is that the 4f electron configuration of Eu^{3+} becomes the $4f^7$ half-filled configuration by accepting one electron from ligands [11, 80–83]. Therefore, the γ value has the largest value for Eu due to the smallest energy gap between the final state and LMCT states. As discussed before, γ has a larger value for a transition whose final state has a larger LMCT character.

Next, we discuss the relation between TDMs and LMCT weights in the final states of hypersensitive transitions by using their indexes τ_2(ab) and γ. As shown in Table 2, τ_2(ab) decreases as γ increases. To establish the origin of this relationship, we remind the character of LMCT in the LnX_3 molecule. Among the LMCT CSFs, the major CSFs, whose CI coefficients are the largest, are those for LMCT from occupied orbitals of the X_3 portion to 4f empty orbitals of Ln. The TDMs between these LMCT CSFs and the $4f^N$ reference CSFs can be expressed as one-electron matrix elements between the 4f orbitals and the occupied orbitals of the X_3 portion, and their values critically depend on their orbital size. The radial expectation values of 4f orbital of Pr^{3+}, 3p orbital of Cl^-, 4p orbital of Br^-, and 5p orbital of I^- were calculated to be 1.0, 1.7, 1.9, and 2.3 bohr, respectively, in reasonable agreement with the relativistic ones by Desclaux [84], 1.1, 1.8, 2.1, and 2.5 bohr. Considering the distances between Ln and the halogens that are about 4.8–5.5 bohr, these 4f on Pr and the valence p orbitals on X have little overlap. Therefore, TDMs between these orbitals are negligibly small compared to those for intra-ligand excitations within X_3. By taking these factors into account, the reason why τ_2(ab) of EuX_3 has a smaller value than that of TbX_3 can be explained by the fact that the relative importance of the intra-ligand excitation CSFs is reduced by the increase of the "dark" LMCT CSFs, especially those from the occupied orbitals of the ligands to the 4f orbitals of Ln.

Table 2 Relation between the weights of LMCT configurations γ in the final states of hypersensitive transitions, and τ_2(ab) × 10^8 (cm) of $LnBr_3$

Ln	$4f^N$	Transitions	γ $LnCl_3$	τ_2(ab)	γ $LnBr_3$	τ_2(ab)	γ LnI_3	τ_2(ab)
Pr	$4f^2$	$^3H_4 \rightarrow {}^3F_2$	0.20	9.0	0.22	11.7	0.25	24.1
Nd	$4f^3$	$^4I_{9/2} \rightarrow {}^4G_{5/2}$	0.25	6.9	0.28	10.0	0.32	22.7
Pm	$4f^4$	$^5I_4 \rightarrow {}^5G_2$	0.25	5.7	0.29	6.8	0.32	18.8
Sm	$4f^5$	$^6H_{5/2} \rightarrow {}^6F_{1/2}$	0.30	4.5	0.33	6.7	0.35	13.2
Eu	$4f^6$	$^7F_0 \rightarrow {}^7F_2$	0.42	4.9	0.43	5.9	0.46	6.8
		$^7F_0 \rightarrow {}^5D_2$	0.64	1.8	0.69	2.6	0.72	4.7
Tb	$4f^8$	$^7F_6 \rightarrow {}^7F_5$	0.16	6.2	0.17	7.7	0.20	12.8
Dy	$4f^9$	$^6H_{15/2} \rightarrow {}^6F_{11/2}$	0.24	6.8	0.25	8.6	0.28	16.8
Ho	$4f^{10}$	$^5I_8 \rightarrow {}^5G_6$	0.35	5.9	0.37	7.4	0.43	15.2
Er	$4f^{11}$	$^4I_{15/2} \rightarrow {}^2H_{11/2}$	0.32	1.9	0.35	2.9	0.40	6.8
		$^4I_{15/2} \rightarrow {}^4G_{11/2}$	0.35	4.4	0.38	5.6	0.43	13.4
Tm	$4f^{12}$	$^3H_6 \rightarrow {}^3H_4$	0.18	7.8	0.21	9.1	0.22	13.8
		$^3H_6 \rightarrow {}^3F_4$	0.19	2.5	0.29	2.9	0.24	6.1

As seen above, the values of TDMs and τ_2(ab) cannot be explained by using only the DC model; the mixing of LMCT CSFs to $4f^N$ states should be considered simultaneously, even though the latter does not contribute to intensity. Our suggestion is consistent with previous assumptions in the JO theory and the DC model that the orbital overlaps are neglected. However, several previous studies explained that the intensities of hypersensitive transitions increased with the magnitude of covalency [20–22, 71–76]. In fact, their reasoning seems to be consistent with the experimental facts, as long as the halogen anions X^- are concerned because both the covalency and the polarizability increase as X becomes heavier. From the above discussion, it is now clear that the genuine covalent interaction between the orbitals of Ln and X is negligibly small and only the polarizability increases the magnitude of τ_2(dc) directory. Therefore, the previous explanation [20–22, 71–76] is probably not "the right answers for the right reasons," but just connected unrelated characteristics of the hypersensitivity.

To summarize the analysis given so far, the relative phase between each contribution to the TDM is expressed as

$$\left|\mathbf{M}^{FI}(ab)\right| = \left|\mathbf{M}^{FI}(DC)\right| - \left|\mathbf{M}^{FI}(JO)\right| + \cdots \tag{14}$$

and the CSFs that represent LMCT from X_3 to empty 4f orbitals of Ln do not contribute to the intensity, but play a role as "dark" states. This relation is applicable for most f–f transitions of LnX_3, which obey $|\Delta J| \leq 2$ and $|\Delta L| \leq 2$, because the τ_2 value is unique for the compound as explained in Sect. 2.1. However, some transitions have different picture due to the larger contributions from τ_4 and τ_6. It can be confirmed by our previous analysis in Fig. 5 of Ref. [4]. The ratios of contributions from the JO theory, the DC model, and the overlap region between Ln^{3+} and X^- are quite similar for most of the f–f transitions from the ground state to lower-laying excited states. On the other hands, the f–f transitions to higher excited states of $PrBr_3$, such as ^3P_J (J = 0, 1, 2), had large contributions from the JO theory which means 4f–5d mixing. The transitions to higher excited states of $TmBr_3$, such as ^3P_J (J = 1, 2), had large contributions from the overlap region between Tm^{3+} and Br^-. The different behaviors of these two kinds of higher excited states can be understood by considering the larger configuration mixing of their $4f^{N-1}5d$ and LMCT excited states, respectively. Here, it is important to note the relation of the energy levels between the LMCT and $4f^{N-1}5d$ states. As explained before, the excitation energies of the LMCT states become decreased from Pr to Eu and once increased between Eu and Tb and then lowered again from Tb to Tm [11, 80–83]. On the other hand, those of the $4f^{N-1}5d$ states are known to display an opposite behavior and become increased from Pr to Eu and once lowered from Eu to Tb and increased again from Tb to Tm [11, 80–85]. Therefore, the excitation levels of the $4f^{N-1}5d$ states are lower in $PrBr_3$, and those of the LMCT states are lower in $TmBr_3$, which results in the large contribution from each effect to the higher $4f^N$ excited states. Going back to the hypersensitive transitions, the difference in the LMCT excitation energies across Ln has a more strong impact to TDMs than that of $4f^{N-1}5d$ states, considering the behavior of τ_2(ab). This phenomenon can be understood because the LMCT excitation energies are lower than those of $4f^{N-1}5d$ states, especially around EuX_3. Taken altogether, the behaviors of τ_2(ab) depend on the individual Ln systems because the difference in the excitation energies between the $4f^{N-1}5d$ and LMCT states heavily depends on the systems. Therefore, it is clear that the mixing of LMCT and intra-Ln excitation configurations also affects the intensities, and they must be considered simultaneously.

Additionally, we would like to point out the limitation of the crystal field treatment from another aspect. Any f–f transitions whose $\mathbf{U}^{(2)}$ matrix elements are zero should not have contributions form the DC model because their intensities do not contain the contributions of τ_2. However, as shown in Fig. 5 of Ref. [4], ${}^3H_4 \rightarrow {}^3P_{0,1}$ in $PrBr_3$ and ${}^3H_6 \rightarrow {}^3F_{2,3}$, ${}^1D_2$, ${}^3P_{0-2}$ in $TmBr_3$ have nonzero contributions from the DC model even though their $\mathbf{U}^{(2)}$ matrix elements are zero [45]. It indicates that the crystal field treatment is not enough to represent all the f–f transition intensities quantitatively, and the electrons in both Ln and at least the nearby ligands have to be considered explicitly through ab initio calculations.

5 Conclusion

In this work, we have studied the oscillator strengths of the f–f hypersensitive transitions of LnX_3 (Ln = Pr, Nd, Pm, Sm, Eu, Tb, Dy, Ho, Er, Tm; X = Cl, Br, I) based on the multi-reference spin–orbit configuration interaction (MRSOCI) method. To compare the dynamic-coupling (DC) model and our ab initio results, we calculated two kinds of the Judd–Ofelt (JO) intensity parameters τ_2(dc), which is derived from the DC model, and τ_2(ab), which reflects all the effects considered in our ab initio calculations. Although these two parameters are in the same order of magnitude and their halogen dependence is similar, their Ln dependence is different, especially between Eu and Tb. The reason of the similarity between τ_2(dc) and τ_2(ab) in terms of the ligand dependence is that the dominant mechanism is the dynamical polarization effects within the ligands, and the reason of the different Ln dependence between Eu and Tb is explained by the different amount of the charge transfer from the occupied ligand orbitals to the vacant 4f orbitals of Ln. Additionally, the JO mechanism has a lesser and opposite sign effect relative to the dynamical coupling mechanism, and this relation is common to all the LnX_3 series.

As is clear from the formula of the JO intensity parameter in the DC model (Eq. 7), the effect of the DC model, in other words, the Coulomb correlation between transition quadrupole moment of Ln and induced dipole moment of ligands must be weakened in a system in which the distances between Ln and the ligands are long. It can be understood that this effect is one of the examples of near-field effects, which have recently gained much attention in the field of surface chemistry.

These findings will be applied for the future theoretical calculations of realistic large systems containing Ln atoms, for example, with the hybrid QM/MM approaches [86–88]. For such calculations, atoms or fragments in the nearby ligands around Ln must be treated as the QM region and cannot be treated as point charges because the intensities of the f–f transitions are determined by the crystal field not from point charges but from induced dipole moments on the ligands. In addition, we should keep in mind that the artificial extraction of the QM region sometimes causes large errors in the amount of charge transfer which might play a role as the dark state to the f–f transition intensities.

Acknowledgments M.H. acknowledges Hayashi Memorial Foundation for Female Natural Scientists. This work was supported in part by Grants-in-Aid for Scientific Research and by the MEXT-Supported Program for the Strategic Research Foundation at Private Universities, 2009–2013. The computations were partly carried out using the computer facilities at the Research Center for Computational Science, Okazaki National Institutes.

Appendix: Definition of weight of LMCT configurations γ

In this appendix, we describe the detailed definition of γ. The $N_{\text{occ}} \times (N_{\text{act}} + N_{\text{vir}})$ rectangular block transition density matrix $\mathbf{T}$ is translated to "diagonalized" rectangular matrix by using the "corresponding orbital" [77] or "natural transition orbital" [78, 79] method as follows,

$$[\mathbf{U}^{+}\mathbf{T}\mathbf{V}]_{ij} = \lambda_i \delta_{ij} \tag{15}$$

where $\mathbf{U} = (\mathbf{u}_1, \mathbf{u}_2, \ldots, \mathbf{u}_{N_{occ}})$ and $\mathbf{V} = (\mathbf{v}_1, \mathbf{v}_2, \ldots, \mathbf{v}_{N_{act}+N_{vir}})$ are the unitary matrices determined by solving the following eigenvalue equations,

$$\begin{aligned} (\mathbf{T}\mathbf{T}^{+})\mathbf{u}_i &= \lambda_i^2 \mathbf{u}_i \quad (i = 1, \ldots, N_{occ}) \\ (\mathbf{T}^{+}\mathbf{T})\mathbf{v}_i &= \lambda_i^2 \mathbf{v}_i \quad (i = 1, \ldots, N_{act} + N_{vir}) \end{aligned} \tag{16}$$

Here, the new set of the occupied and (active + virtual) orbitals is obtained by using the unitary transformations, and a pair of thc orbitals whosc matrix elements with $(\mathbf{U}^{+}\mathbf{T}\mathbf{V})_{ii} = \lambda_i$ are called the i-th hole-particle pair orbitals. The importance of a particular hole-particle transition to the overall rectangular matrix $\mathbf{T}$ is reflected in the magnitude of the associated eigenvalue λ_i^2.

Next, we define a quantity of charge transfer from occupied orbitals of X_3 in Ψ_I to active and virtual orbitals of Ln in Ψ_F. This quantity, called γ, is evaluated by averaging the change of the Mulliken charge population on Ln Δp_i^{Ln} with the associated weight factor of λ_i^2 as follows,

$$\gamma = \frac{\sum_i^{N_{occ}} \lambda_i^2 \Delta p_i^{Ln}}{\sum_i^{N_{occ}} \lambda_i^2} \tag{17}$$

This parameter γ of LnX_3 takes a positive value because the amount of mixing of LMCT CSFs is much larger than that of MLCT CSFs. This weight of λ_i^2 is a "diagonalized" transition density matrix element and has the information of the products of the CI coefficients, i.e., those for the reference CSFs in Ψ_I times and those for one electron excitation CSFs in Ψ_F. With these weight factors, parameter γ represents the component of LMCT mixing in the final state of the target transition.

References

1. Eliseeva SV, Bünzli J-CG (2010) Chem Soc Rev 39:189–227
2. Binnemans K (2009) Chem Rev 109:4283–4374
3. Bünzli J-CG, Comby S, Chauvin A-S, Vandevy DB (2007) J Rare Earths 25:257–274
4. Hatanaka M, Yabushita S (2009) J Phys Chem A 113:12615–12625
5. Hatanaka M, Yabushita S (2011) Chem Phys Lett 504:193–198
6. Yang Y, Pitzer RM (2010) J Phys Chem A 114:7117–7120
7. Borowska L, Fritzsche S, Kik PG, Masunov AE (2011) J Mol Model 17:423–428
8. Judd BR (1962) Phys Rev 127:750–761
9. Ofelt GS (1962) J Chem Phys 37:511–520
10. Jørgensen CK, Judd BR (1964) Mol Phys 8:281–290
11. Bünzli J-CG, Eliseeva SV (2010) In: Hänninen P, Härmä H (ed) Lanthanide luminescence: photophysical, analytical and biological aspects. Springer Ser Fluoresc: Amsterdam, 7: 1–46
12. Gruen DM, DeKock CW (1966) J Chem Phys 45:455–460
13. Gruen DM, DeKock CW, McBeth RL (1967) Adv Chem Ser 71:102–121
14. Hasegawa Y, Wada Y, Yanagida S (2004) J Photochem Photobiol C Photochem Rev 5:183–202
15. Görller-Varland C, Binnemans K (1998) In: Gschneider KA, Eyring L (ed) Handbook on the physics and chemistry of rare-earths. Elsevier: Amsterdam, 25: 101–264
16. Wybourne BG, Smentek L (2007) Optical spectroscope of lanthanides: magnetic and hyperfine interactions. CRC Press, Boca Raton
17. Peacock RD (1975) Struct Bond 22:83–122
18. Judd BR (1966) J Chem Phys 44:839–840
19. Nieuwport WC, Blasse G (1966) Solid State Commun 4:227–229
20. Henrie DE, Choppin GR (1968) J Chem Phys 49:477–481
21. Henrie DE, Fellows RL, Choppin GR (1976) Coord Chem Rev 18:199–224
22. Peacock RD (1977) Mol Phys 33:1239–1246
23. Smentek L (1998) Phys Rep 297:155–237
24. Smentek L, Wybourne BG, Hess BA Jr (2001) J Alloys Compd 323–324:645–648
25. Smentek L (2003) Mol Phys 101:893–897
26. Wybourne BG (1968) J Chem Phys 48:2596–2611
27. Downer MC, Burdick GW, Sardar DK (1988) J Chem Phys 89:1787–1797
28. Smentek L, Wybourne BG (2000) J Phys B 33:3647–3651
29. Smentek L, Wybourne BG (2001) J Phys B 34:625–630
30. Kędziorski A, Smentek L (2007) J Lumin 127:552–560
31. Mason SF, Peacock RD, Stewart B (1975) Mol Phys 30:1829–1841
32. Mason SF, Peacock RD, Stewart B (1974) Chem Phys Lett 29:149–153
33. Peacock RD (1978) J Mol Struct 46:203–207
34. Reid MF, Dallara JJ, Richardson FS (1983) J Chem Phys 79:5743–5751
35. Reid M, Richardson FS (1983) Chem Phys Lett 95:501–507
36. Reid MF (1992) J Alloys Compd 180:93–103
37. Reid MF (1993) J Alloys Compd 193:160–164
38. Höhn EG, Weigang OE Jr (1968) J Chem Phys 48:1127–1137
39. Mason SF, Vane GW (1966) J Chem Soc B 370–374. doi:10.1039/J29660000370
40. Malta OL, Couto dos Santos MA, Thompson LC, Ito NK (1996) J Lumin 69:77–84
41. Malta OL, Carlos LD (2003) Quim Nova 26:889–895
42. Reid MF (2005) In: Liu G, Jacquier B (ed) Spectroscopic properties of rare earth in optical materials. Tsuinghua University Press and Springer-Verlag: Beijing, Heidelberg 95–129
43. Ng B, Newman DJ (1985) J Chem Phys 83:1758–1768
44. Ng B, Newman DJ (1987) J Chem Phys 87:7096–7109
45. Carnall WT, Firlds PR, Rajnak K (1968) J Chem Phys 49:4424–4442
46. Carnall WT, Firlds PR, Rajnak K (1968) J Chem Phys 49:4447–4449
47. Carnall WT, Firlds PR, Rajnak K (1968) J Chem Phys 49:4450–4455
48. Carnall WT, Firlds PR, Wybourne BG (1965) J Chem Phys 42:3797–3806
49. Wybourne BG, Smentek L, Kędziorski A (2004) Mol Phys 102:1105–1111
50. Blasse G, Bril A, Nieuwpoort WC (1966) J Phys Chem Solids 27:1587–1592
51. Kisliuk P, Krupke WF, Gruber JB (1964) J Chem Phys 40:3606–3610
52. Faulker TR, Richardson FS (1978) Mol Phys 35:1141–1161
53. Reid MF, Richardson FS (1984) Mol Phys 51:1077–1094
54. Crooks SM, Reid MF, Tanner PA, Zhao YY (1997) J Alloys Compd 220:297–301
55. Judd BR (1979) J Chem Phys 70:4830–4833
56. Malta OL, Riberio SJL, Faucher M, Porcher P (1991) J Chem Phys Solids 52:587–593
57. Shepard R, Shavitt I, Pitzer RM, Comeau DC, Pepper M, Lischka H, Szalay PG, Ahlrichs R, Brown FB, Zhao J-G (1988) Int J Quantum Chem S22:149–165
58. Yabushita S, Zhang Z, Pitzer RM (1999) J Phys Chem A 103:5791–5800
59. Sakai Y, Miyoshi E, Tatewaki H (1998) J Mol Struct Theochem 451:143–150
60. MCP online library: http://meg.cube.kyushu-u.ac.jp/~meg/MCP1.html
61. Miyoshi E, Sakai Y, Tanaka K, Masamura M (1998) J Mol Struct Theochem 451:73–79
62. MCP online library: http://setani.sci.hokudai.ac.jp/sapporo/
63. Hargittai M (1998) Coord Chem Rev 91:35–88
64. Wadt WR (1982) Chem Phys Lett 89:245–248
65. Itoh S, Saito R, Kimura T, Yabushita S (1993) J Phys Soc Jpn 62:2924–2933
66. Sanoyama E, Kobayashi H, Yabushita S (1998) J Mol Struct Theochem 451:189–204
67. Wolfarm Research Inc (2008) Mathematica Version 7.0. Wolfarm Research, Champaign, IL
68. Dalgarno A (1962) Adv Phys 11:281–315
69. Xia S-D, Reid MF (1993) J Chem Phys Solids 54:777–778
70. Smentek-Mielczarek L (1993) Phys Rev B 48:9273–9278
71. Bukietynska K, Choppin GR (1970) J Chem Phys 52:2875–2880
72. Hoshina S, Imanaga S, Yokono S (1977) J Lumin 15:455–471
73. Khan AA, Iftikhar K (1997) Polyhedron 16:4153–4161
74. Choppin GR (2002) J Alloys Compd 344:55–59
75. Hussain HA, Ansari AA, Iftikhar K (2004) Spectrochim Acta Part A 60:873–884
76. Ansari AA, Ilmi R, Iftikhar K (2012) J Lumin 132:51–60
77. Amos A, Hall GG (1961) Proc R Soc A 263:483–493
78. Martin RL (2003) J Chem Phys 118:4775–4777
79. Mayer I (2007) Chem Phys Lett 437:284–286
80. Krupa JC (1995) J Alloys Compd 225:1–10
81. Dorenbos P (2003) J Phys Condens Matter 15:8417–8434

82. Dorenbos P (2009) J Alloys Compd 488:568–573
83. Dorenbos P (2005) J Lumin 111:89–104
84. Desclaux JP (1973) At Data Nucl Data Tables 12:311–406
85. Dorenbos P (2000) J Lumin 91:91–106
86. Warshel A, Karplus M (1972) J Am Chem Soc 94:5612–5625
87. Maseras F, Morokuma K (1995) J Comput Chem 16:1170–1179
88. Lin H, Truhlar DG (2007) Theor Chem Acc 117:185–199

Theor Chem Acc (2014) 133:1486
DOI 10.1007/s00214-014-1486-5

REGULAR ARTICLE

A Hirshfeld interpretation of the charge, spin distribution and polarity of the dipole moment of the open shell $\left({}^{3}\Sigma^{-}\right)$ phosphorus halides: PF and PCl

James F. Harrison

Received: 18 December 2013 / Accepted: 26 March 2014 / Published online: 17 April 2014

Abstract The charge, spin distribution and dipole moments of the open shell molecules PF and PCl have been analyzed using two variants of the Hirshfeld partitioning of the electronic density. In the HI or iterative Hirshfeld approach, one keeps the number of electrons on a given atom in the molecule and proto molecule equal and does not constrain the spin distribution in any way. In the spin-adapted approach, one constrains both the charges and spins on the *in situ* and proatoms to be equal. We find that while allowing for both spin and charge equalization results in a spin distribution that is different from that of the conventional HI method, the behavior of the atomic spin populations as a function of internuclear separation is similar. Both methods predict that as the bond is formed, the halogen gains α and looses β electrons with the converse for P. These electron shifts are further broken down into their σ and π components and we find that while the α electrons gained by the halogen are essentially all in the σ system, they come from both the σ and π system of P. The β electrons gained by P occupy both σ and π densities, but come essentially from the π system on the halogen. The dipole moment curves are partitioned into their α and β components and showing that the dipole due to the σ spin density has the polarity $P^{+}X^{-}$ while that due to the β spin density has the polarity $P^{-}X^{+}$, while the net dipole polarity at equilibrium is $P^{+}X^{-}$, consistent with the spin-dependent charge shifts described above.

Dedicated to the memory of Professor Isaiah Shavitt and published as part of the special collection of articles celebrating his many contributions.

J. F. Harrison (✉)
Department of Chemistry, Michigan State University,
East Lansing, MI 48823, USA
e-mail: harrison@chemistry.msu.edu

Keywords Hirshfeld · Electron density · Phosphorous halides · Dipole moments

1 Introduction

Recently we reported [1] on a study of the spin, charge distribution and dipole moments of the open shell molecules NF, NCl and NBr in their ground electronic ${}^{3}\Sigma^{-}(m_s = +1)$ state using the iterative Hirshfeld (HI) method [2, 3] to partition the spin density between the two centers and examine the change in the number of α and β electrons on the two centers as the chemical bond forms. Subsequent to this study, Geldof et al. [4] have suggested that while the HI method is appropriate for closed shell systems, it leaves some ambiguity in the treatment of open shell systems. In particular, they note that while the HI method keeps the number of electrons on a given atom in the molecule and promolecule equal, it provides no constraints on the spin of the atoms. To address this issue, they proposed an extension to the HI method in which the charges and spins on the in situ and proatoms are equal and implemented this using the fractional occupation Hirshfeld-I (FOHI). Following up on this theme, we report results on the open shell molecules PF and PCl in their ground electronic state ${}^{3}\Sigma^{-}(m_s = +1)$, which allow for both charge and spin equalization between the atoms in the molecule and promolecule using a slightly different implementation which we call the open shell HI or HIOS approach and compare the results with the original HI approach. We find that while allowing for both spin and charge equalization (HIOS) results in a spin distribution that is different from that of the conventional HI method, the behavior of the atomic spin populations as a function of internuclear separation is similar. We find that as the separated atoms come together, both the HI and HIOS partitioning predict that the halogen gains α and looses β electrons, and since the number of α and β electrons is conserved, P gains β and looses α electrons. As a function of internuclear separation, the changes are monotonic for PF and essentially so for PCl. These electron shifts are further broken down into their σ and π components, and we find that while the α electrons gained by the halogen are essentially all in the σ system, they come from both the σ and π system of P. The β electrons gained by P occupy both σ and π densities but come essentially from the π system on the halogen. We find that the spin density calculated with the HI method compares more favorably with experiment than does those calculated with the HIOS partitioning scheme.

Partitioning the charge and spin density as described above allows one to interpret the molecular dipole moment in terms of the charge and spin shifts on the constituent atoms. The dipole moment curves are calculated for internuclear separations ranging from the separated atoms to equilibrium and are partitioned into their α and β components. We find that the dipole due to the α spin density has the polarity P^+X^-, while that due to the β spin density has the polarity P^-X^+, while the net dipole polarity at equilibrium is P^+X^-, consistent with the spin-dependent charge shifts described above. Partitioning these spin-dependent dipole moments into their charge and induced atomic dipole components confirms that they are largely, but not exclusively, determined by the charge distribution.

2 Previous work

Douglas and Frackowiak [5] first observed the electronic spectra of PF and deduced the spectroscopic constants ($R_e = 1.5897$ Å, $\omega_e = 847$ cm^{-1}) of the ground $X^3\Sigma^-$ state. The bond length of PCl ($X^3\Sigma^-$) has been determined to be 2.0146 Å by Minowa et al. [6] from the microwave spectrum, while the vibrational frequency (551.4 cm^{-1}) was determined by Kanamori et al. [7] from its vibration rotation spectrum. The spin density on P and the halogens has been determined by Minowa et al. (PCl) [8] and by Saito et al. (PF) [6] from the hyperfine coupling constants obtained from the microwave spectra of the radicals.

While there are no experimental values for the dipole moment of either radicals, there are several theoretical predictions. Early SCF calculations by O'Hare and Wahl [9] and O'hare [10] predicted 0.24 ea_0 for PF with the polarity P^+F^-. Subsequently, Nguyen [11] reported dipole moments for both PF and PCl of 0.583 ea_0 and 0.543 ea_0, respectively, using the UHF-MP2 method with a 6-31G** basis. This was followed by calculations by Peterson and Woods [12] using the CISD, CEPA-1 and CASSCF

Table 1 Calculated dipole moments of PF and PCl in the $^3\Sigma^-$ state

Molecules	Method	Basis	R (a_0)	μ (ea_0) P^+X^-
PF	SCF[a]	STO	3.0039	0.460
	SCF[b]	STO	2.9858	0.24
	MP2[c]	6-31G**	3.0614	0.583
	CASSCF[d]	11*s*8*p*3*d*1*f*/P 9*s*6*p*2*d*1*f*/F	3.0209	0.2903
	CEPA[d]	11*s*8*p*3*d*1*f*/P 9*s*6*p*2*d*1*f*/F	3.0115	0.3350
	SOCI[e]	7*s*6*p*3*d*/P 4*s*3*p*2*d*/F	3.109	0.409
	MRCI[f]	aug-cc-pvqz	3.0041	0.3163
	SCF[g]	aug-cc-pv5z (*spdf*)	2.9603	0.3263
	CASSCF[g]	aug-cc-pv5z (*spdf*)	3.0064	0.2948
	MRCI[g]	aug-cc-pv5z (*spdf*)	3.0283	0.3284
	RCCSD(t)[g]	aug-cc-pv5z (*spdf*)	3.0173	0.3135
PCl	SCF[h]	cc-pvtz	3.866	0.4162
	CISD[h]	cc-pvtz	3.847	0.3134
	MRCI[i]	cc-pvtz	3.9	0.326
	MRCI[i]	cc-pvqz	3.9	0.331
	MRCI[j]	aug-cc-pvqz no g plus extra diffuse	3.830	0.338
	MP2[c]	6-31G**	3.849	0.543
	SCF[g]	aug-cc-pv5z (*spdf*)	3.845	0.3649
	CASSCF[g]	aug-cc-pv5z (*spdf*)	3.890	0.2715
	MRCI[g]	aug-cc-pv5z (*spdf*)	3.844	0.2361
	RCCSD(t)[g]	aug-cc-pv5z (*spdf*)	3.832	0.2205

[a] Reference [9], [b] Reference [10], [c] Reference [11], [d] Reference [12], [e] Reference [13], [f] Reference [18], [g] This work, [h] Reference [14], [i] Reference [17], [j] Reference [19]

methods and the contracted Gaussian basis sets 11*s*8*p*3*d*1*f* on P and 9*s*6*p*2*d*1*f* on F. They report dipole moment functions and dipole moments at the experimental equilibrium separation of 0.3479 ea_0, 0.3350 ea_0 and 0.2903 ea_0, respectively. Latifzadeh and Balasubramanian [13] used the SOCI approach with the contracted basis sets 7*s*6*p*3*d* on P and 4*s*3*p*2*d* on F and reported a dipole moment of 0.409 ea_0 at their calculated internuclear separation of 1.647 Å. Papakondylis et al. [14] studied several states of PCl using a variety of ab initio methods and the cc-pVTZ basis of Dunning et al. [15, 16] and reported dipole moments for the SCF (0.416 ea_0) and CISD (0.3134ea_0) calculations. Bravo and Machado [17] constructed dipole moment functions for PCl($X^3\Sigma^-$) using a MRSDCI function and reported dipole moments of 0.326ea_0 and 0.331ea_0 at $R = 3.9a_0$ with the cc-pVTZ and cc-pVQZ basis sets, respectively. de Brouckere [18] used a MRCI approach and the aug-cc-pVQZ basis and reported the dipole moment function between $R = 2.4372$ a_0 and

Table 2 Calculated bond length and harmonic vibration frequency for PF and PCl in the $^3\Sigma^-$ state compared with experiment

Methods	PF ($^3\Sigma^-$)		PCl ($^3\Sigma^-$)	
	R_e (Å)	ω_e (cm^{-1})	R_e (Å)	ω_e (cm^{-1})
scf	1.5665	904	2.0349	551
casscf	1.5909	844	2.0586	515
mrci	1.6025	845	2.0342	543
rccsd(t)	1.5967	839	2.0279	544
experiment	1.5897[a]	846.8[a]	2.0146[b]	551.4[c]

[a] Reference [5], [b] Reference [6], [c] Reference [7]

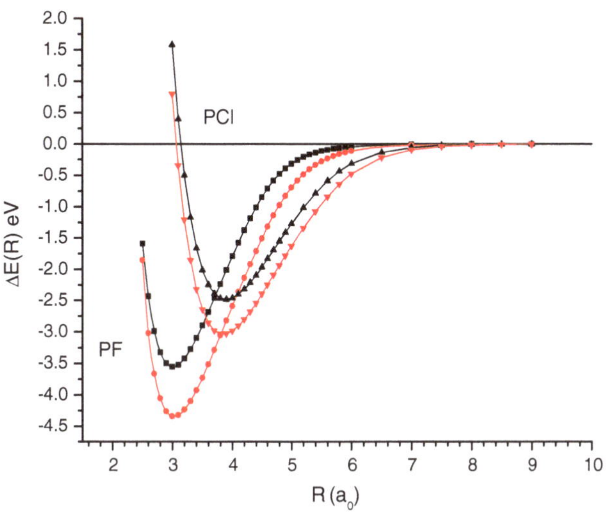

Fig. 1 Potential energy curves for the $^3\Sigma^-$ state of PF and PCl calculated with the CASSCF (*black curves*) (12,8) and MRCI (*red curves*) methods using the av5z basis

4.50615 a_0 and $\mu = 0.3163\ ea_0$ at the experimental equilibrium separation. Da Silva-Neto et al. [19] reported CASSCF/MRCI calculations on several states of PCl using the aug-cc-pVQZ-g basis and reported a dipole moment function for the ground state and an equilibrium dipole moment of 0.338 ea_0. These dipole moments are collected in Table 1 along with our results (vide infra).

3 Computational details

We used the aug-cc-pv(5 + d)z basis for P [20] and the aug-cc-pv5z basis set for F and Cl [21], and in all calculations, we restricted the angular momentum on the basis of *spdf*. This was primarily because the locally written codes that performed the subsequent Hirshfeld analysis are limited to basis functions with these symmetries. We characterized the molecules using the SCF, CASSCF, the internally contracted mrci and the RCCSD(t) methods as implemented in MOLPRO [22], and in all cases, we only correlated the valence electrons. Table 2 compares our calculated internuclear separations and vibrational frequencies for each method with experiment, and Fig. 1 shows the potential energy curves for both PF and PCl at the CASSCF and MRCI levels. In balance, the agreement with the limited experimental data is satisfactory. We anticipate that the small discrepancies between the RCCSD(t) results and experiment are due to the omission of core–valence correlation. In what follows, we will use the charge and spin density matrices available from a CASSCF calculation in MOLPRO to generate the charge and spin densities used in the Hirshfeld analysis.

4 Preliminaries

The Hirshfeld analysis has been discussed previously [1, 2], so we will outline the process as it applies to the current study. Hirshfeld [2] suggested that if one wanted to partition the molecular electron density among the constituent atoms in a molecule, one does so by allotting density to each atom in proportion to the density that atom would have in the promolecule. For PCl, one would define the density of the *insitu* P atom as $\eta_{\text{P}} = \left(\frac{\eta_{\text{P}}^0}{\eta_{\text{proto}}}\right)\eta = W_{\text{P}}\eta$, where η is the calculated electron density in PCl and the promolecule density, $\eta_{\text{pro}} = \eta_{\text{P}}^0 + \eta_{\text{X}}^0$, is the sum of the separated neutral atom densities placed at the appropriate nuclear positions in the molecule. The *in situ* number of electrons on P is $N_{\text{P}} = \int W_{\text{P}}\eta dV$. The arbitrariness of this choice of promolecule has been noted by several authors [23] and has been addressed by Bultinck et al. [3] who described an algorithm for determining a unique promolecule self-consistently. The algorithm is based on the demonstration by Nalewajski et al. [24] that the Hirshfeld atoms are the best in the Kulback-Lieber [25] information theoretical sense if the density of an in situ atom integrates to the number of electrons on the same atom in the promolecule, which means the proatoms will be charged. See also the earlier work of Nalewajski et al. [26] as well as Ayers [27]. The in situ atomic densities obtained from this algorithm are referred to as the Hirshfeld-I or HI densities.

As an example, consider PCl. At large separations, we expect Cl to be negative and so we write the number density for Cl and P in the proatoms as

$$\eta_{\text{Cl}}^{pa} = \eta_{\text{Cl}}^0 + \chi(\eta_{\text{Cl}}^{-1} - \eta_{\text{Cl}}^0) \text{ and } \eta_{\text{P}}^{pa} = \eta_{\text{P}}^0 + \chi(\eta_{\text{P}}^{+1} - \eta_{\text{P}}^0) \tag{1}$$

where η_{Cl}^0 and η_{Cl}^{-1} are the number densities for Cl^0 and Cl^{-1} with the corresponding interpretation for P, and χ is a positive number to be determined. The promolecule density is then

$$\eta^{pm} = \eta^{pa}_{\mathrm{Cl}} + \eta^{pa}_{\mathrm{P}} = \eta^0_{\mathrm{Cl}} + \eta^0_{\mathrm{P}} + \chi(\eta^{-1}_{\mathrm{Cl}} - \eta^0_{\mathrm{Cl}} + \eta^{+1}_{\mathrm{P}} - \eta^0_{\mathrm{P}}). \tag{2}$$

Note that the charge on the promolecule is conserved, while the number of electrons on the Cl and P atoms in the promolecule are

$$N^{pa}_{\mathrm{Cl}} = \int \left(\eta^0_{\mathrm{Cl}} + \chi(\eta^{-1}_{\mathrm{Cl}} - \eta^0_{\mathrm{Cl}})\right)\mathrm{d}V = N^0_{\mathrm{Cl}} + \chi \tag{3}$$

and

$$N^{pa}_{\mathrm{P}} = \int \left(\eta^0_{\mathrm{P}} + \chi(\eta^{+1}_{\mathrm{P}} - \eta^0_{\mathrm{P}})\right)\mathrm{d}V = N^0_{\mathrm{P}} - \chi \tag{4}$$

and differ from those in the neutral free atoms by $\pm\chi$. One forms the weight factor

$$W_{\mathrm{Cl}}(\chi) = \frac{\eta^0_{\mathrm{Cl}} + \chi(\eta^{-1}_{\mathrm{Cl}} - \eta^0_{\mathrm{Cl}})}{\eta^0_{\mathrm{Cl}} + \eta^0_{\mathrm{P}} + \chi(\eta^{-1}_{\mathrm{Cl}} - \eta^0_{\mathrm{Cl}} + \eta^{+1}_{\mathrm{P}} - \eta^0_{\mathrm{P}})} \tag{5}$$

with the corresponding form for $W_{\mathrm{P}}(\chi)$ and initially one chooses $\chi = 0$, the traditional Hirshfeld selection, and determines the atomic charges in the usual way. One then updates the weight factor and proatom densities with χ equal to the absolute value of the atomic charges. Using the updated weights and proatom densities, one determines a new set of atomic charges and therefore an updated χ and so on. In our implementation, this happens to within 0.0001 electrons in fewer than nine iterations. The number of electrons on the ith atom is then

$$N(i) = \int \eta_i \mathrm{d}V = \int W_i(\chi)\eta \mathrm{d}V. \tag{6}$$

When we partition the spin between the two atom (vide infra), we define the density of α electrons on atom i as

$$N_\alpha(i) = \int \eta_{\alpha i}\mathrm{d}V = \int W_i(\chi)\eta_\alpha \mathrm{d}V \tag{7}$$

where η_α is the computed α spin density, and $W_i(\chi)$ is the weight factor defined above. In what follows the electron distributions determined as described above will be referred to as the HI results.

The HIOS approach is very similar. One simply replaces the weight factors $W_{\mathrm{Cl}}(\chi)$ and $W_{\mathrm{P}}(\chi)$ by the spin-dependent factors $W_{\alpha\mathrm{Cl}}(\chi_\alpha), W_{\alpha\mathrm{P}}(\chi_\alpha), W_{\beta\mathrm{Cl}}(\chi_\beta)\text{and}W_{\beta\mathrm{P}}(\chi_\beta)$ and proceeds as above.

Using PCl as an example, one defines the α and β densities of the proto atoms as

$$\eta^{pa}_{\alpha\mathrm{Cl}} = \eta^0_{\alpha\mathrm{Cl}} + \chi_\alpha\left(\eta^0_{\alpha\mathrm{Cl}} - \eta^{-1}_{\alpha\mathrm{Cl}}\right) \tag{8}$$

$$\eta^{pa}_{\beta\mathrm{Cl}} = \eta^0_{\beta\mathrm{Cl}} + \chi_\beta\left(\eta^0_{\beta\mathrm{Cl}} - \eta^{-1}_{\beta\mathrm{Cl}}\right) \tag{9}$$

$$\eta^{pa}_{\alpha\mathrm{P}} = \eta^0_{\alpha\mathrm{P}} + \chi_\alpha\left(\eta^0_{\alpha\mathrm{P}} - \eta^{+1}_{\alpha\mathrm{P}}\right) \tag{10}$$

$$\eta^{pa}_{\beta\mathrm{P}} = \eta^0_{\beta\mathrm{P}} + \chi_\beta\left(\eta^0_{\beta\mathrm{P}} - \eta^{+1}_{\beta\mathrm{P}}\right) \tag{11}$$

Then define the α and β densities of the promolecule as

$$\eta^{pm}_\alpha = \eta^{pa}_{\alpha\mathrm{Cl}} + \eta^{pa}_{\alpha\mathrm{P}} \quad \text{and} \quad \eta^{pm}_\beta = \eta^{pa}_{\beta\mathrm{Cl}} + \eta^{pa}_{\beta\mathrm{P}} \tag{12}$$

The α and β weight factors for Cl are

$$W_{\alpha\mathrm{Cl}} = \frac{\eta^{pa}_{\alpha\mathrm{Cl}}}{\eta^{pm}_\alpha} \text{ and } W_{\beta\mathrm{Cl}} = \frac{\eta^{pa}_{\beta\mathrm{Cl}}}{\eta^{pm}_\beta} \tag{13}$$

with similar definitions for P. The *in situ* density of the α spin electrons on Cl is then

$$\eta^{\mathrm{mol}}_{\alpha\mathrm{Cl}} = \eta^{\mathrm{mol}}_\alpha W_{\alpha\mathrm{Cl}} \tag{14}$$

One determines χ_α by requiring that the number of α spin electrons on the *insitu* Cl equals those in the promolecule

$$\begin{aligned} N^{\mathrm{mol}}_{\alpha\mathrm{Cl}} &= \int \eta^{\mathrm{mol}}_\alpha W_{\alpha\mathrm{Cl}}\mathrm{d}V = N^{\mathrm{pa}}_{\alpha\mathrm{Cl}} = \int \eta^{\mathrm{pa}}_{\alpha\mathrm{Cl}}\mathrm{d}V \\ &= N^0_{\alpha\mathrm{Cl}} + \chi_\alpha\left(N^0_{\alpha\mathrm{Cl}} - N^{-1}_{\alpha\mathrm{Cl}}\right) \end{aligned} \tag{15}$$

and so

$$\chi_\alpha = \frac{N^{\mathrm{mol}}_{\alpha\mathrm{Cl}} - N^0_{\alpha\mathrm{Cl}}}{N^0_{\alpha\mathrm{Cl}} - N^{-1}_{\alpha\mathrm{Cl}}} \tag{16}$$

As with the HI method, one initially chooses $\chi_\alpha = 0$, forms $W_{\alpha\mathrm{Cl}}(\chi_\alpha = 0)$, calculates $N^{\mathrm{mol}}_{\alpha\mathrm{Cl}}$, updates $W_{\alpha\mathrm{Cl}}(\chi_\alpha)$ and iterates until $N^{\mathrm{mol}}_{\alpha\mathrm{Cl}}$ converges, to 10^{-4} which usually happens within 10 or so iterations. This procedure insures that the number of electron and their spins are equal on the atoms in the molecule and the promolecule.

These integrals must be evaluated numerically and we do so using the Euler–Maclaurin [26] method for the radial integrations and the Gauss–Legendre [27] method for the angular integration.

5 Charge and spin distribution

The wave functions for PF and PCl were calculated in C_{2v} symmetry with z as the C_2 axis, and since the charge and spin density matrices are symmetry blocked, we can write the molecular density as a sum of a_1, a_2, b_1 and b_2 symmetries. We will define the b_1 and b_2 symmetries as π and the a_1 and a_2 symmetries as σ. Allotting the a_2 symmetry as σ is arbitrary but has little effect on the σ, π composition of the valence orbitals. Using the spin and charge density matrices available from the CASSCF calculations, we constructed the α and β electron densities, and partitioned these into their σ and π components.

$$\eta = \eta_\alpha + \eta_\beta = \eta_{\alpha\sigma} + \eta_{\beta\sigma} + \eta_{\alpha\pi} + \eta_{\beta\pi} \tag{17}$$

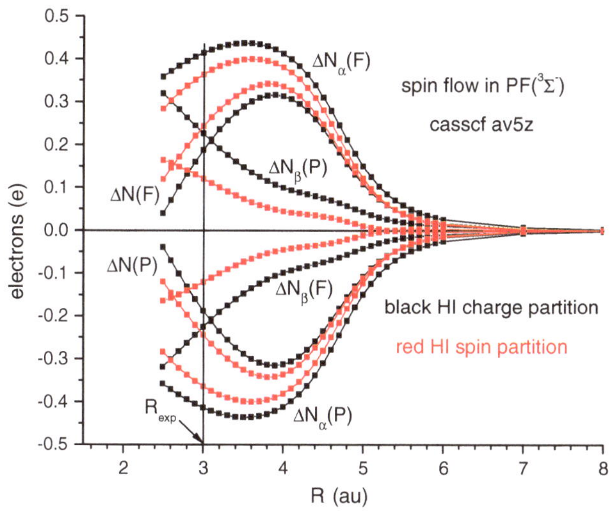

Fig. 2 Comparison of the HI and HIOS partitioning of the number of α and β spin electrons transferred between P and F in the $^3\Sigma^-$ ($M_s = +1$) state of PF calculated as a function of internuclear separation with the CASSCF(12,8) method and the av5z basis

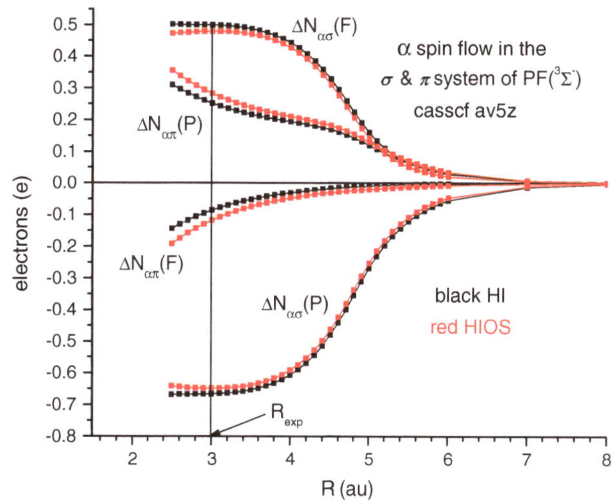

Fig. 4 Comparison of the HI and HIOS partitioning of the number of α spin electrons transferred between P and F in the σ and π systems in the $^3\Sigma^-$ ($M_s = +1$) state of PF calculated as a function of internuclear separation with the CASSCF(12,8) method and the av5z basis

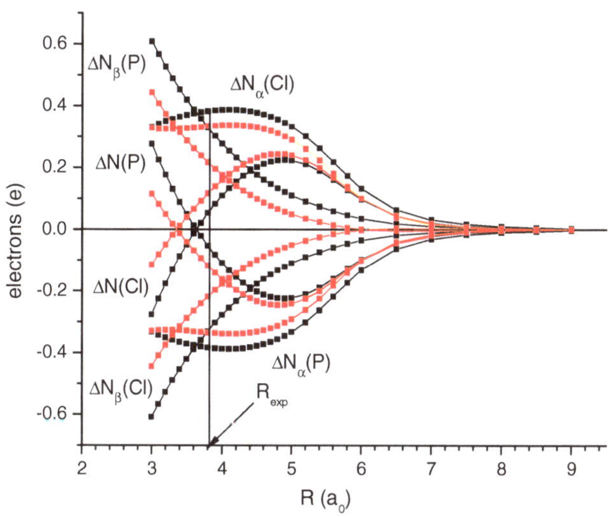

Fig. 3 Comparison of the HI (*black curves*) and HIOS (*red curves*) partitioning of the number of α and β spin electrons transferred between P and Cl in the $^3\Sigma^-$ $M_s = +1$) state of PCl calculated as a function of internuclear separation with the CASSCF(12,8) method and the av5z basis

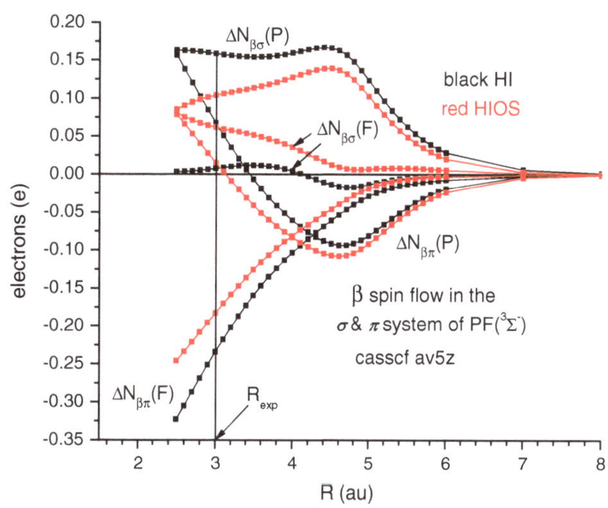

Fig. 5 Comparison of the HI and HIOS partitioning of the number of β spin electrons transferred between P and F in the σ and π systems in the $^3\Sigma^-$ ($M_s = +1$) state of PF calculated as a function of internuclear separation with the CASSCF(12,8) method and the av5z basis

Using the HI and HIOS weight factors, we partition the electron density between the two centers. So, for example, in the HI method, the number of β electrons of σ symmetry on P is given by $N_{\beta\sigma}(\mathrm{P}) = \int W_{\mathrm{P}}\eta_{\beta\sigma}\mathrm{d}V$, while in the HIOS method its $N_{\beta\sigma}(\mathrm{P}) = \int W_{\beta\mathrm{P}}\eta_{\beta\sigma}\mathrm{d}V$.

The $^3\Sigma^-$ state of PF or PCl separates to P in its ground ^4S_u state and the halogen in its ^2P_u state. At ∞ internuclear separation, the $^3\Sigma^-(m_s = +1)$ state of PX obtains when the $m_s = +3/2$ and $+1/2$ spin states of P couple with the $m_s = -1/2$ and $+1/2$ spin states of the halogen, resulting in the asymptotic CASSCF function

$$\left|^3\Sigma^-(R = \infty)\right\rangle = \left| z_X z_P x_P y_P \left(\frac{\sqrt{3}}{2} \beta\alpha\alpha\alpha - \frac{1}{2\sqrt{3}} (\alpha\beta\alpha\alpha + \alpha\alpha\beta\alpha + \alpha\alpha\alpha\beta) \right) \right\rangle \quad (18)$$

where we only show the singly occupied atomic orbitals. The resulting α spin (valence orbital) occupancy at ∞ is then

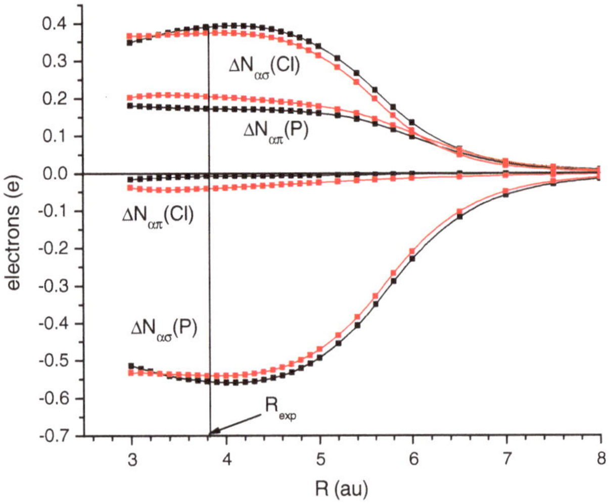

Fig. 6 Comparison of the HI (*black curves*) and HIOS (*red curves*) partitioning of the number of α spin electrons transferred between P and Cl in the σ and π systems in the $^3\Sigma^-$ ($M_s = +1$) state of PCl calculated as a function of internuclear separation with the CASSCF(12,8) method and the av5z basis

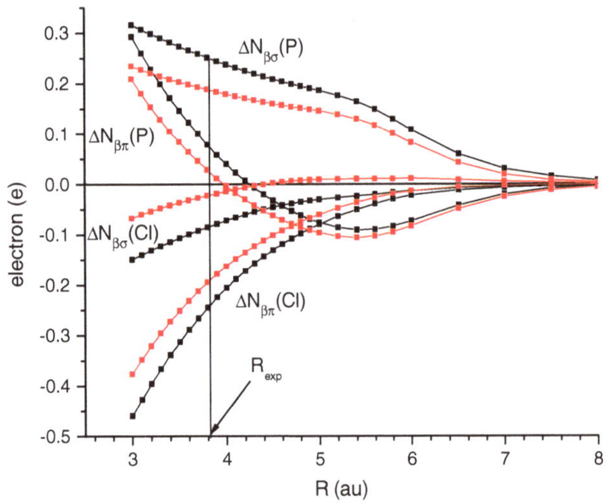

Fig. 7 Comparison of the HI (*black curves*) and HIOS (*red curves*) partitioning of the number of β spin electrons transferred between P and Cl in the σ and π systems in the $^3\Sigma^-$ ($M_s = +1$) state of PCl calculated as a function of internuclear separation with the CASSCF(12,8) method and the av5z basis

$$s_P^1 x_P^{11/12} y_P^{11/12} z_P^{11/12} s_X^1 x_X^1 y_X^1 z_X^{0.25} \tag{19}$$

with the corresponding β spin occupancy

$$s_P^1 x_P^{1/12} y_P^{1/12} z_P^{1/12} s_X^1 x_X^1 y_X^1 z_X^{0.75} \tag{20}$$

Note that the spin composition of the single electron in the p_z orbital on the halogen is 0.75β and 0.25α, while that on the P atom is $11/12\alpha$ and $1/12\beta$. Clearly, at ∞, the halogen is rich in β spins, while P is rich in α spins.

With these asymptotic occupations, we can define the shift in the number of electrons of a specific spin and spatial symmetry on a given center via

$$N_{\alpha\sigma}(\mathrm{X}) = 0.25 + \Delta N_{\alpha\sigma}(\mathrm{X}) \tag{21}$$

$$N_{\beta\sigma}(\mathrm{X}) = 0.75 + \Delta N_{\beta\sigma}(\mathrm{X}) \tag{22}$$

$$N_{\alpha\pi}(\mathrm{X}) = 2.0 + \Delta N_{\alpha\pi}(\mathrm{X}) \tag{23}$$

$$N_{\beta\pi}(\mathrm{X}) = 2.0 + \Delta N_{\beta\pi}(\mathrm{X}) \tag{24}$$

$$N_{\alpha\sigma}(\mathrm{P}) = 11/12 + \Delta N_{\alpha\sigma}(\mathrm{P}) \tag{25}$$

$$N_{\beta\sigma}(\mathrm{P}) = 1/12 + \Delta N_{\beta\sigma}(\mathrm{P}) \tag{26}$$

$$N_{\alpha\pi}(\mathrm{P}) = 11/6 + \Delta N_{\alpha\pi}(\mathrm{P}) \tag{27}$$

$$N_{\beta\pi}(\mathrm{P}) = 1/6 + \Delta N_{\beta\pi}(\mathrm{P}) \tag{28}$$

where, for example, $\Delta N_{\beta\pi}(\mathrm{X})$ represents the change in the number of β electrons of π symmetry on the halogen relative to the number at $R = \infty$.

6 Comparison of the HI and HIOS electron distributions

Figures 2 and 3 compares the distribution of α and β spins as a function of the internuclear separation, R for PF and PCl. The distance dependence for both methods is similar in kind but does differ in degree. Both methods show that as the internuclear separation decreases, the halogen (X) gains α and looses β electrons with the converse happening on P and since the number of α and β electrons is constant, $\Delta N_\alpha(\mathrm{P}) = -\Delta N_\alpha(\mathrm{X})$ and $\Delta N_\beta(\mathrm{P}) = -\Delta N_\beta(\mathrm{X})$ for all R. Both methods predict that the α electrons are transferred from P to the halogen much earlier than the corresponding β are transferred from the halogen to P. Additionally, both predict that the halogen is negative with the HIOS predicting it to be more so.

Figures 4, 5, 6 and 7 compare the spin flow for the HI and HIOS methods broken down into their σ and π components. The σ and π breakdown of the α spin distribution in both molecules (Figs. 4, 6) for the two methods is remarkably similar. Both methods predict that α electrons P looses in the σ system are going into the σ system of the halogen, an interatomic transfer, and also into the π system of P, an intra-atomic transfer. This is most apparent at large R where $\Delta N_{\alpha\pi}(\mathrm{P}) \sim \Delta N_{\alpha\sigma}(\mathrm{X})$ and each is $\sim -\Delta N_{\alpha\sigma}(\mathrm{P})/2$. Note that all of the α electrons gained by the halogen are going into the σ system, which one might expect given the asymptotic spin distribution (vide supra). Figures 4 and 6 compare the corresponding β spin distribution.

Turning to Fig. 5, we see that at large R, the increase in the number of β electrons in the σ system of P in PF is due to an intra-atomic transfer from the Pπ system which continues until around $R = 4.5$ a_0 at which point the

Table 3 Equilibrium electron distribution in the $^3\Sigma^-$ state of PF and PCl calculated using the HI and HIOS methods

	PF				PCl			
	F HI	F HIOS	P HI	P HIOS	Cl HI	Cl HIOS	P HI	P HIOS
$\Delta N_{\alpha\sigma}$	0.500	0.481	−0.666	−0.646	0.388	0.375	−0.556	−0.538
$\Delta N_{\alpha\pi}$	−0.086	−0.120	0.253	0.286	−0.008	−0.039	0.175	0.202
ΔN_{α}	0.414	0.361	−0.413	−0.360	0.380	0.336	−0.379	−0.336
$\Delta N_{\beta\sigma}$	0.009	0.063	0.159	0.107	−0.083	−0.023	0.250	0.188
$\Delta N_{\beta\pi}$	−0.234	−0.183	0.066	0.014	−0.239	−0.190	0.073	0.026
ΔN_{β}	−0.225	−0.120	0.225	0.121	−0.322	−0.213	0.323	0.214
ΔN	0.189	0.241	−0.188	−0.239	0.058	0.123	−0.056	−0.122

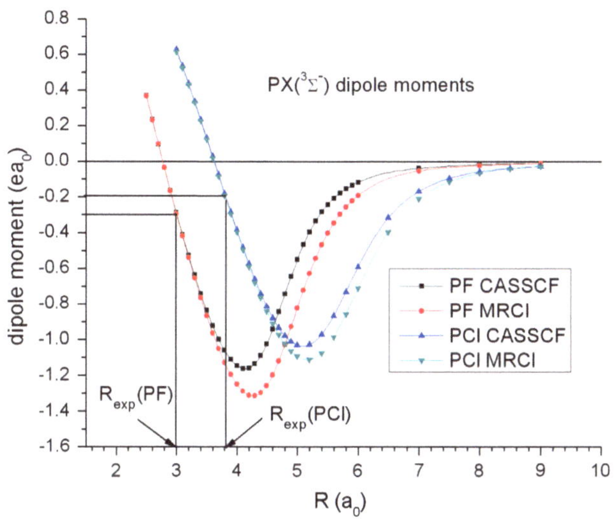

Fig. 8 Dipole moment curves of PF and PCl in the $^3\Sigma^-$ state calculated with the CASSCF and MRCI methods using the av5z basis

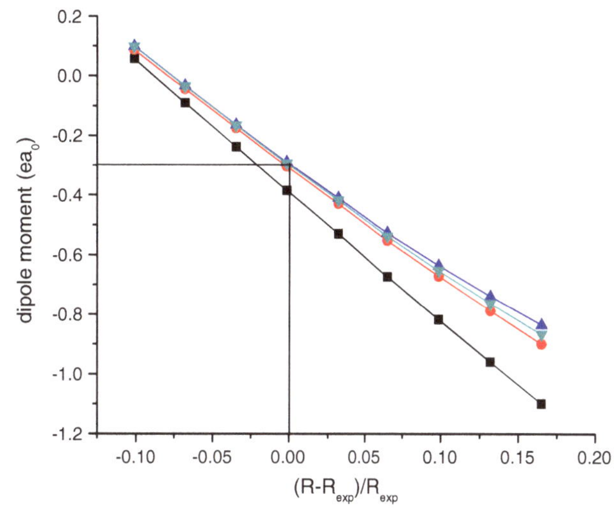

Fig. 9 Dipole moment of PF ($^3\Sigma^-$) calculated with the RHF [*square symbol*], CASSCF [*triangle symbol*], MRCI [*inverted triangle symbol*], RCCSD(t) [*dot symbol*] methods using the av5z basis. Rexp = 3.0041a0

number of β electrons in the σ system of P levels off and the $\beta\pi$ electrons on F start to populate the $\beta\pi$ system of P, i.e., traditional back donation. Note that virtually all of the β electrons lost from the halogen are from the π system.

Figure 7 shows the β electron shifts in PCl and they are similar to those in PF, differing in degree rather than kind. Table 3 summarizes the various electron distributions at equilibrium for both systems.

7 Comparison with experiment

Saito et al. [8] observed the microwave spectrum of PF in the $X^3\Sigma^-$ state and from the observed magnetic coupling constants calculated that the unpaired spin density was 91.3 % on P and 9.1 % on F. In a similar study of PCl, Minowa et al. [6] calculated that 87.2 % of the unpaired spin density was on P, while 14.1 % was on Cl. For PF, the HI method predicts 93.1 % for P and 6.9 % for F and for PCl, 89.9 % on P and 10.1 % on Cl, in reasonable agreement with the experimental values. Interestingly the HIOS

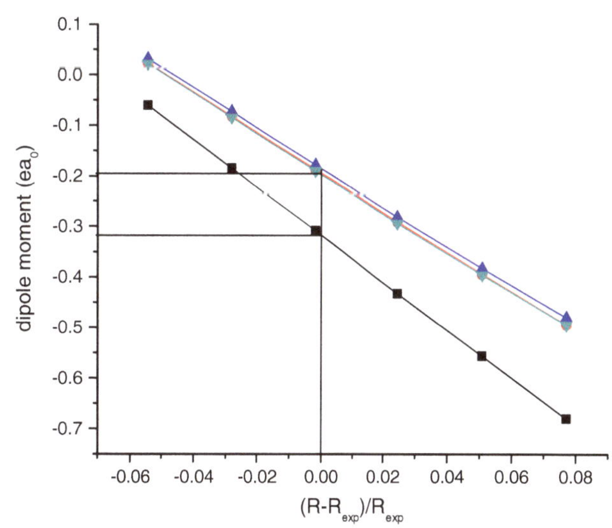

Fig. 10 Dipole moment of PCl ($^3\Sigma^-$) calculated with the RHF [*square symbol*], CASSCF [*triangle symbol*], MRCI [*inverted triangle symbol*] and RCCSD(t) [*dot symbol*] methods using the av5z basis. Rexp = 3.8070 *a*0

Table 4 Equilibrium dipole moments of PF and PCl in the $^3\Sigma^-$ state, calculated at the SCF, CASSCF, MRCI and RCCSD(t) levels with the av5z basis and fit to the polynomial $\sum_{i=0}^{4} c_i[(R-R_e)/R_e]^i$

	PF ($^3\Sigma^-$)				PCl ($^3\Sigma^-$)			
	scf	casscf	mrci	rccsd(t)	scf	casscf	mrci	rccsd(t)
μ_0	−0.3905	−0.2938	−0.2980	−0.3079	−0.3172	−0.1850	−0.1972	−0.1941
μ_1	−1.4602	−1.2364	−1.2623	−1.2732	−4.7294	−3.9917	−4.0011	−3.9890
μ_2	0.0489	0.2269	0.1763	0.1225	0.0196	1.2111	1.1071	0.6073
μ_3	0.0432	0.1274	0.1154	0.0919	3.4848	17.2568	11.1129	7.8403
μ_4	−0.0343	0.0190	0.0089	−0.0097	−8.8157	−72.4101	−29.359	−6.0653
$\mu(R_e)$	−0.3263	−0.2948	−0.3284	−0.3135	−0.3649	−0.2715	−0.2361	−0.2205

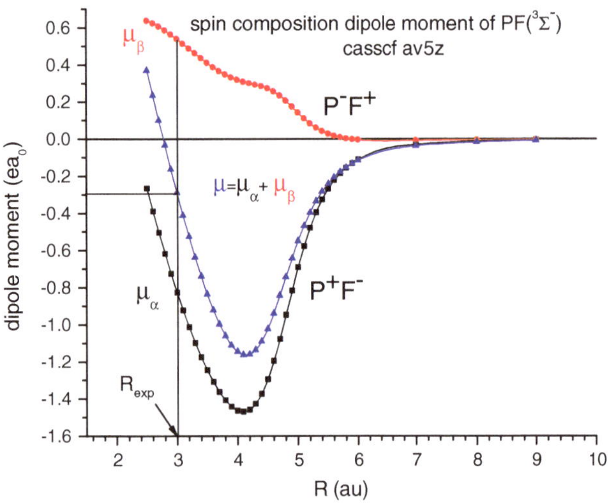

Fig. 11 Dipole moment curve of PF ($^3\Sigma^-$) calculated with the CASSCF(12,8) method and the av5z basis and partitioned into its α and β spin components

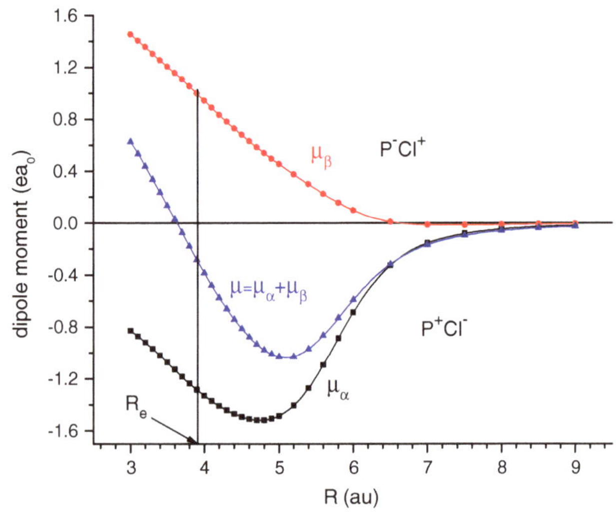

Fig. 12 Dipole moment curve of PCl ($^3\Sigma^-$) calculated with the CASSCF(12,8) method and the av5z basis and partitioned into its α and β spin components

partitioning predicts P(100.8 %) and F(−0.80 %) in PF and (97.9 %) and Cl(2.1 %) in PCl; i.e., much more β electron density on the halogen. Nguyen [11] estimated the spin densities in these molecules using the UHF method with a 6-31++G** basis and found for PF, P(106.5 %), F(−7.0 %) and for PCl, P(99 %) and Cl(1 %), also considerably different from the experimental results and also with a larger β electron density on the halogens. We are exploring the predictions of the HI and HIOS partitioning for spin densities in other open shell systems where experimental data are known.

8 Dipole moment

The dipole moment functions for PF and PCl, calculated at the CASSCF and MRCI levels, are shown in Fig. 8. There is a significant difference between the CASSCF and MRCI functions at large R but little difference around equilibrium. This is illustrated in Figs. 9 and 10, which show the functions around the experimental equilibrium bond length for a variety of methods. These curves are fit to a fourth-order polynomial in $(R-R_{\mathrm{exp}})/R_{\mathrm{exp}}$, and the results are summarized in Table 4. In this table, μ_0 is the dipole moment calculated at the experimental internuclear separation, while $\mu(R_e)$ is the dipole moment at the calculated R_e for each method. To the best of our knowledge, the dipole moment of either molecule has not been determined experimentally but there have been a few theoretical estimates all of which predict the polarity P^+X^- and these are collected in Table 1. Also included for ease of comparison are our $\mu(R_e)$ from Table 4. Comparing the results is difficult because no two calculations are identical, and as we can see from Table 1, the dipole moments are very sensitive to the bond length. It looks like the correlated wave functions with a large flexible basis predict a dipole moment for PF of ~0.31 ea_0. The results for PCl are a bit more confusing. The MRCI calculations of da Siva-Neto et al. [19] use the aug-cc-pvqz basis augmented with several diffuse functions and predict a dipole moment of 0.338 ea_0 at an Re of 3.830 a_0, while our comparable calculations using the aug-cc-pv5z (spdf) basis and the MRCI and RCCSD(t) predict dipoles that are much smaller. We redid our RCCSD(t) calculations using the da Siva-Neto basis

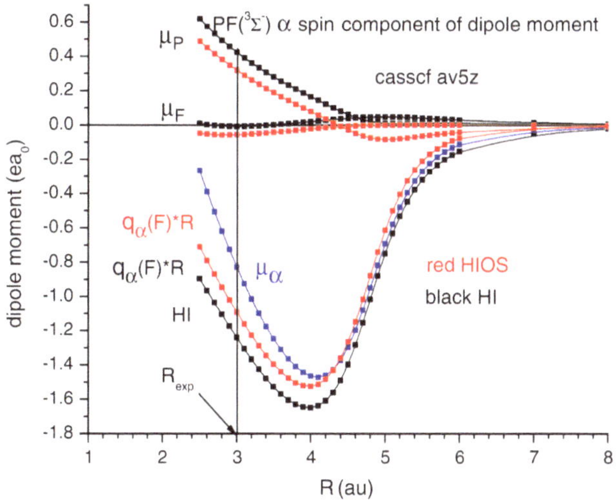

Fig. 13 Comparison of the HI and HIOS α spin component of the dipole moment of PF ($^3\Sigma^-$) calculated with the CASSCF(12,8) method using the av5z basis and partitioned into the charge transfer contribution $q(\mathrm{F})R$ and the induced atomic dipoles $\mu(\mathrm{F})$ and $\mu(\mathrm{P})$

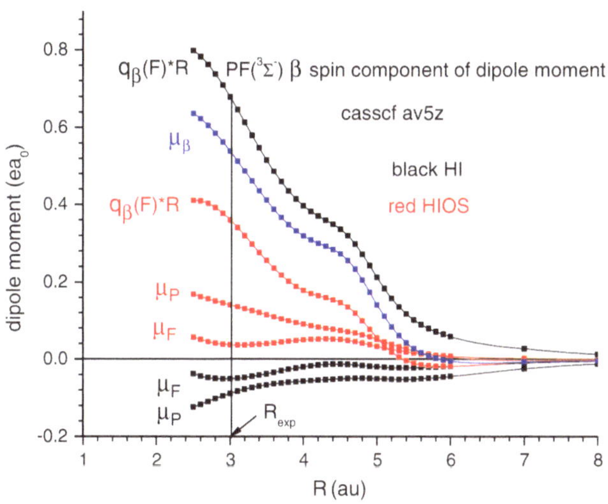

Fig. 14 Comparison of the HI and HIOS β spin component of the dipole moment of PF ($^3\Sigma^-$) calculated with the CASSCF(12,8) method using the av5z basis and partitioned into the charge transfer contribution $q(\mathrm{F})R$ and the induced atomic dipoles $\mu(\mathrm{F})$ and $\mu(\mathrm{P})$

and found $R_e = 3.847\ a_0$ and $\mu = 0.2374\ ea_0$, which are similar to our tabulated results of 3.832 a_0 and 0.2205 ea_0, still somewhat different from their reported values.

Figures 11 and 12 show the CASSCF dipole moment functions of PF and PCl decomposed into their α and β spin components. In both molecules, the long-range dipole moment is entirely due to shifts in the α spin distribution with P loosing α electrons to the halogen. At shorter internuclear distances, the intermolecular shift of β electrons begins and the P^-X^+ polarity of μ_β begins to offset the P^+X^- polarity of μ_α.

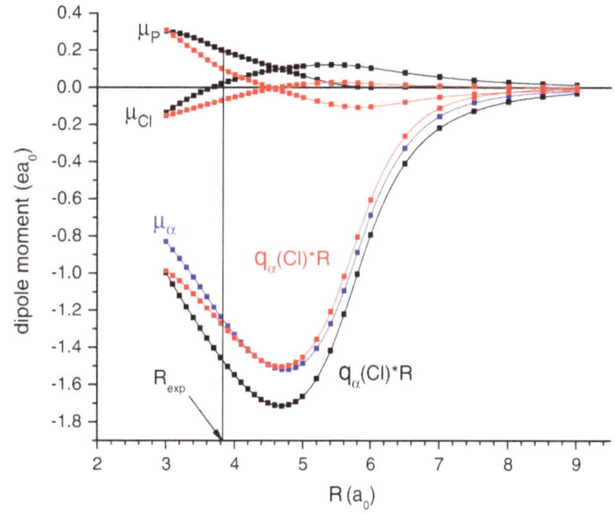

Fig. 15 Comparison of the HI (*black curves*) and HIOS (*red curves*) α spin component of the dipole moment of PCl ($^3\Sigma^-$) calculated with the CASSCF(12,8) method using the av5z basis and partitioned into the charge transfer contribution $q(\mathrm{Cl})R$ and the induced atomic dipoles $\mu(\mathrm{Cl})$ and $\mu(\mathrm{P})$

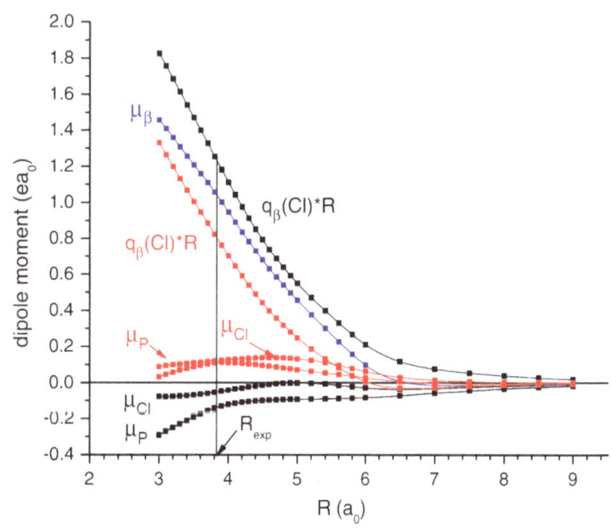

Fig. 16 Comparison of the HI (*black curves*) and HIOS (*red curves*) β spin component of the dipole moment of PCl ($^3\Sigma^-$) calculated with the CASSCF(12,8) method using the av5z basis and partitioned into the charge transfer contribution $q(\mathrm{Cl})R$ and the induced atomic dipoles $\mu(\mathrm{Cl})$ and $\mu(\mathrm{P})$

Figures 13, 14, 15 and 16 compare the HI and HIOS decomposition of the α and β spin dipoles into their *in situ* atomic dipole and charge component, calculated via $\mu_i = \mu_i(X) + \mu_i(P) + q_i(X)R$ where $i = \alpha$ or β and $X =$ F or Cl. From the decomposition of the α spins in Figs. 13 and 15, we see that a consequence of the halogens gaining fewer α spins in the HIOS partitioning is that the induced atomic dipoles due to the α spins must also be smaller since both the HI and HIOS partitioning must sum to the same μ_α. Figures 14 and 16 compare the partitioning of the β spins in the two methods and there are fewer β spins lost by the halogen in the HIOS partitioning, resulting in induced atomic dipoles that are much larger than predicted by the HI partitioning.

9 Summary

The charge, spin distribution and dipole moments of the open shell $^3\Sigma^-(m_s = +1)$ molecules PF and PCl have been analyzed using two variants of the Hirshfeld partitioning of the calculated electronic density: the iterative Hirshfeld method (HI) and the iterative open shell method (HIOS). We find the distance dependence of the charge and spin exchange as the molecules form is similar in both methods but differs in degree. Interestingly, the HI method is in better agreement with the experimental spin densities. Both methods predict that the dipole moment associated with the α spin density has the polarity $P^+X,^-$while the polarity associated with the β spin density is P^-X^+. The net polarity at equilibrium is P^+X^-.

References

1. Harrison JF (2009) J Chem Phys 131:044117
2. Hirshfeld FL (1977) Theor Chim Acta 44:129
3. Bultinck P, VanAlsenoy C, Ayers PW, Carbo-Dorca R (2007) J Chem Phys 126:144111
4. Geldof D, Krishtal A, Blockhuys F, Van Asenoy C (2011) J Chem Theory Comput 7:1328
5. Douglas AE, Frackowiak M (1962) Can J Phys 40:832
6. Minowa T, Saito S, Hirota E (1985) J Chem Phys 83:4939
7. Kanamori H, Yamada C, Butler JE, Kawaguchi K, Hirota E (1985) J Chem Phys 83:4945
8. Saito S, Endo Y, Hirota E (1985) J Chem Phys 82:2947
9. O'Hare PAG, Wahl AC (1971) J Chem Phys 54:4563
10. O'Hare PAG (1973) J Chem Phys 59:3842
11. Nguyen MT (1986) Mol Phys 59:547
12. Peterson KA, Woods C (1990) J Chem Phys 93:1876
13. Latifzadeh L, Balasubramanian K (1995) Chem Phys Lett 243:243
14. Papakondylis A, Mavridis A, Metropoulos A (1995) J Phys Chem 99:10759
15. Dunning TH Jr (1989) J Chem Phys 90:1007
16. Woon DE, Dunning TH Jr (1993) J Chem Phys 98:1358
17. Bravo R, Machado BC (1999) Chem Phys Lett 307:511
18. de Brouckere G (2000) Chem Phys 262:229
19. da Silva-Neto AG, Roberto-Neto O, Ornellas FR, Machado FBC (2004) Chem Phys Lett 395:239
20. Dunning TH Jr, Peterson KA, Wilson AK (2001) J Chem Phys 114:9244
21. Kendall RH, Dunning TH Jr, Harrison RJ (1992) J Chem Phys 96:6796
22. Werner HJ, Knowles PJ, Almof J, Amos RD, Deegan MJO, Elbert ST, Hampel C, Meyer W, Peterson WK, Pitzer R, Stone AJ, Taylor PR, Lindh R, Mura ME, Thorsteinsson T *Molpro*, A package of ab initio programs
23. Davidson ER, Chakravorty S (1992) Theor Chim Acta 83:319
24. Parr RG, Ayers PW, Nalewajski RF (2005) J Phys Chem A 109:3957
25. Kulback S (1977) Information theory and statistics. Dover, Mineola
26. Nalewajski RF, Parr RG (2000) Proc Natl Acad Sci 97:8879
27. Ayers PW (2000) J Chem Phys 113:10886
28. Martin JML, Bauschlicher CW Jr, Ricca A (2001) Comput Phys Commun 133:189
29. Abramowitz M, Stegun IA (1970) Handbook of mathematical functions. Dover Publications Inc, New York

Theor Chem Acc (2014) 133:1479
DOI 10.1007/s00214-014-1479-4

REGULAR ARTICLE

Loss of a C_2H_n fragment from pyrene and circumcoronene

Charles W. Bauschlicher Jr. · Alessandra Ricca

Received: 23 December 2013 / Accepted: 6 March 2014 / Published online: 8 April 2014

Abstract Reactions at the edge of pyrene and circumcoronene are studied using the B3LYP approach in conjunction with the 4-31G and 6-31G** basis sets. The loss of a C_2H_n fragment from the edge of either molecule requires more than 7 eV, which is much larger than for the loss of an H atom. Some paths can be broken down into a series of less energetic steps, but this does not change the overall endothermicity of the process. The exception is a path where a hydrogen atom adds to pyrene or circumcoronene. The resulting molecule rearranges to have a C_2H_2 side group, which is subsequently lost. This process has an overall endothermicity of only about 2.5 eV. This path is actually less endothermic than the loss of an H atom from the same species.

Keywords DFT · Polycyclic aromatic hydrocarbon · Loss of C_2H_n

1 Introduction

Phenolic polymer $(-C_6H_3OHCH_2-)_n$ is an important component of most ablative heat shields. While much is known about phenolic polymer, namely its curing and pyrolysis, we are still lacking a detailed molecular-level understanding of its pyrolysis. Such an understanding would be very useful in attempts to improve the properties of phenolic polymer for thermal protection systems. With this in mind, molecular dynamics (MD) simulations and ab initio calculations have been performed [1–4] to obtain a molecular-level understanding of the pyrolysis of this important system. These studies have yielded some useful information about several important pyrolysis reactions, namely those that lead to CO, H_2, and H_2O formation. One limitation of the MD studies is that temperatures much higher than those observed in experiment must be used in order to observe any significant amount of chemical activity, but temperatures that are too large must be avoided or mostly fragmentation occurs instead of polymer cross-linking followed by the formation of fused rings. In the MD studies where fragmentation is not a problem, the growth of the fused-ring structures competes with ring opening and loss of a C_2H_n fragment, which reduces the number of rings connected together. In essentially all of the ring-opening reactions observed in the MD studies that we investigated in detail, the ring opening was proceeded by the addition or transfer of a hydrogen to the ring before it opened.

Many pyrolysis reactions are similar in spirit to those observed in combustion chemistry, where hydrogen atoms can add to species and significantly change the chemistry. The importance of H atom reactions in combustion has been demonstrated by Melius and coworkers [5, 6]. For example, they showed that the barrier for the fulvene to benzene isomerization was reduced from 3.2 to 0.08 eV by the addition of an H atom.

In this work, we perform a detailed study of the ring opening and loss of C_2H_n using the density functional

Dedicated to the memory of Professor Isaiah Shavitt and published as part of the special collection of articles celebrating his many contributions.

C. W. Bauschlicher Jr. (✉)
Entry Systems and Technology Division, NASA Ames Research Center, Mail Stop 230-3, Moffett Field, CA 94035, USA
e-mail: Charles.W.Bauschlicher@nasa.gov

A. Ricca
Carl Sagan Center, SETI Institute, 189 Bernardo Ave., Suite 100, Mountain View, CA 94043, USA

theory (DFT) approach and a pyrene model ($C_{16}H_{10}$). We do not include the OH group or the bridging CH_2 group present in phenolic polymer to simplify the model because these groups are commonly lost from the polymer when the ring opening occurs. The one exception is the path for loss of CO, but this path has been reported previously [4]. We fully characterize the paths by finding both minima and transition states. The paths investigated are inspired by the MD studies and chemical intuition.

Before reporting on our results, we should note that there is an entirely different application where ring opening and loss of C_2H_n fragments are of interest, namely in the destruction of polycyclic aromatic hydrocarbons (PAHs) in the interstellar medium (ISM). From PAH mid-infrared emission, it is known that they are common in our Galaxy as well as other galaxies [7]. PAHs are believed to be formed principally in the outflows of carbon rich stars by a mechanism similar to their formation in oxygen-poor flames on earth. However, much less is known about their processing and destruction. PAHs are good absorbers of UV radiation, and from their IR emission, it is known that the excited state undergoes a rapid intersystem crossing to the ground electronic state with a high level of vibrational excitation [8]. MD simulations of a hot $C_{54}H_{18}$ molecule show [9] that it can loose C_2H_n fragments in a manner similar to that found in our polymer simulations. Since the average size of PAHs in the ISM are believed to be much larger than pyrene, we extend our path studies to include $C_{54}H_{18}$ as well as pyrene.

2 Models and methods

We used pyrene ($C_{16}H_{10}$) and circumcoronene ($C_{54}H_{18}$) as our model systems. Pyrene is similar in size to the species observed in our polymer pyrolysis studies and is sufficiently small that it is possible to use a sizeable basis set. $C_{54}H_{18}$ is consistent with the size of PAHs believed to exist in the ISM, but its size requires the use of a smaller basis set. The pyrene calculations in the small basis set are used to calibrate the $C_{54}H_{18}$ results.

We used the hybrid B3LYP [10, 11] functional in conjunction with the 4-31G or 6-31G** basis sets [12]. In our previous study of polymer reaction paths [4], we found this level of theory to be in good agreement with the more accurate CCSD(T) approach. In that study, the average absolute and maximum difference between the B3LYP/6-31G** and CCSD(T)/cc-pVTZ calculations were 0.16 and 0.37 eV, respectively. We further note that calibration calculations using the cc-pVQZ basis set in the CCSD(T) calculations reduced the difference between the B3LYP and CCSD(T) calculations. Thus, our B3LYP/6-31G** calculations are expected to have an uncertainty of approximately 0.5 eV. The agreement between the 6-31G** and 4-31G basis sets suggests that the B3LYP/4-31G results have an uncertainty of approximately 0.8 eV. However, as we show below, the difference between the most favorable path and the other paths is several eV; thus, our level of approximation is sufficiently accurate to illustrate the difference between the possible reaction paths.

The structures were fully optimized and the harmonic frequencies computed using analytic derivatives. The harmonic frequencies were used to characterize the stationary points as minima or transition states. Since the paths that we found were consistent with those observed in MD studies, we did not check the paths using the intrinsic reaction coordinate method, but rather displaced the transition state geometry in both directions along the imaginary mode and performed a normal geometry optimization. In all cases, the two optimizations recovered the reactants and products associated with the transition state. The harmonic frequencies, without scaling, were also used to compute the zero-point energies. All of the DFT calculations were performed using Gaussian 09 [13] or previous versions. The interactive molecular graphics tool MOLEKEL [14] was used for the visualization of the molecular structures and the vibrational modes.

To minimize the computational effort, the paths were first fully characterized using pyrene at the B3LYP/4-31G level. Starting from the B3LYP/4-31G minima and transition states, the calculations were repeated using the larger 6-31G** basis set. The starting geometries for $C_{54}H_{18}$ were created by modifying the $C_{54}H_{18}$ edge structure to correspond to the pyrene B3LYP/4-31G minima and transition states. This approach reduced the time required to optimize the $C_{54}H_{18}$ structures and especially for locating the transition states, as all of the experimentation was performed on the smaller system with the smaller basis set.

3 Computational results

The reaction pyrene$\rightarrow C_{14}H_8 + C_2H_2$ is endothermic by 7.3 eV, while the loss of an H atom is endothermic by only 4.8 eV. The energetics are very similar for $C_{54}H_{18}$. Given the large energy difference between the loss of an H atom and the loss of a C_2H_2 molecule, it seems unlikely that the direct loss of a C_2H_2 can compete with the loss of an H atom. The MD studies of both the polymer and $C_{54}H_{18}$ confirm that the loss of H is much more common than any ring opening. Therefore, the ring opening and the edge erosion by loss of C_nH_m groups from a PAH probably occur through a sequence of events where the energy in a given step is comparable with the loss of an H atom or by some reaction that reduces the endothermicity.

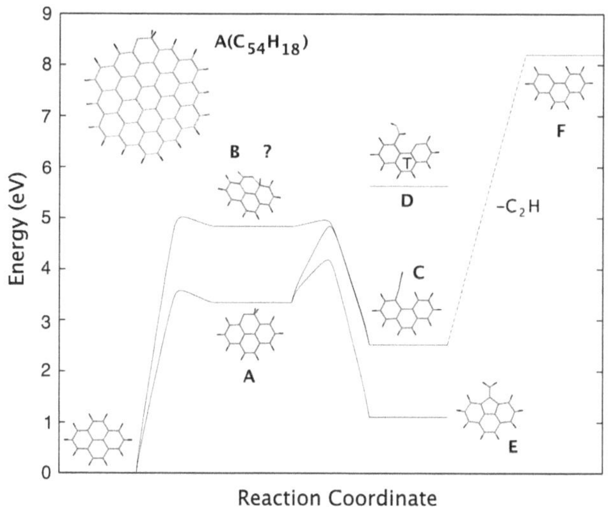

Fig. 1 B3LYP/6-31G** reaction paths for the decomposition of pyrene. The "T" indicates a triplet state. The "?" indicates that this path could only be found at the B3LYP/4-31G level and therefore is probably an artifact. The A structure for $C_{54}H_{18}$ is shown at the *top left* and illustrates the relationship between the two molecules

The first processes that we consider involve a series of unimolecular reactions. They are shown in Fig. 1 and their energetics are summarized in Table 1. Since the effect of the zero-point energy (ZPE) is similar for both molecules and both basis sets, we only report the results without ZPE for the B3LYP/6-31G** approach. We give the energetics to two places after the decimal to better show the small changes between the pyrene and circumcoronene and between two basis sets, but as noted above, our energetics are probably uncertain to about 0.5 eV. The figure shows the energetics for pyrene in the 6-31G** basis set. The first step is the shift of a hydrogen atom from one carbon to the adjacent carbon to produce A. The analogous product for circumcoronene is shown in Fig. 1; this illustrates the correspondence between the two molecules. This reaction is endothermic by 3.35 eV and has a barrier of 3.50 eV. Molecule A can undergo two possible reactions. The path with the lower barrier of the two leads to the formation of a five-membered ring with a CH_2 side group (molecule E). This species is 1.11 eV energetically higher than the starting material. The other path involves a transfer of one of the hydrogens in the CH_2 group of molecule A to the adjacent carbon with three C–C bonds along with the breaking of the C–C bond, which opens the six-membered ring and leads to the formation of a C_2H side group (C). The loss of the C_2H group to form $F + C_2H$ requires the addition of 5.68 eV of energy. This path is observed in our MD studies.

In addition to these two paths, we searched for a path to break a C–C bond on the starting molecule to form a CHCH side group. We were unable to find a molecule with this structure that has a singlet ground state, but we did find

Table 1 Energetics (in eV) for paths shown in Fig. 1

Structure[a]	$C_{16}H_{10}$			$C_{54}H_{18}$
Basis set	6-31G**		4-31G	4-31G
	+ZPE	No ZPE	+ZPE	+ZPE
Path for the loss of C_2H				
Start	0.00	0.00	0.00	0.00
TS Start to A	3.50	3.65	3.86	3.86
A	3.35	3.45	3.55	3.54
TS A to C	4.84	5.03	5.19	5.17
C	2.52	2.63	2.46	2.53
$F + C_2H$	8.20	8.55	8.21	8.23
Path to form 5-membered ring				
Start	0.00	0.00	0.00	0.00
TS Start to A	3.50	3.65	3.86	3.86
A	3.35	3.45	3.55	3.54
TS A to E	4.15	4.26	4.49	4.66
E	1.11	1.15	1.19	1.61
Possible alternative path for loss of C_2H				
Start	0.00	0.00	0.00	0.00
TS Start to B	NA[b]	NA	4.88	4.86
B	NA	NA	4.84	4.78
TS B to C	NA	NA	4.90	4.96
C	2.52	2.63	2.46	2.53
$F + C_2H$	8.20	8.55	8.21	8.23

[a] "TS" denotes transition state and the structures are labeled as in the figure

[b] We are unable to find these structures using the larger basis set, which suggests that they are an artifact of the smaller basis set

a triplet state with this structure (D). It is quite high in energy, as is shown on the figure. The failure to find any singlet structure of interest and the high energy of the triplet state suggests that this path is not an important process. Indeed we did not observe it in the MD studies.

The last path in this series that we found involved first the transfer of a hydrogen to a carbon with three C–C bonds to form structure B. This structure can convert to structure C, with the C_2H side group, through a transition state involving a hydrogen transfer and C–C bond breaking. In addition to being less favorable than the other ring opening process, we were unable to find this path using the large basis set, but were able to find it for both the pyrene and $C_{54}H_{18}$ models using the small basis set. This suggests that this process is an artifact of the small basis set.

In Table 1, we explore the similarities of the pyrene and $C_{54}H_{18}$ reactions for all three paths. We first note that the pyrene results obtained using the 4-31G and 6-31G** basis sets agree reasonably well except for the third path, where as noted above, the first three stationary points from the small basis set do not exist for the bigger basis set. A comparison of the $C_{54}H_{18}$ and pyrene results obtained using the small basis

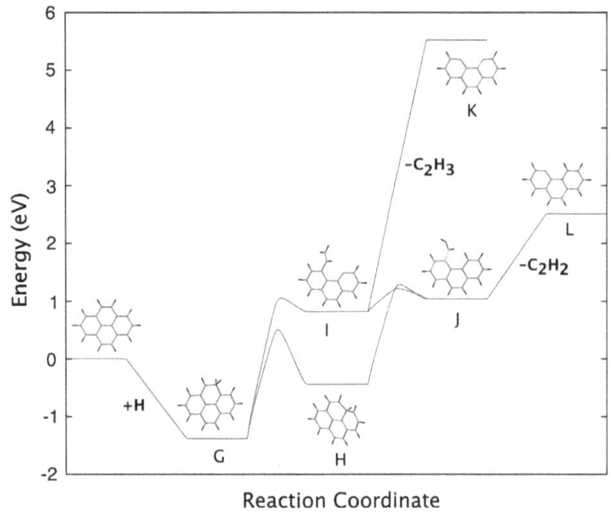

Fig. 2 B3LYP/6-31G** reaction paths for the addition of one hydrogen atom to pyrene and its subsequent decomposition

set show that they agree very well except for structure E (i.e., the one containing the 5-membered ring) and the transition state to form it, where the differences are much larger. The pyrene structures are more stable because the smaller number of rings allow them to distort more and hence have a larger stabilization energy.

The next series of reactions are summarized in Fig. 2 and Table 2. These involve the addition of a hydrogen atom as the first step to form molecule G. This was motivated by previous polymer MD studies, where H addition preceded some ring opening reactions. This is a very plausible step for PAHs in the ISM because of the large excess of atomic hydrogen population and observations that suggest the existence of PAH molecules with excess hydrogens [15]. We have characterized two paths that can occur after hydrogen addition. We should note that the barriers for both of these processes are larger that the energy required to remove the hydrogen and restore the starting materials. Thus, most of the time that G is formed, it dissociates back to the starting material, and this is what we observe in our MD studies. However, some of the time the reaction continues as shown in Fig. 2. The lower energy path involves the H transfer from the CH_2 group in molecule G to the adjacent carbon that has three C–C bonds to form molecule H with a barrier of 1.88 eV. This structure can undergo a ring opening reaction to form molecule J, with a barrier only slightly larger than its endothermicity. The resulting product (J) has a C_2H_2 side group that requires only 1.5 eV to be removed forming $L + C_2H_2$. This is the final step in a reaction path that overall requires only 2.51 eV to remove a C_2H_2 fragment from pyrene plus a H atom.

The other path from the hydrogen addition product (molecule G) opens the ring to form a C_2H_3 side group (I).

Table 2 Energetics (in eV) for paths shown in Fig. 2

Structure	$C_{16}H_{10}$			$C_{54}H_{18}$
Basis set	6-31G**		4-31G	4-31G
	+ZPE	No ZPE	+ZPE	+ZPE
Path for the loss of C_2H_2				
Start	0.00	0.00	0.00	0.00
G ($C_{54}H_{19}$)	−1.38	−1.62	−1.29	−1.30
TS G to H	0.50	0.41	0.66	0.68
H	−0.44	−0.67	−0.34	−0.34
TS H to J	1.26	1.12	1.32	1.32
J	1.04	0.86	1.08	1.12
$L + C_2H_2$	2.51	2.48	2.49	2.51
Alternative path for the loss of C_2H_2				
Start	0.00	0.00	0.00	0.00
G	−1.38	−1.62	−1.29	−1.30
TS G to I	1.00	0.84	1.06	1.08
I	0.82	0.64	0.87	0.89
TS I to J	1.22	1.20	1.29	1.39
J	1.04	0.86	1.08	1.12
$L + C_2H_2$	2.51	2.48	2.49	2.51
Path for the loss of C_2H_3				
Start	0.00	0.00	0.00	0.00
G	−1.38	−1.62	−1.29	−1.30
TS G to I	1.00	0.84	1.06	1.08
I	0.82	0.64	0.87	0.89
$K + C_2H_3$	5.52	5.58	5.62	5.64

This path is only slightly less favorable. This species can either loose a C_2H_3 group ($K + C_2H_3$), which is a very endothermic reaction, or can transfer one of the terminal hydrogen atoms back to the ring to form the structure with the C_2H_2 side group (J). The path then continues as before.

As shown in Table 2, there is good agreement between the large and small basis set results for pyrene and between the pyrene and $C_{54}H_{18}$ models. In fact, the agreement is better than observed for the reaction in Table 1. The agreement between pyrene and circumcoronene suggests that the very favorable energetics for this path will be common for most PAHs.

The last set of pathways that we consider are shown in Fig. 3, and the energetics are summarized in Table 3. Since the C–H bond is the weakest in the PAHs, the loss of a hydrogen is expected to occur before any ring opening reactions occur. The loss of H is observed in the MD simulations and believed to occur in the ISM. These paths begin with the loss of an H atom forming molecule M, which is endothermic by almost 5 eV. While this is a sizeable energy, it is less than the cost of removing C_2H_2. The next step is the transfer of a hydrogen atom to a carbon bonded to three other carbons (O). This has a sizeable

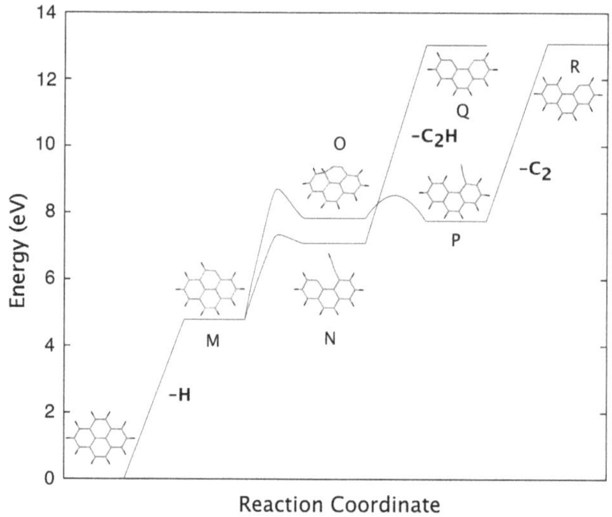

Fig. 3 B3LYP/6-31G** reaction paths for the loss of one hydrogen from pyrene and its subsequent decomposition

Table 3 Energetics (in eV) for paths shown in Fig. 3

Structure	$C_{16}H_{10}$			$C_{54}H_{18}$
Basis set	6-31G**		4-31G	4-31G
	+ZPE	No ZPE	+ZPE	+ZPE
Path for the loss of C_2				
Start	0.00	0.00	0.00	0.00
M	4.79	5.14	4.79	4.79
TS M to O	8.66	9.21	9.01	8.99
O	7.83	8.28	7.90	7.92
TS O to P	8.52	9.00	8.64	8.99
P	7.75	8.20	7.86	7.83
R + C_2	13.06	13.66	13.23	13.24
Path for the loss of C_2H				
Start	0.00	0.00	0.00	0.00
M	4.79	5.14	4.79	4.79
TS M to N	7.28	7.77	7.22	7.24
N	7.08	7.56	7.01	7.03
Q + C_2H	13.02	13.73	13.04	14.01

barrier. This species can open the ring to form a C_2 side group with only a small barrier (P). The loss of the C_2 side group from P which forms R + C_2 requires considerable energy.

An alternative path is to open the ring, forming a C_2H side group (N). This has a lower barrier and the product is more stable than the product of shifting the hydrogen. The loss of the C_2H fragment from N yields Q + C_2H which is even more endothermic than the loss of the C_2 group. As for the other reactions, the pyrene and $C_{54}H_{18}$ models agree reasonably well as do the small and large basis set results.

4 Conclusions

We have considered several paths for pyrene to loose some edge carbon atoms on the ground electronic state potential. All of the reaction mechanisms are very endothermic except for a path that first adds a hydrogen atom. These species can rearrange by two paths to form a C_2H_2 side group. The overall energetics for the loss of the C_2H_2 are about 2.5 eV. This path is essentially driven by the strong H-C bond energy in C_2H_2. We find that the results in the small basis set are very similar to those in the large basis set, thus supporting its use in the larger $C_{54}H_{18}$ model.

The lowest path is consistent with polymer MD simulations where many of the ring opening events are proceeded by the addition of an H atom that was lost from the same or another polymer molecule. This reaction should be important at high temperatures where free H atoms are found. At lower temperatures, this mechanism should be less efficient as few H atoms will exist; however, reactions where the H atom transfers from another molecule are still possible, but the first step is no longer as exothermic as in case of a free H atom. In case of PAH molecules in the ISM, this H addition erosion mechanism should be important in regions where there is a significant hydrogen atom concentration, such as in PDRs.

Acknowledgments AR thanks NASA's Astrophysics Theory and Fundamental Physics (ATFP) (NNX09AD18G) program for its generous support of this work.

References

1. Jiang D, van Duin ACT, Goddard WAG, Dai S (2009) J Phys Chem A 113:6891
2. Desai TG, Lawson JW, Keblinski P (2011) Polymer 52:577
3. Qi T, Bauschlicher CW, Lawson JW, Desai TG, Reed EJ (2013) J Phys Chem A 17:11115
4. Bauschlicher CW, Qi T, Reed EJ, Lefant A, Lawson JW, Desai TG (2013) J Phys Chem A 117:11126
5. Madden LK, Mebel AM, Melus CF (1996) J Phys Organ Chem 9:801
6. Melius CF, Colvin ME, Marinov NM, Pit WJ, Senkan SM (1996) Proc Combust Inst 26:685
7. Tielens AGGM (2005) The physics and chemistry of the interstellar medium. Cambridge University Press, Cambridge
8. Allamandola LJ, Tielens AGGM, Barker JR (1989) Astrophys J Suppl 71:733
9. Ricca A, unpublished results
10. Becke AD (1993) J Chem Phys 98:5648
11. Stephens PJ, Devlin FJ, Chabalowski CF, Frisch MJ (1994) J Phys Chem 98:11623
12. Frisch MJ, Pople JA, Binkley JS (1984) J Chem Phys 80:3265 and references therein
13. Frisch MJ et al (2013) Gaussian 09, Revision D.01. Gaussian Inc, Wallingford
14. Flükiger P, Lüthi HP, Portmann S, Weber J (2000) MOLEKEL 4.2. Swiss Center for Scientific Computing, Manno
15. Bernstein MP, Sanford SA, Allamandola LJ (1996) Astrophys J 472:L127

Theor Chem Acc (2014) 133:1483
DOI 10.1007/s00214-014-1483-8

REGULAR ARTICLE

Heats of formation of the amino acids re-examined by means of W1-F12 and W2-F12 theories

Amir Karton · Li-Juan Yu · Manoj K. Kesharwani · Jan M. L. Martin

Received: 20 January 2014 / Accepted: 15 March 2014 / Published online: 16 April 2014

Abstract We have obtained accurate heats of formation for the twenty natural amino acids by means of explicitly correlated high-level thermochemical procedures. Our best theoretical heats of formation, obtained by means of the ab initio W1-F12 and W2-F12 thermochemical protocols, differ significantly (RMSD = 2.3 kcal/mol, maximum deviation 4.6 kcal/mol) from recently reported values using the lower-cost G3(MP2) method. With the more recent G4(MP2) procedure, RMSD drops slightly to 1.8 kcal/mol, while full G4 theory offers a more significant improvement to 0.72 kcal/mol (max. dev. 1.4 kcal/mol for glutamine). The economical G4(MP2)-6X protocol performs equivalently at RMSD = 0.71 kcal/mol (max. dev. 1.6 kcal/mol for arginine and glutamine). Our calculations are in excellent agreement with experiment for glycine, alanine and are in excellent agreement with the recent revised value for methionine, but suggest revisions by several kcal/mol for valine, proline, phenylalanine, and cysteine, in the latter case confirming a recent proposed revision. Our best heats of formation at 298 K ($\Delta H^{\circ}_{f,298}$) are as follows: at the W2-F12 level: glycine −94.1, alanine −101.5, serine −139.2, cysteine −94.5, and methionine −102.4 kcal/mol, and at the W1-F12 level: arginine −98.8, asparagine −146.5, aspartic acid −189.6, glutamine −151.0, glutamic acid −195.5, histidine −69.8, isoleucine −118.3, leucine −118.8, lysine −110.0, phenylalanine −76.9, proline −92.8, threonine −149.0, and valine −113.6 kcal/mol. For the two largest amino acids, an average over G4, G4(MP2)-6X, and CBS-QB3 yields best estimates of −58.4 kcal/mol for tryptophan, and of −117.5 kcal/mol for tyrosine. For glycine, we were able to obtain a "quasi-W4" result corresponding to $TAE_e = 968.1$, $TAE_0 = 918.6$, $\Delta H^{\circ}_{f,298} = -90.0$, and $\Delta H^{\circ}_{f,298} = -94.0$ kcal/mol.

Keywords Thermochemistry · Amino acids · Explicitly correlated methods · Density functional theory · Ab initio

Dedicated to the memory of Professor Isaiah Shavitt and published as part of the special collection of articles celebrating his many contributions.

Electronic supplementary material The online version of this article (doi:10.1007/s00214-014-1483-8) contains supplementary material, which is available to authorized users.

A. Karton (✉) · L.-J. Yu
School of Chemistry and Biochemistry, The University of Western Australia, Perth, Australia
e-mail: amir.karton@uwa.edu.au

M. K. Kesharwani · J. M. L. Martin (✉)
Department of Organic Chemistry, Weizmann Institute of Science, 76100 Reḥovot, Israel
e-mail: gershom@weizmann.ac.il

J. M. L. Martin
Department of Chemistry and Center for Advanced Scientific Computing and Modeling (CASCaM), University of North Texas, Denton, TX 76201, USA

1 Introduction

Due to the increasing computational power provided by supercomputers and recent advances in the development of economical ab initio methods (e.g., advances in explicitly correlated techniques [1–4]), high-level ab initio methods have now been refined to the point where they are applicable to biologically relevant systems (see Refs. [5–13] for some recent studies). Proteinogenic amino acids are the most basic building blocks of proteins and play key roles in protein structure and function. They also serve as precursors of many biologically relevant molecules, such as polypeptides, nucleotides, hormones, neurotransmitters, and antioxidants [14]. Despite their importance, the experimental gas-phase heats of formation of most of the natural amino acids are not accurately known. Determination of these fundamental thermochemical quantities may be important in understanding why nature chose these molecules as fundamental biological building blocks—for example, by comparing the relative stabilities of α- versus β-amino acids [15, 16]. Accurate heats of formation for the amino acids are also important from a theoretical point of view, e.g., for the validation and parameterization of computationally cost-effective procedures such as density functional theory, semiempirical molecular orbital theory, and molecular mechanics. In recent years, a large number of theoretical studies were dedicated to obtaining thermochemical properties of amino acids using high-level theoretical procedures [15–25].

In the present work, we obtain accurate theoretical heats of formation for the lowest-energy conformers for the 18 proteinogenic amino acids using the high-level, ab initio W1-F12 and W2-F12 thermochemical procedures [31]. These thermochemical procedures represent layered extrapolations to the all-electron, relativistic CCSD(T)/CBS energy (complete basis set limit coupled cluster with singles, doubles, and quasiperturbative triple excitations) and can achieve an accuracy in the sub-kcal/mol range for molecules whose wave functions are dominated by dynamical correlation [31, 32]. We use these benchmark values to evaluate the performance of a variety of Gn-type procedures [33] that were recently used for obtaining accurate thermochemical properties of amino acids [15–20].

The present paper also seeks to pay tribute to the scientific achievements of Prof. Isaiah Shavitt (OBM) and specifically to his seminal contributions to coupled cluster theory [26], to the theory and development of Gaussian basis sets [27], to accurate applied quantum chemistry [28–30], and to computational biochemistry [5].

2 Computational details

Most calculations were run on the CRUNTCh (Computational Research at UNT in Chemistry) Linux farm at the University of North Texas, on the high-performance computing National Computational Infrastructure (NCI) National Facility at Canberra, and on the iVEC@UWA facilities. Some additional calculations were carried out on the Faculty of Chemistry Linux farm at the Weizmann Institute of Science.

The geometries have been optimized at the B3LYP/A'VTZ level of theory [34–36] (where A'VTZ indicates the combination of the standard correlation-consistent cc-pVTZ basis set on hydrogen, [37] the aug-cc-pVTZ basis set on first-row elements, [38] and the aug-cc-pV(T+d)Z basis set on sulfur) [39]. All geometry optimizations and frequency calculations were performed using the Gaussian 09 program suite [40]. Benchmark relativistic, all-electron CCSD(T)/CBS energies were then obtained by means of our recently developed W1-F12 and W2-F12 thermochemical protocols [31] using the Molpro 2012.1 program suite [41]. The computational protocols of W1-F12 and W2-F12 theories have been specified and rationalized in reference [31].

In W1-F12 theory, the Hartree–Fock component is extrapolated from the VDZ-F12 and VTZ-F12 basis sets, using the $E(L) = E_\infty + A/L^\alpha$ two-point extrapolation formula with $\alpha = 5$ (where L is the highest angular momentum represented in the basis set, and VnZ-F12 denotes the cc-pVnZ-F12 basis sets of Peterson et al. [42] which were developed for explicitly correlated calculations). Optimal values for the geminal Slater exponents (β) used in conjunction with the VnZ-F12 basis sets were taken from reference [43]. The valence CCSD-F12 correlation energy is extrapolated from the same basis sets, using the said two-point extrapolation formula. Extrapolation exponents (α) were taken from references [31, 43]. In all of the explicitly correlated coupled cluster calculations the diagonal, fixed-amplitude 3C(FIX) ansatz [45–47] and the CCSD-F12b approximation [48, 49] are employed. The (T) valence correlation energy is obtained in the same way as in the original Weizmann-1 (W1) theory, [50] i.e., extrapolated from the A'VDZ and A'VTZ basis sets using the above two-point extrapolation formula with $\alpha = 3.22$. The CCSD inner-shell contribution is calculated with the core-valence weighted correlation-consistent A'PWCVTZ basis set of Peterson and Dunning, [51] while the (T) inner-shell contribution is calculated with the PWCVTZ(no f) basis set (where A'PWCVTZ indicates the combination of the cc-pVTZ basis set on hydrogen and the aug-cc-pwCVTZ basis set on carbon, and PWCVTZ(no f) indicates the cc-pwCVTZ basis set without the f functions). The scalar relativistic contribution (in the second-order Douglas–Kroll–Hess approximation [52, 53]) is obtained as the difference between non-relativistic CCSD(T)/A'VDZ and relativistic CCSD(T)/A'VDZ-DK calculations [54] (where A'VDZ-DK indicates the combination of the cc-pVDZ-DK basis set on H and aug-cc-pV(D+d)Z-DK basis set on heavier elements). The atomic spin–orbit coupling terms are taken from the experimental fine structure, and the diagonal Born–Oppenheimer correction (DBOC) is calculated at the HF/A'VTZ level of theory. The zero-point vibrational energies (ZPVEs) are derived from B2PLYP/def2-TZVPP harmonic frequencies (and scaled by 0.9833, see Sect. 3.3).

In W2-F12, the Hartree–Fock component is calculated with the VQZ-F12 basis set. The valence CCSD-F12 correlation energy is extrapolated from the VTZ-F12 and VQZ-F12 basis sets, using the above two-point extrapolation formula with $\alpha = 5.94$. The quasiperturbative triples, (T), corrections are obtained from standard CCSD(T)/VTZ-F12 calculations (i.e., without inclusion of F12 terms) and scaled by the factor $f = 0.987 \times E_{MP2-F12}/E_{MP2}$. This approach has been shown to accelerate the basis set convergence [31, 49]. The CCSD inner-shell contribution is calculated with the core-valence weighted correlation-consistent A'PWCVTZ basis set, while the (T) inner-shell contribution is calculated with the PWCVTZ(no f) basis set. The scalar relativistic, spin–orbit coupling, DBOC, and ZPVE corrections are obtained in the same way as in W1-F12 theory.

The total atomization energies at 0 K (TAE_0) are converted to heats of formation at 298 K using the Active Thermochemical Tables (ATcT) [55–59] atomic heats of formation at 0 K (H 51.633 ± 0.000, C 170.024 ± 0.014, N 112.469 ± 0.007, O 58.997 ± 0.000, and S 65.709 ± 0.036 kcal/mol), and the CODATA [60] enthalpy functions, $H_{298} - H_0$, for the elemental reference states (H_2(g) = 2.024 ± 0.000, C(cr,graphite) = 0.251 ± 0.005, N_2(g) = 2.072 ± 0.000, O_2(g) = 2.075 ± 0.000, and S(cr,rhombic) = 1.054 ± 0.001 kcal/mol), while the enthalpy functions for the amino acids are obtained within the rigid rotor harmonic oscillator (RRHO) approximation from B3LYP/A'VTZ geometries and harmonic frequencies.

W1-F12 shows excellent performance for systems containing first-row elements (and H). Specifically, for the 97 first-row atomization energies in the W4-11 dataset, [32] W1-F12 attains a root mean square deviation (RMSD) of 0.19 kcal/mol relative to all-electron, relativistic CCSD(T) reference atomization energies at the infinite basis set limit. However, for second-row systems, it was found that the performance of W1-F12 is significantly degraded owing to shortcomings of the cc-pVDZ and cc-pVDZ-F12 basis sets for second-row elements (see Ref. [31] for details): for the 40 second-row atomization energies in the W4-11 dataset, RMSD actually exceeds 1 kcal/mol. W2-F12 does not suffer from this problem and yields similar RMSDs of 0.18 kcal/mol for first-row and 0.24 kcal/mol for second-row systems. (For further details, see reference [31]). Thus, for the sulfur-containing amino acids (cysteine and methionine) and for the small amino acids (alanine, glycine, and serine), the heats of formation are also obtained using W2-F12 theory.

The case of glycine is small enough (especially considering the C_s symmetry) that the result can be independently verified using accurate thermochemical procedures based on layered extrapolation of orbital basis sets, specifically the high-accuracy W4 method [68]. Full details of the method are given in that reference and will not be repeated here: suffice to say that for a set of molecules where accurate experimental atomization energies are available via ATcT, the RMSD from experiment is 0.10 kcal/mol [32, 68]. The largest-scale calculation involved here, CCSD/aug'-cc-pV6Z, entails 1400 basis functions and required 3 terabyte of scratch space, yet ran to completion within a day on a machine with a large solid-state disk array. The CCSD(T)/aug'-cc-pV5Z calculation, involving 910 basis functions, ran in under a day on 32 cores and 512 GB of RAM.

The heats of formation have also been obtained using computationally more economical composite procedures, namely the Gaussian-4 (G4) protocol [33, 63] and its computationally more economical G4(MP2) and G4(MP2)-6X variants [64, 65]. These calculations were performed using the Gaussian 09 program suite [40]. The G4 and G4(MP2) protocols are widely used for the calculation of thermochemical properties and are applicable to relatively large systems (of up to 20–30 non-hydrogen atoms). They, generally, give RMSDs from experimental or high-accuracy theoretical thermochemical data of 1–2 kcal/mol [32, 63, 64]. For example, for the 454 experimental thermochemical determinations of the G3/05 test set (including heats of formation, ionization energies, and electron affinities), [66] G4 and G4(MP2) attain RMSDs of 1.2 and 1.5 kcal/mol, respectively [63, 64]. For the set of 137 very accurate theoretical atomization energies in the W4-11 set, both procedures attain an RMSD of 2.0 kcal/mol [32]. Finally, we have also considered the performance of the CBS-QB3 procedure [67] using Gaussian 09.

3 Results and discussion

3.1 Computational cost of the W1-F12 calculations

For systems consisting of more than eight non-hydrogen atoms (with C_1 symmetry), W1 theory [50] becomes prohibitively expensive with current commodity server hardware. W1-F12 theory is an explicitly correlated version of the W1 method, [50] which combines explicitly correlated F12 methods [1–4] with extrapolation techniques in order to approximate the CCSD(T)/CBS energy. Because of the drastically accelerated basis set convergence of the F12 methods [42, 43], W1-F12 is superior to the original W1 method, not only in terms of performance but also in terms of computational cost [31]. For example, the cpu times for calculating W1 and W1-F12 energies for a system containing 8 non-hydrogen atoms (with C_1 symmetry) are 595 and 163 h, respectively (both calculations ran on 8 Intel Xeon Sandy Bridge cores at 2.6 GHz). In terms of disk space requirements, the W1 calculation used about five times the amount of scratch disk (660 GB) that the W1-F12 calculation required (126 GB).

In the present work, we obtain W1-F12 energies for the 18 amino acids with up to 12 non-hydrogen atoms. Of these, the largest amino acids are glutamic acid, glutamine, and lysine (10 non-hydrogen atoms); histidine (11 non-hydrogen atoms); arginine and phenylalanine (12 non-hydrogen atoms). Considering the fact that none of the amino acids (apart from glycine) have any spatial symmetry, these represent the largest W1-F12 calculations reported to date. For example, the W1-F12 calculation for arginine ran for 51 days on 6 Intel Nehalem 8837 cores at 2.67 GHz and used 253 GB of RAM and 1.1 TB of scratch disk. Due to this very steep computational cost, we obtain our best heats of formation for the two amino acids with more than 12 non-hydrogen atoms (i.e., tryptophan and tyrosine) with the G*n* and CBS-QB3 methods [33, 67], which have a significantly reduced computational cost. In Sect. 3.4, we show that, relative to W1-F12 and W2-F12 heats of formation, G4, G4(MP2)-6X, and CBS-QB3 result in RMSDs of 0.72, 0.71, and 1.01 kcal/mol, respectively, i.e., near or below the threshold of "chemical accuracy" (traditionally arbitrarily defined as 1 kcal/mol).

3.2 W1-F12 and W2-F12 benchmark heats of formation

Since W1-F12 and W2-F12 theories represent a layered extrapolations to the CCSD(T) basis set limit energy, it is of interest to estimate whether the contributions from post-CCSD(T) excitations are likely to be significant for the atomization energies of the amino acids. The percentage of the total atomization energy accounted for by parenthetical connected triple excitations, $\%TAE_e[(T)]$, has been shown to be a reliable energy-based diagnostic for the importance of non-dynamical correlation effects [68, 74]. It has been suggested that $\%TAE_e[(T)] < 2\ \%$ indicates systems that are dominated by dynamical correlation, while $2\ \% < \%TAE_e[(T)] < 5\ \%$ indicates systems that include mild non-dynamical correlation. $\%TAE_e[(T)]$ values for the amino acids are gathered in Table 1. The amino acids are characterized by $\%TAE_e[(T)]$ values ranging from 1.7 (leucine) to 2.5 % (histidine). Note also that in all cases, the SCF component accounts for 69–77 % of the total atomization energy. These values suggest that our all-electron, non-relativistic, vibrationless benchmark atomization energies should, in principle, be considerably closer than 1 kcal/mol of the atomization energies at the full configuration interaction (FCI) basis set limit. For example, for systems that are associated with similar $\%TAE_e[(T)]$ values in the W4-11 dataset [32], post-CCSD(T) contributions to the atomization energy are 0.2 kcal/mol or less, although somewhat larger values were found for benzene [75, 76].

Table 2 gives an overview of basis set convergence of the CCSD-F12 component of the total atomization energy.

Table 1 Diagnostics indicating the importance of post-CCSD(T) contributions for the amino acids

Isomer	$\%TAE_e[SCF]$[a]	$\%TAE_e[(T)]$[b]
Alanine	71.4	1.99
Arginine	70.1	2.12
Asparagine	69.4	2.34
Aspartic acid	69.7	2.37
Cysteine	70.1	2.23
Glutamine	70.5	2.19
Glutamic acid	70.9	2.20
Glycine	69.6	2.17
Histidine	69.6	2.45
Isoleucine	73.7	1.73
Leucine	73.7	1.72
Lysine	72.3	1.83
Methionine	72.4	1.94
Phenylalanine	73.4	2.17
Proline	72.4	1.98
Serine	69.8	2.18
Threonine	71.0	2.05
Tryptophan[c]	76.6	1.85
Tyrosine[c]	76.5	1.76
Valine	73.1	1.79

Obtained from W1-F12 theory, unless otherwise indicated

[a] Percentages of the valence CCSD(T)/CBS atomization energy accounted for by the SCF component

[b] Percentages of the valence CCSD(T)/CBS atomization energy accounted for by the (T) component

[c] Obtained at the CCSD(T)/cc-pVDZ level of theory, see Ref. [32] for further details

The magnitude of the valence CCSD-F12 correlation component spans a relatively large range. For example, the CCSD-F12/V{D,T}Z-F12 results extrapolated with $\alpha = 3.67$ (which was optimized to minimize the RMSD over 137 first- and second-row systems in the W4-11 dataset [31]) extend from 272.48 (glycine) up to 701.76 (arginine) kcal/mol. The differences between the CCSD-F12/V{D,T}Z-F12 results obtained with $\alpha = 3.67$ (optimized over the entire W4-11 set of small molecules) and $\alpha = 3.38$ (optimized over the subset of 97 first-row species only) can get quite significant for these medium-sized species, ranging from 0.25 kcal/mol for glycine to 0.71 kcal/mol for arginine. Note that these differences still only correspond to about 0.1 % of the valence CCSD correlation component. For comparison, for the systems in the W4-11 dataset, the absolute differences between the CCSD-F12/V{D,T}Z-F12 component extrapolated with $\alpha =$ 3.67 and 3.38 are reduced to just 0.00–0.22 kcal/mol, or 0.08 kcal/mol mean absolute—likewise, about 0.1 % of the valence CCSD correlation component of the atomization energy. Finally, using instead the extrapolation

Table 2 Overview of the basis set convergence of the CCSD-F12 component of the total atomization energies for the amino acids (kcal/mol)

	VDZ-F12	VTZ-F12	VQZ-F12	V{D,T}Z-F12				V{T,Q}Z-F12	
				3.67[a]	3.38[b]	3.144[c]	Best[d]	5.94[a]	4.596[c]
Alanine	327.84	334.23	335.98	336.10	336.41	336.71	336.56 ± 0.30	336.36	336.61
Arginine	683.07	697.54		701.76	702.46	703.15	702.80 ± 0.68		
Asparagine	472.43	481.98		484.76	485.23	485.68	485.46 ± 0.45		
Aspartic acid	448.56	457.12		459.61	460.03	460.44	460.23 ± 0.41		
Cysteine	357.89	363.68	365.51	365.38	365.66	365.93	365.80 ± 0.27	365.91	366.17
Glutamine	534.06	544.83		547.97	548.50	549.01	548.75 ± 0.51		
Glutamic acid	509.77	519.52		522.36	522.84	523.30	523.07 ± 0.46		
Glycine	265.77	270.96	272.46	272.48	272.73	272.97	272.85 ± 0.25	272.80[e]	273.00
Histidine	570.49	582.21		585.63	586.20	586.76	586.48 ± 0.55		
Isoleucine	514.53	524.49		527.40	527.88	528.35	528.12 ± 0.47		
Leucine	513.54	523.49		526.39	526.87	527.34	527.11 ± 0.47		
Lysine	581.62	593.25		596.65	597.21	597.76	597.49 ± 0.55		
Methionine	478.88	486.87	489.21	489.20	489.59	489.97	489.78 ± 0.38	489.72	490.06
Phenylalanine	596.04	607.63		611.01	611.58	612.13	611.85 ± 0.55		
Proline	428.78	437.25		439.72	440.14	440.54	440.34 ± 0.40		
Serine	372.23	379.39	381.45	381.48	381.83	382.17	382.00 ± 0.34	381.90	382.20
Threonine	436.04	444.43		446.87	447.28	447.68	447.48 ± 0.40		
Valine	452.54	461.33		463.89	464.32	464.73	464.53 ± 0.42		
RMSD[f]				0.43	0.14	0.24	0.12		
MAD[f]				0.41	0.11	0.21	0.11		
MSD[f]				−0.41	−0.10	0.21	0.06		

[a] Extrapolation exponent (α) from Ref. [31], optimized to minimize the RMSD over the entire W4-11 dataset

[b] Extrapolation exponent (α) from Ref. [31], optimized to minimize the RMSD over the first-row systems in the W4-11 dataset

[c] Extrapolation exponent (α) from Ref. [43]

[d] The values are the average between the CCSD-F12 components extrapolated with $\alpha = 3.38$ and 3.144, the difference between these two CCSD-F12 components may be regarded as a conservative error bar

[e] Extrapolating from the A'VQZ and A'V5Z basis sets with $\alpha = 3.0$ results in a CCSD-F12 component of 272.94 kcal/mol

[f] Root mean square deviation (RMSD), mean absolute deviation (MAD), and mean signed deviation (MSD) relative to the CCSD-F12/V{T,Q}Z-F12 results for 5 systems

exponent optimized by Hill et al. [43] ($\alpha = 3.144$), which was optimized over a smaller set of 14 absolute correlation energies, results in atomization energies increased by 0.24 (glycine) up to 0.69 (arginine) kcal/mol over the values with $\alpha = 3.38$ (Table 2).

For five smaller amino acids (alanine, cystine, glycine, methionine, and serine), we were able to obtain CCSD-F12/VQZ-F12 energies. Table 2 gives the CCSD-F12/V{T,Q}Z-F12 results extrapolated with $\alpha = 5.94$ (used in W2-F12 theory [31]) and 4.596 (from Ref. [43]). For these systems, the difference between the CCSD-F12/V{T,Q}Z-F12 contributions extrapolated with $\alpha = 5.94$ and 4.596 ranges between 0.20 (glycine) and 0.34 (methionine) (Table 2). We note that the error statistics over the 137 systems in the W4-11 dataset are as follows: RMSD = 0.13, MAD = 0.10, and MSD = 0.01 for $\alpha = 5.94$, and RMSD = 0.15, MAD = 0.11, and MSD = 0.08 kcal/mol for $\alpha = 4.596$. Peterson and Feller [44] obtained benchmarks extrapolated from basis sets as large as aug-cc-pV8Z for a fairly large sample of molecules that overlaps W4-11 and found that CCSD-F12b/V{T,Q}Z-F12 tends to overestimate the valence CCSD component on average: as they were using $\alpha = 4.596$, this is consistent with the present finding. (They also report difficulties reaching 0.1 kcal/mol convergence for CCSD-F12b energies with aug-cc-pV5Z basis sets: We were only able to apply this basis set to glycine, and in any case 0.1 kcal/mol is smaller than other potential error sources in the present work).

For the five W2-F12 amino acids, the RMSDs for CCSD-F12/V{D,T}Z-F12 with various choices of extrapolation exponent are 0.43 ($\alpha = 3.67$), 0.14 ($\alpha = 3.38$), and 0.24 ($\alpha = 3.144$) kcal/mol. Taking the average between the CCSD-F12/V{D,T}Z-F12 components extrapolated with $\alpha = 3.38$ and 3.144 results in an RMSD of 0.12 kcal/mol and a mean signed deviation of only +0.06 kcal/mol. We thus use this averaged CCSD-F12/V{D,T}Z-F12 component in our final W1-F12 atomization energies. The spread between the $\alpha = 3.38$ and 3.144 values can be considered a crude gauge of the uncertainty in the basis set limit.

The component breakdowns of the W1-F12 and W2-F12 atomization energies are gathered in Table 3. The following general observations may be noted:

- As pointed out above, the magnitude of the valence CCSD-F12 correlation component runs a large gamut, extending from 272.85 (glycine) up to 702.80 (arginine) kcal/mol.
- The magnitude of the valence (T) correlation component can be rather large, reaching 54.28 kcal/mol for phenylalanine.
- The core–valence contribution approaches or exceeds 10 kcal/mol for the largest systems. Namely, it is 9.88 (arginine) and 11.68 (phenylalanine) kcal/mol.
- The DBOC contribution ranges from 0.28 (glycine) up to as much as 0.72 (arginine) kcal/mol.

Comparison of the W1-F12 and W2-F12 results for alanine, cystine, glycine, methionine, and serine reveals the following:

- The HF/V{D,T}Z-F12 component systematically underestimates the HF/VQZ-F12 basis set limit, namely by 0.03 (glycine), 0.04 (alanine and cysteine), 0.05 (serine), and 0.08 (methionine) kcal/mol.
- Our best CCSD-F12/V{D,T}Z-F12 component overestimates the CCSD-F12/V{T,Q}Z-F12 component by 0.05 (glycine), 0.06 (methionine), 0.10 (serine), 0.20 (alanine) kcal/mol, and underestimates it by 0.11 kcal/mol for cysteine.
- The valence (T) contribution from W1-F12 theory systematically overestimates the W2-F12 results, specifically by 0.06 (cysteine), 0.13 (methionine), 0.17 (glycine), 0.20 (alanine), and 0.25 (serine) kcal/mol.
- The core–valence contribution from W1-F12 systematically underestimates the W2-F12 result, namely by 0.09 (glycine), 0.12 (alanine), 0.14 (serine), and 0.16 (cysteine) kcal/mol (we were not able to obtain the core–valence contribution for methionine from W2-F12 theory).
- Overall, the TAE_e from W1-F12 theory overestimates the TAE_e from W2-F12 theory by 0.11 (glycine and methionine), 0.16 (serine), and 0.23 (alanine) kcal/mol, and underestimates it by 0.30 kcal/mol for cysteine.

As noted in the "Methods" section, we were able to "cross-check" the result for glycine at the W4 level: the lower-cost W2.2 level is obtained as a by-product. As seen in Table 3, the SCF, CCSD, (T), core-valence, and relativistic components of the W2-F12 calculation are all in excellent agreement with the W4 calculation, the cumulative difference being just 0.04 kcal/mol. The higher-order correlation steps, CCSDT(Q)/cc-pVTZ, and CCSDTQ/cc-pVDZ are more problematic from a computational point of view, but their importance is typically quite small for molecules dominated by a single reference configuration (due to error compensation between "antibonding" higher-order T_3 and "bonding" T_4 contributions [68–73]). Absent a direct calculation, their importance can be estimated by assuming that their contribution to the following isodesmic reaction energy will be approximately zero:

$$CH_3COOH + CH_3NH_2 \rightarrow \text{glycine} + CH_4 \qquad (1)$$

From Table SI-II of Ref. [32], we find the post-CCSD(T) contributions to the TAEs to be −0.05 kcal/mol for acetic acid, −0.09 kcal/mol for methyl amine, and +0.01 kcal/mol for methane, leading to an estimated post-CCSD(T) correction of −0.15 kcal/mol for glycine.

3.3 A note on zero-point vibrational energies (ZPVEs)

In view of the magnitude of the zero-point vibrational energies (50–140 kcal/mol, see Table 4), some remarks are due concerning their calculation. Ideally, one should obtain them from accurate anharmonic force fields, and for small molecules, this is indeed a practical option [68, 85, 91]. In the present case, however, the computational cost would be prohibitive with the computational resources at hand, and multiplication of calculated harmonic frequencies with a scaling factor λ(ZPVE) appropriate for zero-point vibrational energies [50, 83, 84, 86, 90] is the only practical option. As shown in Ref. [84], ZPVEs are typically almost exact averages of one-half the sum of the harmonics and one-half the sum of the fundamentals, the difference being just $\text{ZPVE} - (1/4)\sum_i \omega_i + \nu_i = G_0 - \sum_i X_{ii}/4$, where the X_{ii} are the diagonal anharmonicity constants and G_0 is the polyatomic counterpart of the small Y_{00} Dunham constant [82] in diatomics. Consequently [50, 84, 90], the optimal scaling factor for ZPVEs is almost exactly midway between a $\lambda(\omega)$ suitable for harmonic frequencies (as an approximate correction for systematic bias in the calculated frequencies) and a $\lambda(\nu)$ suitable for fundamental frequencies (which additionally seeks to approximately corrects for anharmonicity). In fact, Alecu et al. [86] found for a large variety of basis sets and ab initio and DFT methods that $\lambda(\omega)/\lambda(\text{ZPVE}) = 1.014 \pm 0.002$, which is almost exactly the ratio of 1.0143 found by Perdew and coworkers [87] between harmonic frequencies and ZPVEs derived from experimental anharmonic force fields. Note that the "small" uncertainty of 0.002 on a ZPVE of 140 kcal/mol still would translate to about 0.3 kcal/mol, and even that is probably optimistic for the uncertainty in an individual

Table 3 Component breakdown of the W1-F12 and W2-F12 atomization energies and final gas-phase heats of formation at 0 and 298 K for the lowest-energy conformers of the amino acids (kcal/mol)

	SCF	CCSD	(T)	CV	Rel.	DBOC	SO	ZPVE	TAE_e	TAE_0	$\Delta_f H^\circ_{0K}$	$hcf_{298}{}^{RRHO}H$	$H_{conf,298}$	$\Delta_f H^\circ_{298K}$
Alanine	900.58	336.56	25.06	4.51	−1.25	0.34	−0.70	66.93	1265.10	1198.17	−96.20	4.97	0.45[d]	−101.73
Ditto W2-F12	900.62	336.36	24.86	4.63	−1.25	0.34	−0.70	66.93	1264.87	1197.94	−95.97	4.97	0.45[d]	−101.49
Arginine	1771.89	702.80	53.45	9.88	−2.61	0.72	−0.95	138.18	2535.20	2397.02	−86.14	8.92	0.33[e]	−98.78
Asparagine	1189.40	485.46	40.11	6.6	−1.91	0.52	−1.01	84.81	1719.16	1634.35	−139.26	6.39	0.61[f]	−146.54
Aspartic acid	1147.26	460.23	38.93	6.12	−1.85	0.46	−1.23	76.55	1649.91	1573.37	−183.38	6.47	0.61[g]	−189.57
Cysteine	926.11	365.80	29.45	4.79	−1.67	0.37	−1.26	67.37	1323.58	1256.21	−88.53	5.52	0.81[h]	−94.21
Ditto W2-F12	926.15	365.91	29.39	4.95	−1.65	0.37	−1.26	67.37	1323.88	1256.50	−88.83	5.52	0.81[h]	−94.50
Glutamine	1415.13	548.75	43.96	7.77	−2.11	0.58	−1.09	102.14	2012.99	1910.85	−142.47	7.40	0.64[i]	−150.99
Glutamic acid	1375.14	523.07	42.74	7.29	−2.04	0.52	−1.32	94.19	1945.40	1851.22	−187.94	7.32	0.68[j]	−195.50
Glycine	672.51	272.85	20.98	3.36	−1.06	0.28	−0.62	49.44	968.31	918.87	−90.19	4.16	0.49[k]	−94.21
Ditto W2-F12	672.54	272.80	20.81	3.45	−1.07	0.28	−0.62	49.44	968.20	918.76	−90.08	4.16	0.49[k]	−94.10
Ditto W2.2	672.50	272.72	20.92	3.43	−1.06	0.28	−0.62	49.44	968.18	918.74	−90.06	4.16	0.49[k]	−94.08
Ditto W4[b]	672.53	272.80	20.85	3.43	−1.06	0.28	−0.62	[49.438][a]	968.07[b]	918.63	−89.95	4.16	0.49[k]	−93.97
Histidine	1456.87	586.48	51.29	9.33	−2.24	0.56	−0.95	100.02	2101.33	2001.31	−61.06	6.80	0.22[l]	−69.84
Isoleucine	1579.21	528.12	37.11	8.00	−1.83	0.52	−0.95	119.69	2150.17	2030.48	−108.64	7.51	0.59[m]	−118.31
Leucine	1580.74	527.11	36.91	8.00	−1.82	0.52	−0.95	119.68	2150.49	2030.81	−108.97	7.50	0.48[m]	−118.76
Lysine	1670.47	597.49	42.32	8.54	−2.10	0.60	−0.95	132.05	2316.37	2184.32	−98.38	7.65	0.60[m]	−109.95
Methionine	1380.29	489.78	36.92	7.08	−2.06	0.47	−1.43	102.81	1911.06	1808.24	−93.98	7.66	0.37[o]	−102.50
Ditto W2-F12	1380.37	489.72	36.79	7.08[c]	−2.06[c]	0.47	−1.43	102.81	1910.95	1808.14	−93.88	7.66	0.37[o]	−102.40

Table 3 continued

	SCF	CCSD	(T)	CV	Rel.	DBOC	SO	ZPVE	TAE_e	TAE_0	$\Delta_f H^\circ_{0K}$	$hcf_{298}{}^{RRHO}H$	$H_{conf,298}$	$\Delta_f H^\circ_{298K}$
Phenylalanine	1839.81	611.85	54.28	11.68	−2.25	0.56	−1.21	117.86	2514.73	2396.87	−68.22	7.32	0.47[p]	−76.93
	SCF	CCSD	(T)	CV	Rel.	DBOC	SO	ZPVE	TAE_e	TAE_0	$\Delta_f H^\circ_{0K}$	$hcf_{298}{}^{RRHO}$	$H_{conf,298}$	$\Delta_f H^\circ_{298K}$
Proline	1241.14	440.34	33.97	6.75	−1.63	0.40	−0.87	89.92	1720.11	1630.19	−84.91	5.20	0.43[q]	−92.75
Serine	950.49	382.00	29.62	4.73	−1.52	0.40	−0.92	70.49	1364.80	1294.31	−133.34	5.45	0.56[r]	−139.32
Ditto W2-F12	950.54	381.90	29.37	4.87	−1.52	0.40	−0.92	70.49	1364.64	1294.15	−133.19	5.45	0.56[r]	−139.16
Threonine	1178.57	447.48	34.03	5.93	−1.71	0.48	−1.01	87.98	1663.77	1575.79	−141.53	6.13	0.71[s]	−148.95
Valine	1353.27	464.53	33.16	6.84	−1.63	0.46	−0.87	102.04	1855.76	1753.72	−105.17	6.64	0.42[m]	−113.60

hcf stands for the heat content function (a.k.a., enthalpy function), $H_T - H_0$

[a] anharmonic CCSD(T)/CBS ZPVE, Ref. [81] [b] values include −0.15 kcal/mol isodesmic correction for higher-order correlation effects (see text)

[c] Taken from W1-F12 theory

[d] CCSD(T)/CBS including zero-point corrections for 12 conformers from Ref. [24]

[e] Ref. [100]. Lowest 12 conformers, CCSD/6-31++G(d,p)

[f] Ref. [101]. B3LYP/6-311++G(d,p) relative energies and zero-point corrections for 62 lowest conformers taken from Table 1

[g] Ref. [102]. CCSD/6-311++G(d,p) total energies for 15 lowest conformers with B3LYP/6-311++G(d,p) zero-point energies. To be regarded as a lower bound. If MP2 is substituted for CCSD, drops to 0.55 kcal/mol

[h] MP2/cc-pVTZ structures for 71 lowest conformers taken from Ref. [23]. CCSD(T*)-F12b/cc-pVDZ-F12 relative energies, and MP2/cc-pVTZ zero-point corrections, calculated in present work

[i] Ref. [104]. Lowest 21 conformers, MP2/6-311G(2df,2pd)//B3LYP/6-311++G(d,p) with zero-point energy from BHandHLYP/6-311++G(d,p)

[j] Ref. [103]. MP2/6-311++G(2df,p)//B3LYP/6-311++G(d,p) relative energies at the bottom of the well for the lowest 35 conformers, combined with ZPVEs at the B3LYP/6-311++G(d,p) level, both taken from Table 1

[k] Ref. [81]. Best CCSD(T)/CBS values including ZPVE taken for all 6 conformers

[l] Ref. [105]. Lowest 42 conformers, MP2/6-311++G(d,p) relative energies with B3LYP/6-311G(d) zero-point corrections

[m] Ref. [106], relative energies at 0 K taken from Table 1. MP2/6-311++G(2d,2p) with scaled MP2(full)/6-31G* ZPVEs. As only lowest 6, 8, and 12 conformers are included for VAL, LEU, and ILE, respectively, these values are to be regarded as lower limits

[n] Ref. [107]. Relative energies at 0 K taken from Table 1. MP2/6-31++G**//B3LYP/6-31++G** relative energies for 18 lowest conformers

[o] From approximate CCSD(T)/6-311++G(3df,2p) conformer energies in Table 7 of Ref. [21]

[p] Ref. [108]. Lowest 37 conformers at MP2/6-311++G(d,p) level combined with B3LYP/6-311G(d) zero-point energies. Substituting B3LYP/6-311++G(d,p) reduces value to 0.31 kcal/mol

[q] Ref. [109]. Focal point analysis for lowest 4 conformers. To be regarded as lower limit

[r] Present work. MP2/cc-pVTZ geometries from cysteine conformer set used as starting structures. For the 58 unique minima obtained, CCSD(T*)-F12b/cc-pVDZ-F12 relative energies and MP2/cc-pVTZ zero-point corrections

[s] Ref. [110]. Focal point analysis on lowest 5 conformers. To be regarded as lower limit on account of the small number of conformers

Table 4 Dependence of computed ZPVEs (kcal/mol) on the level of theory

	HF/ 6-31G(d) G3,G3MP2	B3LYP/ 6-31G(d) G3B3, G3MP2B3	B3LYP/ 6-31G(2df,p) G4,G4MP2	B3LYP/ 6-311G (2d,d,p) CBS-QB3	B3LYP/ cc- pV(T+d)Z W1	B3LYP/ aug'-cc-pV(T+d)Z	B3LYP/ def2- TZVPP	B2PLYP/ def2- TZVPP
Scaling factor[a]	0.8929[b]	0.9600[c]	0.9854[d]	0.9900[e]	0.9850[f]	0.9896[g]	0.9884[g]	0.9833[g]
HFREQ27 RMSD	88.1	74.2	39.1	34.0	34.6	30.9	31.9	16.2
(cm^{-1}) w/o F_2	86.1	73.7	36.1	33.4	32.2	27.9	29.2	13.2
Alanine	65.49	65.33	66.67	66.95	66.52	66.73	66.82	66.93
Arginine	135.38	134.90	137.67	138.21	137.38	137.92	137.99	138.18
Asparagine	83.29	82.91	84.67	85.04	84.43	84.61	84.77	84.81
Aspartic	75.42	74.74	76.40	76.78	76.19	76.41	76.51	76.55
Cysteine	66.21	65.78	67.07	67.43	66.97	67.16	67.25	67.37
Glutamic	92.52	91.94	93.99	94.41	93.75	94.01	94.14	94.20
Glutamine	100.08	99.83	101.85	102.35	101.63	101.99	102.07	102.13
Glycine	48.56	48.25	49.27	49.51	49.15	49.29	49.35	49.44
Histidine	98.15	97.65	99.83	100.23	99.61	99.97	100.07	100.02
Isoleucine	116.70	116.95	119.27	119.67	118.97	119.38	119.51	119.69
Leucine	116.63	116.91	119.14	119.60	118.90	119.30	119.45	119.68
Lysine	128.89	129.00	131.51	132.13	131.20	131.72	131.76	132.05
Methionine	100.66	100.42	102.35	102.79	102.12	102.54	102.59	102.81
Phenylalanine	115.16	115.18	117.60	118.10	117.47	117.94	117.97	117.86
Proline	87.94	87.91	89.70	90.09	89.54	89.86	89.95	89.92
Serine	69.18	68.74	70.29	70.65	70.07	70.22	70.37	70.49
Threonine	86.26	85.97	87.75	88.09	87.41	87.71	87.81	87.98
Valine	99.57	99.69	101.69	102.04	101.42	101.75	101.86	102.04

[a] Scaling factor λ(ZPVE) appropriate for ZPVEs. All scaling factors for harmonic frequencies $\lambda(\omega) = 1.014\lambda$(ZPVE)

[b] As specified in Ref. [61]

[c] As specified in Ref. [62]

[d] As specified in Ref. [63, 64]

[e] As specified in Ref. [67]

[f] As specified in Ref. [50]. Sometimes in molecules with ionic bonding character, the aug′-cc-pV(T+d)Z basis set is employed with the same scaling factor: RMSD(HFREQ27) for that is 32.5 cm^{-1} including, and 29.9 cm^{-1} excluding F_2

[g] Obtained in this work by minimizing the RMSD over the HFREQ27 [95] dataset. For B3LYP/cc-pV(T+d)Z, the same procedure leads λ(ZPVE) = 0.9892. For B2PLYP near the basis set limit, Ref. [95] has $\lambda(\omega) = 0.997$, which corresponds to λ(ZPVE) = 0.983

molecule [88]. It has been argued earlier [91] (see also Ref. [92]) that for organic and bio organic molecules that are "well-behaved" from an electronic structure point of view, the main factor limiting accuracy in computational thermochemistry may well be the treatment of the nuclear motion, rather than the electronic problem as such.

Computed zero-point vibrational energies for the amino acids at various levels of theory (including those used in the composite thermochemistry schemes compared in this work) are listed in Table 4. In search of an alternative that was more accurate than B3LYP yet still comparatively affordable, we considered the B2PLYP double hybrid functional [93] in conjunction with the def2-TZVPP basis set [94] and optimized a $\lambda(\omega)$ scaling factor by minimizing the RMSD for the HFREQ27 dataset [95] of accurately known harmonic frequencies. As can be seen in Table 4, the RMSD over the HFREQ27 set is only half that of B3LYP and drops to 13.2 cm^{-1} if the anomalous F_2 molecule is eliminated. (For comparison, the HFREQ27 RMSD for CCSD(T)/cc-pV(Q+d)Z is still 8.4 cm^{-1}.) The optimum scale factor $\lambda(\omega) = 0.9971$ is very close to unity, and in conjunction with the "universal" ratio of 1.014 translated into $\lambda(\text{ZPVE}) = 0.9833$. As a sanity check on our procedure, we re-evaluated the λ(ZPVE) for B3LYP/6-31G(2df,p) and B3LYP/6-311G(2d,d,p) and obtained 0.9858 and 0.9896, respectively, which agree to better than 3 decimal places with the "official" values used in G4 theory and CBS-QB3, respectively [63, 67].

It can be seen in Table 4 that the lower levels of theory used for ZPVEs in G3(MP2) [61] and G3(MP2)B3, [62] can yield values several kcal/mol lower than the highest-level method: the RMSD from B2PLYP/def2-TZVPP are 2.12 and 2.29 kcal/mol, respectively, compared to 0.33 and 0.14 kcal/mol, respectively, for B3LYP/6-31G(2df,p) (scaled by 0.9854) as used by the G4 variants, and B3LYP/6-311G(2d,d,p) as used by CBS-QB3. (The "2d" refers to the use of an extra d function on second-row elements.) But also B3LYP/cc-pV(T+d)Z scaled by 0.985, as used in W1- and W1-F12 theory, appears to yield values that are too low, and indeed λ(ZPVE) as obtained from the HFREQ27 set is 0.9892. For B3LYP with a basis set that is effectively at the Kohn-Sham limit, $\lambda(\text{ZPVE}) = 1.004$ was found, which corresponds to $\lambda(\text{ZPVE}) = 0.99$, and the database of Radom and coworkers [90] likewise lists scaling factors near 0.99 for B3LYP with large basis sets. While a scaling factor of 0.985 vs. 0.990 may rightly be considered a distinction without a difference for small molecules (where anybody concerned about 0.1 kcal/mol in a ZPVE should seriously consider an accurate anharmonic ZPVE), the problem is much more obvious in larger systems such as presently considered.

For one system, glycine, an anharmonic value of 49.438 kcal/mol is available due to Puzzarini and coworkers [81], who combined CCSD(T)/CBS harmonic frequencies with a DFT anharmonic force field. Fortuitously, our scaled B2PLYP/def2-TZVPP value agrees to two decimal places. As an additional observation, for ethane, the accurate anharmonic ZPVE is 46.29 kcal/mol, [91] compared to 45.97 kcal/mol B3LYP/cc-pVTZ scaled by 0.985, 46.20 with a revised scaling factor of 0.99, and 46.33 kcal/mol at the B2PLYP/def2-TZVPP level scaled by 0.9833.

3.4 Performance of G*n* methods for the heats of formation of the amino acids

In this Section we use our best heats of formation from W1-F12 and W2-F12 theories (given in Table 3) to evaluate the performance of a variety of composite thermochemical Gaussian-*n* (G*n*) procedures including G3(MP2), [61] G3(MP2)B3, [62], G4, [63] G4(MP2), [64] and G4(MP2)-6X [65]. Table 5 presents the deviations (G*n*–W*n*-F12) from our benchmark W*n*-F12 results, as well as the RMSD, mean absolute deviations (MAD), and mean signed deviations (MSD) for the G*n* methods. Stover et al. [17] obtained G3(MP2) heats of formation for the amino acids: except for phenylalanine, cysteine, and methionine, the deviations between their heats of formation and our reference values exceed 1 kcal/mol. The mean signed deviation (MSD) of 1.90 kcal/mol being nearly equal to the RMSD of 2.25 kcal/mol indicates a very systematic error. Simply switching to G3(MP2)B3 cuts the MSD to 0.78 kcal/mol and the RMSD to 1.13 kcal/mol, while "upgrading" to G3B3 lowers these numbers even further to 0.45 and 0.60 kcal/mol, respectively. While both methods use MP2 rather than B3LYP reference geometries, the entire G3 family suffers from underestimated ZPVEs for the amino acids (Table 4), so apparently some of that issue is absorbed by the empirical correction. Stover et al. [17] also obtained G3(MP2) heats of formation via isodesmic bond separation reactions. As expected this improves the performance, with RMSD = 1.48 kcal/mol and a maximum deviation of 2.40 kcal/mol for phenylalanine. We note, however, that their CCSD(T)/CBS anchor value for the heat of formation at room temperature of glycine, −92.6 kcal/mol, is 1.5 kcal/mol lower than our W2-F12 value. If we substitute the latter in their isodesmic reactions, their RMSD plunges to just 0.47 kcal/mol.

We now turn our attention to the performance of the Gaussian-4 family: G4, [63] G4(MP2), [64] and G4(MP2)-6X [65]. The G4(MP2) procedure exhibits somewhat disappointing performance, its RMSD = 1.80 kcal/mol placing intermediately between G3(MP2) and G3(MP2)B3. The largest deviations are obtained for asparagine (2.48), lysine (2.32), glutamine (3.15), and arginine (3.34 kcal/mol), but all other deviations exceed 1 kcal/mol apart from phenylalanine, cysteine, and methionine. The computationally more expensive "full" G4 procedure yields much better performance with an RMSD of 0.72 kcal/mol, and just three cases exceeding 1 kcal/mol (glutamine 1.39, arginine 1.21, and lysine 1.37 kcal/mol). However, an essentially identical RMSD = 0.71 kcal/mol is afforded by the G4(MP2)-6X procedure, which involves the same computational steps and cost as G4(MP2) but entails six additional empirical scaling factors. Deviations larger than 1 kcal/mol are obtained for just four systems, namely arginine (1.63), glutamine (1.63), asparagine (1.10), and methionine (−1.02 kcal/mol). Finally, we note that the CBS-QB3 method clocks in at RMSD = 1.01 kcal/mol.

Very recently, Ramabhadran et al. [21] determined the enthalpies of formation of cysteine and methionine using their connectivity-based hierarchy (CBH-*n*) approach [77, 78]. From their Table 3, the best enthalpies of formation obtained for the lowest-energy conformer at the CBH-2 (isoatomic) rung using experimental heats of formation for the reference species and CCSD(T)/6-311++G(3df,2p) reaction energies are −96.1 (cysteine) and −104.3 (methionine) kcal/mol. From their Table 7, we calculate conformer corrections of +0.77 kcal/mol for cysteine and +0.37 kcal/mol for methionine: The latter we actually use in the present work, while the former is slightly less than our own calculation of 0.81 kcal/mol. According to their Table 9, the heats of formation after conformer correction are −95.3 and −104.0 kcal/mol (the latter value presumably after roundoff), both more exothermic than our W2-F12 values (Table 3) of −94.5 and −102.4 kcal/mol. We do note that some of the experimental data for reference species used in Ref. [21] carry non-trivial uncertainties, which could account for at least some of the discrepancy.

Table 5 Performance of a selection of composite procedures of the G*n* family for the calculation of heats of formation ($\Delta_f H^\circ_{298K}$, exclusive of conformer correction) of the 18 amino acids in Table 3

	CBS-QB3	G3(MP2)[a]	G3(MP2)[b]	G3(MP2)B3[c]	G3B3[c]	G4(MP2)[c]	G4(MP2)-6X[c]	G4[c]
Alanine	−0.41	1.74	1.24	0.48	−0.44	1.04	−0.02	0.25
Arginine	0.99	N/A	4.61	2.74	0.32	3.34	1.63	1.21
Asparagine	−0.64	1.35	2.75	1.44	−0.55	2.48	1.10	0.67
Aspartic acid	−1.70	0.58	1.28	−0.01	−1.55	1.64	0.09	0.20
Cysteine	−0.75	1.71	0.91	0.05	0.03	0.39	0.15	0.47
Glutamine	0.20	N/A	3.11	2.04	0.06	3.15	1.63	1.39
Glutamic acid	−0.91	1.08	1.58	0.29	−1.17	1.92	0.19	0.54
Glycine	−0.46	N/A	2.69	0.76	−0.09	1.22	0.38	0.51
Histidine	0.24	N/A	4.26	1.27	0.06	1.60	0.44	0.29
Isoleucine	0.86	1.30	1.00	0.52	−0.43	1.21	−0.46	0.37
Leucine	1.02	1.54	1.14	0.69	−0.18	1.26	−0.41	0.49
Lysine	1.50	N/A	2.65	1.74	0.53	2.32	0.61	1.37
Methionine	−0.54	1.17	0.27	−0.37	−0.37	−0.26	−1.02	−0.23
Phenylalanine	2.23	2.40	0.30	−0.47	0.29	0.68	−0.13	0.85
Proline	0.87	0.68	2.08	1.18	0.13	1.94	0.46	0.95
Serine	−1.16	1.52	1.52	0.50	−0.75	1.47	0.11	0.39
Threonine	−0.81	1.76	1.76	0.74	−0.73	1.86	0.34	0.58
Valine	0.39	1.52	1.12	0.47	−0.49	1.14	−0.32	0.31
RMSD[c]	1.01	1.48	2.25	1.13	0.60	1.80	0.71	0.72
MAD[c]	0.87	1.41	1.90	0.88	0.45	1.61	0.53	0.62
MSD[c]	0.05	1.41	1.90	0.78	−0.30	1.58	0.27	0.59
LD[c]	2.23	2.40	4.61	2.74	−1.55	3.34	1.63	1.39
	Phe	Phe	Arg	Arg	Asp	Arg	Arg, glu	Glu
ND[c]	6	11[e]	14	6	2	15	4	3

The values listed are deviations (G*n*–W*n*-F12) from our benchmark W1-F12 and W2-F12 heats of formation (kcal/mol)

[a] From reference [17], obtained from isodesmic reactions

[b] From reference [17], obtained from atomization reactions

[c] This work

[d] RMSD = root mean square deviation, MAD = mean absolute deviation, MSD = mean signed deviation, LD = largest deviation, ND = number of deviations (in absolute value) exceeding 1 kcal/mol

[e] Out of a total of 13 determinations

3.5 Comparison with experiment

Comparison with experiment obviously entails thermal corrections. The RRHO approximation will cause some errors, the largest of which will be neglect of the population of the various low-energy conformers. If we neglect the difference between the rovibrational partition functions of the different conformers, then the conformer contribution to the enthalpy function $hcf_{298} \equiv H_{T=298} - E_0$ is easily found as [96]

$$hcf_{298}conf = \frac{\sum_i x_i \exp(-x_i)}{\sum_i \exp(-x_i)} \quad \text{where} \quad x_i \equiv \frac{E_i - E_0}{RT} \qquad (2)$$

where the index *i* runs over the conformers. The effect of accounting for different rovibrational partition functions in the different conformers was considered in Ref. [96] for the alkane conformers and is negligible compared to other potential error sources in the present calculation, such as the neglect of anharmonicity and the uncertainty in the basis set extrapolation. Conformer energies were gathered from published calculations in the literature [21–24, 81, 100–112]: these range from complete basis set CCSD(T) studies for glycine [81] and alanine [24] to relatively low-level MP2 or DFT calculations for some other species. Details are given in the footnotes to Table 3.

Table 6 lists the available experimental gas-phase heats of formation at 298 K ($\Delta H^\circ_{f,298}$). Our W2-F12 value for alanine (−101.5 kcal/mol) is spot on the experimental value of Dorofeeva and Ryzhova [97] (−101.5 ± 0.5 kcal/mol) and still agrees to within mutual uncertainties with that of da Silva et al. [15] (−101.9 ± 0.7). However, the NIST chemistry WebBook [79] value (−99.1 ± 1.0 kcal/mol) is clearly incompatible with our calculations.

Table 6 Experimental gas-phase heats of formation at 298 K for the amino acids (kcal/mol)

$\Delta_f H^\circ_{298K}$	Expt.[a]	Uncert.	Best theor.[b]
Alanine	−99.1	1.0	−101.5
	−101.5[c]	0.5	
	−101.9[d]	0.7	
	−101.3[e]		
Cysteine	−91.4[f]	0.4	−94.2
Glycine	−93.3	1.1	−94.1, −93.97[h]
	−94.1[c]	0.4	
	−93.7[e]		
Methionine	−98.8	1.0	−102.4
	−102.8[g]	2.4	
Phenylalanine	−74.8[d,e]		−76.9
Proline	−87.5	1.0	−92.8
Valine	−108.8[e]		−113.6

[a] Values are taken from the NIST chemistry WebBook, [79] unless otherwise indicated

[b] From Table 3

[c] From Dorofeeva and Ryzhova [97]

[d] From da Silva et al. [15]

[e] From CRC Handbook [80]

[f] From Roux et al. 2010 [19]

[g] From Roux et al. 2012 [18]

[h] Quasi-W4 value including anharmonic ZPVE and estimated higher-order correlation correction (see Table 3 and text)

[i] Note that in the CRC Tables, the value for L-alanine (111.4 kcal/mol) seems to have been transposed with the value for β-alanine (101.3 kcal/mol)

Our W2-F12 heat of formation for cysteine (−94.2 kcal/mol) suggests that the experimental value of Roux et al. [19] should be revised downward by about 2.8 kcal/mol; the recent study of Ramabhadran et al. [21] suggests even further downward revision (*vide supra*). As for glycine, the W2-F12 heat of formation (−94.1 kcal/mol from W2-F12, −94.0 from quasi-W4) and the available experimental values agree to within overlapping uncertainties. Specifically, our calculations are spot on the experimental value of Dorofeeva and Ryzhova [97] (−94.1 ± 0.4 kcal/mol), just slightly below the experimental value from the CRC Handbook (−93.7 kcal/mol), and in the upper end of the uncertainty band of the NIST WebBook value (−93.3 ± 1.1 kcal/mol). Our W2-F12 value for methionine (−102.4 kcal/mol) agrees well with the new measurement of Roux et al. [18] (−102.8 ± 2.4 kcal/mol), and both imply a downward revision of the NIST Chemistry Webbook value (−98.8 ± 1.0 kcal/mol) by about 3–4 kcal/mol. As for phenylalanine, our W1-F12 value (−76.9 kcal/mol) suggests that the experimental value from the CRC Handbook (−74.8 kcal/mol) should be revised downward by about 2 kcal/mol. The W1-F12 values for proline (−92.8 kcal/mol) and valine (−113.6 kcal/mol) suggest that the experimental values should be revised downward by about 5 kcal/mol (Table 6).

For the two largest amino acids, tryptophan and tyrosine, we were unable to calculate W1-F12 atomization energies. At the G4, CBS-QB3, and G4(MP2)-6X levels, respectively, we obtain heats of formation at 0 K for tryptophan of −49.60, −47.87, and −48.77 kcal/mol, and for tyrosine of −109.12, −108.58, and −108.49 kcal/mol. At room temperature, the corresponding values are −59.98, −58.27, and −58.98 kcal/mol for tryptophan and −118.56, −118.03, and −117.78 kcal/mol for tyrosine. Averaging over all three levels of theory, and adding in conformers corrections for tryptophan of 0.71 kcal/mol [111] and for tyrosine of 0.65 kcal/mol, we finally obtain estimated heats of formation at 298 K of −58.37 kcal/mol for tryptophan, and of −117.47 kcal/mol for tyrosine.

4 Conclusions

We have obtained benchmark heats of formation at the CCSD(T)/CBS limit for the 20 natural amino acids. Our best heats of formation at 298 K ($\Delta H^\circ_{f,298}$) are −101.5 (alanine), −98.8 (arginine), −146.5 (asparagine), −189.6 (aspartic acid), −94.5 (cysteine), −151.0 (glutamine), −195.5 (glutamic acid), −94.0 (glycine, quasi-W4) or −94.1 (glycine, W2-F12), −69.8 (histidine), −118.3 (isoleucine), −118.8 (leucine), −110.0 (lysine), −102.4 (methionine), −76.9 (phenylalanine), −92.8 (proline), −139.2 (serine), −149.0 (threonine), and −113.6 (valine) kcal/mol. These heats of formation are obtained at the W2-F12 level for alanine, cysteine, glycine, methionine, and serine, and at the W1-F12 level for all of the rest. For the two largest amino acids, an average over G4, G4(MP2)-6X, and CBS-QB3 yields best estimates of −58.4 kcal/mol for tryptophan, and of −117.5 kcal/mol for tyrosine.

Uncertainties caused by issues with the zero-point vibrational energy and the conformer corrections rival, and probably exceed, those directly related to the electronic structure treatment. The overall uncertainty is somewhat difficult to quantify, but a semi-quantitative estimate would range from about ±0.5 kcal/mol for the smaller, to about ±1 kcal/mol for the larger, amino acids.

For glycine, by way of validation, we were able to obtain a "quasi-W4" result corresponding to $\mathrm{TAE}_e = 968.1, \mathrm{TAE}_0 = 918.6, \Delta H^\circ_{f,298} = -90.0$, and $\Delta H^\circ_{f,298} = -94.0$ kcal/mol.

Our best theoretical values suggest that the experimental gas-phase heats of formation from the NIST WebBook should be revised downward by 2.4 (alanine), 0.7–0.8 (glycine), 3.2 (methionine), and 5.3 (proline) kcal/mol. Similarly, we suggest that the experimental values from the

CRC Handbook should be revised downward by 0.4 (glycine), 2.0 (phenylalanine), and 4.8 (valine) kcal/mol. Our best theoretical values are in good agreement with the recently reported experimental values of Roux and coworkers for alanine [15] and methionine, [18] but suggest that their experimental value for cysteine should be revised downward by 2.8 kcal/mol. Finally, our best theoretical values for alanine and glycine are in excellent agreement with the recent values of Dorofeeva and Ryzhova [97].

Using our W1-F12 and W2-F12 benchmark heats of formation, we benchmark the performance of the empirical composite G*n* procedures. We obtain the following RMSDs: 2.25 (G3(MP2)), 1.13 (G3(MP2)B3), 0.60 (G3B3), 1.80 (G4(MP2)), 0.71 (G4(MP2)-6X), and 0.72 (G4) kcal/mol. Particularly G4(MP2)-6X appears to offer an excellent performance-to-computational cost ratio.

Finally, it appears that for W1- and W1-F12, the scaling factor for the B3LYP/cc-pV(T+dZ)Z or B3LYP/aug'-cc-pV(T+d)Z zero-point vibrational energy should be revised upward to 0.990.

5 Supporting information

B3LYP/A'VTZ optimized geometries for the species considered in the present work (Table S1). Full references for ref [40] (Gaussian 09) and ref [41] (Molpro 2010) (Table S2). B2PLYP/def2-TZVPP harmonic frequencies for all amino acids except tryptophan and tyrosine, and B3LYP/aug'-cc-pV(T+d)Z frequencies for all amino acids.

Acknowledgments JMLM is the Baroness Thatcher Professor of Chemistry at the Weizmann Institute of Science and acknowledges partial financial support from the Lise Meitner-Minerva Center for Computational Quantum Chemistry and the Helen and Martin Kimmel Center for Molecular Design. This research was supported in part by the Weizmann AERI (Alternative Energy Research Initiative) and by a startup grant from the University of North Texas from which the Martin group Linux cluster was purchased. The authors would like to thank Dr. David Hrovat for assistance with procurement and management of the latter. A.K. is the recipient of an Australian Research Council (ARC) Discovery Early Career Researcher Award (project number: DE140100311). We also acknowledge the generous allocation of computing time from the National Computational Infrastructure (NCI) National Facility and the support of iVEC through the use of advanced computing resources located at iVEC@UWA.

Note added in revision For nine of the amino acids, we were able to compute correlation corrections to the DBOC at the CCSD/cc-pVDZ level using the CFOUR package [98, 99]. (We have shown earlier [91] that the correlation contribution to DBOCs converges very rapidly with the basis set.) They uniformly reduce the DBOCs: values are Gly −0.10, Ala −0.13, Ser −0.14, Cys −0.13, Asn −0.16, Asp −0.15, Pro −0.16, Thr −0.17, and Val −0.18 kcal/mol. Linear regression through the origin reveals that DBOC[CCSD] ~0.642 DBOC[HF]: This suggests TAE reductions of up to 0.26 kcal/mol (arginine) for the remaining amino acids if correlation were included in the DBOC. Once again, this highlights that the nuclear motion, rather than the clamped-nuclei electronic structure, is the accuracy-limiting factor here.

References

1. Hättig C, Klopper W, Köhn A, Tew DP (2012) Chem Rev 112:4
2. Kong L, Bischoff FA, Valeev E (2012) Chem Rev 112:75
3. Ten-no S, Noga J (2012) WIREs Comput Mol Sci 2:114
4. Peterson KA, Feller D, Dixon DA (2012) Theor Chem Acc 131:1079
5. Quinn JR, Zimmerman SC, Del Bene JE, Shavitt I (2007) J Am Chem Soc 129:934
6. Distasio RA Jr, Steele RP, Rhee YM, Shao Y, Head-Gordon M (2007) J Comp Chem 28:839
7. Šponer J, Riley KE, Hobza P (2008) Phys Chem Chem Phys 10:2595
8. Valdes H, Pluháčková K, Pitonák M, Řezáč J, Hobza P (2008) Phys Chem Chem Phys 10:2747
9. Valdes H, Pluháčková K, Hobza P (2009) J Chem Theory Comput 5:2248
10. Jiang J, Wu Y, Wang Z-X, Wu C (2010) J Chem Theory Comput 6:1199
11. Tkatchenko A, Rossi M, Blum V, Ireta J, Scheffler M (2011) Phys Rev Lett 106:118102
12. Bokatzian-Johnson SS, Stover ML, Dixon DA (2012) J Phys Chem B 116:14844
13. Goerigk L, Karton A, Martin JML, Radom L (2013) Phys Chem Chem Phys 15:7028
14. Nelson DL, Cox MM (2008) Lehninger principles of biochemistry. Palgrave-Macmillan, New York
15. Ribeiro da Silva MAV, Ribeiro da Silva MDMC, Santos AFLOM, Roux MV, Foces-Foces C, Notario R, Guzman-Majia R, Juaristi E (2010) J Phys Chem B 114:16471
16. Notario R, Roux MV, Foces-Foces C, da Silva MAVR, da Silva MDMCR, Santos AFLOM, Guzman-Meja R, Juaristi E (2011) J Phys Chem B 115:9401
17. Stover ML, Jackson VE, Matus MH, Adams MA, Cassady CJ, Dixon DA (2012) J Phys Chem B 116:2905
18. Roux MV, Notario R, Segura M, Chickos JS, Liebman JF (2012) J Phys Org Chem 25:916
19. Roux MV, Foces-Foces C, Notario R, da Silva MAVR, da Silva MDMC, Santos AFLOM, Juaristi E (2010) J Phys Chem B 114:10530
20. Brás NF, Perez MAS, Fernandes PA, Silva PJ, Ramos MJ (2011) J Chem Theory Comput 7:3898
21. Ramabhadran RO, Sengupta A, Raghavachari K (2013) J Phys Chem A 117:4973
22. Jaeger HM, Schaefer HF III, Demaison J, Császár AG, Allen WD (2010) J Chem Theory Comput 6:3066
23. Wilke JJ, Lind MC, Schaefer HF III, Császár AG, Allen WD (2009) J Chem Theory Comput 5:1511
24. Balabin RM (2011) Comp Theor Chem 965:15
25. Balabin RM (2009) Chem Phys Lett 479:195
26. Shavitt I, Bartlett RJ (2009) Many-body methods in chemistry and physics: MBPT and coupled-cluster theory. Cambridge Molecular Science, Cambridge
27. Shavitt I (1993) The history and evolution of Gaussian basis sets. Isr J Chem 33:357
28. Bartlett RJ, Cole SJ, Purvis GD, Ermler WC, Hsieh HC, Shavitt I (1987) J Chem Phys 87:6579
29. Shavitt I (1985) Tetrahedron 41:1531
30. Comeau DC, Shavitt I, Jensen P, Bunker PR (1989) J Chem Phys 90:6491

31. Karton A, Martin JML (2012) J Chem Phys 136:124114
32. Karton A, Daon S, Martin JML (2011) Chem Phys Lett 510:165
33. Curtiss LA, Redfern PC, Raghavachari K (2011) WIREs Comput Mol Sci 1:810
34. Lee C, Yang W, Parr RG (1988) Phys Rev B 37:785
35. Becke AD (1993) J Chem Phys 98:5648
36. Stephens PJ, Devlin FJ, Chabalowski CF, Frisch MJ (1994) J Phys Chem 98:11623
37. Dunning TH (1989) J Chem Phys 90:1007
38. Kendall RA, Dunning TH, Harrison RJ (1992) J Chem Phys 96:6796
39. Dunning TH Jr, Peterson KA, Wilson AK (2001) J Chem Phys 114:9244
40. Frisch MJ, Trucks GW, Schlegel HB, Scuseria GE, Robb MA, Cheeseman JR, Scalmani G, Barone V, Mennucci B, Petersson GA, et al (2009) Gaussian 09, Revision D01, Gaussian Inc, Wallingford CT. See also: URL: http://www.gaussian.com
41. Werner H-J, Knowles PJ, Manby FR, Schütz M, Celani P, Knizia G, Korona T, Lindh R, Mitrushenkov A, Rauhut G et al (2012) Molpro 2012.1, University College Cardiff Consultants Limited: Cardiff U.K. See also: http://www.molpro.net.
42. Peterson KA, Adler TB, Werner H-J (2008) J Chem Phys 128:084102
43. Hill JG, Peterson KA, Knizia G, Werner H-J (2009) J Chem Phys 131:194105
44. Feller D, Peterson KA (2013) J Chem Phys 139:084110
45. Ten-no S (2004) Chem Phys Lett 398:56
46. Werner H-J, Adler TB, Manby FR (2007) J Chem Phys 126:164102
47. Knizia G, Werner H-J (2008) J Chem Phys 128:154103
48. Adler TB, Knizia G, Werner H-J (2007) J Chem Phys 127:221106
49. Knizia G, Adler TB, Werner H-J (2009) J Chem Phys 130:054104
50. Martin JML, de Oliveira G (1999) J Chem Phys 111:1843
51. Peterson KA, Dunning TH (2002) J Chem Phys 117:10548
52. Douglas M, Kroll NM (1974) Ann Phys 82:89
53. Heß BA (1986) Phys Rev A 33:3742
54. de Jong WA, Harrison RJ, Dixon DA (2001) J Chem Phys 114:48
55. Ruscic B, Pinzon RE, Morton ML, von Laszewski G, Bittner S, Nijsure SG, Amin KA, Minkoff M, Wagner AF (2004) J Phys Chem A 108:9979
56. Ruscic B (2004) Encyclopedia of Science and Technology (2005 Yearbook of Science and Technology). McGraw-Hill, New York, p 3
57. Ruscic B, Pinzon RE, Morton ML, Srinivasan NK, Su M-C, Sutherland JW, Michael JV (2006) J Phys Chem A 110:6592
58. Stevens WR, Ruscic B, Baer T (2010) J Phys Chem A 114:13134
59. Ruscic B, Feller D, Peterson KA (2014) Theor Chem Acc 133:1415. doi:10.1007/s00214-013-1415-z
60. Cox JD, Wagman DD, Medvedev VA (1989) CODATA key values for thermodynamics, Hemisphere Publishing Corp.: New York. http://www.codata.org/resources/databases/key1.html
61. Curtiss LA, Redfern PC, Raghavachari K, Rassolov V, Pople JA (1999) J Chem Phys 110:4703
62. Baboul AG, Curtiss LA, Redfern PC, Raghavachari K (1999) J Chem Phys 110:7650
63. Curtiss LA, Redfern PC, Raghavachari K (2007) J Chem Phys 126:84108
64. Curtiss LA, Redfern PC, Raghavachari K (2007) J Chem Phys 127:124105
65. Chan B, Deng J, Radom L (2011) J Chem Theory Comput 7:112
66. Curtiss LA, Redfern PC, Raghavachari K (2005) J Chem Phys 123:124107
67. Montgomery JA, Frisch MJ, Ochterski JW, Petersson GA (1999) J Chem Phys 110:2822
68. Karton A, Rabinovich E, Martin JML, Ruscic B (2006) J Chem Phys 125:144108
69. Bak KL, Jørgensen P, Olsen J, Helgaker T, Gauss J (2000) Chem Phys Lett 317:116
70. Boese AD, Oren M, Atasoylu O, Martin JML, Kallay M, Gauss J (2004) J Chem Phys 120:4129
71. Stanton JF (1997) Chem Phys Lett 281:130
72. Karton A, Taylor PR, Martin JML (2007) J Chem Phys 127:064104
73. Harding ME, Vázquez J, Ruscic B, Wilson AK, Gauss J, Stanton JF (2008) J Chem Phys 128:114111 and references therein
74. Fogueri UR, Kozuch S, Karton A, Martin JML (2013) Theor Chem Acc 132:1291
75. Karton A, Kaminker I, Martin JML (2009) J Phys Chem A 113:7610
76. Harding ME, Vázquez J, Gauss J, Stanton JF, Kállay M (2011) J Chem Phys 135:044513
77. Ramabhadran RO, Raghavachari K (2011) J Chem Theory Comput 7:2094
78. Ramabhadran RO, Raghavachari K (2012) J Phys Chem A 116:7531
79. Afeefy HY, Liebman JF, Stein SE "Neutral Thermochemical Data" in NIST Chemistry WebBook, NIST Standard Reference Database Number 69, Eds Linstrom PJ, Mallard WG, National Institute of Standards and Technology, Gaithersburg MD, 20899. http://webbook.nist.gov. Retrieved October 22, 2013
80. (2012) CRC handbook of chemistry and physics, 93rd edn. CRC Press, Boca Raton 2013
81. Barone V, Biczysko M, Bloino J, Puzzarini C (2013) Phys Chem Chem Phys 15:10094–10111. The absolute ZPVE for the lowest-energy conformer is not given explicitly in the paper, but is 49.438 kcal/mol: Puzzarini C, personal communication to authors (January, 2014).
82. Dunham JL (1932) Phys Rev 41:721
83. Del Bene JE, Aue DH (1992) Shavitt I 114:1631
84. Grev RS, Janssen CL, Schaefer HF (1991) J Chem Phys 95:5128
85. Schuurman MS, Allen WD, Schaefer HF (2005) J Comput Chem 26:1106
86. Alecu IM, Zheng J, Zhao Y, Truhlar DG (2010) J Chem Theor Comput 6:2872. http://comp.chem.umn.edu/freqscale/index.html
87. Csonka G, Ruzsinszky A, Perdew JP (2005) J Phys Chem A 109:6779
88. Irikura KK, Johnson RD, Kacker RN, Kessel R (2009) J Chem Phys 130:114102
89. Sinha P, Boesch SE, Gu C, Wheeler RA, Wilson AK (2004) J Phys Chem A 108:9213
90. Merrick JP, Moran D, Radom L (2007) J Phys Chem A 111:11683. http://groups.chem.usyd.edu.au/radom/More/ScaleFactor.html
91. Karton A, Ruscic B, Martin JML (2007) J Mol Struct Theochem 811:345
92. Pfeiffer F, Rauhut G, Feller D, Peterson KA (2013) J Chem Phys 138:044311
93. Grimme S (2006) J Chem Phys 124:034108
94. Weigend F, Ahlrichs R (2005) Phys Chem Chem Phys 7:3297
95. Kozuch S, Gruzman D, Martin JML (2011) J Phys Chem C 114:20801, Table S-16.
96. Gruzman D, Karton A, Martin JML (2009) J Phys Chem A 113:11974
97. Dorofeeva OV, Ryzhova ON (2009) J Chem Thermodyn 41:433
98. CFOUR, Coupled-Cluster techniques for Computational Chemistry, a quantum-chemical program package by Stanton JF,

Gauss J, Harding ME, Szalay PG with contributions from Auer AA, Bartlett RJ, Benedikt U, Berger C, Bernholdt DE, Bomble YJ, Cheng L, Christiansen O, Heckert M, Heun O, Huber C, Jagau TC, Jonsson D, Juslius J, Klein K, Lauderdale WJ, Matthews DA, Metzroth T, Mück LA, O'Neill DP, Price DR, Prochnow E, Puzzarini C, Ruud K, Schiffmann F, Schwalbach W, Simmons C, Stopkowicz S, Tajti A, Vázquez J, Wang F, Watts JD and the integral packages MOLECULE (Almlöf J, Taylor PR), PROPS (Taylor PR), ABACUS (Helgaker T, Jensen HJA, Jørgensen P, Olsen J), and ECP routines by Mitin AV, van Wüllen C. For the current version. http://www.cfour.de
99. Gauss J, Tajti A, Kállay M, Stanton JF, Szalay PG (2006) J Chem Phys 125:144111
100. Ling S, Yu W, Huang Z, Lin Z, Haranczyk M, Gutowski M (2006) J Phys Chem A 110:12282–12291
101. Chen M, Huang Z, Lin Z (2005) J Mol Struct Theochem 719:153–158
102. Chen M, Lin Z (2007) J Chem Phys 127:154314
103. Meng L, Lin Z (2011) Comp Theor Chem 976:42–50
104. Pang R, Guo M, Ling S, Lin Z (2013) Comp Theory Chem 1020:14–21
105. Huang Z, Yu W, Lin Z (2006) J Mol Struct Theochem 801:7–20
106. Dokmaisrijan S, Lee VS, Nimmanpipug P (2010) J Mol Struct Theochem 953:28–38
107. Boeckx B, Maes G (2012) J Phys Chem B 116:12441–12449
108. Huang Z, Yu W, Lin Z (2006) J Mol Struct Theochem 758:195–202
109. Czinki E, Császár AG (2003) Chem Eur J 9:1008–1019
110. Szidarovszky T, Czakó G, Császár AG (2009) Mol Phys 107:761–775
111. Huang Z, Lin Z (2005) J. Phys. Chem. A 109:2656. MP2/6-311++G(d, p) total energies for 45 lowest conformers, with B3LYP/6-311G* zero-point energies.
112. Zhang M, Huang Z, Lin Z (2005) J Chem Phys 122: 134313. Lowest 36 conformers, MP2/6-311G(2df, p)//B3LYP/6-311++G(d, p) with zero-point energy from B3LYP/6-311++G(d, p).

Theor Chem Acc (2014) 133:1474
DOI 10.1007/s00214-014-1474-9

REGULAR ARTICLE

Quantum chemical characterization of the $\tilde{X}(^1A_1)$, $\tilde{a}(^3B_1)$, $\tilde{A}(^1B_1)$ and $\tilde{B}(2^1A_1)$ states of diiodomethylene and the enthalpies of formation of diiodomethylene, iodomethylene and iodomethylidyne

George B. Bacskay

Received: 19 December 2013 / Accepted: 25 February 2014 / Published online: 26 March 2014

Abstract The equilibrium energies of the $\tilde{X}(^1A_1)$, $\tilde{a}(^3B_1)$, $\tilde{A}(^1B_1)$ and $\tilde{B}(2^1A_1)$ states of diiodomethylene (CI_2) and its atomization and dissociation energies in the complete basis limit were determined by extrapolating valence correlated (R/U)CCSD(T) and Davidson corrected multi-reference configuration interaction energies calculated with the aug–cc–pVxZ ($x = T, Q, 5$) basis sets and the ECP28MDF pseudopotential of iodine plus corrections for core and core-valence correlation, scalar relativity, spin–orbit coupling and zero-point energies. The geometries and vibrational frequencies were obtained at the CCSD and complete active space second-order perturbation theory levels of theory with the cc–pVTZ basis. Spin–orbit energies were computed in a large basis of configurations chosen so as to accurately describe dissociation to the ^3P and ^2P states of C and I, respectively. These computations were extended to iodomethylene (CHI) and iodomethylidyne (CI), resulting in small corrections to the thermochemistry and the singlet–triplet gap of CHI computed previously. The onset (T_{00}) of the $\tilde{A} \leftarrow \tilde{X}$ excitations in CI_2 is predicted to be 12,680 cm^{-1}. The Renner–Teller intersection is computed to have a substantially lower energy (6.5 kcal mol^{-1}) than the dissociation barrier on the $\tilde{A}$ surface, thus internal relaxation via Renner–Teller coupling is expected to be the dominant photochemical channel. The predicted enthalpies of formation of CI_2, CHI and CI in their ground states at 0 K are 109.1 ± 1, 102.8 ± 1 and 132.9 ± 1 kcal mol^{-1}, respectively. The computed singlet–triplet gaps in CI_2 and CHI are 11.1 and 4.4 kcal mol^{-1}, respectively.

Keywords Iodocarbene · Enthalpy of formation · MRCI · Singlet–triplet splitting · Renner–Teller effect · Electronic excitation

Dedicated to the memory of Professor Isaiah Shavitt and published as part of the special collection of articles celebrating his many contributions.

G. B. Bacskay (✉)
School of Chemistry, The University of Sydney, Sydney, NSW 2006, Australia
e-mail: george.bacskay@sydney.edu.au

1 Introduction

Carbenes are a fascinating class of chemical species! My own interest in this area was first inspired by Shi Shavitt's seminar on methylene in 1979—the material presented was subsequently published as a review article in Tetrahedron [1]. It is a comprehensive summary of the most important experimental and theoretical determinations of the geometry and singlet–triplet splitting in methylene to date, i.e. 1984. I have read that paper many times. Its clarity is typical of Shavitt's publications, where pioneering research was beautifully presented and discussed. It is an honour to contribute a piece of work on carbenes to this memorial issue that honours the life and scientific contributions of Isaiah Shavitt.

Carbenes are prominent in synthetic and polymer chemistry as well as in interstellar and atmospheric chemistry [2–5]. The great interest stems largely from the chemical differences between singlet and triplet carbenes, particularly in their reactivity and rich spectroscopy and photochemistry. The latter were recently discussed and summarized in a review by Kable et al. [6]. Laser-induced fluorescence (LIF) studies in particular have yielded a wealth of structural and spectroscopic data as well as useful information on the nature of the potential energy surface of the excited $\tilde{A}$ state, such as barriers to dissociation and

Renner–Teller (RT) intersections. The latter occur at linear geometries where the singlet $\tilde{X}$ and $\tilde{A}$ states become degenerate (in the Born–Oppenheimer approximation). The RT coupling of these states provides a relaxation mechanism for carbenes in their excited singlet states via internal conversion. The relative heights of the barriers to dissociation and linearity will therefore determine whether photodissociation would occur.

The interesting and unusual properties of carbenes stem from their open-shell electronic configurations, whereby two non-bonding valence electrons occupy two near-degenerate orbitals. The latter are effectively an in-plane sp^2 hybrid and an out-of-plane $2p$ atomic orbital on the carbon atom. Thus, several near-degenerate electronic states can arise. In the case of CI_2 (as in all halocarbenes) in the $\tilde{X}(^1A_1)$ ground state, the two electrons occupy the same a_1 molecular orbital, giving rise to a closed shell configuration, whereas the excited $\tilde{a}(^3B_1)$ and $\tilde{A}(^1B_1)$ states correspond to triplet and singlet spin-coupled a_1b_1 configurations, respectively. The $\tilde{B}$ state, observed recently in a number of carbenes [6–9], is dominated by the b_1^2 configuration. Although in many carbenes the singlet is the ground state, there are exceptions. Most notably, in methylene, the $^3B_1$ state lies 9.0 kcal mol^{-1} lower in energy than the $^1A_1$ state [10]. (In general, electronegative substituents give rise to singlet ground states, while electropositive substituents produce triplet ground states.)

Our group has already studied all the fluoro-, chloro- and bromo-carbenes [11–16]. We used the complete active space self-consistent field (CASSCF) method as well as complete active space second-order perturbation theory (CASPT2) and multi-reference configuration interaction (MRCI) approaches to compute the geometries, force constants, and vibrational frequencies of the (singlet) $\tilde{X}$ and $\tilde{A}$ states as well as the (triplet) $\tilde{a}$ states. Our theoretical studies of most of these carbenes were carried out specifically to complement LIF studies that were pursued in our laboratories by Kable et al. [6]. In addition to the determination of spectroscopic constants, the spectroscopic and theoretical studies considered dynamics on the $\tilde{A}$ surfaces, i.e. whether photodissociation or internal conversion to the ground state would occur.

Iodocarbenes represent an additional computational challenge, inasmuch as spin–orbit coupling effects are expected to be substantially more important than even in bromocarbenes, given that the spin–orbit splitting in the iodine atom is 21.7 kcal mol^{-1}, roughly twice what it is in Br (10.5 kcal mol^{-1}) [17]. While in molecules there is substantial orbital angular momentum quenching, as found in our previous work [18] on CBr, CI, CHBr and CHI too, the large spin–orbit stabilization that occurs, e.g. on C–I bond breaking may actually result in a barrier to dissociation that could lie below the energy of the RT intersection. Thus, spin–orbit coupling may not only affect the energy of the barrier, but the qualitative photochemical behaviour of an iodocarbene.

There is very little experimental information on iodocarbenes, thus high quality quantum chemical studies are all the more important. The only spectroscopic data on iodocarbenes come from the negative ion photoelectron work of the Lineberger group [19, 20] and more recently the fluorescence excitation and emission spectroscopic studies of CHI and CDI by Tao et al. [21, 22]. In particular, a value of -1 ± 3 kcal mol^{-1} was deduced for the singlet–triplet gap in CI_2 from the photoelectron experiments [20], the negative value implying a triplet ground state. Subsequent quantum chemical calculations by Hajgató et al. [23] and by Hargittai et al. [24, 25], however, contradicted this, predicting that in CI_2 the ground state is singlet, lying by $\sim$9 kcal mol^{-1} below the triplet, in qualitative agreement with earlier calculations [26–28]. CHI also has a singlet ground state. The most recent computed value of the singlet–triplet splitting is 3.70 kcal mol^{-1} [18], in good agreement with the experimental lower bounds of 3.76 and 4.1 kcal mol^{-1}, as determined by Tao et al. [21, 22]. Similar values were obtained by Hajgató et al. [23] as well as in several earlier computations [28–31].

This paper reports the results of quantum chemical computations on the $\tilde{X}(^1A_1)$, $\tilde{a}(^3B_1)$, $\tilde{A}(^1B_1)$ and $\tilde{B}(2^1A_1)$ states of CI_2, including characterization of the barriers to dissociation and to linearity and its atomization energy that in turn enables the prediction of its enthalpy of formation. Several additional excited states ($^3A_2$, $^1A_2$, $^3B_2$, $^1B_2$, $^3\Pi_u$ and $^1\Pi_u$) that lie in the same region of energy as the $\tilde{A}$ and $\tilde{B}$ states are also characterized, although at a lower level of theory than the latter. As in previous work of ours [16, 18] and of other workers [32–34], the strategy is to obtain accurate valence-correlated energies in a hypothetical complete basis set (CBS) limit, which are further corrected for core and core-valence correlation, scalar relativistic, spin–orbit coupling, zero-point vibration and thermal effects, as described below. This approach, when applied to the heats of formation of CHBr and CBr [18], was found to be of comparable in accuracy to the earlier work of Dixon et al. [32] and of Oren et al. [33].

2 Theory and computational methods

The highest correlated level of theory employed for the $\tilde{X}(^1A_1)$ and $\tilde{a}(^3B_1)$ states is the singles and doubles coupled cluster theory with perturbational treatment of triples, viz.

(R/U)CCSD(T) [35–37], where (R/U) indicates that for the triplet state, a restricted open-shell reference state is used, but the coupled cluster equations are solved using the unrestricted formalism [37–39]. For the $\tilde{A}(^1B_1)$, $\tilde{B}(2\,^1A_1)$ and higher states the (singles plus doubles) MRCI method [40, 41] was used with Davidson's correction [42] for quadruple excitations, denoted MRCI + Dav. The correlation consistent basis sets of Dunning and co workers [43–47], ranging from cc–pVTZ to aug–cc–pV5Z and cc–pwCVQZ, have been employed in these calculations. For iodine, the ECP28MDF small-core relativistic pseudopotential (PP) was used (which accounts for the $1s^2 \ldots 3d^{10}$ core), along with the aug–cc–pVxZ-PP ($x = T$, Q, 5) and cc–pwCVQZ–PP basis sets [47].

The geometries and frequencies of the singlet ground and triplet $\tilde{a}(^3B_1)$ states were computed using the CCSD method, while for all other states, including transition states, CASPT2 [48–50] (with a full-valence CASSCF reference) was employed, all using the cc–pVTZ basis sets, along with the ECP28MDF PP for I with the appropriate cc–pVTZ–PP basis [51]. No scaling was applied to the computed frequencies.

The $\tilde{a}(^3B_1) - \tilde{X}(^1A_1)$ adiabatic energy gap was calculated at the valence correlated (R/U)CCSD(T) as well the MRCI + Dav levels of theory, using the aug–cc–pVxZ basis sets, where $x = T, Q, 5$. All other excitation energies, including that of the $\tilde{A}(^1B_1) \leftarrow \tilde{X}(^1A_1)$ transition, were evaluated at the MRCI + Dav level of theory. The internally contracted MRCI calculations are based on full-valence (18 active electrons in 12 active orbitals) CASSCF reference states and include all single and double excitations from the valence orbitals. The total valence-correlated energies of the various molecular states and of their constituent atoms in their ground states were extrapolated to the complete basis set (CBS, $x = \infty$) limit, so CBS limits of the adiabatic excitation and atomization energies could be deduced. The x^{-3} extrapolation scheme [52] was used, i.e. application of the equation

$$E(x) = A + Bx^{-3} \quad (1)$$

where the constants A [$=E(\infty)$] and B are determined by fitting the last two energies in a given sequence, e.g. $x = 4$, 5. The full CBS energies contain further corrections for core–core and core-valence (CV) correlation, scalar relativistic effects and spin–orbit coupling.

The CV correlation contributions were computed at the MRCI + Dav level of theory using the cc–pwCVQZ basis sets, where the CASSCF reference states were obtained by employing an active space of six electrons in six active orbitals. Unfortunately, MRCI calculations with a full-valence CASSCF reference state with all electrons correlated proved to be too demanding computationally, hence the truncation of the active space. CV correlation corrections for the $\tilde{X}$ and $\tilde{a}$ states were computed also at the (R/U)CCSD(T) level of theory.

The scalar relativistic (SR) corrections were calculated by the second-order Douglas–Kroll–Hess (DKH2) method [53–57] at the (U/R)CCSD(T) or MRCI level of theory in conjunction with the all-electron aug–cc–pVQZ–DK2 basis sets that had been recently developed for iodine [58]. The SR contributions, as computed here, account for the scalar relativistic effects on carbon as well as corrections for the PP approximation for iodine. Note, however, that the Stuttgart–Köln PPs that are used in this work include Breit corrections that are absent in the Douglas–Kroll–Hess approach [58].

Spin–orbit (SO) interaction energies were calculated by diagonalization of the SO Hamiltonian (containing both electronic and SO coupling contributions) in a basis of selected CASSCF or MRCI configurations [59]. In this work, the SO matrix elements were computed using the ECP28MDF PP for iodine and via the Breit–Pauli operator for carbon. The SO energies will of course depend on the number and type of basis states (configurations) that are used to construct the SO matrix.

By way of illustration, let us consider iodomethylidyne (CI) that has a $^2\Pi$ ground state. At the minimal level, the SO splitting is described by considering the interaction between the degenerate set of $^2\Pi(M_S = \pm 1/2)$ configurations. At the internuclear distance of 3.86 a_0 (that is approximately the equilibrium bond length), the doubly degenerate eigenvalues are $\pm$409 cm^{-1} relative to the unperturbed ground state MRCI energy, i.e. corresponding to a ΔE_{LS} of -409 cm^{-1} and a SO splitting of 818 cm^{-1}. Enlarging the set of basis states so that it includes all doublet states that can be constructed from the ^3P states of C and ^2P states of I, i.e. a total of nine distinct configurations, each with a spin multiplicity of 2, results in a SO energy lowering of -460 cm^{-1} and a splitting of 857 cm^{-1}. Adding the nine quartet states to this set (i.e. a total of 18 states) yields a SO lowering of -600 cm^{-1} and a splitting of 808 cm^{-1}. However, as the data in Fig. 1 shows, ΔE_{LS} now has the correct asymptotic behaviour, i.e. at large R, the SO energy shift is that of an I atom: $\sim -2{,}300$ cm^{-1}. The inclusion of the extra states that arise from the interaction of the ^1S and ^1D states of C with the ^2P state of I, i.e. a total of 36 states, results in only a relatively small change to the energy which is now 659 cm^{-1} below the unperturbed energy. The corresponding SO splitting is 841 cm^{-1}. Except at short distances, the MRCI-18 and MRCI-36 values are essentially the same, as shown in Fig. 1. Note that the CASSCF results agree closely with the MRCI values. At the equilibrium geometry in particular, the SO energy

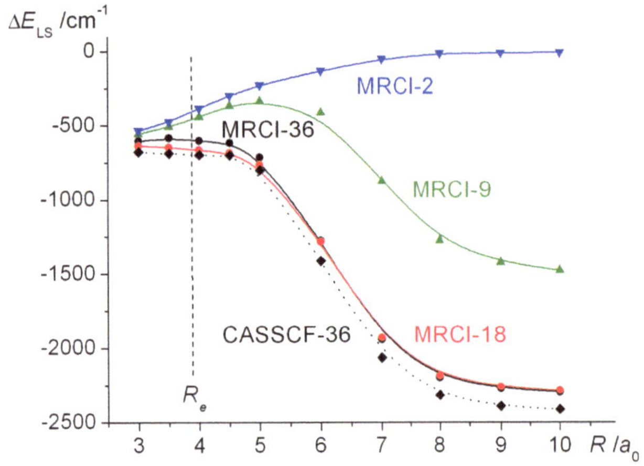

Fig. 1 Distance dependence of SO correction in iodomethylidyne (CI) computed at MRCI/cc–pVTZ level of theory. The SO Hamiltonians were constructed in bases of 2, 9, 18 and 36 electronic states, as indicated and as discussed in text. One set of results obtained at the CASSCF level is shown for comparison (*broken line*)

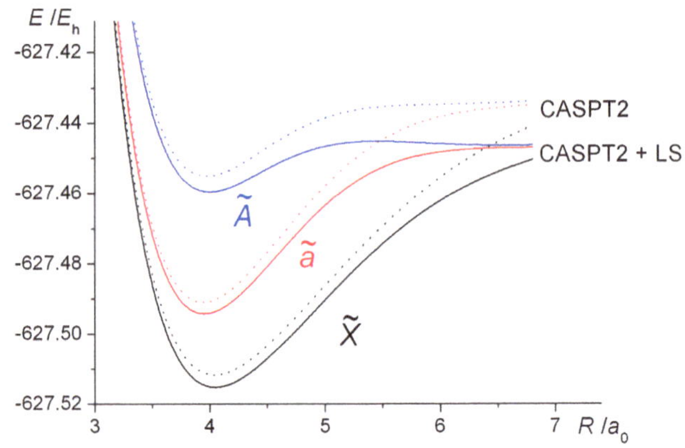

Fig. 2 CI_2: C–I distance dependence of total energies of lowest electronic states with SO coupling (*full lines*) and without it (*dotted lines*) computed at CASPT2/cc–pVTZ level of theory. The SO energies were computed using CASSCF. The other C–I distance and bond angle are kept fixed at their ground state equilibrium values, as given in Table 1

predicted by CASSCF-36 is $-698\ \mathrm{cm}^{-1}$, just $29\ \mathrm{cm}^{-1}$ lower than the MRCI-36 value. It is worth noting that the SO splitting is very stable, the maximum variation is just $23\ \mathrm{cm}^{-1}$. The analogous treatment of iodine oxide IO at its equilibrium geometry has resulted in a SO splitting of $1{,}808\ \mathrm{cm}^{-1}$, in good agreement with the experimental value of $2{,}091\ \mathrm{cm}^{-1}$ [60], as well as the computed values of $1{,}683\ \mathrm{cm}^{-1}$ [61] and $1{,}827\ \mathrm{cm}^{-1}$ [47].

In the case of CI_2, the aim was to include all singlet, triplet and quintet states that arise from the interaction of the ^3P states of C and ^2P states of the I atoms. In C_{2v} symmetry, this would correspond to 27 singlets, 54 triplets and 27 quintets that would lead to a SO matrix of dimensions $27 + 54 \times 3 + 27 \times 5 = 324$. As MOLPRO currently has an upper limit of 200, the number of states was reduced to include 27 singlets, 31 triplets and 16 quintets by leaving out the high energy triplets and quintets. Test calculations have shown that convergence is reasonably rapid, so neglecting the high energy triplets and quintets is unlikely to degrade the quality of the results. Furthermore, all SO calculations on CI_2 were carried out using CASSCF wave functions, but replacing the CASSCF unperturbed diagonal energies with CASPT2 energies, although the effects of such a procedure on the SO energies are fairly minor, $\sim 10\ \%$.

For a clear and comprehensive discussion of SO coupling in C_{2v} molecules (in particular in methylene and dihydrides of other group IV atoms), the reader is referred to the paper by Matsunaga et al. [62]. As explained there, the SO-coupled components of a $^3B_1$ state transform as A_1, A_2 and B_2, since the three components of the triplet spin functions ($\alpha\alpha + \beta\beta$, $\alpha\alpha - \beta\beta$ and $\alpha\beta + \beta\alpha$) in a C_{2v} molecule transform as B_1, B_2 and A_2, respectively. The SO-coupled states are thus denoted $^3B_1(A_1)$, $^3B_1(A_2)$, $^3B_1(B_2)$, where the symbols in parentheses denote the total symmetry. Thus, SO coupling gives rise to mixing of the $^3B_1(A_1)$ and $^1A_1(A_1)$ states and an avoided crossing, as demonstrated in the next section. (The singlet $\alpha\beta - \beta\alpha$ spin function transforms as A_1.)

The heat of formation of CI_2 in its ground state at 0 and 298 K was determined from computed enthalpies of atomization and experimental heats of formation of the atoms [17]. The atomization energies and enthalpies at the specified temperatures were obtained from the full (R/U)CCSD(T)/CBS energies of CI_2 and its constituent atoms, which include CV, SR, SO, zero-point and thermal corrections.

All computations were performed using the MOLPRO 2009 [63] and MOLPRO 2012 [64] codes on the supercomputer facility of the National Computational Infrastructure (NCI) National Facility at the Australian National University, Canberra.

3 Results and discussion

3.1 The $\tilde{X}(^1A_1)$, $\tilde{a}(^3B_1)$ and $\tilde{A}(^1B_1)$ states of CI_2

Qualitative potential energy surfaces representing the dissociation

$$CI_2(^1A_1, ^3B_1, ^1B_1) \rightarrow CI(^2\Pi) + I(^2P) \qquad (2)$$

are shown in Fig. 2. These were computed at the CASPT2/cc-pVTZ level of theory, with and without SO coupling (with one C–I distance and the bond angle frozen at the ground state equilibrium values, as given in Table 1). The importance of SO coupling is obvious, especially close to

Table 1 Computed geometries, harmonic vibrational frequencies, excitation and dissociation energies of CI_2 (based on CBS energies from Table 2)

	$\tilde{X}(^1A_1)$ Eq^m	$\tilde{a}(^3B_1)$ Eq^m	$\tilde{A}(^1B_1)$ Eq^m	$\tilde{A}(^1\Delta_g)$ Linear	$\tilde{A}(^1B_1)$ Barrier to dissociation	$\tilde{B}(2\,^1A_1)$ Linear
$R_{C\text{–}I}$/Å	2.104	2.040	2.032	1.958	2.050, 2.871	1.959
θ/°	112.1	132.6	133.5	180.0	119.2	180.0
ω_1/cm^{-1}	512.9	427.6	367.5	221.7	97.9i	207.4
ω_2/cm^{-1}	135.6	128.6	129.8	286.4i	69.1	32.1i
ω_3/cm^{-1}	617.6	814.9	770.2	1,030.6	598.5	1,033.0
T_{00}/cm^{-1}		3,882	12,680	15,578	17,836	20,991
D_0/kcal mol^{-1}	49.7	38.6	13.5	5.2	−1.3	−10.3

Geometries and frequencies obtained at CCSD/cc–pVTZ ($\tilde{X}$ and $\tilde{a}$ states) and CASPT2/cc–pVTZ (all other systems) levels of theory

Dissociation to CI + I

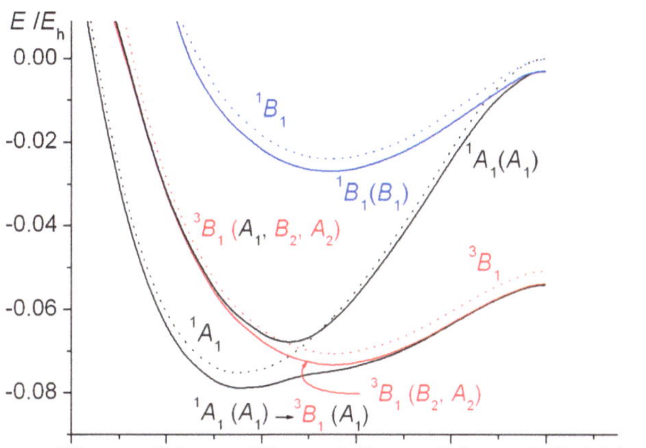

Fig. 3 CI_2: bond angle dependence of total energies of lowest electronic states with SO coupling (*full lines*) and without it (*dotted lines*) computed at CASSCF/cc–pVTZ level of theory. The C–I distances are kept fixed at their triplet state equilibrium values

dissociation, as noted indeed for CI (see Fig. 1). Clearly, as foreshadowed in the Sect. 1, SO coupling is responsible for the presence of a (small) barrier to recombination on the $\tilde{A}(^1B_1)$ surface. In the equilibrium regions, the distance dependence of the SO energy in both CI_2 and CI is quite weak, so neglecting its effects on geometries and frequencies is justified.

The bond angle dependence of the energies in CI_2 is illustrated in Fig. 3, computed at the CASSCF/cc–pVTZ level for the lowest three electronic states (with the C–I distances frozen at the triplet equilibrium values). SO effects play a critical role in the singlet–triplet crossing region. The lowest energy $^1A_1(A_1)$ surface, obviously corresponding to the singlet ground state at angles smaller than about 120°, morphs into the $^3B_1(A_1)$ component of the triplet as the bond angle increases, with no minimum in the triplet region. The doubly degenerate $^3B_1(B_2, A_2)$ surface has a minimum (essentially at the same geometry as the triplet surface with no SO coupling) and this is then taken as the equilibrium structure and energy of the triplet state. The higher energy surface with A_1 total symmetry corresponds to a triplet to singlet morphing, which has a well-defined minimum at ~127°. This state has not been fully characterized in this work.

The computed geometries, harmonic vibrational frequencies of the $\tilde{X}(^1A_1)$, $\tilde{a}(^3B_1)$ and $\tilde{A}(^1B_1)$ states of CI_2 as well as of the two saddle points of interest on the $\tilde{A}$ surface are summarized in Table 1, where the $\tilde{a}$ state is taken to be the $^3B_1(B_2, A_2)$ state. A breakdown of the individual energetic effects of the CBS extrapolation process, CV correlation, scalar relativity, SO coupling and zero-point energy are given in Table 2. The computed SO energies for CI_2 (along with CI and CHI) are listed in Table 3. The energies of all stationary points on the potential energy surfaces that are studied in this work are summarized in Fig. 4.

As for carbenes in general, the most noticeable differences in the geometries of CI_2 are the shorter bond lengths and larger bond angles in the excited states than in the ground state. These differences are reflected in the larger stretching and lower bending force constants and frequencies in the excited states. The computed geometries and frequencies of the ground and lowest triplet states are consistent with the theoretical literature values [23–25].

3.2 Singlet–triplet splitting in CI_2 and CHI

The computed singlet–triplet splitting of 11.1 kcal mol^{-1} in CI_2 is higher by ~2.5 kcal mol^{-1} than what had been calculated before [23–25]. This is partly due to the more thorough treatment of basis set enlargement and CV cor-

Table 2 Computed contributions to atomization and adiabatic excitation energies, heats of formation and zero-point-corrected excitation energies (in kcal mol^{-1}) in CI_2

	Atomization from $\tilde{X}$ state (R/U)CCSD(T)	$^3B_1 \leftarrow {}^1A_1$ (R/U)CCSD(T)	$^1B_1 \leftarrow {}^1A_1$ MRCI + Dav	$^1\Delta_g - {}^1B_1$ linear MRCI + Dav	$^1A'' - {}^1B_1$ barrier MRCI + Dav	$2^1A - {}^1A_1$ MRCI + Dav
aug–cc–pVTZ	118.52	9.91	33.61	11.07	16.51	56.12
aug–cc–pVQZ	121.57	9.98	33.36	10.41	17.96	55.18
aug–cc–V5Z	122.68	10.06	33.32	10.22	18.41	54.93
Extrap (aVT, aVQ)	123.80	10.02	33.18	9.93	19.01	54.49
Extrap (aVQ,aV5)	123.85	10.15	33.27	10.02	18.89	54.67
CV correlation	2.15	0.14	2.51	−2.07	0.38	4.16
Scalar relativity	0.05	0.07	0.09	0.03	−0.09	0.13
Spin–orbit coupl.[a]	−12.12	0.59	0.39	0.32	−3.59	1.28
Zero point energy	−1.81	0.15	0.00	−0.02	−0.86	−0.22
Total (CBS)	112.12	11.10	36.26	8.28	14.74	60.02
$\Delta_f H_0^0$	109.08 ± 1.0					
$\Delta_f H_{298}^0$	109.27 ± 1.0					

[a] Using experimental data [17] for C and I, and computed values for molecules (Table 3)

Table 3 Computed spin–orbit energies (cm^{-1})

	ΔE_{LS}		ΔE_{LS}
C ^3P	−24.7, −32.8[a]	I ^2P	−2,299.9, −2,534.4[a]
CHI $^1A'$ (A')	−467.8	CI_2 $^1A_1$ (A_1)	−863.3
CHI $^3A''(A')$	−358.1	CI_2 $^3B_1$ (A_1)	−1,151.4
CHI $^3A''$ (A', A'')	−264.8	CI_2 $^3B_1$ (B_2, A_2)	−658.4
CHI $^1A''$ (A'')	−306.7	CI_2 $^1B_1$ (B_1)	−727.7
CHI $^1\Delta$ linear	−324.1	CI_2 $^1\Delta_g$ linear	−615.8
CHI $^1A''$ barrier	−292.1	CI_2 $^1A''$ barrier	−1,981.8
CI $^2\Pi$	−658.8	CI_2 2^1A_1	−416.9

CI_2 calculations as described in text

CHI data computed at CASSCF/cc–pVTZ level of theory with CASPT2 unperturbed energies using all 36 states (singlets, triplets and quintets) arising from C ^3P, I ^2P and H ^2S atomic states

CI data computed at MRCI/cc–pVTZ level of theory using all 36 states (doublets and quartets) arising from C ^3P, ^1D, ^1S and I ^2P atomic states

C and I data computed using C ^3P, ^1D, ^1S and I ^2P atomic states, respectively

The symmetry labels in parentheses for the electronic states of CI_2 and CI represent the overall symmetries of the SO-coupled states

[a] Experiment, Ref. [17]

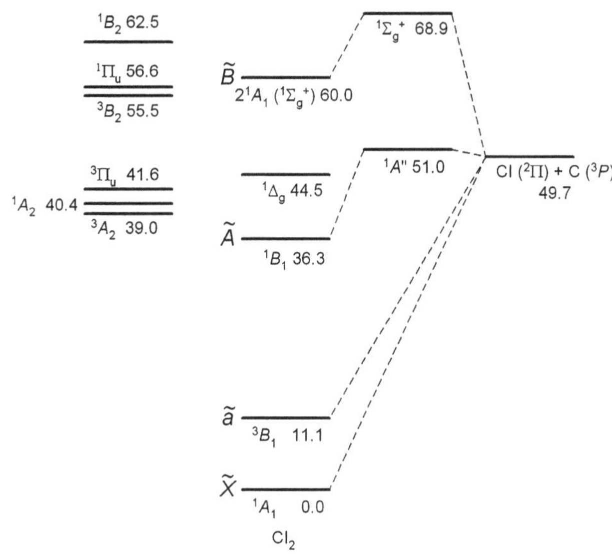

Fig. 4 Single point energies (in kcal mol^{-1}) of important stationary points on the CI_2 potential energy surfaces. The relative energies on the $^1A_1$, $^3B_1$, $^1B_1$and 2^1A_1 surfaces are CBS values as summarized in Tables 1, 2 and 4 (and include zero-point correction). The equilibrium energies of the $^{1,3}A_2$, $^{1,3}B_2$ and $^{1,3}\Pi_u$ states (on *left* of the diagram) are from Table 4, computed at the MRCI + Dav/cc–pVTZ level of theory and do not contain zero-point correction

relation in this work, that collectively result in a singlet–triplet gap of 10.5 kcal mol^{-1}, that is 1.3 kcal mol^{-1} higher than the CCSD(T) value of Hargittai et al. [24] before SO correction. Now, depending on which value we choose as the appropriate SO correction in the case of the triplet state, the singlet–triplet splitting will increase by ~0.6 kcal mol^{-1} [$^3B_1(B_2, A_2)$] or decrease by ~0.8 kcal mol^{-1} [$^3B_1(A_1)$] according to the data in Table 3. However, as noted above, a stable structure exists on the $^3B_1(B_2, A_2)$ surface, but not on the lowest $^3B_1(A_1)$ surface, except the one identified as the $\tilde{X}[^1A_1(A_1)]$ state.

In the author's previous work [18] on CHI and CI, the SO energy was computed using only the $\tilde{X}$, $\tilde{a}$ and $\tilde{A}$ states of CHI and the two ground state doublets for CI. As discussed in Sect. 2, such a procedure for CI results in an error of ~250 cm^{-1}. Thus, in addition to CI and CI_2, the SO energies of CHI have been also recomputed and are listed in Table 3. As for CI_2, in the case of triplet CHI, the lowest local minimum is on the $^3A''(A', A'')$ surface; thus, the appropriate SO correction is −264.8 cm^{-1}, and therefore, the SO component of the singlet–triplet gap is 467.8 − 264.8 = 202.9 cm^{-1}, rather than −13.5 cm^{-1} that was previously calculated [18]. Also, with the recent publication of the aug–cc–pVQZ–DK2 basis set for iodine [58], it has become possible to calculate the DKH2 SR correction that was previously neglected [18]. The resulting SR correction is 22.7 cm^{-1}. With these corrections, the singlet–triplet splitting becomes 4.4 kcal mol^{-1}, which agrees well with the experimental lower bound of 4.1 kcal mol^{-1} [22].

Until recently iodocarbenes such as CHI and CI_2 were believed to have triplet ground states, largely on the basis of negative ion photoelectron spectroscopic studies [19, 20, 65]. Recent theoretical studies [18, 20, 23–25], including the current one, have clearly contradicted this, as discussed above. Moreover, the spectroscopic work of Tao et al. [21, 22] on CHI has confirmed that its ground state is indeed singlet.

3.3 The $\tilde{A}(^1B_1) \leftarrow \tilde{X}(^1A_1)$ excitation in CI_2 and CHI

The $\tilde{A}(^1B_1) \leftarrow \tilde{X}(^1A_1)$ excitation energy (T_{00}) in CI_2 is predicted to be 12,680 cm^{-1}. It is, as expected, between that of CHI (10,500 cm^{-1} [22]) and CBr_2 (15,093 cm^{-1} [66]), very close in fact to CHBr (11,972 cm^{-1} [67]). The previous theoretical study on CHI and CHBr (using the same methods as in this work) predicted excitation energies that are 352 and 371 cm^{-1}, respectively, higher than experiment. (In the case of CHI, the computed excitation energy needs to be increased by 64 + 18 cm^{-1}, because of the revised SO and SR corrections, respectively.) Consequently, we can confidently expect that the current prediction for CI_2 will not be in error by more than ~430 cm^{-1}, i.e. ~1.2 kcal mol^{-1}.

The barrier to linearity on the $\tilde{A}(^1B_1)$ surface (including zero point corrections) has been computed to be 2,898 cm^{-1} that is 2,258 cm^{-1} below the barrier to dissociation. Therefore, we may confidently expect CI_2 to undergo RT relaxation, as predicted also for CHI [18]. (The SO corrections as computed here for CHI do not change that conclusion.)

3.4 The $\tilde{B}(2\,^1A_1)$ state of CI_2

The $\tilde{B}(2\,^1A_1)$ state of CI_2, as in CHF [7, 8] and CHCl [9], has a quasi-linear equilibrium geometry, whereby the vibrationally averaged structure in the double minimum bending potential of the molecule is linear. The computed

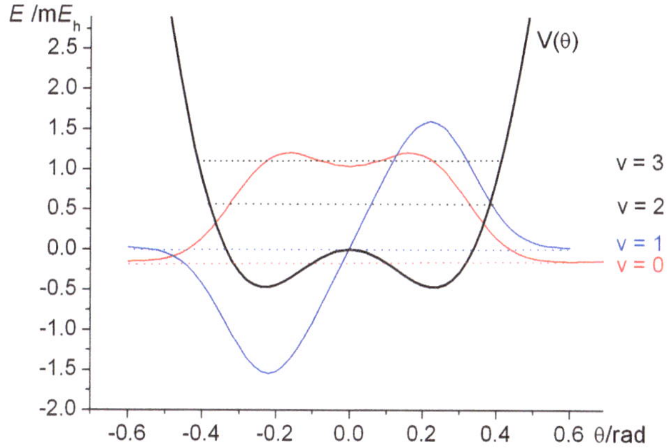

Fig. 5 Bending potential of CI_2 in $\tilde{B}(2^1A_1)$ state, showing computed anharmonic energy levels and wave functions of lowest two states. (The amplitudes of the wave functions are indicated by the unscaled y axis units, i.e. −2.0 to 2.5 au.)

CASPT2/cc–pVTZ potential generated with bond lengths fixed at the linear geometry values, along with anharmonic energy levels and wave functions (computed by the generalized finite element method [68]) are shown in Fig. 5. The zero-point energy level lies just below the central barrier height, but the corresponding wave function is fully delocalized with only a small dip over the barrier. This metastable state dissociates to CI + I via a linear symmetric transition state that lies ~9 kcal mol^{-1} above the energy of the $\tilde{B}$ state. The imaginary frequency of this transition state corresponds to the symmetric stretch. The top of the barrier is a bifurcation point as there are two equivalent pathways for bond breaking to occur. In CHF and CHCl, an alternative dissociation pathway corresponds to the formation of HF or HCl plus a C atom [7–9]. The possibility of such a process in CI_2 was not investigated in this work.

3.5 The $^{1,3}A_2$, $^{1,3}B_2$ and $^{1,3}\Pi_u$ states of CI_2

The work of Hargittai et al. [24] suggested that in addition to the $\tilde{a}$ state, there may be several low-lying triplet states, $^3A_2$, $^3B_2$ and $^3\Pi_u$, although no energies were given. A study of these states, along with the corresponding singlets, was therefore carried out. The results, obtained at the MRCI + Dav/cc–pVTZ level of theory, are summarized in Table 4 and in Fig. 4. The six states appear to lie in the energy range of 39–63 kcal mol^{-1} above the ground state, i.e. energetically overlapping with the $\tilde{A}$ and $\tilde{B}$ surfaces. (Since the $^1A_2$ state actually lies below the 2^1A_1 state, the label $\tilde{B}$ should really be attached to the former.) The $^{1,3}A_2$ and $^{1,3}B_2$ states are dominated by $\ldots a_1^2b_1^1b_2^1a_2^2$ and $\ldots a_1^2b_1^1b_2^2a_2^1$ type configurations where a_1, b_1, b_2 and a_2 denote the highest occupied orbitals of the specified symmetry. Relative to the $^{1,3}B_1$ ($\ldots a_1^1b_1^1b_2^2a_2^2$) states the above configurations thus represent $b_2 \rightarrow a_1$ and $a_2 \rightarrow a_1$ excitations of nominally iodine

Table 4 Computed geometries (CASPT2/cc-pVTZ) and energies of the $^{1,3}A_2$, $^{1,3}B_2$, $^{1,3}\Pi_u$ equilibrium and $^1A_1$ transition state of CI_2

	R_{C-I}/Å	θ/°	ΔE/kcal mol^{-1}		R_{C-I}/Å	θ/°	ΔE/kcal mol^{-1}
$^1A_2$	2.337	86.0	40.4	$^3A_2$	2.326	86.1	39.0
$^1B_2$	2.392	101.0	62.5	$^3B_2$	2.269	101.6	55.5
$^1\Pi_u$	2.214	180.0	56.6	$^3\Pi_u$	2.166	180.0	41.6
2^1A_1 (TS)	2.076	180.0	68.9[a]				

Relative electronic energies at MRCI + Dav/cc–pVTZ level of theory (without zero-point corrections) unless indicated otherwise

[a] CBS level of theory with all corrections as in Tables 1 and 2

lone pair electrons. The $^{1,3}\Pi_u$ states are different, inasmuch as they correspond to the excitation of a carbon π electron (a_1 and b_1 in C_{2v} symmetry) to a low-lying antibonding σ_g molecular orbital. It remains to be seen if any of these 'new' states could be observed spectroscopically or in photochemical experiments.

3.6 The enthalpies of formation of CI_2, CHI and CI

The computed atomization energy of CI_2, as given in Table 2, is 112.12 kcal mol^{-1}, from which the heats of formation at 0 and 298 K were calculated to be 109.1 and 109.3 kcal mol^{-1}, respectively, with an estimated error of $\pm$1 kcal mol^{-1}. Spin–orbit coupling contributes -12.1 to the atomization energy, which includes a molecular contribution of 2.5 kcal mol^{-1}, a definitely non-negligible quantity. The above enthalpy of formation at 298 K is consistent with the experimental value of 112.0 $\pm$ 14.3 kcal mol^{-1} [69] as well as with the G2 value of 107.6 kcal mol^{-1} [31].

As discussed in Sect. 2, computing the SO energy of CI using only the ground state doublets for CI results in an error of ~250 cm^{-1}. With the best value computed in this work (-659 cm^{-1}) to replace that used previously (-373 cm^{-1}), the atomization energy that had been computed [18] needs to be corrected by 0.82 kcal mol^{-1}. With this revision, plus the SR correction of 0.05 kcal mol^{-1}, also computed in this work, the enthalpies of formation of CI become 132.9 $\pm$ 1 and 133.6 $\pm$ 1 kcal mol^{-1} at 0 and 298 K, respectively.

In the case of CHI, analogous corrections need to be applied, due to the previous underestimation of the SO energy [18] and the SR correction (which in this case is only 0.01 kcal mol^{-1}). The revised enthalpies of formation of CHI are 102.8 kcal mol^{-1} at both 0 and 298 K.

4 Conclusion

The thermochemistry of CI_2 and its spectral properties, as for a range of other halocarbenes studied already, have been characterized by quantum chemical methods using a CBS approach in conjunction with the CCSD(T) and MRCI + Dav methods. Particular emphasis was placed on the accurate prediction of SO corrections which are especially important in iodine-containing molecules. The predicted enthalpies of formation of CI_2 in its singlet ground state at 0 and 298 K are 109.1 $\pm$ 1 and 109.3 $\pm$ 1 kcal mol^{-1}, respectively. The singlet–triplet splitting is predicted to 11.1 kcal mol^{-1}, which is substantially larger than what had been computed for CHI (4.4 kcal mol^{-1}). The $\tilde{A}(^1B_1) \leftarrow \tilde{X}(^1A_1)$ excitation energy (T_{00}) in CI_2 is predicted to be 12,680 cm^{-1}. In the $\tilde{A}$ state, there is a small barrier to dissociation, estimated to be 1.3 kcal mol^{-1} above the dissociation products CI + I, whose existence is entirely due to SO effects. The RT intersection is computed to be 5.2 kcal mol^{-1} below the dissociation products; thus, internal $\tilde{A} \rightarrow \tilde{X}$ conversion via RT coupling is expected to be the dominant photochemical channel.

Acknowledgments I wish to express my thanks to the National Computational Infrastructure (NCI) National Facility and Intersect Australia Ltd for their generous allocation of computer time.

References

1. Shavitt I (1985) Tetrahedron 41:1531–1542
2. Wentrup C (1962) Reactive molecules. Wiley, New York
3. Liebman JF, Simons J (1986) Molecular structure and energetics, vol 1. VCH, Deerfield Beach
4. Finlayson-Pitts BJ, Pitts JN (1986) Atmospheric chemistry: fundamentals and experimental techniques. Wiley, New York
5. Moss RA, Platz MS, Jones M (eds) (2004) Reactive intermediate chemistry. Wiley-Interscience, Hoboken
6. Kable SH, Reid SA, Sears TJ (2009) Int Rev Phys Chem 28:435–480
7. Richmond C, Tao C, Mukarakate C, Dawes R, Brown EC, Kable SH, Reid SA (2011) J Chem Phys 135:104316
8. Tao C, Richmond C, Mukarakate C, Dawes R, Kable SH, Reid SA (2011) J Chem Phys 135:104315
9. Tao C, Richmond C, Mukarakate C, Kable SH, Bacskay GB, Brown EC, Dawes R, Lolur P, Reid SA (2012) J Chem Phys 137:104307
10. Gu J-P, Hirsch G, Buenker RJ, Brumm M, Osmann G, Bunker PR, Jensen PA (2000) J Mol Struct 517–518:247–264
11. Cameron MR, Kable SH, Bacskay GB (1995) J Chem Phys 103:4476–4483

12. Knepp PT, Scalley CK, Bacskay GB, Kable SH (1998) J Chem Phys 109:2220–2232
13. Schmidt TW, Bacskay GB, Kable SH (1998) Chem Phys Lett 292:80–86
14. Sendt K, Bacskay GB (2000) J Chem Phys 112:2227–2238
15. Sendt K, Schmidt TW, Bacskay GB (2000) Int J Quantum Chem 76:297–305
16. Guss JS, Bacskay GB, Kable SH (2005) Chem Phys Lett 405:258–264
17. Chase MW (ed) (1998) In: Journal of Physical Chemistry Reference Data Monograph 9. NIST-JANAF thermochemical tables 4th Ed. American Chemical Society, Washington
18. Bacskay GB (2010) J Phys Chem A 114:8625–8630
19. Gilles MK, Ervin KM, Ho J, Lineberger WC (1992) J Phys Chem 96:1130–1141
20. Schwartz RL, Davico GE, Ramond TM, Lineberger WC (1999) J Phys Chem A 103:8213–8221
21. Tao C, Ebben C, Ko H-T, Reid SA (2008) Phys Chem Chem Phys 10:6090–6092
22. Tao C, Ebben C, Reid SA (2009) J Phys Chem A 113:13407–13412
23. Hajgató B, Nguyen HMT, Veszprémi T, Nguyen MT (2000) Chem Phys Phys Chem 2:5041–5045
24. Hargittai M, Schultz G, Schwerdtfeger P, Seth M (2001) Struct Chem 12:377–391
25. Szabados A, Hargittai M (2003) J Phys Chem A 107:4314–4321
26. Russo N, Sicilia E, Toscano M (1992) J Chem Phys 97:5031–5036
27. Gobbi A, Frenking G (1993) J Chem Soc Chem Commun, issue no:14, 1162–1164
28. Garcia VM, Castell O, Reguero M, Caballo R (1996) Mol Phys 87:1395–1404
29. Russo N, Sicilia E, Toscano M (1993) Chem Phys Lett 213:245–249
30. Worthington SE, Cramer CJ (1997) J Phys Org Chem 10:755–767
31. Schwartz M, Marshall P (1999) J Phys Chem A 103:7900–7906
32. Dixon DA, de Jong WA, Peterson KA, Francisco JS (2002) J Phys Chem A 106:4725–4728
33. Oren M, Iron MA, Burcat A, Martin JML (2004) J Phys Chem A 108:7752–7761
34. Peterson KA, Feller D, Dixon DA (2012) Theor Chem Acc 131:1079
35. Raghavachari K, Trucks GW, Pople JA, Head-Gordon M (1989) Chem Phys Lett 157:479–483
36. Hampel C, Peterson KA, Werner H-J (1992) Chem Phys Lett 190:1–12
37. Knowles PJ, Hampel C, Werner H-J (1993) J Chem Phys 99:5219–5227
38. Rittby M, Bartlett RJ (1988) J Phys Chem 92:3033–3036
39. Deegan MJO, Knowles PJ (1994) Chem Phys Lett 227:321–326
40. Werner H-J, Knowles PJ (1988) J Chem Phys 89:5803–5814
41. Knowles PJ, Werner H-J (1988) Chem Phys Lett 145:514–522
42. Langhoff SR, Davidson ER (1974) Int J Quantum Chem 8:61–72
43. Dunning TH Jr (1989) J Chem Phys 90:1007–1023
44. Wilson AK, Woon DE, Peterson KA, Dunning TH Jr (1999) J Chem Phys 110:7667–7676 and references therein
45. Feller D (1996) The role of databases in support of computational chemistry calculations. J Comp Chem 17:1571–1586
46. Schuchardt KL, Didier BT, Elsethagen T, Sun L, Gurumoorthi V, Chase J, Li J, Windus TL (2007) Basis set exchange: a community database for computational sciences. J Chem Inf Model 47:1045–1052. doi:10.1021/ci600510j
47. Peterson KA, Shepler BC, Figgen D, Stoll H (2006) J Phys Chem A 110:13877–13883
48. Andersson K, Malmqvist P-A, Roos BO (1992) J Chem Phys 96:1218–1226 and references therein
49. Werner H-J (1996) Mol Phys 89:645–661
50. Celani P, Werner H-J (2000) J Chem Phys 112:5546–5557
51. Peterson KA, Figgen D, Goll E, Stoll H, Dolg M (2003) J Chem Phys 119:11113–11123
52. Helgaker T, Klopper W, Koch H, Noga J (1997) J Chem Phys 106:9639–9646
53. Douglas M, Kroll NM (1974) Ann Phys 82:89–155
54. Hess BA (1986) Phys Rev A 33:3742
55. Wolf A, Reiher M, Hess BA (2002) J Chem Phys 117:9215–9226
56. Reiher M, Wolf A (2004) J Chem Phys 121:2037–2047
57. Reiher M, Wolf A (2004) J Chem Phys 121:10945–10956
58. Bross DH, Peterson KA (2014) Theor Chem Acc 133:1434
59. Berning A, Schweizer M, Werner H-J, Knowles PJ, Palmieri P (2000) Mol Phys 98:1823–1833
60. Gilles MK, Polak ML, Lineberger WC (1991) J Chem Phys 95:4723–4724
61. Roszak S, Krauss M, Alekseyev AB, Liebermann HP, Buenker RJ (2000) J Phys Chem A 104:2999–3003
62. Matsunaga N, Koseki S, Gordon MS (1996) J Chem Phys 104:7988–7996
63. Werner HJ, Knowles PJ, Lindh R, Manby FR, Schütz M, Celani P, Korona T, Mitrushenkov A, Rauhut G, Adler TB, Amos RD, Bernhardsson A, Berning A, Cooper DL, Deegan MJO, Dobbyn AJ, Eckert F, Goll E, Hampel C, Hetzer G, Hrenar T, Knizia G, Köppl C, Liu Y, Lloyd AW, Mata RA, May AJ, McNicholas SJ, Meyer W, Mura ME, Nicklass A, Palmieri P, Pflüger K, Pitzer R, Reiher M, Stoll H, Stone AJ, Tarroni R, Thorsteinsson T, Wang M, Wolf A (2009) MOLPRO, version 2009.1, a package of ab initio programs. See http://www.molpro.net
64. MOLPRO, version 2012.1, a package of ab initio programs, Werner H-J, Knowles PJ, Knizia G, Manby FR, Schütz M, and others, see http://www.molpro.net
65. Murray KK, Leopold DG, Miller TM, Lineberger WC (1988) J Chem Phys 89:5442–5453
66. Xu S, Harmony MD (1993) J Phys Chem 97:7465–7470
67. Yu H-G, Gonzalez-Lezana T, Marr AJ, James T, Muckerman JT, Sears TJ (2001) J Chem Phys 115:5433–5444
68. Nordholm S, Bacskay G (1976) Chem Phys Lett 42:253–258
69. Gurvich L, Veyts IV, Alcock CB (eds) (1992) Thermodynamic properties of individual substances, vol 2. Hemisphere, New York

Theor Chem Acc (2014) 133:1497
DOI 10.1007/s00214-014-1497-2

REGULAR ARTICLE

Theoretical studies of the excited states of p-cyanophenylalanine and comparisons with the natural amino acids phenylalanine and tyrosine

Stephen L. Meloni · Spiridoula Matsika

Received: 26 February 2014 / Accepted: 3 May 2014 / Published online: 21 May 2014

Abstract The absorption and fluorescence properties of the non-natural amino acid p-cyanophenylalanine (PheCN) were examined using high-level ab initio methods and were compared to those of natural amino acids phenylalanine and tyrosine. Single-reference and multireference methods were surveyed for their accuracy in predicting the excited state energies and transition dipole moments of the chromophores and the corresponding amino acids. The excitation energies were found to be very similar between the chromophores and the various conformers of the amino acids for all three amino acids studied here, but the transition dipole moments and consequently the radiative lifetimes were very sensitive to conformation. In agreement with experimental data, PheCN is predicted to have increased fluorescence intensity compared to phenylalanine and the amino acid group is partly responsible for this effect.

Keywords Excited states · Fluorescence · Amino acids · Cyanophenylalanine · Tyrosine · Phenylalanine · Radiative lifetime

Dedicated to the memory of Professor Isaiah Shavitt and published as part of the special collection of articles celebrating his many contributions.

Electronic supplementary material The online version of this article (doi:10.1007/s00214-014-1497-2) contains supplementary material, which is available to authorized users.

S. L. Meloni · S. Matsika (✉)
Department of Chemistry, Temple University, Philadelphia, PA 19122, USA
e-mail: smatsika@temple.edu

1 Introduction

Fluorescence spectroscopy is an important tool for studying proteins [23]. It has sensitivity and time-scale features that can make it more attractive than absorbance spectroscopy [40], and it is one of the most common methods used to determine native protein structure, folding dynamics, and other conformational changes [33]. Tryptophan is the most popular amino acid to study with fluorescence because it has the largest fluorescence quantum yield of the natural fluorescent amino acids, and its fluorescence is sensitive to its environment [5]. The use of natural amino acid fluorophores is limited, however, because tryptophan is not very common, and phenylalanine has low fluorescence quantum yield.

Because of this limitation, non-natural amino acids which have useful spectroscopic properties are utilized as well. The use of non-natural amino acids has been shown to be effective enough that several methods of incorporating them into proteins have been developed [25]. Non-natural amino acid analogs have been shown to be useful when incorporated into natural proteins for the study of dynamics, conformation, and binding [48]. The properties for these synthetic amino acids can be tuned more specifically than is possible if only considering natural amino acids. Synthetic amino acids can also participate in Förster resonance energy transfer (FRET) as donors to tryptophan [32]. FRET efficiency is inversely proportional to the sixth power of the distance between donor and acceptor, so this is a sensitive technique for measuring conformational changes in a protein. Nitrile derivatives have also been shown to have potential applications as probes of local environment [13, 15].

p-Cyanophenylalanine (PheCN) is a fluorescent probe shown to be a suitable spectroscopic tool for studying

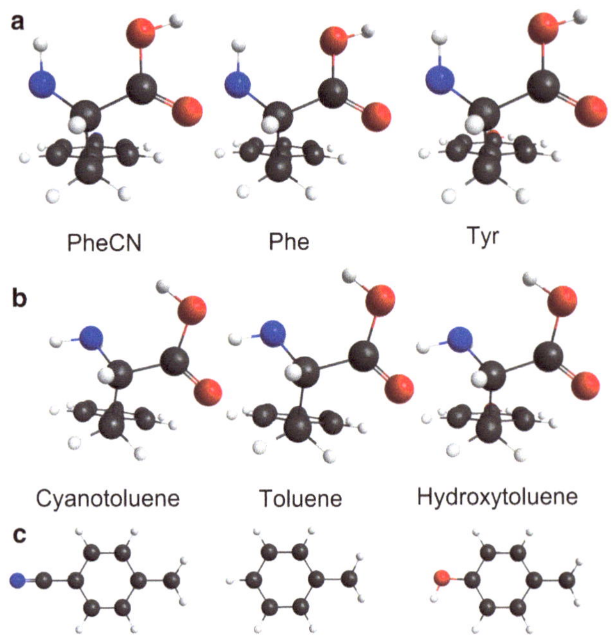

Fig. 1 Amino acids and their respective chromophores. **a** Lowest energy conformers of amino acids at the MP2 level. **b** Lowest energy conformers of amino acids at the AM1 level. **c** Chromophores of each amino acid studied

protein binding and formation [27, 36, 46]. It has three absorption maxima in the ultra violet region at 233, 274, and 280 nm, and has significantly increased molar absorptivity compared to phenylalanine [47]. Along with increased absorptivity, PheCN can be selectively excited at 240 nm in the presence of tryptophan [45]. Other advantages of PheCN are that it has a fluorescence quantum yield five times that of phenylalanine, it does not have a photoinduced proton transfer which can complicate interpretation, and it does not disturb the natural protein structure significantly [36, 47]. It shows an emission peak at 295 nm, which is insensitive to excitation wavelength and environment [35]. Its fluorescence quantum yield has been shown to depend on the hydrogen bonding ability of its environment, and lifetimes have been correlated with the Kamlet-Taft α parameter of its solvent, a measure of the ability of the solvent to be a hydrogen bond donor [35]. However, the radiative lifetimes are not correlated in the same way, which suggests that non-radiative processes are important for understanding the fluorescence behavior in various solvents.

We have used high-level theoretical methods to better understand the excited state properties of PheCN, and compare to tyrosine (Tyr), and phenylalanine (Phe), as well as their corresponding chromophores 4-cyanotoluene, 4-hydroxytoluene, and toluene. The systems studied in this work are seen in Fig. 1. The objective of this work is both to examine how structure affects the photophysical properties and to examine the performance of different levels of theory on the description of these properties. The absorption and emission energies as well as the radiative lifetimes were examined. The sensitivity of the excitation energies and transition dipole moments on the inclusion of the amino acid, the solvent, as well as the level of theory used was also examined.

2 Methods

The geometries for the chromophores benzene, toluene, 4-hydroxytoluene, and 4-cyanotoluene were optimized using second order Møller-Plesset perturbation theory (MP2) and the cc-pVDZ basis set. Amino acid conformations were obtained using a Monte Carlo conformation search and the semi-empirical AM1 method [9]. The obtained geometries at the AM1 level were reoptimized at the MP2/cc-pVDZ level. Geometries of solvated, zwitterionic amino acids were optimized at the MP2/cc-pVDZ level using a polarizable continuum model [8]. Excited state geometries were optimized using second order perturbation theory corrections to configuration interaction singles (CIS), specifically the CIS(2) method [22].

Excited state energies were calculated for the chromophores using a variety of methods. The following single-reference methods were used: CIS [11]; second order perturbation theory corrections to CIS with either the resolution of identity approximation CIS(D) method (RI-CIS(D)) [2] or CIS(2) [22]; the equation of motion coupled cluster singles and doubles (EOM-CCSD) method [19, 44]; and time-dependent density functional theory (TDDFT) with the B3LYP functional. In order to explore the effect of non-dynamical correlation, multiconfigurational methods were explored as well. In particular, the quasidegenerate multiconfigurational second order perturbation theory (MCQDPT) method [30] was used. Orbitals were obtained using a two-state averaged complete active space self-consistent field (CASSCF) approach with an active space of (6,6) for toluene and hydroxytoluene and (8,8) for 4-cyanotoluene. Basis set effects were explored using a variety of correlation consistent basis sets (cc-pVDZ, cc-pVTZ, aug-cc-pVDZ, aug-cc-pVTZ). cc-pVDZ+diff indicates the modified aug-cc-pVDZ basis set where diffuse functions are only added to the heavy atoms and not hydrogen.

The "surface and simulation of volume polarization for electrostatics" [SS(V)PE] solvation model [6] was used to calculate the excited states of solvated zwitterionic amino acids in combination with the RI-CIS(D) method.

Excited state radiative lifetimes τ_{01} for fluorescence were calculated using the relationship $\tau_{01} = 1/A_{01} = \frac{\epsilon_0 c^3 \omega_{01}}{2\pi^2 m_e^2 f_{01} n^3}$, where A_{01} is the Einstein coefficient, m_e is the

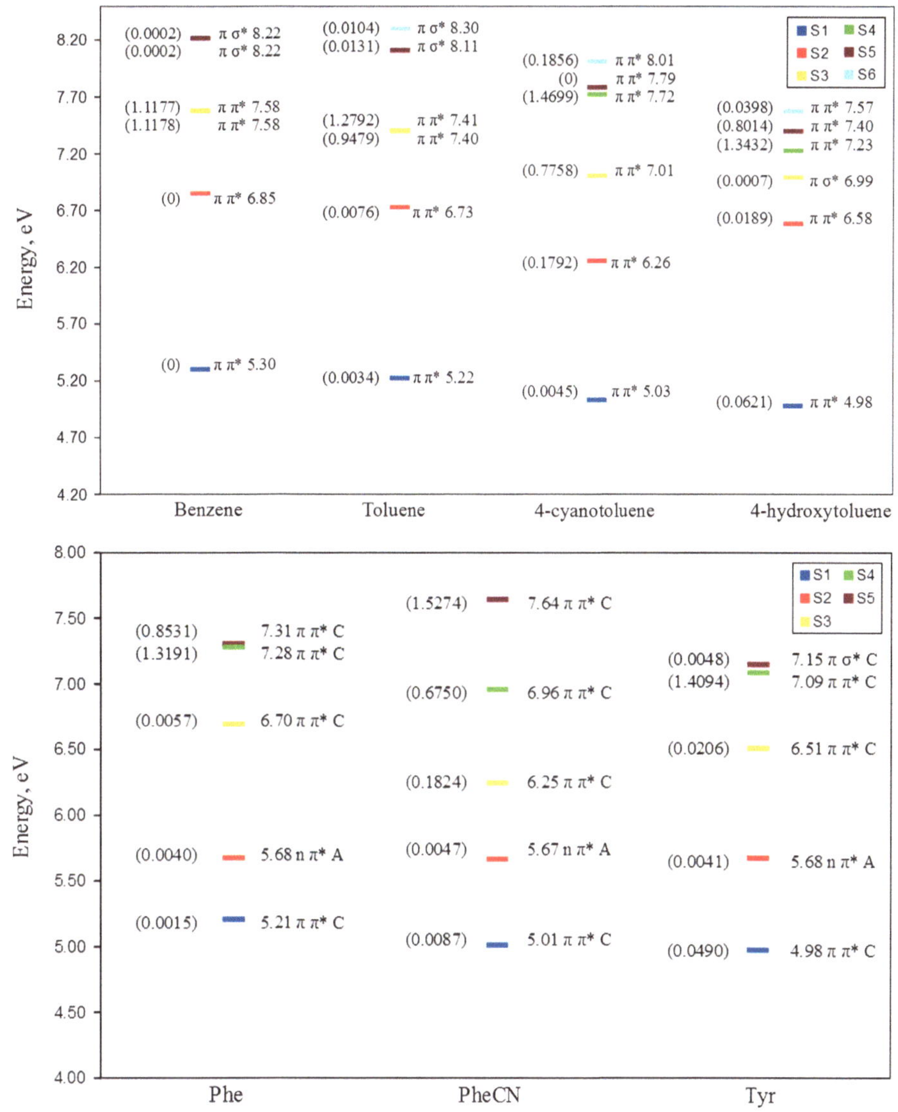

Fig. 2 Vertical excitation energies and oscillator strengths of the lowest states of (*top panel*) benzene, toluene, 4-cyanotoluene, and 4-hydroxytoluene and (*bottom panel*) phenylalanine, cyanophenylalanine, and tyrosine at the CIS(D)/cc-pVDZ level with transitions assigned with natural orbitals at the CIS/cc-pVDZ level and oscillator strengths in parentheses. A denotes an excited state localized primarily on the amino acid chain, while C denotes an excited state localized on the chromophore

mass of the electron, n is the index of refraction which is taken to be 1 in a vacuum, ϵ_0 is the permittivity of free space, f_{01} is the oscillator strength, and ω_{01} is the frequency of the transition [23]. The frequencies of transitions for absorption were determined at the CCSD/aug-cc-pVDZ or MCQDPT/cc-pVDZ levels for the chromophores, while for fluorescence, the Stokes shift found at the CIS(2)/cc-pVDZ+diff level for the amino acids was subtracted from the absorption frequencies. The oscillator strengths were calculated from the transition dipole moments at the CIS(2)/cc-pVDZ+diff level using the relationship $f_{01} = \frac{2}{3}\mu_{01}^2 E$, where E is the vertical excitation used in ω_{01} and μ_{01}.

The conformational search was performed using Spartan 10 [49] and MP2 geometry optimizations using GAMESS [34]. CASSCF transition dipole moments and MCQDPT energies were also calculated with GAMESS. CIS(2) calculations were performed with the Priroda 7 computational package [21]. Gaussian 03 was used to compute the solvated cyanophenylalanine geometry [12]. RI-CIS(D) and EOM-CCSD calculations were performed using Q-Chem [37]. MacMolPlt [3] was used for visualization.

3 Results and discussion

3.1 Comparisons with benzene

The chromophores of the amino acids studied here are substituted benzene molecules. In order to get a better qualitative idea of how the lower excited states originate, we compare the excited states of benzene to the other chromophores and discuss the effects of the substituents. A methyl group has been added to each chromophore in the

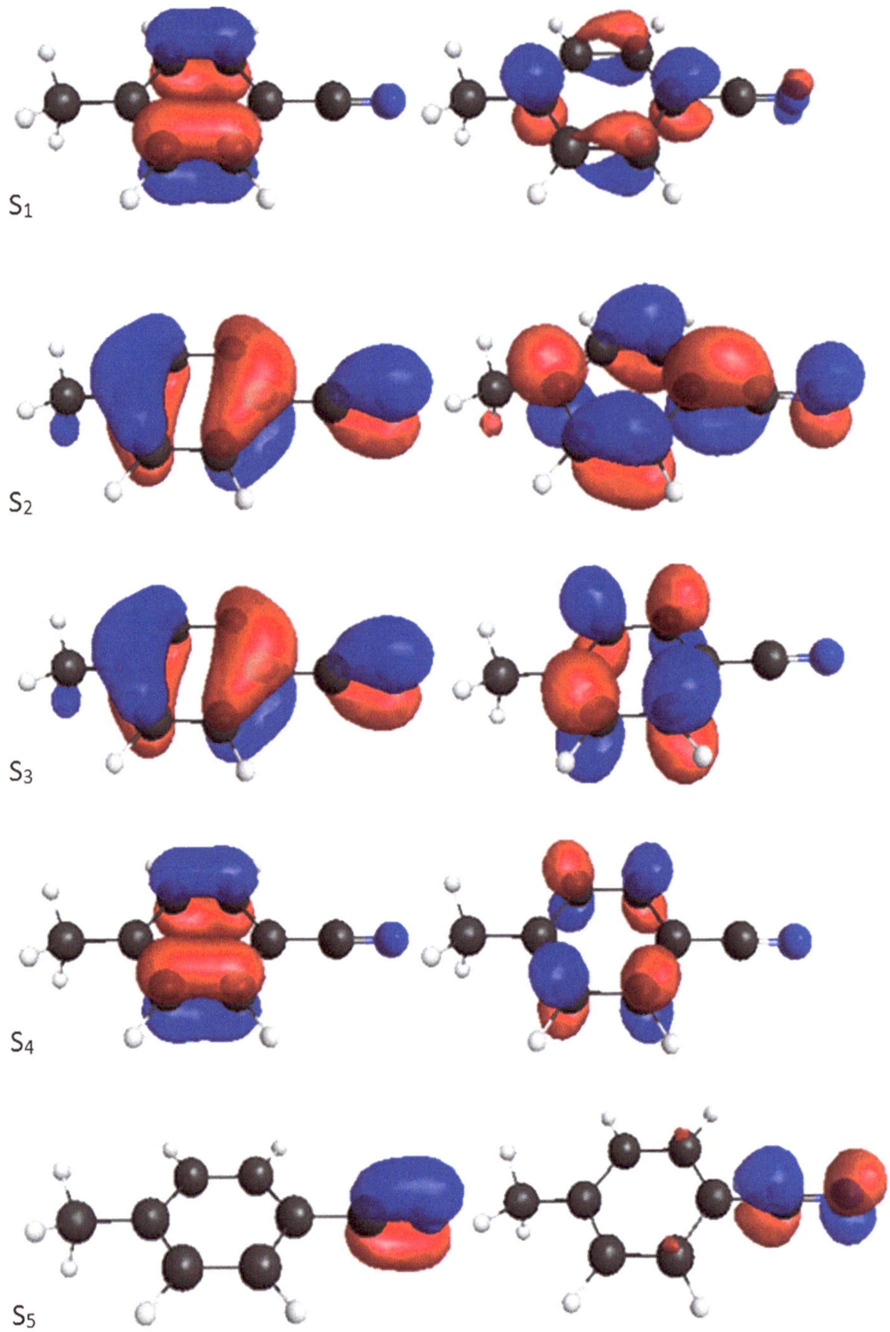

Fig. 3 Natural orbitals of the excited states of cyanotoluene obtained at the CIS/cc-pVDZ level

place of the amino acid. The HOMO and LUMO of benzene are both degenerate orbitals, so the excitation HOMO→LUMO leads to four states with two of them being degenerate. The first two excited states of benzene are dark non-degenerate B_{1u}, B_{2u} states with $\pi\pi^*$ character, while the third is a bright and doubly degenerate E_{1u} state. The effect of the substituents is examined in Fig. 2 (top panel) which shows that all chromophores have red shifted absorption to the first excited state compared to benzene. This can be explained by frontier molecular orbital (FMO) theory [10]. The hydroxy group is an electron donating substituent, while the cyano group is an electron withdrawing group. FMO theory predicts that both the HOMO and the LUMO are destabilized by the substituent which creates a node between the ring and the substituent [10]. However, the relative stability of the HOMO versus the LUMO determines the final shift in the excitation energy, which in this case decreases. Substituent effects for the methyl group are smaller than for the OH and CN substituents. The natural orbitals describing the excited states in 4-cyanotoluene are shown in Fig. 3, where it is seen that for the first four excited states, the majority of the density is

Table 1 Relative energies and Boltzmann distributions at room temperature of Phe conformers at the AM1 and MP2 levels

Conformer	AM1 (eV)	AM1 distr. %	MP2 (eV)	MP2 distr. %
M01	0.000	30.111	0.000	3.672
M02	0.002	27.781	−0.039	17.1067
M03	0.024	11.894	0.047	0.583
M04	0.032	8.702	0.062	0.334
M05	0.034	8.031	0.103	0.067
M06	0.038	6.897	0.047	0.593
M07	0.063	2.551	0.047	0.593
M08	0.075	1.645	0.079	0.170
M09	0.097	0.689	0.086	0.131
M10	0.104	0.533	0.092	0.102
M11	0.104	0.527	0.083	0.143
M12	0.114	0.354	0.125	0.029
M13	0.127	0.213	0.070	0.242
M14	0.167	0.045	0.134	0.020
N01	0.204	0.011	−0.006	4.610
M15	0.224	0.005	0.178	0.004
N02	0.233	0.003	0.185	0.003
N03	0.234	0.003	−0.073	62.121
N04	0.242	0.002	−0.007	4.759
N05	0.260	0.001	−0.007	4.757

Conformer names beginning with M have hydrogens from the carboxylic acid group pointed away from the amino group, while conformers beginning with N have carboxylic acid hydrogens pointed toward the amino group. Energies are relative to the lowest lying AM1 conformer, M01

Table 2 Relative energies and Boltzmann distributions at room temperature of Tyr conformers at AM1 and MP2 levels

Conformer	AM1 (eV)	AM1 distr. %	MP2 (eV)	MP2 distr %
M01	0.000	34.285	0.000	2.687
M02	0.005	27.779	−0.039	12.168
M03	0.025	13.063	0.048	0.422
M04	0.039	7.475	0.113	0.033
M05	0.040	7.172	0.062	0.243
M06	0.043	6.382	0.104	0.046
M07	0.067	2.505	0.049	0.398
M08	0.102	0.641	0.049	0.398
M09	0.110	0.474	0.085	0.097
M10	0.132	0.203	0.054	0.168
N01	0.214	0.008	−0.002	2.950
N02	0.232	0.004	−0.085	72.737
N03	0.241	0.003	0.183	0.002
N04	0.250	0.002	−0.009	3.826
N05	0.251	0.002	−0.009	3.826

Conformer names beginning with M have hydrogens from the carboxylic acid group pointed away from the amino group, while conformers beginning with N have carboxylic acid hydrogens pointed toward the amino group. Energies are relative to the lowest AM1 conformer, M01

localized on the ring (these are the states mainly coming from the benzene states), although there is some density on the substituent. On the other hand, the fifth excited state is primarily localized on the substituent. 4-Hydroxytoluene also has states localized on the substituent. In this molecule, S_3 is a $\pi\sigma^*$ transition localized on the hydroxyl group.

There are also effects observed from the symmetry being reduced. The degenerate benzene states split in all chromophores, with the gap being smallest for toluene. The oscillator strength of the lowest dark state increases when the symmetry of the benzene molecule is lost due to substituents. The effect is strongest in 4-hydroxytoluene. Quite interestingly the lowest dark state in benzene becomes the absorbing/fluorescent state in the chromophores present in the amino acids.

The bottom panel of Fig. 2 shows the excited states for the corresponding amino acids at the same level of theory. Comparison between the chromophore and amino acid excited states shows that the states localized on the chromophore are not affected much by the addition of the carbonyl and amino groups. However, a new state appears as the second excited state which is localized on the amino acid. This is the second excited state and is only about half an eV above S_1. It should be pointed out, however, that these results are in the gas phase, and inclusion of the solvent and the zwiterionic form of the amino acid moves that state higher by about 0.5 eV according to our calculations (using the [SS(V)PE] solvation model).

In the following discussion, we focus on the S_1 state since this is the state from which fluorescence is observed.

3.2 Distribution of conformers of amino acids

Tables 1, 2 and 3 show energies and abundances at room temperature of conformers of Phe, Tyr, and PheCN at the AM1 and MP2 levels. The natural abundances of conformers can be important for characterizing fluorescence properties if these properties are sensitive to the conformation. Conformational distributions in the gas phase have been studied theoretically and experimentally before for Phe and Tyr [1, 7, 16, 17, 24, 28, 39, 41, 50], but not for PheCN. For all three amino acids examined here, the lowest energy conformers at the AM1 level favor hydrogen bonding involving the O from the acid group and NH from the amino group, while the lowest energy conformers at the MP2 level have hydrogen bonding involving the OH from the acid group and the N from the amino group. The conformer predicted to be the lowest at the MP2 level has less than 0.01% abundance at the AM1 level. The dissimilarity between AM1 and MP2 populations shows that the accuracy of ground state energies is important to

Table 3 Relative energies and Boltzmann distributions at room temperature of PheCN conformers at AM1 and MP2 levels

Conformer	AM1 (eV)	AM1 distr. %	MP2 (eV)	MP2 distr. %
M01	0.000	35.146	0.000	8.949
M02	0.008	25.651	−0.033	32.266
M03	0.015	19.338	0.030	2.797
M04	0.028	11.702	0.091	0.255
M05	0.050	5.054	0.090	0.266
M06	0.076	1.825	0.132	0.053
M07	0.105	0.588	0.333	0.000
M08	0.110	0.489	0.094	0.231
M09	0.134	0.190	0.094	0.231
N01	0.225	0.005	0.033	2.481
M10	0.226	0.005	0.091	0.262
N02	0.244	0.003	−0.044	50.419
N03	0.267	0.001	0.041	1.788
N04	0.267	0.001	0.242	0.001
N05	0.276	0.001	0.293	0.000

determine the accurate distribution of conformers. The lowest MP2 conformers found here agree with previous theoretical studies of Phe by Huang et al. [16] and of Tyr by Zhang et al. [50] which were both done using DFT at the B3LYP/6-311++G** level. The Phe lowest conformer also agrees with the experimentally observed one in the gas phase using resonant two-photon ionization [24]. The number of conformers detected experimentally varies by experimental conditions, particularly the temperature. The main focus of this work is not to thoroughly discuss conformation distribution, but rather to examine the sensitivity of the excited state properties on the different conformers, and we discuss this topic next.

3.3 Sensitivity of excitation energies and transition dipole moments on conformers

The dependence of the excited state properties on the different conformers was examined by calculating the excited states of several conformers at the CIS(D)/cc-pVDZ level. Table 4 shows the excitation energies and magnitude of the transition dipole moments for several conformers of PheCN, Phe, and Tyr. The data indicate that the excitation energy to the S_1 state is not sensitive to the rotation around the single bonds of the conformers analyzed since the conformers differ in energy by at most 0.015 eV (approximately 0.3 % effect). This small effect agrees with experimental studies which also have shown that different conformers of Phe and Tyr have origins varying by 0.02 eV in the gas phase [28, 39]. On the other hand, the magnitudes of the transition dipole moments (μ_{01}) differ significantly between conformers. The largest relative variation

Table 4 CIS(D)/cc-pVDZ excitation energies in eV with magnitudes of transition dipole moments in atomic units in parenthesis at the CIS level for MP2 optimized geometries

Conformer	E (μ_{01})
Phe	
N03 (MP2)	5.210 (0.1008)
M01 (AM1)	5.210 (0.0638)
Tyr	
N02 (MP2)	4.977 (0.5849)
M01 (AM1)	4.984 (0.5777)
PheCN	
N02 (MP2)	5.015 (0.2467)
N03	5.011 (0.2100)
M01 (AM1)	5.019 (0.2617)
M02	5.009 (0.2409)
M03	5.019 (0.2747)
M05	5.004 (0.2048)
M06	5.007 (0.2081)

between transition dipole moments of AM1 and MP2 conformers is observed with Phe which has the lowest magnitude overall (ca. 50 % effect), although the differences between PheCN and Tyr conformers are also significant (1–20 %). These results indicate that the excitation energies (and thus absorption maxima and possibly fluorescence maxima) are not sensitive to the amino acid conformation, but the intensity of the absorption or emission is Snoek et al. [41] have also seen great variations in the orientation of the transition dipole moment of the first excited state of tyrosine depending on the conformer using CIS calculations.

3.4 Vertical excitation energies and transition dipole moments

3.4.1 Isolated chromophores

In the following discussion, we first focus on isolated chromophores. Very accurate methods can be used for these systems since they are smaller, and the effect of the remaining amino acid will be examined later. The first column of Tables 5, 6, and 7 lists the various methods used, while the second column shows the vertical excitation energies to S_1 and the magnitudes of the corresponding transition dipole moments found for each chromophore. The effects of dynamical correlation are examined by comparing CIS, CIS(D), CIS(2), and EOM-CCSD methods. The performance of TDDFT has also been examined. CIS as expected predicts excitation energies that are in general too high. Perturbation theory corrections at the CIS(D) level lower the energies considerably, by ca. 0.8 eV when compared to CIS using the same basis set. CIS(2) does not perform as well. The effect of including dynamical correlation at the EOM-CCSD level only lowers the

Table 5 Vertical excitation energies and transition dipole moments (in parenthesis) of toluene and Phe

Method	Toluene S_0	Phe S_0	Toluene S_1	Phe S_1
CIS	6.027 (0.1517)	6.022 (0.1008)	5.712 (0.1878)	5.712 (0.0909)
CIS(D)	5.224 (0.1517)[a]	5.210 (0.1008)[a]		4.860 (0.0909)[a]
CIS(D)/cc-pVTZ	5.144 (0.1640)[a]		4.797 (0.2019)[a]	
CIS(D)/cc-pVQZ	5.117 (0.1667)[a]			
CIS(D)/aug-cc-pVTZ	5.099 (0.1497)[a]	5.097 (0.0988)[a]		
B3LYP/aug–cc-pVTZ	5.160 (0.1125)		4.859 (0.1478)	
CIS(2)	5.589 (0.0992)	5.677 (0.0620)	5.327 (0.1231)	5.324 (0.0869)
CIS(2)/cc-pVDZ+diff	5.686 (0.0975)	5.600 (0.0722)	5.242 (0.1292)	5.259 (0.0939)
CCSD/aug-cc-pVDZ	5.013 (0.0793)	(0.0576)[c]	4.691 (0.1022)	(0.0742)[c]
MCQDPT	4.717 (0.0525)[b]	(0.0382)[c]	4.357 (0.0644)	(0.0468)[c]

Energies are in eV and transition dipole moments are in atomic units. Basis set used is cc-pVDZ unless otherwise specified. S_0 corresponds to the ground state equilibrium and the corresponding energies represent absorption maxima. S_1 corresponds to the minimum on S_1 and the energies represent fluorescence maxima

[a] Transition dipole moments at the CIS level. [b] Transition dipole moments at the CASSCF level. [c] Extrapolated values from the chromophore using the ratio of the transition dipole moments between chromophore and amino acid at the CIS(2)/cc-pVDZ+diff level

Table 6 S_1 vertical excitation energies and transition dipole moments (in parenthesis) of 4-hydroxytoluene and Tyr

Method	OH-toluene S_0	Tyr S_0	OH-toluene S_1	Tyr S_1
CIS	5.838 (0.6590)	5.846 (0.5849)	5.431 (0.7879)	5.383 (0.7445)
CIS(D)	4.982 (0.6590)[a]	4.976 (0.5849)[a]		4.511 (0.7445)[a]
CIS(D)/cc-pVTZ	4.884 (0.6613)[a]			
CIS(D)/cc-pVQZ	4.849 (0.6557)[a]			
CIS(D)/aug-cc-pVTZ	4.808 (0.6371)[a]	4.823 (0.5506)[a]		
B3LYP/aug-cc-pVTZ	4.773 (0.5304)		4.401 (0.6106)	
CIS(2)	5.366 (0.5822)	5.365 (0.5217)	4.927 (0.6751)	4.87 (0.6363)
CIS(2)/cc-pVDZ+diff	5.236 (0.5853)	5.263 (0.5106)	4.805 (0.6822)	4.78 (0.6328)
CCSD/aug-cc-pVDZ	4.754 (0.4435)	(0.3869)[c]	4.368 (0.5304)	(0.4920)[c]
MCQDPT	4.579 (0.2971)[b]	(0.2591)[c]	4.280 (0.3654)	(0.3390)[c]

Energies are in eV and transition dipole moments are in atomic units. Basis set used is cc-pVDZ unless otherwise specified. S_0 corresponds to the ground state equilibrium and the corresponding energies represent absorption maxima. S_1 corresponds to the minimum on S_1 and the energies represent fluorescence maximal

[a] Transition dipole moments at the CIS level. [b] Transition dipole moments at the CASSCF level. [c] Extrapolated values from the chromophore using the ratio the transition dipole moments between chromophore and amino acid at the CIS(2)/cc-pVDZ+diff leve

energies by less than 0.15 eV compared to CIS(D). The TDDFT calculations using the B3LYP functional and the aug-cc-pVTZ basis set predict energies which are within 0.02–0.06 eV of the CIS(D) energies when using the same basis set. The effect of non-dynamical correlation measured through the use of MCQDPT is to decrease the excitation energy by about 0.2–0.3 eV compared to CCSD for almost all cases. The comparison, however, is not direct since different basis sets have been used for the two methods.

Basis set effects have also been examined systematically using CIS(D) and a serious of the cc-pVnZ basis sets. The results are plotted in Supporting Information. When the energy is plotted as a function of 1/n, an extrapolation limit can be reached. This is estimated to drop the energy from the cc-pVDZ results by 0.13 eV, 0.18 eV, and 0.12 eV for toluene, OH-toluene and CN-toluene, respectively. The inclusion of diffuse functions is also very important for calculating the S_1 energy for all chromophores. Adding diffuse functions to cc-pVDZ decreases the energy by 0.08 eV while adding them to cc-pVTZ decreases the energy by 0.03–0.04 eV.

Transition dipole moments are a lot more sensitive to the level of theory. Basis sets are important for the transition dipole moments; however, the effect is more difficult to monitor. Plotting the transition dipole moments versus 1/n in most cases gives non-monotonic behavior, so an extrapolation is not possible. Diffuse functions are also

Table 7 Vertical excitation energies with transition dipole moments (in parenthesis) of cyanotoluene and the lowest energy conformer of cyanophenylalanine (PheCN) at the MP2 level

Method	CN-toluene S_0	PheCN S_0	PheCN $S_{0_{aq}}$	CN-toluene S_1	PheCN S_1
CIS	5.852 (0.1772)	5.835 (0.2467)	5.807 (0.3815)	5.542 (0.2103)	5.521 (0.2954)
CIS(D)	5.034 (0.1772)[a]	5.015 (0.2467)[a]	5.011 (0.3815)[a]		4.684 (0.2954)[a]
CIS(D)/cc-pVTZ	4.954 (0.1683)[a]				
CIS(D)/cc-pVQZ	4.929 (0.1709)[a]				
CIS(D)/aug-cc-pVDZ	4.947 (0.1928)[a]	4.941 (0.2803)[a]			
CIS(D)/aug-cc-pVTZ	4.919 (0.1854)[a]				
B3LYP/aug-cc-pVTZ	4.893 (0.1191)			4.601 (0.1332)	
CIS(2)	5.504 (0.1328)	5.484 (0.1812)		5.150 (0.1335)	5.132 (0.2117)
CIS(2)/cc-pVDZ+diff	5.433 (0.1566)	5.424 (0.2157)		5.087 (0.1573)	5.079 (0.2472)
CCSD/aug-cc-pVDZ	4.869 (0.0846)	(0.1164)[c]		4.553 (0.0966)	(0.1519)[c]
MCQDPT	4.591 (0.0618)[b]	(0.0851)[c]		4.255 (0.0739)	(0.1161)[c]

Energies are given in eV and transition dipole moments are in atomic units. Basis set used is cc-pVDZ unless otherwise specified. S_0 corresponds to the ground state equilibrium and the corresponding energies represent absorption maxima. S_1 corresponds to the minimum on S_1 and the energies represent fluorescence maxima.

[a] Transition dipole moments at the CIS level. [b] Transition dipole moments at the CASSCF level. [c] Extrapolated values from the chromophore using the ratio the transition dipole moments between chromophore and amino acid at the CIS(2)/cc-pVDZ+diff level

very important for the transition dipole moments. The transition dipole moment magnitudes at the CCSD level are much smaller than the values obtained with more approximate methods, often by a factor of 2 or more. Transition dipole moments at the CASSCF level are even lower than the CCSD ones. Transition dipole moments at the B3LYP level are closer to those obtained with the CIS(2) method. Diffuse functions cause an increase in transition dipole moment, the opposite of what is seen with adding correlation.

Our results compare well with previous CASPT2 calculations on the chromophores of tyrosine and phenylalanine using a larger basis set [20, 31]. Previously reported CC2 results on the amino acids phenylalanine and tyrosine [18] are close to our CIS(D) values.

3.4.2 Effect of adding the amino acid side

The effect of the amino acid group is examined next. The third column of Tables 5, 6, and 7 shows excitation energies and transition dipole moments of Phe, Tyr, and PheCN at various levels of theory using the MP2 lowest energy conformer. The energies of all amino acids are within 0.02 eV of their chromophores at every level of theory showing a very small effect. This agrees with the discussion in Sect. 3.3.

The effect of the amino acid on the transition dipole moments is not consistent for the various chromophores. Table 7 shows that PheCN has a larger transition dipole moment than cyanotoluene at all levels of theory. The transition dipole moment for the chromophore is about 70 % that of PheCN. But the amino acid has a smaller transition dipole moment for Phe and Tyr. All methods surveyed showed similar effects of the amino acids compared to the chromophores, so the shifts predicted are consistent at the different levels of theory. The increased transition dipole moment of PheCN compared to cyanotoluene contributes to its fluorescence properties as will be discussed in more detail in Sect. 3.7.

3.4.3 Solvation effect

The effect of solvation on PheCN was investigated by using a polarizable continuum model. The zwiterionic form of the amino acid is used in these calculations. The results are shown on the fourth column of Table 7, where it can be seen that the energy is not affected by solvation, in agreement with experimental findings from Serrano et al. [35]. The solvated amino acid is shifted 0.02 eV from the gas phase at the CIS level and is not shifted at all at the CIS(D) level. So, interestingly even the very different charge distribution of the zwiterion does not have a big effect on the energies. The CIS transition dipole moment was increased by 54.6 % to 0.38145 a.u.

3.5 Fluorescence properties

In order to investigate fluorescence properties, we have to examine the energies and transition dipole moments at the S_1 minimum of each chromophore and amino acid. The geometries at the S_0 and S_1 minima can be seen in Fig. 4. Distortions in geometry are similar in all molecules, which is consistent with the fact that the character of the excited

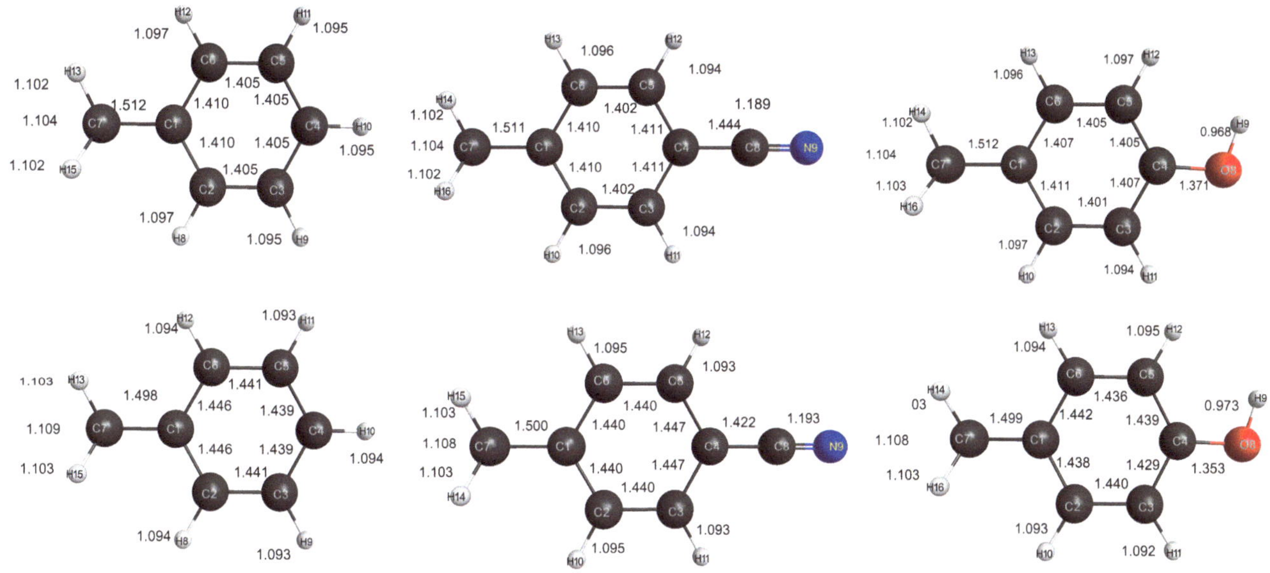

Fig. 4 Geometries of chromophores toluene, cyanotoluene, and hydroxytoluene obtained at the MP2/cc-pVDZ level for the S_0 minimum (*top*), and at the CIS(2)/cc-pVDZ for the S_1 minimum (*bottom*). Bond lengths are shown in Å

state is similar for all three molecules and is mostly localized on the ring. The aromatic ring remains planar for all molecules and expands at the S_1 minimum as all the bond lengths increase by ca. 0.02–0.04 Å. The bond between the methyl group and the C1 carbon shortens by 0.01 Å in each S_1 minimum. For hydroxytoluene and cyanotoluene the bonds between the C4 carbon and functional group shorten by 0.02 Å.

The S_1 minima were used for single point excited state calculations using various levels of theory shown in Tables 5, 6, and 7. The Stokes shift for toluene is between 0.30 eV and 0.36 eV for all the methods used. For Phe, the Stokes shift is predicted to be similar to toluene. The experimental Stokes shift in solution is 0.37 eV. The best value is at the MCQDPT level which is 0.36 eV. PheCN follows trends similar to Phe. Stokes shifts are between 0.29 and 0.35 eV with the chromophore and amino acid being very similar. The experimental shift is 0.3 eV in aqueous solution. Tyrosine and hydroxytoluene Stokes shifts differ slightly as shown in Table 6 but are within 0.05 eV with the amino acid Stokes shift being larger. The CCSD/aug-cc-pVDZ shift is 0.39 eV, while the MCQDPT is 0.3 eV. The experimental value is 0.43 eV, in better agreement with the CCSD value.

The magnitude of the transition dipole moments is larger at the S_1 minimum compared to that at the S_0 minimum for all chromophores and amino acids. As in the S_0 geometries μ_{01} are affected by both method and basis set for all chromophores, especially for cyanotoluene. Table 7 shows that PheCN has a larger transition dipole moment than cyanotoluene at all levels of theory.

Table 8 Experimental origins of Phe and Tyr from Martinez et al. [28] in eV compared to theoretical origins at the MCQDPT/cc-pVDZ and CCSD/aug-cc-pVDZ levels for the chromophores

Species	Exp.	MCQDPT	MCQDPT diff (%)	CCSD	CCSD diff (%)
Phe	4.65	4.35	6.45	4.70	1.1
Tyr	4.40	4.38	0.45	4.42	0.39
PheCN		4.25		4.61	

3.6 Comparison with experimental results

We initially compare our results to gas phase electronic spectra measured in a supersonic jet using laser-induced fluorescence spectroscopy [28]. A comparison with gas phase results is more applicable to our results. Furthermore, we use the origins obtained from the chromophores at the highest levels of theory used here, since we have demonstrated that the amino acid has negligible effect on the energies. The theoretical adiabatic energies including Zero Point Energy (ZPE) corrections and the experimental origins corresponding to 0–0 transitions are shown in Table 8. We use the adiabatic energies at the MCQDPT/cc-pVDZ and CCSD/aug-cc-pVDZ levels with ZPE corrections taken from MP2 and CIS(2) frequency calculations at the S_0 and S_1 minima, respectively. The results agree well with experimental origins for both Phe and Tyr. There are no published gas phase spectra of PheCN to obtain origin information. Table 8 shows that the errors for Tyr are less than 0.1% for both methods. Errors for Phe are somewhat

Table 9 Experimental absorbance and fluorescence energies with quantum yield (QY) and lifetimes from Tucker and Chen [4, 23, 35]

Exp.	$E_{abs}(f)$	$E_{fluor}(f)$	QY	τ_0	τ_{rad}
Phe	4.77	4.40	0.02	6.8	340
Tyr	4.51	4.08	0.14	3.6	26
PheCN	4.43	4.13	0.11	7.0	63
MCQDPT					
Phe	4.72 (0.00017)	4.36 (0.00023)			263.4
Tyr	4.58 (0.00753)	4.28 (0.01205)			5.3
PheCN	4.59 (0.00081)	4.25 (0.00142)			45.5
CCSD					
Phe	5.01 (0.00041)	4.69 (0.00063)			83.8
Tyr	4.75 (0.01743)	4.37 (0.02590)			2.4
PheCN	4.87 (0.00162)	4.55 (0.00257)			21.9

Theoretical results were calculated as described in the text. Energies are in eV, and lifetimes in ns. Oscillator strengths (*f*) are given in parenthesis

larger when MCQDPT is used. The ZPE corrections lower the origin by about 0.14–0.17 eV. The ZPE for the excited state is somewhat lower than those for the ground state so the overall effect is that adding ZPE lowers the adiabatic energies.

We next compare absorption and fluorescence data in aqueous solutions, where more data exist. Absorption and emission data for all three systems are shown in Table 9. The energies shown in Table 9 are those of the chromophore at the MCQDPT and CCSD levels using again the fact that the energies are not sensitive to the amino acid group. The oscillator strengths, however, are sensitive to the amino acid, so we tried to use the information we obtained in our benchmark studies to estimate their value more accurately. The oscillator strengths were calculated using the equation $f_{01} = \frac{2}{3}\mu_{01}^2 E_{01}$ with E_{01} being the vertical excitation energy used in Table 9 in hartree, μ_{01} is the transition dipole moments in atomic units shown in Tables 5, 6, and 7 and indicated with a superscript *c*. This is obtained by scaling the chromophore transition dipole moment at the CASSCF or CCSD level with the ratio of amino acid to chromophore μ at the CIS(2)/cc-pVDZ+diff level. This approximation is most reasonable for PheCN where this ratio does not vary much with the method. These estimates were made to obtain our best theoretical numbers and to use them to calculate fluorescence lifetimes using the equation $A_{01} = \frac{2\pi^2 m_e^2 f_{01} n^3}{\epsilon_0 c^3 \omega_{01}}$. The lifetimes are then compared to experimental radiative lifetimes that are extracted from the quantum yields and the total lifetimes using the relationship $QY = \frac{k_{rad}}{k_{tot}}$ where k_{rad} and k_{tot} are the rate constants for radiative and total decay.

The theoretical values for absorption and emission maxima are very close to the experimental ones, especially at the MCQDPT level. The difference between theoretical and experimental values is on average 0.1 eV. The difference at the CCSD level is larger, around 0.3 eV on average. Both theory and experiment predict that Phe has the highest absorption and emission energy, while Tyr and PheCN have similar absorption and emission values. The radiative lifetimes have larger deviations between theory and experiment. Again, MCQDPT performs better than CCSD with the most accurate value being that of Phe which is 23 % different from experiment. Even if quantitative accuracy is not found, qualitative the values agree with experiment, with Phe having the longest lifetime among the three amino acids and Tyr having the shortest.

It should be kept in mind that comparison with absorption and fluorescence maxima in aqueous solution is not as straightforward as comparing to gas phase origins which we did in Table 8. Even though our results indicate that the energies in the gas phase are very similar to aqueous solution, there are still other uncertainties present. Specifically, a maximum on the spectrum does not necessarily correspond exactly to the vertical excitation. With that in mind, we can conclude that both MCQDPT and CCSD describe these properties quite well.

There are more reasons why there are discrepancies in radiative lifetimes. As we discussed earlier, these values are very sensitive to the methods used, as well as the environment. Furthermore, the transition dipole moments are significantly different for various conformers of the amino acids. The experimental findings are a weighted average of the lifetimes for each conformer present, but only one conformer for each amino acid was used in our results. The conformational distributions are in the gas phase and also very sensitive with respect to method. It is likely that this distribution will be different in aqueous solution. The transition dipole moment changed significantly between gas phase and solution for PheCN. All these facts indicate that getting an accurate value for μ_{01} is very challenging. As already mentioned, other workers have also seen great variations in the transition dipole moment of the first excited state of tyrosine depending on the conformer [41]. Hashimoto et al. have found in 1+1 REMPI and LIF spectra that the lifetimes for gas phase phenylalanine vary by up to a factor of three [14]. The variation in transition dipole moment found here suggests that these differences in conformer lifetimes observed by Hashimoto et al. could be attributed to differences in radiative lifetimes instead of variation in accessibility of non-radiative decay pathways which is what they proposed in their work.

3.7 What makes PheCN more fluorescent

The theoretical results reported here predict the experimental trend between the amino acids. Addition of an OH or CN substituent on the benzene ring increases the radiative decay rate and decreases the radiative lifetime. Specifically, incorporation of the CN group in phenylalanine decreases the radiative lifetime by a factor of 5–6 and this is reproduced by the theoretical results. Tyrosine has an even shorter radiative lifetime. It is interesting to examine the source of these changes, as this is partly responsible of the development of the non-natural amino acid as a fluorescent probe. At the chromophore level, toluene and cyanotoluene have similar transition dipole moment magnitudes as can be seen in Tables 5 and 7 with cyanotoluene being only 1.2 times higher than toluene. On the other hand, hydroxytoluene has a transition dipole moment magnitude 6 times higher than that of toluene. So, the OH substituent has a much stronger effect on the transition dipole moment of the chromophore than the CN substituent. Quite interestingly, the effect of the amino acid is quite important but has opposite trends for phenylalanine compared to cyanophenylalanine. In phenylalanine, the transition dipole moment decreases compared to the isolated chromophore, while in PheCN, it increases. The net effect is that the ratio of the transition dipole moments for PheCN versus Phe is 2.5. Since the oscillator strength and the radiative lifetimes depend on the square of the transition dipole moment, this leads to a ratio of about 6, which is the same as the experimental observation. It seems that small mixing of the wavefunctions of the $n\pi^*$ state (S_2 state of the amino acids localized on the amino group) to the wavefunction of the S_1 $\pi\pi^*$ state can lead to such changes in the transition dipole moments.

3.8 Conclusions

In this study, we have discussed excited state properties of the two natural amino acids Phe, and Tyr and compared them to non-natural amino acid PheCN. Both EOM-CCSD and MCQDPT methods give quite good absorption and emission energies, as well as origins, when compared to experimental data available for Phe and Tyr. The origin for PheCN is expected to be near 4.43 eV if differences between experiment and theory are consistent.

Our results also show that the chromophores toluene, 4-cyanotoluene, and 4-hydrotoluene may be reliable used to calculate excitation energies for the amino acids phenylalanine, p-cyanophenylalanine, and tyrosine, respectively, since the amino acid has minimum effect on these energies. Comparison of benzene versus toluene in Fig. 2 indicates that these are more accurate chromophores to use than benzene, phenol, and benzonitrile because the methyl group affects excitation energies for most states.

The transition dipole moment, and therefore the oscillator strengths, are sensitive to the inclusion of the amino acid, conformation, method, and basis sets. This makes accurate calculations involving the oscillator strengths difficult to obtain. The oscillator strengths of the amino acids qualitatively show their order of absorptivity and lifetimes, but do not agree well with the experimental lifetimes [4, 5, 35]. The sensitivity of these properties on the conformers and the method contributes to the poor agreement with experiment.

The calculations predict that the radiative rate of fluorescence of PheCN is about six times larger than that of Phe in agreement with experimental observations. This agrees with the fact that PheCN is an improved fluorescent probe compared to Phe. Interestingly, the amino acid contributes to this increase in radiative rate. Even though here we have examined the radiative decay of PheCN, more theoretical work should be dedicated to non-radiative decay channels on PheCN in order to obtain a more complete picture of its properties. The non-radiative decay is very sensitive to the solvent, and this should be taken into account in any studies. Radiationless decay pathways for the natural amino acids, dipeptides, and their complexes with water have been investigated theoretically, and efficient pathways for decay often involving proton transfer have been identified [26, 29, 38, 42, 43]. It is very likely that similar pathways are involved in the deactivation of PheCN.

Acknowledgments Support by the National Science Foundation under grant CHE-1213614 is acknowledged. SM thanks the Alexander von Humboldt Foundation for support during a visit to Germany where part of this paper was written. SLM thanks the Undergraduate Research Program of the College of Science and Technology at Temple University for partial support.

References

1. Abo-Riziq A, Grace L, Crews B, Callahan MP, van Mourik T, de Vries MS (2011) Conformational structure of tyrosine, tyrosylglycine, and 2 tyrosyl-glycyl-glycine by double resonance spectroscopy. J. Phys. Chem. A 115:6077–6087
2. Berkeley L, Field M (1997) A diagnostic for the applicability of the CIS and CIS(D) excitation energy methods. Chem. Phys. Lett. 279:151–157
3. Bode BM, Gordon MS (1998) MacMolPlt: a graphical user interface for GAMESS. J. Mol. Graphics Modell. 16(133—8):164
4. Chen RF (1967) Fluorescence Quantum Yields of Tryptophan and Tyrosine. Anal. Lett. 1:35–42
5. Chen Y, Berkley MD (1998) Toward understanding tryptophan fluorescence in proteins. Biochemistry 37:9976–82
6. Chipman DM (2000) Reaction field treatment of charge penetration. J. Chem. Phys. 112:5558

7. Cohen R, Brauer B, Nir E, Grace L, de Vries MS (2000) Resonance-enhanced multiphoton ionization spectroscopy of dipeptides. J. Phys. Chem. A 104:6351–6355
8. Cossi M, Barone V, Cammi R, Tomasi J (1996) Ab initio study of solvated molecules: a new implementation of the polarizable continuum model. Chem. Phys. Lett. 255:327–335
9. Dewar MJS, Zoebisch EG, Healy EF, Stewart JJP (1985) AM1: A New General Purpose Quantum Mechanical Molecular Model. J. Am. Chem. Soc. 107:3902–3909
10. Fleming I (1976) Frontier Orbitals and Organic Chemical Reactions. John Wiley & Sons Ltd, New York
11. Foreman JB, Frisch MJ (1992) Toward a Systematic Molecular Orbital Theory for Excited States. J. Phys. Chem. 96:135–149
12. Frisch MJ, Trucks GW, Schlegel HB, Scuseria GE, Robb MA, Cheeseman JR, Montgomery Jr. JA, Vreven T, Kudin KN, Burant JC, Millam JM, Iyengar SS, Tomasi J, Barone V, Mennucci B, Cossi M, Scalmani G, Rega N, Peter GA, Pople JA (2003) Gaussian 03
13. Getahun Z, Huang CY, Wang T, De León B, DeGrado WF, Gai F (2003) Using nitrile-derivatized amino acids as infrared probes of local environment. J. Am. Chem. Soc. 125:405–11
14. Hashimoto T, Takasu Y, Yamada Y, Ebata T (2006) Anomalous conformer dependent S1 lifetime of l-phenylalanine. Chem. Phys. Lett. 421:227–231
15. Huang CY, Wang T, Gai F (2003) Temperature dependence of the CN stretching vibration of a nitrile-derivatized phenylalanine in water. Chem. Phys. Lett. 371:731–738
16. Huang Z, Yu W, Lin Z (2006) Exploration of the full conformational landscapes of gaseous aromatic amino acid phenylalanine: An ab initio study. Journal of Molecular Structure: THEOCHEM 758:195–202
17. Inokuchi Y, Kobayashi Y, Ito T, Ebata T (2007) Conformation of L-tyrosine studied by fluorescence-detected UV-UV and IR-UV double-resonance spectroscopy. J. Phys. Chem. A 111:3209–3215
18. Kawashima Y, Nakano H, Jung J, Ten-no S (2011) A combined quantum mechanical and molecular mechanical method using modified generalized hybrid orbitals: implementation for electronic excited states. Phys. Chem. Chem. Phys. 13:11,731–11,738
19. Krylov AI (2008) Equation-of-Motion Coupled-Cluster Methods for Open-Shell and Electronically Excited Species: The Hitchhiker's Guide to Fock Space. Annu. Rev. Phys. Chem. 59:433–462
20. Kushwaha P, Mishra P (2000) Electronic spectra, excited-state geometries and molecular electrostatic potentials of aromatic amino acids. J. Photochem. Photobiol., A 137:79–86
21. Laikov D Priroda: An Electronic Structure Code
22. Laikov D, Matsika S (2007) Inclusion of second-order correlation effects for the ground and singly-excited states suitable for the study of conical intersections: The CIS(2) model. Chem. Phys. Lett. 448:132–137
23. Lakowicz JR (2006) Principles of Fluorescence Spectroscopy, 6th edn. Springer, Baltimore, USA
24. Lee KT, Sung J, Lee KJ, Park YD, Kim SK (2002) Conformation-dependent ionization energies of L-phenylalanine. Angew. Chem. 41:4114–7
25. Link A, Mock M, Tirrell D (2003) Non-canonical amino acids in protein engineering. Current Opinion in Biotechnology 14:603–609
26. Marazzi M, Sancho U, Castano O, Domcke W (2010) Manuel Frutos, L.: Photoinduced Proton Transfer as a Possible Mechanism for Highly Efficient Excited-State Deactivation in Proteins. J. Phys. Chem. Lett. 1(1):425–428
27. Marek P, Gupta R, Raleigh DP (2008) The fluorescent amino acid p-cyanophenylalanine provides an intrinsic probe of amyloid formation. ChemBioChem 9:1372–1374
28. Martinez SJ, Alfano JC, Levy DH (1992) The Electronic Spectroscopy of the Amino Acids Tyrosine and Phenylalanine in a Supersonic Jet. J. Mol. Spectrosc. 430:421–430
29. Matsika S, Krause P (2011) Nonadiabatic Events and Conical Intersections. Annu. Rev. Phys. Chem. 62:621–643
30. Nakano H (1993) Quasidegenerate perturbation theory with multiconfigurational self-consistent-field reference functions. J. Chem. Phys. 99:7983. doi:10.1063/1.465674
31. Rogers D, Hirst J (2003) Ab initio study of aromatic side chains of amino acids in gas phase and solution. J. Phys. Chem. A **107**
32. Rogers JMG, Lippert LG, Gai F (2010) Non-natural amino acid fluorophores for one- and two-step fluorescence resonance energy transfer applications. Anal. Biochem. 399:182–9
33. Royer CA (2006) Probing protein folding and conformational transitions with fluorescence. Chem. Rev. 106:1769–84
34. Schmidt MW, Baldridge KK, Boatz JA, Elbert ST, Gordon MS, Jensen JH, Koseki S, Matsunaga N, Nguyen KA, Shyjun SU, Dupuis M, Montgomery JA (1993) General Atomic and Molecular Electronic Structure System. J. Comput. Chem. 14:1347–1363
35. Serrano AL, Troxler T, Tucker MJ, Gai F (2010) Photophysics of a Fluorescent Non-natural Amino Acid: p-Cyanophenylalanine. Chem. Phys. Lett. 487:303–306
36. Serrano AL, Waegele MM, Gai F (2012) Spectroscopic studies of protein folding: linear and nonlinear methods. Protein Sci. 21:157–170
37. Shao Y, Molnar LF, Jung Y, Kussmann J, Ochsenfeld C, Brown ST, Gilbert ATB, Slipchenko LV, Levchenko SV, O'Neill DP (2006) DiStasio, R.a., Lochan, R.C., Wang, T., Beran, G.J.O., Besley, N.a., Herbert, J.M., Lin, C.Y., Van Voorhis, T., Chien, S.H., Sodt, A., Steele, R.P., Rassolov, V.a., Maslen, P.E., Korambath, P.P., Adamson, R.D., Austin, B., Baker, J., Byrd, E.F.C., Dachsel, H., Doerksen, R.J., Dreuw, A., Dunietz, B.D., Dutoi, A.D., FOPTurlani, T.R., Gwaltney, S.R., Heyden, A., Hirata, S., Hsu, C.P., Kedziora, G., Khalliulin, R.Z., Klunzinger, P., Lee, A.M., Lee, M.S., Liang, W., Lotan, I., Nair, N., Peters, B., Proynov, E.I., Pieniazek, P.a., Rhee, Y.M., Ritchie, J., Rosta, E., Sherrill, C.D., Simmonett, A.C., Subotnik, J.E., Woodcock, H.L., Zhang, W., Bell, A.T., Chakraborty, A.K., Chipman, D.M., Keil, F.J., Warshel, A., Hehre, W.J., Schaefer, H.F., Kong, J., Krylov, A.I., Gill, P.M.W., Head-Gordon, M.: Advances in methods and algorithms in a modern quantum chemistry program package. Phys. Chem. Chem. Phys. 8:3172–91
38. Shemesh D, Domcke W (2011) Effect of the Chirality of Residues and gamma-Turns on the Electronic Excitation Spectra, Excited-State Reaction Paths and Conical Intersections of Capped Phenylalanine-Alanine Dipeptides. ChemPhysChem 12:1833–1840
39. Shimozono Y, Yamada K (2013) Ishiuchi, S.i., Tsukiyama, K., Fujii, M.: Revised conformational assignments and conformational evolution of tyrosine by laser desorption supersonic jet laser spectroscopy. Phys. Chem. Chem. Phys. 15(14):5163–5175
40. Sinkeldam RW, Greco NJ, Tor Y (2010) Fluorescent analogs of biomolecular building blocks: design, properties, and applications. Chem. Rev. 110:2579–619
41. Snoek L, Robertson E, Kroemer R, Simons J (2000) Conformational landscapes in amino acids: infrared and ultraviolet ion-dip spectroscopy of phenylalanine in the gas phase. Chem. Phys. Lett. 321:49
42. Sobolewski AL, Domcke W, Dedonder-Lardeux C, Jouvet C (2002) Excited-state hydrogen detachment and hydrogen transfer driven by repulsive $^1\pi\sigma^*$ states: a new paradigm for nonradiative decay in aromatic biomolecules. Phys. Chem. Chem. Phys. 4:1093–1100
43. Sobolewski AL, Shemesh D, Domcke W (2009) Computational Studies of the Photophysics of Neutral and Zwitterionic Amino

Acids in an Aqueous Environment: Tyrosine-(H2O)(2) and Tryptophan-(H2O)(2) Clusters. J. Phys. Chem. A 113(3):542–550
44. Stanton JF, Bartlett RJ (1993) The equation of motion coupled-cluster method. A systematic biorthogonal approach to molecular excitation energies, transition probabilities, and excited state properties. J. Chem. Phys. 98:7029
45. Taskent-Sezgin H, Chung J, Patsalo V, Miyake-Stoner SJ, Miller AM, Brewer SH (2009) Mehl, R.a., Green, D.F., Raleigh, D.P., Carrico, I.: Interpretation of p-cyanophenylalanine fluorescence in proteins in terms of solvent exposure and contribution of side-chain quenchers: a combined fluorescence, IR and molecular dynamics study. Biochemistry 48:9040–9046
46. Taskent-Sezgin H, Marek P, Thomas R, Goldberg D, Chung J, Carrico I, Raleigh DP (2010) Modulation of p-cyanophenylalanine fluorescence by amino acid side chains and rational design of fluorescence probes of alpha-helix formation. Biochemistry 49:6290–6295
47. Tucker MJ, Oyola R (2006) A Novel Fluorescent Probe for Protein Binding and Folding Studies: p-Cyano-Phenylalanine. Biopolymers 83:571–576
48. Twine SM, Szabo AG (2003) Fluorescent amino acid analogs. Methods Enzymol. 360:104–127
49. Wavefunction (2010) spartan 10; wavefunction, inc
50. Zhang M, Huang Z, Lin Z (2005) Systematic ab initio studies of the conformers and conformational distribution of gas-phase tyrosine. J. Chem. Phys. 122:134, 313

Theor Chem Acc (2014) 133:1482
DOI 10.1007/s00214-014-1482-9

REGULAR ARTICLE

Trihalide cations MF_3^+, MCl_3^+ and MBr_3^+, M = B, Al, Ga: pseudo Jahn–Teller coupling, electronic spectra and ionization potentials of MX_3

Friedrich Grein

Received: 6 January 2014 / Accepted: 15 March 2014 / Published online: 23 April 2014

Abstract Optimized geometries and energies, vertical excitation energies and vibrational frequencies are reported for nine cations MX_3^+, with M = B, Al, Ga, and X = F, Cl, Br. Density functional theory using long-range corrected functionals, coupled cluster and multireference configuration interaction methods were applied with triple-zeta polarized basis sets. All cations were shown to be distorted from the high D_{3h} to the lower C_{2v} symmetry due to a strong pseudo Jahn–Teller effect. Geometry optimizations lead to two $^2B_2$ states, one with an X(axial)–M–X angle above 120°, to give a structure with one short and two long bonds (1S2L), the other having such angle below 120°, resulting in a structure with two short and one long bond (2S1L). In most cases, the 1S2L structure was found to be more stable than 2S1L, but the stabilization energies of 1S2L and 2S1L differ by no more than 0.2 eV. There is a saddle point at D_{3h} symmetry. Adiabatic and vertical ionization energies of MX_3 are also reported. Good to excellent agreement with available experimental data was found.

Keywords Trihalide cations · Pseudo Jahn–Teller effect · Double minima in C_{2v} symmetry · Coupled cluster and density functional calculations · Excitation energies · Ionization potentials

Dedicated to the memory of Professor Isaiah Shavitt and published as part of the special collection of articles celebrating his many contributions.

Electronic supplementary material The online version of this article (doi:10.1007/s00214-014-1482-9) contains supplementary material, which is available to authorized users.

F. Grein (✉)
Department of Chemistry, University of New Brunswick, PO Box 4400, Fredericton, NB E3B5A3, Canada
e-mail: fritz@unb.ca

1 Introduction

In the present series of studies, AX_3 molecules and ions are chosen which may be subject to the pseudo Jahn–Teller effect [1], whereby the symmetry of a non-degenerate state is lowered due to vibronic coupling with low-lying excited states. With the highest symmetry of (nonlinear) AX_3 systems being D_{3h}, molecules are selected for which the lowest state in D_{3h} symmetry is not degenerate. If degenerate, first-order Jahn–Teller splitting [2] will occur. Molecules with 22 valence electrons (VE), such as CO_3, have in D_{3h} symmetry the electron configuration … e'^4 e''^4, giving rise to a non-degenerate $^1A_1'$ state. Systems with 19–21 VEs have a degenerate ground state, with e'' partially occupied. With 23 VEs, examples are NO_3 and CO_3^-, and 24 VEs, examples are NO_3^- and CO_3^{2-}, an a_2' orbital is occupied, leading to a non-degenerate lowest state. For systems with 25 (example SO_3^-) and 26 VEs (example SO_3^{2-}), the next higher orbital a_1' is filled, and with 27 and 28 VE's an a_2'' orbital is occupied, in all cases leading to non-degenerate lowest states in D_{3h} symmetry. For more VEs, degeneracies will occur. As demonstrated, AX_3 systems with 22–28 VEs have non-degenerate lowest D_{3h} states and are therefore candidates for studying the PJTE.

In the first paper of this series [3], trioxides with 22 and 23 VEs were studied. It was shown that the 22 VE trioxides CO_3, NO_3^+, BO_3^- and isovalent systems with the central atom being Si, P and Al, have minima at both D_{3h} and C_{2v} symmetry. The C_{2v} minimum, at lower energy, results from vibronic coupling of the $1^1A_1'$ ground state with $^1E'$ excited states via the e' vibrational mode. For GeO_3, AsO_3^+ and

GaO_3^-, however, only the C_{2v} minimum remains. The $^1A_1$ Jahn–Teller component of the excited $1^1E'$ state has a minimum in C_{2v} symmetry at an O(axial)–A–O angle well in excess of 120°, causing the C_{2v} minimum.

The situation is quite different for 23 VE systems [3]. The trioxides with the lighter atoms, CO_3^-, SiO_3^-, NO_3, PO_3 and SO_3^+, have a $^2A_2'$ ground state in D_{3h} symmetry and are not subject to the PJTE. However, GeO_3^-, AsO_3 and SeO_3^+ have a double-well potential in C_{2v} symmetry, with a saddle point at D_{3h} symmetry. A strong PJTE interaction between the $^2A_2'$ lowest state and the $^2E'$ exited states is encountered, due to lower excitation energies and favorable ∂/∂q matrix elements. In all cases investigated, the lowest excited $^2E'$ state has a minimum at D_{3h} symmetry, not at C_{2v} symmetry as for the 22 VE systems.

In the second paper of this series [4], the structure of BF_3^+ was investigated. The ground state of BF_3^+, with 23 VEs, might be expected to be most stable in D_{3h} symmetry, such as in NO_3 or CO_3^-. While both BF_3^+ and NO_3 have in D_{3h} symmetry a $1^2A_2'$ ground state and low-lying excited $^2E'$ states (at 1.16 eV for $1^2E'$ of BF_3^+, and 2.10 eV for NO_3; MRCI results), their structures are in fact quite different. As Haller et al. [5] have shown, due to strong PJT coupling the lowest state of BF_3^+ has C_{2v} symmetry with a minimum at an F(axial)-B–F angle of $\sim 108°$. According to ref [4], a strong PJTE actually leads to a double-well potential, with minima at 134° (1S2L) and 106° (2S1L), the latter being slightly lower in energy. There is a saddle point at D_{3h} symmetry. Astonishingly, density functional theory (DFT) methods with common density functionals, such as B3LYP and B3PW91, only obtain the 1S2L minimum at the larger angle. Optimizations with non-hybrid functionals such as BLYP and BPW91 settle on a D_{3h} minimum only. Long-range corrected functionals are required to cover both minima, and to obtain results similar to Møller–Plesset (MP) and coupled cluster (CC) values.

In the present paper, DFT, coupled cluster and multireference configuration interaction (MRCI) results will be presented for MF_3^+, MCl_3^+ and MBr_3^+, where M = B, Al, Ga, all cations having 23 VEs. Results for BF_3^+ from ref 4 will be included for comparison.

DFT, second-order MP (MP2) and coupled cluster calculations with single and double substitutions and non-iterative triple excitations CCSD(T) studies were performed by Kaltsoyannis and Price [6] for the D_{3h} and 2S1L minima of BF_3^+ and AlF_3^+. Using the B3LYP method, only a D_{3h} state was reported. Vibrational frequencies were not presented. In combination with experimental studies, CCSD(T) calculations by Yang et al. [7] obtained a global 1S2L minimum for BCl_3^+. Transition structures were found at D_{3h} and C_{2v} (2S1L) symmetries. They also reported adiabatic (IE_{ad}) and vertical (IE_{vert}) ionization energies for BCl_3^+, as well as vibrational frequencies for the 1S2L structure.

No theoretical results were found for the other systems. Photoelectron and photoionization studies are available for MF_3, MCl_3 and MBr_3, providing ionization energies as well as vertical excitation energies and some vibrational frequencies for the corresponding cations in D_{3h} symmetry.

2 Methods

The highest symmetry of (nonlinear) $\mathbf{MX_3^+}$ systems is D_{3h}. Due to the pseudo Jahn–Teller effect, distortion to C_{2v} symmetry occurs. In C_{2v} symmetry, the bond angle α is defined as X^1–M–X^2, where X^1 lies on the symmetry axis. DFT, time-dependent DFT (TD-DFT) and equation of motion CCSD (EOM-CCSD) calculations were performed with the Gaussian09 programs [8], using the long-range corrected LC-B97D functional [9, 10] for DFT. With common functionals, such as B3LYP, the 2S1L minimum of BF_3^+ could not be obtained [4]. The basis set is 6-311+G(3df) for DFT, TD-DFT and EOM-CCSD.

Closed-shell and restricted open-shell CCSD (RCCSD) and RCCSD(T) calculations were carried out with the MOLPRO programs [11, 12], using the cc-pVTZ basis set.

Multi-reference configuration interaction (MRCI) calculations [13, 14] with single and double excitations were conducted at optimized LC-B95D geometries, using the valence triple-zeta plus double polarization basis set TZVPP [15]. Calculations were done with ground state orbitals for eight roots of each C_{2v} irreducible representation (IR), at configuration selection thresholds of 0.5, 0.2 and 0.1 μh. The core orbitals (1*s* for B, 1*s*, 2*s*, 2*p* for Al, etc.) were frozen. Starting at 0.5 μh with CAS(13,15) reference functions, at 0.2 μh and 0.1 μh, the reference space consisting of all configurations with a coefficient $c^2 \geq 0.001$ produced in the previous step was used. In the final 0.1 μh step, about 1.3×10^6 configurations per IR were included for diagonalization, using about 180 reference functions per IR. The final 'estimated full CI energies,' as given in the tables, were obtained by applying to the diagonalized CI energies a Davidson-type correction to account for higher excitations [16, 17], and by extrapolating to zero threshold, using the sum of multi-reference MP2 energies of discarded configurations. MRCI energies for Fig. 1 were obtained at angles α varying from 90° to 150°, in spaces of 10°, using the respective LC-B95D optimized geometries.

3 Results

3.1 Geometries and energies of ground state

RCCSD(T) and LC-B97D results for geometries and energies of the MX_3^+ cations are shown in Table 1 (see Tables

S1 and S2 in the supplementary information for more details). For each system, geometries were optimized at D_{3h} symmetry for the lowest $1^2A_2'$ state, and at C_{2v} symmetry for the lowest 1^2B_2 state, starting with angles α below and above 120°. Structures with $\alpha < 120°$ have two short and one long bonds (2S1L) and $r_1 > r_2$, whereas structures with $\alpha > 120°$ have one short and two long bonds (1S2L) with $r_1 < r_2$. This is a consistent finding, also seen for the CO_3 and CO_3^- series [3]. The reason is eventual dissociation of MX_3^+ to MX_2^+ (r_2) + X (r_1) for $\alpha < 120°$, and to MX^+ (r_1) + X_2 (r_2) for $\alpha > 120°$. MX_2^+ is linear. In all cases, two C_{2v} minima were found. In several instances, one of them is associated with an imaginary frequency. For most systems, the lowest frequency of the D_{3h} structure was calculated to be imaginary, indicating a saddle point.

The angles of the 2S1L states decrease (move further below 120°) as the central atom gets heavier. For the MF_3^+ series, the angles go from 106.18° to 98.26° to 96.30° (RCCSD(T) values). Understandably, for a given X, the bond distances increase as one goes from BX_3^+ to GaX_3^+. The angles of the 1S2L states increase from 134.08° for BF_3^+ to 144.40° for AlF_3^+ to 146.30° for GaF_3^+. The same trends of decreasing/increasing angles and increasing bond distances can be seen for the other series.

In most cases, both C_{2v} states are more stable than the D_{3h} state. Stabilization energies (with respect to the D_{3h} energy) increase within each series, e.g., from 0.49 eV for BF_3^+ to 0.90 eV for AlF_3^+ to 1.32 eV for GaF_3^+. The stabilization energies of the two C_{2v} minima are usually very similar, differing by no more than 0.2 eV. In general, the MX_3^+ cations are calculated to be more stable in the 1S2L structure. However, 2S1L is more stable for BF_3^+ and GaF_3^+. According to the BF_3^+ paper [4], all ab initio methods (HF, MP2, CCSD, CCSD(T)) lead to BF_3^+ being more stable in 2S1L. In the case of BCl_3^+ and BBr_3^+, the 2S1L structure is calculated to be less stable than the D_{3h} isomer (positive sign). Such optimizations should have led to D_{3h} symmetry.

The optimized bond distances and angles obtained by LC-B97D are very similar to those found by RCCSD(T). Stabilization energies differ by up to 0.3 eV from RCCSD(T) values, but again the differences between 1S2L and 2S1L lie within 0.2 or even 0.1 eV. In most cases, 1S21 is calculated to be more stable than 2S1L.

The CCSD(T)/6-311G* results of Kaltsoyannis and Price [6] for BF_3^+ show a C_{2v} minimum at 104.80°, with B–F bond distances of 1.729 and 1.250 Å, lying 56.5 kJ/mol (0.59 eV) lower than the D_{3h} energy. The CCSD(T)/6-311G* minimum of AlF_3^+ lies at 98.01°, being 91.5 kJ/mol (0.95 eV) lower than the D_{3h} value, whereas the B3LYP/6-311G* C_{2v} minimum was found at 109.13°, 0.90 kJ/mol (0.01 eV) lower than D_{3h}. CCSD(T) calculations on BCl_3^+ by Yang et al. [7] found a global C_{2v} minimum at 132.2° (1S2L), with bond distances of 1.670 and 1.776 Å, lying 0.21 eV below the D_{3h} state.

Table 1 RCCSD(T)/cc-pVTZ (first line) and LC-B97D/6-311+G(3df) (second line) geometries and energies for MF_3^+ to MBr_3^+ systems with 23 valence electrons, at D_{3h} and C_{2v} (2S1L and 1S2L) symmetries

System	D_{3h} ($1^2A_2'$)	2S1L (1^2B_2)			1S2L (1^2B_2)			ΔE 2S1L	ΔE 1S2L
	r	r_1	r_2	α	r_1	r_2	α		
BF_3^+	1.313	1.660	1.254	106.18	1.250	1.365	134.08	−0.49	−0.38
	1.298	1.638	1.245	105.94	1.240	1.355	134.39	−0.33	−0.38
AlF_3^+	1.667	2.004	1.608	98.26	1.602	1.722	144.40	−0.90	−0.90
	1.635	1.911	1.589	99.56	1.582	1.695	144.07	−0.59	−0.84
GaF_3^+	1.752	2.173	1.673	96.30	1.671	1.801	146.30	−1.32	−1.17
	1.728	2.110	1.660	95.94	1.658	1.789	146.24	−1.23	−1.25
BCl_3^+	1.727	1.845	1.695	112.38	1.670	1.776	132.10	+0.44	−0.19
	1.709	1.874	1.675	112.04	1.656	1.761	131.72	−0.21	−0.31
$AlCl_3^+$	2.087	2.321	2.035	104.83	2.023	2.143	140.07	−0.31	−0.50
	2.044	2.285	2.003	105.29	1.989	2.107	138.78	−0.53	−0.58
$GaCl_3^+$	2.163	2.475	2.092	102.63	2.082	2.219	141.59	−0.51	−0.63
	2.092	2.418	2.037	103.06	2.026	2.158	139.70	−0.85	−0.78
BBr_3^+	1.886	1.973	1.960	113.20	1.831	1.935	131.61	+0.05	−0.16
	1.871	2.012	1.843	112.79	1.820	1.923	131.22	−0.16	−0.28
$AlBr_3^+$	2.244	2.457	2.195	106.11	2.180	2.301	138.91	−0.21	−0.42
	2.200	2.431	2.160	106.43	2.144	2.265	137.34	−0.43	−0.48
$GaBr_3^+$	2.299	2.577	2.236	104.70	2.222	2.356	139.99	−0.34	−0.51
	2.237	2.543	2.188	104.62	2.174	2.305	137.96	−0.68	−0.64

Distances r in Å, angles α in degree, energy differences $\Delta E = E(C_{2v}) - E(D_{3h})$ in eV

3.2 Vertical excitation energies at D_{3h} symmetry and the pseudo Jahn–Teller effect

At D_{3h} symmetry, the ground state of the MX_3^+ systems is $^2A_2'$ (with a ....e'^4 e''^4 $a_2'^1$ configuration). The lowest excited state is $1^2E''$ ($e'' \rightarrow a_2'$), followed by $1^2E'$ ($e' \rightarrow a_2'$), $1^2A_2''$ ($a_2'' \rightarrow a_2'$) and at higher energies by $2^2E'$ (lower $e' \rightarrow a_2'$).

TD-DFT, EOM-CCSD and MRCI vertical excitation energies (VEE) at D_{3h} symmetry are presented in Table 2. Energies calculated by different methods are similar. For example, for BF_3^+ VEEs for $1^2E''$ range from 0.88 eV (MRCI) to 1.02 eV (EOM), for $1^2E'$ from 1.16 eV (MRCI) to 1.30 eV (TD), and for $2^2E'$ from 4.20 eV (MRCI) to 4.44 eV (EOM). From photoelectron spectra, experimental VEEs are available for BF_3^+ [18], AlF_3^+ [19], BCl_3^+ [20], $AlCl_3^+$ [21], BBr_3^+ [20, 22] and $AlBr_3^+$ [21], as shown in Table 2. Experimental values for the Ga compounds could not be found. As expected, the MRCI values come closest to the experimental ones, with good to excellent agreement. The MRCI values of BF_3^+ lie within 0.15 eV of the experimental values obtained by Romero et al. [18]. Good agreement is obtained for AlF_3^+, except that the experimental 4.64 eV could not be matched. The calculated VEEs of BCl_3^+ also lie within 0.15 eV of the experimental numbers, except for $1^2E'$ (0.79 MRCI, 1.15 expt.). Good agreement is obtained for $AlCl_3^+$ and $AlBr_3^+$ (maximal 0.2 eV deviation). For BBr_3^+, there is a discrepancy in $1^2E'$, with 0.78 eV calculated by MRCI, and 1.20 eV experimental, similar to the BCl_3^+ deviation. The excitation energies of the lowest state are very low, getting lower as the central atom gets heavier. The lowest energy is obtained for the $1^2E'$ state GaF_3^+, with a value of 0.07 eV.

The distortion of the MX_3^+ cations from the highest symmetry D_{3h} to C_{2v} is caused by the pseudo Jahn–Teller effect. Vibronic coupling of the $^2A_2'$ ground state with $^2E'$ excited states via an e' vibrational mode causes a change from D_{3h} to C_{2v} symmetry ($e' * E' = A_1', A_2', E'$). Distortion to C_{3v} symmetry via the a_2'' vibrational mode would in principle be possible by coupling of the ground state with $^2A_1''$ excited states ($a_2'' * A_1'' = A_2'$), but there are no low-lying $^2A_1''$ excited states, as Table 2 shows. One of the criteria for strong vibronic coupling is a low excitation energy.

MRCI potential curves for BF_3^+, as function of the F–B–F angle α ($\alpha = 120°$: D_{3h}, $\alpha \neq 120°$:C_{2v}) [Fig. 2 of ref [4]], show shallow minima around 105° and 135° in the potential curve of the ground state, in good agreement with the calculated minima (Table 1), and a maximum (saddle point) at 120°. They also display the Jahn–Teller splitting of $1^2E'$ into 1^2A_1, with a minimum below 90°, and 2^2B_2, with a minimum close to 120°. At higher energies, $2^2E'$ splits into 2^2A_1, with a minimum around 110°, and 3^2B_2, with a minimum around 120°. With 2^2B_2 and 3^2B_2 minimizing around 120°, the double minimum in the ground state potential is caused by strong PJT interaction.

Table 2 Vertical excitation energies (eV) at D_{3h} symmetry for trihalides MF_3^+ to MBr_3^+ with 23 valence electrons

	$1^2E''$	$1^2E'$	$1^2A_2''$	$2^2E'$	$1^2A_1'$ ($2^2E''$)	Next state
BF_3^+						
TD-DFT	0.94	1.30	3.18	4.26	5.19	11.36
EOM	1.02	1.20	3.34	4.44	5.66	11.60
MRCI	0.88	1.16	3.10	4.20	5.40	11.15
Expt. Ref. [18]	0.75	1.12	3.20	4.16	5.55	
AlF_3^+						
TD-DFT	0.59	0.82	1.18	1.76	2.93	8.74
EOM	0.63	0.67	1.16	1.81	3.08	8.51
MRCI	0.39	0.60	1.12	1.62	2.83	8.22
Expt. Ref. [19]		0.65		1.62	4.64	
GaF_3^+						
TD-DFT	0.23	0.12	0.96	1.45	3.25	6.69
EOM	0.34	0.06	0.92	1.51	3.30	6.60
MRCI	0.13	0.07	1.01	1.44	3.11	6.75
BCl_3^+						
TD-DFT	0.76	0.39	2.93	4.02	5.93	6.22
EOM	0.85	0.92	2.91	4.08	6.16	6.30
MRCI	0.71	0.79	2.66	3.82	5.94	6.22
Expt. Ref. [20]	0.63	1.15	2.59	3.87	6.11	
$AlCl_3^+$						
TD-DFT	0.56	0.29	1.37	2.02	4.05	6.69
EOM	0.65	0.70	1.43	2.26	4.09	7.42
MRCI	0.50	0.68	1.29	2.07	3.82	6.66
Expt. Ref. [21]	0.46	0.72	1.32	2.03	3.96	
$GaCl_3^+$						
TD-DFT	0.34	0.48	1.24	1.83	4.56	5.43
EOM	0.26	0.41	1.13	1.95	4.32	5.80
MRCI	0.34	0.35	1.08	1.89	4.11	5.39
BBr_3^+						
TD-DFT	0.65	0.18	2.64	3.70	4.91	5.85
EOM	0.79	0.83	2.70	3.83	5.05	5.45
MRCI	0.71	0.78	2.64	3.81	5.82	5.94
Expt. Ref. [20, 22]	0.62	1.20	2.43	3.58	6.12	6.63
$AlBr_3^+$						
TD-DFT	0.50	0.34	1.33	2.05	4.30	5.79
EOM	0.57	0.58	1.30	2.16	4.25	6.00
MRCI	0.48	0.63	1.27	2.08	4.02	5.75
Expt. Ref. [21]	0.62	0.83	1.46	2.26	4.32	
$GaBr_3^+$						
TD-DFT	0.31	0.47	1.23	1.80	4.65	4.89
EOM	0.41	0.30	1.17	2.00	4.61	5.12
MRCI	0.31	0.35	1.13	1.92	4.34	4.67

TD-DFT, EOM-CCSD and MRCI results. Comparison with photoelectron results. The lowest state in D_{3h} symmetry is $1^2A_2'$

The PJT interaction between the ground state and the excited states is much stronger for GaF_3^+, as seen in the MRCI potential curves, Fig. 1. Again, the ground state has two C_{2v} minima, around 96° and 146°, and a maximum at 120°. The energy difference between the maximum at 120° and the minimum at 146° is 1.25 eV in LC-B97D and 1.17 eV in RCCSD(T), whereas the energy difference to the minimum around 96° is 1.23 eV in LC-B97D and 1.32 eV in RCCSD(T). As in the case of BF_3^+, the Jahn–Teller states 2^2B_2 and 3^2B_2 have their minima around 120°. There is an avoided crossing between 2^2B_2 and 3^2B_2 around 130°, as confirmed by an exchange of leading configurations.

3.3 Vertical excitation energies at C_{2v} symmetry

Vertical excitation energies for the 1S2L and 2S1L structures, calculated by TD-DFT and EOM-CCSD, using the respective geometries are shown in Tables 3 and 4, respectively. (A definite decision on which C_{2v} isomer is more stable cannot be made due to the small difference in stabilization energies.) VEEs for the lowest excited state

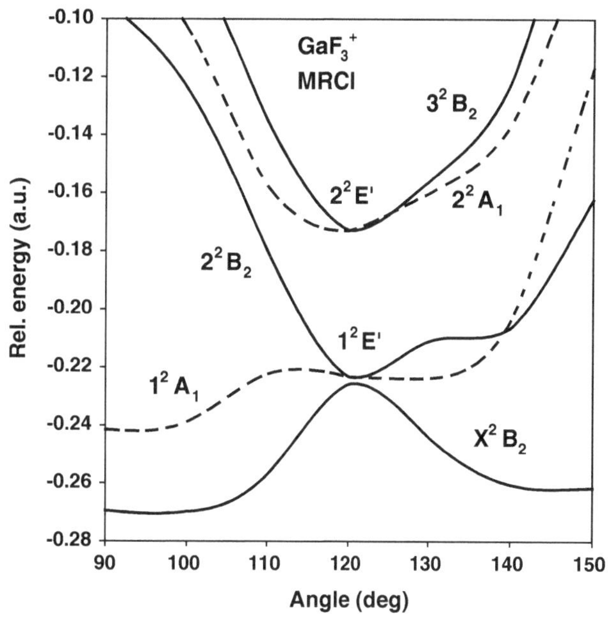

Fig. 1 MRCI potential curves for GaF_3^+ as function of the F^1–Ga–F^2 angle

Table 3 Vertical excitation energies (eV) for trihalides MF_3^+ to MBr_3^+ with 23 valence electrons at optimized 1S2L geometry

	1^2A_2	2^2B_2	1^2B_1	1^2A_1	2^2B_1	3^2B_2	2^2A_1	3^2A_1
BF_3^+								
TD	1.37	2.84	2.86	3.06	5.34	5.61	5.97	7.51
EOM	1.40	3.12	3.11	2.97	6.02	6.27	6.19	8.51
AlF_3^+								
TD	1.57	2.70	2.93	3.31	5.52	5.50	4.60	6.84
EOM	1.60	2.75	3.01	3.19	–	6.31	4.67	7.72
GaF_3^+								
TD	1.51	2.31	2.83	3.12	6.27	6.29	4.77	6.88
EOM	1.52	2.32	2.75	2.85	–	6.54	4.67	7.16
BCl_3^+								
TD	1.09	2.42	2.37	1.90	4.55	5.00	5.52	6.12(B_1)
EOM	1.03	2.03	2.06	1.97	3.94	4.57	5.22	6.21(B_1)
$AlCl_3^+$								
TD	1.18	2.43	2.32	2.08	4.50	4.53	4.21	6.52
EOM	1.14	2.37	2.16	2.18	3.84	3.87	4.10	6.11
$GaCl_3^+$								
TD	1.15	2.15	2.24	1.85	4.81	4.87	4.31	5.56
EOM	1.14	2.16	2.11	2.09	4.03	4.08	4.20	5.82
BBr_3^+								
TD	0.92	2.12	2.08	1.68	4.05	4.55	5.10	5.10 (B_1)
EOM	1.05	1.94	2.03	2.01	3.73	4.33	5.02	5.07 (B_1)
$AlBr_3^+$								
TD	0.98	2.17	2.01	1.78	4.01	4.05	3.97	5.66
EOM	1.13	2.38	2.19	2.24	3.56	3.65	4.25	5.61
$GaBr_3^+$								
TD	0.95	1.92	1.94	1.58	4.24	4.31	4.00	4.75
EOM	1.13	2.21	2.12	2.14	3.66	3.75	4.24	4.96 (B_1)

The ground state is X^2B_2. TD-DFT and EOM-CCSD results

Table 4 Vertical excitation energies (eV) for trihalides MF_3^+ to MBr_3^+ with 23 valence electrons at optimized 2S1L geometry

	1^2B_1	1^2A_1	1^2A_2	2^2B_2	2^2B_1	2^2A_1	3^2B_2	3^2A_1
BF_3^+								
TD	0.47	1.32	4.31	4.43	6.00	6.01	7.46	7.93
EOM	0.25	1.30	5.83	5.79	7.44	7.31	9.06	9.72
AlF_3^+								
TD	0.43	1.19	3.27	3.37	3.55	3.58	4.22	5.21
EOM	0.19	1.05	5.00	4.93	5.20	5.14	5.99	6.92
GaF_3^+								
TD	0.28	0.85	4.47	4.47	5.06	5.01	5.77	6.84
EOM	0.14	0.81	5.39	5.34	5.80	5.71	6.52	7.39
BCl_3^+								
TD	0.47	1.21	2.37	2.46	4.02	4.37	5.57	6.07
EOM	0.54	1.14	1.79	1.83	3.45	4.12	–	6.12
$AlCl_3^+$								
TD	0.26	1.12	2.89	2.97	3.33	3.38	4.35	5.86
EOM	0.28	1.19	2.61	2.57	3.01	3.10	4.08	5.56
$GaCl_3^+$								
TD	0.07	0.77	3.49	3.55	4.08	4.09	5.11	5.69
EOM	0.19	0.94	3.19	3.16	3.66	3.67	4.73	5.93
BBr_3^+								
TD	0.16	0.92	2.16	2.25	3.61	3.97	5.10	5.04
EOM	0.55	0.99	1.47	1.56	3.04	3.94	–	5.16
$AlBr_3^+$								
TD	0.15	0.89	2.43	2.54	2.90	2.99	3.93	5.71
EOM	0.28	1.14	2.14	2.12	2.61	2.83	3.80	5.56
$GaBr_3^+$								
TD	0.23	0.61	2.67	2.74	3.10	3.14	3.95	4.12
EOM	0.21	0.95	2.55	2.54	3.05	3.16	4.24	5.06

The ground state is X^2B_2. TD-DFT and EOM-CCSD results

1^2A_2 of the 1S2L structures are in the 1–2 eV range, followed by 2^2B_2, 1^2B_1 and 1^2A_1 in the 2–3 eV range. Oscillator strengths (not reported) are very low for all but the $^2A_1$ states.

The lowest excited state in 2S1L is 1^2B_1, followed by 1^2A_1. Both states have very low VEEs. The energy of 1^2B_1 (transition from ground state is forbidden) ranges from 0.07 eV ($GaCl_3^+$) to 0.47 eV (BF_3^+, BCl_3^+) in LC-B97D, and from 0.14 eV (GaF_3^+) to 0.64 eV (BBr_3^+) in EOM-CCSD. VEEs of 1^2A_1 are still in the 1 eV range. Oscillator strengths are very low, except for $^2B_2$ excited states.

Comparison of the C_{2v} excitation energies with D_{3h} values is quite interesting. $1^2E''$ in D_{3h} correlates with $1^2B_1 + 1^2A_2$ in C_{2v}; $1^2E'$ with $1^2A_1 + 2^2B_2$, and $1^2A_2''$ with 2^2B_1. For 1S2L, $1^2E''$ in D_{3h} splits into $1^2A_2 + 1^2B_1$, with 1^2A_2 having the lower energy, and 1^2B_1 lying 1–2 eV higher. Both states have energies above $1^2E''$. On the other hand, $1^2E'$ splits into $1^2A_1 + 2^2B_2$, both having energies higher than $1^2E'$, with 1^2A_1 and 2^2B_2 energies differing by about 0.5 eV or less. One or the other may be at lower energy. The $1^2A_2''$ state in D_{3h} correlates with 2^2B_1 in C_{2v}. The 2^2B_1 energies are much higher than the corresponding $1^2A_2''$ values.

The situation is somewhat different for the 2S1L VEEs. The 1^2B_1 component of $1^2E''$ has VEE below that of $1^2E''$, whereas the 1^2A_2 energy lies well above the $1^2E''$ energy, with differences between the 1^2B_1 and 1^2A_2 energies amounting to 2–4 eV. The splitting between the 1^2A_1 and 2^2B_2 components of $1^2E'$ is also quite high, 1^2A_1 always having the lower VEE. Due to such splitting, there are two low-lying states 1^2B_1 and 1^2A_1, followed by a large gap in excitation energies.

3.4 Vibrational frequencies at D_{3h} and C_{2v} symmetries

In Table 5, harmonic frequencies for the 1S2L and 2S1L structures, calculated by LC-B97D/6-311+G(3df), are displayed. Some experimental frequencies are known for BF_3^+ [23], BCl_3^+ [7] and BBr_3^+ [24] from photoelectron and photoionization spectra, as listed in Table 5. For BF_3^+, the frequencies are better assigned to 2S1L than 1S2L, but for BCl_3^+ and BBr_3^+ to 1S2L, since in these cases, the

Table 5 Harmonic frequencies ν (cm^{-1}) and IR intensities (km/mol) for trihalides MF_3^+ to MBr_3^+ at 1S2L (first line) and 2S1L (second line) geometry

	b_2^a	a_1	b_1	b_2	a_1	a_1
BF_3^+ expt.[b]	364	439	**529**	**539**	880	**1,741**
	129	282	**537**	**1,852**	**455**	**1,084**
			509.6	**1,790.8**		**(1,790.8)**
AlF_3^+	**166**	362	**190**	**500**	685	**1,103**
	−64	**205**	**200**	**1,143**	**395**	776
GaF_3^+	73	349	**149**	**366**	670	**887**
	54	**174**	**155**	**933**	278	750
BCl_3^+ expt.[c]	238	247	399	**524**	488	**1,137**
	−145	233	418	**1,097**	381	**699**
	194	209		**1,090**		**(1,090)**
$AlCl_3^+$	103	185	**153**	**328**	400	**753**
	14	**137**	**163**	**758**	**306**	**473**
$GaCl_3^+$	−39	172	114	**213**	386	**558**
	53	**119**	121	**584**	**214**	431
BBr_3^+ expt.[d]	146	151	346	**496**	291	**958**
	−121	141	364	**892**	252	**583**
				930		(930)
$AlBr_3^+$	71	109	**131**	**256**	240	**616**
	−22	89	**140**	**614**	215	**348**
$GaBr_3^+$	17	101	90	**148**	234	**423**
	18	**80**	94	**437**	**160**	**268**

LC-B97D results. High IR intensities indicated in bold. Comparison with experimental values

[a] Minus sign for imaginary frequency

[b] Experimental values, Ref. [23]

[c] Experimental values, Ref. [7]

[d] Experimental values, Ref. [24]

harmonic frequencies are higher than the experimental values, as is usually the case due to neglect of anharmonic contributions. The most intense frequency b_2 for the 2S1L isomer of BF_3^+ is calculated to be 1,852 cm^{-1}, experimentally 1,791 cm^{-1}; a_1 for the 1S2L isomer of BCl_3^+ is calculated as 1,137 cm^{-1}, experimentally 1,090 cm^{-1}, and a_1 for the 1S2L isomer of BBr_3^+ is calculated as 958 cm^{-1}, experimentally 930 cm^{-1}. The suggestion that 1S2L is the more stable isomer of BCl_3^+ and BBr_3^+ agrees with the lower energy calculated for 1S2L, but contrasts the finding for BF_3^+, where the LC method predicts 1S2L to be more stable (by only 0.05 eV). Imaginary frequencies (Table 5) were obtained for the 2S1L isomers of AlF_3^+, BCl_3^+, BBr_3^+ and $AlBr_3^+$, and for the 1S2L isomer of $GaCl_3^+$, in complete agreement with the higher stabilization energies calculated in the LC method for these isomers. For the remaining four cations, both 1S2L and 2S1L have real frequencies, although stabilization energies prefer one isomer over the other, however, by very small amounts (0.05 eV or less). Possible reasons for this outcome are insufficient accuracy of the calculations or a double minimum.

Harmonic frequencies at D_{3h} symmetry, calculated by the same method, are shown in Table S3 in the supplementary information. In most, but not all cases, the lowest frequency is imaginary, indicating that the cations are unstable in D_{3h} symmetry. As this is expected for all cations covered in this paper, it is assumed that the accuracy of the calculations is not sufficient to properly resolve shallow minima (see above). The frequencies decrease as one goes from B to Al to Ga within a given group.

3.5 Adiabatic and vertical ionization energies

In Table 6, optimized geometries of neutral AX_3 molecules, as obtained by the RCCSD(T) method, are shown. The bond distances for BF_3 and AlF_3 agree closely with values obtained by Gutsev et al. [25], using the CCSD(T) method with a 6-311+G(2d) basis set for B and F, and the 6-311+G(3d) basis set for Al. RCCSD(T) values for adiabatic (IE_{ad}) and vertical (IE_{vert}) ionization energies of BF_3 to $GaBr_3$ are also presented in Table 6. The calculated values are in good to excellent agreement with available experimental data [19, 21, 26–31], which were taken from the compilation in NIST [32]. It is seen that theoretical IE_{ad} values differ by no more than 0.2 eV from the experimental ones, with the exception of $AlCl_3$. However, the (older) experimental value for $AlCl_3$ of 12.8 ± 0.5 eV is obviously in error, as it is higher than IE_{vert} (12.01 eV). The correct IE_{ad} for $AlCl_3$ should be close to 11.4 eV. Excellent agreement is seen for IE_{vert}, with several theoretical values differing by 0.03 eV or less from the experimental ones. The largest deviation of 0.08 eV occurs for $AlCl_3$. As mentioned in the Introduction, Yang et al. [7] obtained 11.61 eV for IE_{ad} of BCl_3, and 11.71 eV for IE_{vert}.

As Table 6 shows, the IE values of MX_3 are nearly independent of the central atom, and closely reflect the ionization energies of the halogen atoms. (The ionization energy of F is about 17.4 eV, of Cl 13.0 eV and of Br 11.8 eV). This is understandable as the HOMO of MX_3 is a halogen lone pair orbital.

4 Discussion

In Table 7, optimized C_{2v} bond angles and stabilization energies with respect to the D_{3h} saddle point are summarized for the nine cations. In both the LC-B97D and RCCSD(T) methods, two C_{2v} geometries were optimized for each system, with a potential maximum at D_{3h} symmetry. Such results are indicative of strong pseudo Jahn–

Table 6 Optimized geometries (distances in Å) and energies *E* (hartree) for neutral MX_3 molecules at D_{3h} symmetry

	MX_3 D_{3h} ($1^1A_1'$)		IE_{ad}			IE_{vert}		
MX_3	*r*	*E*	Theor	Exptl	Ref.	Theor	Exptl	Ref.
BF_3	1.313	−324.190962	15.44	15.7 ± 0.3	[26]	15.94	15.96	[27]
AlF_3	1.642	−541.443745	14.38	–		15.31	15.45 ± 0.02	[19]
GaF_3	1.719	−2,222.489206	14.00	–		15.41	–	
BCl_3	1.747	−1,404.108654	11.44	11.64	[28]	11.65	11.62	[29]
$AlCl_3$	2.083	−1,621.417917	11.43	12.8 ± 0.5	[30]	11.93	12.01	[21]
$GaCl_3$	2.150	−3,302.726189	11.31	11.52	[31]	11.94	11.96	[21]
BBr_3	1.909	−7,742.826068	10.40	10.51 ± 0.02	[28]	10.58	–	
$AlBr_3$	2.245	−7,960.146915	10.49	10.4	[21]	10.91	10.91	[21]
$GaBr_3$	2.293	−9,641.470778	10.41	10.4	[31]	10.92	10.94	[21]

Adiabatic (IE_{ad}) and vertical (IE_{vert}) ionization energies in eV. RCCSD(T)/cc-pVTZ results

Table 7 Comparison of optimized angles and energy differences $\Delta E = E(C_{2v}) - E(D_{3h})$ obtained by LC-B97D, RCCSD(T) and B3PW91 methods

	Optimized angles (deg)			ΔE (eV)		
	LC	CCSD(T)	B3PW91	LC	CCSD(T)	B3PW91
BF_3^+	134.39	134.08	132.45	−0.38	−0.38	−0.07
	105.94	106.18	120.00	−0.33	−0.49	0.00
AlF_3^+	144.07	144.40	142.83	−0.84	−0.90	−0.27
	99.56	98.26	109.77	−0.59	−0.90	−0.01
GaF_3^+	146.24	146.30	145.27	−1.25	−1.17	−0.54
	95.94	96.30	96.58	−1.23	−1.32	−0.27
BCl_3^+	131.72	132.10	129.58	−0.31	−0.19	0.01
	112.04	112.38	120.00	−0.21	+0.44	0.00
$AlCl_3^+$	138.78	140.07	138.66	−0.58	−0.50	−0.15
	105.29	104.83	115.51	−0.53	−0.31	0.00
$GaCl_3^+$	139.70	141.59	139.96	−0.78	−0.63	−0.27
	103.06	102.63	108.44	−0.85	−0.51	−0.02
BBr_3^+	131.22	131.61	129.02	−0.28	−0.16	0.01
	112.79	113.20	120.00	−0.16	+0.05	0.00
$AlBr_3^+$	137.34	138.91	137.56	−0.48	−0.42	−0.11
	106.43	106.11	120.00	−0.43	−0.21	0.00
$GaBr_3^+$	139.40	139.99	137.96	−0.64	−0.51	−0.20
	104.62	104.70	112.75	−0.68	−0.34	−0.01

The first line for each system describes the 1S2L, the second line the 2S1L structure

Teller coupling, aided in part by very low excitation energies. Table 7 shows that RCCSD(T) geometries (as represented by the bond angle) are very similar to LC-B978D values. Stabilization energies may be larger in one or the other method, but they are again of similar magnitude. To the LC-B97D and RCCSD(T) results, taken from Table 1, B3PW91 values have been added. As stated before, and described in [4], methods with commonly used density functionals such as B3PW91 and B3LYP completely miss the 2S1L minimum of BF_3^+. Optimizations starting at a low angle, for example 100°, end in T_d symmetry. According to Table 7, the B3PW91 method also fails to obtain the 2S1L minima of BCl_3^+, BBr_3^+ and $AlBr_3^+$. For the other 2S1L isomers, B3PW91 shows a minimum at an angle below 120°, which, however, is in most cases higher than the RCCSD(T) or LC angle. For example, in AlF_3^+, the 2S1L angle is 98.26° in RCCSD(T), 99.56° in LC and 109.77° in B3PW91. Even in cases of close geometry, such as in GaF_3^+, the B3PW91 stabilization energies are usually well below the values obtained with the LC and RCCSD(T) methods. For the 1S2L isomers, the B3PW91 geometries are similar to the LC and RCCSD(T) results, but the stabilization energies are again below the reference values. This shows that the failure of common density functionals is not restricted to BF_3^+, but extends to all isovalent molecules covered in this paper.

In several, but not all cases, imaginary frequencies were calculated for the less stable C_{2v} isomer. According to the LC-B97D results of Table 5, for BF_3^+, GaF_3^+, $AlCl_3^+$ and $GaBr_3^+$, none of the two C_{2v} isomers has an imaginary frequency, and one may therefore conclude that these cations have a double minimum in C2v symmetry. However, these conclusions have to be taken with caution, as the appearance of imaginary frequencies differs with method and basis set.

The difference between vertical and adiabatic ionization energies should be close to the stabilization received by distortion from D_{3h} to C_{2v} symmetry (since in D_{3h} symmetry, the geometry of MX_3 is about the same as that of MX_3^+). In cases of good agreement between calculated and experimental IEs, experimental confirmation is given to the calculated C_{2v} structures of the MX_3^+ cations. For example, $GaBr_3^+$ has a theoretical stabilization energy of 0.51 eV (RCCSD(T) value for 1S2L), compared with an experimental difference between IE_{ad} and IE_{vert} of 0.54 eV. As the other isomer of $GaBr_3^+$, 2S1L, has a calculated stabilization energy of 0.34 eV, experimental data appear to confirm that 1S2L is more stable than 2S1L.

Experimentally known vibrational frequencies indicate that the 1S2L isomer of BCl_3^+ and BBr_3^+ is more stable than the 2S1L isomer and that 2S1L of BF_3^+ is more stable than 1S2L, in agreement with the calculated stabilization energies.

The findings for MX_3^+ species with 23 VEs are to be contrasted to those reported for XO_3 molecules with 23 VEs, which were found to have a D_{3h} structure for central atoms from the first and second row, such as NO_3 and PO_3, but a C_{2v} double-well potential for central atoms from the third row, such as AsO_3, mainly due to lower excitation energies [3].

5 Summary and conclusion

Nine cations MX_3^+, with M = B, Al, Ga, and X = F, Cl, Br, have been studied. All were shown to be distorted from the high D_{3h} to the lower C_{2v} symmetry due to the pseudo Jahn–Teller effect. In D_{3h} symmetry, the lowest state $1^2A_2'$ couples with low-lying excited $^2E'$ states via e′ vibrational modes, causing distortion to C_{2v} symmetry. Strong PJTE coupling is aided by very low excitation energies to the lowest $^2E'$ state, calculated—and in several cases experimentally confirmed—to lie between 0.07 and 1.2 eV. Geometry optimizations lead to two C_{2v} states, both having $^2B_2$ symmetry, one with an X(axial)–M–X angle above 120°, to give an 1S2L structure, the other having such angle below 120°, resulting in an 2S1L structure. In all cases, the stabilization energies between 1S2L and 2S1L differ by no more than 0.2 eV, such that, within the accuracy of the calculations, a clear decision on which isomer is more stable cannot always be made. In some cases, one or the other structure has an imaginary frequency; however, again there is some ambiguity. Despite this, based on the calculations and on some experimental evidence, preference is given to the 1S2L structure as being the more stable one for the majority of cations. An exception is BF_3^+, where all ab initio methods point to 2S1L being more stable [4].

Vertical excitation energies and harmonic vibrational frequencies at D_{3h} and C_{2v} symmetries, as well as adiabatic and vertical ionization energies of the neutral molecules, are reported. Good to excellent agreement with a multitude of experimental values was found.

The data supplied in this paper should help with experimental verification of the appropriate structures. In addition, ESR measurements might also be of use. Spin densities for 1S2L isomers are mainly located on F^2 and F^3, whereas they are concentrated on the axial F^1 for 2S1L isomers. Similarly, NMR studies might be able to distinguish between the two possible C_{2v} structures.

Other MX_3 systems, such as NF_3 with 26 VEs, are known to distort to C_{3v} symmetry, due to coupling of $1^1A_1'$ with a low-lying excited $1^1A_2''$ state. More detailed studies on NF_3 and related systems are in progress.

Acknowledgments Financial support by NSERC (Canada) in form of a Discovery Grant is gratefully acknowledged. Thanks to Dr. Pablo J. Bruna for helpful discussions and valuable suggestions, and to ACEnet for providing adequate computing facilities.

References

1. Oepik U, Pryce MLH (1957) Proc R Soc A238:425
2. Jahn A, Teller E (1937) Proc R Soc A 161:220
3. Grein F (2013) J Chem Phys 138:204305
4. Grein F (2013) Chem Phys 423:151
5. Haller E, Koeppel H, Cederbaum LS, von Niessen W, Bieri G (1983) J Chem Phys 78:1359
6. Kaltsoyannis N, Price SD (1999) Chem Phys Lett 313:679
7. Yang J, Mo Y, Lau KC, Song Y, Qian XM, Ng CY (2005) Phys Chem Chem Phys 7:1518
8. Frisch MJ, Trucks GW, Schlegel HB et al (2009) Gaussian 09, Revision A.02. Gaussian, Inc., Wallingford CT
9. Ikura H, Tsuneda T, Yanai T, Hirao K (2001) J Chem Phys 115:3540
10. Grimme S (2006) J Comp Chem 27:1787
11. Werner H-J, Knowles PJ, Knizia G, Manby FR, Schuetz M (2012) WIREs Comput Mol Sci 2:242
12. Werner H-J, Knowles PJ (2012) Molpro: a package of ab initio programs (see http://www.molpro.net for more details)
13. Grimme S, Waletzke M (1999) J Chem Phys 111:5645
14. Grimme S, Waletzke M (2000) Phys Chem Chem Phys 2:2075
15. Schäfer A, Huber C, Ahlrichs R (1994) J Chem Phys 100:5829
16. Langhoff SR, Davidson ER (1974) Int J Quant Chem 8:61
17. Bruna PJ, Peyerimhoff SD, Buenker RJ (1980) Chem Phys Lett 72:278
18. Romero J, Lane IC, Powis IJ (1993) Chem Soc Faraday Trans 89:1179
19. Dyke JM, Kirby C, Morris A, Gravenor BWJ, Klein R, Rosmus P (1984) Chem Phys 88:289

20. Mackie RA, Shpinkova LG, Holland DMP, Shaw DA (2003) Chem Phys 288:211
21. Barker GK, Lappert MF, Pedley JB, Sharp GJ, Westwood NPC (1975) J Chem Soc Dalton Trans 1975:1765
22. Bassett PJ, Lloyd DR (1971) J Chem Soc A 641:641–645
23. Jacox ME, Thompson WE (2011) J Chem Phys 134:194306
24. Miller JH, Andrews L (1980) J Am Chem Soc 102:4900
25. Gutsev GL, Jena P, Bartlett RJ (1998) Chem Phys Lett 292:289
26. Farber M, Srivastava RD (1984) J Chem Phys 81:241
27. Dehmer JL, Parr AC, Southworth SH, Holland DMP (1984) Phys Rev A 30:1783
28. Bassett PJ, Lloyd DR (1971) J Chem Soc A 1971:1551
29. King GH, Krishnamurthy SS, Lappert MF, Pedley JB (1972) Faraday Discuss Chem Soc 54:70
30. Porter RF, Zeller EE (1960) J Chem Phys 33:858
31. Dehmer JL, Berkowitz J, Cusachs LC, Aldrich HS (1974) J Chem Phys 61:594
32. NIST Chemistry WebBook, NIST Standard Reference Database Number 69, Eds. Linstrom PJ, Mallard WG

Theor Chem Acc (2014) 133:1491
DOI 10.1007/s00214-014-1491-8

REGULAR ARTICLE

Anomeric effects in fluoro and trifluoromethyl piperidines: a computational study of conformational preferences and hydration

Nathan D. Erxleben · Gary S. Kedziora · Joseph J. Urban

Received: 11 December 2013 / Accepted: 15 April 2014 / Published online: 14 May 2014

Abstract A computational investigation of anomeric effects in piperidine rings bearing fluoro and trifluoromethyl substituents shows for both compounds the most pronounced evidence of the anomeric effect, as expressed as hyperconjugative delocalization of the nitrogen lone pair, in structures with the substituent in the axial position and the N–H bond in the equatorial position. This structure is the lowest-energy structure in the fluoro case but not in the trifluoromethyl case where there is an increased axial penalty associated with the CF_3 group. The anomeric effect is characterized via geometrical evidence, natural bond orbital analysis, electrostatic effects, and energetic criteria. Computational results from a variety of levels of theory are presented including CCSD(T) with complete basis set extrapolation, B2PLYP-D, ωB97XD, B97-D, M06-2X, B3LYP, and MP2 allowing for a comparison of performance. The CCSD(T)/CBS results are very well represented by either B2PLYP-D or ωB97XD with moderate to large basis sets (aug-cc-pVTZ or aug-cc-pVDZ). Hyperconjugation, electrostatic effects, and steric effects play a role in the relative energetic ordering of the isomers considered.

Keywords Stereoelectronic effect · Conformation · Fluorine · Computational · Density functional theory

Dedicated to the memory of Professor Isaiah Shavitt and published as part of the special collection of articles celebrating his many contributions.

Electronic supplementary material The online version of this article (doi:10.1007/s00214-014-1491-8) contains supplementary material, which is available to authorized users.

N. D. Erxleben · J. J. Urban (✉)
Chemistry Department, United States Naval Academy,
572 Holloway Road, Annapolis, MD 21402, USA
e-mail: urban@usna.edu

G. S. Kedziora
Engility Corporation, Wright-Patterson AFB, OH 45433, USA

1 Introduction

Fluorine has long attracted attention as a useful substituent for the modification of the properties of organic compounds. Its small size, high electronegativity, and high carbon–fluorine bond strength make possible significant changes in polarity, solubility, reactivity, and metabolic stability with minimal steric penalties [1]. While the introduction of fluorine may not change the size of a molecule appreciably, it can have a profound impact on its shape and interactions with neighboring groups [2].

Fluorine participates in a number of subtle effects that impact the conformational preferences of organic compounds suggesting that the C–F bond has potential utility as a molecular design tool by allowing for the systematic modification of a compound's folding, self-assembly, and molecular recognition tendencies [3]. This can potentially be exploited in the design of peptidomimetics, novel materials, polymers, bio-inspired materials, engineered proteins, and therapeutic agents [4–10]. The effects impacting conformational properties are stereoelectronic effects [11, 12] that result in the stabilization of a particular geometry as in the gauche effect [13–15] and the anomeric effect [16–21] (Fig. 1). Beyond conformational equilibria, stereoelectronic effects can play a role in determining the course of reactions through stabilization of particular transition state geometries [22, 23].

The anomeric effect has been the topic of much research and has been reviewed extensively in the literature [11, 17–

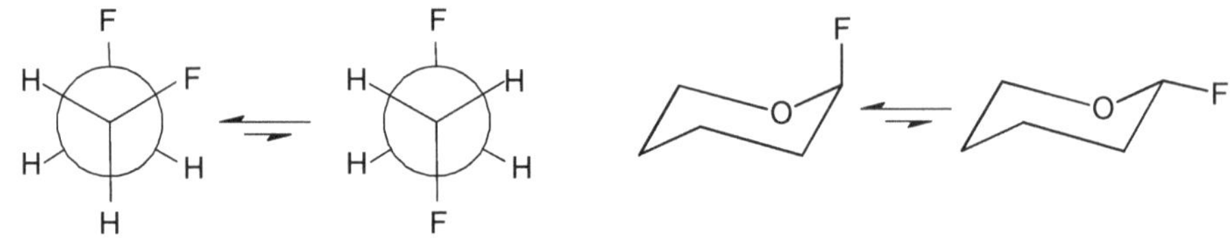

Fig. 1 Gauche effect in 1,2-difluoroethane (*left*) and the anomeric effect in substituted tetrahydropyrans (*right*)

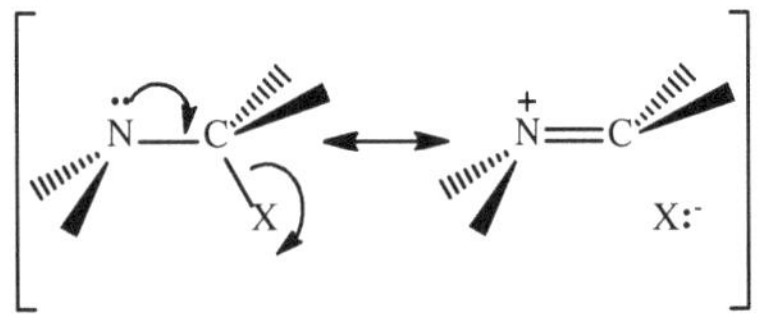

Fig. 2 Double bond-no-bond resonance depiction of hyperconjugative delocalization of nitrogen lone pair density into C–X σ*

19, 21, 24–26]. Commonly offered explanations include the concept of an energetically favorable participation of the C–F σ* orbital as an electron acceptor in a particular geometry, for example, corresponding to an anti-periplanar orientation of bonds, resulting in a preferred conformation, although many alternative explanations exist [17]. Such interactions have been described as a type of hyperconjugation. The IUPAC Compendium of Chemical Terminology [27] states "In the formalism that separates bonds into σ and π types, hyperconjugation is the interaction of σ-bonds with a π-network" and that "the interaction between filled π- or *p*-orbitals and adjacent antibonding σ* orbitals is referred to as negative hyperconjugation." The Compendium also uses what it terms "double bond-no-bond resonance" structures of the type shown in Fig. 2, which are prevalent in the literature on stereoelectronic effects [11, 28, 29] to illustrate the definition of hyperconjugation.

The concept of hyperconjugation relies on a local orbital picture of quantum chemistry that is consistent with the common chemist's view of Lewis structures. Natural bond orbital (NBO) analysis [30] has been used to explain the anomeric effect by showing a favorable interaction between the heteroatom lone pair and the σ* orbital of the substituent bonded to the anomeric carbon [16, 31, 32]. This was done most recently by Freitas [33] for a series of 2-substituted tetrahydropyrans, who found that NBO analysis provides a coherent framework in which to analyze the results, as we do in this report. Freitas points out that steric, electrostatic, and hyperconjugative effects play a role whether or not the anomeric effect is observed in his calculations, and sometimes the hypercongjugative effect is not the dominate factor in the preference of one isomer over the other.

There has been an ongoing controversy regarding the origin of the anomeric effect. Most recently, Mo denied the importance of hyperconjugation based on block-localized wavefunction (BLW) methods [24], arguing that delocalization between different fragments containing the heteroatom atoms does not lead to an energy lowering. The underlying assumption is that delocalization is equivalent to hyperconjugation, and to prove that hyperconjugation is not important only requires one to enforce localization and show that the anomeric effect is still reproduced. However, the localization enforced in the BLW calculations, based on molecular fragments, is not the same as using the very specific types of localized orbitals in NBO theory or other local bonding theories based on atomic orbitals and bonding orbitals. The BLW calculations show that electron transfer between the fragments is not necessary to account for the anomeric effect, which seemingly calls into question the basic underlying assumptions of exactly what is the anomeric effect in terms of physical properties or energy decomposition and has spurred some interest in the recent literature. The conclusion that hyperconjugation does not play a role in the anomeric effect in 2-fluoro tetrahydrofuran is contrary to findings of Freitas based on NBO energy analysis [33].

In the BLW approach, deleting off-diagonal blocks of the molecular orbital coefficient matrix enforces localization, where the basis set expansion for each fragment includes only those basis functions from the atoms in that fragment. The interaction between the electrons in the different blocks or fragments is included via the effective Fock matrix [34]. Gianninetti et al. [35] used the same type of block localizing method to eliminate basis set superposition error in intermolecular interaction calculations and found that there is an electron transfer between nucleic acid pairs using Bader charges [35], which indicates that localization may not be fully enforced with this method. Clearly, different conclusions may be drawn depending on which definition of localization one uses. Other recent work attempting to explain the origin of the anomeric effect includes studies that do not employ localization but rather decompose the energy of the total electron distribution into various contributions [36–38]. These various points of view often have conflicting conclusions about how to explain the anomeric effect. Our work shows that different points of view are not necessarily mutually exclusive.

The purpose of this study is to examine the anomeric effect in 2-fluoropiperidine (FP) and 2-trifluoromethylpiperidine

Fig. 3 Investigated isomers of FP with X = F, and TP with X = CF_3

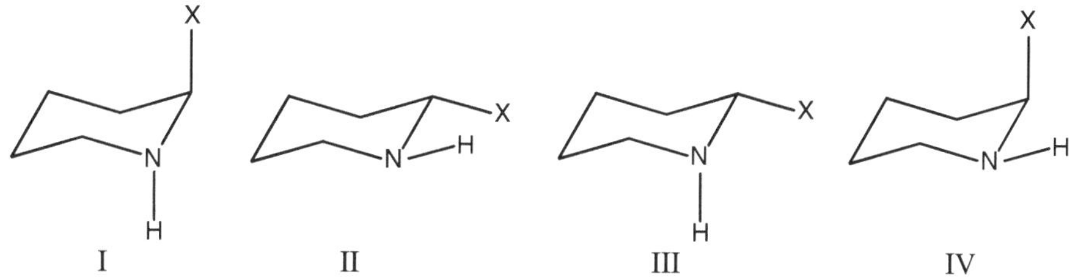

(TP) (see Fig. 3). A significant portion of the previous work on the anomeric effect has focused on oxygen-based heterocycles with less attention given to nitrogen-containing rings [25, 31, 32, 39]. We are interested in ascertaining the significance of stereoelectronic effects in these systems and in making a comparison between the fluorine and trifluoromethyl substituents in the role of electron acceptor. We have chosen to incorporate the –CH(F)–NH– and –CH(CF_3)–NH– moieties into a piperidine ring because of the conformational simplicity that results and because it coincides with the well-studied anomeric problem in the related pyranose systems. The nitrogen atom in the piperidine ring also has an NH bond that can be used to choose the direction of the nitrogen lone pair, which allows investigation of the effect of the position on stereoelectronic effects. We are also interested in establishing the performance record for a variety of computational protocols, including several density functional methods that were not yet available or widely used when much of the work on the anomeric effect was originally conducted.

1.1 Computational methods

The highest-cost and most reliable results reported here are obtained from coupled-cluster theory including the popular perturbative triples correction, CCSD(T) [40]. Also at the higher end of the computational spectrum, but much less costly than CCSD(T), three methods were chosen. These include Møller-Plesset perturbation theory [41] at the MP2 level; Grimme's double-hybrid method, B2PLYP-D [42], which empirically combines MP2 with DFT and includes a dispersion correction; and Head-Gordon's ωB97X-D [43] long-range corrected non-local exchange functional. On the more computationally efficient end of the spectrum, density functional theory calculations have been performed using Grimme's B97-D [44] functional, Zhao and Truhlar's M06-2X functional [45], and the popular B3LYP functional [46]. Both Pople basis sets [6-31+G(d,p) and 6-311++G(d,p)] and Dunning correlation consistent basis sets (cc-pVXZ [47] and aug-cc-pVXZ [48], where X = D, T, Q, and 6) were employed.

Complete geometry optimization of structures I–IV (Fig. 3) were conducted with several computational protocols including B2PLYP-D/aug-cc-pVDZ, B2PLYP-D/aug-cc-pVTZ, ωB97X-D/aug-cc-pVDZ, ωB97X-D/aug-cc-pVTZ, B97-D/6-31+G(d,p), B97-D/6-311++G(d,p), M06-2X/6-31+G(d,p), M06-2X/6-311++G(d,p), B3LYP/6-31+G(d,p), B3LYP/6-311++G(d,p), MP2/6-31+G(d,p), MP2/aug-cc-pVDZ, MP2/aug-cc-pVTZ, using the Gaussian 09 [49] program. Diffuse functions were included for accurate non-bonded interactions and to reduce the basis set superposition error. Frequency calculations were also performed to characterize the stationary points found as minima except for the 2-trifluoropiperidine system at B2PLYP-D/aug-cc-pVTZ and MP2/aug-cc-pVTZ due to the computationally intensive nature of the calculations. Since no imaginary frequencies were seen in the corresponding calculations employing the aug-cc-pVDZ basis set with nearly identical geometries, the B2PLYP-D and MP2 optimized structures with aug-cc-pVDZ were assumed to also be true minima. Natural bond orbital (NBO) analysis was performed using the NBO program version 3.1 [50] built into Gaussian 09 [49].

The CCSD(T) calculations were performed with the basis sets indicated in Tables 1 and 2 with NWChem 6.0 [51], GAMESS [52], and Gaussian 09 [49]. Energies at the

Table 1 CCSD(T) relative energies, in kcal/mol, for structures[a] I–IV of FP

	I	II	III	IV
CCSD(T)/CBS	4.90	4.01	4.96	0.00
CCSD(T)/cc-pVQZ	4.91	3.97	4.91	0.00
CCSD(T)/cc-pVTZ	4.87	3.87	4.72	0.00
CCSD(T)/aug-cc-pVTZ	4.97	4.05	4.94	0.00

[a] Geometries optimized at B2PLYP-D/aug-cc-pVTZ

Table 2 CCSD(T) relative energies, in kcal/mol, for structures[a] I–IV of TP

	I	II	III	IV
CCSD(T)/CBS	3.34	0.00	1.10	1.22
CCSD(T)/cc-pVQZ	3.37	0.00	1.06	1.29
CCSD(T)/cc-pVTZ	3.40	0.00	1.01	1.39

[a] Geometries optimized at B2PLYP-D/aug-cc-pVTZ

complete basis set (CBS) limit (CCSD(T)/CBS) were estimated by extrapolation of the correlation energy using the two-point extrapolation scheme of Halkier and coworkers [53] with the CCSD(T)/cc-pVTZ and CCSD(T)/cc-pVQZ values. For the CBS estimate of the Hartree–Fock energy, the cc-pV6Z energy was used.

2 Results

2.1 Relative energies of structures I–IV

The CCSD(T) relative energies of structures I–IV for FP and TP are provided in Tables 1 and 2, respectively. The relative energies in Tables 1 and 2 are taken from B2PLYP-D/aug-cc-pVTZ geometries and very similar results are obtained from the other geometries tested here (see Supplemental Information). In addition to the CCSD(T) results, a large variety of more approximate computational protocols have been examined and the results are detailed in Tables S1–S4 of the Supplemental Information. All levels of theory find the lowest-energy structures to be IV for fluoropiperidine and II for trifluoromethylpiperidine.

For FP, at the CCSD(T)/CBS level (Table 1), structure IV is stabilized by 4.01 kcal/mol over structure II followed in stability by structure I at 4.90 kcal/mol and structure III at 4.96 kcal/mol. Structures I and III, which both have the nitrogen lone pair in the equatorial position, are nearly identical in energy relative to structure IV. Comparison of the CCSD(T)/cc-pVTZ and CCSD(T)/aug-cc-pVTZ results in Table 1 reveals that inclusion of diffuse basis set functions add about 0.1–0.2 kcal/mol to the relative energies of isomers I–III, which indicates that the diffuse function contributes to a further stabilization of structure IV relative to the others. Increasing the number of polarization functions from TZ to QZ increases the relative energies of I–III by 0.1–0.2 kcal/mol.

For TP, at the CCSD(T)/CBS level (Table 2), structure II with the nitrogen lone pair in the axial position and the CF_3 group in the equatorial position is the global minimum. Structures III and IV are calculated to be very similar in energy and roughly 1 kcal/mol above the ground state.

2.2 Comparison of computational methods

The deviations with respect to the CCSD(T)/CBS standard energies in Table 1 for various methods for structures I–III are summarized in Fig. 4 for FP. Mean unsigned errors (MUE) for each approximate method and the individual energy differences appearing in Fig. 4 are tabulated in the Supplemental Information (Table S3). As expected, the best agreement is seen with B2PLYP-D and ωB97X-D with the large aug-cc-pVTZ and aug-cc-pVDZ basis sets with MUE values ranging from 0.07 to 0.31 kcal/mol. Among the more approximate levels of theory that offer greater computational efficiency, M06-2X provides an MUE of only 0.22 kcal/mol with 6-311++G(d,p) or 6-31+G(d,p). B97-D produces an MUE with either basis set of 0.74 kcal/mol. At the MP2 level, MUE of 1.14, 0.67,

Fig. 4 Errors in ΔE values relative to structure IV, in kcal/mol, compared to CCSD(T)/CBS//B2PLYP-D/aug-cc-pVTZ for structures I–III of FP

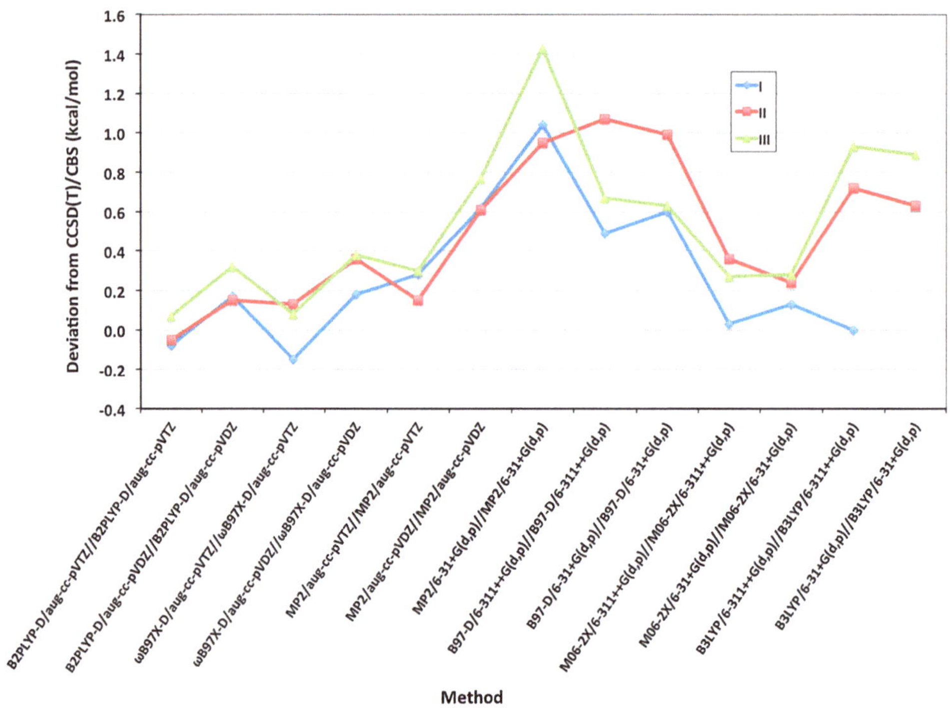

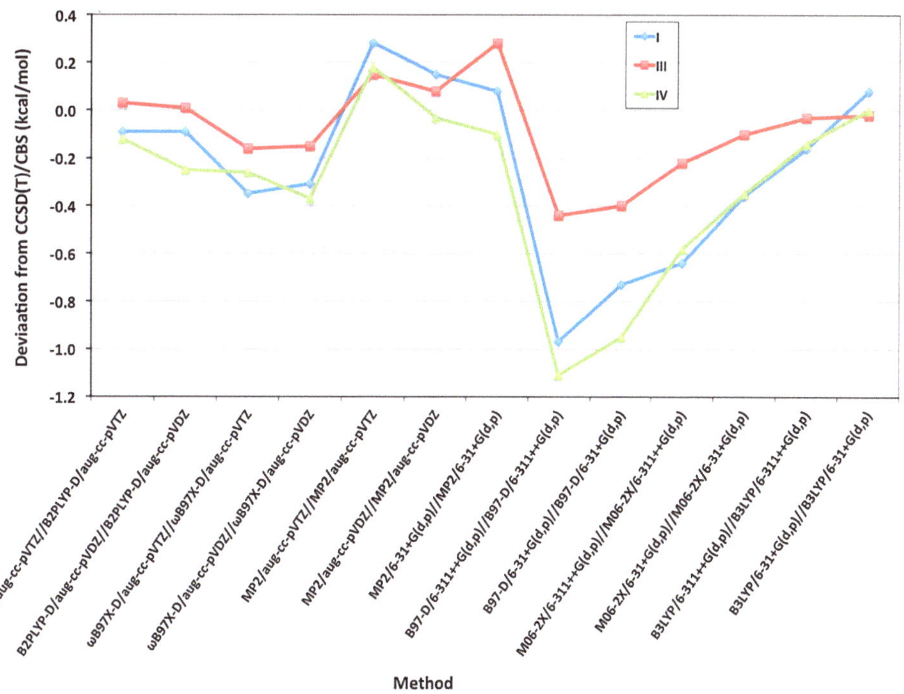

Fig. 5 Errors in ΔE values relative to ground-state structure II, in kcal/mol, compared to CCSD(T)/CBS for structures I, II, IV of TP

and 0.24 kcal/mol are found as the basis set is increased from 6-31+G(d,p) to aug-cc-pVDZ and then aug-cc-pVTZ, respectively. It should be noted that a local minimum corresponding to structure I is not found on the potential energy surface with B3LYP at either 6-31+G(d,p) or 6-311++G(d,p) but is found with all other methods.

Figure 5 provides a plot of errors relative to the CCSD(T)/CBS standard, across the set of three local minima (I, III, and IV) for TP. The B2PLYP-D results with either aug-cc-pVTZ or aug-cc-pVDZ are in good agreement with the CCSD(T)/CBS standard with MUE values (see Table S4 of the Supplemental Information) of only 0.08 and 0.12 kcal/mol, respectively. With ωB97X-D, the MUE increase slightly to 0.26 and 0.28 kcal/mol with aug-cc-pVTZ and aug-cc-pVDZ, respectively. MP2 produces good results with MUE of 0.20, 0.09, and 0.15 kcal/mol with aug-cc-pVTZ, aug-cc-pVDZ, and 6-31+G(d,p), respectively. B97-D results in MUE values of 0.84 and 0.69 kcal/mol with 6-311++G(d,p) and 6-31+G(d,p), respectively. The MUE with M06-2X are 0.48 and 0.27 kcal/mol at 6-311++G(d,p) and 6-31+G(d,p), respectively. B3LYP produces MUE of only 0.11 and 0.03 kcal/mol with 6-311++G(d,p) and 6-31+G(d,p), respectively. Remarkably, B3LYP, which missed one local minimum and produced relatively large errors for fluoropiperidine, produces *ΔE*s for the trifluoromethyl system that are in close agreement with the CCSD(T)/CBS values.

For FP, the largest deviations from the CCSD(T) results are seen with MP2/6-31+G(d,p), B97-D/6-311+G(d,p), and B3LYP with either basis set. All of these methods overestimate the relative energies of FP structures I–III. For TP, the largest deviations are the underestimation of the relative energies of I, III, and IV by B97-D with either basis set and M06-2X with 6-311+G(d,p).

Table 3 contains thermochemical energies relative to the lowest-energy isomer for structures I–IV using frequency calculations performed at the B2PLYP-D level of theory with aug-cc-pVTZ and aug-cc-pVDZ basis sets for fluoropiperidine and with aug-cc-pVDZ for

Table 3 Thermochemical energies, in kcal/mol, for structures I–IV of fluoropiperidine (FP) and trifluoromethylpiperidine (TP)

	Δ(*E* + ZPE)	Δ*E*(298)	Δ*H*(298)	Δ*G*(298)
F-I	5.06[a]	5.13	5.13	4.93
	4.74[b]	*4.84*	*4.84*	*4.58*
F-II	4.48	4.48	4.48	4.44
	4.19	*4.19*	*4.19*	*4.16*
F-III	5.49	5.43	5.43	5.44
	5.12	*5.10*	*5.10*	*5.11*
F-IV	0.00	0.00	0.00	0.00
	0.00	*0.00*	*0.00*	*0.00*
CF_3-I	3.10	3.09	3.09	3.14
CF_3-II	0.00	0.00	0.00	0.00
CF_3-III	0.99	1.01	1.01	0.97
CF_3-IV	0.93	0.89	0.89	1.04

[a] Geometry optimization and frequency calculation at B2PLYP-D/aug-cc-pVDZ

[b] Italicized values derived from geometry optimization and frequency calculations at B2PLYP-D/aug-cc-pVTZ

trifluoromethylpiperidine. B2PLYP-D has been shown to perform well for vibrational frequencies using B2PLYP-D/aug-cc-pVDZ [54]. The thermochemical results from the fluoro compound show little difference, on the order of 0.3 kcal/mol, between the larger and smaller basis sets, where the larger basis set results are specified in the italicized rows, and, for both compounds, the thermochemical corrections result in corrections to the relative energies that are small in magnitude.

2.3 Assessment of the anomeric effect: energetic criteria

Several authors [11, 17, 25] have used a scheme of the type $\Delta\Delta E = \Delta E_N - \Delta E_C$ to provide an energetic characterization of anomeric effect magnitude where the ΔE_N term is the axial-equatorial energy difference in the heterocyclic system ($\Delta E_N = E_{Neq} - E_{Nax}$), and the ΔE_C term is the value for the corresponding monosubstituted cyclohexane ($\Delta E_C = E_{Ceq} - E_{Cax}$). The magnitude of the anomeric effect is thus expressed as $\Delta\Delta E$ with a positive value indicating that the relative stability of the axial conformer is greater in the heterocyclic case (piperidine, in this work) than in cyclohexane. The experimental energetic terms used are often Gibbs' free energies or enthalpies. But, ΔE values from electronic structure calculations are also commonly used [17, 25]. The data in Table 3 indicate that the magnitude of thermochemical corrections is small in these systems indicating ΔE values should provide meaningful results. $\Delta\Delta E$ values for fluoro and trifluromethylpiperidine at selected levels of theory are summarized in Table 4. The $\Delta\Delta E$ values obtained from additional levels of theory can be found in the Supplemental Information (Table S5).

The penalty, in free energy terms, for placing a substituent in the axial position of cyclohexane versus the equatorial position is commonly termed the "A value" or the "conformational energy" [55]. These are available in the literature for a large variety of substituents with values for the fluoro group reported to be from 0.25 to 0.42 kcal/mol and from 2.4 to 2.5 kcal/mol for the CF_3 group [56]. The ΔE_C value corresponds to the negative of the A value. A values derived from ΔE_C values calculated at the various levels of theory are also provided in Table 4 and in the Supplemental Information. Thus, the data in Table 4 (and S5) provide the magnitudes of the axial steric penalty in cyclohexane for the fluoro and trifluoromethyl groups in the form of the A values and the magnitude of the anomeric benefit to axial occupation in the piperidine systems in the form of the $\Delta\Delta E$ values. Values for the latter are presented for the two cases where the N–H bond is equatorial (i.e., the nitrogen lone pair is axial) and the N–H bond is axial.

Table 4 Cyclohexane A values and FP and TP anomeric energy ($\Delta\Delta E$) values, in kcal/mol

	FP			TP		
	A[a]	$\Delta\Delta E$ NHeq[b]	$\Delta\Delta E$ NHax[c]	A	$\Delta\Delta E$ NHeq	$\Delta\Delta E$ NHax
CCSD(T)/CBS//B2PLYP-D/aug-cc-pVTZ	0.13	4.14	0.19	2.26	1.05	0.02
CCSD(T)/cc-pVQZ//B2PLYP-D/aug-cc-pVTZ	0.14	4.07	0.10	2.30	1.01	−0.01
B2PLYP-D/aug-cc-pVTZ//B2PLYP-D/aug-cc-pVTZ	0.17	4.13	0.38	2.24	1.14	0.12
ωB97X-D/aug-cc-pVTZ//ωB97X-D/aug-cc-pVTZ	0.28	4.42	0.57	2.13	1.17	0.08

[a] Axial–equatorial for cyclohexane bearing fluoro in FP case and trifluoromethyl in TP case

[b] $\Delta\Delta E = \Delta E_N - \Delta E_C$ with N–H bond equatorial (see text for further explanation)

[c] $\Delta\Delta E = \Delta E_N - \Delta E_C$ with N–H bond axial (see text for further explanation)

All of the computational results in Table 4 are in good agreement with the experimental A values of 0.25–0.42 kcal/mol for F and 2.4–2.5 kcal/mol for CF_3, especially considering that ΔE values are being compared to ΔG values. The A value for the fluoro group has near zero magnitude, and in two cases, MP2/6-31+G(d,p) and M06-2X/6-311++G(d,p), a small negative value is obtained (Table S5). The steric penalty in cyclohexane for a CF_3 group is far more significant. The highest levels of theory employed here give results within a few tenths of a kcal/mol of the experimental range of 2.4–2.5 kcal/mol. The largest deviations are seen with B97D calculations and the M06-2X/6-311++G(d,p) which produce values just under 2 kcal/mol (Table S5).

For both the fluoropiperidine and trifluoromethylpiperidine systems, the magnitude of the anomeric effect ($\Delta\Delta E$ value) is significantly larger in the case where the N–H bond is in the equatorial position. This is consistent with the expectation that the nitrogen lone pair can best delocalize into the axial C–F (or C–CF_3) bond when it occupies the axial position. At the highest levels of theory, the $\Delta\Delta E$ values range from 4.07 to 4.60 kcal/mol with NH bond equatorial, but are only 0.10–0.57 kcal/mol when the NH bond is axial (and the nitrogen lone pair is equatorial). Given the small magnitude of the axial penalty for a fluoro group, the conformational landscape of the fluoropiperidine system is dominated by the anomeric effect. B2PLYP-D/aug-cc-pVTZ is in excellent agreement with the CCSD(T)/CBS $\Delta\Delta E$ NHeq value, and ωB97X-D/aug-cc-pVTZ produces a slightly overestimated value.

Table 5 Dipole moments and selected geometrical features of FP and TP calculated at B2PLYP-D/aug-cc-pVTZ

	μ[a]	C–F[b]	C–N[b]	N–C–F[c]	α: H–(C–C)–N[d]	β: F–(N–C)–C[d]
Fluoropiperidine						
I	2.90	1.408	1.432	107.6	−26.4	36.6
II	2.69	1.395	1.441	107.4	29.7	−35.4
III	3.06	1.392	1.443	107.8	−30.8	−35.1
IV	1.40	1.440	1.413	111.3	26.9	35.1
	μ[a]	C–CF_3[b]	C–N[b]	N–C–<u>C</u>F_3[c]	α: H–(C–C)–N[d]	β: <u>C</u>F_3–(N–C)–C[d]
Trifluoromethylpiperidine						
I	3.44	1.525	1.459	109.7	−29.2	32.0
II	3.06	1.517	1.455	109.3	29.3	−34.4
III	3.44	1.517	1.458	108.8	−30.3	−33.5
IV	1.76	1.529	1.451	113.8	27.2	30.2

[a] Dipole moment, in Debeye

[b] Bond length, in Ångstroms

[c] Bond angle, in degrees

[d] Improper dihedral angle, in degrees. See text for definition

The results in Table 4 for the TP system illuminate the underlying causes for the relative energies of structures I through IV. In FP, structure IV is the lowest-energy structure driven largely by the magnitude of the anomeric effect. Although structure II is the lowest-energy structure for TP (Table 2), there is still an anomeric effect involving the CF_3 group at work. When the nitrogen lone pair is axial, the magnitude of the anomeric effect as indicated by the $\Delta\Delta E$ NHeq values is 1.01–1.05 kcal/mol with CCSD(T) and 1.14–1.17 kcal/mol with B2PLYP-D and ωB97X-D. However, it is outweighed by the magnitude of the steric penalty for placing the large CF_3 group in the axial position which is 2.13–2.30 kcal/mol for cyclohexane at those levels of theory. Thus, modeling larger systems containing a trifluoromethyl group, such as trifluoroethylamine peptidomimetics, requires computational techniques that adequately capture the anomeric effect.

It should also be noted that it has been postulated that the A values for cyclohexane provide an underestimation of the axial penalty in oxygen or nitrogen-containing rings because the latter involve a C–N (or C–O) bond that is shorter than a C–C bond producing even greater 1,3-diaxial strain in the heterocycle [11]. As with the FP system, the magnitude of the anomeric effect is much smaller in TP when the nitrogen lone pair is in the equatorial position ($\Delta\Delta E$ NHax values).

2.4 Assessment of the anomeric effect: geometric criteria

Table 5 contains geometrical parameters calculated at B2PLYP-D/aug-cc-pVTZ for structures I through IV of both compounds under study here. Analysis of this data reveals trends that are consistent with what would be expected due to a stereoelectronic effect involving delocalization of nitrogen lone pair electron density into the σ* of the C–F bond, or the corresponding C–CF_3 bond in the case of TP, as depicted in double bond-no-bond resonance structures of Fig. 2. Structures where this effect is greatest are expected to exhibit the longest C–F (or, C–CF_3) bond lengths and the shortest C–N bonds. For FP, structure IV displays the longest C–F bond at 1.440 Å, and structure III displays the shortest at 1.392 Å. Structure I has a longer C–F bond than structure II with values of 1.408 and 1.395 Å, respectively. The C–N bond lengths show the inverse correlation that is expected from Fig. 2. Structure IV shows the shortest C–N bond at 1.413 Å with the values of the other structures, in increasing length, at 1.432 Å for I, 1.441 Å for II, and 1.443 Å for III. These bond length changes are consistent with the notion that the lone pair hyperconjugative delocalization is more pronounced when the C–X bond is axial rather than equatorial and when the nitrogen lone pair is axial versus equatorial. This leads to the trend of IV > I > II > III in decreasing stereoelectronic effect due to an exchange of charge between the nitrogen atom and the fluorine atom. Similar conclusions can be drawn from the geometry data obtained at the other levels of theory investigated here (see Supplemental Information, Tables S6 and S7).

The double bond-no-bond resonance scheme invoked in Fig. 2 implies geometrical changes beyond those noted above in bond lengths. The delocalization of the nitrogen lone pair into the C–X σ* is expected to also involve a

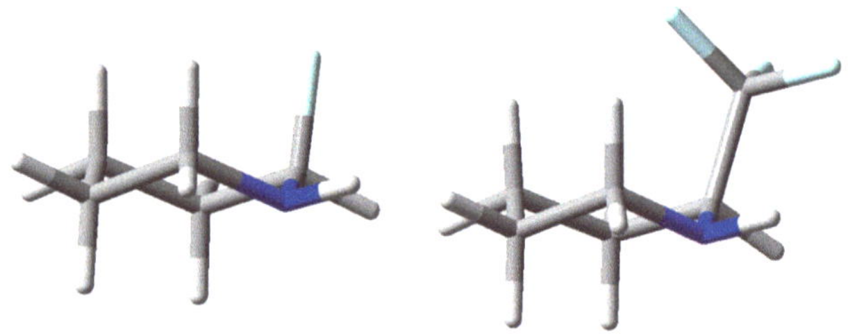

Fig. 6 B2PLYP-D/aug-cc-pVTZ optimized structure IV for fluoropiperidine (*left*) and trifluoropiperidine (*right*) viewed along the N–C bond vector

decrease in pyramidality (increase in planarity) at the N and C sites shown in Fig. 2, consistent with increasing s-character. Structure IV exhibits the widest N–C–X bond angle with values of 111.3° for X = F and 113.8° for X = CF_3. Improper dihedral angles (i.e., those defined by four points that are not contiguously chemically bonded) have been defined to illustrate the degree of planarity at the C and N centers of interest. For example, the degree of planarity at a trigonal center A, surrounded with ligands, B, C, and D, can be expressed by dihedral angle made by the points B–(C–D)–A. The parentheses are meant to highlight the fact that the central two atoms are not chemically bonded to each other. A trigonal planar center has a value of 0° for this angle. A typical pyramidal center such as the nitrogen of ammonia has a value on the order of ±35° for this angle. The values for two such angles, α at the N atom and β at the C atom bearing the F or CF_3, are in Table 5 where α is defined as the H–(C–C)–N improper dihedral angle and β as the improper dihedral X–(N–C)–C with X = F, CF_3. The α values are smallest, indicating highest degree of planarity at nitrogen, for structure IV and increase in magnitude following the trend of IV < I < II < III in correspondence with the C–X and C–N bond length variations. The β value is also of smallest magnitude in structure IV for both compounds indicating the greatest degree of planarity at the C bearing the F or CF_3. The expected pattern in β is followed for the fluoro structures (i.e., IV < I < II < III) but not for the CF_3 compounds with structure II showing the largest (in magnitude) β value.

Figure 6 shows the B2PLYP-D/aug-cc-pVTZ optimized structure IV for both FP and TP. For both compounds, the dipole moment for structure IV is significantly lower than that of the other structures with a value of 1.40 D compared to a range of 2.90–3.60 D for the fluoro case. With trifluoromethylpiperidine, the dipole moment of structure IV is 1.76 D and those for the other structures range from 3.06 to 3.44 D. Inspection of the 3D structures and dipole moment vectors (Supplemental Information, Figure S1.) suggests that a reasonable approximation of the molecular dipole moment can be obtained from the sum of the C–F (or C–CF_3) and N lone pair dipole vectors. The charge on the nitrogen atom is directional depending on where the lone pair is pointing. Plots of the electrostatic potential (ESP) mapped onto an electron density isosurface (Supplemental Information, Figure S2) show the directionality of the negative charge of the nitrogen lone pair in the various structures. In structure IV, there is a negative charge cloud pointing away from the fluorine lone pairs, which reduces the dipole moment and leads to less electrostatic repulsion. Structure I has a more negative charge on the side of the piperidine ring where the fluorine is situated, contributing to a longer dipole moment. Comparing the ESP for structures II and III also shows the variation in the direction of the dipole moment is largely based on the position and direction of the N lone pair and the F lone pairs. Based on a detailed analysis of the geometrical parameters taking into account the resonance interaction in Fig. 2 and an analysis of the dipole moments taking into account the ESP, we can conclude that there is indeed an electrostatic explanation and a hyperconjugative explanation for the anomeric effect and other stereoelectronic effects in these substituted piperidines.

2.5 Assessment of the anomeric effect: natural orbital analysis

To further examine hyperconjugation, NBO analysis [30, 57] was carried out on structures I–IV of both fluoropiperidine and trifluoromethylpiperidine. The key results at the ωB97X-D/aug-cc-pVTZ level are summarized in Table 6. The nitrogen lone pair occupancy, n_N, is lowest for structure IV at 1.872 electrons for FP and 1.908 electrons for TP. Since the nitrogen lone pair population is less than two, we conclude that the endo-anomeric effect is operative for structure IV, and we will show below that this is due to hyperconjugation in the NBO analysis framework.

The NBO analysis provides second-order perturbative estimates of specific donor–acceptor interactions in the NBO basis. These are expressed in kcal/mol and represent the stabilizing effect due to the interaction between filled Lewis-type NBOs and non-Lewis (i.e., antibonding or Rydberg) NBOs. The most significant interactions are listed as $E^{(2)}$ values for each structure of each compound in Table 6. We are particularly interested in the delocalization of the nitrogen lone pair into the CF σ^* of FP and the corresponding C–CF_3 σ^* of TP as they represent the antiperiplanar interaction (see, for example, Alabugin et al. [58] or Weinhold [30]). These two interactions are shown in bold in Table 6. For FP, the structures with an axial fluorine I and IV show the most significant amount of nitrogen lone pair to C–F σ^* delocalization ($n_N \rightarrow \sigma_{CF}^*$). This interaction ranks first in structures I and IV, but is a much stronger interaction in structure IV where the $E^{(2)}$

Table 6 Natural bond orbital analysis results ωB97X-D/aug-cc-pVTZ. See text for details

	I	II	III	IV
Fluoropiperidine				
n_N occupancy	1.891	1.910	1.914	1.872
$E^{(2)}$	**(1)15.05 $n_N \rightarrow \sigma_{CF}$***	(1)10.34 $n_N \rightarrow \sigma_{CH}$*	(1) 9.63 $n_F \rightarrow \sigma_{CN}$*	**(1) 29.01 $n_N \rightarrow \sigma_{CF}$***
	(2)9.89 $n_F \rightarrow \sigma_{CC}$*	(2) 9.84 $n_N \rightarrow \sigma_{CH}$*	(2) 9.51 $n_N \rightarrow \sigma_{CC}$*	(2) 9.50 $n_N \rightarrow \sigma_{CH}$*
	(3)9.05 $n_N \rightarrow \sigma_{CC}$*	(3) 9.77 $n_F \rightarrow \sigma_{CN}$*	(3) 9.37 $n_F \rightarrow \sigma_{CC}$*	(3) 9.39 $n_F \rightarrow \sigma_{CN}$*
	(4)7.65 $n_F \rightarrow \sigma_{CN}$*	(4) 7.67 $n_F \rightarrow$ C*	(4) 8.31 $n_F \rightarrow$ C*	(4) 5.80 $\sigma_{CH} \rightarrow \sigma_{CN}$*
	(5)6.61 $n_F \rightarrow$ C*	(5) 7.19 $n_F \rightarrow \sigma_{CH}$*	(5) 6.49 $n_F \rightarrow \sigma_{CC}$*	(5) 5.79 $\sigma_{CH} \rightarrow \sigma_{CN}$*
	(6)6.27 $n_F \rightarrow \sigma_{CC}$*	(6) 4.72 $n_F \rightarrow \sigma_{CC}$*	(6) 5.06 $n_F \rightarrow \sigma_{CH}$*	(6) 5.22 $n_F \rightarrow \sigma_{CC}$*
	(7)5.80 $n_F \rightarrow \sigma_{CH}$*	**(7) 4.36 $n_N \rightarrow \sigma_{CF}$***	**(10) 4.06 $n_N \rightarrow \sigma_{CF}$***	(7) 5.21 $n_F \rightarrow$ C*
ΔE	4.75	4.14	5.04	0
ΔE^{NoCF}	7.35	4.55	0	13.32
ΔE^{NoStar}	5.09	0	2.94	4.3
Trifluoropiperidine				
n_N occupancy	1.916	1.916	1.921	1.908
$E^{(2)}$	(1) 14.39 $n_F \rightarrow \sigma_{CF}$*	(1) 14.53 $n_F \rightarrow \sigma_{CF}$*	(1) 14.23 $n_F \rightarrow \sigma_{CF}$*	**(1) 14.26 $n_N \rightarrow \sigma_{CC}$***
	(2) 14.25 $n_F \rightarrow \sigma_{CF}$*	(2) 14.34 $n_F \rightarrow \sigma_{CF}$*	(2) 14.10 $n_F \rightarrow \sigma_{CF}$*	(2) 14.01 $n_F \rightarrow \sigma_{CF}$*
	(3) 13.39 $n_F \rightarrow \sigma_{CF}$*	(3) 13.76 $n_F \rightarrow \sigma_{CF}$*	(3) 14.02 $n_F \rightarrow \sigma_{CF}$*	(3) 13.61 $n_F \rightarrow \sigma_{CF}$*
	(4) 12.81 $n_F \rightarrow \sigma_{CF}$*	(4) 13.37 $n_F \rightarrow \sigma_{CF}$*	(4) 13.28 $n_F \rightarrow \sigma_{CF}$*	(4) 13.40 $n_F \rightarrow \sigma_{CF}$*
	(5) 11.85 $n_F \rightarrow \sigma_{CF}$*	(5) 11.28 $n_N \rightarrow \sigma_{CH}$*	(5) 11.45 $n_F \rightarrow \sigma_{CF}$*	(5) 12.43 $n_F \rightarrow \sigma_{CF}$*
	(6) 11.12 $n_F \rightarrow \sigma_{CF}$*	(6) 11.20 $n_F \rightarrow \sigma_{CF}$*	(6) 10.92 $n_F \rightarrow \sigma_{CF}$*	(6) 11.89 $n_F \rightarrow \sigma_{CF}$*
	(7) 10.14 $n_N \rightarrow \sigma_{CC}$*	(7) 10.65 $n_F \rightarrow \sigma_{CF}$*	(7) 10.45 $n_N \rightarrow \sigma_{CC}$*	(7) 11.83 $n_F \rightarrow \sigma_{CF}$*
	(8) 9.20 $n_N \rightarrow \sigma_{CC}$*	(8) 9.85 $n_N \rightarrow \sigma_{CH}$*	(8) 9.29 $n_N \rightarrow \sigma_{CC}$*	(8) 10.35 $n_N \rightarrow \sigma_{CH}$*
	(13) 7.35 $n_N \rightarrow \sigma_{CC}$*	**(59) 1.82 $n_N \rightarrow \sigma_{CC}$***	**(60) 2.00 $n_N \rightarrow \sigma_{CC}$***	
ΔE	2.99	0	0.94	0.96
ΔE^{NoCC}	6.95	0	1.07	10.13
ΔE^{NoStar}	6.31	0	2.4	6.23

value is 29.01 versus 15.05 kcal/mol in structure I. In structures II and III, with the C–F bond in the equatorial position, the $n_N \rightarrow \sigma_{CF}$* interaction drops in ranking to 7th and 10th, respectively. Inspection of Table 6 reveals that a number of delocalization interactions other than $n_N \rightarrow \sigma_{CF}$* are also important, especially in structures II and III. Nitrogen lone pair delocalization into a C–H σ* ranks first in structure II, and fluorine lone pair lone pair delocalization into the C–N σ* ranks first for structure III. The large $E^{(2)}$ value for the $n_N \rightarrow \sigma_{CF}$* in structure IV indicates that the hyperconjugation effect is most significant in this structure as expected with the nitrogen lone pair and the C–F bond are anti-periplanar to each other.

The ramifications of this stereoelectronic effect can be further elucidated by examining the ΔE, ΔE^{noCF}, and ΔE^{nostar} values. The ΔE value is simply the difference in the ωB97X-D/aug-cc-pVTZ energies for each isomer, whereas the ΔE^{nostar} and ΔE^{noCF} incorporate NBO deletions in the calculations. The ΔE^{nostar} set of relative energies is the difference in energies obtained when all non-Lewis-type interactions are removed from the calculation. Thus, this value represents the conformational energy differences for the perfectly localized (each Lewis NBO doubly occupied) system. In other words, the stabilizing effect due to delocalization is deleted from the calculation. The ΔE^{noCF*} values are arrived at in a similar manner except that only the specific $n_N \rightarrow \sigma_{CF}$* interaction has been deleted. In the absence of the $n_N \rightarrow \sigma_{CF}$* interaction, structure III, with an equatorial C–F bond and an equatorial nitrogen lone pair, becomes the global minimum. The relative energy of structure IV of FP rises significantly when $n_N \rightarrow \sigma_{CF}$* hyperconjugation is ignored resulting in a ΔE^{NoCF} value of 13.32 kcal/mol above the global minimum. The relative energy of structure I also increases upon removal of the $n_N \rightarrow \sigma_{CF}$* interaction.

When all hyperconjugative interactions are ignored (ΔE^{NoStar} values), structure II becomes the lowest-energy structure for FP. It is interesting to note that the ordering of the structures based on their ΔE^{NoStar} values corresponds to how we would order them based on our intuitive notion of steric hinderance, where II and III are less hindered that I and IV, since I and IV have the lone-pair-rich substituent in the more sterically hindered axial position. It appears that III would be more sterically hindered than II, since in structure III the nitrogen lone pairs and fluorine lone pairs are near each other in the equatorial position, and we are assuming based on VSEPR theory [59] that lone pairs have a larger steric effect than hydrogen atoms. Isomer I is more hindered than IV, because in I both sets of lone pairs are in the axial position. Thus, based on this analysis, if steric effects were the only ones contributing to the ordering of the isomers, the order would be II < III < IV < I, which is the ordering based on $\Delta E^{\mathrm{NoStar}}$.

Table 6 also contains the $E^{(2)}$ values for donor–acceptor interactions in trifluoromethylpiperidine (TP). In the case of TP, the delocalization interaction that corresponds to the $n_N \rightarrow \sigma_{CF}^*$ stereoelectronic effect in FP is the $n_N \rightarrow \sigma_{CC}^*$ where the C–C bond of interest is between the ring carbon and the carbon of the trifluoromethyl group. There are other $n_N \rightarrow \sigma_{CC}^*$ interactions possible involving carbon–carbon bonds within the ring. The $n_N \rightarrow \sigma_{CC}^*$ involving the C–CF_3 bond is shown in bold in Table 6. In structure IV, this $n_N \rightarrow \sigma_{CC}^*$ interaction is the most significant with an $E^{(2)}$ value of 14.26 kcal/mol. It is ranked 13th for structure I and drops to 59th and 60th in structures II and III, respectively. Thus, in NBO terms, as with FP, hyperconjugation in TP is most significant in structure IV and least significant in structures II and III with the CF_3 group in the equatorial position.

The results of the NBO deletion studies also mirror those seen for FP. Deletion of the delocalization of the nitrogen lone pair into the C–CF_3 σ^* results in structure IV rising significantly in relative energy as indicated by the ΔE^{NoCC} value of 10.13 kcal/mol. Structure I also rises significantly in energy, to 6.95 kcal/mol, upon removal of this delocalization. When all hyperconjugative interactions are ignored, the relative energies for TP are 6.31, 0.00, 2.40, and 6.23 kcal/mol for structures I–IV, respectively. Similar to our analysis of FP, the order of these energies corresponds to the order based on the common understanding of steric hindrance as typically presented in chemistry textbooks.

Structure II is the lowest-energy structure for TP in contrast to FP where structure IV is lowest in energy. However, the NBO analysis reveals that there is stabilization due to hyperconjugation in TP as well. Structure IV, the one favored by the anomeric effect, is not the lowest-energy structure in TP. As stated above, the large steric demand of the CF_3 group creates too great a penalty for its occupation of the axial position to outweigh the anomeric effect. Structure IV has a relative energy of 0.96 kcal/mol in TP with ωB97X-D/aug-cc-pVTZ (Supplemental Info, Table S2). But, the NBO deletion studies reveal that, in the absence of the hyperconjugative delocalization of the nitrogen lone pair into the C–CF_3 σ^* orbital, it would be significantly higher in relative energy.

Table 7 Selected parameters derived from NBO analysis partial atomic charge ωB97X-D/aug-cc-pVTZ

	I	II	III	IV
Fluoropiperidine				
F Charge[a]	−0.401	−0.383	−0.380	−0.429
C–F polarity[b]	0.784	0.783	0.777	0.803
Trifluoropiperidine				
CF_3 charge[c]	−0.012	−0.000	−0.001	−0.034
C–CF_3 polarity[d]	1.031	1.025	1.024	1.042

[a] Charge on F atom in electron units

[b] Difference in charge across C–F bond ($q_C - q_F$) in electron units

[c] Sum of atomic charges in CF_3 group in electron units

[d] Difference in charge across C–CF_3 bond ($q_C - q_{CF3}$) in electron units

Table 7 shows selected parameters derived from the NBO partial atomic charges. For fluoropiperidine, the negative charge on fluorine is greatest in structure IV. Also, the enhancement in negative charge upon placing the fluoro substituent into the axial position is greater when the nitrogen lone pair is axial than when it is equatorial. In going from structure II to structure IV, fluorine's charge is increased by 0.046 versus an increase of 0.021 in going from structure III to I. The charge difference across the C–F bond is also greatest in structure IV. These characteristics are consistent with enhanced charge flow from nitrogen to fluorine in structure IV. The charges on the atoms in the CF_3 group were summed in Table 7 to allow for an analogous analysis for trifluoropiperidine. The same trends are seen but the magnitude is much smaller. The CF_3 group is most negative in structure IV (−0.034), and the CF_3 group exhibits a greater increase in charge in going from II to IV than III to I. The charge difference across the C-CF_3 bond is also greatest in structure IV.

2.6 Aqueous solvation

The M06-2X/6-311++G(d,p) relative energies of structures I–IV for FP and TP in an aqueous environment, ΔE_{aq}, were determined by use of the polarized continuum solvent model using the integral equation formalism variant (IE-FPCM) as implemented in Gaussian 09 [60]. The results are summarized in Table 8. The differences between the total energies in the aqueous medium and in isolation are reported in Table 8 as the hydration free energies, ΔG_{Hyd}, for each structure. Comparison of the ΔE_{aq} values to the M06-2X/6-311++G(d,p) gas-phase relative energies, ΔE_{gas}, reveals that the same structures are found to be the global minima in the aqueous phase as were seen in the gas phase; structure IV for FP and structure II for TP. For both compounds, structure IV is the least-well solvated with ΔG_{hyd} values that are smaller in magnitude than for the

Table 8 Aqueous relative energies and hydration free energies at M06-2X/6-311++G(d,p)

	Structure			
	I	II	III	IV
Fluoropiperidine (FP)				
ΔE_{gas}	4.93	4.37	5.23	0.00
ΔE_{aq}	3.75	3.57	4.06	0.00
ΔG_{Hyd}	−4.70	−4.31	−4.67	−3.51
Trifluoropiperidine (TP)				
ΔE_{gas}	2.70	0.00	0.88	0.64
ΔE_{aq}	2.11	0.00	0.44	1.44
ΔG_{Hyd}	−4.19	−3.61	−4.04	−2.81

other structures. This is consistent with the dipole moment results reported above in Table 5. For both FP and TP, structure IV has a significantly smaller molecular dipole moment, which in turn results in a smaller aqueous hydration free energy. There may also be local hydration effects at work, such as explicit hydrogen bonding, which is not considered in this work.

The fact that the stereoelectronic effect is most pronounced in structure IV should result in greater negative charge on the fluorine in that structure than in structures I–III, and, correspondingly, greater positive charge on nitrogen. Organic fluorine is a notoriously poor hydrogen bond acceptor [61]. But, electron flow via the anomeric effect could potentially increase the hydrogen bond acceptor ability of fluorine, and the hydrogen bond donor ability of the NH bond, in structure IV. Such electronic rearrangement is accounted for the IEFPCM solvent model via the underlying quantum mechanical description of the solute. In addition, the results in Table 8 are derived from complete geometry optimizations in the presence of the continuum solvent model, and thus, any structural rearrangement in the aqueous medium is accounted for. However, it is possible that structure IV is the least-well solvated by a bulk aqueous medium but has local interactions with individual water molecules that are enhanced due to the flow of charge expected via the anomeric effect. With the exception of the drop in magnitude of the ΔG_{Hyd} values noted above for structure IV, the hydration free energies among the remaining structures are very consistent with values ranging from −4.31 to −4.70 kcal/mol for FP and −3.61 to −4.04 for TP.

3 Conclusions

The anomeric effect in piperidine rings bearing a fluoro or trifluoromethyl substituent at the anomeric carbon has been examined with a variety of computational chemistry approaches including benchmark-level CCSD(T)/CBS calculations and several far more computationally economical approaches. The anomeric effect in these systems has been characterized by energetic, electrostatic, and geometric criteria as well as through NBO analysis. The results indicate a pronounced anomeric bias in both systems. The anomeric effect in the trifluoromethylpiperidine system is masked by offsetting steric effects.

The results obtained here for the nitrogen heterocycle supplement those available in the literature for the well-studied oxygen-containing ring systems, provide a performance assessment for several more recently developed DFT approaches, allow for an assessment of participation of the trifluoromethyl group in the anomeric effect, and examine the impact of aqueous hydration on fluoro, and trifluoromethyl-mediated anomeric effects. An advantage of the piperidine systems is that the lone pair in the ring atom can be in either the axial or equatorial position, which allows us to test refined hypotheses about stereoelectronic effects relating to the direction of the lone pair. NBO analysis provides a coherent description of the anomeric effect in these molecules that includes an important contribution from hyperconjugation, which supports our common understanding of stereoelectronic effects based on Lewis structures. We observe that the position of the nitrogen lone pair relative to the fluorine atom lone pairs is an important factor in the electrostatic and steric interactions in the isomers.

With the aug-cc-pVTZ and aug-cc-pVDZ basis sets, very good agreement is seen between ωB97X-D and the B2PLYP-D double-hybrid approach and with CCSD(T)/CBS extrapolated results. For fluoropiperidine, structure IV, with the nitrogen lone pair anti-periplanar to the axial fluorine, is found to be the lowest-energy structure. By energetic, geometric, and NBO criteria, the anomeric effect is greatest in structure IV. Qualitatively similar results are obtained for the trifluoromethylpiperidine system but the increased axial steric penalty for the larger CF_3 group results in structure II as the lowest-energy structure. Among the less computationally demanding approaches investigated here, M06-2X with 6-31+G(d,p) or 6-311++G(d,p) generally captures the salient results seen with the higher levels of theory (see Supplemental Information).

Calculations employing a continuum hydration model indicate that structure IV, which shows the most pronounced stereoelectronic effect, is also the least-well-solvated structure. While this causes the other structures to be lower in energy relative to the gas-phase results, it does not change the identity of the lowest-energy structure in aqueous media, which remains IV for fluoropiperidine and II for trifluoromethylpiperidine as in the gas phase.

Acknowledgments Support from the Defense Threat Reduction Agency, the Office of Naval Research (via midshipmen research funds administered by the USNA Research Office), the DoD High Performance Computing Modernization Program, and the Air Force Research Laboratory DoD Supercomputing Resource Center is gratefully acknowledged. GSK wishes to acknowledge the kind mentorship of Isaiah Shavitt, who taught him many-body methods in electronic structure theory, helped him sharpen his thinking and writing, and demonstrated the importance of attention to detail.

References

1. Smart BE (2001) Fluorine substituent effects (on bioactivity). J Fluorine Chem 109(1):3–11
2. O'Hagan D (2008) Understanding organofluorine chemistry. An introduction to the C–F bond. Chem Soc Rev 37:308–319. doi:10.1039/b711844a
3. Hunter L (2010) The C–F bond as a conformational tool in organic and biological chemistry. Beilstein J Org Chem 6:38. doi:10.3762/bjoc.6.38
4. McKinney BE, Urban JJ (2010) Fluoroolefins as peptide mimetics. 2. A computational study of the conformational ramifications of peptide bond replacement. J Phys Chem A 114(2):1123–1133. doi:10.1021/Jp9094535
5. Urban JJ, Tillman BG, Cronin WA (2006) Fluoroolefins as peptide mimetics: a computational study of structure, charge distribution, hydration, and hydrogen bonding. J Phys Chem A 110(38):11120–11129
6. Molteni M, Bellucci MC, Bigotti S, Mazzini S, Volonterio A, Zanda M (2009) Ψ[CH(CF_3)NH]gly-peptides: synthesis and conformation analysis. Org Biomol Chem 7(11):2286–2296
7. Sinisi R, Ghilardi A, Ruiu S, Lazzari P, Malpezzi L, Sani M, Pani L, Zanda M (2009) Synthesis and in vitro evaluation of trifluoroethylamine analogues of enkephalins. ChemMedChem 4(9):1416–1420
8. Bigotti S, Meille SV, Volonterio A, Zanda M (2008) Synthesis of ψ[CH(RF)NH]gly-peptides: the dramatic effect of a single fluorine atom on the diastereocontrol of the key aza-michael reaction. J Fluorine Chem 129(9):767–774
9. Molteni M, Pesenti C, Sani M, Volonterio A, Zanda M (2004) Fluorinated peptidomimetics: synthesis, conformational and biological features. J Fluorine Chem 125(11):1735–1743
10. Zanda M (2004) Trifluoromethyl group: an effective xenobiotic function for peptide backbone modification. New J Chem 28(12):1401–1411
11. Kirby AJ (1996) Stereoelectronic effects, vol 36. Oxford, New York
12. Deslongchamps P (1983) Stereoelectronic effects in organic chemistry. Oxford, Pergamon
13. O'Hagan D (2012) Organofluorine chemistry: synthesis and conformation of vicinal fluoromethylene motifs. J Org Chem 77:3689–3699. doi:10.1021/jo300044q
14. Wiberg KB, Murcko MA, Laidig KE, MacDougall PJ (1990) Origin of the gauche effect in substituted ethanes and ethenes. J Phys Chem 94:6956–6959
15. Wolfe S (1972) The gauche effect. Some stereochemical consequences of adjacent electron pairs and polar bonds. Acc Chem Res 5:102
16. Trapp ML, Watts JK, Weinberg N, Pinto BM (2006) Component analysis of the X–C–Y anomeric effect (X=O, S; Y=F, OMe, NHMe) by dft molecular orbital calculations and natural bond orbital analysis. Can J Chem 84(4):692–701. doi:10.1139/v06-048
17. Salzner U, PvR Schleyer (1994) Ab initio examination of anomeric effects in tetrahydropyrans, 1,3-dioxanes, and glucose. J Org Chem 59(8):2138–2155
18. Thatcher GR (1993) The anomeric effect and associated stereoelectronic effects. ACS symposium series, vol 539. Wiley, New York
19. Juaristi E, Cuevas G (1992) Recent studies of the anomeric effect. Tetrahedron 48(24):5019–5087
20. Wiberg KB, Murcko MA (1989) Rotational barriers 4. Dimethoxymethane—the anomeric effect revisited. J Am Chem Soc 111(13):4821–4828
21. Wolfe S, Rauk A, Tel LM, Csizmadia IG (1971) Theoretical study of the edward-lemieux effect (the anomeric effect). Stereochemical requirements of adjacent electron pairs and polar bonds. J Chem Soc Sect B Phys Organ (1):136–145
22. Gold B, Dudley GB, Alabugin IV (2013) Moderating strain without sacrificing reactivity: design of fast and tunable noncatalyzed alkyne-azide cycloadditions via stereoelectronically controlled transition state stabilization. J Am Chem Soc 135(4):1558–1569. doi:10.1021/ja3114196
23. Gold B, Shevchenko NE, Bonus N, Dudley GB, Alabugin IV (2012) Selective transition state stabilization via hyperconjugative and conjugative assistance: stereoelectronic concept for copper-free click chemistry. J Org Chem 77(1):75–89. doi:10.1021/jo201434w
24. Mo Y (2010) Computational evidence that hyperconjugative interactions are not responsible for the anomeric effect. Nat Chem 2:666–671. doi:10.1038/nchem.721
25. Weldon AJ, Vickrey TL, Tschumper GS (2005) Intrinsic conformational preferences of substituted cyclohexanes and tetrahydropyrans evaluated at the ccsd(t) complete basis set limit: implications for the anomeric effect. J Phys Chem A 109(48):11073–11079. doi:10.1021/jp0550311
26. Alabugin IV (2000) Stereoelectronic interactions in cyclohexane, 1,3-dioxane, 1,3-oxathiane, and 1,3-dithiane: W-effect, sigma(C–X)—sigma*(C–H) interactions, anomeric effect—what is really important? J Org Chem 65(13):3910–3919
27. Iupac. Compendium of chemical terminology, 2nd ed. (the "Gold Book"). Compiled by A. D. Mcnaught and A. Wilkinson, Blackwell Scientific Publications, oxford (1997). Xml on-line corrected version: http://goldbook.Iupac.Org (2006-) created by M. Nic, J. Jirat, B. Kosata; updates compiled by A. D. Jenkins
28. Deslongchamps G, Deslongchamps P (2011) Bent bonds, the antiperiplanar hypothesis and the theory of resonance. A simple model to understand reactivity in organic chemistry. Org Biomol Chem 9(15):5321–5333. doi:10.1039/C1ob05393k
29. Reed AE, PvR Schleyer (1987) The anomeric effect with central atoms other than carbon. 1. Strong interactions between nonbonded substituents in polyfluorinated first- and second-row hydrides. J Am Chem Soc 109(24):7362–7373
30. Weinhold F (1998) Natural bond orbital methods. In: Schleyer PV, Allinger NL, Clark T et al (eds) Encylopedia of computational chemistry, vol 3. Wiley, Chichester, pp 1792–1811
31. Urban JJ (2005) Computational study of stereoelectronic effects in fluorinated alkylamines. J Phys Org Chem 18(11):1061–1071
32. Senderowitz H, Aped P, Fuchs B (1993) Computation of O–C–F and N–C–F systems—ab initio calculations and a mm2 parameterization study—theory vs experiment. Tetrahedron 49(18):3879–3898
33. Freitas MP (2013) The anomeric effect on the basis of natural bond orbital analysis. Org Biomol Chem 11(17):2885–2890. doi:10.1039/C3OB40187A
34. Mo Y, Song L, Lin Y (2007) Block-localized wavefunction (blw) method at the density functional theory (dft) level. J Phys Chem A 111(34):8291–8301. doi:10.1021/jp0724065

35. Gianinetti E, Raimondi M, Tornaghi E (1996) Modification of the roothaan equations to exclude bsse from molecular interactions calculations. Int J Quantum Chem 60(1):157–166. doi:10.1002/(SICI)1097-461X(1996)60:1<157:AID-QUA17>3.0.CO;2-C
36. Bauerfeldt GF, Cardozo TM, Pereira MS, da Silva CO (2013) The anomeric effect: the dominance of exchange effects in closed-shell systems. Org Biomol Chem 11(2):299–308. doi:10.1039/C2OB26818C
37. Huang Y, Zhong A-G, Yang Q, Liu S (2011) Origin of anomeric effect: a density functional steric analysis. J Chem Phys 134(8):084103/084101–084103/084109. doi:10.1063/1.3555760
38. Liu S (2007) Steric effect: a quantitative description from density functional theory. J Chem Phys 126(24):244103. doi:10.1063/1.2747247
39. Bjornsson R, Arnason I (2009) Conformational properties of six-membered heterocycles: accurate relative energy differences with dft, the importance of dispersion interactions and silicon substitution effects. Phys Chem Chem Phys 11(39):8689–8697. doi:10.1039/b910016d
40. Raghavachari K, Trucks GW, Pople JA, Head-Gordon M (1989) A fifth-order perturbation comparison of electron correlation theories. Chem Phys Lett 157:479–483. doi:10.1016/s0009-2614(89)87395-6
41. Møller C, Plesset MS (1934) Phys Rev 46:618
42. Schwabe T, Grimme S (2007) Double-hybrid density functionals with long-range dispersion corrections: higher accuracy and extended applicability. Phys Chem Chem Phys 9:3397–3406. doi:10.1039/b704725h
43. Chai JD, Head-Gordon M (2008) Long-range corrected hybrid density functionals with damped atom-atom dispersion corrections. Phys Chem Chem Phys 10(44):6615–6620. doi:10.1039/B810189b
44. Grimme S (2006) Semiempirical gga-type density functional constructed with a long-range dispersion correction. J Comput Chem 27(15):1787–1799. doi:10.1002/Jcc.20495
45. Zhao Y, Truhlar DG (2008) The m06 suite of density functionals for main group thermochemistry, thermochemical kinetics, non-covalent interactions, excited states, and transition elements: two new functionals and systematic testing of four m06-class functionals and 12 other functionals. Theor Chem Acc 120(1–3):215–241. doi:10.1007/s00214-007-0310-x
46. Becke AD (1988) Density-functional exchange-energy approximation with correct asymptotic behavior. Phys Rev A At Mol Opt Phys 38(6):3098–3100
47. Dunning TH Jr (1989) Gaussian basis sets for use in correlated molecular calculations. I. The atoms boron through neon and hydrogen. J Chem Phys 90(2):1007–1023
48. Kendall RA, Dunning TH, Harrison RJ (1992) Electron-affinities of the 1st-row atoms revisited—systematic basis-sets and wave-functions. J Chem Phys 96(9):6796–6806. doi:10.1063/1.462569
49. Frisch MJ, Trucks GW, Schlegel HB, Scuseria GE, Robb MA, Cheeseman JR, Scalmani G, Barone V, Mennucci B, Petersson GA, Nakatsuji H, Caricato M, Li X, Hratchian HP, Izmaylov AF, Bloino J, Zheng G, Sonnenberg JL, Hada M, Ehara M, Toyota K, Fukuda R, Hasegawa J, Ishida M, Nakajima T, Honda Y, Kitao O, Nakai H, Vreven T, Montgomery J, Peralta JE, Ogliaro F, Bearpark M, Heyd JJ, Brothers E, Kudin KN, Staroverov VN, Kobayashi R, Normand J, Raghavachari K, Rendell A, Burant JC, Iyengar SS, Tomasi J, Cossi M, Rega N, Millam JM, Klene M, Knox JE, Cross JB, Bakken V, Adamo C, Jaramillo J, Gomperts R, Stratmann RE, Yazyev O, Austin AJ, Cammi R, Pomelli C, Ochterski JW, Martin RL, Morokuma K, Zakrzewski VG, Voth GA, Salvador P, Dannenberg JJ, Dapprich S, Daniels AD, Farkas Ö, Foresman JB, Ortiz JV, Cioslowski J, Fox DJ (2009) Gaussian09, revision b.01. Gaussian Inc., Wallingford
50. Badenhoop JK, Weinhold F (1997) Natural bond orbital analysis of steric interactions. J Chem Phys 107(14):5406–5421
51. Valiev M, Bylaska EJ, Govind N, Kowalski K, Straatsma TP, Van Dam HJJ, Wang D, Nieplocha J, Apra E, Windus TL, de Jong WA (2010) Nwchem: a comprehensive and scalable open-source solution for large scale molecular simulations. Comput Phys Commun 181(9):1477–1489. doi:10.1016/j.cpc.2010.04.018
52. Schmidt MW, Baldridge KK, Boatz JA, Elbert ST, Gordon MS, Jensen JH, Koseki S, Matsunaga N, Nguyen KA, Su SJ, Windus TL, Dupuis M, Montgomery JA (1993) General atomic and molecular electronic-structure system. J Comput Chem 14(11):1347–1363. doi:10.1002/jcc.540141112
53. Halkier A, Helgaker T, Jorgensen P, Klopper W, Koch H, Olsen J, Wilson AK (1998) Basis-set convergence in correlated calculations on Ne, N_2, and H_2O. Chem Phys Lett 286(3–4):243–252. doi:10.1016/S0009-2614(98)00111-0
54. Biczysko M, Panek P, Scalmani G, Bloino J, Barone V (2010) Harmonic and anharmonic vibrational frequency calculations with the double-hybrid b2plyp method: analytic second derivatives and benchmark studies. J Chem Theory Comput 6:2115–2125. doi:10.1021/ct100212p
55. Winstein S, Holness NJ (1955) Neighboring carbon and hydrogen. Xix. Tert-butylcyclohexyl derivatives. Quantitative conformational analysis. J Am Chem Soc 77:5562–5578
56. Eliel EL, Wilen SH, Mander LN (1994) Stereochemistry of organic compounds. Wiley, Hoboken
57. Reed AE, Curtiss LA, Weinhold F (1988) Intermolecular interactions from a natural bond orbital, donor–acceptor viewpoint. Chem Rev 88(6):899–926. doi:10.1021/cr00088a005
58. Alabugin IV, Gilmore KM, Peterson PW (2011) Hyperconjugation. Wiley Interdiscip Rev Comput Mol Sci 1(1):109–141. doi:10.1002/wcms.6
59. Gillespie RJ (1963) The valence-shell electron-pair repulsion (vsepr) theory of directed valency. J Chem Educ 40:295
60. Tomasi J, Mennucci B, Cammi R (2005) Quantum mechanical continuum solvation models. Chem Rev (Washington, DC, US) 105(8):2999–3093. doi:10.1021/Cr9904009
61. Dunitz JD, Taylor R (1997) Organic fluorine hardly ever accepts hydrogen bonds. Chem Eur J 3(1):89–98

Theor Chem Acc (2014) 133:1464
DOI 10.1007/s00214-014-1464-y

REGULAR ARTICLE

σ–σ and σ–π pnicogen bonds in complexes H_2XP:PCX, for X = F, Cl, OH, NC, CN, CCH, CH_3, and H

Janet E. Del Bene · Ibon Alkorta · José Elguero

Received: 21 November 2013 / Accepted: 11 February 2014 / Published online: 5 March 2014

Abstract Ab initio MP2/aug'-cc-pVTZ calculations have been carried out on complexes H_2XP_s:P_tCX, for X = F, Cl, OH, NC, CN, CCH, CH_3, and H, in search of complexes stabilized by P···P pnicogen bonds. These intermolecular bonds arise when a pnicogen atom acts as a Lewis acid for complex formation. Three sets of equilibrium structures have been found on the H_2XP_s:P_tCX potential surfaces. Conformation A complexes have P···P σ–σ pnicogen bonds, which involve the σ systems of both P atoms. Conformations B and C are stabilized by σ–π pnicogen bonds, which involve the σ system of H_2XP and the π system of PCX. Binding energies of B and C complexes are similar and are greater than the binding energies of the A conformers. Charge transfer stabilizes A, B, and C conformers. In A complexes, the dominant charge transfer is from the lone pair of PCX to the antibonding σ*P–A orbital of PH_2X, with A the atom of X directly bonded to P. For conformations B and C, the dominant charge transfer is from the P=C π orbital to the σ*P–A orbital of H_2XP. Although the binding energies of these complexes do not correlate with the intermolecular P–P distances, both the charge-transfer energies and the equation-of-motion coupled cluster singles and doubles one-bond ^{31}P–^{31}P spin–spin coupling constants do correlate with the P–P distances. The largest coupling constants ^{1p}J(P–P) are found for complexes with conformation A, due to the nature of the σ–σ pnicogen bond and the dominance of the Fermi contact term. For a given X, ^{1p}J(P–P) values are ordered A > C > B.

Keywords Structures and binding energies · Intermolecular interactions · σ–σ and σ–π pnicogen bonds · Charge-transfer energies · ^{31}P–^{31}P EOM-CCSD spin–spin coupling constants ^{1p}J(P–P)

Dedicated to the memory of Professor Isaiah Shavitt and published as part of the special collection of articles celebrating his many contributions.

Electronic supplementary material The online version of this article (doi:10.1007/s00214-014-1464-y) contains supplementary material, which is available to authorized users.

J. E. Del Bene (✉)
Department of Chemistry, Youngstown State University, Youngstown, OH 44555, USA
e-mail: jedelbene@ysu.edu

I. Alkorta (✉) · J. Elguero
Instituto de Química Médica (IQM–CSIC), Juan de la Cierva, 28006 Madrid, Spain
e-mail: ibon@iqm.csic.es

1 Introduction

Subsequent to the landmark 2011 paper of Hey-Hawkins et al. [1], there have been many papers published on the subject of intermolecular interactions through the formation of pnicogen bonds [2–32]. This bond arises when a pnicogen atom (N, P, As) acts as a Lewis acid by accepting a pair of electrons from a Lewis base. When two pnicogen atoms participate in forming a bond, each acts as both an electron-pair acceptor and an electron-pair donor. Most studies of pnicogen bonds have involved the PH_3 molecule and its derivatives.

Recently, we asked to what extent a formally sp^2-hybridized P atom could participate in a P···P pnicogen bond in complexes $(H_2C{=}PX)_2$ and $H_2C{=}(X)P{:}PXH_2$ for a variety of substituents X [30, 32]. In the latter study, we identified a series of complexes stabilized by pnicogen bonds, which involve π electron donation by $H_2C{=}PX$ to PH_2X through the σ-hole, and donation of the lone pair of

PH_2X to H_2C=PX through the π-hole. To our knowledge, this was the first time that pnicogen-bonded complexes with a π electron donor and π-hole acceptor involving the same π bond had been reported. We referred to this bond as a π–σ pnicogen bond to indicate that it is the π system of one molecule and the σ system of the other that interact. It should be noted, however, that pnicogen bonds involving π-electron donors or π-hole acceptors have been discussed previously in the literature, but the π-donors and π-hole acceptors are not the same π bond [7, 21, 29, 33, 34].

Since PH_2X can form stable σ–π pnicogen-bonded complexes with H_2C=(X)P, we decided to investigate a related series of complexes in which PH_2X interacts with PCX, searching for both σ–π and σ–σ pnicogen bonds. This study yielded three different conformations of complexes H_2XP:PCX, for X = F, Cl, OH, NC, CN, CCH, CH_3, and H. In this paper, we report the equilibrium structures of these complexes, their binding energies, bonding properties, and ^{31}P–^{31}P spin–spin coupling constants.

2 Methods

The structures of the isolated monomers and the complexes H_2XP:PCX were optimized at second-order Møller-Plesset perturbation theory (MP2) [35–38] with the aug'-cc-pVTZ basis set [39]. This basis set is derived from the Dunning aug-cc-pVTZ basis set [40, 41] by removing diffuse functions from H atoms. Frequencies were computed to establish that the optimized structures correspond to equilibrium structures on their potential surfaces. In addition, we determined transition structures which are the barriers to the interconversion of the B and C conformers of H_2FP:PCF and H_3P:PCH. Optimization and frequency calculations were performed using the Gaussian 09 program [42].

The electron densities of these complexes have been analyzed using the atoms in molecules (AIM) methodology [43–46] and the electron localization function (ELF) [47] employing the AIMAll [48] and Topmod [49] programs. The topological analysis of the electron density produces the molecular graph of each complex. This graph identifies the location of electron density features of interest, including the electron density (ρ) maxima associated with the various nuclei, saddle points which correspond to bond critical points (BCPs), and ring critical points which indicate a minimum electron density within a ring. The zero gradient line which connects a BCP with two nuclei is the bond path. The ELF function illustrates those regions of space at which the electron density is high. MP2/aug'-cc-pVTZ atomic and molecular charges have been obtained using the natural bond orbital (NBO) method [50]. The stabilizing charge-transfer interactions in these binary complexes have been computed using the NBO-6 program [51].

Spin–spin coupling constants were evaluated using the equation-of-motion coupled cluster singles and doubles (EOM-CCSD) method in the CI (configuration interaction)-like approximation [52, 53], with all electrons correlated. For these calculations, the Ahlrichs [54] qzp basis set was placed on ^{13}C, ^{15}N, ^{17}O, and ^{19}F, and the qz2p basis set on ^{31}P and ^{35}Cl. The Dunning cc-pVDZ basis set was placed on all ^{1}H atoms. Only ^{1P}J(P–P) coupling constants are reported in this paper. The EOM-CCSD calculations were performed using ACES II [55] on the IBM Cluster 1350 (Glenn) at the Ohio Supercomputer Center.

3 Results and discussion

In order to differentiate the two P atoms, we will refer to them as P_s, the P atom that forms single covalent bonds in P_sH_2X, and P_t for the triply bonded P in P_tCX. Although many minima may exist on the H_2XP_s:P_tCX surfaces, we have restricted our searches to regions that have structural characteristics associated previously with complexes stabilized by pnicogen bonds. We began by searching the region in which the interaction involves the formation of a σ–σ pnicogen bond. In the resulting conformation A complexes, A–P_s···P_t–C approaches linearity, with A being the atom of X directly bonded to P_s, and C the carbon of PCX. We then investigated regions in which PCX interacts through its π-electron system with PXH_2 to form σ–π bonds. In conformation B complexes, a σ–π pnicogen bond forms in which A–P_s···C approaches linearity. In conformation C, the σ–π pnicogen bond has A–P_s···P_t approaching linearity. These three conformations are illustrated in Fig. 1. We did not search regions in which one H of PH_2X

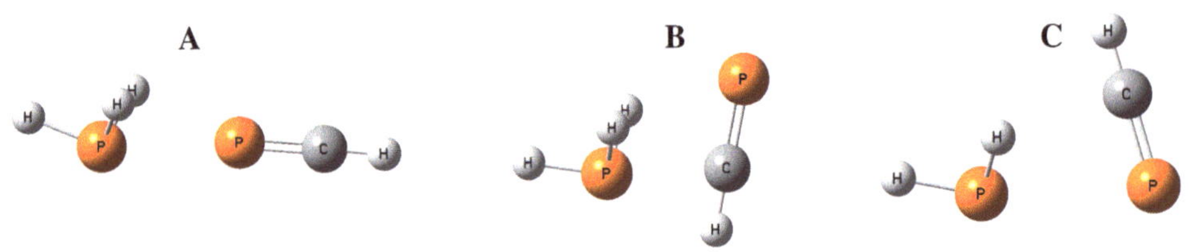

Fig. 1 H_3P:PCH complexes with conformations **A**, **B**, and **C**. All complexes have C_s symmetry except for $H_2(CN)P$:PCCN and $H_2(CCH)P$:PCCCH B which have C_1 symmetry

assumes a linear arrangement, since $(PH_2X)_2$ complexes with H–P···P–A linear have significantly reduced binding energies relative to the same complexes with A–P···P–A linear [18, 23].

The binding energies of conformation A, B, and C complexes are reported in Table 1. These are ordered according to decreasing binding energies of conformation C. All substituents X form B and C complexes stabilized by σ–π bonds. Conformation A complexes have P···P σ–σ pnicogen bonds and are formed by all molecules except those containing the more electronegative substituents F and Cl. $H_2(OH)P$:PCOH is a planar complex, but it has not been included in this study since it is stabilized primarily by an O–H···P_s hydrogen bond. Its binding energy of −34.2 kJ/mol is significantly greater than the binding energies of the pnicogen-bonded complexes.

3.1 Conformation A complexes

The structures, total energies, and bond paths connecting the two P atoms of conformation A complexes are reported in Table S1 of the Electronic Supporting Material. Table 2 reports the binding energies, P–P distances, and the angles P_t–P_s–A and P_s–P_t–C in these complexes. Of the three conformations, the A complexes are the most weakly bound, with binding energies ranging from −3.1 kJ/mol for $H_2(CN)P$:PCCN to −7.4 kJ/mol for $H_2(CCH)P$:PCCCH. Moreover, these binding energies are lower than the binding energies of the corresponding complexes $(PH_2X)_2$ and H_2C=(X)P:PXH_2. For a given X, the intermolecular P–P distances in A conformers decrease in the order H_2XP:PCX > H_2C=(X)P:PXH_2 > $(PH_2X)_2$. As noted previously for complexes with σ–σ pnicogen bonds, A–P···P–A' arrangements tend to approach linearity. The P_s–P_t–C alignment in conformation A complexes closely approaches linearity, with the P_s–P_t–C angle varying between 175 and 179°. The P_t–P_s–A angle deviates to some extent from linearity, with values between 158 and 167°.

Complexes with pnicogen bonds are stabilized by charge transfer. The more favorable charge-transfer interaction involves donation of the P_t lone pair of PCX to the σ*P–A orbital of PH_2X. Charge-transfer energies range from 3.4 kJ/mol for X = CH_3 to 9.4 kJ/mol when X = NC. In contrast, charge transfer from the P_s lone pair of PH_2X to the σ*P=C orbital of PCX is 2.5 kJ/mol when X = NC, and 1.4 or 1.5 kJ/mol for the remaining complexes. The preference for charge transfer from H_2C=PX to PH_2X was observed previously in conformation A complexes [32]. The P_t(lp) → σ*P–A charge-transfer energies do not correlate with the binding energies of these complexes, but do correlate with the P–P distances, as can be seen in Fig. 2. The data of Table 3 also indicate that the PH_2X molecules become slightly negatively charged in the complexes, except for $P(CH_3)H_2$ which is uncharged. Both P_s and P_t are positively charged in the monomers, and that positive charge is reduced upon complexation. The positive charge on P_s in conformation A complexes varies from 0.043e in H_3P:PCH to 0.576e in $H_2(NC)P$:PCNC. The positive charge on P_s decreases in the order:

Table 1 Binding energies (ΔE, kJ/mol) of complexes H_2XP:PCX with conformations A, B, and C

H_2XP:PCX, X =	ΔE(A)	ΔE(B)	ΔE(C)
Cl		−16.4	−17.6
F		−16.6	−15.6
CCH	−7.4	−13.3	−14.7
OH[a]	[a]	−13.2	−14.2
NC	−4.2	−12.0	−13.5
CN	−3.1	−9.6	−10.6
CH_3	−5.7	−12.6	−10.0
H	−4.7	−8.7	−7.5

Ordered according to decreasing binding energy of conformation C complexes

[a] This complex is stabilized primarily by an O–H···P_s hydrogen bond, with a binding energy of −34.2 kJ/mol

Table 2 Binding energies (ΔE, kJ/mol), P–P distances (R, Å), and P_t–P_s–A and P_s–P_t–C angles (<, deg) for conformation A complexes

H_2XP:PCX, X =	ΔE	R(P–P)	<P_t–P_s–A	<P_s–P_t–C
CCH	−7.4	3.594	163	179
NC	−4.2	3.521	166	179
CN	−3.1	3.649	167	176
CH_3	−5.7	3.705	160	176
H	−4.7	3.772	158	175

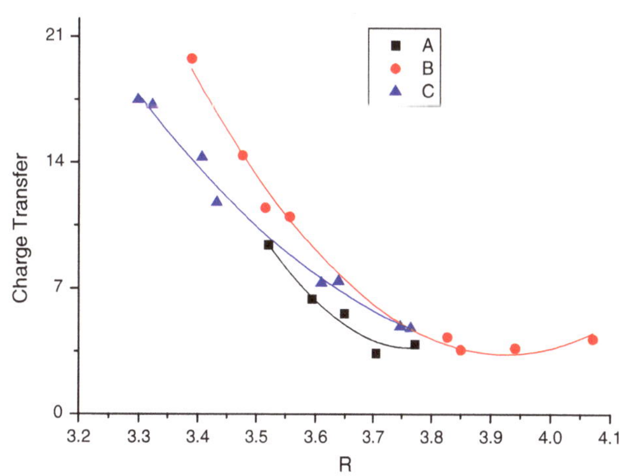

Fig. 2 Charge-transfer energies from PCX to PH_2X (kJ/mol) versus the P–P distance (R, Å) for conformation *A*, *B*, and *C* complexes. Correlation coefficients R^2 are 0.963, 0.992, and 0.990, respectively

Table 3 Charges on PH_2X, changes in the charges on the P atoms (δe, au), and charge-transfer energies (kJ/mol) for conformation A complexes H_2XP_s:P_tCX

H_2XP:PCX, X =	Charge on PH_2X	δe(P_s)[a]	δe(P_t)[a]	P_t(lp) → σ*P_s–A	P_s(lp) → π*P_t=C
CCH	−0.003	−0.011	−0.014	6.4	1.4
NC	−0.006	−0.017	−0.033	9.4	2.5
CN	−0.003	−0.016	−0.028	5.6	1.5
CH_3	0.000	−0.003	−0.003	3.4	1.5
H	−0.001	−0.005	−0.007	3.9	1.4

[a] Both P_s and P_t are positively charged in the isolated monomers

Table 4 Binding energies (ΔE, kJ/mol), P–P and P_s–C distances (R, Å), and C–P_s–A angles (<, deg) for conformation B complexes

H_2XP:PCH, X =	ΔE	R(P–P)	R(P_s–C)	<C–P_s–A
Cl	−16.4	3.476	3.326	175
F	−16.6	3.389	3.070	172
CCH	−13.3	3.827	3.404	161
OH	−13.2	3.516	3.251	175
NC	−12.0	3.557	3.241	177
CN	−9.6	3.941	3.426	165
CH_3	−12.6	3.850	3.439	177
H	−8.7	4.071	3.413	171

These complexes have C_s symmetry, except for H_2(CN)P:PCCN and H_2(CCH)P:PCCCH which have C_1 symmetry

$$NC \approx CN > CCH > H > CH_3.$$

The positive charge on P_t varies from 0.479e in $H_2(CH_3)$-P:$PCCH_3$ to 0.617e in H_2(CN)P:PCCN and decreases in a similar order:

$$NC > CN > CCH > H > CH_3.$$

For complexes with X = NC, CN, and CCH, the decrease in the positive charge on P_t is noticeably greater than the decrease in the positive charge on P_s.

3.2 Conformation B complexes

The structures, total energies, and bond paths of conformation B complexes are given in Table S1 of the Electronic Supporting Material. These bond paths connect P_s with PCX through the π system at the P≡C C atom. The classification as conformation B is based on the C–P_s–A angles which range from 161 to 177° and are closer to linearity than the corresponding P_t–P_s–A angles. Conformation B pnicogen bonds are σ–π bonds, which involve the σ system of PH_2X and the P≡C π bond. The binding energies, P–P and P_s–C distances, and C–P_s–A angles are reported in Table 4. The binding energies of these complexes range from −8.7 kJ/mol for H_3P:PCH to −16.6 kJ/mol for H_2FP:PCF. The ordering is consistent with the ordering of conformation C complexes, except for H_2-(CH_3)P:$PCCH_3$, which has a noticeably higher binding energy than its C counterpart. The P_s–C distances are always shorter than the corresponding P–P distances, but once again, there is no correlation between the binding energies and either the P–P or the P_s–C intermolecular distances. Both linear and quadratic trendlines have correlation coefficients R^2 of 0.73.

Conformation B complexes are also stabilized by charge-transfer interactions. Since the P≡C bond is polarized toward C, charge transfer occurs from the π bond at C through the σ-hole to P_s, and from the lone pair on P_s to P_t through the π hole. In all complexes but one, the dominant charge transfer is from the πP=C orbital of PCX to the σ*P–A antibonding orbital of PH_2X, as can be seen from the data of Table 5. The single exception is X = H, for which charge transfer from the πP=C orbital of PCH to the σ*P–H orbital is 0.2 kJ/mol less stabilizing. Having the C–P_s–A angle approach linearity is favorable for charge transfer from PCX to PH_2X. In addition, the P_s–P_t–C angles are acute, ranging from 55 to 68°, thereby leading to shorter distances between P_s and C, and facilitating charge transfer from the P=C π bond, which is polarized toward C. The πP=C→σ*P–A charge-transfer energies vary from 3.6 kJ/mol for X = CH_3 to 19.8 kJ/mol for X = F. Charge-transfer energies from the lone pair on P_s to the π*P=C orbital range from 1.3 kJ/mol for X = CCH to 10.0 kJ/mol for X = F. The πP=C→σ*P–A charge-transfer energies do not correlate with the binding energies of conformation B complexes or with the P_s–C distances, but do correlate with the intermolecular P–P distances, as shown in Fig. 2. The net result of charge transfer is to produce a slightly negatively charged PH_2X molecule, except for PH_2CH_3 and PH_3, and to reduce the positive charge on P_s. The positive charge on P_t may increase or decrease, as seen from the data of Table 5. Figure 3 illustrates the regions of high electron density involved in charge transfer in H_2FP:PCF conformation B.

3.3 Conformation C complexes

Table S1 of the Electronic Supporting Material reports the structures, total energies, and molecular graphs for

Table 5 Charges on PH_2X, changes in the charges on the P atoms (δe, au), and charge-transfer energies (kJ/mol) for conformation B complexes $H_2XP_s{:}P_t{\equiv}CX$

$H_2XP{:}PCX$, X =	Charge on PH_2X	$\delta e(P_s)$	$\delta e(P_t)$	$\pi P{=}C \rightarrow \sigma^*P_s{-}A$	$P_s(lp) \rightarrow \pi^*P{=}C$
Cl	−0.011	−0.017	−0.003	14.4	5.1
F	−0.005	−0.020	0.006	19.8	10.0
CCH	−0.003	−0.005	0.005	4.3	1.3
OH	−0.002	−0.007	−0.002	11.5	5.1
NC	−0.007	−0.014	−0.002	11.0	5.6
CN	−0.003	−0.010	0.014	3.7	1.5
CH_3	0.003	−0.011	−0.010	3.6	3.4
H	0.002	−0.008	−0.002	4.2	4.4

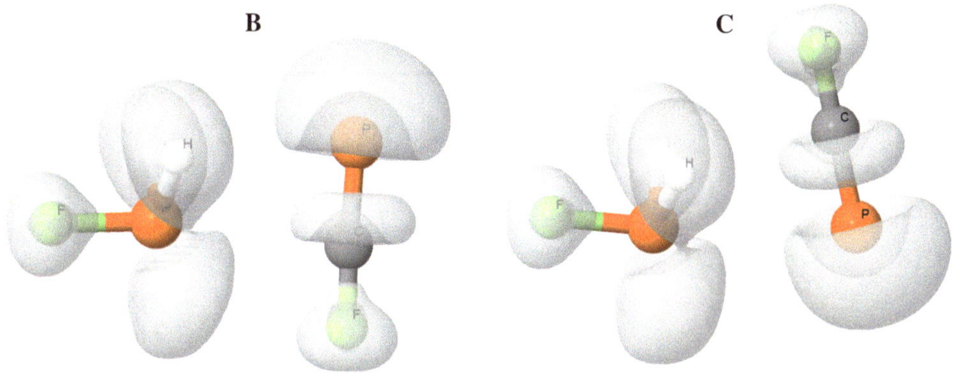

Fig. 3 The ELF representations of the regions of high electron density in $H_2FP{:}PCF$ B and C

Table 6 Binding energies (ΔE, kJ/mol), P–P and P–C distances (R, Å), and P_t–P_s–A angles (<, deg) for conformation C complexes

$H_2XP{:}PCH$, X =	ΔE	R(P–P)	$R(P_s{-}C)$	$<P_t{-}P_s{-}A$
Cl	−17.6	3.323	3.313	174
F	−15.6	3.298	3.270	175
CCH	−14.7	3.610	3.431	172
OH	−14.2	3.432	3.360	171
NC	−13.5	3.407	3.317	175
CN	−10.6	3.639	3.428	176
CH_3	−10.0	3.747	3.571	166
H	−7.5	3.765	3.702	168

conformation C complexes. The bond paths connect P_s to the π system of PCX, usually but not always at P_t. These complexes are differentiated from the conformation B complexes in so far as the P_t–P_s–A angles approach closer to linearity than the corresponding C–P_s–A angles. The values of the P_t–P_s–A angles are reported in Table 6 and can be seen to vary between 166 and 176°. Table 6 also shows that the P_s–C distances are still shorter than the P–P distances, although the difference between them is much less than found for conformation B complexes due to the values of the P_s–P_t–C angles. These angles vary between 70 and 76°, and are therefore larger than the corresponding angles in conformation B.

Table 6 also reports the binding energies of conformation C complexes. These energies range from −7.5 kJ/mol for X = H to −17.6 kJ/mol for X = Cl. Conformation C complexes are more stable than conformation B for 5 of 8 complexes, but the binding energies of B and C are similar, differing by 1 to 1.5 kJ/mol. The single exception is conformation C of $H_2(CH_3)P{:}PCCH_3$, which is 2.5 kJ/mol less stable than B. The binding energies do not correlate with the P–P distances.

Figure 3 illustrates the regions of high electron density in $H_2FP{:}PCF$ conformation C. The regions associated with the lone pair on P_s and the P=C π bond give rise to the charge-transfer interactions. A charge-transfer pattern similar to that observed for conformation B complexes is found for conformation C. Charge transfer involves electron donation by πP=C to the σ*P_s–A antibonding orbital through the σ-hole at P_s, and lone-pair donation by P_s to the π*P=C orbital through the π-hole at P_t. As can be seen from the data of Table 7, the πP=C→σ*P_s–A charge-transfer energies are significantly greater than the P_s(lp) → π*P=C energies and are also greater than the corresponding πP=C→σ*P_s–A energies of conformation B complexes, except for $H_2FP{:}PCF$ which has the largest charge-transfer energy among all complexes. Once again, the πP=C→σ*P_s–A charge-transfer energies correlate with the P–P distances, as seen in Fig. 2. The net result of charge transfer is to make PH_2X negatively charged in the

Table 7 Charges on PH_2X, changes in the charges on the P atoms (δe, au), and charge-transfer energies (kJ/mol) for conformation C complexes $H_2XP_s{:}P_t{\equiv}CX$

$H_2XP{:}PCX$, X =	Charge on PH_2X	$\delta e(P_s)$	$\delta e(P_t)$	$\pi P{=}C \rightarrow \sigma^*P_s{-}A$	$P_s(lp) \rightarrow \pi^*P{=}C$
Cl	−0.018	−0.019	0.020	17.2	5.1
F	−0.012	−0.017	0.025	17.5	6.2
CCH	−0.004	−0.013	0.018	7.3	1.0
OH	−0.007	−0.010	0.019	11.8	4.1
NC	−0.012	−0.023	0.021	14.3	2.7
CN	−0.005	−0.023	0.022	7.4	0.7
CH_3	−0.002	0.010	0.015	4.9	0.9
H	−0.003	−0.004	0.012	4.8	1.5

Table 8 Intermolecular P–P distances (R, Å) and ^{31}P–^{31}P spin–spin coupling constants (Hz) for $H_2XP{:}PCX$ complexes with conformations A, B, and C

$H_2XP{:}PCX$	A		B		C	
	R(P–P)	^{1p}J(P–P)	R(P–P)	^{1p}J(P–P)	R(P–P)	^{1p}J(P–P)
X = Cl			3.476	41.1	3.323	116.7
F			3.389	54.3	3.298	119.7
CCH	3.594	157.6	3.827	16.9	3.610	49.4
OH			3.516	41.9	3.432	83.3
NC	3.521	209.4	3.557	31.9	3.407	86.8
CN	3.649	150.4	3.941	14.9	3.639	48.0
CH_3	3.705	100.8	3.850	11.9	3.747	32.6
H	3.772	88.3	4.071	7.3	3.765	34.5

complex, decrease the positive charge on P_s, and increase the positive charge on P_t.

B and C complexes are energetically and structurally similar, since both are stabilized by pnicogen bonds involving P_s of PH_2X and the PCX π system. What is the barrier for interconversion of these conformers? To answer this question, we have optimized transition structures on the $H_3P{:}PCH$ and $H_2FP{:}PCF$ surfaces. These structures have C_1 symmetry, with the PCH and PCF molecules nearly perpendicular to the PH_3 and PH_2F symmetry planes, respectively. The structures suggest that B and C are interconverted by rotation of the PCH or PCF molecules about an axis which connects P_s to the P=C π bond, and by accompanying changes in the P_s–C and P–P bond lengths. For $H_3P{:}PCH$, the P_s–C bond length in the transition structure is similar to that in C, while the P–P bond length increases. Both P_s–C and P–P bonds are longer in the $H_2FP{:}PCF$ transition structure that they are in both B and C. The interconversion of B and C via rotation indicates that these complexes remain intact in the transition state, with similar binding energies of –6.1 and –6.9 kJ/mol, respectively. Relative to the less stable conformer C, the barriers to converting C to B are 2.6 and 9.7 kJ/mol for $H_3P{:}PCH$ and $H_2FP{:}PCF$, respectively.

3.4 ^{31}P–^{31}P spin–spin coupling constants

^{31}P–^{31}P spin–spin coupling constants across pnicogen bonds have been computed for complexes with conformations A, B, and C. Table S2 of the Electronic Supporting Material reports components of ^{1p}J(P–P). Although previous investigations of ^{31}P–^{31}P coupling across pnicogen bonds suggest that the Fermi contact term is an excellent approximation to total ^{1p}J(P–P), we computed all terms, the paramagnetic spin orbit (PSO), diamagnetic spin orbit (DSO), Fermi contact (FC), and spin dipole (SD) terms for 12 $H_2XP{:}PCX$ complexes. The largest differences between the FC terms and total ^{1p}J(P–P) are found for complexes in which X is one of the more electronegative substituents, namely F, Cl, and OH. For these, the difference between the FC term and ^{1p}J(P–P) ranges from 2.0 to 5.8 Hz and arises primarily from the contribution of the PSO term. For the remaining 9 complexes with C_s symmetry, PSO, DSO, and FC terms were computed. The largest PSO contribution is 1.5 Hz for X = NC, with the PSO terms for the remaining complexes having absolute values less than 0.7 Hz. For the two conformation B complexes of C_1 symmetry with X = CCH and CN, only the FC terms were evaluated due to computational expense. This can be justified by noting that the PSO terms for the A and C complexes with these same substituents have absolute values no greater than 0.1 Hz. In Table 8, total ^{1p}J(P–P) values are reported for 12 complexes, and the FC terms have been used to approximate ^{1p}J(P–P) for the remaining complexes.

Table 8 reports the P–P distances and values of ^{1p}J(P–P) for complexes $H_2XP{:}PCX$ with conformations A, B, and C. For a given substituent X, the order of decreasing ^{1p}J(P–P)

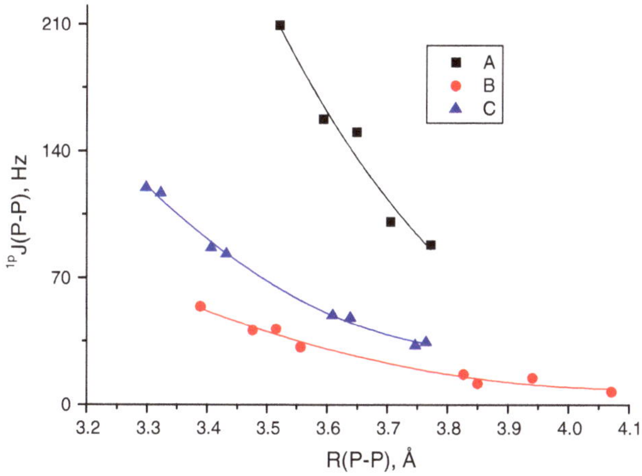

Fig. 4 ^{1p}J(P–P) versus the P–P distance for complexes with conformations *A*, *B*, and *C*

is A > C > B. The large values of ^{1p}J(P–P) for conformation A complexes may be attributed primarily to the nature of the pnicogen bond, which is a σ–σ bond, and the dependence of the dominant FC term on s-electron densities in both ground and excited states. The nature of the FC term is also consistent with the reduced values of ^{1p}J(P–P) for conformations B and C, since they are stabilized by σ–π pnicogen bonds which involve the π electrons of PCX. That ^{1p}J(P–P) for a given X is greater for the conformation C complex is consistent with the shorter P–P distances in C, and with the A–P_s–P_t arrangement which approaches linearity. Figure 4 presents plots of ^{1p}J(P–P) versus the P–P distance for complexes with conformations A, B, and C. The good correlation between these two variables is evident, with the second-order trendlines having correlation coefficients R^2 of 0.961, 0.995, and 0.976, respectively.

In a previous study, we compared coupling constants ^{1p}J(P–P) for conformation A complexes $(PH_2X)_2$, H_2C=(X)P:PXH_2, and $(H_2C{=}PX)_2$ with σ–σ pnicogen bonds. We can now compare ^{1p}J(P–P) for the σ–π pnicogen bonds in conformations B and C of H_2XP:PCX with ^{1p}J(P–P) σ–π bonds for conformation C complexes of H_2C=(X)P:PXH_2. That comparison is illustrated in Fig. 5. The exponential trendline for the entire set of points has a correlation coefficient of 0.859. However, it is apparent from Fig. 5 that all of the points for conformation C complexes of H_2XP:PCX with A–$P_s\cdots P_t$ approaching linearity lie above the trendline and have larger values of ^{1p}J(P–P) at each P–P distance. Values for H_2XP:PCX conformation B with A–$P_s\cdots$C linear and conformation C of H_2C=(X)P_d:P_sXH_2 with A–$P_s\cdots P_d$ linear lie either on or below the trendline, and have similar values at similar distances. The exponential trendline for ^{1p}J(P–P) versus R(P–P) for these two sets treated together has a correlation coefficient R^2 of 0.958.

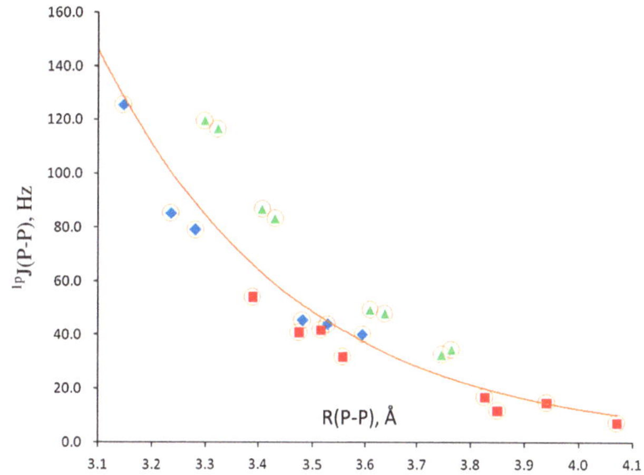

Fig. 5 Coupling constants ^{1p}J(P–P) versus the P–P distances for σ–π pnicogen bonds in H_2XP:PCX conformations B (*red square*) and C (*green triangle*) and H_2C=(X)P:PXH_2 conformation C (*blue diamond*). The symbol *circle* includes all complexes for which the best-fit trendline is an exponential

4 Conclusions

Ab initio MP2/aug'-cc-pVTZ calculations have been carried out on a series of complexes H_2XP:PCX, for X = F, Cl, OH, NC, CN, CCH, CH_3, and H, to identify and characterize σ–σ and σ–π pnicogen bonds. Three sets of complexes have been identified.

1. Conformation A complexes are stabilized by P···P σ–σ bonds. Of the three sets of complexes, the A conformers are the most weakly bound, with binding energies ranging from −3.1 to −7.4 kJ/mol. A conformers are stabilized by charge-transfer interactions, the more stabilizing of which arises from transfer of the P_t lone pair of P_tCH to the σ*P_s-A orbital of P_sH_2X, where A is the atom of X directly bonded to P. Charge transfer from the lone pair of P_s to the σ*P=C orbital is a less stabilizing interaction.
2. Conformation B and C complexes are stabilized by σ–π bonds, that is, bonds that involve interaction between the σ system of PH_2X and the π system of PCH. In both sets of conformers, the preferred direction of charge transfer is from the π bond of PCH through the σ-hole to the σ*P–A orbital of PH_2X and secondarily from the lone pair on P_s through the π-hole to the π*P=C orbital.
3. Conformation B and C complexes have similar binding energies, which range from −7.5 to −17.6 kJ/mol. They are differentiated structurally in so far as B complexes have A-$P_s\cdots$C approaching linearity, whereas C complexes have A-$P_s\cdots P_t$ approaching linearity.

4. The binding energies of conformation A, B, and C complexes do not correlate with the intermolecular P–P distances. However, the dominant charge-transfer energies in each set do correlate with the intermolecular P–P distance.
5. EOM-CCSD ^{31}P–^{31}P spin–spin coupling constants $^{1p}J(P–P)$ correlate with the P–P distances in conformations A, B, and C. The largest coupling constants are found for conformation A complexes, a result of the nature of the σ–σ pnicogen bond and the dominance of the Fermi contact term. For a given substituent X, the ordering of $^{1p}J(P–P)$ is A > C > B >.

Acknowledgments This work was carried out with financial support from the Ministerio de Educación y Ciencia (Project No. CTQ2012-35513-C02-02) and Comunidad Autónoma de Madrid (Project MADRISOLAR2, ref S2009/PPQ1533). Thanks are given to the Ohio Supercomputer Center for its continued support and to the CTI (CSIC). This paper is dedicated in memory of Shi Shavitt, mentor, colleague, and friend.

References

1. Zahn S, Frank R, Hey-Hawkins E, Kirchner B (2011) Pnicogen bonds: a new molecular linker? Chem Eur J 17:6034–6038
2. Solimannejad M, Gharabaghi M, Scheiner S (2011) SH···N and SH···P blue-shifting H-bonds and N···P interactions in complexes pairing HSN with amines and phosphines. J Chem Phys 134(1–6):024312
3. Scheiner SA (2011) New noncovalent force: comparison of P···N interaction with hydrogen and halogen bonds. J Chem Phys 134(1–9):094315
4. Scheiner S (2011) Effects of substituents upon the P···N noncovalent interaction: the limits of its strength. J Phys Chem A 115:11202–11209
5. Adhikari U, Scheiner S (2012) Substituent effects on Cl···N, S···N and P···N noncovalent bonds. J Phys Chem A 116:3487–3497
6. Adhikari U, Scheiner S (2012) Sensitivity of pnicogen chalcogen halogen and H-bonds to angular distortions. Chem Phys Lett 532:31–35
7. Scheiner S (2011) Can two trivalent N atoms engage in a direct N···N noncovalent interaction? Chem Phys Lett 514:32–35
8. Scheiner S (2011) Effects of multiple substitution upon the P···N noncovalent interaction. Chem Phys 387:79–84
9. Scheiner S (2011) On the properties of X···N noncovalent interactions for first- second- and third-row x atoms. J Chem Phys 134(1–9):164313
10. Adhikari U, Scheiner S (2011) Comparison of P···D (D = P, N) with other noncovalent bonds in molecular aggregates. J Chem Phys 135(1–10):184306
11. Scheiner S, Adhikari U (2011) Abilities of different electron donors (d) to engage in a p d noncovalent interaction. J Phys Chem A 115:11101–11110
12. Scheiner S (2011) Weak H-bonds: comparisons of CH···O to NH···O in proteins and PH···N to direct P···N interactions. Phys Chem Chem Phys 13:13860–13872
13. Del Bene JE, Alkorta I, Sánchez-Sanz G, Elguero J (2011) ^{31}P-^{31}P spin–spin coupling constants for pnicogen homodimers. Chem Phys Lett 512:184–187
14. Del Bene JE, Alkorta I, Sánchez-Sanz G, Elguero J (2014) Structures, energies, bonding, and NMR properties of pnicogen complexes $H_2XP:NXH_2$ (X = H, CH_3, NH_2, OH, F, Cl). J Phys Chem A 115:13724–13731
15. Adhikari U, Scheiner S (2012) Effects of carbon chain substituents on the P···N noncovalent bond. Chem Phys Lett 536:30–33
16. Li Q-Z, Li R, Liu X-F, Li W-Z, Cheng J-B (2012) Pnicogen-hydride interaction between FH_2X (X = P and As) and HM (M = ZnH, BeH, MgH, Li and Na). J Phys Chem A 116:2547–2553
17. Li Q-Z, Li R, Liu X-F, Li W-Z, Cheng J-B (2012) Concerted interaction between pnicogen and halogen bonds in XCl-FH_2P-NH_3 (X = F, OH, CN, NC and FCC). ChemPhysChem 13:1205–1212
18. Del Bene JE, Alkorta I, Sánchez-Sanz G, Elguero J (2012) Structures, binding energies, and spin–spin coupling constants of geometric isomers of pnicogen homodimers $(PHFX)_2$ X = F, Cl, CN, CH_3, NC. J Phys Chem A 116:3056–3060
19. Del Bene JE, Alkorta I, Sánchez-Sanz G, Elguero J (2012) Homo- and heterochiral dimers $(PHFX)_2$ X = Cl, CN, CH_3, NC: To what extent do they differ? Chem Phys Lett 538:14–18
20. Alkorta I, Sánchez-Sanz G, Elguero J, Del Bene JE (2012) Influence of hydrogen bonds on the P···P pnicogen bond. J Chem Theor Comp 8:2320–2327
21. An X-L, Li R, Li Q-Z, Liu X-F, Li W-Z, Cheng J-B (2012) Substitution cooperative and solvent effects on π pnicogen bonds in the FH_2P and FH_2As Complexes. J Mol Model 18:4325–4332
22. Bauzá A, Quiñonero D, Deyà PM, Frontera A (2012) Pnicogen-π complexes: theoretical study and biological implications. Phys Chem Chem Phys 14:14061–14066
23. Alkorta I, Sánchez-Sanz G, Elguero J, Del Bene JE (2013) Exploring $(NH_2F)_2$, $H_2FP:NFH_2$ and $(PH_2F)_2$ potential surfaces: hydrogen bonds or pnicogen bonds? J Phys Chem A 117:183–191
24. Sánchez-Sanz G, Alkorta I, Elguero J (2013) Intramolecular pnicogen interactions in PHF-$(CH_2)_n$-PHF (n = 2–6) systems. ChemPhysChem 14:1656–1665
25. Del Bene JE, Alkorta I, Sánchez-Sanz G, Elguero J (2013) Phosphorus as a simultaneous electron-pair acceptor in intermolecular P···N pnicogen bonds and electron-pair donor to Lewis acids. J Phys Chem A 117:3133–3141
26. Grabowski SJ, Alkorta I, Elguero J (2013) Complexes between dihydrogen and amine phosphine and arsine derivatives: hydrogen bond versus pnictogen interaction. J Phys Chem A 117:3243–3325
27. Politzer P, Murray J, Clark T (2013) Halogen bonding and other σ-hole interactions: a perspective. Phys Chem Chem Phys 15:11178–11189
28. Alkorta I, Elguero J, Del Bene JE (2013) Pnicogen-bonded cyclic trimers $(PH_2X)_3$ with X = F, Cl, OH, NC, CN, CH_3, H and BH_2. J Phys Chem A 117:4981–4987
29. Sánchez-Sanz G, Trujillo C, Solimannejad M, Alkorta I, Elguero J (2013) Orthogonal interactions between nitryl derivatives and electron donors: pnictogen bonds. Phys Chem Chem Phys 15:14310–14318
30. Del Bene JE, Alkorta I, Elguero J (2013) Characterizing complexes with pnicogen bonds involving sp^2 hybridized phosphorus atoms: $(H_2C{=}PX)_2$ with X = F, Cl, OH, CN, NC, CCH, H, CH_3 and BH_2. J Phys Chem A 117:6893–6903
31. Alkorta I, Elguero J, Del Bene JE (2013) Pnicogen bonded complexes of PO_2X (X = F, Cl) with nitrogen bases. J Phys Chem A 117:10497–10503
32. Del Bene JE, Alkorta I, Elguero J (2013) Properties of complexes $H_2C{=}(X)P:PXH_2$ for X = F, Cl, OH, CN, NC, CCH, H, CH_3 and BH_2: P···P pnicogen bonding at σ-holes and π-holes. J Phys Chem A 117:11592–11604

33. Solimannejad M, Nassirinia N, Amani SA (2013) Computational study of 1:1 and 1:2 complexes of nitryl halides (O_2NX) with HCN and HNC. Struct Chem 24:651–659
34. Solimannejad M, Ramezani V, Trujillo C, Alkorta I, Sánchez-Sanz G, Elguero J (2012) Competition and interplay between σ-hole and π-hole interactions: a computational study of 1:1 and 1:2 complexes of nitryl halides (O_2NX) with ammonia. J Phys Chem A 116:5199–5206
35. Pople JA, Binkley JS, Seeger R (1976) Theoretical models incorporating electron correlation. Int J Quantum Chem Quantum Chem Symp 10:1–19
36. Krishnan R, Pople JA (1978) Approximate fourth-order perturbation theory of the electron correlation energy. Int J Quantum Chem 14:91–100
37. Bartlett RJ, Silver DM (1975) Many-body perturbation theory applied to electron pair correlation energies in closed-shell first-row diatomic hydrides. J Chem Phys 62:3258–3268
38. Bartlett RJ, Purvis GD (1978) Many-body perturbation theory coupled-pair many-electron theory and the importance of quadruple excitations for the correlation problem. Int J Quantum Chem 14:561–581
39. Del Bene JE (1993) Proton affinities of ammonia, water, and hydrogen fluoride and their anions: a quest for the basis-set limit using the Dunning augmented correlation-consistent basis sets. J Phys Chem 97:107–110
40. Dunning TH (1989) Gaussian basis sets for use in correlated molecular calculations: i. the atoms boron through neon and hydrogen. J Chem Phys 90:1007–1023
41. Woon DE, Dunning TH (1995) Gaussian basis sets for use in correlated molecular calculations: v. core-valence basis sets for boron through neon. J Chem Phys 103:4572–4585
42. Frisch MJ, Trucks GW, Schlegel HB, Scuseria GE, Robb MA, Cheeseman JR, Scalmani G, Barone V, Mennucci B, Petersson GA et al. Gaussian Inc: Wallingford CT 2009 Gaussian-09 Revision A01
43. Bader RFW (1991) A quantum theory of molecular structure and its applications. Chem Rev 91:893–928
44. Bader RFW (1990) Atoms in molecules a quantum theory. Oxford University Press, Oxford, England
45. Popelier PLA (2000) Atoms in molecules: an introduction. Prentice Hall, Harlow, England
46. Matta CF, Boyd RJ (2007) The quantum theory of atoms in molecules: from solid state to DNA and drug design. Wiley-VCH, Weinheim, Germany
47. Silvi B, Savin A (1994) Classification of chemical bonds based on topological analysis of electron localization functions. Nature 371:683
48. AIMAll (Version 110823) Todd A, Keith TK Gristmill Software Overland Park KS USA 2011. http://aim.tkgristmill.com. Accessed August 1 2013
49. Noury S, Krokidis X, Fuster F, Silvi B (1997) TopMod Package. Universite Pierre et Marie Curie, Paris, France
50. Reed AE, Curtiss LA, Weinhold F (1988) Intermolecular interactions from a natural bond orbital donor-acceptor viewpoint. Chem Rev 88:899–926
51. Glendening ED, Badenhoop JK, Reed AE, Carpenter JE, Bohmann JA, Morales CM, Landis CR, Weinhold F (2013) NBO 60. University of Wisconsin, Madison, WI
52. Perera SA, Nooijen M, Bartlett RJ (1996) Electron correlation effects on the theoretical calculation of nuclear magnetic resonance spin–spin coupling constants. J Chem Phys 104:3290–3305
53. Perera SA, Sekino H, Bartlett RJ (1994) Coupled-cluster calculations of indirect nuclear coupling constants: the importance of non-fermi contact contributions. J Chem Phys 101:2186–2196
54. Schäfer A, Horn H, Ahlrichs R (1992) Fully optimized contracted gaussian basis sets for atoms Li to Kr. J Chem Phys 97:2571–2577
55. Stanton JF, Gauss J, Watts JD, Nooijen M, Oliphant N, Perera SA, Szalay PS, Lauderdale WJ, Gwaltney SR, Beck S, et al. ACES II, University of Florida Gainesville, Fl

Theor Chem Acc (2014) 133:1505
DOI 10.1007/s00214-014-1505-6

REGULAR ARTICLE

The second-order Ehrenfest method

A practical CASSCF approach to coupled electron–nuclear dynamics

**Morgane Vacher · David Mendive-Tapia ·
Michael J. Bearpark · Michael A. Robb**

Received: 1 April 2014 / Accepted: 14 May 2014 / Published online: 4 June 2014

Abstract This article describes the Ehrenfest method and our second-order implementation (with approximate gradient and Hessian) within a CASSCF formalism. We demonstrate that the second-order implementation with the predictor–corrector integration method improves the accuracy of the simulation significantly in terms of energy conservation. Although the method is general and can be used to study any coupled electron–nuclear dynamics, we apply it to investigate charge migration upon ionization of small organic molecules, focusing on benzene cation. Using this approach, we can study the evolution of a non-stationary electronic wavefunction for fixed atomic nuclei, and where the nuclei are allowed to move, to investigate the interplay between them for the first time. Analysis methods for the interpretation of the electronic and nuclear dynamics are suggested: we monitor the electronic dynamics by calculating the spin density of the system as a function of time.

Keywords Ehrenfest method · CASSCF · Coupled electron–nuclear dynamics · Charge migration · Charge transfer

Dedicated to the memory of Professor Isaiah Shavitt and published as part of the special collection of articles celebrating his many contributions.

M. Vacher · M. J. Bearpark · M. A. Robb (✉)
Department of Chemistry, Imperial College London,
London SW7 2AZ, UK
e-mail: mike.robb@imperial.ac.uk

D. Mendive-Tapia
Laboratoire CEISAM - UMR CNR 6230,
Université de Nantes, 2 Rue de la Houssinière, BP 92208,
44322 Nantes Cedex 3, France

1 Introduction

Photoionization can create a coherent superposition of electronic states and therefore initiates electronic dynamics in atoms and molecules. Experiments on the latter are particularly difficult to interpret as change in the nuclear geometry is also expected. Indeed, the equilibrium geometry of the ionized and neutral species are unlikely to be the same. Therefore, the initial electron dynamics, that may last up to a few femtoseconds, is then followed by the onset of nuclear dynamics [1]. Theoretical methods are needed to help understand the effects seen in attosecond laser experiments (see, e.g., the reviews of Kling [2] and Ivanov [3]).

Methods for non-adiabatic dynamics were the subject of a recent special issue of J. Chem. Phys.; in particular, the lead article of Tully [4] provides a current summary of the state of the art. Quantum mechanical simulations are expensive computationally. Reducing the number of nuclear degrees of freedom of the system is sometimes done to make the calculation feasible, but the validity of this approximation is limited [5–7]. Conventional molecular dynamics (MD) only allows one to simulate nuclear motion on a single potential energy surface and therefore does not describe non-adiabatic processes involving non-radiative electronic transitions. Mixed quantum-classical dynamics methods have been developed to address this issue. In the Ehrenfest method, one propagates quantum mechanically an electronic wavepacket consisting of a superposition of electronic eigenstates by solving the time-dependent Schrödinger equation; and one moves classically the nuclei integrating Newton's equation of motion. The feedback between the quantum electronic and classical nuclear degrees of freedom is described in a mean-field manner. This simplification allows one to study the

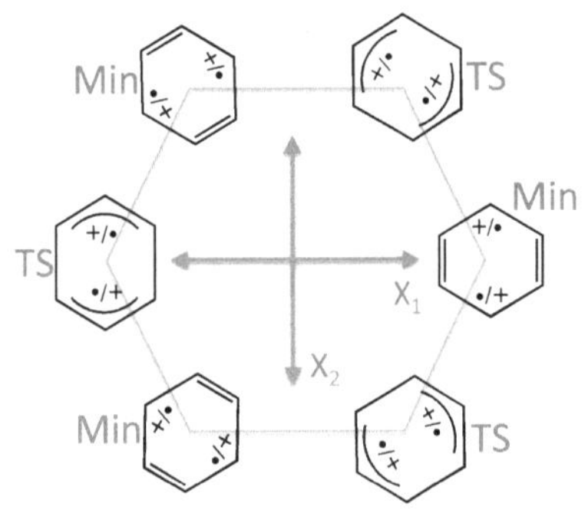

Fig. 1 Benzene radical cation resonance structures. The directions X_1 and X_2 are the gradient difference and the gradient of the interstate coupling vector, respectively. The gradient difference direction X_1 connects a pair of quinoid/antiquinoid forms by lowering the symmetry from D_{6h} to D_{2h}. The motion along the gradient of the interstate coupling vector X_2 preserves only C_{2h} and allows one to move from one quinoid (or antiquinoid) structure to a "60° rotated" antiquinoid (or quinoid) structure. Note that for each quinoid or antiquinoid structure, there are actually two VB resonance structures with the unpaired electron and the positive charge exchanged (this is indicated by •/+ interchange in each case)

electronic and nuclear dynamics independently and discover whether the motions of the electrons and the nuclei are synchronous or asynchronous.

The Ehrenfest method is general and we have previously used it for photochemistry [8] and electron transfer [9]. In this article, we focus on the application of the Ehrenfest method to the simulation of electron dynamics (and the coupled nuclear dynamics) upon ionization in molecules. We have recently shown [10] how charge migration and charge transfer in benzene and 2-phenylethylamine cations can be studied using the Ehrenfest method within a CASSCF [11, 12]-like formulation. Charge migration is defined at a fixed nuclear geometry and corresponds to oscillations in the electronic density due the non-stationarity of the electronic state. Charge transfer is a change in the electronic density due to a change in the nuclear geometry. In this work, after a review of the theory, we present our approximate second-order CASSCF implementation of the Ehrenfest method. We show that a second-order method augmented with a predictor–corrector integration method (devised by Schlegel [13] and implemented in the Gaussian program [14]), permits the use of larger step sizes while conserving the total energy.

We apply our second-order Ehrenfest method to a model system: benzene radical cation. Ionization of the neutral from the degenerate HOMO/HOMO-1 leads to the Jahn–Teller [15] effect in the cation. There is a peaked conical intersection between the two lowest-energy eigenstates D_0 and D_1 at geometries with D_{6h} symmetry. Figure 1 represents the surrounding "moat" of the conical intersection seen from above. It contains several valence bond (VB) resonance structures: three equivalent quinoid structures that are minima (Min) and three antiquinoid structures that are transition structures (TS). The degeneracy is lifted along two directions: the gradient difference X_1 and the interstate coupling X_2. In this system, there is the possibility of charge migration/charge transfer around the ring.

2 The Ehrenfest approach: general theoretical development

The Ehrenfest method has been extensively discussed in the literature [16–26]. In this section, we review the Ehrenfest formalism following the elegant derivation of Tully [27]. Our aim is to explicitly state the approximations underlying the method and to discuss their implications.

2.1 Separation of nuclear and electronic variables

We shall start with the non-relativistic time-dependent Schrödinger equation, where r and R refer to the electronic (fast) and nuclear (slow) variables, respectively

$$i\hbar\frac{\partial}{\partial t}\Phi(r,R,t) = \mathscr{H}\Phi(r,R,t) \tag{1}$$

In order to derive mixed quantum-classical dynamics, the nuclear and electronic variables have to be separated. The simplest possible form is a product ansatz:

$$\Phi(r,R,t) = \Psi(r,t)\cdot\chi(R,t) \tag{2}$$

The first approximation made in the Ehrenfest method is thus the factorisation of the total wavefunction into a product of electronic and nuclear parts. One deficiency of the ansatz (2) is the fact that the electronic wavefunction does not have the possibility to decohere: the populated electronic states in $\Psi(r,t)$ share the same nuclear wavepacket $\chi(R,t)$ by definition of the total wavefunction. Decoherence here is defined as the tendency of the time-evolved electronic wavefunction to behave as a statistical ensemble of electronic states rather than a coherent superposition of them [26]. The neglect of electronic decoherence could lead to non-physical asymptotic behaviors in case of bifurcating paths. It is not expected to be a problem here as we are interested in relatively short timescale dynamics.

In order to simplify the appearance of the expressions at a later stage of the derivation, a phase factor is introduced for the total wavefunction and also some internal phase factors for the two individual wavefunctions. More details can be found here [27]. Inserting the ansatz (2) with the additional phase factors into Eq. (1), multiplying on the left by $\chi^*(R,t)$ and $\Psi^*(r,t)$ and integrating over R and r gives, respectively:

$$i\hbar\frac{\partial\Psi(r,t)}{\partial t} = -\sum_i \frac{\hbar^2}{2m_e}\nabla_i^2\Psi(r,t) + \langle\chi(R,t)|\mathscr{V}_{n-e}(r,R)|\chi(R,t)\rangle_R\cdot\Psi(r,t) \quad (3)$$

$$i\hbar\frac{\partial\chi(R,t)}{\partial t} = -\sum_I \frac{\hbar^2}{2M_I}\nabla_I^2\chi(R,t) + \langle\Psi(r,t)|\mathscr{H}_e(r;R)|\Psi(r,t)\rangle_r\cdot\chi(R,t) \quad (4)$$

The indices i and I refer, respectively, to the electrons and the nuclei; m_e is used to denote the mass of an electron, and M_I is used to denote the mass of the nucleus I. $\mathscr{V}_{n-e}(r,R)$ includes all inter-particle interactions (electron–electron, nucleus–nucleus and electron–nucleus), and $\mathscr{H}_e(r;R)$ is the electronic Hamiltonian for the nuclei fixed at positions R.

To understand the consequence of using the ansatz (2), let us, for instance, look closer at the second term on the right hand side of Eq. (3). The interaction between electrons at points r in space and nuclei at points R is weighted by the probability that the nuclei are at these particular points R. This is the effective potential experienced by the electrons due to the nuclei. The corresponding remark can be made about the second term on the right hand side of Eq. (4). According to the set of coupled Eqs. (3) and (4), the feedback between electronic and nuclear degrees of freedom is described in a mean-field manner, in both directions. In other words, both electrons and nuclei move in time-dependent effective potentials obtained from appropriate expectation values of the nuclear and electronic wavefunctions, respectively.

2.2 Classical limit for nuclear motion

The Ehrenfest method is obtained by taking the classical limit of Eqs. (3) and (4). To do that in Eq. (4), the nuclear wavefunction is (exactly) rewritten in a polar coordinate system in terms of an amplitude A and a phase S which are both considered to be real and positive:

$$\chi(R,t) = A(R,t)\cdot\exp\left(\frac{i}{\hbar}S(R,t)\right) \quad (5)$$

After inserting Eq. (5) in (4), the real parts on each side can be equal:

$$\frac{\partial S}{\partial t} + \sum_I \frac{1}{2M_I}(\nabla_I S)^2 + \langle\Psi(r,t)|\mathscr{H}_e(r;R)|\Psi(r,t)\rangle_r = \hbar^2\sum_I \frac{1}{2M_I}\frac{\nabla_I^2 A}{A} \quad (6)$$

The classical limit is taken as $\hbar\to 0$:

$$\frac{\partial S}{\partial t} + \sum_I \frac{1}{2M_I}(\nabla_I S)^2 + \langle\Psi(r,t)|\mathscr{H}_e(r;R)|\Psi(r,t)\rangle_r = 0 \quad (7)$$

The resulting equation is isomorphic to the Hamilton–Jacobi equation and is thus equivalent to Newton's equation of motion, where $P_I = \nabla_I S$ is the classical momentum of nucleus I:

$$\frac{dP_I}{dt} = -\nabla_I\langle\Psi(r,t)|\mathscr{H}_e(r;R)|\Psi(r,t)\rangle_r \quad (8)$$

In Eq. (3), we can replace $\chi(R,t)$ by a delta function at the classical trajectory $R(t)$:

$$i\hbar\frac{\partial\Psi(r,t;R)}{\partial t} = \left(-\sum_i \frac{\hbar^2}{2m_e}\nabla_i^2 + \mathscr{V}_{n-e}(r,R(t))\right)\Psi(r,t;R) = \mathscr{H}_e(r;R(t))\cdot\Psi(r,t;R) \quad (9)$$

Note that now the electronic wavefunction Ψ depends parametrically on $R(t)$ through $\mathscr{V}_{n-e}(r,R(t))$ and thus $\mathscr{H}_e(r;R(t))$. By treating the nuclear motion classically, we lose the spatial delocalisation of the nuclei and their motion is now described by a classical trajectory.

Equations (8) and (9) define the Ehrenfest method. It is important to keep in mind that it succeeds in describing nuclear motion if the potential energy surfaces of the various electronic states are similar in topology and energies [20]. However, in the case of weakly coupled electronic states, the nuclear motion will be dominated by the potential corresponding to the highly populated electronic state and regions of space accessible only on the sparsely populated electronic state may not be explored properly [28, 29]. One advantage of the Ehrenfest method is that its applications and results do not depend on the choice of basis functions (if complete) and can, in principle, be applied without choosing basis functions by numerical integration of Eq. (9).

2.3 Non-adiabatic couplings

Expanding the electronic wavefunction in a basis of orthonormal configurations $\{\phi_l\}$ or eigenstates $\{\tilde{\phi}_l\}$ gives:

$$\Psi(r,t;R) - \sum_l a_l(t)\phi_l(r;R) - \sum_l c_l(t)\tilde{\phi}_l(r;R) \quad (10)$$

In order to prove the presence of the non-adiabatic couplings, it is convenient to use the eigenstate expansion. Substituting expansion (10) into Eq. (9), multiplying on the left by $\tilde{\phi}_k^*(r;R)$ and integrating over r gives:

$$i\hbar\frac{\partial c_k(t)}{\partial t} = c_k(t)E_k(R) - i\hbar\sum_l c_l(t)\left\langle\tilde{\phi}_k\middle|\frac{\partial}{\partial t}\tilde{\phi}_l\right\rangle = c_k(t)E_k(R) - i\hbar\sum_{l,J} c_l(t)d_{kl}^J(R)\cdot\dot{R}_J \quad (11)$$

with the non-adiabatic couplings $d_{kl}^J(R) = \int\tilde{\phi}_k^*(r;R)\nabla_{R_J}\tilde{\phi}_l(r;R)dr$ and the eigenvalues E_k. Thus, if the electronic wavefunction is expanded in the adiabatic basis, Eq. (11) gives the time-varying amplitudes along the classical trajectory. The latter depends on the non-adiabatic coupling terms. Note that the same equations define the time-dependence of the electronic basis amplitudes for the surface hoping method. Note that a similar expression can be derived for the diabatic basis [27].

3 Implementation within a CASSCF formalism

We now describe our CASSCF implementation of the Ehrenfest method that allows the electronic wavefunction to be made of multiple states. An approximate second-order method with a numerical fit procedure is used for the propagation of the nuclei.

3.1 Quantum propagation of a CASSCF electronic wavefunction

In practice, time is discretized and when integrating Eq. (9) assuming a constant Hamiltonian over the time step, we obtain:

$$\Psi(r, t_n; R(t_n)) = \exp\left(-\frac{i}{\hbar}\mathscr{H}_e(r; R(t_n)) \cdot (t_n - t_{n-1})\right) \cdot \Psi(r, t_{n-1}; R(t_{n-1})) \tag{12}$$

The time-dependent electronic wavefunction is expanded in the basis of configurations, here $\mathbf{A}(t_n)$ is the vector gathering the expansion coefficients at time t_n defined in Eq. (10). We use bold font to signify vectors and matrices.

$$\mathbf{A}(t) = \begin{pmatrix} a_1(t) \\ \vdots \\ a_k(t) \\ \vdots \end{pmatrix} \tag{13}$$

Using matrix notation, Eq. (12) reads as:

$$\mathbf{A}(t_n) = \exp\left(-\frac{i}{\hbar}\boldsymbol{H}_e(t_n) \cdot (t_n - t_{n-1})\right)\mathbf{A}(t_{n-1}) \tag{14}$$

$\boldsymbol{H}_e(t_n)$ is the matrix representation of the electronic Hamiltonian in the basis of configurations at time t_n. Using its spectral resolution, it becomes:

$$\mathbf{A}(t_n) = \mathbf{U}(t_n)\exp\left(-\frac{i}{\hbar}\mathbf{D}_e(t_n) \cdot (t_n - t_{n-1})\right)\mathbf{U}^{\dagger}(t_n) \cdot \mathbf{A}(t_{n-1}) \tag{15}$$

$\mathbf{U}$ is the matrix containing the eigenvectors arranged as columns. $\mathbf{D}_e$ is the matrix representation of the electronic Hamiltonian in the basis of eigenstates: it thus contains the eigenvalues $\{E_k\}$ on the diagonal and is zero elsewhere. Both $\mathbf{U}$ and $\mathbf{D}_e$ are assumed to be constant over a time step, and their value at time t_n is used. If one expands $\mathbf{D}_e(t_n)$ as a Taylor series around its value at time t_{n-1}, one obtains to first order:

$$\mathbf{D}_e(t_n) = \mathbf{D}_e(t_{n-1}) + \frac{d\mathbf{D}_e(t_{n-1})}{dt} \cdot (t_n - t_{n-1}) \tag{16}$$

The electronic Hamiltonian is time-independent (no external electric field), but it changes with time if the nuclei are allowed to move. Hence, we have the following relation (if it obeys the Hellmann–Feynman theorem):

$$\frac{d\mathbf{D}_e}{dt} = \nabla_R \mathbf{D}_e \cdot \dot{R} = \langle \tilde{\phi}_k | \nabla_{R_I}(\mathscr{H}_e) | \tilde{\phi}_l \rangle \cdot \dot{R} \tag{17}$$

We can now identify the derivative coupling (see Sect. 2.3):

$$\begin{aligned} \langle \tilde{\phi}_k | \nabla_{R_I}(\mathscr{H}_e) | \tilde{\phi}_l \rangle &= \langle \tilde{\phi}_k | \nabla_{R_I} \tilde{\phi}_l \rangle \cdot (E_l - E_k) \\ &= d_{kl}^I(R) \cdot (E_l - E_k) \end{aligned} \tag{18}$$

The non-adiabatic couplings are, therefore, included in our propagation of the electronic wavefunction by finite differentiation.

We thus obtain a sequence of vectors corresponding to the different steps:

$$\mathbf{A}(t_0) \to \mathbf{A}(t_1) \to \cdots \to \mathbf{A}(t_n) \to \cdots \tag{19}$$

The sequence of vectors may be obtained keeping the nuclei fixed, in which case the basis of configurations $\{\tilde{\phi}_l\}$ and also the basis of CASSCF eigenvectors do not change with time. The matrices $\mathbf{U}$ and $\mathbf{D}_e$ are time-independent.

Alternatively, the sequence of vectors $\{\mathbf{A}\}$ may be obtained in concert with nuclear motion. In the latter case, at each step t_n of the dynamics (or at each geometry of the classical trajectory), a state-averaged (SA) CASSCF calculation is done to update the electronic Hamiltonian matrix $\mathbf{H}_e$. One obtains a basis of CASSCF eigenvectors $\{\mathbf{U}_j(t_n)\}$ from its diagonalization and a set of SA optimized orbitals from the solution of the SA-MCSCF equations. Although we assume the configurations $\{\phi_l\}$ do not change for the propagation of the time-dependent wavefunction, the orbitals are in practice optimized at each step. This strategy is reasonable if the active space is large enough so that only small relaxations of the inactive orbitals occur.

The energy of the system is computed as the expectation value of the electronic Hamiltonian:

$$E(t_n) = \langle \Psi(r, t_n) | \mathscr{H}_e | \Psi(r, t_n) \rangle_r = \mathbf{A}^{\dagger}(t_n) \cdot \boldsymbol{H}_e(t_n) \cdot \mathbf{A}(t_n) \tag{20}$$

More conveniently, we can define a real time-dependent vector $\mathbf{M}$, by rotating the complex expansion coefficients into real coefficients. The rotation must be performed in the eigenstates basis for the energy to be preserved (see

"Appendix" for details). The energy can then be computed without the need for complex algebra:

$$E(t_n) = \mathbf{M}^{\dagger}(t_n) \cdot \boldsymbol{H}_e(t_n) \cdot \mathbf{M}(t_n) \tag{21}$$

3.2 Classical nuclear trajectory

The nuclear geometry is updated at each time step, if wanted, by integrating the equation of motion (8). This is done using the Hessian-based predictor–corrector algorithm designed by Hase and Schlegel [13]. We use the first and second derivatives of the energy to make a local quadratic approximation of the energy:

$$E(R) = E^0 + \mathbf{G}^0 \cdot (R - R^0) + \frac{1}{2}(R - R^0)^{\dagger} \cdot \mathbf{H}^0 \cdot (R - R^0) \tag{22}$$

where E^0, $\mathbf{G}^0$ and $\mathbf{H}^0$ are the energy, the gradient and the Hessian evaluated at R^0, respectively. Newton's equation of motion on a quadratic surface is as follows:

$$\frac{dP_I}{dt} = -\mathbf{G}_I^0 - \sum_J \mathbf{H}_{IJ}^0 (R_J - R_J^0) \tag{23}$$

Note that unlike the energy, the gradient and Hessian cannot be calculated using a real TD vector: They depend intrinsically on the relative complex phase of the electronic eigenstates occupied (see "Appendix" for details on the gradient).

We now review the general approach to the gradient and Hessian computations using the methods of Almlöf and Taylor [30] and apply it specifically to the Ehrenfest wavefunction. We also explain the approximations used in computing the gradient and Hessian.

3.2.1 Gradient computation

The energy of a configuration interaction (CI)-like wavefunction depends on the large number of parameters that define the wavefunction. Among them, the optimized molecular orbital (MO) and CI coefficients depend on the nuclear geometry. One constraint is that the MO coefficients remain orthonormal under the change in geometry, which is possible using unitary transformations (a simpler alternative to Lagrange multipliers). We introduce the following notation: Superscripts denote partial derivatives, while subscripts denote total derivatives. The expression for the gradient is obtained by applying the chain rule to the energy derivative with respect to a change in geometry [10, 30]:

$$E_{R_I} = E^{R_I} + \mathbf{E}^C \mathbf{C}_{R_I} + \mathbf{E}^X \mathbf{X}_{R_I} - \frac{1}{2}\mathbf{E}^Y \mathbf{S}_{R_I} \tag{24}$$

E^{R_I} is the Hellmann–Feynman term. It represents the intrinsic dependence of the energy with respect to a change in nuclear geometry. As the energy depends on parameters that depend themselves on the nuclear geometry, one obtains the non-Hellmann–Feynman terms: $\mathbf{C}_{R_I}$ and $\mathbf{X}_{R_I}$ are the CI vector and MO rotation derivatives and $\mathbf{E}^C$ and $\mathbf{E}^X$ are the corresponding energy derivatives. $\mathbf{S}_{R_I}$ is the atomic orbital overlap matrix, and $\mathbf{Y}$ is the symmetric matrix arising from re-orthogonalization of the MO.

When the derivative is computed for an eigenvector $\mathbf{U}_i$, the $\mathbf{C}$ matrix corresponds to the rotation of the eigenvector $\mathbf{U}_i$ with the remaining orthogonal vectors $\mathbf{U}_{j \neq i}$. Here, the reference vector is chosen to be the TD vector of the previous step $\mathbf{A}(t_{n-1})$ and the matrix $\mathbf{C}$ expresses $\mathbf{A}(t_n)$ by a rotation between $\mathbf{A}(t_{n-1})$ and its orthogonal complements. As we propagate the time-dependent wavefunction assuming that the MO do not change, $\mathbf{X}_{R_I}$ is neglected and the gradient reads as:

$$E_{R_I} = E^{R_I} + \mathbf{E}^C \mathbf{C}_{R_I} - \frac{1}{2}\mathbf{E}^Y \mathbf{S}_{R_I} \tag{25}$$

By differentiating the Newton–Raphson equation with respect to nuclear motion, the derivative CI coefficients can be written as:

$$\mathbf{C}_{R_I} = -\left(\mathbf{E}^{CC}\right)^{-1} \mathbf{E}_{R_I}^C = -\left(\mathbf{E}^{CC}\right)^{-1} \left(\mathbf{E}^{R_I C} - \frac{1}{2}\mathbf{E}^{CY}\mathbf{S}_{R_I}\right) \tag{26}$$

Note that Eq. (26) assumes a quadratic expansion of the energy as a function of the CI parameters $\mathbf{E}(\mathbf{C})$ about a minimum. This approximation results in some error in the gradient. In practice, we also neglect the derivative due to the complex phase of the CI expansion coefficients, i.e., we calculate the matrix $\mathbf{C}$ by expressing the real vector $\mathbf{M}(t_n)$ by a rotation between $\mathbf{M}(t_{n-1})$ and its orthogonal complements. As we show in Sect. 4.2, these errors are corrected via numerical fitting of the hypersurface along the trajectory.

3.2.2 Hessian computation

In the present work, at each step, a CASSCF calculation is done with state-averaged orbitals over two states. The Hessian used in the integration of Newton's equation of motion is calculated for the highest of the two roots of the SA calculation.

$$\begin{aligned} E_{R_I R_J} = {} & E^{R_I R_J} + \mathbf{E}^C \mathbf{C}_{R_I R_J} + \mathbf{E}^{CR_I} \mathbf{C}_{R_J} + \mathbf{E}^{CR_J} \mathbf{C}_{R_I} + \mathbf{E}^{CC} \mathbf{C}_{R_I} \mathbf{C}_{R_J} + \mathbf{E}^{CX} \mathbf{C}_{R_I} \mathbf{X}_{R_J} \\ & - \frac{1}{2}\mathbf{E}^{CY} \mathbf{C}_{R_I} \mathbf{S}_{R_J} + \mathbf{E}^X \mathbf{X}_{R_I R_J} + \mathbf{E}^{XR_I} \mathbf{X}_{R_J} + \mathbf{E}^{XR_J} \mathbf{X}_{R_I} + \mathbf{E}^{XX} \mathbf{X}_{R_I} \mathbf{X}_{R_J} \\ & + \mathbf{E}^{XC} \mathbf{X}_{R_I} \mathbf{C}_{R_J} - \frac{1}{2}\mathbf{E}^{XY} \mathbf{X}_{R_I} \mathbf{S}_{R_J} - \frac{1}{2}\mathbf{E}^{YR_I} \mathbf{S}_{R_J} - \frac{1}{2}\mathbf{E}^{YR_J} \mathbf{S}_{R_I} + \frac{1}{2}\mathbf{E}^Y \mathbf{S}_{R_I} \mathbf{S}_{R_J} \\ & - \frac{1}{2}\mathbf{E}^Y \mathbf{S}_{R_I R_J} + \frac{1}{4}\mathbf{E}^{YY} \mathbf{S}_{R_I} \mathbf{S}_{R_J} - \frac{1}{2}\mathbf{E}^{YC} \mathbf{S}_{R_I} \mathbf{C}_{R_J} - \frac{1}{2}\mathbf{E}^{YX} \mathbf{S}_{R_I} \mathbf{X}_{R_J} \end{aligned} \tag{27}$$

For an optimized SA-CASSCF wavefunction, $\mathbf{E}^C = 0$ but only the weighted average of the anti-symmetric

Lagrangians will vanish: $\sum_{i=1}^{2} \omega_i \mathbf{E}^X(i) = 0$. In addition, we neglect the second derivative of the MO rotation matrix with respect to nuclear distortion $\mathbf{X}_{R_I R_J}$. With these few simplifications, the Hessian becomes:

$$\begin{aligned} E_{R_I R_J}(i) = {} & E^{R_I R_J}(i) + \mathbf{E}^{CR_J}(i)\mathbf{C}_{R_I}(i) + \mathbf{E}^{CX}(i)\mathbf{C}_{R_I}(i)\mathbf{X}_{R_J} \\ & - \frac{1}{2}\mathbf{E}^{CY}(i)\mathbf{C}_{R_I}(i)\mathbf{S}_{R_J} \\ & + \mathbf{E}^{XR_I}(i)\mathbf{X}_{R_J} + \mathbf{E}^{XR_J}(i)\mathbf{X}_{R_I} + \mathbf{E}^{XX}(i)\mathbf{X}_{R_I}\mathbf{X}_{R_J} \\ & - \frac{1}{2}\mathbf{E}^{XY}(i)\mathbf{X}_{R_I}\mathbf{S}_{R_J} - \frac{1}{2}\mathbf{E}^{YR_I}(i)\mathbf{S}_{R_J} \\ & - \frac{1}{2}\mathbf{E}^{YR_J}(i)\mathbf{S}_{R_I} + \frac{1}{2}\mathbf{E}^{Y}(i)\mathbf{S}_{R_I}\mathbf{S}_{R_J} - \frac{1}{2}\mathbf{E}^{Y}(i)\mathbf{S}_{R_I R_J} \\ & + \frac{1}{4}\mathbf{E}^{YY}(i)\mathbf{S}_{R_I}\mathbf{S}_{R_J} - \frac{1}{2}\mathbf{E}^{YX}(i)\mathbf{S}_{R_I}\mathbf{X}_{R_J} \end{aligned} \quad (28)$$

The terms $\mathbf{C}_{R_I}(i)$ and $\mathbf{X}_{R_I}$ are obtained by solving the standard coupled-perturbed SA-MCSCF equations. These are obtained by differentiating Newton–Raphson equations with respect to a nuclear distortion:

$$\begin{pmatrix} \omega_1\mathbf{E}^{XX}(1) + \omega_2\mathbf{E}^{XX}(2) & \omega_1\mathbf{E}^{XC}(1) & \omega_2\mathbf{E}^{XC}(2) \\ \mathbf{E}^{CX}(1) & \mathbf{E}^{CC}(1) & 0 \\ \mathbf{E}^{CX}(2) & 0 & \mathbf{E}^{CC}(2) \end{pmatrix} \begin{pmatrix} \mathbf{X}_{R_I} \\ \mathbf{C}_{R_I}(1) \\ \mathbf{C}_{R_I}(2) \end{pmatrix} = - \begin{pmatrix} \omega_1\left\{\mathbf{E}^{R_I X}(1) - \frac{1}{2}\mathbf{E}^{XY}(1)\mathbf{S}_{R_I}\right\} + \omega_2\left\{\mathbf{E}^{R_I X}(2) - \frac{1}{2}\mathbf{E}^{XY}(2)\mathbf{S}_{R_I}\right\} \\ \mathbf{E}^{R_I C}(1) - \frac{1}{2}\mathbf{E}^{CY}(1)\mathbf{S}_{R_I} \\ \mathbf{E}^{R_I C}(2) - \frac{1}{2}\mathbf{E}^{CY}(2)\mathbf{S}_{R_I} \end{pmatrix} \quad (29)$$

Using an optimized SA-CASSCF wavefunction to calculate the Hessian for the time-dependent vector is a rough approximation; again, we show in Sect. 4.2 that the numerical fitting of the hypersurface along the trajectory corrects most of the error.

3.2.3 Fifth order polynomial fit

The second-order method is used with the fifth order predictor–corrector integration scheme of Schlegel [13] in the Gaussian program [14]. In this method, the quadratic approximation (gradient, Hessian) at the current point R^0 is used in a predictor step to a predicted geometry R^p. Then the energies, gradients and Hessians at geometries R^0 and R^p are fitted by a fifth order polynomial. The equations of motion on this fitted surface are then integrated to give the corrector step to the geometry R^c.

4 Application to coupled electron–nuclear dynamics upon ionization

We apply our second-order Ehrenfest method to study the coupled electron–nuclear dynamics of benzene upon ionization. Because the nuclear geometry will be a non-stationary point on the cation potential energy surfaces and the electronic wavefunction will be a non-stationary state, we expect some electron and nuclear dynamics to occur after ionization. We use the Ehrenfest method to investigate the interplay between the electronic and nuclear dynamics.

The electronic structure is computed using the CASSCF method. Using the standard 6-31G* basis set, we choose the 6 π orbitals as active. The degenerate HOMO, HOMO-1 and matching degenerate LUMO, LUMO+1 are needed to recover the non-dynamic electron correlation. The remaining pair of benzene π orbitals contributes to dynamic correlation and has to be included for stability (because of a large dynamic correlation effect).

4.1 Initial conditions

The initial conditions of the dynamics calculation (the nuclear geometry and the electronic wavefunction) depend on the state of the neutral species before ionization. The absorption of a photon leading to ionization is itself instantaneous so neither the electrons or the nuclei have time to relax. We assume that the system was in its nuclear and electronic ground state before ionization. In principle, to mimic the initial nuclear wavepacket distribution and obtain a realistic dynamics of the system, one must simulate many trajectories starting with sampled positions and momentum of the nuclei. Here, our aim is to illustrate the method using a single simulation with the initial nuclear geometry close to the equilibrium geometry of the neutral species, i.e., the minimum of its ground state, without initial kinetic energy. The initial electronic wavefunction will be a non-stationary state, i.e., a superposition of several cationic eigenstates. We neglect the interaction with the electric field and the interaction between the outgoing electron and the cation. These approximations are reasonable if a high-energy ultrashort pulse is used for the ionization so that the outgoing electron has a high kinetic energy and moves rapidly away from the cation. In other words, we assume the "sudden" removal of an electron.

In many studies [31–36] of the electronic dynamics upon ionization, the initial electronic wavefunction is created using the so-called single-channel sudden approximation [37–39]. This assumes the sudden removal of an electron *from a particular orbital*. The initial superposition of electronic eigenstates results therefore from electron correlation only (in this case, from the CI expansion used) [31]. This single-channel sudden approximation was used in our previous work [10]. However, one can investigate a particular ionization channel independently of the others only if they do not interfere: The valence ionization channels are not well separated in energy so they will interfere. We therefore choose to stay general: We assume the “sudden” removal of an electron, but do not aim at studying electronic dynamics following a single-channel ionization. Relative weights and phases of the eigenstates in the initial superposition are parameters that can be investigated in numerical simulations. Experimentally, they depend on the photon energy but also on the field polarization for example.

We study the valence ionization of benzene. An instantaneous photoionization experiment would be carried out using a pulse of broad bandwidth which means that several electronic states may be populated. However, considering the energy gap between the first and second excited states (more than 2 eV), we can assume that the dynamics of the two lowest-energy electronic states will not interfere with the dynamics on the second excited state within the first few tens of femtoseconds. Our aim is to model the dynamics resulting from populating the lowest two eigenstates which would be part of any observed dynamics. We thus choose the initial electronic wavefunction to be an equal mixture of the two lowest eigenstates: $|\Psi(r, t=0)\rangle = \frac{1}{\sqrt{2}}(|D_0\rangle + |D_1\rangle)$. As we observed in our first study [10], when the two eigenstates are exactly degenerate (which is the case at the equilibrium geometry of the neutral species), there is no charge migration. Indeed, at exact degeneracy, any combination of the two eigenstates is also an eigenstate, and therefore, it is a stationary electronic wavefunction. An alternative explanation is the fact that the period of oscillation in charge migration is inversely proportional to the energy gap; thus, if the energy gap is zero, the period of oscillation is infinite, which means there is no charge migration. We suggested in our first study [10] to enhance charge migration by distorting the initial geometry away from the conical intersection in order to lift the degeneracy. We decide to start the ionization at a distorted geometry along the derivative coupling vector. The same test calculation is used throughout the rest of the article.

4.2 Conservation of energy

The accuracy in the integration of the equations of motion is monitored by the conservation of energy. To illustrate this, we run test calculations using two different integration algorithms involving: approximate gradient computation (Sect. 3.2.1) only, or approximate gradient and approximate Hessian computations with the fifth order polynomial fit (Sects. 3.2.2 and 3.2.3). The results with approximate gradient and approximate Hessian are not shown here as we do not expect any improvement of the conservation of energy (the Hessian is calculated for the upper eigenstate of the SA-CASSCF calculation and not for the TD vector). In

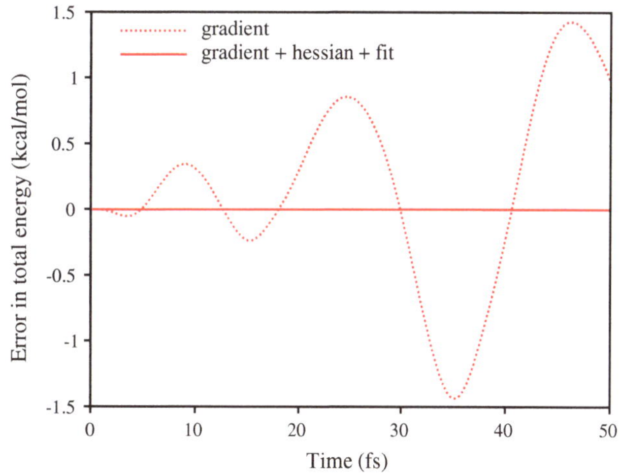

Fig. 2 Error in conservation of energy. The error in the total energy is compared for two different integration algorithms involving the approximate gradient computation only, or the approximate gradient and approximate Hessian computations using the fifth order polynomial fit. The mass-weighted step size is 0.03 amu$^{1/2}$bohr (about 0.3 fs). This illustrates how the polynomial fit performs significantly better (error below 2×10^{-4} kcal/mol for 50 fs)

addition, the computational cost of the Hessian-based integrator using the polynomial fit is the same as without the polynomial fit. We use a mass-weighted step size of 0.03 amu$^{1/2}$bohr (about 0.3 fs). The trajectories are started with no kinetic energy.

Figure 2 shows the error in the total energy as a function of time for the different integration methods. We see oscillations that become bigger and bigger with time when using the approximate gradient only. Varying the step size does not improve the conservation of energy (not shown) which suggests that the error is more likely to come from the approximations in the analytical gradient rather than the integrator. The combination of approximate analytical gradient and Hessian with a numerical fit decreases the error significantly (below 2×10^{-4} kcal/mol for 50 fs). Indeed, the error in the conservation of energy due to the approximations made in the analytical expressions for gradient and Hessian is corrected by fitting numerically the potential energy surface. The difference in cost between the two methods is essentially the computation of the Hessian at each step: This can partly be eliminated if a Hessian updating procedure is used (not explored here).

Figure 3 shows the total energy, the kinetic energy and the potential energy during the 50 fs trajectory using the fifth order polynomial fit (same simulation as in Fig. 2). The total and potential energies have been shifted by the initial potential energy so that they are plotted on the same vertical axis. The variations in kinetic and potential energies exactly cancel each other. Therefore, we can safely use step sizes as large as we use here (about 0.3 fs) using the

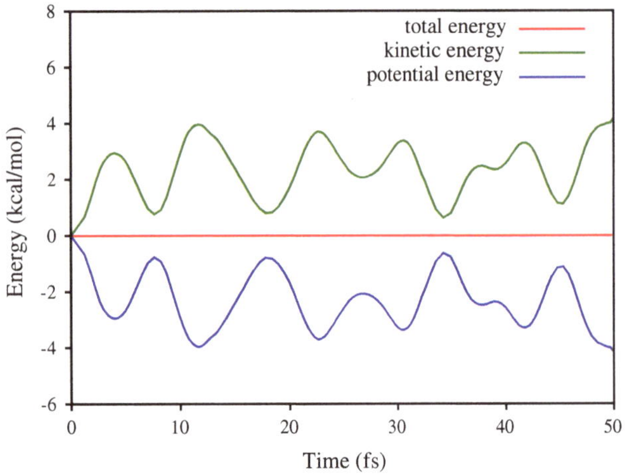

Fig. 3 Evolution of total, kinetic and potential energies during the trajectory. The fifth order polynomial fit procedure is used with a mass-weighted step size of 0.03 amu$^{1/2}$bohr (about 0.3 fs). The decrease of potential energy is compensated by a gain in kinetic energy so that the total energy is perfectly conserved for over 50 fs (error below 2×10^{-4} kcal/mol for 50 fs)

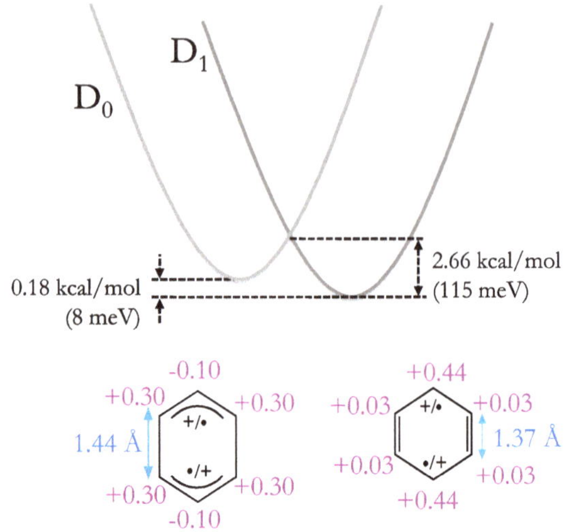

Fig. 4 Schematic representation of the benzene radical cation D_1/D_0 crossing. This is a cross-section through the conical intersection along X_1 shown in Fig. 1. The quinoid structure (on the *right*) is a minimum whereas the antiquinoid structure (on the *left*) is a transition structure in the moat of the conical intersection. Mulliken atomic spin densities are indicated in *pink* next to each carbon atom. Characteristic bond lengths are indicated in *blue*. Note that for the quinoid or antiquinoid structure, there are actually two resonance structures with the unpaired electron and the positive charge exchanged (this is indicated by +/•)

fifth order polynomial fit and this is the integration algorithm we choose for applications.

4.3 Analysis tools

The number of degrees of freedom increases with the size of the system, and it becomes difficult to extract qualitative trends from large amounts of data. Some efforts are necessary in analyzing results and getting a physical picture. Here, we suggest some ways to monitor electron and nuclear dynamics during such a simulated trajectory. We illustrate them with our model system, benzene radical cation, but they could in principle be applied to other systems where charge migration is initiated near a conical intersection.

We suggest using Fig. 1 to represent both the electronic and the nuclear dynamics. Indeed, the structures can refer both to the nature of the electronic density and to the nuclear geometry. Figure 4 is a schematic cross-section of the conical intersection along the X_1 direction shown in Fig. 1. A set of optimized quinoid/antiquinoid structures is presented with their respective spin densities and important bond lengths. One can imagine the electronic character and the nuclear geometry evolving synchronously in equilibrium or asynchronously during the trajectory.

4.3.1 Monitoring changes in the electronic structure

Benzene radical cation can adopt different electronic characters during the simulation. The electronic wavefunction is, in theory, a superposition of several VB structures and we would like to monitor its time-dependence. One way to do it is to calculate its spin density as a function of time. The spin density is defined as the difference between the alpha density (density of electrons with spin up) and the beta density (density of electrons with spin down): $|\Psi^S(t)|^2 = |\Psi^\alpha(t)|^2 - |\Psi^\beta(t)|^2$. It allows one to locate the unpaired electron within the molecule. The spin density can be partitioned onto atomic sites using the standard Mulliken population analysis. Figure 4 gives the partitioned spin densities for the quinoid and antiquinoid VB structures: The unpaired electron is located on the top and bottom carbon atoms in the quinoid VB structure shown in the figure, whereas it is delocalized onto the four carbons on the sides in the antiquinoid VB structure. One can thus assign a different spin density pattern to each VB structure.

To follow the evolution of the electronic wavefunction, its spin density is computed, partitioned onto the atoms and can be decomposed in the space of the VB structures of the moat at each step of the simulation. One can then plot the electron dynamics trajectory on the moat diagram (Fig. 1) where the structures represent exclusively the nature of the electronic wavefunction in this case. Because each structure in Fig. 1 is a superposition of two resonance structures where the unpaired electron and the positive charge are interchanged, following the unpaired electron is equivalent to following the positive charge. This is how we monitor the "hole" dynamics.

4.3.2 Monitoring changes in nuclear geometry

The question we would like to answer is whether nuclear geometry and electronic wavefunction evolve in equilibrium or not; how synchronous or asynchronous the electron and nuclear dynamics are. We are therefore interested in the nuclear motion in the branching space of the conical intersection. The nuclear trajectory can be plotted on the same moat diagram (Fig. 1), but here, the structures represent the nuclear geometry exclusively (i.e., the pattern of single bonds corresponding to longer bond lengths, versus double bonds corresponding to shorter bond lengths, etc.). By comparing the electronic dynamics with the nuclear motion when both are represented in terms of VB structures, we can study how they differ.

4.4 Simulations with fixed nuclei and nuclei moving

In the first calculation, we propagate only the electronic wavefunction without allowing the nuclei to move. This is to study first "pure" charge migration. In the second calculation, the nuclei are allowed to move along with the propagation of the electronic wavefunction. In Fig. 5, the spin densities partitioned onto the atoms are plotted as a function of time in each case. The bottom half aims to represent the corresponding evolution of both the electronic character (in pink) and the nuclear geometry (in blue) within the moat.

With fixed nuclei, we observe oscillation between a quinoid and an antiquinoid VB structures with a total period of about 10 fs. With nuclei moving, the electronic dynamics is the same as with fixed nuclei during the first 2 fs but after that, the effect of the nuclear motion is not negligible. Indeed, the pink arrow representing the electronic "trajectory" has the same initial direction but then, it deviates because of the nuclear motion. Now, let us look at the nuclear trajectory. The first blue arrow shows that the initial direction of the nuclear trajectory is "pulled" by the initial electronic character: The nuclear motion is driven by the effective electronic potential so they move in a way that minimizes the potential energy. In a classical picture, it means adopting the geometry that is in equilibrium with the electronic character. Then, the nuclear motion becomes more difficult to interpret because it is driven by the electronic potential but also by the kinetic energy built up. Note that the initial dynamics of the electronic wavefunction and the nuclei are asynchronous (the pink and blue arrows are not superimposed). At about 17 fs, both the electronic and nuclear trajectories are in the bottom left corner of the diagram and they both evolve toward the right. Their dynamics become more synchronous. Indeed, asynchronous dynamics of electrons and nuclei arise from

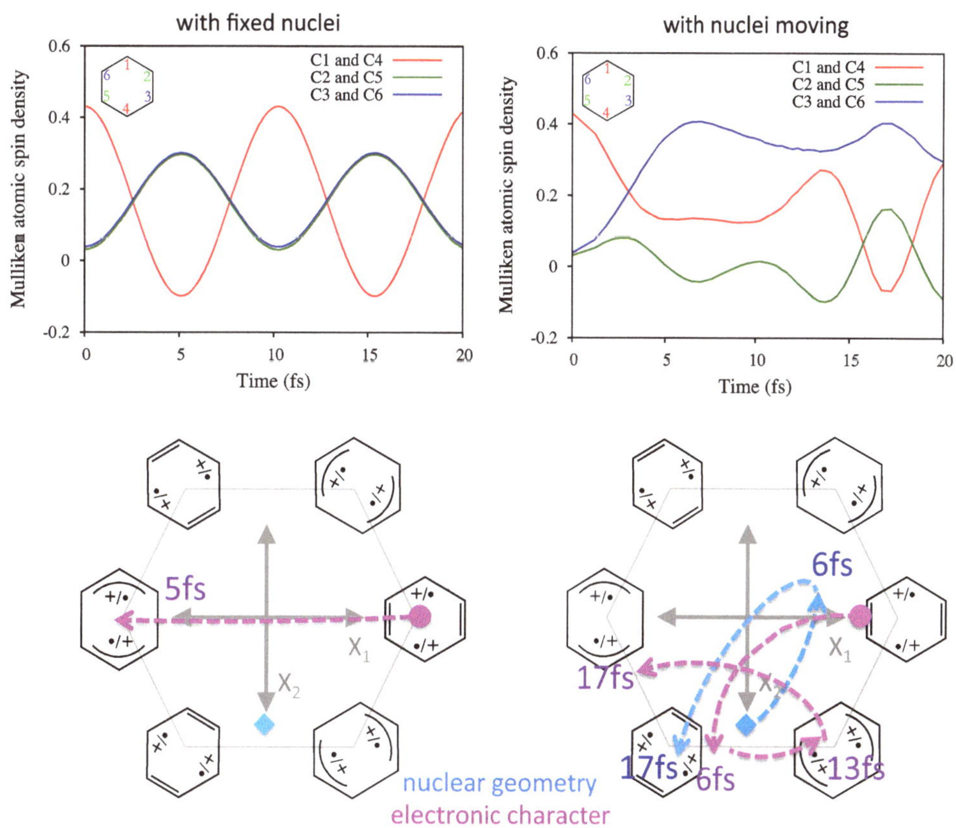

Fig. 5 Results of two Ehrenfest simulations on benzene cation: with fixed nuclei (*left side*) and with nuclei moving (*right side*). The *top figures plot* the evolution of the Mulliken spin densities as the function of time. The electron and nuclear motions are represented on the *bottom moat diagrams* (the nuclear geometry in *blue* and the electronic character in *pink*). With nuclei fixed, the electronic character of the system between a set of quinoid/antiquinoid structures. With moving nuclei, the oscillations in the electronic character seem damped until the nuclear geometry slowly "catches" the electronic character

the population of several electronic eigenstates. As the trajectory decays onto the ground state, the residual electronic dynamics will be charge transfer, where the electronic and nuclear dynamics are synchronous.

5 Conclusion

The Ehrenfest method allows one to study coupled electron–nuclear dynamics by treating the feedback between electronic and nuclear degrees of freedom in a mean-field manner. The nuclear state (expressed as a single classical trajectory) experiences an effective potential due to a time-dependent superposition of electronic eigenstates. Using this approach, we can study the evolution of a non-stationary electronic wavefunction for fixed atomic nuclei, and where the nuclei are allowed to move, to investigate the differences. So far, we have used it to study both charge migration and charge transfer upon ionization of small organic molecules. Our CASSCF implementation with a second-order integration algorithm and an additional fifth order polynomial fit allows us step sizes of the order of 0.3 fs (or 0.03amu$^{1/2}$bohr) while keeping the error in energy conservation below 0.0002 kcal/mol for dynamics runs longer than 50 fs.

We choose benzene cation as a prototype because ionizing the neutral species leads to a Jahn–Teller degeneracy between ground and first excited states of the cation. Note that only illustrative simulations were presented here; one must sample the positions and momentum of nuclei to obtain a realistic dynamics of the system. Taking advantage of the approximation of the Ehrenfest method, we can analyze nuclear and electronic dynamics independently. We monitor the electronic dynamics by calculating the spin density of the system as a function of time. With nuclei fixed, there is no electron dynamics in this case. However, if we distort the geometry [10] away from the exact degeneracy, we see "pure" charge migration: oscillations in the spin density that we can correlate with particular localized electronic structures, with a period depending on the gap between the states initially populated. Close to a conical intersection, the energy gap will be small and the resulting electron dynamics will be on a femtosecond timescale. Here, the oscillations in the spin density have a period of 10 fs (see Fig. 5), so we can use relatively big step sizes (0.3 fs). If we allow nuclear motion, we see changes in the period of the electronic dynamics as the nuclei start to couple.

Acknowledgments This work was supported by UK-EPSRC Grant EP/I032517/1. All calculations were run using the Imperial College High Performance Computing service. The original SA CP-MCSCF programs were written by Thom Vreven. The work on the Ehrenfest programs was initiated by Patricia Hunt (see the supplementary information of reference [9]).

Appendix: Computation of the energy and the gradient of a complex wavefunction

Energy computation

Let us consider the expansion of the TD wavefunction in the eigenstate basis set as defined in Eq. (10) with complex coefficients $\{c_k\}$. $\mathscr{H}_e$ is the electronic Hamiltonian operator and $\boldsymbol{H}$ its matrix representation in the eigenstate basis with elements $\boldsymbol{H}_{kl} = \langle\tilde{\phi}_k|\mathscr{H}_e|\tilde{\phi}_l\rangle$. The energy of the TD wavefunction is computed as the expectation value of this operator:

$$E = \langle\Psi|\mathscr{H}_e|\Psi\rangle = \sum_k \sum_l c_k^* c_l \boldsymbol{H}_{kl} \tag{30}$$

Note that in the eigenstate basis, $\mathbf{H}_{kl} = 0$ for $k \neq l$ so the double sum reduces to one. The energy expression then reads:

$$E = \sum_k |c_k|^2 E_k \tag{31}$$

Here, we see that the energy depends only on the weight of each eigenstate and not on their relative phase. From an implementation point of view, instead of repeating the operations for the real and imaginary components, we create a real wavefunction that has the same energy and using directly the machinery already programmed. For that, one needs to rotate all the complex coefficients in the TD vector expansion so that they are all real but conserving their magnitude. So the energy evaluated with the vector rotated to real is equal to the energy of the complex TD vector. Note this is only true because the rotation is done in the eigenstate basis. In general, we can compute the expectation value of an operator with the wavefunction rotated to real only if the operator is diagonal in the basis set we do the rotation in.

Gradient computation

The Hellmann–Feynman term of the gradient is defined as the partial derivative of the energy with respect to a nuclear distortion R_I. To consider the intrinsic dependence of the energy, we assume an expansion in exact eigenstates. By applying the product rule, we obtain:

$$E^{R_I} = \sum_k \nabla_{R_I}(|c_k|^2) \cdot E_k + \sum_k |c_k|^2 \cdot \nabla_{R_I}(E_k) \tag{32}$$

The second term on the right hand side is the average of the gradient of each electronic eigenstate weighted by their occupation. It represents the change in potential energy staying on the same potential, i.e. keeping the same occupation on each electronic eigenstate. The first term however is the change in potential energy due to change in occupation of the electronic eigenstates because of non-adiabatic transitions. To calculate the derivative of $|c_k|^2$ with respect to a nuclear distortion, we can invoke the time derivative by applying the chain rule:

$$E^{R_I} = \sum_k \frac{\partial |c_k|^2}{\partial t} \frac{1}{\dot{R_I}} \cdot E_k + \sum_k |c_k|^2 \cdot \nabla_{R_I}(E_k) \tag{33}$$

The time derivative of the norm squared of the expansion coefficient $|c_k|^2$ with respect to time can be obtained using Eq. (11):

$$\frac{\partial |c_k(t)|^2}{\partial t} = -\sum_{l,J} \left(c_l(t) c_k^*(t) + c_l^*(t) c_k(t) \right) \cdot d_{kl}^J(R) \cdot \dot{R_J} \tag{34}$$

By inserting this in Eq. (33), it reads:

$$\begin{aligned} E^{R_I} = &-\sum_{k \neq l} \left(c_l(t) c_k^*(t) + c_l^*(t) c_k(t) \right) \cdot d_{kl}^J(R) \cdot E_k \\ &+ \sum_k |c_k|^2 \cdot \nabla_{R_I}(E_k) \end{aligned} \tag{35}$$

We see that the non-adiabatic coupling $d_{kl}^J(R)$ is present in the term representing the change in energy due to electronic transitions. Using the relation $d_{kl}^J(R) = -d_{lk}^J(R)$ gives:

$$\begin{aligned} E^{R_I} = &-\sum_{k \neq l} c_l^*(t) c_k(t) \cdot d_{kl}^J(R) \cdot (E_k - E_l) \\ &+ \sum_k |c_k|^2 \cdot \nabla_{R_I}(E_k) \end{aligned} \tag{36}$$

Using Eq. (18), it becomes:

$$\begin{aligned} E^{R_I} = &\sum_{k \neq l} c_l^*(t) c_k(t) \cdot \langle \tilde{\phi}_l | \nabla_{R_I}(\mathscr{H}_e) | \tilde{\phi}_k \rangle \\ &+ \sum_k |c_k|^2 \cdot \langle \tilde{\phi}_k | \nabla_{R_I}(\mathscr{H}_e) | \tilde{\phi}_k \rangle \end{aligned} \tag{37}$$

The Hellmann–Feynman term of the gradient is the expectation value of the derivative of the Hamiltonian operator. On one hand, the diagonal terms represent the weighted average potential and they depend only on the norms of the expansion coefficients. On the other hand, the off-diagonal terms represent the change in energy due to non-adiabatic transitions and they do depend on the relative complex phase of the expansion coefficients. For this reason, one can not construct a real wavefunction whose gradient would be equal to the gradient of a complex wavefunction.

References

1. Breidbach J, Cederbaum LS (2005) Universal attosecond response to the removal of an electron. Phys Rev Lett 94:033901
2. Kling MF, Vrakking MJJ (2008) Attosecond electron dynamics. Annu Rev Phys Chem 59:463–492
3. Krausz F, Ivanov M (2009) Attosecond physics. Rev Mod Phys 81:163–234
4. Tully JC (2012) Perspective: nonadiabatic dynamics theory. J Chem Phys 137(22):22A301
5. Joubert-Doriol L, Lasorne B, Gatti F, Schröder M, Vendrell O, Meyer H-D (2012) Suitable coordinates for quantum dynamics: applications using the multiconfiguration time-dependent Hartree (MCTDH) algorithm. Comput Theor Chem 990:75–89 Chemical reactivity, from accurate theories to simple models, in honor of Professor Jean-Claude Rayez
6. Vibók Á, Csehi A, Gindensperger E, Köppel H, Halász GJ (2012) Quantum dynamics through conical intersections: combining effective modes and quadratic couplings. J Phys Chem A 116(11):2629–2635
7. Worth GA, Meyer H-D, Köppel H, Cederbaum LS, Burghardt I (2008) Using the MCTDH wavepacket propagation method to describe multimode non-adiabatic dynamics. Int Rev Phys Chem 27(3):569–606
8. Klein S, Bearpark MJ, Smith BR, Robb MA, Olivucci M, Bernardi F (1998) Mixed state 'on the fly' non-adiabatic dynamics: the role of the conical intersection topology. Chem Phys Lett 292(3):259–266
9. Blancafort L, Hunt P, Robb MA (2005) Intramolecular electron transfer in bis(methylene) adamantyl radical cation: a case study of diabatic trapping. J Am Chem Soc 127(10):3391–3399 PMID: 15755157
10. Mendive-Tapia D, Vacher M, Bearpark MJ, Robb MA (2013) Coupled electron–nuclear dynamics: charge migration and charge transfer initiated near a conical intersection. J Chem Phys 139(4):044110
11. Roos BO, Taylor PR, Siegbahn PE (1980) A complete active space SCF method (CASSCF) using a density matrix formulated super-CI approach. Chem Phys 48(2):157–173
12. Roos BO (1987) The complete active space self-consistent field method and its applications in electronic structure calculations. Adv Chem Phys 69:399–446
13. Millam JM, Bakken V, Chen W, Hase WL, Schlegel HB (1999) Ab initio classical trajectories on the Born–Oppenheimer surface: Hessian-based integrators using fifth-order polynomial and rational function fits. J Chem Phys 111(9):3800–3805
14. Frisch MJ, Trucks GW, Schlegel HB, Scuseria GE, Robb MA, Cheeseman JR, Scalmani G, Barone V, Mennucci B, Petersson GA, Nakatsuji H, Caricato M, Li X, Hratchian HP, Izmaylov AF, Bloino J, Zheng G, Sonnenberg JL, Hada M, Ehara M, Toyota K, Fukuda R, Hasegawa J, Ishida M, Nakajima T, Honda Y, Kitao O, Nakai H, Vreven T, Montgomery JA Jr, Peralta JE, Ogliaro F, Bearpark M, Heyd JJ, Brothers E, Kudin KN, Staroverov VN, Kobayashi R, Normand J, Raghavachari K, Rendell A, Burant JC, Iyengar SS, Tomasi J, Cossi M, Rega N, Millam JM, Klene M, Knox JE, Cross JB, Bakken V, Adamo C, Jaramillo J, Gomperts R, Stratmann RE, Yazyev O, Austin AJ, Cammi R, Pomelli C, Ochterski JW, Martin RL, Morokuma K, Zakrzewski VG, Voth GA, Salvador P, Dannenberg JJ, Dapprich S, Daniels AD, Farkas Ö, Foresman JB, Ortiz JV, Cioslowski J, Fox DJ (2010) Gaussian development version, revision h.10. Gaussian Inc., Wallingford, CT
15. Teller E (1937) The crossing of potential surfaces. J Phys Chem 41(1):109–116

16. Amarouche M, Gadea F, Durup J (1989) A proposal for the theoretical treatment of multi-electronic-state molecular dynamics: hemiquantal dynamics with the whole dim basis (HWD). A test on the evolution of excited ar3+ cluster ions. Chem Phys 130:145–157
17. Delos JB, Thorson WR, Knudson SK (1972) Semiclassical theory of inelastic collisions. I. Classical picture and semiclassical formulation. Phys Rev A 6:709–720
18. Delos JB, Thorson WR (1972) Semiclassical theory of inelastic collisions. II. Momentum-space formulation. Phys Rev A 6:720–727
19. Herman MF (1994) Dynamics by semiclassical methods. Annu Rev Phys Chem 45(1):83–111
20. Jasper AW, Nangia S, Zhu C, Truhlar DG (2006) Non-Born–Oppenheimer molecular dynamics. Acc Chem Res 39(2):101–108 PMID: 16489729
21. Li X, Tully JC, Schlegel HB, Frisch MJ (2005) Ab initio ehrenfest dynamics. J Chem Phys 123(8):084106
22. Saita K, Shalashilin DV (2012) On-the-fly ab initio molecular dynamics with multiconfigurational ehrenfest method. J Chem Phys 137(22):22A506
23. Shalashilin DV (2010) Nonadiabatic dynamics with the help of multiconfigurational ehrenfest method: improved theory and fully quantum 24d simulation of pyrazine. J Chem Phys 132(24):244111
24. Takatsuka K, Yonehara T (2011) Exploring dynamical electron theory beyond the Born–Oppenheimer framework: from chemical reactivity to non-adiabatically coupled electronic and nuclear wavepackets on-the-fly under laser field. Phys Chem Chem Phys 13(11):4987–5016
25. Yonehara T, Hanasaki K, Takatsuka K (2012) Fundamental approaches to nonadiabaticity: toward a chemical theory beyond the Born–Oppenheimer paradigm. Chem Rev 112(1):499–542
26. Zhu C, Nangia S, Jasper AW, Truhlar DG (2004) Coherent switching with decay of mixing: an improved treatment of electronic coherence for non-Born–Oppenheimer trajectories. J Chem Phys 121(16):7658–7670
27. Tully JC (1998) Mixed quantum–classical dynamics. Faraday Discuss 110:407–419
28. Hack MD, Wensmann AM, Truhlar DG, Ben-Nun M, Martinez TJ (2001) Comparison of full multiple spawning, trajectory surface hopping, and converged quantum mechanics for electronically nonadiabatic dynamics. J Chem Phys 115(3):1172–1186
29. Tully JC (1990) Molecular dynamics with electronic transitions. J Chem Phys 93(2):1061–1071
30. Almlöf J, Taylor PR (1985) Molecular properties from perturbation theory: a unified treatment of energy derivatives. Int J Quantum Chem 27(6):743–768
31. Cederbaum L, Zobeley J (1999) Ultrafast charge migration by electron correlation. Chem Phys Lett 307(3–4):205–210
32. Kuleff AI, Cederbaum LS (2007) Charge migration in different conformers of glycine: the role of nuclear geometry. Chem Phys 338:320–328
33. Kuleff AI, Lünnemann S, Cederbaum LS (2010) Ultrafast charge migration following valence ionization of 4-methylphenol: jumping over the aromatic ring. J Phys Chem A 114(33):8676–8679
34. Kuleff AI, Lünnemann S, Cederbaum LS (2012) Ultrafast reorganization of the hole charge created upon outer-valence ionization of porphyrins. Chem Phys 399:245–251
35. Lünnemann S, Kuleff AI, Cederbaum LS (2008) Ultrafast charge migration in 2-phenylethyl-n, n-dimethylamine. Chem Phys Lett 450:232–235
36. Periyasamy G, Levine R, Remacle F (2009) Electronic wave packet motion in water dimer cation: a many electron description. Chem Phys 366:129–138
37. Aberg T (1967) Theory of X-ray satellites. Phys Rev 156:35–41
38. Aberg T (1970) Asymptotic double-photoexcitation cross sections of the helium atom. Phys Rev A 2:1726–1729
39. Manne R, Åberg T (1970) Koopmans' theorem for inner-shell ionization. Chem Phys Lett 7(2):282–284

Theor Chem Acc (2014) 133:1506
DOI 10.1007/s00214-014-1506-5

REGULAR ARTICLE

Steric and electrostatic effects on photoisomerization dynamics using QM/MM ab initio multiple spawning

Aaron M. Virshup · Benjamin G. Levine · Todd J. Martínez

Received: 3 March 2014 / Accepted: 15 May 2014 / Published online: 5 June 2014

Abstract Photoisomerization of conjugated systems is a common pathway for photomechanical energy conversion in biological chromophores. There are many examples where the local environment of the chromophore plays an important role in determining the outcome of photoisomerization. We have investigated the effect of simple steric and electrostatic environments on the excited-state photodynamics of ethylene, a simple model for larger conjugated systems. Ab initio electronic structure methods were combined with molecular mechanical force fields to describe the ground and excited-state potential energy surfaces of ethylene embedded in electrostatic and steric environments. The time evolution of the system following photoabsorption was modeled using the ab initio multiple spawning (AIMS) method for quantum dynamics. We introduce a new method for integration of the equations of motion in AIMS, which detects conical intersections automatically and then decreases the timestep adaptively around them. Neither steric hindrance nor electrostatics have a large effect on the excited-state lifetime, even at effective pressures as large as 2 GPa. However, a nearby point charge creates an electric field that stabilizes one of two symmetry-related conical intersections, biasing the reaction toward a particular photoisomerization pathway. For the larger tetramethylethylene, where steric hindrance is expected to be more pronounced, we also see no effect on the excited-state lifetime. Our results suggest that electrostatic interactions are more effective than steric hindrance in modifying the course of excited-state reactions.

Keywords Isomerization · Nonadiabatic dynamics · Surface hopping

Dedicated to the memory of Professor Isaiah Shavitt and published as part of the special collection of articles celebrating his many contributions.

A. M. Virshup
Department of Chemistry, Duke University,
Durham, NC 27708, USA

B. G. Levine
Department of Chemistry, Michigan State University,
East Lansing, MI 48824, USA

T. J. Martínez (✉)
Department of Chemistry and the PULSE Institute,
Stanford University, Stanford, CA 94305, USA
e-mail: toddjmartinez@gmail.com

T. J. Martínez
SLAC National Accelerator Laboratory, Menlo Park,
CA 94025, USA

1 Introduction

Conical intersections (CIs), points of true degeneracy between different electronic potential energy surfaces (PESs), are now understood to play a fundamental role in photochemical dynamics [1–5]. These molecular geometries act as doorways between different electronic PESs, efficiently funneling population between different electronic states. In particular, inter-state population transfer through CIs is a crucial step in many photoinduced isomerization reactions. In biological systems, chromophores containing conjugated π-bond systems are a common site for such processes. It is also increasingly apparent that interaction of such chromophores with their local environment plays an important role in determining their photochemistry [6–14]. In this work, we investigate the effects of local environment on photoisomerization in ethylene to better understand the mechanisms of this environment-dependent tuning.

Interactions of a chromophore with its local environment can be broadly decomposed into two categories: steric and electrostatic effects. The precise interplay of these effects in biological systems has been the subject of a great deal of research and some controversy. Over the past 30 years, for instance, many broad mechanisms of protein activity have been ascribed to steric strain [15–17]. However, later theoretical studies have disputed many of these conclusions, finding that electrostatic interactions, in particular, play a more important role [6–14, 18–20]. Electrostatic interactions are increasingly recognized as an important factor in many realms of biochemistry [20–22].

As the basic building block for many conjugated molecules, ethylene has long been of interest as a model for photoisomerization dynamics. The photodynamics of isolated ethylene after photoexcitation of an electron from the HOMO (π) to the LUMO (π*) orbital have been the subject of numerous theoretical [23–28], pump probe [29–31], and Raman spectroscopy [32] studies. While ethylene is planar in its ground state, the $\pi \rightarrow \pi^*$ excited state favors a twisted D_{2d} geometry. Interaction of this $\pi\pi^*$ state with a charge transfer state of doubly excited character leads to the phenomenon of "sudden polarization," [33–36] in which one of the methylene groups pyramidalizes and the molecule acquires a large dipole moment. This pyramidalization distortion also leads to a set of CI geometries, resulting in rapid quenching to the ground electronic state. In competition with this quenching, a hydrogen atom can migrate from one side of the molecule to the other, yielding ethylidene ($HCCH_3$). Thus, the excited-state reaction path can be roughly characterized by a brief period of twisting motion about the C=C bond, followed by pyramidalization and/or hydrogen migration, as depicted in Fig. 1. There are CI geometries near both of the twisted/pyramidalized and ethylidene regions, leading to very fast quenching to the ground state. In fact, previous studies provide estimates of the excited-state lifetime ranging from 30 to 250 fs [5, 23, 27, 28, 31, 37, 38].

2 Methods

2.1 Hybrid QM/MM potential energy surfaces

While certain quantum chemical techniques can, in principle, describe the full electronic structure of a system in both ground and excited states, in practice these methods are far too computationally expensive to describe a molecule in a complex environment. Fortunately, in excited-state photochemical problems, it is often true that the electronically perturbed region is localized to a small, well-defined chromophore. This separation of the system into perturbed and unperturbed subsystems makes it amenable to simulation by multi-resolution techniques. In particular, while expensive quantum chemical simulations are necessary to model the photochemically active subsystem, the larger system can be modeled using much more efficient molecular mechanical force fields, which provide reasonable and inexpensive descriptions of the PESs of systems near equilibrium. These two levels of theory can be combined to generate a hybrid quantum mechanical/molecular mechanical (QM/MM) PES. These QM/MM techniques were initially developed by Warshel and Levitt to treat bond rearrangement in calculations of enzyme docking kinetics, [39] but QM/MM methods have subsequently been the subject of much development and have found use in a variety of chemical fields, including dynamics and free energy calculations [40–44].

Here, the QM/MM Hamiltonian is written as an approximation to the full electronic Hamiltonian:

$$\hat{H}_{\mathrm{el}} \cong \hat{H}_{\mathrm{QM}} + \hat{H}_{\mathrm{MM}} + \hat{H}_{\mathrm{QM/MM}}, \tag{1}$$

where $\hat{H}_{\mathrm{QM}}$ represents the exact Hamiltonian for the QM system, $\hat{H}_{\mathrm{MM}}$ the force field of the MM system, and $\hat{H}_{\mathrm{QM/MM}}$ the interaction between the two systems. The interaction term consists of Lennard-Jones and electrostatic interactions between the two systems:

$$\hat{H}_{\mathrm{QM/MM}} = \hat{H}_{\mathrm{LJ}} + \hat{H}_{\mathrm{ES}}. \tag{2}$$

Dispersion and Pauli repulsion between the two systems is modeled with a Lennard-Jones potential,

$$\hat{H}_{\mathrm{LJ}} = \sum_{i \in \mathrm{MM}, j \in \mathrm{QM}} \varepsilon_{ij} \left[\frac{\sigma_{ij}^{12}}{R_{ij}^{12}} - \frac{\sigma_{ij}^{6}}{R_{ij}^{6}} \right], \tag{3}$$

where ε_{ij} and σ_{ij} are empirical parameters drawn from the MM force field and R_{ij} is the distance between the ith and jth nuclear centers. The origin of this force is electronic; the repulsive R^{-12} term, for instance, arises from Pauli repulsion between electrons on the two atoms. However,

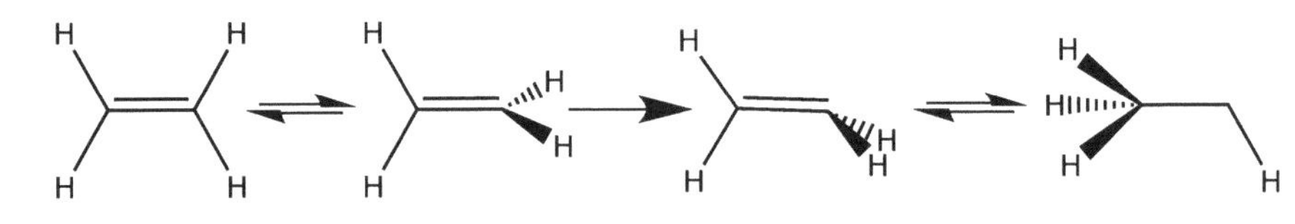

Fig. 1 Schematic mechanism for photodynamics of ethylene on the $\pi \rightarrow \pi^*$ excited state. After photoexcitation, twisting and pyramidalization brings the molecule to an S1/S0 CI, and further hydrogen migration creates ethylidene

Table 1 Lennard-Jones parameters for the Ar-solvated simulations

Interaction	ε/kcal/mol	σ/Å
C–Ar	0.143	1.91
H–Ar	0.061	1.70
Ar–Ar	0.239	1.91

the form of the Lennard-Jones potential depends only upon atomic positions and *not* electronic coordinates [45, 46] From a practical standpoint, this means that the electronic structure is not coupled to the MM system through the Lennard-Jones potential.

The electrostatic interaction term, in contrast, does include the electronic coordinates, and thus must be included in the electronic structure calculation. This interaction is given by

$$\hat{H}_{\text{ES}} = \sum_{i \in \text{MM}, j \in \text{QM}} \frac{q_i Q_j}{R_{ij}} + \sum_p \sum_{i \in \text{MM}} \frac{q_i}{|\mathbf{r}_p - \mathbf{R}_i|}, \tag{4}$$

where q_i and Q_j are the charges of the ith MM and jth QM nuclei, respectively, and p indexes the electrons in the QM system (with $\mathbf{r}_\text{p}$ being the position of the pth electron).

For the MM system and QM/MM interaction, Lennard-Jones parameters (Table 1) for argon were drawn from Maitland et al. [47] and those for the ethylene atoms were drawn from the AMBER ff99 force field [48]. For interactions between distinct atom types, the Lennard-Jones σ parameter was chosen to be the arithmetic average of the values for the two atom types, while the ε parameter was the geometric mean of the two values.

2.2 Ab initio electronic structure

The electronic wavefunction was evaluated using the state averaged complete active space self-consistent field (SA-CASSCF) method [49, 50], as implemented in the Molpro [51] electronic structure package. We used the 6-31G** basis set [52] and an active space of two electrons in two orbitals with equally weighted averaging over the lowest three singlet states, i.e., SA-3-CAS(2/2)/6-31G**.

We caution that this level of theory was chosen in part for its computational efficiency; CASSCF does have some deficiencies, and in particular is unable to describe dynamic electron correlation. The PESs for ethylene from SA-3-CAS(2/2) are qualitatively similar to the PESs from higher level calculations, but dynamics calculations with higher levels of theory (e.g., including perturbation theory corrections with MS-CASPT2 [27] and/or including diffuse orbitals to represent the Rydberg states [28]) suggest that there is a tendency for SA-3-CAS(2/2) dynamics to overestimate the importance of the ethylidene conical intersection.

2.3 Ab initio multiple spawning (AIMS) dynamics

The time evolution of the system was modeled using the ab initio multiple spawning (AIMS) method for quantum molecular dynamics. This method has been extensively described elsewhere [53, 54]; we present only a brief overview here. AIMS takes as its starting point a linear combination of time-dependent Born–Oppenheimer states:

$$\Psi_{\text{AIMS}}(\mathbf{R}, \mathbf{r}, t) = \sum_{I=1}^{N(t)} c_I(t) \chi_I(\mathbf{R}, t) \phi_{N_I}(\mathbf{r}; \mathbf{R}), \tag{5}$$

where $\chi_I(\mathbf{R}, t)$ is a nuclear basis function and $\phi_{N_I}(\mathbf{r}; \mathbf{R})$ is the electronic wavefunction on state N_I. The nuclear basis functions are minimum uncertainty wave packets, i.e., frozen Gaussians [55], in each of the nuclear coordinates, and their average positions and momenta are chosen to evolve classically on their given potential energy surfaces. The time evolution of the amplitudes of the basis functions $\{c_I(t)\}$ is determined by direct substitution of this wavefunction into the time-dependent Schrödinger equation. For the results presented here, simulations begin with a single basis function, with classical parameters sampled from some initial distribution (typically a Wigner distribution for the desired initial state, see below). However, when a wave packet on a given PES enters a region of high nonadiabatic coupling to a different PES, the number of basis functions $N(t)$ is expanded, and a basis function [with $c_I(t) = 0$] is placed on the complementary PES. This dynamic "spawning" of new basis functions allows description of population transfer between different PESs without a priori assumptions about the nature of the dynamics. All dynamical simulations were performed in our implementation of AIMS with the Molpro [electronic structure package [56].

At each timestep, the classical positions and momenta for each trajectory basis function (TBF) are evolved from time t to $t + \Delta t$ using the velocity Verlet integrator, while the quantum mechanical amplitudes $\{c_I(t)\}$ are evolved from time t to $t + \Delta t$ using a second-order Runge–Kutta integrator to solve the time-dependent Schrödinger equation. Computational efficiency dictates the choice of a large timestep Δt, while accurate integration demands a timestep small enough to capture the variation of the independent variables (positions, momenta, and complex amplitudes). Near conical intersections, the curvature of the PES can be very large (in fact, the derivatives of the PES are discontinuous at a CI). Furthermore, at conical intersections and very weakly avoided crossings, the time-derivative coupling exhibits a narrow and sharp spike, as described in recent works on the accurate treatment of trivially unavoided crossings in surface hopping simulations [57–59]. To avoid "jumping over" this spike in the coupling, which would cause large errors in the predicted population

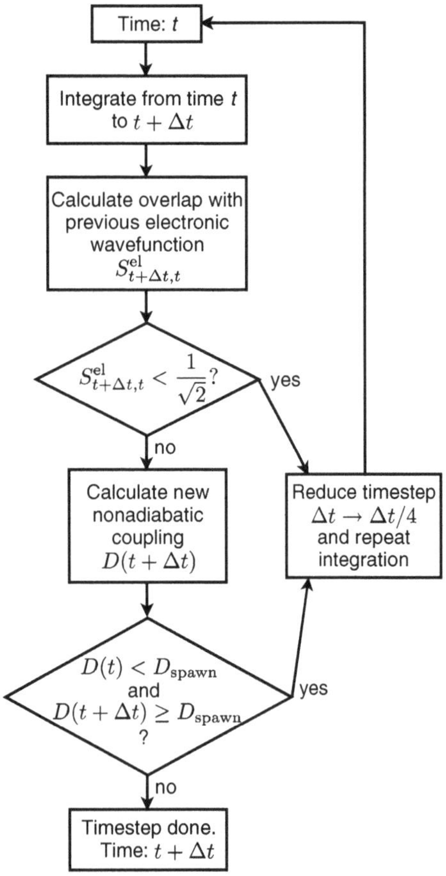

Fig. 2 Acceptance scheme for adaptive timestep integration

transfer, a very small timestep may be needed near CIs, but in regions far from CIs such a small timestep would waste computational effort.

To address this conflict, we developed an adaptive timestep integration scheme. Simulations generally begin with a larger timestep (for this system, 0.5 fs), suitable for efficient integration of dynamics on typical regions of the PES. The simulation adaptively reduces the size of the timestep (we have found that reducing the timestep by a factor of four works well) in regions near conical intersections, as detailed in Fig. 2. Two tests are used to detect the presence of conical intersections and weakly avoided crossings: large derivative coupling values, and rapid changes in the character of the electronic wavefunction.

If the derivative coupling at the center of a TBF, $D(t)$, rises above the predetermined spawning threshold D_{spawn}, or the off-diagonal inter-state matrix elements of the Hamiltonian are large, the smaller timestep is used to accurately capture the expected spikes in the derivative coupling. While this was found to improve our description of nonadiabatic events, it nevertheless neglects situations in which a TBF "jumps over" a high-coupling region in a single timestep. Such situations can be corrected by examining changes in the electronic structure.

In the vicinity of a CI (or weakly avoided crossing), the electronic structure (and related quantities such as potential energy and derivative coupling) varies rapidly with molecular geometry. These rapid changes can be detected by comparing the overlap of the electronic wavefunctions at the current and previous timesteps [56]:

$$S^{\text{el}}_{t,t-\Delta t} = \langle \phi_J(\mathbf{r}; \bar{\mathbf{R}}(t - \Delta t)) \mid \phi_J(\mathbf{r}; \bar{\mathbf{R}}(t)) \rangle_{\mathbf{r}}, \quad (6)$$

where $|\phi_J(\mathbf{r}; \mathbf{R})\rangle$ is the electronic wavefunction for the Jth electronic state at nuclear position $\mathbf{R}$, and $\bar{\mathbf{R}}(t)$ is the centroid of the TBF at time t. A small overlap $S^{el}_{t,t-\Delta t}$ suggests both the presence of a CI and that the integration timestep must be reduced to compensate for the rapid variation of the electronic wavefunction. Note that, although the PES, derivative coupling, and electronic wavefunction are discontinuous at a point of true degeneracy between two states, the CI seam has dimensionality 2 less than the full dimensionality of the system. In practice, a TBF will therefore never exactly encounter a point of degeneracy and thus, given a small enough timestep, each TBF's electronic structure is expected to be continuous from one timestep to the next. Therefore, if $S^{el}_{t,t-\Delta t}$ falls below a threshold (by default, $2^{-1/2}$, indicating that the normalized overlap with the previous timestep is <50 %), the timestep is discarded, and dynamics are restarted from time $t - \Delta t$ with a smaller timestep. This procedure allows us to detect CI regions that would otherwise have been "jumped over" by a larger timestep.

This adaptive integration scheme also permits recovery from numerical integration errors. Specifically, the program will reject a timestep if the total wavefunction norm changes by more than 5 % or if the *classical* energy of any single TBF changes by more than 5 millihartree. In these situations, as when a TBF jumps over an intersection, the simulation is restarted from an earlier time and reintegrated with the smaller timestep. Note that this scheme permits recovery only from errors in integration of the classical and quantum mechanical equations of motion; it cannot correct, e.g., discontinuities in the calculated PESs or nonconvergence of the electronic structure. Note too that the quantum mechanical energy (that is, the expectation value of the Hamiltonian) is not monitored, and in fact is not necessarily conserved for small basis sets [60].

In parallel to the spawning of new TBFs when necessary, we also terminate TBFs which are no longer necessary to describe population transfer. A given TBF is marked for potential termination if it (a) has too little population to trigger new spawning events or (b) has reached the user-specified "target" electronic state, usually the ground state. Because the TBFs are all coupled in the solution of the complex amplitudes, it is only possible to remove a TBF from the simulation (without ill effects) if

Fig. 3 Cross section of C_2H_4/Ar-cluster simulation geometry

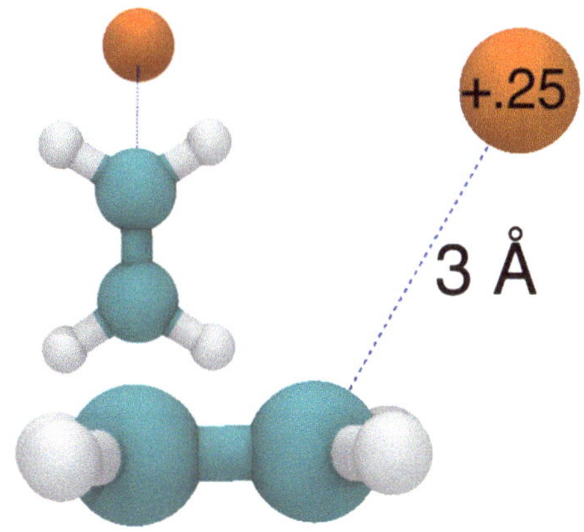

Fig. 4 Top and side views of geometry used for C_2H_4 with a surrounding point charge

Table 2 Excited-state lifetimes (with standard error from bootstrapping) for ethylene in various environments

	No of runs	t_0/fs	τ/fs
Gas-phase ethylene	200	24 ± 3	116 ± 9
Argon-solvated ethylene	80	29 ± 5	93 ± 11
Gas-phase TME	25	34 ± 11	276 ± 100
Argon-solvated TME	25	22 ± 10	237 ± 82
Ethylene w/point charge	200	16 ± 3	103 ± 8

The timescale for IVR is given by t_0 and the time constant for nonadiabatic transitions from S_1 to S_0 is given by τ, as described in Eq. (7)

the TBF has negligible matrix elements (both overlap and Hamiltonian) with all TBFs which will remain in the simulation. Therefore, we continue to propagate the TBFs that are marked for potential deletion for an additional time period (chosen to be 5 fs in the cases discussed here). A TBF is removed from the simulation if and only if it is both marked for deletion *and* has negligible overlap with all other TBFs during this time period. Otherwise, it continues to propagate with the other TBFs. If further propagation of a terminated TBF is desired (for instance, to investigate the dynamics on the target state), it can later be propagated, decoupled from the original simulation. This dynamic pruning of unimportant trajectories leads to significant computational savings, as well as automatic termination of simulations once all population has been transferred to the target state.

2.4 Simulation details

As a simple model for steric interactions, we simulated the photodynamics of ethylene microsolvated in a cluster of 1,200 argon atoms at 0 K (Fig. 3). The system was prepared by randomly generating a cluster of 1,200 argon atoms around an ethylene molecule at its ground state minimum geometry, then optimizing the energy of the MM system with the QM (ethylene) atoms held fixed. In order to improve Lennard-Jones contact and exaggerate the effects of steric hindrance, the system was confined with half-harmonic restraints to a sphere of radius 19.7 Å, corresponding to an external pressure of approximately 2 GPa (calculated using the internal virial $P = \frac{NkT}{V} + \frac{\mathbf{F}_{\text{int}} \cdot \mathbf{R}}{3V}$, where $\mathbf{F}_{\text{int}}$ are forces excluding those from the walls of the system [61]). These simulations were also repeated with tetramethylethylene (TME), again in an attempt to exaggerate any steric effects on the reaction. For comparison, the gas-phase dynamics of ethylene and TME were also simulated.

To investigate the effect of an external electric field on the photodynamics, a single point charge of magnitude $+0.25\ q_e$ was placed 3 Å from one of the carbon atoms (Fig. 4). Both the ethylene molecule's center of mass and the position of the point charge were held fixed using the RATTLE algorithm [62] to constrain their positions relative to one another.

The initial positions and momenta of the ethylene or TME molecule in each simulation were sampled from a Wigner transform of the ground state vibrational wavefunction in the harmonic approximation. Each set of initial conditions was run as a single, independent simulation, i.e., using the independent first-generation approximation [63] The number of independent simulations for each system is shown in Table 2.

To determine the excited-state lifetime of each simulated system, the time dependence of the average excited-state population was fit to a "delayed exponential" of the form

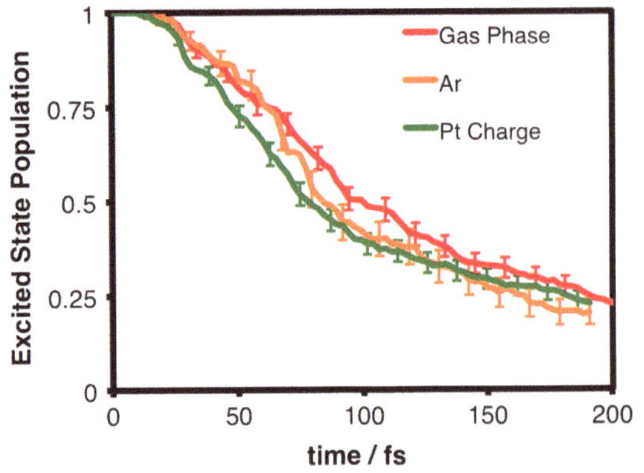

Fig. 5 Average excited-state (S_1) population as a function of time after photoexcitation, from AIMS SA-3-CAS(2/2) dynamics. The three traces correspond to ethylene in a gas-phase (*red*, $N = 200$), argon-solvated (*orange*, $N = 80$), or point charge (*green*, $N = 200$) environment at 0 K. *Error bars* indicate estimated standard error. The surrounding environment has little effect on the excited-state lifetime for ethylene under the chosen conditions

$$P_{S_1}(t;\tau,t_0) = \begin{cases} 1, & t < t_0 \\ \exp\left(\frac{t_0 - t}{\tau}\right), & t \geq t_0 \end{cases}, \quad (7)$$

with fitting parameters t_0 and τ. Here, t_0 represents the time for ethylene to move from the Franck–Condon point to the CI region, and τ is the decay constant for population transfer to the ground state. The time dependence of the excited-state populations (averaged over all initial conditions) for ethylene in the gas phase, Ar cluster, and with surrounding point charge are shown in Fig. 5. The lifetimes obtained from this data are summarized in Table 2, along with standard error estimates obtained from a nonparametric bootstrap over 10,000 samples [64].

3 Results

3.1 Argon cluster: steric influence of the surroundings

Previously published results from an initial study of ethylene solvated in a 300-argon cluster suggested that the excited-state dynamics were not perturbed by the argon environment [56]. In this study, we have increased the size of the argon cluster to 1,200 atoms, which more accurately replicates the bulk properties of argon at the ethylene solvation site and have investigated a high pressure of 2 GPa. Contrary to initial expectations, but in line with previous results, solvation in an argon cluster did not have any significant effects on the excited-state dynamics of ethylene. We note that reorganization of the argon cluster does not explain the lack of an effect; the simulations were rerun with the argon atoms fixed in place, which again produced no significant change in either the lifetime or the branching ratios.

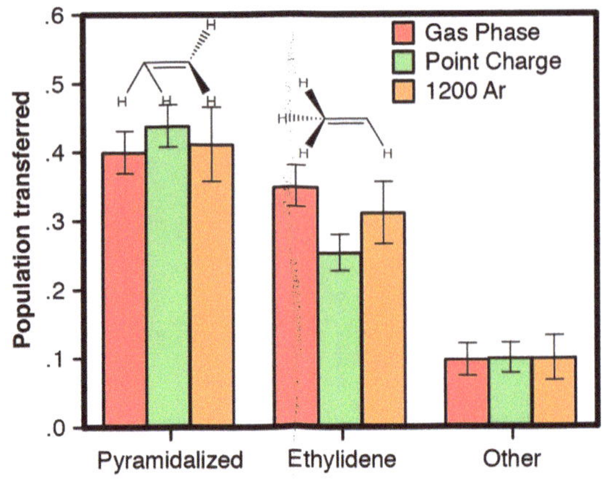

Fig. 6 Amount of population transfer through different ethylene conical intersection geometries, as classified by Lischka et al. *Error bars* are bootstrap-estimated standard error. Solvation in highly pressurized argon (*orange*, 1,200 Ar) shows no statistically significant effect on the quenching mechanism compared to isolated ethylene (*red*, Gas Phase). In contrast, the electric field from a surrounding point charge (*green*, Point Charge) leads to suppression of population transfer through ethylidene intersections and enhancement of population transfer through pyramidalized intersections

From frictional and steric hindrance considerations, one might reasonably expect that solvation in a pressurized argon cluster would slow or block the excited-state dynamics, resulting in a longer electronically excited lifetime. As shown in Fig. 5, the lifetime is not significantly affected for the photoisomerization of ethylene. Additionally, we have characterized the excited-state reaction mechanism by classifying the population transferred according to the molecular geometry when nonadiabatic effects are occurring. We use the scheme introduced by Lischka et al. [26] to classify the molecular geometries as pyramidalized or ethylidene-like. As shown in Fig. 6, the introduction of the argon environment does not significantly alter the amount of population transferred through pyramidalized and ethylidene CIs, compared to the excited-state dynamics of isolated ethylene.

In order to better understand how the excited-state dynamics can be unaffected by the surrounding pressurized argon environment, we have explored the volume swept out by photoisomerization and its relation to the cavity in the solvent which houses the chromophore. We define a cavity volume in analogy to conventional solvent-accessible volume calculations [65]. It is calculated as the union of all points that can be occupied by a hydrogen atom's van der Waals (vdW) sphere without overlapping the vdW spheres of the surrounding argon atoms, as shown in Fig. 7. The chromophore-accessible volume of the ethylene cavities was calculated for each of the 100 initial conditions generated. This gives a mean cavity size of 26 ± 5 Å^3 at the beginning of dynamics. We also calculated the total volume swept out by isomerization by carrying out the same procedure for

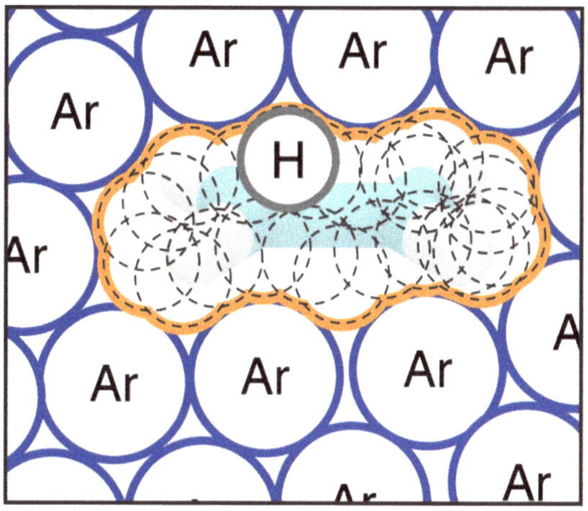

Fig. 7 Depiction of method used to define cavity volume. The cavity is defined as the union of volumes of a probe sphere (corresponding to a H atom in this case) such that the distance between the probe sphere and any surrounding Ar atoms never exceeds the sum of the van der Waals radii for the probe and Ar atoms

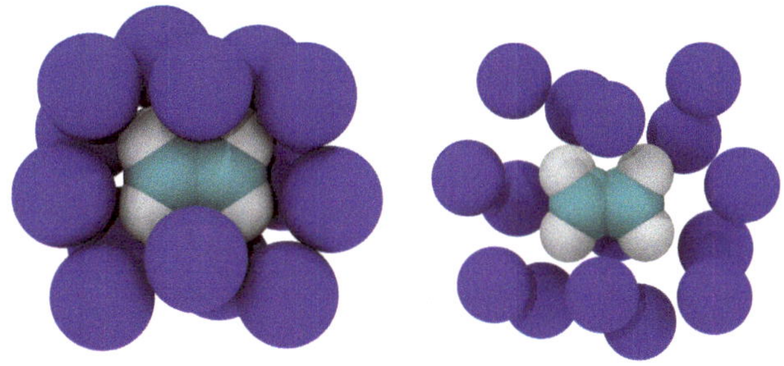

Fig. 8 *Left* cross section rendering of a representative geometry of ethylene and nearest surrounding argon atoms using spheres corresponding to vdW radii. Note that there appears to be little space for large amplitude motion of the ethylene molecule. *Right* as above, but with effective vdW radii scaled to incorporate the kinetic energy accumulated by relaxation of the chromophore on the excited state. Note that the cavity now appears quite open and there is ample space for isomerization and hydrogen atom migration

each timestep and defining the total volume as the union of all solute cavities during the first 200 fs of the dynamics. This leads to a total reaction volume of 96 ± 12 Å^3, which is significantly larger than the initial mean cavity size.

We monitored the solute–solvent distances during the photodynamics and found that these often fall below the vdW contact distance. This occurs because of a disparity in energy scales; the energetic penalty from significant interpenetration of the vdW spheres is much smaller than the energy scale of the photodynamics. The excited-state dynamics involve conversion of over 4 eV of potential into kinetic energy, or about 0.2 eV per ethylene degree of freedom—an order of magnitude greater than thermal energy at room temperature. This change in energy scale opens up much of the previously inaccessible volume. To illustrate this, we define a set of effective vdW radii at which the Lennard-Jones repulsion energy (Eq. 3) is 0.2 eV; these effective radii for the carbon and hydrogen atoms are 30 % less than their conventional values. Using these effective vdW radii leads to a reaction volume of only 59 ± 7 Å^3 swept out by the excited-state dynamics of the isolated ethylene molecule. The effective cavity volume given the amount of kinetic energy which will be accumulated during the dynamics is thus large enough to easily accommodate the reaction. In Fig. 8, we compare cross sections of the ethylene molecule and nearest surrounding Ar atoms using the usual vdW radii (left) and the effective vdW radii (right). The openness of the solvent cavity is quite clear in the latter case, which is the appropriate measure for an excited-state reaction.

Of course, the dynamics of ethylene have a limited spatial extent, as they consist mostly of hydrogen motion. To increase the spatial extent of the reaction, the hydrogens were replaced with methyl groups, creating tetramethylethylene (TME). The S_1 photodynamics and S_1/S_0 MECIs for TME are similar to those of ethylene (see Fig. 11)—hydrogen migration, in which a methyl hydrogen transfers to an ethylenic carbon and twisted-pyramidalized MECIs are observed (3.2 and 3.8 eV below the Franck–Condon point, respectively). However, population is also transferred to the ground state through a pyramidalized C–C stretched intersection. We could not find a minimal energy CI of this type; thus, these are higher energy points on the intersection seam. Typically, the C–C stretched geometries leading to significant population transfer had potential energies <1 eV below the Franck–Condon point. Decay through this stretched intersection appears to be kinetically favored, as the positions of the methyl groups are essentially unchanged from their Frank–Condon point positions.

As with ethylene, Fig. 9 shows that we observe no significant change in the excited-state lifetime of argon-solvated TME (relative to isolated TME). Furthermore, even when the argon atoms are held fixed, the excited-state lifetime is negligibly affected. However, the insensitivity of the excited-state lifetime says little about the excited-state mechanism. In Fig. 10, we show the amount of population transferred through three different types of intersections—hydrogen migration, twisted-pyramidalized, and stretched C–C. A slight increase in the propensity for population transfer through H migration CIs is observed for TME in pressurized argon compared to isolated TME. However, when the surrounding argon atoms are held fixed, the change in the mechanism is dramatic and nonadiabatic transitions through the stretched intersection are strongly enhanced. This is consistent with the expectation that motion of the methyl groups will be highly restricted by the surrounding Ar.

In Fig. 11, we characterize the excited-state mechanism in more detail, depicting the geometries associated with nonadiabatic transitions. For each such transition leading to population transfer to the ground electronic state, we

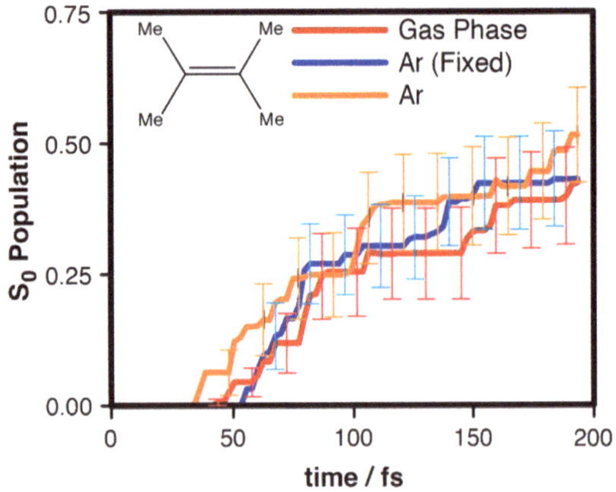

Fig. 9 Ground state population recovery after photoexcitation of tetramethylethylene (TME) in isolation (*red*, $N = 25$), pressurized argon cluster (*orange*, $N = 25$), and in an argon cluster with Ar atoms held fixed (*blue*, $N = 25$). Within the error bars, the excited-state lifetime is not significantly altered

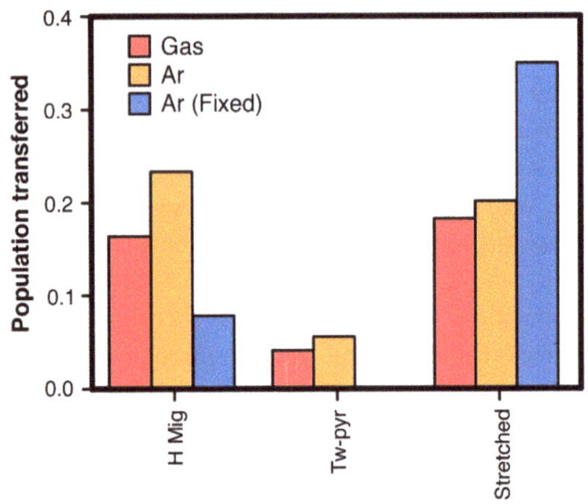

Fig. 10 Characterization of the excited-state quenching mechanism in TME for three different environments. The pressurized argon cluster (*orange*, Ar) leads to enhancement of population transfer through hydrogen migration (H Mig) intersections, compared to isolated TME (*red*, Gas). Fixing the argon atoms in the cluster (*blue*, Ar-Fixed) leads to a dramatic enhancement of population transfer through C–C stretched intersections, which require little motion of the methyl groups

determine the distance of the ethylenic C–C bond and the pyramidalization angle for the most pyramidalized ethylenic carbon atom. Each geometry is then plotted as a dot (colored according to the simulation it corresponds to: isolated TME, pressurized argon cluster, or environment of fixed argon atoms). Three clusters are easily distinguished and circled, with representative geometries shown in the insets. As discussed above, the presence of an argon environment leads to more population decay through the H migration intersections. Fixing the argon atoms further enhances the decay through C–C stretched intersections (which require little motion of the methyl groups).

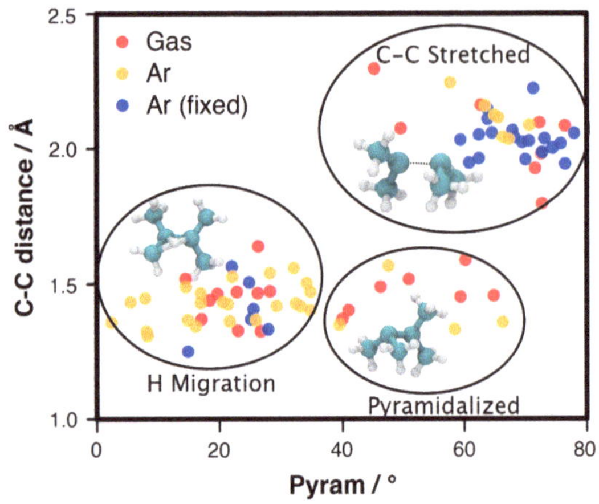

Fig. 11 Depiction of molecular geometries where population transfer occurs (i.e., "spawning geometries") for isolated TME (*red*, Gas), TME in a pressurized argon cluster (*orange*, Ar), and for TME in a pressurized argon cluster where the argon atoms are fixed (*blue*, Ar-Fixed). Each dot corresponds to the ethylenic C–C distance and pyramidalization angle around the most pyramidalized carbon atom at the spawning geometry. Three clusters can be recognized (*circled*): one corresponding to geometries that involve hydrogen migration, a second corresponding to geometries that involve pyramidalization about one of the ethylenic C atoms, and a third corresponding to pyramidalization and stretching of the ethylenic C–C bond. The presence of the argon environment leads to more nonadiabatic events corresponding to H migration or C–C stretching (compared to isolated TME). Fixing the argon atoms in the environment leads to an even more dramatic shift toward electronic state quenching through stretching of the C–C bond

3.2 Point charge: electrostatic influence of the surroundings

The phenomenon of "sudden polarization" in the photoisomerization of ethylene provides a possible mechanism for external control of the reaction through electrostatic fields. Electrostatic control is already thought to play an important role in the photoisomerization of larger conjugated chromophores; for instance, the retinal chromophore demonstrates a marked bond selectivity in its photoisomerization processes in vivo, while showing no such bias in solution or the gas phase. It has been proposed [14] that this bond selectivity is a product of a charged protein residue stabilizing the charge transfer state around a particular bond. Theoretical studies in our group and elsewhere have also suggested that external stabilization of charge transfer plays an important role in the in vivo photoisomerization of the PYP chromophore [7, 8].

The location and magnitude of the point charge we use here is similar to that of the charged residues in the binding pockets of the mentioned biological chromophores. We have previously shown that a positive charge stabilizes the pyramidalized intersection about the carbon *adjacent* to the point charge, while destabilizing the *opposite* intersection (pyramidalized about the carbon furthest from the point

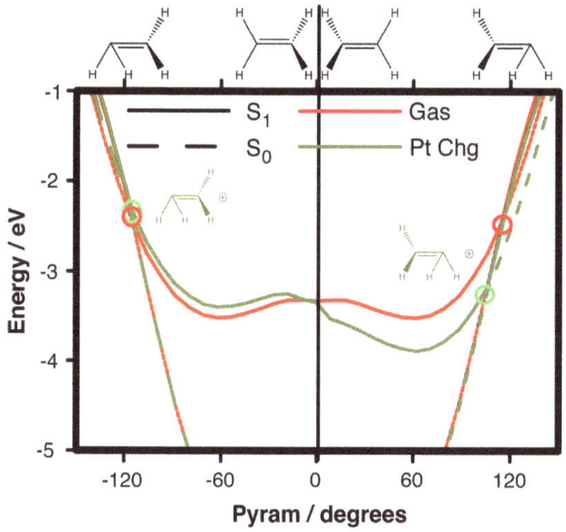

Fig. 12 Ground (*dashed line*) and excited-state (*solid line*) potential energy curves for pyramidalization of twisted ethylene in isolation (*red*) or in the presence of a positive point charge (*green*). Negative/positive pyramidalization angles refer to pyramidalization about the left/right carbon atom, which is furthest/closest to the point charge, respectively. Conical intersections are *circled*, showing that the intersections for pyramidalization about the left/right carbon atoms are degenerate in isolation, but this degeneracy is broken in the presence of the point charge

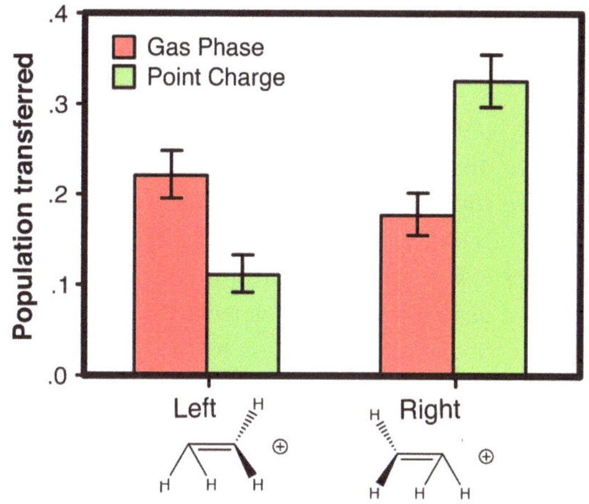

Fig. 13 Amount of population transferred (after 250 fs) from the excited state to the ground state through conical intersections corresponding to pyramidalization about the left/right carbon atoms in ethylene. In the isolated molecule (*red*), these would be equivalent by symmetry (and are indeed equivalent within the error bars which are depicted). In the presence of a positive point charge (*green*), the intersection which has pyramidalization about the carbon nearest the point charge (*right*) dominates

charge) [14]. In Fig. 12, we show the potential energy curves for ethylene which is pyramidalized about the carbon atom closest to (positive pyramidalization angles) or furthest from (negative pyramidalization angles) the point charge. To generate this figure, the molecule is first twisted by 90° about the C=C bond. These potential energy curves are compared for isolated ethylene and ethylene surrounded by a single point charge (as shown in Fig. 4). Two intersections can be seen in each case (highlighted with circles). In isolation, these two intersections are degenerate by symmetry. In contrast, the surrounding point charge stabilizes one of these and breaks the degeneracy.

Ab initio multiple spawning (AIMS) dynamics show that the electric field creates a strong bias toward the adjacent intersection, as shown in Fig. 13. More than twice as much population flows through this intersection than through the opposite intersection over the 250 fs of AIMS dynamics. The presence of the point charge also leads to a slightly smaller (but statistically significant) lag time before quenching to the ground state begins (t_0 in Table 2)—a result of the stabilization of the intersection (and thus increased gradient of the path leading to it after twisting, as shown in Fig. 12).

4 Discussion

We have used the AIMS method with ab initio QM/MM potential energy surfaces to investigate the role of environmental effects in the photoisomerization reaction of ethylene. This serves as a useful prototype for unraveling the general principles governing environmental effects in photoisomerization, which is an important process in many common biological chromophores. Steric hindrance was found not to have any significant effects on the S_1 dynamics; the photodynamical energy scale reduces the effective distance of the vdW spheres, resulting in an effective cavity size that easily permits the motions required for isomerization and hydrogen migration. Replacing the hydrogen atoms with bulky methyl groups leads to somewhat more noticeable steric effects, but these are still quite minor. In contrast, the charge transfer character of the reaction makes it amenable to control through external electrostatic fields. Our simulations suggest that a surrounding electric field can have a significant effect on the outcome of photoisomerization. This is especially relevant for larger conjugated molecules like retinal protonated Schiff base, where the surrounding electric field could play a significant role in the bond selectivity of photoisomerization [9, 11–14].

It is interesting that steric hindrance, normally an important consideration in biological reactions, plays almost no role in this case. This is due to the large amount of kinetic energy which is accumulated on the excited state, i.e., the energy of S_1 at the Franck–Condon point is almost 4 eV above the energy at the S_1/S_0 minimal energy conical intersection. For ethylene and TME, there is sufficient kinetic energy liberated during the excited-state reaction to overcome significant amounts of vdW repulsion (in the ethylene case) and reorganize the surrounding solvent to increase the cavity size (in the case of TME). The excitation energy generally decreases as molecules become more highly conjugated. Thus, the amount of kinetic energy accumulated on the excited state will also decrease and the role of steric effects may be more pronounced in longer conjugated molecules such as polyenes and retinal protonated Schiff base.

We have also introduced a new method to adaptively decrease the timestep used for the AIMS simulations in the vicinity of a conical intersection. This allows one to use larger timesteps in regions far from conical intersections. The method also automatically detects when the molecule has traversed a conical intersection by monitoring the overlap of the electronic wavefunctions at successive timesteps.

Acknowledgments This work has been supported by the NSF (OCI-10-47577 and CHE-11-24515) with computational support through DOE from the AMOS program within the Chemical Sciences, Geosciences and Biosciences Division of the Office of Basic Energy Sciences, Office of Science, US Department of Energy. We are pleased to dedicate this article to the memory of Isaiah Shavitt, who all the authors interacted with many times at UIUC.

References

1. Yarkony DR (1996) Diabolical conical intersections. Rev Mod Phys 68:985
2. Yarkony DR (1998) Conical intersections: diabolical and often misunderstood. Acc Chem Res 31:511
3. Bernardi F, Olivucci M, Robb MA (1996) Potential energy surface crossings in organic photochemistry. Chem Soc Rev 25:321
4. Klessinger M, Michl J (1995) Excited states and photochemistry of organic molecules. VCH Publishers Inc, New York
5. Levine BG, Martinez TJ (2007) Isomerization through conical intersections. Ann Rev Phys Chem 58:613
6. Virshup AM, Punwong C, Pogorelov TV, Lindquist B, Ko C, Martínez TJ (2009) Photodynamics in complex environments: ab initio multiple spwaning quantum mechanical/molecular mechanical dynamics. J Phys Chem B 113:3280
7. Groenhof G, Schäfer LV, Boggio-Pasqua M, Grubmüller H, Robb MA (2008) Arginine52 controls the photoisomerization process in photoactive yellow protein. J Am Chem Soc 130:3250
8. Ko C, Virshup AM, Martínez TJ (2008) Electrostatic control of the photoisomerization in the photoactive yellow protein chromophore: hybrid QM/MM ab initio multiple spawning simulation. Chem Phys Lett 460:272
9. Cembran A, Bernardi F, Olivucci M, Garavelli M (2004) Counterion controlled photoisomerization of retinal chromophore models: a computational investigation. J Am Chem Soc 126:16018
10. Altoe P, Bernardi F, Garavelli M, Orlandi G, Negri F (2005) Solvent effects on the vibrational activity and photodynamics of the green fluorescent protein chromophore: a quantum chemical study. J Am Chem Soc 127:3952
11. Cembran A, Bernardi F, Olivucci M, Garavelli M (2005) The retinal chromophore/chloride ion pair: structure of the photoisomerization path and interplay of charge transfer and covalent states. Proc Natl Acad Sci 102:6255
12. Olsen S, Toniolo A, Ko C, Manohar L, Lamothe K, Martinez TJ (2005) Computation of reaction mechanisms and dynamics in photobiology. In: Olivucci M (ed) Computational photochemistry. Elsevier, Amsterdam
13. Toniolo A, Olsen S, Manohar L, Martinez TJ (2004) Conical intersection dynamics in solution: the chromophore of green fluorescent protein. Faraday Discus 127:149
14. Martínez TJ (2006) Insights for light-driven molecular devices from ab initio multiple spawning excited-state dynamics of organic and biological chromophores. Acc Chem Res 39:119
15. Vallee BL, Williams RJP (1968) Metalloenzymes: the entatic nature of their active sites. Proc Natl Acad Sci 59:498
16. Lightstone FC, Bruice TC (1996) Ground state conformations and entropic and enthalpic factors in the efficiency of intramolecular and enzymatic reactions. 1. Cyclic anhydride formation by substituted glutarates, succinate, and 3,6-Endoxo-Δ4-tetrahydrophthalate monophenyl ester. J Am Chem Soc 118:2595
17. Ford L, Johnson L, Machin P, Phillips D, Tijian R (1974) Crystal structure of lysozyme-tetrasaccharide lactone complex. J Mol Bio 88:349
18. Ryde U, Olsson MHM, Pierloot K, Roos BO (1996) The cupric geometry of blue copper proteins is not strained. J Mol Bio 261:586
19. Shurki A, Štrajbl M, Villa J, Warshel A (2002) How much do enzymes really gain by restraining fragments? J Am Chem Soc 124:4097
20. Warshel A (2003) Computer simulations of enzyme catalysis: methods, progress and insights. Ann Rev Biophys Biomol Struct 32:425
21. Warshel A, Sharma PK, Kato M, Parson WW (2006) Modeling electrostatic effects in proteins. Biochim Biophys Acta 1764:1647
22. Suydam IT, Snow CD, Pande VS, Boxer SG (2006) Electric fields at the active site of an enzyme: direct comparison of experiment with theory. Science 313:200
23. Ben-Nun M, Martínez TJ (2000) Photodynamics of ethylene: ab initio studies of conical intersections. Chem Phys 259:237
24. Quenneville J, Martínez T (2003) Ab initio study of cis-trans photoisomerization in stilbene and ethylene. J Phys Chem, 107A
25. Barbatti M, Paier J, Lischka H (2004) Photochemistry of ethylene: a multireference configuration interaction investigation of the excited-state energy surfaces. J Chem Phys 121:11614
26. Barbatti M, Ruckenbauer M, Lischka H (2005) The photodynamics of ethylene: a surface-hopping study on structural aspects. J Chem Phys 122:174307
27. Tao H, Levine BG, Martinez TJ (2009) Ab initio multiple spawning dynamics using multi-state second-order perturbation theory. J Phys Chem A 113:13656
28. Mori T, Glover WJ, Schuurman MS, Martinez TJ (2012) Role of Rydberg States in the photochemical dynamics of ethylene. J Phys Chem A 116:2808
29. Mestdagh JM, Visticot JP, Elhanine M, Soep B (2000) Prereactive evolution of monoalkenes excited in the 6 eV region. J Chem Phys 113:237
30. Farmanara P, Stert V, Radloff W (1998) Ultrafast internal conversion and fragmentation in electronically excited C2H4 and C2H3Cl molecules. Chem Phys Lett 288:518
31. Tao H, Allison TK, Wright TW, Stooke AM, Khurmi C, Tilborg JV, Liu Y, Falcone RW, Belkacem A, Martinez TJ (2011) Ultrafast internal conversion in ethylene. I. The excited state lifetime. J Chem Phys 134:244306
32. Sension RJ, Hudson BS (1989) Vacuum ultraviolet resonance Raman studies of the excited electronic states of ethylene. J Chem Phys 90:1377
33. Brooks BR, Schaefer IHF (1979) Sudden polarization: pyramidalization of twisted ethylene. J Am Chem Soc 101:307
34. Bonacic-Koutecky V, Bruckmann P, Hiberty P, Koutecky J, Leforestier C, Salem L (1975) Sudden Polarization in the Zwitterionic Z1 excited states of organic intermediates Photochemical implications. Ang Chem 14:575
35. Bonacic-Koutecky V (1978) Sudden polarization in zwitterionic excited states of organic intermediates in photochemical reactions. On a possible mechanism for bicyclo[3.1.0]hex-2-ene formation. J Am Chem Soc 100:396

36. Bonacic-Koutecky V, Koutecky J, Michl J (1987) Neutral and charged Biradicals, Zwitterions, funnels in S1, and proton translocation: their role in photochemistry, photophysics, and vision. Ang Chem 26:170
37. Barbatti M, Granucci G, Persico M, Lischka H (2004) Semiempirical molecular dynamics investigation of the excited state lifetime of ethylene. Chem Phys Lett 401:276
38. Ben-Nun M, Martínez TJ (1998) Ab initio molecular dynamics study of cis-trans photoisomerization in ethylene. Chem Phys Lett 298:57
39. Warshel A, Levitt M (1976) Theoretical studies of enzymic reactions: dielectric, electrostatic and steric stabilization of the carbonium ion in the reaction of lysozyme. J Mol Biol 103:227
40. Hu H, Yang W (2008) Free energies of chemical reactions in solution and in enzymes with ab initio quantum mechanics/molecular mechanics methods. Ann Rev Phys Chem 59:573
41. Zhang Y, Liu H, Yang W (2000) Free energy calculation on enzyme reactions with an efficient iterative procedure to determine minimum energy paths on a combined ab initio QM/MM potential energy surface. J Chem Phys 112:3483
42. Field M, Bash P, Karplus M (1990) A combined quantum mechanical and molecular mechanical potential for molecular dynamics simulations. J Comp Chem 11:700
43. Friesner RA, Guallar V (2005) Ab initio quantum chemical and mixed quantum mechanics/molecular mechanics (QM/MM) methods for studying enzymatic catalysis. Ann Rev Phys Chem 56:389
44. Senn HM, Thiel W (2009) QM/MM methods for biomolecular systems. Ang Chem Int Ed 48:1198
45. Slavicek P, Martinez TJ (2006) Multicentered valence electron effective potentials: solution to the link atom problem for ground and excited electronic states. J Chem Phys 124:084107
46. Ben-Nun M, Martinez TJ (1998) Direct evaluation of the Pauli repulsion energy using 'classical' wavefunction in hybrid quantum/classical potential energy surfaces. Chem Phys Lett 290:289
47. Maitland GC, Smith EB (1971) The intermolecular pair potential of argon. Mol Phys 22:861
48. Cornell W, Cieplak P, Bayly C, Gould I, Merz JK, Ferguson D, Spellmeyer D, Fox T, Caldwell J, Kollman P (1995) A second generation force field for the simulation of proteins, nucleic acids, and organic molecules. J Am Chem Soc 117:5179
49. Roos BO (1987) The complete active space self-consistent field method and its applications in electronic structure calculations. In: Lawley KP (ed) Advances in chemical physics: ab initio methods in quantum chemistry II. Wiley, New York, p 399
50. Docken K, Hinze J (1972) LiH potential curves and wave functions. J Chem Phys 57:4928
51. Werner H-J, Knowles PJ, Lindh R, Schuetz M, Celani P, Korona T, Manby FR, Rauhut G, Amos RD, Bernhardsson A, Berning A, Cooper DL, Deegan MJO, Dobbyn AJ, Eckert F, Hampel C, Hetzer G, Lloyd AW, McNicholas SJ, Meyer W, Mura ME, Nicklass A, Palmieri P, Pitzer R, Schumann U, Stoll H, Stone AJ, Tarroni R, Thorsteinsson T MOLPRO, version, a package of ab initio programs
52. Frisch M, Pople J, Binkley J (1984) Self-consistent molecular orbital methods. J Chem Phys 80:3265
53. Ben-Nun M, Martínez T (2002) Ab initio quantum molecular dynamics. Adv Chem Phys, 121
54. Ben-Nun M, Quenneville J, Martínez T (2000) Ab initio multiple spawning: photochemistry from first principles quantum molecular dynamics. J Phys Chem 104A:5161
55. Heller E (1981) Frozen gaussians: a very simple semiclassical approximation. J Chem Phys 75:2923
56. Levine BG, Coe JD, Virshup AM, Martínez TJ (2008) Implementation of ab initio multiple spawning in the MolPro quantum chemistry package. Chem Phys 347:3
57. Fernandez-Alberti S, Roitberg AE, Nelson T, Tretiak S (2012) Identification of unavoided crossings in nonadiabatic photoexcited dynamics involving multiple electronic states in polyatomic conjugated molecules. J Chem Phys 137:014512
58. Nelson T, Fernandez-Alberti S, Roitberg AE, Tretiak S (2013) Artifacts due to trivial unavoided crossings in the modeling of photoinduced energy transfer dynamics in extended conjugated molecules. Chem Phys Lett 590:208
59. Wang L, Prezhdo OV (2014) A simple solution to the trivial crossing problem in surface hopping. J Phys Chem Lett 5:713
60. Habershon S (2012) Linear dependence and energy conservation in Gaussian wave packet basis sets. J Chem Phys 136:014109
61. Allen MP, Tildesley DJ (1987) Computer simulation of liquids. Oxford University Press, Oxford, p 385
62. Kutteh R (1998) RATTLE recipe for general holonomic constraints angle and torsion constraints. CCP5 Newslett, 46:9
63. Hack MD, Wensmann AM, Truhlar DG, Ben-Nun M, Martinez TJ (2001) Comparison of full multiple spawning, trajectory surface hopping and converged quantum mechanics for electronically nonadiabatic dynamics. J Chem Phys 115:1172
64. Efron R, Tibshirani R (1986) Bootstrap methods for standard errors, confidence intervals, and other measures of statistical accuracy. Stat Sci 1:54
65. Connolly ML (1983) Solvent-accessible surfaces of proteins and nucleic acids. Science 221:709

Theor Chem Acc (2014) 133:1480
DOI 10.1007/s00214-014-1480-y

REGULAR ARTICLE

Effects of the second hydration shell on excited-state multiple proton transfer: dynamics simulations of 7-azaindole: $(H_2O)_{1-5}$ clusters in the gas phase

Nawee Kungwan · Khanittha Kerdpol · Rathawat Daengngern · Supa Hannongbua · Mario Barbatti

Received: 23 January 2014 / Accepted: 7 March 2014 / Published online: 22 March 2014

Abstract Dynamics of the multiple excited-state proton transfer (ESPT) in clusters of 7-azaindole with up to five water molecules was investigated with quantum chemical methods. The ultrafast excited-state dynamics triggered by photoexcitation was simulated with the algebraic diagrammatic construction to the second-order scheme. Multiple ESPT through a hydrogen-bonded network is observed in the 100-fs scale. The probability of tautomerization is anti-correlated with the maximum free energy barrier in the excited state. An increasing number of water molecules tends to reduce the barrier by strengthening the hydrogen-bonded network. Barrierless reactions are found already for clusters with four waters. In structures presenting double hydrogen bond circuits, proton transfer happens mostly through the internal circuit by triple proton transfer. The overall role of the second hydration shell is of stabilizing the network, facilitating the proton transfer in the internal circuit. Proton transfers involving the second hydration shell were observed, but with small probability of occurrence. The proton-transfer processes tend to be synchronous, with two of them occurring within 10–15 fs apart.

Keywords On-the-fly dynamics simulation · Excited-state proton transfer · Excited-state tautomerization · Water-assisted proton transfer · Hydrogen bond rearrangement · 7-Azaindole · ADC(2)

Dedicated to the memory of Professor Isaiah Shavitt and published as part of the special collection of articles celebrating his many contributions.

Electronic supplementary material The online version of this article (doi:10.1007/s00214-014-1480-y) contains supplementary material, which is available to authorized users.

N. Kungwan (✉) · K. Kerdpol · R. Daengngern
Department of Chemistry, Faculty of Science, Chiang Mai University, Chiang Mai 50200, Thailand
e-mail: naweekung@gmail.com

S. Hannongbua
Department of Chemistry, Faculty of Science, Kasetsart University, Bangkhen Campus, Bangkok 10930, Thailand

M. Barbatti (✉)
Max-Planck-Institut für Kohlenforschung, Kaiser-Wilhelm-Platz 1, 45470 Mülheim an der Ruhr, Germany
e-mail: barbatti@kofo.mpg.de

1 Introduction

Proton transfer plays crucial roles in a variety of chemical and biological reactions [1–5] such as enol–keto tautomerization [6], proton transport via membrane-spanning proteins [7, 8], and proton relay system in enzymes [8]. The phototropic tautomerism of DNA base pairs, which has attracted much interest due to its relation to UV-induced gene mutation, has the excited-state proton-transfer (ESPT) phenomena as the primary molecular step [9]. Generally, it is specially challenging to monitor ESPT in natural nucleobases and nucleobase pairs due to their conformational and spectroscopic complexity [10, 11]. In this context, 7-azaindole (7AI, Fig. 1), a spectroscopically simpler prototype for purine nucleobases [9], emerges as a good alternative for basic sciences research. It has also been claimed that 7AI can advantageously replace tryptophan, customarily considered as the standard optical probe of protein structure and dynamics [12–14]. As a result, ESPT in 7AI monomer within small molecular clusters and in bulk solvents has been extensively studied by experimentalists and theorists [9, 11, 15–38].

It is well known that the isolated 7AI molecule cannot undergo ESPT without assistance of solvent molecules [39]. With cooperation of a single water molecule (Fig. 1), the energy barrier required for tautomerization of 7AI is considerably reduced in the ground and excited states as reported by Chaban and Gordon [40, 41]. The barrier heights for ESPT are even lower within clusters with two waters [42].

When 7AI-water clusters are excited into the S_1 band origin, ESPT occurs slow [16, 17, 32]. For $7AI(H_2O)_3$, for instance, while the excited-state lifetime (an upper bound to the ESPT time) is more than 10 ns at 0^0_0, it is reduced to only 15 ps upon an energy excess of 300 cm^{-1} [43]. Ab initio molecular dynamics (AIMD) simulations on $7AI(H_2O)_{1,2}$ reported by Kina et al. [44] showed that the excited-state transfer occurs about 50 fs after photoexcitation for $\sim$12,000 cm^{-1} energy excess.

7AI-water complexes with more than three waters have received less attention because it is believed that such large complexes cannot be directly involved in the tautomerization. Moreover, these complexes are difficult to be spectroscopically assigned due to the complexity of their electronic [11] and vibrational [10] structures. 7AI with four waters was studied by Folmer et al. [27] using ultrafast pump–probe spectroscopy combined with theoretical calculations. Their results revealed that the proton-transfer rate increases compared to that of 7AI with two and three waters. Their deuteration studies provided proof for the occurrence of proton transfer (PT), although it was not conclusively confirmed that the proton transfer resulted in a complete tautomerization of the 7AI monomer. For even bigger clusters of 7AI with five waters, there are no experimental investigations available; only a theoretical study was reported on the second hydration shell effect [45].

Recently, Yu et al. [46] reported a theoretical investigation of ESPT on clusters of 7AI with three water molecules by using time-dependent density functional theory (TDDFT), complete active space perturbation theory to the second order (CASPT2) and coupled cluster with approximated doubles (CC2). Their static calculations on the first excited state suggested the possibility of hydrogen bond rearrangement (HBR) of water molecules from a bridged-planar isomer to a cyclic-nonplanar isomer, followed by triple proton transfer. They also found out that triple PT in the cyclic-nonplanar isomer is energetically more favorable than the quadruple PT in the bridged-planar isomer. It was concluded that all proton-transfer processes follow a synchronous mechanism. (In their paper, they call "concerted" what we call "synchronous" in the present work. This classification is discussed below in Sect. 3.2). Furthermore, the barrier for HBR was found to be less than 1 kcal mol^{-1}, consistent with the missing vibronic bands in the resonance-enhanced multiphoton ionization (REMPI) spectra for $7AI(H_2O)_3$ [43]. Another recent theoretical study using on-the-fly dynamics simulations was carried out by our group for the multiple proton transfer in the first excited state of 7AI in the clusters of methanol [15] and mixed methanol–water [47] employing the algebraic diagrammatic construction to the second-order (ADC(2)) scheme. The results showed that the PT is completed in the 100-fs timescale and tends to be slower in methanol than in water because of the lower polarity of the former.

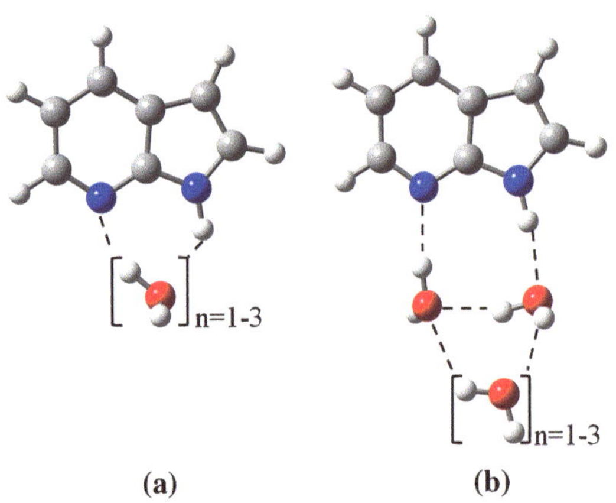

Fig. 1 Structure of 7AI with water molecules showing **a** first and **b** first and second hydration shells

Based on all those previous studies mentioned above, especially the isomerization of $7AI(H_2O)_3$ by Yu et al. [46], it occurred to us that the ab initio molecular dynamics simulations of $7AI(H_2O)_n$ with different isomers and different cluster sizes would be a very interesting way of testing the reaction paths for ultrafast ESPT, specially the role of the second hydration shell. Therefore, the purpose of this work is to carry out a systematic study of ESPT dynamics in $7AI(H_2O)_{1-5}$ complexes. The main goals are to determine the dynamic behavior of different water clusters around 7AI, the dependence of the ultra-fast excited-state tautomerization probability with the cluster size, the effect of hydrogen bond arrangement for complexes with the same number of water but different conformations, and the role played by water in different cluster sizes.

2 Computational details

2.1 Ground-state calculations

Ground-state geometries of $7AI(H_2O)_{n=1-5}$ complexes in the gas phase were optimized with the second-order Møller–Plesset Perturbation Theory (MP2) with the resolution-of-the-identity (RI) approximation for the electron repulsion integrals [48, 49]. The split valence polarized (SVP)

[50] basis set was assigned to heavy atoms and hydrogen atoms involved in the hydrogen-bonded network, whereas the split-valence (SV(P)) [50] basis set was assigned to the remaining hydrogen atoms in the complexes, as implemented in the program package TURBOMOLE 5.10 [51]. The performance of this basis set was tested with comparisons to results computed with the TZVPP [52] basis set. For $7AI(H_2O)_2$, two isomers were investigated. For $7AI(H_2O)_3$, three different isomers were determined based on previous results by Pino et al. [43] and Yu et al. [46]. For $7AI(H_2O)_4$, a cyclic isomer with two circuits was chosen, corresponding to an isomer reported by Folmer et al. [27] and Casadesus et al. [53]. Similarly, for $7AI(H_2O)_5$, the cyclic complex with two circuits was selected. This isomer has also been previously reported by Fernandez-Ramos et al. [45]. All optimized structures were confirmed to be minima on the ground-state surface by normal-mode analysis.

2.2 Excited-state dynamics simulations

On-the-fly Born–Oppenheimer dynamics simulations on the first excited-state (S_1) potential energy surface were carried out for the $7AI(H_2O)_{1-5}$ complexes at RI-ADC(2)/SVP-SV(P) level. The ADC(2) method, originally derived using diagrammatic perturbation theory [54, 55], can be expressed by the symmetric Jacobian $\mathbf{A}^{ADC(2)} = ½(\mathbf{A}^{CIS(D\infty)} + \mathbf{A}^{CIS(D\infty)\dagger})$, where $\mathbf{A}^{CIS(D\infty)}$ is the Jacobian of the CIS(D∞) coupled-cluster approximation [48]. The ADC(2) excited-state energies correspond to the eigenvalues of the Jacobian, while the ground-state energy is given by the MP2 method. ADC(2) has a computational cost similar to that of coupled cluster to approximated second order (CC2), with comparable accuracy [56]. ADC(2), however, possesses the distinct advantage over CC2 of deriving from a Hermitian eigenvalue problem. This increases its numerical stability in the case of quasi-degenerate excited states and reduces the computational effort required for the computation of molecular properties and gradients. The current implementation of ADC(2) is explained in Ref. [57], and the computation of its analytical gradients is explained in Ref. [49]. Initial conditions were generated using a harmonic oscillator Wigner distribution for each vibrational normal mode, as implemented in the NEWTON-X program package [58, 59] interfaced with the TURBOMOLE program. Excitation into the entire S_1 band was allowed to provide enough energy to activate ultrafast processes. Dynamics simulations in the first excited state were carried out with the NEWTON-X program in microcanonical ensembles using Born–Oppenheimer energies and gradients provided by RI-ADC(2) in the TURBOMOLE program. The nuclear motion of all atoms in the complexes was treated classically and computed by numerical integration of Newton's equation by the Velocity Verlet algorithm [60, 61]. Fifty trajectories for each complex and each different isomer—totalizing 400 trajectories—were simulated with a time step of 1 fs. Trajectories were propagated for 300 fs, enough to cover the pre- and post-PT regimes. A subset of five trajectories for one of the three $7AI(H_2O)_3$ isomers was extended up to 1,000 fs to check the possibility of hydrogen bond rearrangement. Furthermore, a statistical analysis was carried out to deliver detailed geometric and energetic properties, which were used to describe the time evolution of the reactions along the hydrogen-bonded network.

Neither tunneling nor nonadiabatic effects were considered in this work. Tunneling could be discarded as our focus was on the description of the ultrafast proton transfer (sub-picosecond scale). Based on our previous investigations of these systems [15, 47], we did not expect occurrence of surface crossings that would demand a nonadiabatic treatment. This was confirmed by the dynamics results, which showed that the character of the S_1 state was always a $\pi\pi^*$ excitation. The $\pi\pi^*$ character of the S_1 state implies that a proton—rather than a hydrogen atom—is transferred. For having a hydrogen transfer, the character of the S_1 state would be $\pi\sigma^*$ (see, for instance, Ref. [62]). We did not observe diabatic changes between $\pi\pi^*$ and $\pi\sigma^*$ characters along the dynamics. The separation between these states is illustrated for one single trajectory of $7AI(H_2O)_3$ in Fig. S1 of the Supplementary Material.

3 Results and discussion

3.1 Ground-state structures

The ground-state geometries of all complexes were optimized at RI-ADC(2)/SVP-SV(P) level. Cartesian coordinates for all structures discussed below are available in the Supplementary Material. Figure 2 shows the ground-state geometries of all three investigated isomers of $7AI(H_2O)_3$. Intermolecular hydrogen bonds of 7AI with water and between water molecules are indicated by dashed lines. The energies of these three isomers are less than 2.4 kcal mol^{-1} apart (see Table S1 in the Supplementary Material), in good agreement with results reported by Yu et al. [46] and Sakota et al. [18, 43]. The cyclic-nonplanar (Fig. 2b) is the lowest energy structure, followed by the bridged-planar (Fig. 2a) and then by the bicyclic-nonplanar isomer (Fig. 2c). Dynamics starting from the cyclic-nonplanar isomer was specially set to test the HBR proposed by Yu et al. [46].

The ground-state optimized geometry of $7AI(H_2O)_4$ is shown in Fig. 3a. There are other possible ground-state

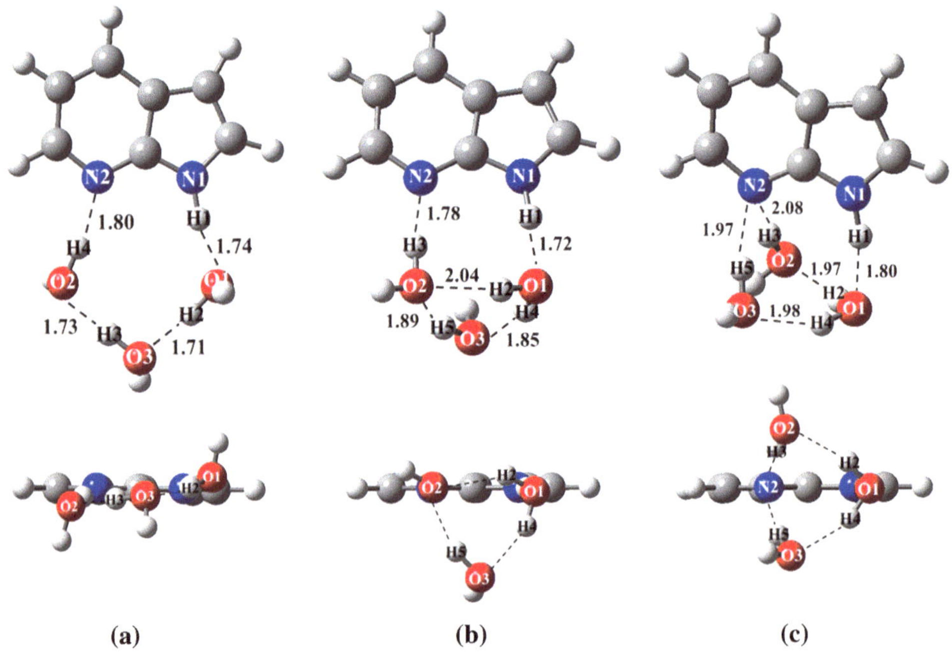

Fig. 2 Three different isomers of $7AI(H_2O)_3$ optimized at RI-ADC(2)/SVP-SV(P) level: **a** bridged-planar, **b** cyclic-nonplanar, and **c** bicyclic-nonplanar. *First row* is top view and *second row* is side view

structures, but only one candidate to exhibit two circuits for the proton transfer was found. The first circuit includes the two nearest waters (first shell), while the second circuit includes all four waters. This structure, which can be viewed as $7AI(H_2O)_{2+2}$, is similar to the B3LYP results reported by Casadesus et al. [53]. Based on Hartree–Fock level optimizations, Folmer et al. [27] also reported a $7AI(H_2O)_{2+2}$ structure. The orientation of the first water in their work, however, is somewhat different from our result and from that of Ref. [53].

The ground-state optimized structure of $7AI(H_2O)_5$ as a cyclic $7AI(H_2O)_2$ with three additional water molecules is illustrated in Fig. 3b. This structure, also previously reported by Fernandez-Ramos et al. [45], was chosen to investigate the effect of the second hydration shell. Like the $7AI(H_2O)_4$ complex, it has two circuits with a well-formed hydrogen-bonded network of water molecules. Proton transfer might occur either through circuit1, involving two waters, or through circuit2, involving all five waters.

The ground-state geometries of $7AI(H_2O)_1$ and two isomers with two waters, $7AI(H_2O)_{1+1}$ and $7AI(H_2O)_2$, were also computed and further employed for initiating dynamics simulation. Their optimized geometries are shown in Fig. S2 of the Supplementary Material.

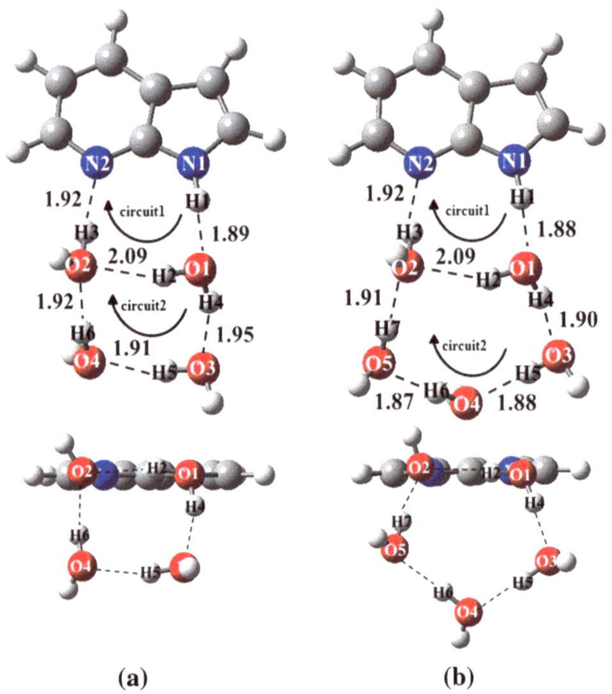

Fig. 3 Ground-state geometries of **a** $7AI(H_2O)_4$ and **b** $7AI(H_2O)_5$ optimized at RI-ADC(2)/SVP-SV(P) level

3.2 Excited-state dynamics simulations

On-the-fly dynamics simulations in the S_1 excited state were carried out for all $7AI(H_2O)_{1-5}$ complexes described above. The simulated trajectories were sorted into two types of reaction: (1) "ESPT" when 7AI tautomerization is complete within the simulation time (300 fs); and (2) "NT" (for "No tautomerization") when tautomerization does not take place within the simulation time. The number of trajectories in each type of reaction, the probability of tautomerization (ratio between the number of ESPT trajectories and the total number of trajectories), and the average time of PT for each complex are summarized in Table 1. The error bars for the statistical uncertainty are

Table 1 Summary of the excited-state dynamics performed at RI-ADC(2)/SVP-SV(P): number of trajectories showing 7AI tautomerization (ESPT) or no tautomerization (NT), tautomerization probability, and average time to complete the proton transfers

Complex	Reaction		Tautom. probability	Time (fs)			
	ESPT	NT		PT1	PT2	PT3	PT4
$7AI(H_2O)_1$	12	38	0.24 ± 0.10	48	59		
				(1.34)	(1.39)		
$7AI(H_2O)_{1+1}$	18	32	0.36 ± 0.11	91	96		
				(1.39)	(1.37)		
$7AI(H_2O)_2$	20	30	0.40 ± 0.11	69	70	78	
				(1.28)	(1.28)	(1.31)	
$7AI(H_2O)_3$-bridged-planar	3	46	0.06 ± 0.06	69	73	79	90
	1[a]			(1.29)	(1.30)	(1.31)	(1.33)
$7AI(H_2O)_3$-cyclic-nonplanar	19	31	0.38 ± 0.11	77	91	99	
				(1.30)	(1.34)	(1.35)	
$7AI(H_2O)_3$-bicyclic-nonplanar	30	20	0.60 ± 0.10	70	105	115	
				(1.27)	(1.30)	(1.34)	
$7AI(H_2O)_4$	40	9	0.82 ± 0.08	74	83	95	
	1[b]			(1.27)	(1.31)	(1.29)	
$7AI(H_2O)_5$	38	9	0.82 ± 0.08	74	85	111	
	3[c]			(1.27)	(1.33)	(1.27)	

Average distances at the proton-transfer time are given in parenthesis (Å)

[a] HBR trajectory was included neither in the probability nor in the PT time analysis

[b] Quintuple ESPT trajectory

[c] Sextuple ESPT trajectories were included in the probability analysis, but not in the PT time analysis

given for 90 % confidence interval. The PT distance, averaged over all ESPT trajectories, is also given in Table 1. The results are discussed in the next sections.

The PT time is defined as the time when the breaking bond length averaged over all ESPT trajectories intersects the average forming bond length. This is the same definition that we have used in our previous investigations [15, 63–65]. The PT mechanism can be assigned as either synchronous, concerted or sequential depending on the time lag between two consecutive PTs [66]. If the delay time is shorter than about 10–15 fs, which corresponds to a vibrational period of N–H and O–H stretching modes, the PT processes are synchronous. Otherwise, they are either concerted (a single kinetic step) or sequential (two distinct kinetic steps via a stable intermediate). It will be concerted if the PTs take place within the time of few N–H stretching modes (roughly, less than 100 fs), and it will be sequential if there is enough time to form a stable intermediate (>100 fs). Average time lags between each PT used to classify the mechanisms are listed in Table S3 of the Supplementary Material.

To test the performance of the SVP-SV(P) basis set, the energy of points along a trajectory was recomputed with ADC(2)/TZVPP. The results show only minor differences between the two levels (<3 kcal mol^{-1}) in the excited-state energy profile (see Figure S3 of the Supplementary Material).

3.2.1 $7AI(H_2O)_3$ complex

3.2.1.1 Bridged-planar isomer From 50 trajectories starting at the bridged-planar isomer, three trajectories showed 7AI tautomerization within 300 fs through a quadruple PT process (see Table 1). Triple ESPT after HBR occurred in only one trajectory, but in later times as discussed below. Tautomerization did not occur in 46 trajectories during the simulation time. Back PT reaction was observed in 1 trajectory.

The structures along the proton-transfer pathway are shown in Fig. 4, and average values of forming bonds, breaking bonds, and energies for the three trajectories following the quadruple ESPT reaction are shown in Fig. 5. The PT process, represented by arrows in Fig. 4, can be visualized as the following sequence of events (atom labels are given in Fig. 2a): a normal (N) form is observed at time 0. The first proton (H1) moves from the pyrrole ring to O1 atom of the nearest water (PT1) at 69 fs (when N1···H1 and O1···H1 bond distances are equal to 1.29 Å see Fig. 5a) and right after that a proton is transferred from this water to the second water (PT2) at 73 fs (O1···H2 and O3···H2 equal to 1.30 Å). The third proton is transferred from the second water to the third water (PT3) at 79 fs (O2···H3 and O3···H3 equal to 1.31 Å). Afterward, the forth proton moves from the third water to N2 in the pyridine ring (PT4) at 90 fs (O2···H4 and N2···H4 equal to 1.33 Å). After completing the reaction, the complex separates. This dynamic behavior is a sequence of synchronous PT processes. In the first 100 fs, the relative energy difference of S_1–S_0 (Fig. 5b) gradually decreases. After the tautomerization is completed, the energy gap is still around 46 kcal mol^{-1}, reflecting the planarity of the 7AI skeleton during the simulation time.

A subset of five trajectories that ended without tautomerization was extended until 1,000 fs. One of these

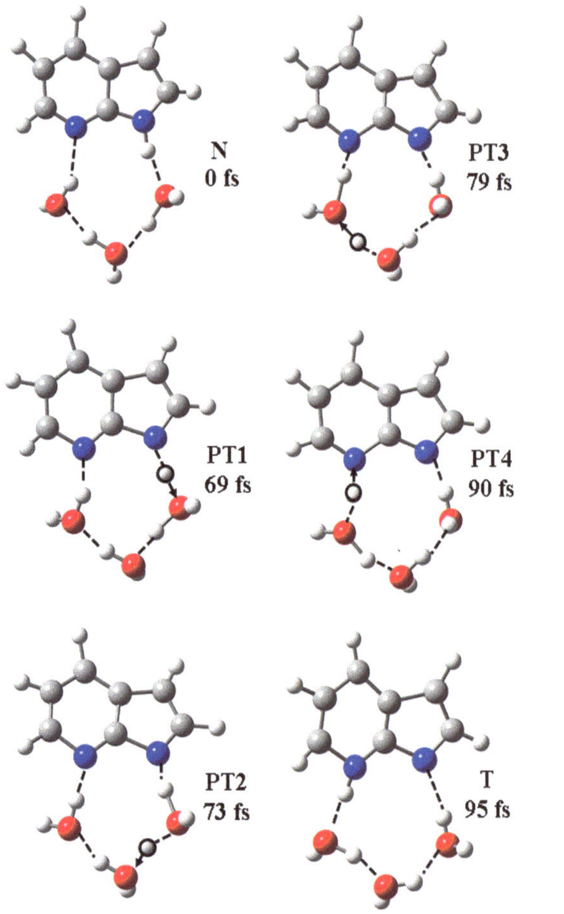

Fig. 4 Snapshots of the $7AI(H_2O)_3$-bridged-planar dynamics showing the time evolution of the quadruple ESPT reaction through the hydrogen-bonded network. Normal (*N*), proton transfer (*PT*), and tautomer (*T*). Values correspond to the average over all quadruple ESPT trajectories

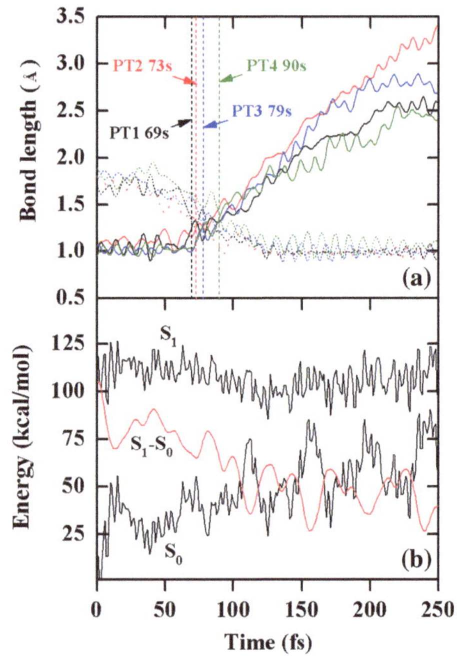

Fig. 5 Average values over quadruple ESPT trajectories of the $7AI(H_2O)_3$-bridged-planar isomer. **a** Average breaking and forming of bonds showing time evolution. N1–H1 and O1···H1 in *black*, O1–H2 and O3···H2 in *red*, O3–H3 and O2···H3 in *blue*, and O2–H4 and N2···H4 in *green*. **b** Average relative energies of excited stated state (S_1), ground state (S_0), and energy difference between S_1 and S_0 state (S_1–S_0)

trajectories exhibited HBR leading the bridged-planar to cyclic-nonplanar isomer. After HBR, triple PT was initiated. Snapshots of this trajectory are shown in Fig. S4. This pathway has been previously proposed by Yu et al. [46]. Considering that the small tautomerization probability for this isomer in the short timescale and the occurrence of HBR in one of the extended trajectories, we may speculate that this rearrangement may be a relevant pathway. Nevertheless, simulation of a larger number of trajectories in the long timescale would be needed to provide a quantitative estimate of its importance.

3.2.1.2 Cyclic-nonplanar isomer 7AI tautomerization through triple ESPT reaction took place in 19 out of 50 trajectories (38 % probability, Table 1), while no proton-transfer reaction was observed in 31 trajectories. Back PT reaction was also observed in 11 trajectories.

Snapshots illustrating the triple ESPT are displayed in Fig. 6, and the atom numbering scheme given in Fig. 2b is used to describe the dynamics. The values of the forming and breaking bonds averaged over all ESPT trajectories are given in Table 1 (see also Fig. S5 of the Supplementary Material). Starting from the normal form (N) at time 0, the PT process is summarized in the following steps: first, the first proton (H1) departs from N1 to O1 (PT1) at 77 fs (N1···H1=O1···H1 at 1.30 Å), and then, the second proton (H2) of the first water moves from O1 to O2 (PT2) at 91 fs (O1···H2=O2···H2 at 1.34 Å). Finally, the third proton (H3) leaves the second water to the N2 in pyridine ring (PT3) at 99 fs (N1···H1=O1···H1 at 1.35 Å). Tautomerization is complete in 105 fs and it is followed by dissociation of the complex.

There are 14 fs time lag between the first and the second PT and 8 fs between the second and the third PT. This indicates two synchronous processes. As in the previous complex, the S_1–S_0 energy gap gradually decreases in the first 100 fs. After that, the average energy difference is always slightly below 50 kcal mol^{-1}, and no intersection between the two states is approached within the simulation time.

3.2.1.3 Bicyclic-nonplanar isomer Thirty trajectories exhibited 7AI tautomerization through triple ESPT reaction (60 %, Table 1), 16 though O2 and 14 though O3 (see numbering in Fig. 2c). Twenty trajectories showed no

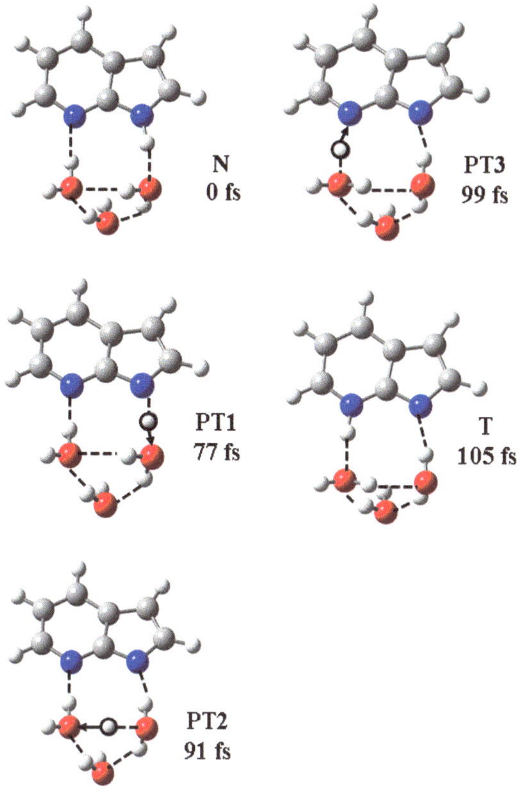

Fig. 6 Snapshots of the $7AI(H_2O)_3$-cyclic-nonplanar dynamics showing the time evolution of the triple ESPT reaction through the hydrogen-bonded network. Normal (*N*), proton transfer (*PT*), and tautomer (*T*). Values correspond to the average over all triple ESPT trajectories

tautomerization within the simulation time. Back PT reaction was also observed in 6 trajectories.

Tautomerization in this complex can occur through two symmetry-equivalent pathways, above and below the 7AI plane. Snapshots of a trajectory are depicted in Fig. 7. The first PT takes place at 70 fs (N1–H1=O1–H1=1.27 Å, Table 1). The second PT occurs at 105 fs for a 1.34 Å O–H distance. The third PT takes place at 115 fs for a 1.35 Å O–H distance (Fig. S6 in the Supplementary Material).

There is a 35 fs time lag between the first and second PT and 10 fs time lag between the second and third PT. The relatively long time delay between the first and the second PT can be attributed to a competition between O2 and O3 to receive the proton. H2 and H4 are pulled back and forth toward O2 and O4, respectively, few times before the transfer actually happens. This behavior implies that the triple proton transfer correspond to a concerted process followed by synchronous process.

3.2.2 $7AI(H_2O)_4$ complex

Tautomerization via circuit1 (Fig. 3a) occurred in 41 out of 50 trajectories (82 %, Table 1). It did not take place in nine

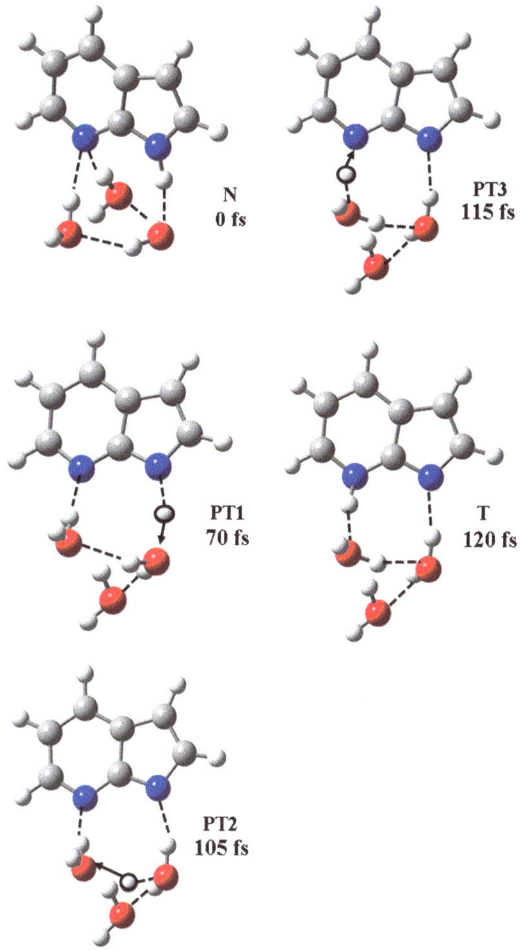

Fig. 7 Snapshots of the $7AI(H_2O)_3$-bicyclic-nonplanar dynamics showing the time evolution of the triple ESPT reaction through the hydrogen-bonded network. Normal (*N*), proton transfer (*PT*), and tautomer (*T*). Values correspond to the average over all triple ESPT trajectories

trajectories during the simulation time. One trajectory exhibited quintuple ESPT reaction through four waters (circuit2). Back PT reaction was also observed in 17 trajectories.

The structures along the reaction pathway are shown in Fig. 8. From our results averaged over the ESPT trajectories (see Fig. S7 in the Supplementary Material; atom numbers given in Fig. 3a), the times for PT1 (N1···H1=O1···H1 at 1.27 Å), PT2 (O1···H2=O2···H2 at 1.31 Å), and PT3 (O2···H3=N2···H3 at 1.29 Å) are 74, 83, and 95 fs, respectively. Time lags of 9 fs (between the PT1 and PT2) and 12 fs (between the PT2 and PT3) were observed, corresponding to two synchronous processes.

For the single trajectory with quintuple ESPT reaction via circuit2, the following events took place: proton transfers from N1 of pyrrole to the first water (O1), then to the second water (O3), then to the third water (O4), then to the forth water (O2) and finally to N2 of pyridine. The transfer times were 64, 78, 90, 95, and 104 fs, respectively.

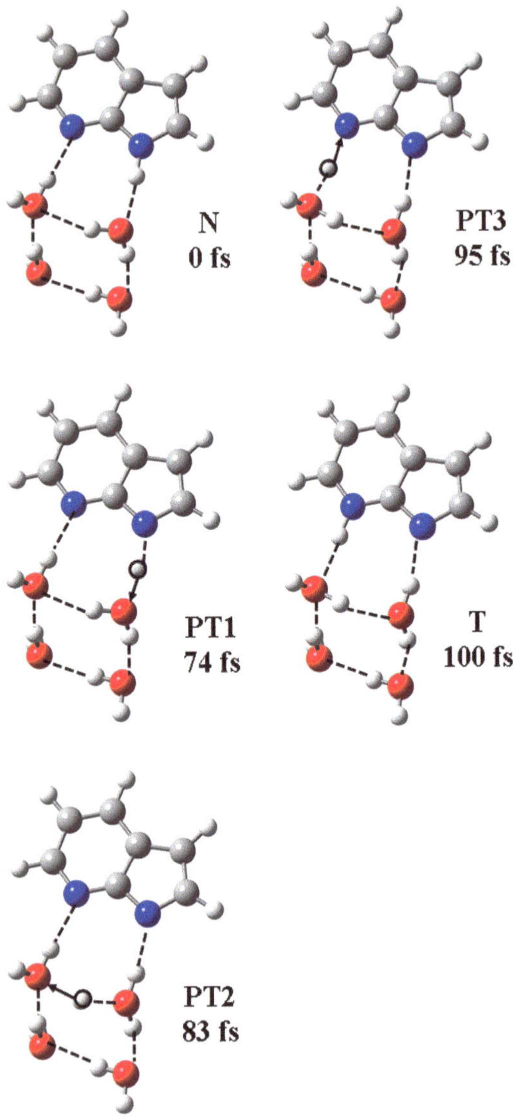

Fig. 8 Snapshots of the 7AI(H_2O)$_4$ dynamics showing the time evolution of the triple ESPT reaction through the hydrogen-bonded network. Normal (*N*), proton transfer (*PT*), and tautomer (*T*). Values correspond to the average over all triple ESPT trajectories

3.2.3 7AI(H_2O)$_5$ complex

From 50 trajectories, 41 trajectories (82 %, Table 1) showed 7AI tautomerization through circuit1 and circuit2 (see Fig. 3b), while no reaction was observed in 9 trajectories within the simulation time. Thirty-eight trajectories showed triple ESPT reaction through two waters (circuit1), whereas only three trajectories exhibited the sextuple ESPT reaction through five waters (circuit2). Back PT reaction was also observed in 18 trajectories.

Details of the triple ESPT process are depicted in Fig. 9 and Table 1 (numbering shown in Fig. 3b; see also Fig. S8 of the Supplementary Material). Transfer times for PT1

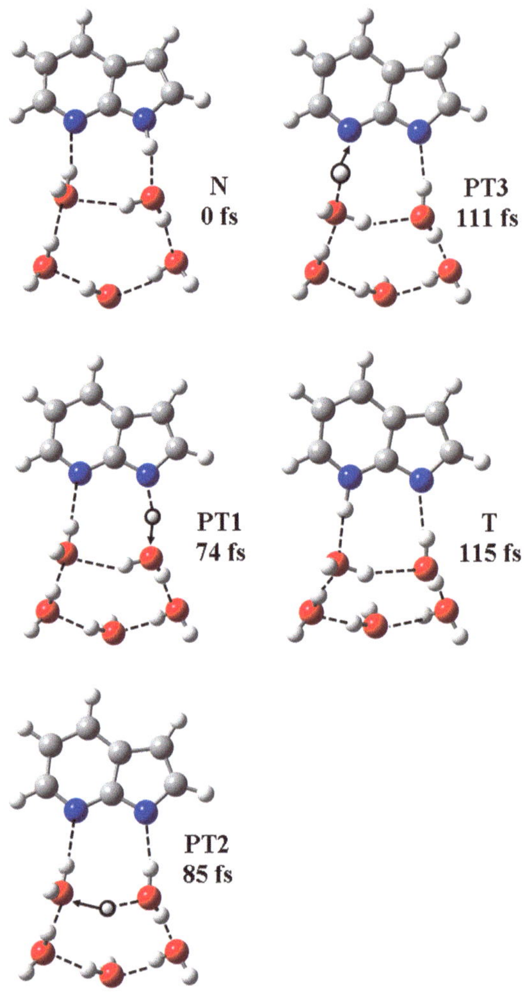

Fig. 9 Snapshots of the 7AI(H_2O)$_5$ dynamics showing the time evolution of the triple ESPT reaction through the hydrogen-bonded network. Normal (*N*), proton transfer (*PT*), and tautomer (*T*). Values correspond to the average over all triple ESPT trajectories

(N1···H1=O1···H1 at 1.27 Å), PT2 (O1···H2=O2···H2 at 1.33 Å), and PT3 (O2···H3=N2···H3 at 1.27 Å), averaged over ESPT trajectories, are 74, 85, and 111 fs, respectively. Time lags of 11 fs (between the PT1 and PT2) and 26 fs (between the PT2 and PT3) were observed, corresponding to a synchronous process followed by a concerted process.

3.2.4 7AI(H_2O)$_1$, 7AI(H_2O)$_{1+1}$, and 7AI(H_2O)$_2$ complexes

The main aim of this work is to investigate the effect of the second hydration shell on the ultrafast ESPT. Nevertheless, for a matter of completeness, we have simulated the dynamics of complexes with one and two water molecules as well. The main results are reported in Tables 1, 2 and Fig. 10b.

For 7AI(H_2O)$_1$, tautomerization occurs through a double PT with 24 % probability. It is completed in less than

Table 2 Tautomerization probabilities and free energy barriers for ESPT reactions in the excited state computed from the dynamics simulations

Initial isomer	Tautom. probability	Barrier (kcal mol^{-1})	ESPT
$7AI(H_2O)_1$	0.24 ± 0.10	6	D
$7AI(H_2O)_{1+1}$	0.36 ± 0.11	3	D
$7AI(H_2O)_2$	0.40 ± 0.11	3	T
$7AI(H_2O)_3$-bridged-planar	0.06 ± 0.06	17	Q
$7AI(H_2O)_3$-cyclic-nonplanar	0.38 ± 0.11	4	T
$7AI(H_2O)_3$-bicyclic-nonplanar	0.60 ± 0.10	2	T
$7AI(H_2O)_4$	0.82 ± 0.08	0	T
$7AI(H_2O)_5$	0.82 ± 0.08	0	T

D double, *T* triple, *Q* quadruple

100 fs (see Figs. S9 and S10 in the Supplementary Material).

With two waters, we investigated the dynamics starting from two isomers: a $7AI(H_2O)_{1+1}$ isomer, where a single water makes the bridge for double PT, while the second water is in the second shell; and a $7AI(H_2O)_2$ isomer, where the two waters form a bridge allowing tautomerization via triple PT.

Dynamics starting from the $7AI(H_2O)_{1+1}$ isomer is very similar to the single-water case. Tautomerization occurs through double PT in less than 100 fs. The tautomerization probability, however, tends to increase due to the network stabilization caused by the external water. More details about the dynamics results are given in Figs. S11 and S12 of the Supplementary Material.

In Ref. [47], we discussed the dynamics of $7AI(H_2O)_2$ in the context of the effect of water–methanol mixing. In the present work, we extended the number of simulated trajectories from 25 to 50, but the results remain, as expected, essentially the same. Tautomerization occurs through triple PT with 40 % probability. It is also completed within 100 fs. More details are given in Figs. S13 and S14 of the Supplementary Material.

3.3 Comparative analysis

For each complex, energies of the ground (S_0) and first excited ($\pi\pi^*$) states averaged over all ESPT trajectories were computed along the reaction pathway for the following structures: initial tautomer (N), intermediary structure for each PT (IS*n*), and final tautomer (T). These average energies for the $7AI(H_2O)_3$-bridged-planar isomer are shown in Fig. 10a. For the other complexes, they are plotted in Fig. S15-S21 of the Supplementary Material. For each complex, the highest average energy barrier is given in Table 2. The barriers computed in this way are, in fact, an approximation to the free energy barrier obtained by

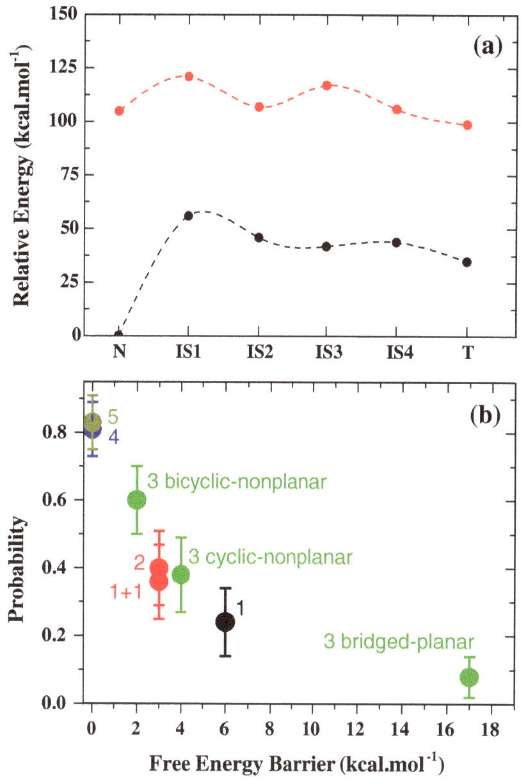

Fig. 10 **a** Average relative energies of the ground (S_0) and the first excited states ($\pi\pi^*$) of $7AI(H_2O)_3$-bridged-planar isomer. **b** Tautomerization probability versus free energy barrier for all complexes. The *labels aside each symbol* indicate the number of water molecules and the isomer

thermodynamic integration of independent trajectories [67] following Gaussian reaction paths. They are in good agreement with relative energies computed on the first excited singlet electronic state using TD-B3LYP by Casadesus et al. [53]. A good agreement is also observed between the barrier for $7AI(H_2O)_1$ computed here (6 kcal mol^{-1}) and the zero-point-corrected energy barrier computed with multi-reference perturbation theory to second order (MRPT2) reported in Ref. [17] (6.39 kcal mol^{-1}).

The free energy barriers are nicely anti-correlated with the tautomerization probabilities (Table 2; Fig. 10b): the case with largest barrier, $7AI(H_2O)_3$-bridged-planar (17 kcal mol^{-1}), has only 6 % probability, while the cases with the smallest barriers, $7AI(H_2O)_4$ and $7AI(H_2O)_5$ (0 kcal mol^{-1}), have 82 % probability of tautomerization.

In the two cases where double ESPT occurs, $7AI(H_2O)_1$ and $7AI(H_2O)_{1+1}$, the addition of one external water to form the latter complex cuts the barrier by half, increasing the tautomerization probability from 24 to 36 % (note, however, that these values are within the error bar for a 90 % confidence interval). In the cases where triple ESPT occurs, the clusters are arranged with $7AI(H_2O)_{2+n}$

($n = 0$–3) structure. As explained above, triple ESPT occurs through the two internal waters (circuit1), while the n other external waters remain as spectators, contributing to the overall stability of the hydrogen-bonded network. Once more, the increase in the number of waters in the second shell tends to reduce the barrier and increase the tautomerization probability. A saturation of this effect is observed already for $n = 2$ (4 waters), when the process becomes barrierless.

Most complexes share a common pattern: after photoexcitation, it takes about 70 fs to trigger the PT process (see Table 1). Exceptions are $7AI(H_2O)_1$ and $7AI(H_2O)_{1+1}$ which have the PT triggered in 48 and 91 fs, respectively. In all cases, as soon as the first PT is initiated, it facilitates a fast sequence of proton transfers through the bridge until the tautomerization is reached within about 100 fs. The PT times (Table 1) indicate that synchronous processes take place in $7AI(H_2O)_1$, $7AI(H_2O)_{1+1}$, $7AI(H_2O)_2$, $7AI(H_2O)_3$-bridged-planar, $7AI(H_2O)_3$-cyclic-nonplanar, and $7AI(H_2O)_4$. A sequence of concerted then-synchronous-processes takes place in $7AI(H_2O)_3$-bicyclic-nonplanar and a sequence of synchronous then concerted processes takes place in $7AI(H_2O)_5$.

As discussed in the introduction, spectroscopic results at the S_1 band origin [16, 32] imply much longer PT times than those that we have described. The main reason for the difference is that in our simulations, the whole S_1 band is excited, providing enough energy to overcome the tautomerization barriers. We have also seen that the tautomerization probability is relatively small for clusters with up to three waters, meaning that even with this excess of energy, substantial fraction of the excited population should tautomerize in longer timescales.

4 Conclusions

On-the-fly molecular dynamics simulations in the first excited state of $7AI(H_2O)_{1-5}$ complexes were carried out at RI-ADC(2)/SVP-SV(P) level. The following conclusions concerning the excited-state proton-transfer process and the effect of the second hydration shell on it can be drawn from our results:

- Multiple excited-state proton transfers through a hydrogen-bonded network are observed in the 100-fs scale.
- The probability of tautomerization is anti-correlated with the maximum free energy barrier in the excited state. This probability can be as low as 8 % and as high as 82 %.
- An increasing number of water molecules tends to reduce the barrier by strengthen the hydrogen-bonded network. Barrierless reactions are found already for clusters with four waters.
- In structures presenting double hydrogen bond circuits, proton transfer happens mostly through the internal circuit by triple proton transfer. Thus, the main role of the second hydration shell is of stabilizing the network, facilitating the proton transfer.
- Proton transfer occurs mostly in the original ground-state conformation that the complex has at the time of photoexcitation. We found, however, qualitative evidence of sub-picosecond photoinduced hydrogen bond reorganization of $7AI(H_2O)_3$ from the bridged-planar to cyclic-nonplanar isomer prior the proton transfer.
- Although rare, quintuple, and sextuple proton transfers were observed. They are completed in about 100 fs.
- The proton transfer tends to be composed of synchronous steps, with two of them occurring within 10–15 fs apart.

Acknowledgments The authors wish to thank the Thailand Research Fund (MRG5480294 and TRG5680098) for financial support. K. Kerdpol and R. Daengngern thank the Science Achievement Scholarship of Thailand (SAST), Faculty of Science, Chiang Mai University, Chiang Mai, Thailand.

References

1. Hynes JT, Klinman JP, Limbach HH, Schowen RL (eds) (2007) Hydrogen-transfer reactions: biological aspects I–II. WILEY-VCH Verlag GmbH & Co. KGaA, Weinheim, pp 1–4
2. Chang SM, Tzeng YJ, Wu SY, Li KY, Hsueh KL (2005) Thin Solid Films 477:38
3. Rey R, Moller KB, Hynes JT (2002) J Phys Chem A 106:11993
4. Arnaut LG, Formosinho SJ (1993) J Photochem Photobiol A 75:1
5. Formosinho SJ, Arnaut LG (1993) J Photochem Photobiol A 75:21
6. Gebicki J, Bally T (1997) Acc Chem Res 30:477
7. Freier E, Wolf S, Gerwert K (2011) Proc Natl Acad Sci USA 108:11435
8. Zewail AH (2000) Pure Appl Chem 72:2219
9. Taylor CA, El-Bayoumi MA, Kasha M (1969) Proc Natl Acad Sci USA 63:253
10. Yokoyama H, Watanabe H, Omi T, Ishiuchi S, Fujii M (2001) J Phys Chem A 105:9366
11. Nakajima A, Hirano M, Hasumi R, Kaya K, Watanabe H, Carter CC, Williamson JM, Miller TA (1997) J Phys Chem A 101:392
12. Smirnov AS, English DS, Rich RL, Lane J, Teyton L, Schwabacher AW, Luo S, Thornburg RW, Petrich JW (1997) J Phys Chem B 101:2758
13. Negrerie M, Gai F, Bellefeuille SM, Petrich JW (1991) J Phys Chem 95:8663
14. Negrerie M, Bellefeuille SM, Whitham S, Petrich JW, Thornburg RW (1990) J Am Chem Soc 112:7419
15. Daengngern R, Kungwan N, Wolschann P, Aquino AJA, Lischka H, Barbatti M (2011) J Phys Chem A 115:14129
16. Young JW, Pratt DW (2011) J Phys Chem 135:084301
17. Duong MPT, Kim Y (2010) J Phys Chem A 114:3403
18. Sakota K, Jouvet C, Dedonder C, Fujii M, Sekiya H (2010) J Phys Chem A 114:11161
19. Sakota K, Kageura Y, Sekiya H (2008) J Chem Phys 129:054303

20. Kwon O-H, Zewail AH (2007) Proc Natl Acad Sci USA 104:8703
21. Sakota K, Inoue N, Komoto Y, Sekiya H (2007) J Phys Chem A 111:4596
22. Sakota K, Komoto Y, Nakagaki M, Ishikawa W, Sekiya H (2007) Chem Phys Lett 435:1
23. Takeuchi S, Tahara T (2007) Proc Natl Acad Sci USA 104:5285
24. Hara A, Sakota K, Nakagaki M, Sekiya H (2005) Chem Phys Lett 407:30
25. Catalan J, Perez P, del Valle JC, de Paz JLG, Kasha M (2004) Proc Natl Acad Sci USA 101:419
26. Waluk J (2003) Acc Chem Res 36:832
27. Folmer DE, Wisniewski ES, Stairs JR, Castleman AW Jr (2000) J Phys Chem A 104:10545
28. Catalán J, del Valle JC, Kasha M (1999) Proc Natl Acad Sci USA 96:8338
29. Folmer DE, Wisniewski ES, Hurley SM, Castleman AW Jr (1999) Proc Natl Acad Sci USA 96:12980
30. Mente S, Maroncelli M (1998) J Phys Chem A 102:3860
31. Takeuchi S, Tahara T (1998) J Phys Chem A 102:7740
32. Huang Y, Arnold S, Sulkes M (1996) J Phys Chem 100:4734
33. Chou P-T, Wei C-Y, Chang C-P, Kuo M-S (1995) J Phys Chem 99:11994
34. Ilich P (1995) J Mol Struct 354:37
35. Chen Y, Gai F, Petrich JW (1994) Chem Phys Lett 222:329
36. Chen Y, Rich RL, Gai F, Petrich JW (1993) J Phys Chem 97:1770
37. Chou PT, Martinez ML, Cooper WC, Collins ST, McMorrow DP, Kasha M (1992) J Phys Chem 96:5203
38. Moog RS, Maroncelli M (1991) J Phys Chem 95:10359
39. Chau P-T (2001) J Chin Chem Soc 48:651
40. Chaban GM, Gordon MS (1999) J Phys Chem A 103:185
41. Gordon MS (1996) J Phys Chem 100:3974
42. Fang H, Kim Y (2011) J Chem Theory Comput 7:642
43. Pino GA, Alata I, Dedonder C, Jouvet C, Sakota K, Sekiya H (2011) Phys Chem Chem Phys 13:6325
44. Kina D, Nakayama A, Noro T, Taketsugu T, Gordon MS (2008) J Phys Chem A 112:9675
45. Fernandez-Ramos A, Smedarchina Z, Siebrand W, Zgierski MZ (2001) J Chem Phys 114:7518
46. Yu X, Yamazaki S, Taketsugu T (2012) J Phys Chem A 116:10566
47. Daengngern R, Kerdpol K, Kungwan N, Hannongbua S, Barbatti M (2013) J Photochem Photobiol A 266:28
48. Hättig C (2005) Adv Quantum Chem 50:37
49. Hättig C (2003) J Chem Phys 118:7751
50. Schäfer A, Horn H, Ahlrichs R (1992) J Chem Phys 97:2571
51. Ahlrichs R, Bär M, Häser M, Horn H, Kölmel C (1989) Chem Phys Lett 162:165
52. Schäfer A, Huber C, Ahlrichs R (1994) J Chem Phys 100:5829
53. Casadesus R, Moreno M, Lluch JM (2003) Chem Phys 290:319
54. Trofimov AB, Schirmer J (1995) J Phys B At Mol Opt Phys 28:2299
55. Schirmer J (1982) Phys Rev A 26:2395
56. Winter NOC, Graf NK, Leutwyler S, Hattig C (2013) Phys Chem Chem Phys 15:6623
57. Hättig C, Weigend F (2000) J Chem Phys 113:5154
58. Barbatti M, Ruckenbauer M, Plasser F, Pittner J, Granucci G, Persico M, Lischka H (2014) WIREs Comput Mol Sci 4:26
59. Barbatti M, Granucci G, Ruckenbauer M, Plasser F, Crespo-Otero R, Pittner J, Persico M, Lischka H (2013) NEWTON-X: a package for Newtonian dynamics close to the crossing seam. www.newtonx.org
60. Swope WC, Andersen HC, Berens PH, Wilson KR (1982) J Chem Phys 76:637
61. Verlet L (1967) Phys Rev 159:98
62. Tanner C, Manca C, Leutwyler S (2003) Science 302:1736
63. Kungwan N, Daengngern R, Piansawan T, Hannongbua S, Barbatti M (2013) Theor Chem Acc 132:1
64. Kungwan N, Plasser F, Aquino AJA, Barbatti M, Wolschann P, Lischka H (2012) Phys Chem Chem Phys 14:9016
65. Barbatti M, Aquino AJA, Lischka H, Schriever C, Lochbrunner S, Riedle E (2009) Phys Chem Chem Phys 11:1406
66. Dewar MJS (1984) J Am Chem Soc 106:209
67. Lawrenz M, Baron R, McCammon JA (2009) J Chem Theory Comput 5:1106

MIX
Papier aus verantwortungsvollen Quellen
Paper from responsible sources
FSC® C105338

If you have any concerns about our products,
you can contact us on
ProductSafety@springernature.com

In case Publisher is established outside the EU,
the EU authorized representative is:
Springer Nature Customer Service Center GmbH
Europaplatz 3, 69115 Heidelberg, Germany

Printed by Libri Plureos GmbH
in Hamburg, Germany